Modern Inorganic Chemistry

Modern Inorganic Chemistry

Editor: Garry Hollis

NY RESEARCH PRESS

New York

Published by NY Research Press
118-35 Queens Blvd., Suite 400,
Forest Hills, NY 11375, USA
www.nyresearchpress.com

Modern Inorganic Chemistry
Edited by Garry Hollis

International Standard Book Number: 978-1-63238-548-2 (Hardback)

The publisher's policy is to use permanent paper from mills that operate a sustainable forestry policy. Furthermore, the publisher ensures that the text paper and cover boards used have met acceptable environmental accreditation standards.

Trademark Notice: Registered trademark of products or corporate names are used only for explanation and identification without intent to infringe.

Cataloging-in-Publication Data

Modern inorganic chemistry / edited by Garry Hollis.
    p. cm.
Includes bibliographical references and index.
ISBN 978-1-63238-548-2
1. Chemistry, Inorganic. 2. Chemistry. 3. Inorganic compounds. I. Hollis, Garry.
QD151.3 .M63 2017
546--dc23

Printed in the United States of America.

Contents

Preface

This book was inspired by the evolution of our times; to answer the curiosity of inquisitive minds. Many developments have occurred across the globe in the recent past which has transformed the progress in the field.

Inorganic chemistry is the study of inorganic compounds and their corresponding chemical reactions. Inorganic chemistry has been implemented in the industrial sector as well the agricultural sector. Bioinorganic compounds, organometallic compounds, cluster compounds are some of the popularly studied chemical compounds in this discipline. This book on modern inorganic chemistry focuses on the latest research that is taking place in this field. Topics are well-explained and follow a systematic approach of analysis. This book is a compilation of the latest studies and emerging trends that have accrued in this field. The book presents chapters that are relevant for the theory as well as practice of inorganic chemistry. It will serve as a vital source of reference to students and scholars alike.

This book was developed from a mere concept to drafts to chapters and finally compiled together as a complete text to benefit the readers across all nations. To ensure the quality of the content we instilled two significant steps in our procedure. The first was to appoint an editorial team that would verify the data and statistics provided in the book and also select the most appropriate and valuable contributions from the plentiful contributions we received from authors worldwide. The next step was to appoint an expert of the topic as the Editor-in-Chief, who would head the project and finally make the necessary amendments and modifications to make the text reader-friendly. I was then commissioned to examine all the material to present the topics in the most comprehensible and productive format.

I would like to take this opportunity to thank all the contributing authors who were supportive enough to contribute their time and knowledge to this project. I also wish to convey my regards to my family who have been extremely supportive during the entire project.

Editor

A molecular catalyst for water oxidation that binds to metal oxide surfaces

Stafford W. Sheehan[1], Julianne M. Thomsen[1], Ulrich Hintermair[1,2], Robert H. Crabtree[1], Gary W. Brudvig[1] & Charles A. Schmuttenmaer[1]

Molecular catalysts are known for their high activity and tunability, but their solubility and limited stability often restrict their use in practical applications. Here we describe how a molecular iridium catalyst for water oxidation directly and robustly binds to oxide surfaces without the need for any external stimulus or additional linking groups. On conductive electrode surfaces, this heterogenized molecular catalyst oxidizes water with low overpotential, high turnover frequency and minimal degradation. Spectroscopic and electrochemical studies show that it does not decompose into iridium oxide, thus preserving its molecular identity, and that it is capable of sustaining high activity towards water oxidation with stability comparable to state-of-the-art bulk metal oxide catalysts.

[1] Department of Chemistry, Yale University, 225 Prospect Street, PO Box 208107, New Haven, Connecticut 06520-8107, USA. [2] Centre for Sustainable Chemical Technologies, University of Bath, Claverton Down BA2 7AY, UK. Correspondence and requests for materials should be addressed to S.W.S. (email: stafford.sheehan@yale.edu) or to U.H. (email: u.hintermair@bath.ac.uk) or to G.W.B. (email: gary.brudvig@yale.edu).

Economic and environmental concerns raised by the extensive use of fossil fuels have made alternative energy sources more attractive[1]. Renewable sources are particularly promising owing to their environmental sustainability and potential for widespread availability. However, there are a number of problems that need to be addressed before renewable energy can be used on a global scale[2]. Chief among these is the lack of reliable methods of concentrating and storing energy on a large scale, since numerous renewable energy sources, including solar and wind, are intermittent and diffuse. Towards this end, the generation of renewable fuels, in which electricity generated by renewable means is stored in the chemical bonds of a suitable fuel, has become a critically important area of research[3]. In one scheme for the formation of renewable fuels that mimics photosynthesis in plants, protons and electrons are extracted from water by electrochemical oxidation to be used for fuel formation, liberating O_2 as a byproduct[4]. The efficient generation of such renewable fuels, therefore, necessitates the development of efficient, fast and stable water-oxidation catalysts (WOCs).

Of the extensive library of available WOCs, molecular species show promise because of their high activity and tunability, as well as their ability to be integrated into sophisticated molecular assemblies[5–12]. Their major drawback is their limited stability, with the best homogeneous systems providing turnover numbers in the thousands[13–15] to tens of thousands[16]. This problem is particularly pronounced for electrode-driven WOCs, which often decompose to less active materials under moderate applied potentials[17,18]. Building on the success of different heterogenization strategies for homogeneous catalysts in organometallic, inorganic and surface chemistry[19–27], immobilization of molecular WOCs on electrode surfaces has been sought to overcome this[28–30]. However, in the case of an electrode-driven WOC, the ligand anchoring the catalyst to the electrode surface must display a high degree of oxidative stability, which is not always the case[31]. Methods that alter the electrode surface, including deposition of coating layers of TiO_2 after catalyst adhesion have been shown to assist in solving this issue[32]. However, to date, no system has fully succeeded in combining the high efficiency and tunability of a molecular catalyst that contains a single, well-defined catalytically active site (single-site)[33] with the durability and stability of a bulk material in a heterogeneous electrocatalyst for water oxidation.

In a recent report[34], we identified highly active homogeneous WOCs that are formed by the oxidative removal of Cp* (pentamethylcyclopentadienyl, $C_5Me_5^-$), an organic placeholder ligand, from well-studied Cp*Ir based precursors[35,36]. The compounds that form from these precatalysts all possess a single bidentate chelate ligand[37] per iridium that is oxidatively stable and prevents the formation of IrO_x-based films[38] or nanoparticles[39] under oxidative conditions. In contrast, Cp*Ir precursors lacking a stable bidentate ligand anodically deposit amorphous IrO_x on electrodes to give a heterogeneous WOC referred to as 'blue layer' (BL)[38]. This BL, hydrated IrO_x nanoclusters[40], and the homogeneous WOCs formed by the oxidative activation of our organometallic iridium precursors[41,42] all display a characteristic deep blue colour in their oxidized form,

owing to an absorption feature near 600 nm. This has caused confusion about the identity of the active species in these systems[34,39]; thus, in the following we refer only to the activated form of the molecular iridium catalyst as the hom-WOC.

In this study, we report the heterogenization of the hom-WOC to form a surface-bound, ligand-modified iridium electrocatalyst for water oxidation in acidic solutions. On self-adhering to the surface of a metal oxide at room temperature, a molecular monolayer of the catalyst is formed, which possesses higher activity than the bulk material analogue, IrO_x. We show that this heterogenized molecular catalyst remains bound to the surface after extended use, eliminating the need for any linking moieties, while retaining its molecular identity and ligand-based tunability.

Results

Catalyst preparation and heterogenization. In the pursuit of combining the high efficiency of the molecular hom-WOC with the stability of bulk metal oxides, we found that when an oxide material is immersed in an aqueous solution of the hom-WOC, the material rapidly and irreversibly chemisorbs some of the blue-coloured complex from the solution (we refer to this supported heterogeneous complex as the het-WOC). The dinuclear structures shown in Fig. 1 are based on the characterization data reported in this paper and on the prior work on the homogeneous analogue (hom-WOC)[34,41]. These structures are consistent with the available data, but are not intended to be definitive.

In contrast, the hom-WOC does not bind to noble metals that do not form a native oxide layer such as Au or Pt[41]. To further probe the electrochemical and spectroscopic properties of the het-WOC, we moved to high surface area transparent conductive electrodes consisting of mesoporous films of tin-doped indium oxide nanoparticles (*nano*ITO)[43] on fluorine-doped tin oxide (FTO)-coated glass slides. A solution of the hom-WOC bearing a 2-(2′pyridyl)-2-propanolate (pyalc) bidentate ligand, previously characterized as $[Ir(pyalc)(H_2O)_2(\mu\text{-}O)]_2^{2+}$, was prepared from [Cp*Ir(pyalc)OH] and $NaIO_4$ using established methods[34]. On immersion, catalyst binding to the *nano*ITO surface is rapid, self-limiting and does not require any external driving force such as photons or an applied potential (Fig. 2a). Control experiments show that the removal of the organometallic placeholder ligand from the precursor is required for surface binding to occur in all the cases (Supplementary Figs 1–4). Formation of a molecular monolayer is complete in 2 h at room temperature, with negligible absorption changes being observed at later times (Fig. 2b). Even after thorough rinsing with deionized water, the catalyst is not washed off the surface. Transmission electron microscope (TEM) and scanning electron microscope (SEM) images of the electrode indicate that no nanoparticulate deposits are formed, and energy-dispersive X-ray spectroscopy (EDX) measurements confirm the presence of iridium on the electrode surface without any traces of iodine or sodium from the $NaIO_4$ used to produce the hom-WOC in solution (Supplementary Figs 5–8).

The hom-WOC absorbance peak at 608 nm in solution blue shifts to 580 nm in the het-WOC formed on binding to *nano*ITO (Fig. 2d), both being distinct from electrodeposited IrO_x[38]. The

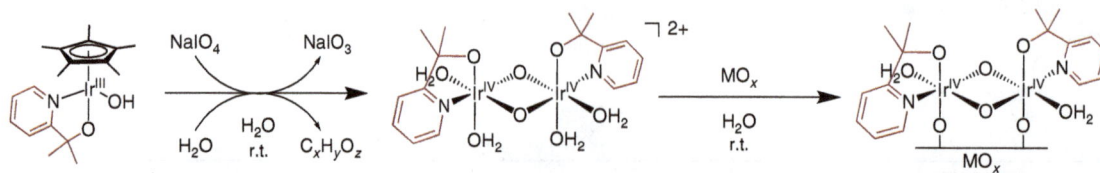

Figure 1 | Formation of the hom-WOC and proposed molecular structure for the adsorbate. Oxidation of the [Cp*Ir(pyalc)OH] precursor (left) to form the $[Ir(pyalc)(H_2O)_2(\mu\text{-}O)]_2^{2+}$ hom-WOC (middle) and heterogenization at room temperature (r.t.) to form the het-WOC (right) is shown.

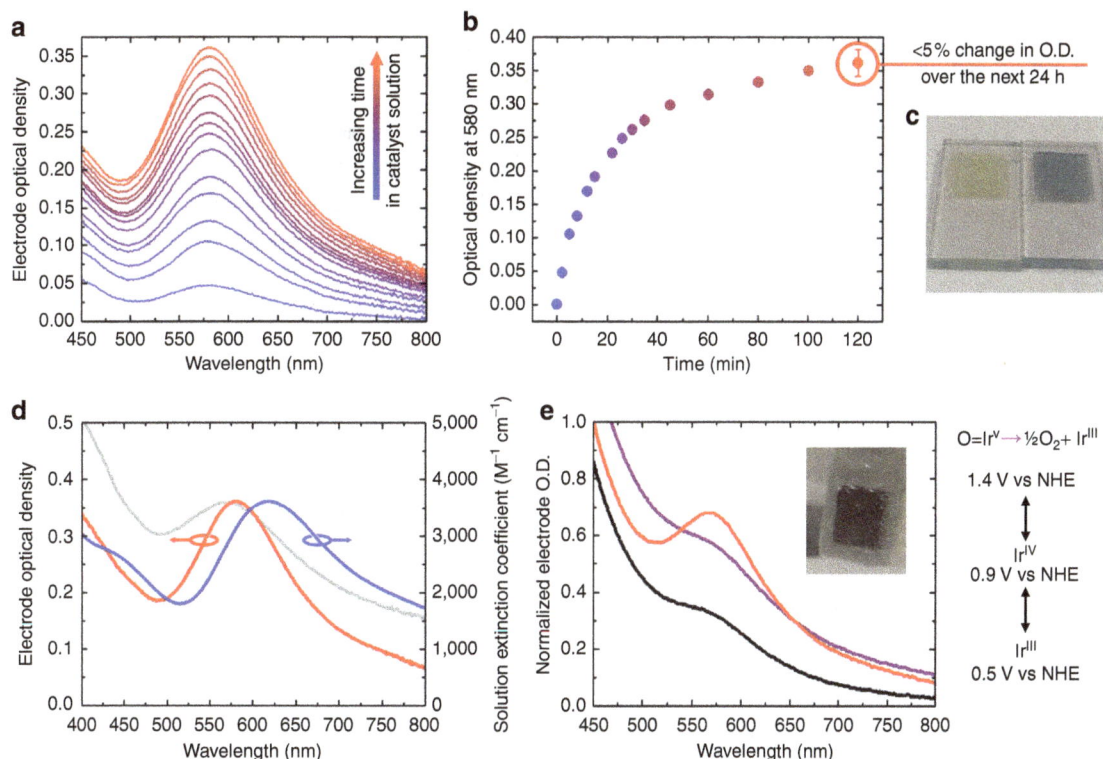

Figure 2 | Spectroscopic characterization of the het-WOC on *nano*ITO electrodes. (**a**) Optical density (O.D.) spectra of an electrode measured after increasing the amounts of time immersed in hom-WOC solution at room temperature. Between each measurement, the electrode was washed thoroughly with deionized water. (**b**) Increase of O.D. at 580 nm for the electrode as a function of immersion time, error bar shows the largest measured deviation in optical density across 3 samples after further immersion for 24 h. (**c**) Photograph of an electrode before (left side of panel) and after (right side of panel) immersion in hom-WOC solution for 2 h. (**d**) Comparison of the spectra of the catalyst on the surface (red) to the catalyst in solution (blue) along with IrO$_x$ electrodeposited on *nano*ITO (grey). (**e**) Spectroelectrochemical response of the electrode showing reversible transitions between Ir^{III} (black) and Ir^{IV} (red) oxidation states, as well as under turnover conditions (purple). The absorption feature at 580 nm is characteristic of Ir^{IV}, and the highly simplified legend on the right shows the potentials applied to reach these oxidation states. Inset is a photograph of the electrode under turnover conditions (1.4 V versus NHE, pH 2.6) corresponding to the spectrum in purple.

shift that occurs during catalyst heterogenization is similar to that previously observed for the reversible deprotonation of bound water ligands[34]. Along with the fact that the catalyst remains bound to the surface after repeated washing, this suggests chemical binding rather than mere physisorption. Both the pH and the concentration of iodate in the hom-WOC solution have pronounced effects on the rate of binding (Supplementary Fig. 9). As the pH of the solution is decreased, the rate of catalyst binding increases. Similarly, when the pH of the solution is increased, the rate of catalyst binding decreases, demonstrating that the catalyst binds faster when retaining aqua ligands rather than more strongly coordinating hydroxo ligands, consistent with a water displacement mechanism. These sites are also rapidly exchanged with the anion present in solution, iodate, which can also act as a ligand[44]. Consistently, we found that an increase in iodate concentration inhibits surface binding; however, a low concentration of iodate is required for heterogenization. When Cp*Ir precatalyst activation to form the hom-WOC is performed electrochemically[41], surface binding does not occur unless iodate is present in the solution. Electron paramagnetic resonance spectroscopy of the het-WOC showed no signals that would be expected for monomeric Ir^{IV} species on the surface[45], while X-ray photoelectron spectroscopy (XPS) proves that iridium is indeed present in the Ir^{IV} state (Supplementary Fig. 10)[46–48]. This is also the case for the hom-WOC in solution, indicating that the catalyst is still in dimer form when bound to the surface.

The absorbance peak at 580 nm is still evident after the electrode is immersed in aqueous solution in a spectroelectrochemical cell (Fig. 2e, Supplementary Figs 11–13). In this experiment, the catalyst-coated *nano*ITO working electrode forms a circuit with an Ag/AgCl reference electrode and a platinum counter electrode. We then vary the potential applied to the working electrode to induce reversible changes in the oxidation state of iridium in the het-WOC while collecting ultraviolet–visible spectra. Importantly, the catalyst remains bound to the electrode not only in its native Ir^{IV} state, but also in its reduced oxidation state, presumably Ir^{III}, as well as the catalytically active state, presumably Ir^V, from which oxygen evolution is observed[49,50].

The stability and versatility of the het-WOC is shown by its irreversible adhesion when exposed both to acidic and basic aqueous solutions (pH values ranging from 1–12) and to numerous organic solvents, including CH$_2$Cl$_2$ and MeCN. Only repeated washing under highly alkaline conditions (pH > 13) was found to remove the het-WOC from *nano*ITO, resulting in a clean electrode surface. The catalyst also adheres rapidly to different nanostructured metal oxides that are commonly used as photoanodes for light-driven water oxidation, including TiO$_2$ and WO$_3$ (Supplementary Fig. 14)[51,52].

Water-oxidation performance and stability. The het-WOC maintains its activity for oxygen evolution with chemical oxidants (Fig. 3a, Supplementary Fig. 15) compared with that previously observed[34] for the hom-WOC in solution. Most importantly, however, the het-WOC shows exceptional activity when driven electrochemically. Cyclic voltammograms (CVs) in an oxygen-saturated solution of 0.1 M KNO$_3$ in water at pH 2.6 (Fig. 3b,

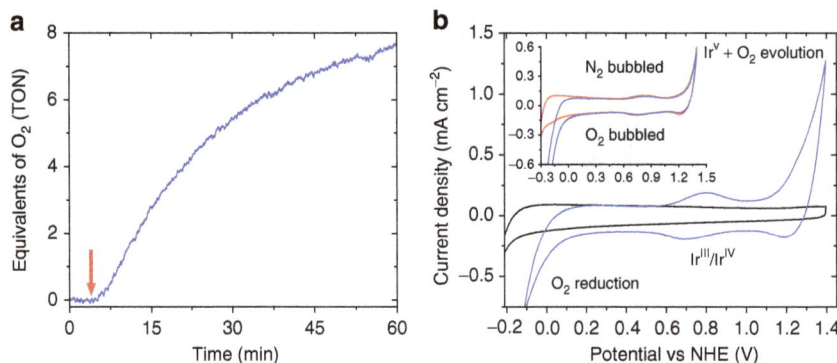

Figure 3 | Water oxidation using the het-WOC. (**a**) Water oxidation using NaIO$_4$ as a sacrificial oxidant with the catalyst bound to *nano*ITO; the red arrow corresponds to injection of NaIO$_4$ solution, initiating catalysis as quantified by the turnover number (TON) of O$_2$ per iridium atom on the mesoporous surface. (**b**) CVs of a catalyst-loaded *nano*ITO electrode (blue) compared with a similar *nano*ITO electrode without catalyst (black) in an oxygen-saturated solution of 0.1 M KNO$_3$ in water at pH 2.6, taken with a 10 mV s^{-1} scan rate. Inset shows the effect of saturating O$_2$ or N$_2$ gas in solution when using thinner electrodes, highlighting the catalytic wave for oxygen reduction at 0 V versus NHE.

Supplementary Fig. 16) show reversible Ir^{III}/Ir^{IV} charging features with E$_{1/2}$ = 0.75 V versus the normal hydrogen electrode (NHE), as well as reversible water oxidation/oxygen reduction similar to traditional iridium oxides, but lacking a redox feature that has been assigned to the oxidation of Ir^{IV} to Ir^V (ref. 53). The onset of the water-oxidation catalytic wave occurs at a distinctively lower potential than the Ir^{IV}/Ir^V redox couple in IrO$_x$ samples prepared by different means[54] and, thereby, obscures the Ir^{IV}/Ir^V charging feature. Drawing a parallel to previously suggested mechanisms for Ir-catalysed water oxidation[36,49,50], we postulate that this is a direct result of this catalyst's highly active Ir^V state, which along with the high electroactivity of the molecular iridium compound on the surface, allows for water oxidation at low overpotentials. Specifically, integration of the Ir^{III}/Ir^{IV} wave and comparison with the total iridium loading derived from ultraviolet–visible measurements demonstrates that >90% of iridium on the electrode is electroactive, as is expected for a molecular monolayer[55].

Previously, we reported that the organometallic precursor complexes used to form the hom-WOCs by oxidative activation do not show any activity for electrode-driven water oxidation[41]. We also find that they do not self-adhere to oxide surfaces. Comparing both the activated forms, the kinetics of water oxidation appear to be different between the hom-WOC and the het-WOC: the H/D kinetic isotope effect (KIE) of 2.01 for the hom-WOC differs significantly from the KIE of 1.0 found for the het-WOC, when run at potentials below the appearance of mass-transport related effects at the electrode surface (Supplementary Fig. 17). This suggests that different rate-limiting steps are applied to each. The KIE of unity for the het-WOC may indicate that the rate-determining step is electron transfer from the Ir centres in the catalyst to the metal oxide scaffold, rather than any step involving water. KIEs that are close to 1 for similar reasons have been seen for iridium oxide colloids and related materials[36,56].

In comparison with bulk IrO$_x$ species, one advantage of using single-site surface-bound molecular catalysts for water oxidation is accurate control of electrode overpotential by tuning the scaffold surface area, thereby changing catalyst loading. By increasing the nanoporous film's thickness, the overpotential of the electrode at specific current densities can be decreased (Fig. 4). For example, typical 3-µm-thick *nano*ITO films require an overpotential of 275 mV to attain a catalytic current of 0.5 mA cm^{-2}, whereas 18-µm-thick films require < 160 mV. Although limited tunability of the number of active sites in bulk or nanostructured WOCs prevents direct comparison, we are not aware of any lower overpotential values reported in the literature

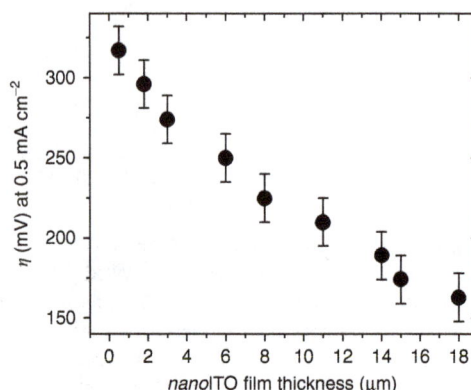

Figure 4 | Electrode overpotential as a function of nanoporous film thickness. The film thickness is directly proportional to catalyst loading. As we increase the catalyst loading on the electrode, the overpotential required to reach 0.5 mA cm^{-2} decreases. Error bars reported represent the largest s.d. (sample size of five electrodes) across the nine film thicknesses measured. Data were gathered using 5-min dwell times to allow the electrodes to adequately stabilize.

for this current density. Standardized benchmarking experiments comparing the het-WOC to IrO$_x$ show that the het-WOC possesses a lower overpotential for water oxidation in all cases (Supplementary Fig. 18)[57]. However, any comparison between single-site molecular species and bulk heterogeneous catalysts is complicated because of the difficulty of determining the turnover rates per metal atom in bulk materials needed to accurately gauge the relative activity on a fair basis.

To further investigate the het-WOC mechanistically, Tafel plots of catalytic currents were made over a range of pH and buffer conditions (Fig. 5a) and with electrodes of varying thickness of the porous *nano*ITO film to increase catalyst loading by increasing the electrode surface area (Fig. 5b). Limitations on proton diffusion through the nanoporous films on electrodes[58,59] cause a decrease in measured activity due to the low buffering capacity of KNO$_3$ at pH 7. As thicker *nano*ITO films are used, the pH gradient formed through the film decreases electrode performance over time regardless of the catalyst's stability, because the locally generated highly acidic conditions etch the ITO support[60]. The use of a buffer may inhibit this effect, and in the presence of phosphate the het-WOC behaves in a manner identical to IrO$_x$ materials[61]. Along with similarities in their pH dependence (Supplementary Fig. 19), CVs and spectro-electrochemical measurements, these results support the

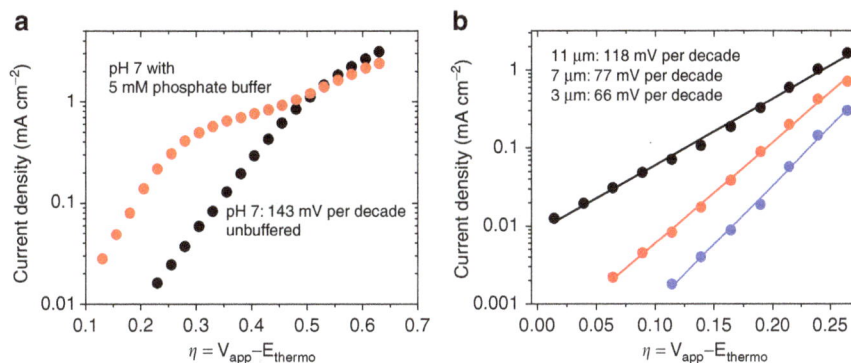

Figure 5 | Tafel plots for the het-WOC on *nano*ITO. (**a**) Tafel plots at pH 7 showing the effect of adding a phosphate buffer to the 0.1 M KNO₃ solution. (**b**) Tafel plots at pH 2.6 without any added buffer. A decrease in Tafel slope ($\Delta\eta\,/\Delta\log(i)$) with decreasing film thickness corresponds to a decrease in diffusion-related pH effects.

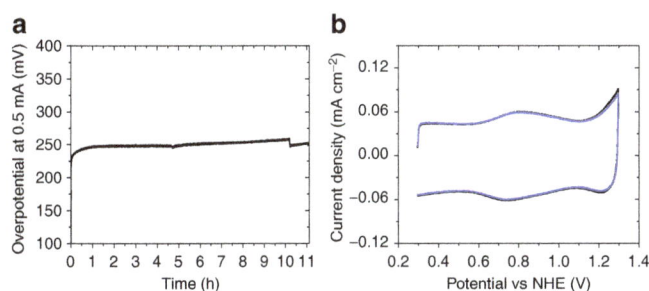

Figure 6 | Stability for electrochemical water oxidation. (**a**) Chronopotentiometry showing stability of the het-WOC during sustained water oxidation for over 11 h (pH 2.6). Small increases and decreases correspond to oxygen bubble build-up and release on the surface, which was minimized by rapidly stirring the solution. (**b**) CVs (pH 2.6, 10 mV s⁻¹ scan rate) after 1 h of electrolysis (black) and after >12 h of electrolysis (blue) using electrodes thin enough to mitigate local pH effects, showing full preservation of the electrode characteristics after sustained use.

hypothesis that the active catalytic sites in iridium oxides such as BL are mechanistically similar to those in the het-WOC. At a pH of 2.6, where pH is less sensitive to proton production from water oxidation, Tafel slopes increase as *nano*ITO film thickness is increased since the protons generated from water oxidation must diffuse through a thicker film. Remarkably, however, as seen in Fig. 5b, 11-μm-thick samples have high enough catalyst loading to allow sustained current densities with an onset of linearity in the Tafel plot beginning at overpotentials as low as 14 mV (where the current density is 11 μA cm⁻²). This is consistent with the onset of water oxidation (E_{cat}) for this catalyst being at nearly the thermodynamic potential, though the current is below the threshold of 0.5 mA cm⁻² required for practical use.

The het-WOC also shows excellent stability, and is capable of sustaining water oxidation for many hours at a 250 mV overpotential without degradation (Fig. 6a, Supplementary Figs 20–22), reaching turnover numbers in excess of 10⁶ O₂ evolved per iridium atom over multiple trials, as calculated by measuring the current passed through the electrode assuming a four-electron oxidative process. CVs of the electrode during and after these stability tests confirm that there is minimal loss in catalyst on the electrode surface (Fig. 6b). Moving to higher applied potentials (+520 mV relative to thermodynamic) we measure a turnover frequency of 7.9 s⁻¹ O₂ molecules evolved per electroactive iridium atom, which is one of the highest values reported to date. In addition, a 99% Faradaic yield is measured for O₂ evolution over 2 h using a phase fluorometric oxygen sensor (Supplementary Figs 23 and 24). To compare this with

benchmark iridium oxide nanomaterials reported in the literature, films of 2 nm IrOₓ clusters with comparable electroactivity require an overpotential of 680 mV to achieve a turnover frequency of 6.0 s⁻¹, while larger 60–100 nm IrO₂ nanoparticles having 16% electroactivity require 580 mV to achieve a rate of 6.6 s⁻¹ O₂ molecules evolved per electroactive iridium atom[53,62]. The observed high performance and atomic efficiency further distinguishes this molecular iridium WOC from traditional bulk iridium oxides.

To further probe the molecular nature of the het-WOC, we compare it with iridium oxide-based materials formed by heating an as-prepared electrode to 500 or 700 °C for 1 h (Fig. 7a). Scanning TEM analysis coupled with EDX mapping (STEM-EDX) displays the nanoscale coverage of iridium on ITO nanoparticles. As deposited, there is a highly conformal coating of iridium on each particle, consistent with a surface-bound molecular monolayer (Fig. 7b, Supplementary Fig. 25). The corresponding CV has a catalytic wave for water oxidation beginning at 1.1 V versus NHE at pH 2.6 (1.25 V versus RHE). Heating an electrode covered with the molecular catalyst at 500 °C in air burns off the pyalc ligand without affecting conformal coating of iridium oxide around each ITO nanoparticle (Fig. 7c, Supplementary Figs 26 and 27). This causes an anodic shift in the catalytic wave for water oxidation to 1.3 V versus NHE, revealing a feature at 1.1 V versus NHE typically assigned to the Ir^{IV}/Ir^{V} redox couple[53,54]; in the unheated sample, the catalytic wave for water oxidation obscures this feature. Heating an electrode coated with the molecular catalyst to 700 °C in air results in the formation of crystalline rutile IrO₂ clusters with ~20 nm diameter (Fig. 7d, Supplementary Fig. 28). In accordance with literature precedent[63], these show even lower activity for water oxidation, in part because most of the iridium is no longer in contact with water, reducing the number of active surface sites.

Molecular nature of the het-WOC. Aside from the monolayer distribution of active iridium, the oxidatively resistant bidentate pyalc ligand bound to iridium in the het-WOC represents another striking difference between single-site molecular catalysts such as this and traditional iridium oxide materials. Therefore, it is important to show that the ligand remains after extended periods of electrolysis, demonstrating that the het-WOC is stable. XPS, thus, serves to confirm the molecular structure and stability of the catalyst on the electrode, showing that the pyalc ligand is still present after ~16 h of electrolysis corresponding to >100,000 turnovers of O₂ per iridium atom (Fig. 8, Supplementary Fig. 29, Supplementary Table 1). From these data, we can also conclude that the resting state for the catalyst on the

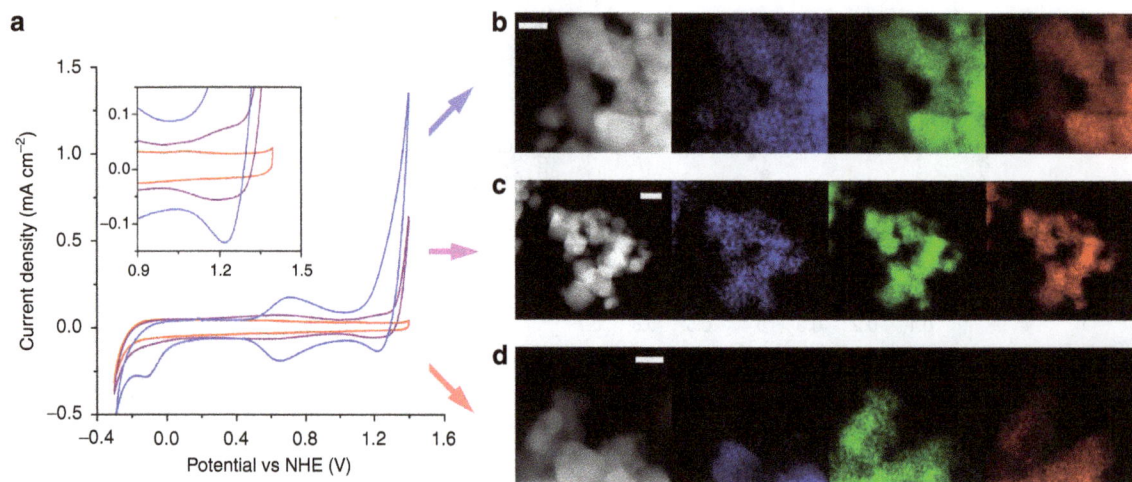

Figure 7 | Comparison of iridium oxide catalysts on *nano*ITO films. (a) CVs taken with a 10 mV s^{-1} scan rate at pH 2.6 of a catalyst-coated *nano*ITO electrode as-prepared (blue), heated to 500 °C (purple) and heated to 700 °C (red). Inset is the 0.5–0.9 V versus NHE region expanded to more easily compare the features in the purple and red traces. **(b)** STEM-EDX maps of electrode material as-prepared: high-angle annular dark field image (grey), iridium detected shown in blue, indium in green and tin in red. **(c)** STEM-EDX map after heating to 500 °C showing iridium remains coated around the ITO nanoparticle scaffold after the pyalc ligand is burned off (X-ray photoelectron spectra shown in Supplementary Fig. 27). **(d)** Corresponding maps after heating to 700 °C, the iridium is now localized in specific regions indicating nanoparticle formation (scale bars, 20 nm).

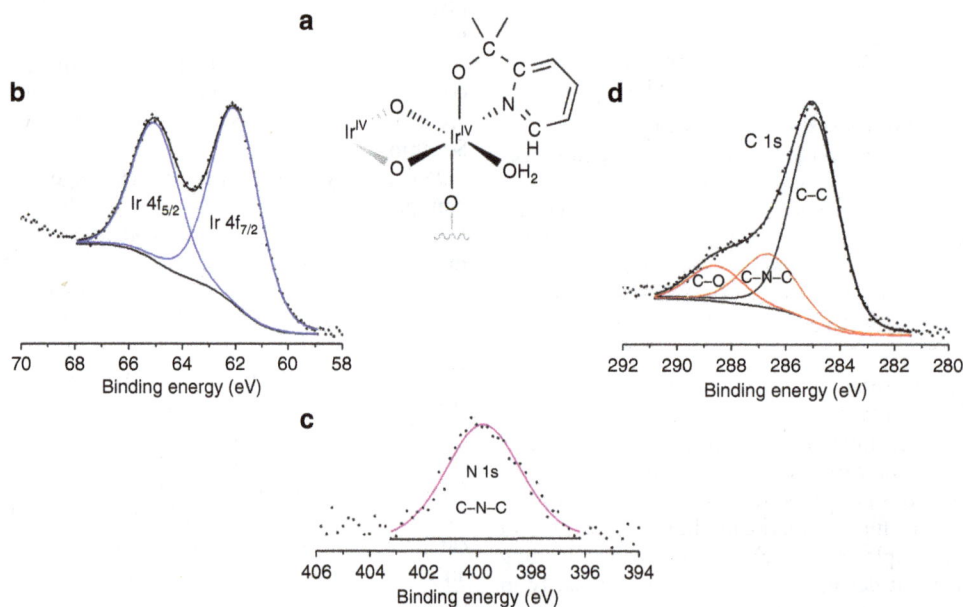

Figure 8 | X-ray photoelectron spectra taken after 16 h of water-oxidation catalysis. Representative colour-coded schematic **(a)** showing elements present. Signals corresponding to the iridium **(b)**, nitrogen **(c)** and carbon **(d)** show that both the metal and ligand are still intact after prolonged electrolysis. No changes are observed on the same film before and after electrolysis.

electrode surface involves Ir^{IV} with a 1:1 ratio of pyalc to Ir as found previously for the hom-WOC in solution[34].

Molecular catalysts are tunable by the variation of their ligands, and the het-WOC behaves in the same way. By changing the bidentate chelating ligand in the hom-WOC before deposition on the electrode surface, the properties of the het-WOC-functionalized electrode can be drastically changed. If, for example, [Cp*Ir(bpy)OH]BF$_4$ bearing the 2,2'-bipyridine (bpy) ligand instead of pyalc is used as a precatalyst to form the active catalyst in solution[34], a semitransparent yellow iridium compound (Supplementary Fig. 30) is deposited, having electrochemical properties that are very distinct from those of the pyalc-derived het-WOC (Supplementary Fig. 31), with much lower oxygen-evolution activity being observed (Supplementary Fig. 32). We find similar differences in activity in both electrochemically and

chemically driven oxygen evolution for these two hom-WOCs in solution[41]. This shows that our heterogenization strategy preserves the ligand-based tunability of the hom-WOC.

Discussion

These results provide a framework for assembling surface-bound molecular catalysts with a variety of direct-binding schemes[64,65]. The mild conditions of deposition are particularly promising with regards to nanostructured electrode materials that are difficult to functionalize by other means[66]. The higher activity and stability of the described materials over previously reported heterogenized molecular WOCs[67,68] shows that direct surface binding is a valid approach to attaching WOCs to electrodes. The present system also outperforms heterogeneous IrO$_x$ materials such as BL as a

WOC, although they do share similar characteristics, and our further studies will determine how this molecular species mechanistically relates to the active sites in bulk iridium oxides[38,69]. The het-WOC requires minimal iridium to efficiently oxidize water relative to IrO_x catalysts, and although water-oxidation catalysis on a global scale as required for solar fuel production will likely require the use of catalysts made from more abundant elements, it is still advantageous to develop WOCs based on rare but highly active metals[70]. Nevertheless, this demonstration of a robust and highly efficient iridium-based molecular heterogeneous catalyst provides a new architecture for molecular heterogeneous catalysts and opens up this field to develop WOCs made from abundant materials using similar design principles.

Methods

General procedures. All the chemicals were purchased from major suppliers and used as received. Synthesis of the precatalyst [(η^5-pentamethylcyclopentadienyl)Ir^{III}(2-(2'pyridyl)-2-propanolate-$\kappa O,\kappa N$)OH] was performed using a published procedure[39] and its activated form, the proposed [Ir(pyalc)(H$_2$O)$_2$(μ-O)]$_2^{2+}$ compound, was synthesized by the oxidation of the precatalyst with 100 equivalents of NaIO$_4$ (Acros Organics, 99%) following previously published methods[34].

Electrode preparation. *Nano*ITO electrodes were prepared by spin coating (Headway PWM32 Spin Coater, Headway Research Inc.) a solution of ITO nanoparticles (Sigma-Aldrich, < 50 nm particle size) suspended in a 5 M acetic acid/ethanol solution onto a 2.2 mm-thick FTO-coated glass slide (FTO, TEC 7, Hartford Glass Co. Inc.), followed by heating at 500 °C in air for 1 h, cooling to room temperature, then heating to 300 °C in a 3% H$_2$/N$_2$ atmosphere for 1 h and cooling back to room temperature. They were immersed in a catalyst solution formed by oxidizing 10 mM [Cp*Ir(pyalc)OH] in 30 ml deionized water with 100 equivalents of NaIO$_4$. The electrodes were removed after 2 h, washed thoroughly with deionized water, and had acquired a visible blue colour. Side-by-side controls were used to measure the relative decrease in absorption of the catalyst solution, to approximately determine the amount of iridium that had adhered to the surface of the electrode. *Nano*ITO film thicknesses were measured using a profilometer (KLA Tencor Alphastep 200). Additional preparatory procedures for electrodes made from TiO$_2$ and WO$_3$, as well as the procedure for determining catalyst loading are detailed in the Supplementary Methods.

Ultraviolet–visible spectroscopy. For optical studies including the data gathered in Fig. 2a,b,d, a Varian Cary 3 spectrophotometer with an integrating sphere attachment in absorption mode was used. A 6.45 cm^2 geometric surface area of *nano*ITO on FTO-coated glass was used to completely cover the aperture of the integrating sphere. A background spectrum was taken before immersion of the substrate into the hom-WOC solution. Ultraviolet–visible spectra for IrO_x on *nano*ITO were taken by electrodepositing IrO_x from a solution containing [Cp*Ir(H$_2$O)$_3$]SO$_4$ using the conditions outlined in ref. 38 with 50 deposition cycles at 50 mV s^{-1} scan rate. For the data gathered in Fig. 2e, an electrochemical cell as described previously was assembled in a 1 cm^2 quartz ultraviolet–visible cuvette attached to an integrating sphere in a Shimadzu UV-2600 spectrophotometer. Standard electrolyte conditions of 0.1 M KNO$_3$ adjusted to pH 2.6 were used. Catalyst-coated *nano*ITO electrodes that were 6.45 cm long and 0.7 cm wide were prepared by cutting an FTO-coated glass slide with a 7 μm thick catalyst-coated *nano*ITO film to the appropriate dimensions. Working electrodes were constructed by attaching a copper wire to an exposed FTO surface on one side of the *nano*ITO-coated FTO slide using conductive epoxy (Fast Setting Conductive Silver Epoxy, SPI). Six hours were allotted for the conductive epoxy to cure, then non-conductive water-resistant marine epoxy (White Marine Epoxy, Loctite, 24 h allotted to cure) was applied on top of the conductive epoxy to prevent electrical contact between the wire leads and electrolyte, so that the only conductive component exposed was the catalyst-coated *nano*ITO. This was placed into the quartz cuvette along with Pt mesh counter and Ag/AgCl reference (Bioanalytical Systems, Inc.) electrodes adjacent to the integrating sphere, connected to a potentiostat (WavenowXV, Pine), and chronoamperometric experiments with the potentials detailed in Fig. 2e were performed. The electrodes were given 3 min to stabilize at that potential, then an ultraviolet–visible spectrum was taken using the integrating sphere in absorption mode. A blank *nano*ITO-coated FTO slide without catalyst was used for a background scan.

Chemical oxidation. Oxygen was detected with a Clark electrode using a custom-made zero-headspace 10 ml glass cell, water jacketed for constant temperature. A Teflon cap through which the Clark electrode membrane contacted deionized water (adjusted to pH 2.5 using nitric acid) also held a catalyst-coated *nano*ITO

film on FTO-coated glass sample that was submerged in the cell. Catalyst-loaded samples were prepared similarly to those used for electrochemistry, except with a 1.6 cm^2 geometric surface area. The Clark electrode was allowed to stabilize while stirring the deionized water solution for 1 h before injection with an oxidizing solution of freshly prepared NaIO$_4$ in deionized water. Catalyst response and O$_2$ generation occurred immediately, without any lag phase. A typical experiment, such as is shown in Fig. 3a, used 25 μl of 0.25 M NaIO$_4$ in deionized water. The oxygen content in the cell was monitored until oxygen evolution ceased, which for a loading of around 50 nmol of iridium (corresponding to a *nano*ITO film 11-μm thick) took ~ 90 min. Data were collected while stirring the solution to ensure steady state oxygen readings.

Electrochemical characterization. Electrolyte pHs were adjusted using 1 M HNO$_3$ or 1 M KOH. Electrochemical data were taken using 0.1 M KNO$_3$ in deionized water as an electrolyte, adjusted to pH 2.6 unless stated otherwise, with an Ag/AgCl reference electrode (Bioanalytical Systems Inc.) and Pt mesh counter electrode. Measurements were taken using a Princeton Applied Research Versastat 4–400 potentiostat in a standard three-electrode configuration. Vigorous stirring was required in unbuffered solutions during long-term experiments to prevent etching of the *nano*ITO electrode under acidic conditions. Long-term stability testing and oxygen detection using phase fluorometry were performed in a two-chamber electrochemical cell, with working and counter electrode chambers separated by a glass frit. For these experiments, such as those shown in Fig. 6, low catalyst loading was achieved with 300–500 nm thickness *nano*ITO films on a 6.45 cm^2 substrate, thereby minimizing local pH effects due to the low buffer capacity of KNO$_3$ over the pH range studied. SEM and TEM data were taken both before and after electrolysis to determine that there were no changes to sample morphology and CVs were taken to ensure the minimal loss of electroactive catalyst over the course of an experiment. Further details on electrochemical methods and additional controls are included in the Supporting Information (Supplementary Figs 33–35).

Electron microscopy. SEM images and SEM-EDX data were taken on a Hitachi SU-70 Analytical Scanning Electron Microscope. Images of the samples were taken both before and after electrolysis. TEM images, EDX data, high-angle annular dark field images and STEM-EDX maps were taken using a FEI Tecnai Osiris TEM operating at 200 kV. Samples were prepared by scraping *nano*ITO off of an electrode into deionized water, then suspending them onto a silicon monoxide coated TEM grid (Ted Pella).

X-ray photoelectron spectroscopy. X-ray photoelectron spectra were collected using an Al anode ($hv = 1486.6$ eV) and a double-pass cylinder mirror analyzer (PHI 15- 255G). Geometric surface area 6.45 cm^2 samples with a ~ 400 nm thick *nano*ITO film on the surface of FTO-coated glass were used for XPS studies. All experiments used a pass energy of 35.75 eV. Spectra were calibrated to an Au standard, and peak fits were performed using XPSPeak (version 4.1). Additional information regarding peak fitting and experimental details for Supplementary Figures can be found in the Supplementary Methods.

References

1. Chu, S. & Majumdar, A. Opportunities and challenges for a sustainable energy future. *Nature* **488**, 294–303 (2012).
2. Lewis, N. S. & Nocera, D. G. Powering the planet: chemical challenges in solar energy utilization. *Proc. Natl Acad. Sci. USA* **103**, 15729–15735 (2006).
3. Faunce, T. *et al.* Artificial photosynthesis as a frontier technology for energy sustainability. *Energy Environ. Sci.* **6**, 1074–1076 (2013).
4. Blankenship, R. E. *et al.* Comparing photosynthetic and photovoltaic efficiencies and recognizing the potential for improvement. *Science* **332**, 805–809 (2011).
5. Llobet, A. *Molecular Water Oxidation Catalysis: A Key Topic for New Sustainable Energy Conversion Schemes* (Wiley, 2014).
6. Ruttinger, W. & Dismukes, G. C. Synthetic water-oxidation catalysts for artificial photosynthetic water oxidation. *Chem. Rev.* **97**, 1–24 (1997).
7. Young, K. J. *et al.* Light-driven water oxidation for solar fuels. *Coord. Chem. Rev.* **256**, 2503–2520 (2012).
8. Wasylenko, D. J., Palmer, R. D. & Berlinguette, C. P. Homogeneous water oxidation catalysts containing a single metal site. *Chem. Commun.* **49**, 218–227 (2013).
9. Cole-Hamilton, D. J. Homogeneous catalysis—new approaches to catalyst separation, recovery, and recycling. *Science* **299**, 1702–1706 (2003).
10. Evangelisti, F., Güttinger, R., Moré, R., Luber, S. & Patzke, G. R. Closer to Photosystem II: A Co$_4$O$_4$ Cubane Catalyst with Flexible Ligand Architecture. *J. Am. Chem. Soc.* **135**, 18734–18737 (2013).
11. Evangelisti, F., Car, P. E., Blacque, O. & Patzke, G. R. Photocatalytic water oxidation with cobalt-containing tungstobismutates: tuning the metal core. *Catal. Sci. Technol.* **3**, 3117–3129 (2013).
12. Eisenberg, R. & Gray, H. B. Preface on Making Oxygen. *Inorg. Chem.* **47**, 1697–1699 (2008).

13. Duan, L. L. *et al.* A molecular ruthenium catalyst with water-oxidation activity comparable to that of photosystem II. *Nat. Chem.* **4**, 418–423 (2012).

14. Fillol, J. L. *et al.* Efficient water oxidation catalysts based on readily available iron coordination complexes. *Nat. Chem.* **3**, 807–813 (2011).

15. Yin, Q. S. *et al.* A fast soluble carbon-free molecular water oxidation catalyst based on abundant metals. *Science* **328**, 342–345 (2010).

16. Lewandowska-Andralojc, A. *et al.* Efficient water oxidation with organometallic iridium complexes as precatalysts. *Phys. Chem. Chem. Phys.* **16**, 11976–11987 (2014).

17. Barnett, S. M., Goldberg, K. I. & Mayer, J. M. A soluble copper-bipyridine water-oxidation electrocatalyst. *Nat. Chem.* **4**, 498–502 (2012).

18. Dau, H. *et al.* The Mechanism of Water Oxidation: From Electrolysis via Homogeneous to Biological Catalysis. *ChemCatChem* **2**, 724–761 (2010).

19. Wegener, S. L., Marks, T. J. & Stair, P. C. Design strategies for the molecular level synthesis of supported catalysts. *Acc. Chem. Res.* **45**, 206–214 (2012).

20. Muratsugu, S. & Tada, M. Molecularly imprinted Ru complex catalysts integrated on oxide surfaces. *Acc. Chem. Res.* **46**, 300–311 (2013).

21. Comas-Vives, A. *et al.* Single-site homogeneous and heterogenized gold(III) hydrogenation catalysts: mechanistic implications. *J. Am. Chem. Soc.* **128**, 4756–4765 (2006).

22. Murray, R. W., Ewing, A. G. & Durst, R. A. Chemically modified electrodes— molecular design for electroanalysis. *Anal. Chem.* **59**, 379A–390A (1987).

23. Beck, J. S. *et al.* A new family of mesoporous molecular-sieves prepared with liquid-crystal templates. *J. Am. Chem. Soc.* **114**, 10834–10843 (1992).

24. Muresan, N. M., Willkomm, J., Mersch, D., Vaynzof, Y. & Reisner, E. Immobilization of a molecular cobaloxime catalyst for hydrogen evolution on a mesoporous metal oxide electrode. *Angew. Chem. Int. Ed.* **51**, 12749–12753 (2012).

25. Soriaga, M. P. Surface Coordination Chemistry of Monometallic and Bimetallic Electrocatalysts. *Chem. Rev.* **90**, 771–793 (1990).

26. Terry, T. J. & Stack, T. D. P. Covalent Heterogenization of a Discrete Mn(II) Bis-Phen Complex by a Metal-Template/Metal-Exchange Method: An Epoxidation Catalyst with Enhanced Reactivity. *J. Am. Chem. Soc.* **130**, 4945–4953 (2008).

27. Somorjai, G., Frei, H. & Park, J. Y. Advancing the Frontiers in Nanocatalysis, Biointerfaces, and Renewable Energy Conversion by Innovations of Surface Techniques. *J. Am. Chem. Soc.* **131**, 16589–16605 (2009).

28. Concepcion, J. J., Binstead, R. A., Alibabaei, L. & Meyer, T. J. Application of the rotating ring-disc-electrode technique to water oxidation by surface-bound molecular catalysts. *Inorg. Chem.* **52**, 10744–10746 (2013).

29. Zhong, D. K., Zhao, S. L., Polyansky, D. E. & Fujita, E. Diminished photoisomerization of active ruthenium water oxidation catalyst by anchoring to metal oxide electrodes. *J. Catal.* **307**, 140–147 (2013).

30. Klepser, B. M. & Bartlett, B. M. Anchoring a molecular iron catalyst to solar-responsive WO$_3$ improves the rate and selectivity of photoelectrochemical water oxidation. *J. Am. Chem. Soc.* **136**, 1694–1697 (2014).

31. Crabtree, R. H. The stability of organometallic ligands in oxidation catalysis. *J. Organomet. Chem.* **751**, 174–180 (2014).

32. Vannucci, A. K. *et al.* Crossing the divide between homogeneous and heterogeneous catalysis in water oxidation. *Proc. Natl Acad. Sci. USA* **110**, 20918–20922 (2013).

33. Thomas, J. M., Raja, R. & Lewis, D. W. Single-site heterogeneous catalysts. *Angew. Chem. Int. Ed.* **44**, 6456–6482 (2005).

34. Hintermair, U. *et al.* Precursor transformation during molecular oxidation catalysis with organometallic iridium complexes. *J. Am. Chem. Soc.* **135**, 10837–10851 (2013).

35. Hull, J. F. *et al.* Highly active and robust (Cp*) iridium complexes for catalytic water oxidation. *J. Am. Chem. Soc.* **131**, 8730–8731 (2009).

36. Blakemore, J. D. *et al.* Half-sandwich iridium complexes for homogeneous water-oxidation catalysis. *J. Am. Chem. Soc.* **132**, 16017–16029 (2010).

37. Schley, N. D. *et al.* Distinguishing homogeneous from heterogeneous catalysis in electrode-driven water oxidation with molecular iridium complexes. *J. Am. Chem. Soc.* **133**, 10473–10481 (2011).

38. Blakemore, J. D. *et al.* Anodic deposition of a robust iridium-based water-oxidation catalyst from organometallic precursors. *Chem. Sci.* **2**, 94–98 (2011).

39. Hintermair, U., Hashmi, S. M., Elimelech, M. & Crabtree, R. H. Particle formation during oxidation catalysis with Cp* iridium complexes. *J. Am. Chem. Soc.* **134**, 9785–9795 (2012).

40. Blakemore, J. D. *et al.* Comparison of amorphous iridium water-oxidation electrocatalysts prepared from soluble precursors. *Inorg. Chem.* **51**, 7749–7763 (2012).

41. Thomsen, J. M. *et al.* Electrochemical activation of Cp* iridium complexes for electrode-driven water-oxidation catalysis. *J. Am. Chem. Soc.* **136**, 13826–13834 (2014).

42. Ingram, A. J. *et al.* Modes of activation of organometallic iridium complexes for catalytic water and C-H oxidation. *Inorg. Chem.* **53**, 423–433 (2014).

43. Chen, Z. F., Concepcion, J. J., Hull, J. F., Hoertz, P. G. & Meyer, T. J. Catalytic water oxidation on derivatized *nano*ITO. *Dalton. Trans.* **39**, 6950–6952 (2010).

44. Rose, D., Lever, F. M., Powell, A. R. & Wilkinson, G. The Nature of iridium(IV) iodate. *J. Chem. Soc. A* **1969**, 1690–1691 (1969).

45. Brewster, T. P. *et al.* An iridium(IV) species, [Cp*Ir(NHC)Cl]$^+$, related to a water-oxidation catalyst. *Organometallics* **30**, 965–973 (2011).

46. Rubel, M. *et al.* Characterization of IrO$_2$-SnO$_2$ thin-layers by electron and ion spectroscopies. *Vacuum* **45**, 423–427 (1994).

47. Wang, C., Wang, J. L. & Lin, W. B. Elucidating molecular iridium water oxidation catalysts using metal-organic frameworks: a comprehensive structural, catalytic, spectroscopic, and kinetic study. *J. Am. Chem. Soc.* **134**, 19895–19908 (2012).

48. Kim, Y. I. & Hatfield, W. E. Electrical, magnetic and spectroscopic properties of tetrathiafulvalene charge-transfer compounds with iron, ruthenium, rhodium and iridium halides. *Inorg. Chim. Acta* **188**, 15–24 (1991).

49. Minguzzi, A. *et al.* Observing the oxidation state turnover in heterogeneous iridium-based water oxidation catalysts. *Chem. Sci.* **5**, 3591–3597 (2014).

50. Sanchez Casalongue, H. G. *et al.* In Situ Observation of Surface Species on Iridium Oxide Nanoparticles during the Oxygen Evolution Reaction. *Angew. Chem. Int. Ed.* **53**, 7169–7172 (2014).

51. Youngblood, W. J. *et al.* Photoassisted overall water splitting in a visible light-absorbing dye-sensitized photoelectrochemical cell. *J. Am. Chem. Soc.* **131**, 926–927 (2009).

52. Lin, Y. J. *et al.* Semiconductor nanostructure-based photoelectrochemical water splitting: a brief review. *Chem. Phys. Lett.* **507**, 209–215 (2011).

53. Nakagawa, T., Beasley, C. A. & Murray, R. W. Efficient electro-oxidation of water near its reversible potential by a mesoporous IrO$_x$ nanoparticle film. *J. Phys. Chem. C* **113**, 12958–12961 (2009).

54. Ouattara, L., Fierro, S., Frey, O., Koudelka, M. & Comninellis, C. Electrochemical comparison of IrO$_2$ prepared by anodic oxidation of pure iridium and IrO$_2$ prepared by thermal decomposition of H$_2$IrCl$_6$ precursor solution. *J. Appl. Electrochem.* **39**, 1361–1367 (2009).

55. Bard, A. J. & Faulkner, L. R. *Electrochemical Methods: Fundamentals and Applications* 2nd edn (Wiley, 2001).

56. Morris, N. D., Suzuki, M. & Mallouk, T. E. Kinetics of electron transfer and oxygen evolution in the reaction of [Ru(bpy)$_3$]$^{3+}$ with colloidal iridium oxide. *J. Phys. Chem. A* **108**, 9115–9119 (2004).

57. McCrory, C. C. L., Jung, S. H., Peters, J. C. & Jaramillo, T. F. Benchmarking heterogeneous electrocatalysts for the oxygen evolution reaction. *J. Am. Chem. Soc.* **135**, 16977–16987 (2013).

58. Bediako, D. K., Costentin, C., Jones, E. C., Nocera, D. G. & Saveant, J. M. Proton-electron transport and transfer in electrocatalytic films. application to a cobalt-based O$_2$-evolution catalyst. *J. Am. Chem. Soc.* **135**, 10492–10502 (2013).

59. Surendranath, Y., Kanan, M. W. & Nocera, D. G. Mechanistic studies of the oxygen evolution reaction by a cobalt-phosphate catalyst at neutral pH. *J. Am. Chem. Soc.* **132**, 16501–16509 (2010).

60. Scholten, M. & Vandenmeerakker, J. E. A. M. On the mechanism of ITO etching—the specificity of halogen acids. *J. Electrochem. Soc.* **140**, 471–475 (1993).

61. Kushner-Lenhoff, M. N., Blakemore, J. D., Schley, N. D., Crabtree, R. H. & Brudvig, G. W. Effects of aqueous buffers on electrocatalytic water oxidation with an iridium oxide material electrodeposited in thin layers from an organometallic precursor. *Dalton Trans.* **42**, 3617–3622 (2013).

62. Yagi, M., Tomita, E., Sakita, S., Kuwabara, T. & Nagai, K. Self-assembly of active IrO$_2$ colloid catalyst on an ITO electrode for efficient electrochemical water oxidation. *J. Phys. Chem. B* **109**, 21489–21491 (2005).

63. Hara, M., Waraksa, C. C., Lean, J. T., Lewis, B. A. & Mallouk, T. E. Photocatalytic water oxidation in a buffered tris(2,2'-bipyridyl)ruthenium complex-colloidal IrO$_2$ system. *J. Phys. Chem. A* **104**, 5275–5280 (2000).

64. Ahn, H. S., Yano, J. & Tilley, T. D. Photocatalytic water oxidation by very small cobalt domains on a silica surface. *Energy Environ. Sci.* **6**, 3080–3087 (2013).

65. Coperet, C., Chabanas, M., Saint-Arroman, R. P. & Basset, J. M. Homogeneous and heterogeneous catalysis: bridging the gap through surface organometallic chemistry. *Angew. Chem. Int. Ed.* **42**, 156–181 (2003).

66. Liu, J. *et al.* Oriented Nanostructures for Energy Conversion and Storage. *ChemSusChem* **1**, 676–697 (2008).

67. Chen, Z. F. *et al.* Nonaqueous catalytic water oxidation. *J. Am. Chem. Soc.* **132**, 17670–17673 (2010).

68. deKrafft, K. E. *et al.* Electrochemical water oxidation with carbon-grafted iridium complexes. *ACS Appl. Mater. Interfaces* **4**, 608–613 (2012).

69. Smith, R. D. L., Sporinova, B., Fagan, R. D., Trudel, S. & Berlinguette, C. P. Facile photochemical preparation of amorphous iridium oxide films for water oxidation catalysis. *Chem. Mater.* **26**, 1654–1659 (2014).

70. McFarland, E. W. Solar energy: setting the economic bar from the top-down. *Energy Environ. Sci.* **7**, 846–854 (2014).

Acknowledgements

This work was partially supported by the National Science Foundation (NSF MRSEC DMR 1119826), an NSF Graduate Research Fellowship (S.W.S.), the U.S. Department of Energy (DE-FG02-07ER15909, S.W.S. and C.A.S.) and both the Center for Catalytic Hydrocarbon Functionalization (CCHF) as well as the Argonne-Northwestern Solar Energy Research

(ANSER) Center, Energy Frontier Research Centers funded by the U.S. Department of Energy, Numbers DE-SC0001298 (R.H.C.) and DE-SC0001059 (J.M.T., R.H.C. and G.W.B.), respectively. U.H. acknowledges the Alexander von Humboldt Foundation for a Feodor Lynen Research Fellowship at Yale and the Centre for Sustainable Chemical Technologies at the University of Bath (EPSRC grant no. EP/G03768X/1) for a Whorrod Research Fellowship. Facilities use was supported by the Yale Institute for Nanoscience and Quantum Engineering (YINQE) and Yale's Center for Research on Interface Structures and Phenomena (CRISP). We thank Dr Lior Kornblum, Marvin Wint, Dr Charles Ahn and Dr Fred Walker for access to XPS and assistance with data collection and analysis. Dr Jesús Campos, Jeffrey Chen, TengHooi Goh and Dr Andre Taylor are acknowledged for their assistance with electrode preparation, Dr Aaron Bloomfield for experimental assistance, Dr Christopher Koenigsmann for comments on electrochemistry and Dr David Vinyard for assistance with electron paramagnetic resonance experiments.

Author contributions

S.W.S. developed the concept of heterogenizing via direct binding the molecular homogeneous WOCs that were discovered previously by U.H., R.H.C. and G.W.B. S.W.S., J.M.T. and U.H. designed the experiments, S.W.S. carried out the electrode preparation, and S.W.S. and J.M.T. performed the experiments. S.W.S. wrote the manuscript and supporting information with contributions from J.M.T. U.H., R.H.C., G.W.B. and C.A.S. revised and edited the manuscript before submission. U.H., R.H.C., G.W.B. and C.A.S. supervised the work.

Additional information

Competing financial interests: US Patent Application Number 14/317,906, a patent application by S.W.S., U.H., J.M.T., G.W.B. and R.H.C, was previously filed for the intellectual property described in this article. C.A.S. declares no competing financial interest.

2

Electrochemical synthesis of mesoporous gold films toward mesospace-stimulated optical properties

Cuiling Li[1], Ömer Dag[2], Thang Duy Dao[1,3,4], Tadaaki Nagao[1,3], Yasuhiro Sakamoto[3,5], Tatsuo Kimura[6], Osamu Terasaki[7,8] & Yusuke Yamauchi[1,9]

Mesoporous gold (Au) films with tunable pores are expected to provide fascinating optical properties stimulated by the mesospaces, but they have not been realized yet because of the difficulty of controlling the Au crystal growth. Here, we report a reliable soft-templating method to fabricate mesoporous Au films using stable micelles of diblock copolymers, with electrochemical deposition advantageous for precise control of Au crystal growth. Strong field enhancement takes place around the center of the uniform mesopores as well as on the walls between the pores, leading to the enhanced light scattering as well as surface-enhanced Raman scattering (SERS), which is understandable, for example, from Babinet principles applied for the reverse system of nanoparticle ensembles.

[1] World Premier International (WPI) Research Center for Materials Nanoarchitectonics (MANA), National Institute for Materials Science (NIMS), 1-1 Namiki, Tsukuba, Ibaraki 305-0044, Japan. [2] Department of Chemistry, Bilkent University, 06800 Ankara, Turkey. [3] PRESTO and CREST, Japan Science and Technology Agency (JST), 4-1-8 Honcho, Kawaguchi, Saitama 332-0012, Japan. [4] Graduate School of Materials Science, Nara Institute of Science and Technology, 8916-5 Takayama, Ikoma, Nara 630-0192, Japan. [5] Department of Physics, Graduate School of Science, Osaka University, 1-1 Machikaneyama-cho, Toyonaka, Osaka 560-0043, Japan. [6] Advanced Manufacturing Research Institute, National Institute of Advanced Industrial Science and Technology (AIST), Shimoshidami, Moriyama, Nagoya 463-8560, Japan. [7] Graduate School of EEWS (BK21Plus), KAIST, Daejeon 305-701, Korea. [8] Department of Materials and Environmental Chemistry, EXSELENT, Stockholm University, 10691 Stockholm, Sweden. [9] Department of Nanoscience and Nanoengineering, Faculty of Science and Engineering, Waseda University, 3-4-1 Okubo, Shinjuku, Tokyo 169-8555, Japan. Correspondence and requests for materials should be addressed to Y.Y. (email: Yamauchi.Yusuke@nims.go.jp).

Nanostructural control is a quite important issue in materials chemistry to bring out unique chemical and physical properties. Mesoporous structures can steadily provide surfaces with many functional sites, which are critical for solving emergent problems. So far, many mesoporous materials with different compositions have been reported through self-assembly of amphiphilic organic molecules and keen interest has been shown because of their wide range of potential applications, including gas storage, separation, catalysis, ion-exchange, sensing, polymerization and drug delivery[1–3]. Especially, metallic mesoporous materials can exhibit rather high carrier density and thus remarkable optical response, which are not attainable by using other compositions of mesoporous materials (for example, silica-[4,5] and carbon-based[6,7] compositions). For example, free space radiation wavelength of the light in the near-infrared (NIR) to visible region are shrunk down by one to two orders of magnitude when they interact with the plasma oscillation of the free electrons at the metal surface[8,9]. The excited surface plasma oscillation, or surface plasmon, subsequently leads to the remarkable property that light can be manipulated flexibly by controlling the surface morphology of metals on the nanometer scale[10–13]. Considering working mechanisms of metal nanoparticles that have often been limited by their tendency to sinter to large-sized aggregates, three-dimensionally (3D) extended metal framework will steadily provide abundant reaction/adsorption sites, which are critical for designing emergently required functions.

Nanostructured gold (Au) materials in the forms of films and particles have not only shown high catalytic activities in several oxidation reactions (for example, CO, glucose, methanol) and O_2 reduction reaction[14–18] but also demonstrated unique optical properties due to their localized surface plasmon resonance (LSPR)[19–21]. Although various mesoporous metals and alloys have been prepared by soft- and hard-templating approaches[22–26], Au thin films with uniform mesopores have not yet been obtained by such approaches, because of the difficulty in controlling the crystal growth of Au around the templates. In almost all the cases using hard-templates, nanostructured Au materials have only been obtained as nanoparticles, nanosheets and nanowires, without well-defined porous structures[27,28]. Although colloidal crystal templating is one of the most common techniques to form 3D ordered macroporous materials[29–32], the use of small-sized particles is failed for templating of Au materials with periodic mesoporous structures (for example, inverse opal similar to 3D hexagonal structure)[33]. As far as we know, only one pathway to prepare mesoporous/nanoporous Au materials has been reported by dealloying treatment[34–36], which is a selective dissolution process of less-noble metals in alloys[37,38]. However, this dealloying approach provides limited control over structural parameters and requires multistep processes under stringent experimental conditions, and nevertheless the resulting pores are irregular in size and shape. Thus, it is really difficult to shape Au crystals into mesoporous architectures.

A mesoporous Au material with tunable and uniform pores can newly be proposed as a reversed system of Au nanocrystals ensemble. Therefore, if we consider Babinet principle in optics, it is expected to show high-performance in light scattering as well as in surface-enhanced Raman scattering (SERS) for molecule detection due to multiple 'hot spots' built up in the mesopores as well as in the vicinity of narrow walls between the pores[39].

Here, we report an effective way to prepare mesoporous Au films with fine tuning of the pore size by utilizing spherical micelles of polystyrene-*block*-poly(oxyethylene) (PS-*b*-PEO) diblock copolymers as soft-templates (Fig. 1a). In the electrolyte solution, $HAuCl_4$ is dissolved into H_3O^+ and $AuCl_4^-$ ions and

then interacts with the EO shells of the micelles through hydrogen bonding. This interaction favours H_3O^+ rather than $AuCl_4^-$, and consequently creates positively charged micelles that can be directed to the working electrode surfaces, where the $AuCl_4^-$ ions are, respectively, reduced to metallic Au with the electrochemical deposition of the micelles. The resultant mesoporous Au films actually exhibit high scattering performance and thus high activity for molecular sensing such as seen in SERS[40] and surface-enhanced infrared absorption (SEIRA)[41]. Significantly, enhanced electric field (E-field) amplitude by excitation of LSPR is clearly seen inside or at the perimeter of the mesopores. The E-field amplitude and LSPR frequency are readily tunable by simply tuning the pore size and we could demonstrate the new methodology for tailoring the optical functionality.

Results

Fabrication of mesoporous Au films. Our approach is based on an electrochemical method, which has been practically utilized for preparation of continuous metallic films (Fig. 1a). For plating Au materials in the presence of PS-*b*-PEO micelles at the electrode surface, precursor solutions were prepared as follows. PS-*b*-PEO diblock copolymer was completely dissolved in tetrahydrofuran (THF), being a good solvent for the PS block, as its unimer. No micelles and/or aggregates were detected in the clear PS-*b*-PEO/THF solution by dynamic light scattering. The use of THF was necessary for the preparation; PS-*b*-PEO cannot be dissolved in water or ethanol without THF. After successive addition of ethanol, aqueous solution of $HAuCl_4$ was slowly dropped to the solution and consequently spherical micelles of PS-*b*-PEO was formed on the basis of less solubility of the PS core in water. As shown in Fig. 1b,c, spherical micelles constructed by the PS cores and the PEO shells were observed by transmission electron microscopy (TEM) and their average diameter was ~25 nm. The Tyndall effect was also observed in the electrolyte (Fig. 1b,c). As illustrated in Fig. 1a, the hydrophilic EO blocks can interact with aqua-$HAuCl_4$ complexes near the outer layer of the micelles, which will be discussed later. To directly confirm the presence of PS-*b*-PEO micelles in the electrolyte solution, we visualized uniformly sized polymeric micelles with the Brownian motion by a confocal laser scanning microscope. By applying potentials, the composite micelles that interacted with Au species were approached to deposit as mesostructured Au materials over the working electrode.

Block copolymers including PS-*b*-PEO and poly(ethylene-*co*-butylene)-*b*-PEO (KLE) have been often utilized as templates to synthesize mesoporous metal oxides on the basis of the sol–gel reactions[42–44]. In these cases, although organization of micelle hybrids (formed through the interaction between the inorganic precursors and the hydrophilic regions) at mesoscale is demonstrated[42–45], the solvent evaporation primarily induces the pre-formed ordered mesophases, where the formation of inorganic frameworks are guided steadily. Our approach is based on 'micelle assembly' induced by electromotive force, in which both the metal deposition (that is, the framework construction) and the micelle assembly occur simultaneously, because the electrochemical deposition process of spherical composite micelles is completely proceeded in liquid phases without evaporation of the solvents. Thus, our approach is conceptually different from the previous works on the fabrication of mesoporous films based on the evaporation-induced self-assembly approach. To confirm the role of micelles more clearly, the mesostructure evolution was investigated by scanning electron microscope (SEM) at the initial stages (that is, 10, 30 and 100 s). After 10 s, a large amount of initial Au seeds

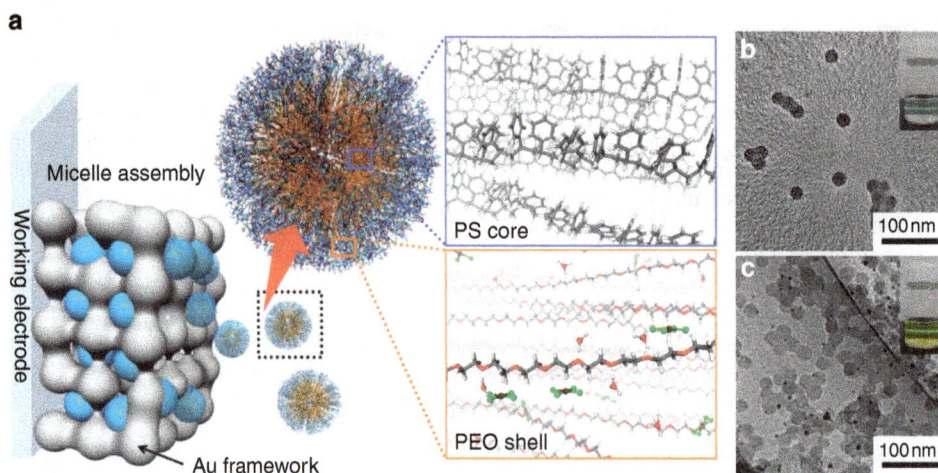

Figure 1 | Synthetic concept of mesoporous Au films. (a) Schematic illustration of formation mechanism of mesoporous Au films through micelle assembly. **(b,c)** TEM images of PS_{18000}-b-PEO_{7500} micelles formed in aqueous solution (with 3 ml THF) **(b)** without and **(c)** with $HAuCl_4$ source. Black dots indicate Au nanoparticles formed on the surface of micelles by irradiation of electron beam. The Tyndall effect is also shown as an inset image.

were formed, as indicated by arrows in Supplementary Fig. 1a. The curved surfaces of the Au branches come from the micelle-directing effect, as indicated by the circle in Supplementary Fig. 1a). When a deposition time was further increased to 30 s, a larger number of mesopores were observed (Supplementary Fig. 1b), though the formation of each mesopore was not perfectly completed. When the deposition proceeded for 100 s, the estimated film thickness reached almost half of the one mesopore size and mesoporous Au framework was well constructed (Supplementary Fig. 1c).

Characterization of mesoporous Au films. After complete removal of the PS-b-PEO micelles, mesoporous Au films were obtained and characterized carefully. Cage-like uniform mesopores were observed in the entire area of the film and connected with one another (Fig. 2a,b). The SEM image of a top surface of the Au film exhibited the presence of uniformly sized mesopores (average 25 ± 5 nm) with wall thickness of 25 nm ± 5 nm (Fig. 2a,b and Supplementary Fig. 2c). The resultant pore size of the Au film (Fig. 2a,b) was almost the same as the size of spherical micelles observed by TEM (Fig. 1b,c) and SEM (Supplementary Fig. 3), meaning that the spherical micelles serve as the porogens. In addition, pore size in Au film was controllable very easily by changing the amount of THF, herein from 1 ml to 2 ml and 3 ml, which is the most important technique related to the tuning of newly proposed optical property arising from concave Au surfaces. Because THF is a good solvent for the PS block, the PS cores in the micelles gradually shrink with the decrease in the amount of THF, thereby reducing pore size of the Au films. Actually, the pore size was controlled down to ~ 19 nm by decreasing the amount of THF without significant change of the wall thickness (Fig. 2c and Supplementary Fig. 2a). While using the minimum amount of THF (1 ml) for complete dissolution of PS-b-PEO, the polymeric micelles, in shrink form, can stably keep the original form and then the resultant mesopores were indeed in spherical form (Fig. 2c). The average pore size and wall thickness are ~ 19 nm and ~ 25 nm, respectively. On the other hand, the pore size can be expanded by adding hydrophobic organic compounds. When 1,3,5-triisopropylbenzene (1,3,5-TIPBz) was used as a swelling agent, the effective pore size was increased from 25 nm (without 1,3,5-TIPBz) to 32 nm (1,3,5-TIPBz; 10 µl), 32 nm (20 µl), 40 nm (30 µl) and 40 nm (40 µl; Supplementary Fig. 4). With the increase of the pore size, the pore

walls started showing the appearance of protrusions. In addition, the molecular weight of the PS block in PS-b-PEO can also be used to control the pore size (Supplementary Fig. 5). As estimated by SEM, the average pore size of a mesoporous Au film prepared using PS_{63000}-b-PEO_{26000} was around 60 nm.

Unlike sol–gel-based fabrication process of ordered mesoporous metal oxide films, electrochemical deposition provides us the advantage of the facile and precise control of the film growth rate, that is, it enables us to fine tune the film thickness by simply adjusting the deposition time. Film thicknesses of the mesoporous Au films prepared using PS_{18000}-b-PEO_{7500} for deposition time of 600, 1,000, 1,800, 2,500 and 3,600 s were, respectively, 70, 140, 170, 230 and 440 nm; in this case the average growth rate was calculated to be 0.11 nm s^{-1} (Supplementary Fig. 6). The electrochemically active surface areas (ECSAs) were also calculated by using cyclic voltammetry in an acidic medium (0.5 M H_2SO_4) (Supplementary Fig. 7). The peak area indicated by a dotted circle is associated with reduction of Au oxide species (Supplementary Fig. 7a). On the basis of the assumption that the charges associated with the reduction of oxide species is 400 µC cm^{-2} for Au surface [46,47], the ECSAs in the films were theoretically calculated from the observed charges. The total surface area of the mesoporous Au films was increased proportionally (Supplementary Fig. 7b). The volume-normalized ECSAs (per film volume, cm^3) were around 49.1 m^2 cm^{-3} (Supplementary Fig. 7c) and almost constant until the film thickness reached 5 µm. Thus, inner parts of the mesoporous Au films can also work as electrochemically active surfaces and it is proved that the mesoporous structures are homogeneously formed inside the films.

To carefully investigate the atomic structure of Au in the pore walls comprehensively, a monolayer of mesoporous Au film was prepared and then its cross-sectional sample was observed by TEM (Fig. 2f–h and Supplementary Fig. 8). The images revealed that spherical mesopores were surrounded by continuous networks of crystalline Au. The lattice fringes associated to Au fcc crystal were clearly observed inside the pore walls. High-resolution TEM image taken along $<110>$ directions (Supplementary Fig. 8) showed (111) lattice fringes with stacking faults related to single crystallinity of the mesoporous Au film. Several crystal domains were connected continuously to create concave surface, which exposed high index planes (Fig. 2f–h). There were also several defects such as (i) stacking fault, (ii) dislocation and (iii) kink band, marked by arrows in

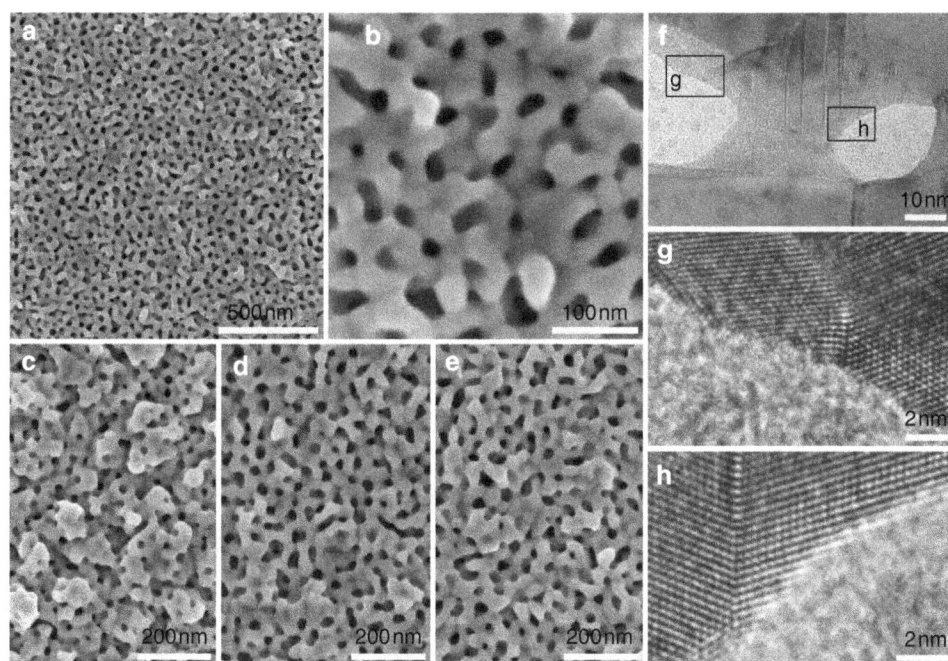

Figure 2 | Microscopic characterization of mesoporous Au films. (**a,b**) Top-surface SEM images of mesoporous Au film prepared with a typical electrolyte containing PS$_{18000}$-b-PEO$_{7500}$ micelles and 3 ml THF as solvent. The deposition time is 1,000 s. (**c–e**) Top-surface SEM images of mesoporous Au films prepared with three electrolytes containing PS$_{18000}$-b-PEO$_{7500}$ micelles and different THF amounts ((**c**) 1 ml, (**d**) 2 ml and (**e**) 3 ml, respectively). (**f–h**) Highly magnified TEM images of mesoporous Au film prepared with a typical electrolyte containing PS$_{18000}$-b-PEO$_{7500}$ micelles and 3 ml THF as solvent. The assignment of crystal lattices is shown in Supplementary Fig. 5.

Supplementary Fig. 8a. These defects could be caused by the localized strains during crystal growth of Au and/or electrochemical deposition of mesostructured films with reduction of soluble Au species. X-ray photoelectronic spectroscopy was used to determine surface composition and oxidation state of the Au film. The high resolution Au 4f scan displayed a doublet at 87.7 and 84.0 eV separated by 3.7 eV due to spin–orbit coupling, confirming the presence of Au(0) species in the film (Supplementary Fig. 9).

To realize well-developed mesoporous structures, the crystal growth speed, that is to say, the resultant crystal size, is the most important factor. Wiesner and colleagues have found that diameter of nanoparticles as building blocks should be below a critical limit relative to the sizes of blocks with which they interact[23]. In case of mesoporous metal oxides, mesoporous structure is well preserved only when the wall thickness is larger than the crystal size, which is controlled by the nucleation and grain growth rates[44]. In our electrochemical process, the control of Au crystal sizes is governed by Au growth speed. The bath temperature for the deposition is one of critical factors for the control of Au growth speed. From the amperometric plots for the deposition of mesoporous Au films under different temperatures, it was proved that the reduction currents increased (that is, the deposition rates increased) with the increase of bath temperature (Supplementary Fig. 10). Obviously, the formation of mesoporous structures was confirmed, only when the temperature was < 25 °C (Supplementary Fig. 11). In contrast, when a temperature was > 40 °C, the Au crystal growth rate was high so that large bulk Au crystals were mainly formed and mesoporous structures were not well developed.

Optical performance of mesoporous Au films. To investigate the fundamental optical properties of the mesoporous Au films, three typical films with different pore sizes were compared. Here we selected mesoporous Au films prepared using PS$_{18000}$-b-PEO$_{7500}$ micelles with different solvent composition (Sample I; 1 ml of THF (average pore size; 19 nm), Sample II; 3 ml of THF (25 nm), Sample III; 3 ml of THF with 40 µl of 1,3,5-TIPBz (40 nm)). The representative scattering spectra were measured by using a custom-made dark-field spectroscopic microscopy (Supplementary Fig. 12), and compared with a spectrum taken from a sputtered Au film without mesopores. As observed in Supplementary Fig. 12b, the scattered light intensity increased systematically with the increase in the pore size, indicating a significant increase of the interaction of the mesopore surfaces with electromagnetic waves in the visible frequency region. The result suggests that the interaction between the Au surface and the light become stronger. This is possibly due to the larger electromagnetic field around the individual mesopores because the charge induced at the surface of the pore becomes larger as the pore size becomes larger as the Au volume exposed to the E-field of the incoming light becomes larger. From another point of view, the reason may be related that expansion of the pores makes the pores respond more efficiently to the light because the pore sizes become closer to the wavelength of the oscillating charge density waves on the Au surface induced by the visible light (Note that the wavelengths of the charge density waves of (localized) surface plasmons are in the sub-100 nm range.). As seen in Supplementary Fig. 12c, each point on the Au surfaces responded in different manner with different scattering intensity as well as different resonance frequency reflecting their various morphologies in nanometer scale. Thus, as shown in Supplementary Fig. 12b, broadband scattering spectra are due to the ensemble effect for the mesopores in the film.

To gain deeper insight into the observed spectra, E-field distribution was assessed on the basis of simulations using finite-difference time domain (FDTD) algorithm. As clearly seen, experimentally observed tendency of the scattering intensity of the three films (Supplementary Fig. 12) had clear correspondence to the strength of the E-field in Fig. 3 and Supplementary Fig. 13. The E-field amplitude increased as the pore size increased

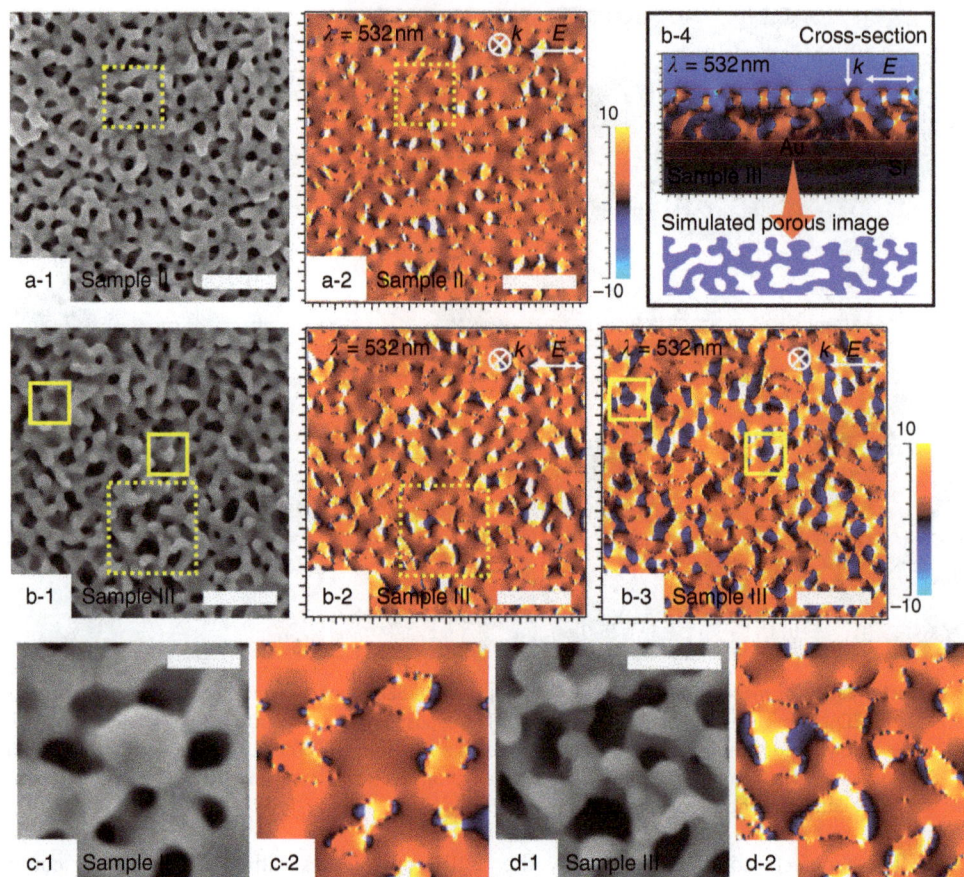

Figure 3 | E-field distributions on mesoporous Au films. (**a-1,b-1**) SEM images and (**a-2,b-2,b-3**) the corresponding E-field distributions on mesoporous Au films prepared with two electrolytes containing PS_{18000}-b-PEO_{7500} micelles with different solvent compositions ((Sample II) 3 ml THF, and (Sample III) 3 ml THF + 40 µl 1,3,5-TIPBz) under 532 nm excitation. The E-field distribution in **a-2** and **b-2** is taken from 10 nm in depth in the films, in which moderate E-field amplitude is clearly observed inside or at the perimeter of the mesopores, as shown in dotted-line squares. The E-field distribution in **b-3** is taken from the film surface, showing strong E-field points (hot spots) at the perimeter of the protruded objects (as marked by solid-line squares). (**b-4**) Cross-sectional E-field distributions on mesoporous Au film (Sample III). (**c,d**) Enlarged SEM images and the corresponding E-field distributions of the areas indicated by dotted-line squares of **a** and **b**. Scale bars, 200 nm (**a,b**), 50 nm (**c**) and 100 nm (**d**), respectively.

(Supplementary Fig. 13). Interestingly, moderate E-field amplitude was clearly seen inside or at the perimeter of the mesopores (Fig. 3c,d). For larger pores on the contrary, there was a clear tendency that strong E-field points (hot spots) emerge at the perimeter of the protruded objects, as clearly seen in Fig. 3b-3. The cross-sectional image also confirmed the above characteristic, which showed the outbreak of the strong E-field inside the mesopores (Fig. 3b-4). Especially for the Sample III, the pores become agglomerated with connected meandering pore walls with prominent nanosized protrusions and the E-field has a tendency to be converged there giving high amplitude (Fig. 3b-3). Such nanoprotrusions on elongated walls show plasmon resonance frequency in the longer wavelength (near infrared) region. In contrast, the other two films (Samples I and II with smaller pores) have less corrugated surface and less protrusions, resulting in slightly higher resonance frequency (that is, shorter wavelength) and smaller E-field as well as scattering.

Stability of the mesoporous Au film in water and strong affinity of the Au surfaces to proteins and amino acids are helpful for utilizing the films as the perfect substrate for biosensing. Most of the state-of-the-art apta-sensing techniques are often based on thiolated aptamers with rationally designed linker molecules for the specific target molecules. Then, it is considered that the present mesoporous Au films are the best substrate suitable for SERS for apta-sensing. Here we examined the SERS performance for the above three typical films (Samples I, II and III) with

different pore size (Fig. 4). A droplet (5 µl) of Nile blue solution (10^{-6} M in ethanol) was spread uniformly on the mesoporous Au films (0.3 cm × 1.5 cm) and the substrates were dried in air at room temperature under a stream of nitrogen gas. The SERS spectra, taken from Nile blue molecule coated on mesoporous Au films, with laser excitation wavelength of 532 nm were recorded. Optical microscope image and corresponding SERS spectral mapping images for Sample III are shown in Fig. 4a–b. It was clearly visualized that the SERS signal was drastically enhanced on mesoporous Au region. The enhancement mechanism of SERS effect is mainly due to the field enhancement effect arising from the mesoporous Au. The sputtered Au surface without mesopores shows far less SERS signal intensity. The most dramatic effect that arises from the nanomorphology is indeed a field enhancement induced by the strong-light Au surface interaction, as can be seen from the enhanced light scattering (Supplementary Fig. 12b) and thus we can assign this SERS effect as electromagnetic origin (Fig. 3 and Supplementary Fig. 13). In addition, the chemical SERS effect induced by the molecule adsorption should be minor, because its contribution to the SERS intensity is in the order of the black spectrum in Fig. 4c where the electromagnetic field amplitude is substantially reduced in comparison with mesoporous Au films. Furthermore, the SERS chemical effect is a local chemical bond effect, which is determined by the local bond nature between the Nile blue molecules and the Au surface. Therefore, we can safely conclude that the difference in SERS

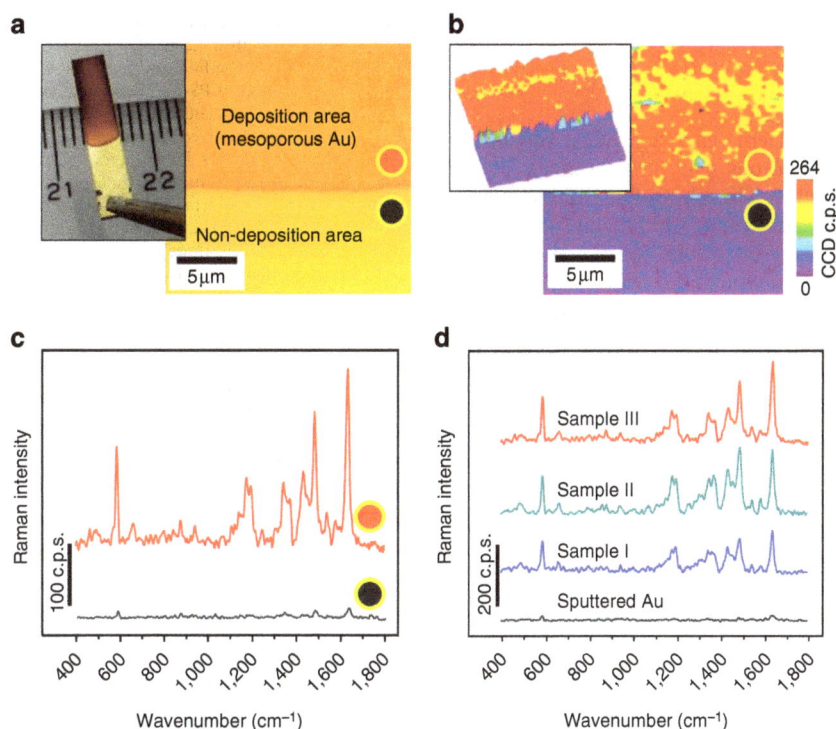

Figure 4 | SERS study of Nile blue-molecules coated on mesoporous Au films. (**a**) Photograph and optical image of mesoporous Au film (Sample III) prepared with an electrolyte containing PS$_{18000}$-b-PEO$_{7500}$ micelles and 3 ml THF + 40 μl 1,3,5-TIPBz as solvent. (**b**) Corresponding SERS spectral mapping with vibrational intensity. (**c**) Representative SERS spectra on mesoporous Au film (Sample III). Mesoporous Au region (that is, deposition area) and non-porous region (that is, non-deposition area) are measured, respectively. (**d**) Representative SERS spectra on mesoporous Au films prepared with three electrolytes containing PS$_{18000}$-b-PEO$_{7500}$ micelles with different solvent compositions ((Sample I) 1 ml THF, (Sample II) 3 ml THF and (Sample III) 3 ml THF + 40 μl 1,3,5-TIPBz). A sputtered Au film without mesopores is also compared. The variability of the SERS spectra is ±5% for each, which is due to the variability of the surface roughness and the uniformity of the adsorbed molecular layers.

intensity is attributed to the surface morphology effect, as the adsorbed Nile blue should be multilayered.

Furthermore, the SERS signal intensity from the molecule was increased with the increase in the mesopore size (Fig. 4d). This result exhibits the same tendency as the scattering intensity shown in Supplementary Fig. 12b, reflecting the difference in the pore architecture of the films. This result is understandable because the stronger light scattering points to the generation of enhanced E-field at the pore surfaces caused by the excitation of localized surface plasmon. Then, this enhanced E-field can subsequently lead to a stronger SERS signal from the adsorbed molecules, as observed in our measurement (Note that a stronger SERS intensity is expected for the laser excitation wavelength around 670 nm close to that of the peak position in the scattering spectra. However, our experiment using a 532 nm laser has already showed enough SERS signal which evidences the high performance of this mesoporous Au substrate.). To estimate the SERS enhancement factor, Au sputtered film (that is, non-SERS substrate) with the same area was coated with a droplet (5 μl) of Nile blue solution with a high concentration of 10^{-3} M in ethanol (to visualize the non-SERS weak signals, Fig. 4c). Using the same measurement conditions, the SESR enhancement factor is simply calculated by the following formula[48].

$$\mathrm{EF} = \frac{I_{\mathrm{SERS}}/C_{\mathrm{SERS}}}{I_{\mathrm{RS}}/C_{\mathrm{RS}}}, \quad (1)$$

where I_{SERS} and I_{RS} are Raman intensities of the SERS and non-SERS signal, whereas C_{SERS} and C_{RS} are concentrations of the dropped molecule SERS and on the SERS and non-SERS substrate, respectively. The intensity of the strongest peak at 1,642 cm^{-1} was used for the calculation, leading to the calculated

enhancement factor of 1.2×10^5. The signal enhancement effect of our materials is in the similar order as other nanostructured samples previously reported[49–51], but the use of mesoporous Au substrates successfully guides to the systematic understanding for correlating the local optical properties and the fine-tuned morphologies by using spectral mapping with high spatial resolution. Such information is definitely useful for elucidating the origin and mechanism of electromagnetic field enhancement.

To further demonstrate the large-sized mesopores, we show selective detection of large biomolecules (protein molecules) on our mesoporous Au film (Supplementary Fig. 14). SEIRA was carried out using Fourier transform infrared (FTIR) configuration in attenuated total reflection geometry. For this measurement, a self-assembled monolayer of bovine serum albumin (BSA, the average diameter of 8.6 nm) adsorbed on the cleaned mesoporous substrate was used as a target protein. Mesoporous Au film (Sample II, 0.3 cm × 1.5 cm) was soaked into BSA solution (10^{-3} M in water, pH = 7.1) for 24 h, then thoroughly rinsed in distilled water to leave a monolayer of BSA on the pore surface, and finally dried under a stream of nitrogen gas. As shown in Supplementary Fig. 14, the protein bands (Amide-I and -II) are clearly observed evidencing the adsorption of monolayer of BSA. Very high signal intensity of self-assembled monolayer-BSA protein is observed on our mesoporous Au film. Note that the enhancement factor scales as the $(E/E_{\mathrm{i}})^2$ for SEIRA, unlike the $(E/E_{\mathrm{i}})^4$ for the SERS, in the visible frequency region (Here E means local electrical field on the mesoporous Au substrate and E_{i} means electrical field of the incident light.). Considering the above difference and also the smaller electromagnetic field enhancement in the infrared compared with the visible region of this sample, the signal enhancement of the SEIRA shown here is high enough and satisfactory.

Discussion

We have developed an efficient electrochemical deposition approach to fabricate mesoporous Au films from a diluted electrolyte containing micelles. The electrolyte solutions were studied by ultraviolet–visible absorption and Raman spectroscopy (Supplementary Fig. 15). In the ultraviolet spectra, the peaks at around 324 nm correspond to a typical d-d transition of $AuCl_4^-$ species. Another band at around 278 nm appears by the presence of benzene rings in the PS block. The Raman spectra display several peaks at 347, 324 and 171 cm^{-1}, due to symmetric (A_{1g}), antisymmetric (B_{2g}) and bending (B_{1g}) modes of $AuCl_4^-$ species in the solution, respectively. All other peaks were related to the solvents and the PS-b-PEO diblock copolymer. Both the ultraviolet–visible absorption and the Raman spectra show that the Au species are considered as Au(III) with Cl$^-$ ligand. Accordingly, in this study, our approach to obtain mesoporous Au films was succeeded through effective interaction between PS-b-PEO micelles and soluble Au species; $HAuCl_4$ are dissolved into H_3O^+ and $AuCl_4^-$ and then interacted with the EO shell domains of the PS-b-PEO micelles by hydrogen bonding (Fig. 1a). Moreover, we consider that $AuCl_4^-$ is free ion in the solution as well as in the hydrophilic domains of the PS-b-PEO micelles. The micelles with $AuCl_4^-$ can be neutral, negatively and positively charged, depending on the $H_3O^+/AuCl_4^-$ ratio in the part of the EO shells. In our experimental conditions, the micelle solution after the addition of aqueous solution of $HAuCl_4$ was slightly positive in zeta-potential. Thus, the H_3O^+ rich micelles were positively charged and then moved to the working (negative) electrode surface. The $AuCl_4^-$ (redox potential of $AuCl_4^-/Au^0$ is 0.93 V versus standard hydrogen electrode) species were reduced to metallic Au from the part of near the electrode surface consecutively under a constant potential of -0.5 V. Our approach shows a fairly high reproducibility (100%) (Supplementary Fig. 16). The electrodeposited mesoporous Au films were well adhered on the substrates (that is, working electrodes) (Supplementary Fig. 17).

As demonstrated above, the present electrochemically deposited mesoporous Au films with tunable pores can provide fascinating optical properties tailored by engineering the soft-templated uniform mesospaces. Although several efforts have been made for preparing mesoporous Au films, continuous Au films with precisely designed mesopores have not been realized yet because of the difficulty in controlling the crystal growth of Au. To overcome this issue, electrochemical synthesis utilizing stable micelles of PS-b-PEO diblock copolymers as pore-directing agents is quite effective. In the resultant films, uniformly sized mesopores are distributed in the entire films and controlled in a wide range from 20 to 60 nm by changing the molecular weight of PS-b-PEO and the electrolyte composition. Our approach not only demonstrates novel optical applications of mesoporous Au films but also sheds new light on the probability of obtaining mesoporous metals through the soft-templating pathway. Such 3D extended metallic frameworks can steadily provide abundant adsorption and reaction sites of target molecules, which are critical to realize emergent functions in the future. Our electrochemical approach is widely applicable to embed uniform mesopores in other metal and alloy systems, which are generally difficult to be synthesized.

Methods

Preparation of mesoporous Au films. Our approach is based on an electrochemical method, which has been practically used for the general preparation of continuous metallic films. In a typical synthesis, 10 mg of polystyrene-b-poly(oxyethylene) (PS-b-PEO) was dissolved in 3 ml of THF completely at 40 °C and then 1.5 ml of ethanol was added to the solution. An aqueous solution of $HAuCl_4$ (the final concentration was 5 mM in 8 ml of electrolyte) was added slowly to the clear PS-b-PEO solution and spherical micelles were formed by the presence of water. Gentle stirring for 30 min at room temperature is used to make sure that the dissolved Au species were well incorporated into the exterior PEO region of the micelles. Finally, a transparent bright-yellow coloured electrolyte (pH = 2.5) was obtained and directly used for electrodeposition. Electrochemical deposition from the precursor solutions was carried out by using an electrochemical machine (CHI 842B electrochemical analyzer, CH Instrument, USA) with a conventional three-electrode system, including a platinum wire as a counter electrode, an Ag/AgCl as a reference electrode and a conducting substrate as a working electrode. The schematic illustration of the typical electrodeposition cell is shown in Supplementary Fig. 18. In this work, the typical conducting substrate used is Au–Si wafer with a representative size of 0.45 cm^2 (0.3 cm × 1.5 cm), which was fabricated by a dicing cutter. The optimal electrodeposition of Au was carried out at a constant potential of -0.5 V (versus Ag/AgCl) for 1,000 s without stirring at room temperature. During the electrodeposition, a stable current was detected for the Au reduction, as displayed in Supplementary Fig. 10. After the Au deposition, the micelles (used as soft-templates) were thoroughly removed by ultraviolet–ozone cleaner or low-powered O_2 plasma treatment, as confirmed by infrared analysis. Calcination of the films in air, which has been commonly used for complete removal of organic templates in mesoporous metal oxide films, led to removal of the templates from the films, but the mesoporous structures collapsed through the grain growth of Au.

Characterization. SEM images were obtained using a Hitachi HR-SEM SU8000 microscope at the accelerating voltage of 5 kV. Transmission electron microscope (TEM) images were taken by a JEOL JEM-2100F microscope at the accelerating voltage of 200 kV. X-ray photoelectronic spectroscopic measurements were conducted by using a JPS-9010TR (JEOL) instrument with an Mg Kα X-ray source. The precursor solutions were studied by a JASCO V-570 ultraviolet–visible–near infrared spectrometer. The Raman spectra were recorded by a Horiba-Jobin-Yvon T64000 Raman spectroscopy system with the laser wavelength of 514.5 nm. The electrodeposition of mesoporous Au films and cyclic voltammetry measurement were performed by using a CHI 842B electrochemical analyzer (CH Instruments). A conventional three-electrode cell was used, including an Ag/AgCl electrode as reference electrode, a platinum wire as counter electrode and a working electrode. The scattering spectra and SERS properties were carried out using a custom-made confocal Raman microscope (WITec Alpha 300S) combined with a monochromator (Action SP2300—Princeton Instruments) and a CCD camera (Andor iDus DU-401A BR-DD-352). Both spatial resolution of dark-field and SERS scanning were estimated to be ~300 nm (64 × 64 pixels on a 20 × 20 μm scanning area) with the integration time of 0.2 s per pixel. An ultraviolet–near infrared light source (halogen lamp HAL 100-Zeiss) and a dark-field lens (× 50/0.55 NA-Zeiss) were used to study the dark-field scattering properties. For the characterization of SERS properties, a second harmonic diode pumped Nd:VO$_4$ laser (Witec) at 532 nm (0.5 mW) was focused on the scanning area of the Nile blue-treated mesoporous Au films by using a × 100 objective (0.9 NA-Olympus). Before each measurement, the intensity calibration was done carefully by measuring the Si Raman peak intensity of the Si wafer. SEIRA was carried out using FTIR configuration in the attenuated total reflection geometry (Nicolet IS50R-FTIR). The electric field distributions of the Au mesoporous structures were calculated using 3D finite-difference time-domain method (Fullwave, Rsoft). The 3D model was performed on the basis of geometries of mesoporous Au films. The excited electric field propagating along the z axis was injected from the top of the Au mesoporous surface with their electric vectors oscillating along the xz plane.

References

1. Kresge, C. T., Leonowicz, M. E., Roth, W. J., Vartuli, J. C. & Beck, J. S. Ordered mesoporous molecular sieves synthesized by a liquid-crystal template mechanism. *Nature* **359**, 710–712 (1992).
2. Zhao, D. et al. Triblock copolymer syntheses of mesoporous silica with periodic 50 to 300 Angstrom Pores. *Science* **279**, 548–552 (1998).
3. MacLachlan, M. J., Coombs, N. & Ozin, G. A. Non-aqueous supramolecular assembly of mesostructured metal germanium sulfides from $(Ge_4S_{10})^{4-}$ clusters. *Nature* **397**, 681–684 (1999).
4. Asefa, T., MacLachlan, M. J., Coombs, N. & Ozin, G. A. Periodic mesoporous organosilicas with organic groups inside the channel walls. *Nature* **402**, 867–871 (1999).
5. Inagaki, S., Guan, S., Ohsuna, T. & Terasaki, O. An ordered mesoporous organosilica hybrid material with a crystal-like wall structure. *Nature* **416**, 304–307 (2002).
6. Liu, J. et al. A facile soft-template synthesis of mesoporous polymeric and carbonaceous nanospheres. *Nat. Commun.* **4**, 2798 (2013).
7. Joo, S. H. et al. Ordered nanoporous arrays of carbon supporting high dispersions of platinum nanoparticles. *Nature* **412**, 169–172 (2001).
8. Ritchie, R. H. Surface plasmons in solids. *Surface Sci.* **34**, 1–19 (1973).

9. Barnes, W. L., Dereux, A. & Ebbesen, T. W. Surface plasmon subwavelength optics. *Nature* **424,** 824–830 (2003).

10. Prodan, E., Radloff, C., Halas, N. J. & Nordlander, P. A hybridization model for the plasmon response of complex nanostructures. *Science* **302,** 419–422 (2003).

11. Atwater, H. A. The promise of plasmonics. *Sci. Am.* **296,** 56–62 (2007).

12. Zhang, X. & Liu, Z. Superlenses to overcome the diffraction limit. *Nat. Mater.* **7,** 435–441 (2008).

13. Cao, L. & Brongersma, M. L. Active plasmonics: ultrafast developments. *Nat. Photonics* **3,** 12–13 (2009).

14. Sun, Y. & Xia, Y. Shape-controlled synthesis of gold and silver nanoparticles. *Science* **298,** 2176–2179 (2002).

15. Zhang, J., Liu, P., Ma, H. & Ding, Y. Nanostructured porous gold for methanol electro-oxidation. *J. Phys. Chem. C* **111,** 10382–10388 (2007).

16. Lang, X.-Y. *et al.* Nanoporous gold supported cobalt oxide microelectrodes as high-performance electrochemical biosensors. *Nat. Commun.* **4,** 2169 (2013).

17. Wittstock, A., Zielasek, V., Biener, J., Friend, C. M. & Bäumer, M. Nanoporous gold catalysts for selective gas-phase oxidative coupling of methanol at low temperature. *Science* **327,** 319–322 (2010).

18. Quaino, P., Luque, N. B., Nazmutdinov, R., Santos, E. & Schmickler, W. Why is gold such a good catalyst for oxygen reduction in alkaline media? *Angew. Chem. Int. Ed.* **51,** 12997–13000 (2012).

19. Li, J. F. *et al.* Shell-isolated nanoparticle-enhanced Raman spectroscopy. *Nature* **464,** 392–395 (2010).

20. Lim, D.-K. *et al.* Highly uniform and reproducible surface-enhanced Raman scattering from DNA-tailorable nanoparticles with 1-nm interior gap. *Nat. Nanotechnol.* **6,** 452–460 (2011).

21. Wi, J.-S., Tominaka, S., Uosaki, K. & Nagao, T. Porous gold nanodisks with multiple internal hot spots. *Phys. Chem. Chem. Phys.* **14,** 9131–9136 (2012).

22. Attard, G. S. *et al.* Mesoporous platinum films from lyotropic liquid crystalline phases. *Science* **278,** 838–840 (1997).

23. Warren, S. C. *et al.* Ordered mesoporous materials from metal nanoparticle-block copolymer self-assembly. *Science* **320,** 1748–1752 (2008).

24. Shin, H. J., Ryoo, R., Liu, Z. & Terasaki, O. Template synthesis of asymmetrically mesostructured platinum networks. *J. Am. Chem. Soc.* **123,** 1246–1247 (2001).

25. Wang, H. *et al.* Shape- and size-controlled synthesis in hard templates: sophisticated chemical reduction for mesoporous monocrystalline platinum nanoparticles. *J. Am. Chem. Soc.* **133,** 14526–14529 (2011).

26. Li, Z. *et al.* Linking experiment and theory for three-dimensional networked binary metal nanoparticle-triblock terpolymer superstructures. *Nat. Commun.* **5,** 3247 (2014).

27. Yang, C.-M., Sheu, H.-S. & Chao, K.-J. Templated synthesis and structural study of densely packed metal nanostructures in MCM-41 and MCM-48. *Adv. Funct. Mater.* **12,** 143–148 (2002).

28. Fukuoka, A. *et al.* Template synthesis of nanoparticle arrays of gold and platinum in mesoporous silica films. *Nano Lett.* **2,** 793–795 (2002).

29. Velev, O. D., Jede, T. A., Lobo, R. F. & Lenhoff, A. M. Porous silica via colloidal crystallization. *Nature* **389,** 447–448 (1997).

30. Fan, W. *et al.* Hierarchical nanofabrication of microporous crystals with ordered mesoporosity. *Nat. Mater.* **7,** 984–991 (2008).

31. Velev, O. D. & Kaler, E. W. Structured porous materials via colloidal crystal templating: from inorganic oxides to metals. *Adv. Mater.* **12,** 531–534 (2000).

32. Jiang, P., Cizeron, J., Bertone, J. F. & Colvin, V. L. Preparation of mesoporous metal films from colloidal crystals. *J. Am. Chem. Soc.* **121,** 7957–7958 (1999).

33. Egan, G. L. *et al.* Nanoscale metal replicas of colloidal crystals. *Adv. Mater.* **12,** 1040–1042 (2000).

34. Erlebacher, J., Aziz, M. J., Karma, A., Dimitrov, N. & Sieradzki, K. Evolution of nanoporosity in dealloying. *Nature* **410,** 450–453 (2001).

35. Ding, Y. & Erlebacher, J. Nanoporous metals with controlled multimodal pore size distribution. *J. Am. Chem. Soc.* **125,** 7772–7773 (2003).

36. Fujita, T. *et al.* Atomic origins of the high catalytic activity of nanoporous gold. *Nat. Mater.* **11,** 775–780 (2012).

37. Tominaka, S., Ohta, S., Obata, H., Momma, T. & Osaka, T. On-chip fuel cell: micro direct methanol fuel cell of an air-breathing, membraneless, and monolithic design. *J. Am. Chem. Soc.* **130,** 10456–10457 (2008).

38. Snyder, J., Fujita, T., Chen, M. W. & Erlebacher, J. Oxygen reduction in nanoporous metal-ionic liquid composite electrocatalysts. *Nat. Mater.* **9,** 904–907 (2010).

39. Born, M. & Wolf, E. *Principles of Optics* (Cambridge Univ. Press, 1999).

40. Stockman, M. I., Shalaev, V. M., Moskovits, M., Botet, R. & George, T. F. Enhanced Raman scattering by fractal clusters: scale-invariant theory. *Phys. Rev. B* **46,** 2821–2830 (1992).

41. Enders, D., Nagao, T., Pucci, A. & Nakayama, T. Surface-enhanced ATR-IR spectroscopy with interface-grown plasmonic gold-island films near the percolation threshold. *Phys. Chem. Chem. Phys.* **13,** 4935–4941 (2011).

42. Cheng, Y.-J. & Gutmann, J. S. Morphology phase diagram of ultrathin anatase TiO_2 films templated by a single PS-b-PEO block copolymer. *J. Am. Chem. Soc.* **128,** 4658–4674 (2006).

43. Fattakhova-Rohlfing, D., Wark, M., Brezesinski, T., Smarsly, B. M. & Rathouský, J. Highly organized mesoporous TiO_2 films with controlled crystallinity: a Li-insertion study. *Adv. Funct. Mater.* **17,** 123–132 (2007).

44. Grosso, D. *et al.* Periodically ordered nanoscale islands and mesoporous films composed of nanocrystalline multimetallic oxides. *Nat. Mater.* **3,** 787–792 (2004).

45. Soler-Illia, G. J. A. A., Angelomé, P. C., Fuertes, M. C., Grosso, D. & Boissiere, C. Critical aspects in the production of periodically ordered mesoporous titania thin films. *Nanoscale* **4,** 2549–2566 (2012).

46. Zhang, L. *et al.* Cu^{2+}-assisted synthesis of hexoctahedral Au-Pd alloy nanocrystals with high-index facets. *J. Am. Chem. Soc.* **133,** 17114–17117 (2011).

47. Asao, N. *et al.* Aerobic oxidation of alcohols in the liquid phase with nanoporous gold catalysts. *Chem. Commun.* **48,** 4540–4542 (2012).

48. Ru, E. C. L., Blackie, E., Meyer, M. & Etchegoin, P. G. Surface enhanced Raman scattering enhancement factors: a comprehensive study. *J. Phys. Chem. C* **111,** 13794–13803 (2007).

49. Malfatti, L. *et al.* Nanocomposite mesoporous ordered films for lab-on-chip intrinsic surface enhanced Raman scattering detection. *Nanoscale* **3,** 3760–3766 (2011).

50. López-Puente, V., Abalde-Cela, S., Angelomé, P. C., Alvarez-Puebla, R. A. & Liz-Marzán, L. M. Plasmonic mesoporous composites as molecular sieves for SERS detection. *J. Phys. Chem. Lett.* **4,** 2715–2720 (2013).

51. Wolosiuk, A. *et al.* Silver nanoparticle-mesoporous oxide nanocomposite thin films: a platform for spatially homogeneous SERS-active substrates with enhanced stability. *ACS Appl. Mater. Interfaces* **6,** 5263–5272 (2014).

Acknowledgements

We acknowledge financial support from the Grant-in-Aid for Young Scientists A (Research Project Number: 26708028) of the Japan Society for the Promotion of Science (JSPS), Japanese-Taiwanese Cooperative Program of the Japan Science and Technology Agency (JST) and The Canon Foundation.

Author contributions

C.L. synthesized and characterized mesoporous Au films, Ö.D. analysed ultraviolet-visible and Raman data, T.D.D. and T.N. studied mesospace-stimulated optical properties, Y.S. and O.T. analysed TEM data, Ö.D. and T.K. contributed to discussion on formation mechanism and Y.Y. designed this work. All the authors discussed the results and participated in writing the manuscript.

Additional information

3

Two-dimensional gold nanostructures with high activity for selective oxidation of carbon–hydrogen bonds

Liang Wang[1,*], Yihan Zhu[2,*], Jian-Qiang Wang[3], Fudong Liu[4,†], Jianfeng Huang[2], Xiangju Meng[1], Jean-Marie Basset[5], Yu Han[2] & Feng-Shou Xiao[1]

Efficient synthesis of stable two-dimensional (2D) noble metal catalysts is a challenging topic. Here we report the facile synthesis of 2D gold nanosheets via a wet chemistry method, by using layered double hydroxide as the template. Detailed characterization with electron microscopy and X-ray photoelectron spectroscopy demonstrates that the nanosheets are negatively charged and [001] oriented with thicknesses varying from single to a few atomic layers. X-ray absorption spectroscopy reveals unusually low gold–gold coordination numbers. These gold nanosheets exhibit high catalytic activity and stability in the solvent-free selective oxidation of carbon–hydrogen bonds with molecular oxygen.

[1] Key Lab of Applied Chemistry of Zhejiang Province, Department of Chemistry, Zhejiang University, Hangzhou 310028, China. [2] Advanced Membranes and Porous Materials Center, Physical Sciences and Engineering Division, King Abdullah University of Science and Technology, Thuwal 23955-6900, Kingdom of Saudi Arabia. [3] Key Laboratory of Interfacial Physics and Technology, Shanghai Institute of Applied Physics, Chinese Academy of Sciences, Shanghai 201800, China. [4] Research Center for Eco-Enviromental Sciences, Chinese Academy of Sciences, Beijing 100085, China. [5] KAUST Catalysis Center, King Abdullah University of Science and Technology, Thuwal 23955-6900, Kingdom of Saudi Arabia. * These authors contributed equally to this work. † Present address: The Materials Sciences Division, Lawrence Berkeley National Laboratory, Berkeley, California 94720, USA. Correspondence and requests for materials should be addressed to Y.H. (email: yu.han@kaust.edu.sa) or to F.-S.X. (email: fsxiao@zju.edu.cn).

Two-dimensional (2D) materials with single to a few atomic layer thicknesses, as represented by graphene, have attracted enormous research interest in the last decade, owing to their fascinating electronic, magnetic, optical and catalytic properties[1-15]. In comparison with lamellar-structured graphite or metal dichalcogenides (for example, MoS_2), which can readily be prepared in the form of 2D materials by physical[16-18] or chemical exfoliation[19-22], noble metals are more difficult to be fabricated into 2D nanostructures, especially by wet chemical synthesis, because of their highly isotropic lattice symmetries. Some anisotropic metallic colloidal nanosheets (NSs) have been synthesized in solution using various capping agents via direct[23-31] or secondary growth[31], but with few exceptions[27], they have thicknesses of several nanometers or even greater and cannot be classified as 2D materials in a strict sense. It is known that with noble metals (for example, Au) extraordinary catalytic activity occurs as a consequence of the quantum size effect when the catalyst contains only two to three atomic layers[32-34]. 2D Au nanostructures have been prepared on well-defined crystal surfaces by physical deposition methods[32,35,36]. They have controllable thicknesses within the range of a few layers of atoms but also limited size (several nanometers) in the other two dimensions, and they are therefore usually referred to as 'islands'[32,35,36]. It was reported that 2D Au islands deposited on the surface of graphene/Ru(0001) could promote CO adsorption and potentially catalyze the CO oxidation at very low temperature[35]; likewise, 2D Au islands prepared on the TiO_2(110) surface showed greatly enhanced molecular binding especially at the edge sites[32]. In general, 2D metallic materials are supposed to show distinct molecular activation ability and catalytic behaviours, considering that they have unusual electronic structures as well as a large proportion of low-coordinated atoms. However, only molecular adsorption properties have thus far been investigated for 2D Au catalysts[32], due to the difficulty in preparing them on the large scale[35,36] and their weak stability[29]. Efficient synthesis of stable 2D Au catalysts remains unattained.

Layered double hydroxides (LDHs) belong to a class of lamellar-structured clay with a general formula of $M(II)_{1-x}M(III)_x(OH^-)_2(A^{n-})_{x/n} \cdot zH_2O$, where the positive charges of the cationic layers made of edge-shared metal (M(II) and M(III)) hydroxide octahedra are balanced by the interlayered anions (A^{n-})[37]. One attractive feature of LDH materials is that the interlayered anions have considerable freedom of movement to be substituted by other types of anions via ion exchange[38-42].

In this work, we exploit the anion-exchange capability of LDH to introduce Au precursors $(AuCl_4^-)$ into the interlayer space of Mg/Al-LDH for subsequent chemical reduction to prepare a Au/LDH hybrid material (Fig. 1). In this way, we successfully synthesize ultra-thin 2D Au NSs between the metal hydroxide layers of LDH, which not only provide a confined-space effect to achieve controllable crystal growth, but also stabilize the resulting Au NSs. A detailed study by high-resolution electron microscopy illustrates that the as-prepared Au NSs are single crystalline with (001) basal planes and {100}-type edges, and that they are ultra-thin down to a few atomic layers. These 2D Au nanostructures exhibit excellent catalytic activity and stability towards the selective and solvent-free oxidation of C–H bonds using molecular oxygen as the oxidant.

Results

Synthesis and characterization. The as-synthesized Au/LDH hybrid contained 0.13 wt% of Au, as determined by inductively coupled plasma optical emission spectrometry (ICP-OES). Such a small loading of Au did not give rise to discernible change in the powder X-ray diffraction pattern (XRD) of LDH (Supplementary Fig. 1). However, the Au species could be clearly distinguished from the LDH substrate in the high-angle annular dark-field (HAADF) scanning transmission electron microscopy (STEM) images by the Z-contrast, which exhibited two kinds of typical morphologies: nanoparticles (NPs) in a size range of 2–10 nm; and irregular NSs with sizes ranging from several nanometers to tens of nanometers (Fig. 2, Supplementary Fig. 2). The NSs show apparently weaker contrast than even the smallest NPs, implying their ultra-thin nature. Assuming that Au NPs are spherical (that is, their 'thickness' is equal to their 'diameter'), the thickness of a Au NS can be approximately determined based on its contrast relative to that of a AuNP, because in HAADF-STEM the image intensity is roughly proportional to the specimen thickness for a given material composition[43,44]. One example is shown in Fig. 2a, in which the thickness of a Au NS was determined to be ~ 1.0 nm from the image intensity by using an adjacent AuNP (4.5 nm) as the reference, and energy dispersive X-ray spectroscopy (EDX) line scanning confirmed that the NSs are made of Au (Fig. 2a,c,e, Supplementary Fig. 3). Figure 2b shows another two NSs with larger lateral dimensions. With the method described above, the NS with brighter contrast was determined to be 1.6-nm thick, and its composition was confirmed by EDX to be Au (Fig. 2d). Although the EDX signal from the NS with lower contrast was too weak to be detected, it is presumed to be Au as well, considering its similarity in morphology to the thicker sheets. Accordingly, its calculated thickness based on the HAADF-STEM image intensity is 0.2–0.4 nm, which corresponds to only 1–2 atomic layers (Fig. 2f). We have randomly examined over 20 NSs and found that the thickness of a large portion of them is <1 nm. These NSs feature sub-nanometer thicknesses and large basal surfaces, which have not been simultaneously achieved in previous 2D Au nanostructures.

Figure 3 shows a high resolution transmission electron microscopy (HRTEM) image of a Au/LDH hybrid, in which lattice fringes are clearly observed over a large area (>100 nm^2). We demonstrated in a control experiment that the crystalline structure of the LDH was immediately amorphized, as evidenced by the rapid disappearance of reflections in the electron diffraction pattern, under similar TEM imaging conditions. Therefore, the observed lattice fringes must be associated with the 2D Au NS. The corresponding fast Fourier transform (FFT) indicates that the Au NS is single crystalline and [001] oriented. Notably, the (110) reflections that are forbidden for bulk Au appear in the FFT (marked with an asterisk, Fig. 3a), which is a characteristic phenomenon for 2D materials due to the loss of symmetry elements in 1D (ref. 27). We simulated HRTEM images for the [001]-oriented Au NSs of different thicknesses starting from a single monolayer. The results show that the (110) reflections appear in the FFTs when the NS contains an odd number of atomic layers and their intensity rapidly decreases as the layer number increases (Supplementary Fig. 4). By comparing the experimental and simulation results, we determined that the observed Au NS has an average thickness of three atomic layers based on the intensity ratio of reflections ($I_{(110)}/I_{(200)}$). As visualized by the Bragg-filtered HRTEM image, the lateral boundaries of the 2D Au NS mainly comprise {100}-type zigzag edges (Fig. 3a).

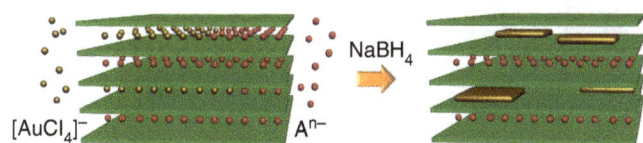

Figure 1 | Schematic illustration of the synthetic procedure of 2D Au NSs. Green slices, golden squares and red/golden spheres represent LDH layers, 2D Au NSs and A^{n-}/$AuCl_4^-$ anions, respectively.

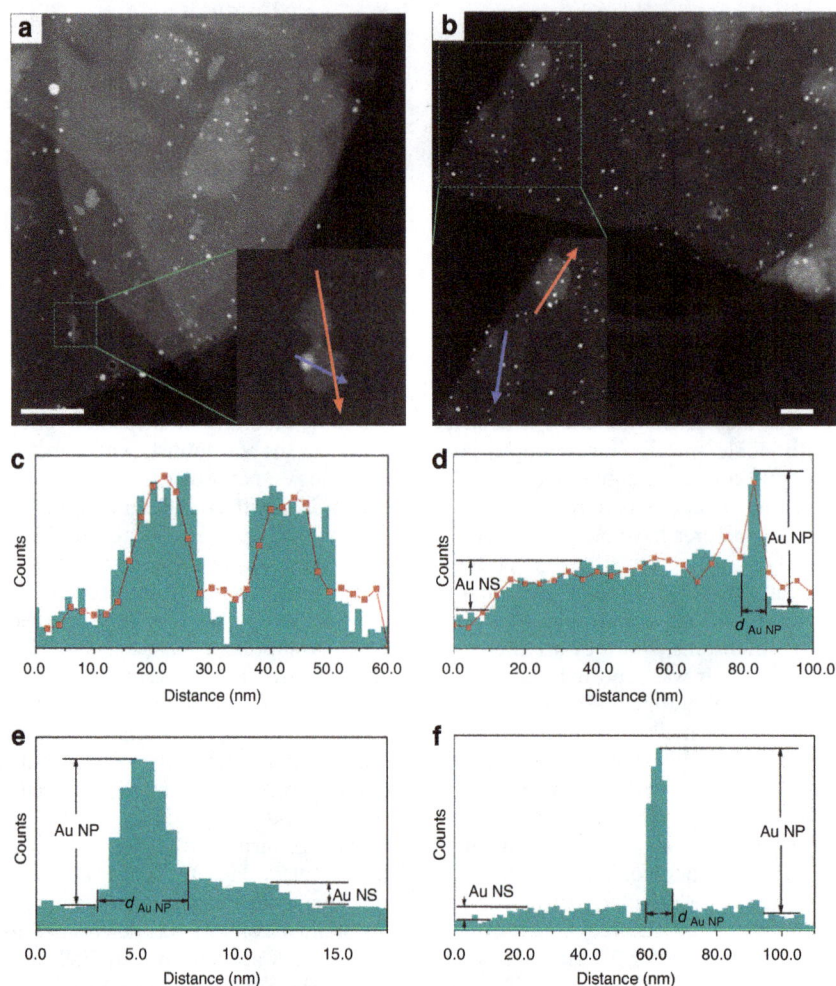

Figure 2 | Two sets of STEM/EDX analysis of the Au/LDH hybrid. (a,b) STEM images in which both Au NPs and Au NSs are observed in the LDH substrate. Scale bar, 100 nm for **a** and 50 nm for **b**. The selected regions are enlarged as the insets, and line scanning was performed over the NSs following the two arrows. (**c,d**) The normalized STEM intensity (green histogram) and the Au M-edge EDX intensity (red curves) line profiles, collected along the red arrows in **a,b**, respectively. (**e,f**) The STEM intensity line profiles collected along the blue arrows in (**a,b**), respectively. The EDX results confirm that the NSs are made of Au; the thickness of the Au NSs can be determined based on the STEM intensity by using Au NPs as a reference.

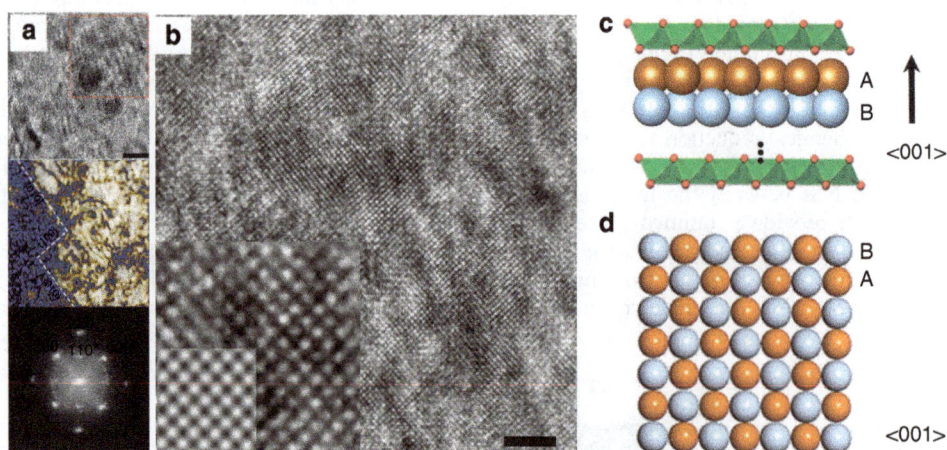

Figure 3 | HRTEM analysis of 2D Au NS. (**a**) HRTEM image taken at the edge of a 2D Au NS with scale bar at 5 nm, and Bragg-filtered image derived by inverse FFT of (200) and (020) reflections in the FFT diffractogram. The boundaries of the Bragg-filtered image were further enhanced by a Sobel filter. (**b**) Enlarged image of the highlighted region in **a** with scale bar at 2 nm. The inset is a further enlarged image to show atomic columns in comparison with a simulated HRTEM image (bottom left corner); the simulation was based on a three-layer [001]-oriented Au structural model (300 kV; Cs: 1.2 mm; Cc = 1.2 mm; $\Delta E = 0.7$ nm; $\Delta f = -65$ nm). Schematic illustrations of (**c**) the AB-stacked [001]-oriented 2D Au NS intercalated in the LDH, and (**d**) 2D Au NSs projected along the [001] axis, where gold atoms in different atomic layers (**a,b**) are represented by spheres in different colours.

It is naturally expected that a lateral image of 2D Au NSs allows a better understanding of their structures and direct measurement of their thicknesses. Therefore, we used focused-ion-beam technique to cut a thin (∼ 50 nm) slice out of a Au/LDH hybrid particle with the cutting direction perpendicular to the basal plane of LDH, to be able to observe the cross-sections of 2D Au NSs with HAADF-STEM. The acquired images indeed show good evidences for the presence of 2D Au NSs intercalated within the LDH layers (Supplementary Fig. 5). However, the ultra-thin nature and the consequent instability (on the irradiation of ion/electron beams) of the Au NSs inhibited more in-depth analysis, as the structure evolved during the focused-ion-beam and STEM imaging processes even with extremely low doses. The HRSTEM images of this direction along with detailed discussions can be found in the Supplementary Information (Supplementary Fig. 5).

(S)TEM characterization explicitly demonstrates the intercalation of sub-nanometer-thick Au NSs in the LDH substrate. It is worth noting that these Au NSs are unique not only in thickness but also in orientation. Unlike previously reported Au NSs that are usually [111] oriented[28,30], the Au NSs in the Au/LDH hybrid are [001] oriented, which may arise from the preferential interaction of the cationic LDH framework and the Au (001) surfaces. We attempted to isolate the Au NSs by treating the hybrid with HCl, but we found that the removal of LDH led to the transformation of Au NSs into NPs (only bulky Au NPs were observed in the final product). This result indicates the crucial stabilization effect of LDH on these anomalously oriented ultra-thin Au NSs. The metal–substrate interaction in the Au/LDH hybrid was characterized using X-ray photoelectron spectroscopy (XPS) by comparison with two reference materials (Supplementary Fig. 6), silica-supported Au NPs and LDH-supported Au NPs (Au NPs are deposited on the external surface of LDH). Silica-supported Au NPs show a binding energy of $Au4f_{7/2}$ at 84.1 eV, which is characteristic of metallic Au species[45]. LDH-supported Au NPs have a smaller $Au4f_{7/2}$ binding energy of 83.8 eV. Remarkably, the $Au4f_{7/2}$ binding energy measured for the Au/LDH hybrid is as low as 83.1 eV, corresponding to a large red shift of 1.0 eV relative to metallic Au. This reveals that the

intercalated 2D Au NSs are highly negatively charged and chemically bonded with the cationic layers of LDH.

The average coordination environment of Au in the Au/LDH hybrid was analysed based on the extended X-ray absorption fine structure (EXAFS) of the $Au-L_{III}$ edge (Fig. 4). In comparison with Au foil (bulk Au) or SiO_2-supported Au NPs, the Au/LDH hybrid has much less intense peaks associated with the Au–Au first-shell coordination. The derived Au–Au coordination number and bond distance are 6.9 ± 0.6 Å and 2.80 ± 0.01 Å, respectively, for the Au/LDH hybrid (Supplementary Table 1); and these values are markedly smaller than those of Au foil (11.2 ± 0.9 and 2.85 ± 0.01) and silica-supported Au NPs (10.1 ± 0.8 and 2.86 ± 0.01). These results provide further evidence for the presence of 2D Au nanostructures with atomic thickness in the Au/LDH hybrid that is responsible for the unusually low coordination number of Au. In addition, the Au–O coordination with a relatively small coordination number (0.7 ± 0.3) was also observed in the hybrid, suggesting a strong interaction between the surface atoms of Au NSs and the LDH cationic layers.

Catalyst evaluation. A series of supported Au catalysts was prepared with comparable Au loading amounts (0.1–0.2 wt%). The statistical analysis based on TEM images shows that, despite slight variations from sample to sample, all the catalysts have a similarly broad size distribution of Au NPs (2–10 nm) with the majority in the range of 3–6 nm (Supplementary Figs 7 and 8). There is no remarkable difference in particles sizes among different catalysts including the Au/LDH hybrid (only taking Au NPs into account). The catalysts were tested for the selective oxidation of ethylbenzene and toluene by molecular oxygen under solvent-free condition, where the ethylbenzene and toluene act as model molecules with secondary and primary C–H bonds, respectively. As summarized in Table 1, the catalysts containing Au NPs on conventional supports exhibit fairly low activity for these reactions: $AuNP/SiO_2$ and $AuNP/FeO_x$ converted <7% of ethylbenzene, while the AuNP/C catalyst was nearly inactive. In contrast, the Au/LDH hybrid exhibited much higher activity,

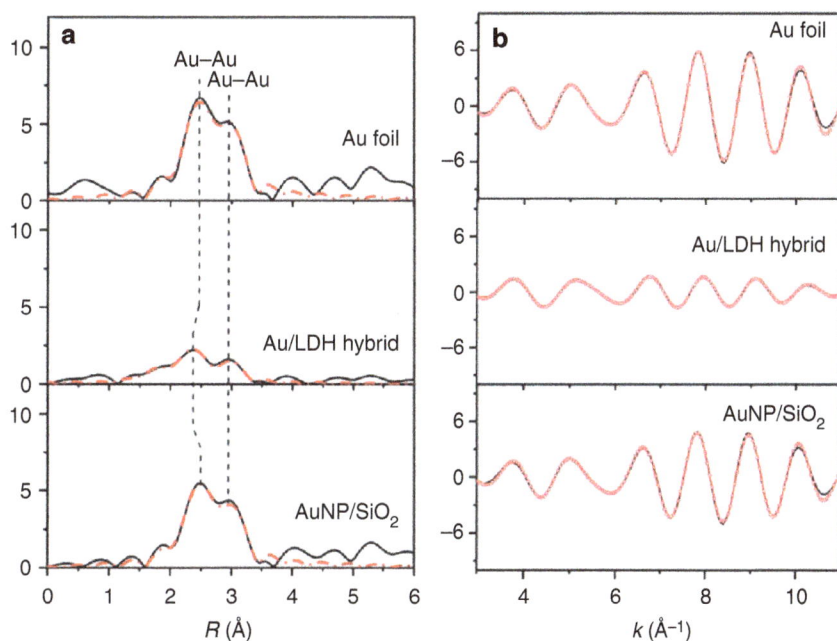

Figure 4 | EXAFS spectra of the Au-L$_{III}$ edge in various Au samples. (**a**) Fourier transforms of filtered $k^3 \cdot \chi(k)$ into the R space, where the red dashed lines correspond to the curve-fitting results; (**b**) filtered $k^3 \cdot \chi(k)$ in the k range of 3-11 Å^{-1}, where the red dotted lines correspond to the curve-fitting results.

Table 1 | Catalytic data in catalytic oxidation of ethylbenzene and toluene.

Entry	Subs.	Catalyst	Au (wt%)	Conv. (%)	Product selectivity (%)			
					P1	P2	P3	Others*
1	PhEt	Au/LDH hybrid	0.13	39.2	91.0	4.0	1.8	3.2
2†		Au/LDH hybrid	0.13	32.5	91.9	5.8	—	2.3
3‡		Au/LDH hybrid	0.14	40.0	90.0	7.0	—	3.0
4		AuNP/FeOx	0.15	6.7	82.7	6.9	4.0	6.4
5		AuNP/C	0.20	Trace	—	—	—	—
6		AuNP/SiO2	0.15	0.5	—	—	—	—
7		AuNP/LDH	0.16	11.0	88.9	3.3	3.5	4.3
8	PhMe	Au/LDH hybrid	0.13	9.2	66.0	34.0	—	—
9		AuNP/LDH	0.16	1.3	90.2	9.8	—	—
10		AuNP/FeOx	0.15	Trace	—	—	—	—
11		AuNP/C	0.20	Trace	—	—	—	—
12		AuNP/SiO2	0.15	Trace	—	—	—	—

Conv., conversion; LDH, layered double hydroxide; Subs., substrate.
Reaction conditions: 47 mmol of substrate, 100 mg of catalyst, 16 h, 140 °C, oxygen pressure at 3 MPa. t-butyl hydroperoxide (TBHP, 3 mol% relative to the substrate) is added as an initiator[49,50]. The results of the reactions without using TBHP are given in Supplementary Table 2.
*2-phenylethanol, benzyl alcohol and others.
†Air was used as the oxidant, reaction time 24 h.
‡The catalyst was treated at 350 °C for 2 h before use.

giving an ethylbenzene conversion of 39.2%. Even when air was used as the oxidant, the conversion was as high as 32.5%. The selectivity of acetophenone was >90% in this system (Table 1). When toluene was used as the reactant, most of the tested catalysts were inactive, which is in good agreement with the results in literature that the sole Au catalysts failed to catalyse the selective oxidation of toluene[46,47]. Very interestingly, the Au/LDH hybrid exhibited a noticeable conversion of 9.2%. Control experiments were carried out using LDH (without loading Au) and Au NP-deposited LDH (AuNP/LDH) as catalysts, during which the former was essentially inactive for both ethylbenzene and toluene oxidation and the latter was performed similarly to the Au NPs on conventional supports (Table 1). These results demonstrated that the high catalytic activity of the Au/LDH hybrid comes mainly from the 2D Au NSs. Considering the configuration of the Au NSs (sandwiched within the LDH layers), the edges of the 2D Au NSs would act as major catalytic active sites for the reactions.

We used molecular probes with different sizes, including ethylbenzene, diphenylmethane and triphenylmethane, to investigate the catalytic roles of the 2D Au NSs in the Au/LDH hybrid. Figure 5 shows the turnover frequency (TOF) of various substrates over the Au/LDH hybrid and AuNP/LDH catalysts, which were calculated based on the overall number of Au atoms in each catalyst. Clearly, the Au/LDH hybrid is much more active than the AuNP/LDH in the catalytic oxidation of both ethylbenzene (TOF: 5240 h^{-1} versus 600 h^{-1}) and diphenylmethane (TOF: 5100 h^{-1} versus 685 h^{-1}). When the bulky triphenylmethane was used as the substrate, however, the Au/LDH hybrid exhibited dramatically lower activity (TOF: 200 h^{-1}), whereas AuNP/LDH essentially retained its activity for small substrates (TOF: 443 h^{-1}). In the AuNP/LDH catalyst, the Au NPs (catalytic active sites) reside on the external surface of

the support with little diffusion limitation, and therefore it exhibits similar reaction rates for different substrates; in the case of the Au/LDH hybrid, however, bulky substrates have difficulty in diffusing into the interlayer regions of LDH and thus have limited contact with the edges of Au NSs, resulting in slower reactions. These results also suggest that the intercalated 2D Au NSs make a greater contribution to the catalytic activity of the Au/LDH hybrid than do the Au NPs on the LDH surface.

To directly probe the activity of 2D Au NSs in the Au/LDH hybrid catalyst, we selectively poisoned the Au NPs on the surfaces of LDH by capping them with polyvinylpyrrolidone (PVP, molecular weight at ~58,000) to inhibit access of the reactant molecules[48]. It is conceivable that the Au NSs would not be poisoned because PVP molecules are too large to enter the interlayer spaces of LDH. This poisoning strategy proved to be effective, as it successfully deactivated the AuNP/LDH catalyst. As shown in Fig. 5, on the poisoning treatment, the TOF value of AuNP/LDH for the oxidation of ethylbenzene decreased by ~80% (from 600 to 105 h^{-1}). In contrast, the Au/LDH hybrid catalyst retained high activity (TOF at 3724 h^{-1}) after the same treatment. This result unambiguously proves that the superior catalytic activity of the Au/LDH hybrid is largely because of the intercalated ultra-thin Au NSs. Since the exact weight percentage of Au NSs in the Au/LDH hybrid cannot be determined due to the presence of Au NPs, the TOF discussed above was calculated based on the total number of Au atoms. The real TOF of the 2D Au NSs should have an even higher value. Our preliminary calculation results based on a simplified system suggest that the negatively charged edge sites of 2D Au NSs synergize with the neighbouring hydroxyl groups of the LDH cationic layer to facilitate the adsorption and activation of oxygen molecules (Supplementary Fig. 9, Supplementary Table 3). However, it

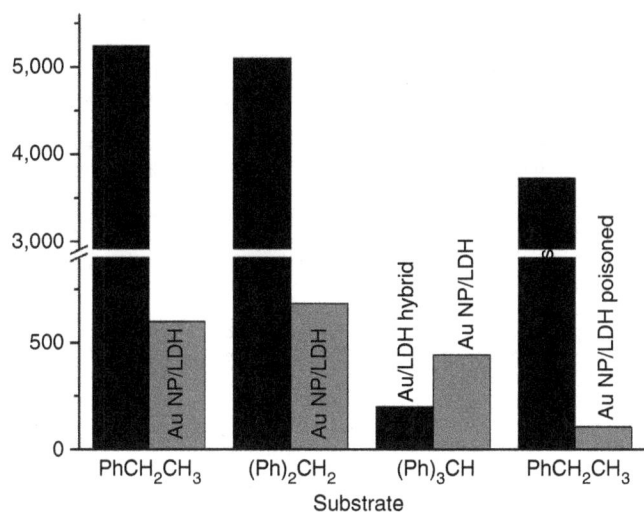

Figure 5 | Turnover frequencies of Au catalysts for the catalytic oxidation of various substrates. The reaction conditions are the same as specified in Table 1. The TOF values were normalized by the total amount of Au in the catalyst during the reaction time of 15 min.

should be pointed out that the adsorption of oxygen molecules only represents one of the several key steps of the reaction. More intensive studies are needed to better understand the origin of the high oxidation activity of the Au/LDH hybrid catalyst.

The Au/LDH hybrid catalyst is reusable. After each reaction run, it can be easily recycled by filtration with negligible Au leaching as confirmed by ICP-OES. Consequently, it exhibits constant catalytic performance during continuous reaction cycles. When used for the aerobic oxidation of ethylbenzene, for example, it gave stable conversion of ethylbenzene ($\sim$38%) and selectivity of acetophenone ($\sim$90%) in five reaction runs (Supplementary Fig. 10a). Furthermore, it is worth noting that the reaction rate of ethylbenzene conversion at the beginning (15 min) of each cycle is similar (Supplementary Fig. 10b), indicating the good recyclability of Au/LDH hybrid catalyst. Moreover, the Au/LDH hybrid catalyst is thermally stable. It showed unchanged catalytic activity for ethylbenzene oxidation after being heated at 350 °C for 2 h (Table 1). We used EXAFS to identify the structural evolution of the Au/LDH hybrid on heating treatment at different temperatures, and we found that the 350 °C-treated sample retained a low average Au–Au first-shell coordination number (7.5 ± 1.3). The 550 °C-treated sample, however, showed markedly increased Au–Au coordination numbers (10.2 ± 1.6), suggesting the transformation of 2D Au NSs to NPs (Supplementary Fig. 11). We characterized a Au/LDH hybrid sample, which was heated at 350 °C for 2 h and then used to catalyse the ethylbenzene oxidation for 16 h, with electron microscopy. The HRTEM image reveals the presence of ultra-thin, single-crystalline and [001]-oriented Au nanostructures (Supplementary Fig. 12), providing more straightforward evidence for the retention of 2D Au NSs during thermal treatment and catalytic use. These results demonstrate that the 2D Au NSs in the hybrid has excellent thermal stability up to at least 350 °C. Further investigation indicated that the Au/LDH hybrid catalyst is generally effective for the selective oxidation of various phenylic alkanes with primary and secondary C–H bonds. In the tested reactions, which were all performed under solvent-free conditions using molecular oxygen, it consistently gave the desired products in good activities and selectivities (Supplementary Table 4). The good stability combined with the general applicability makes the Au/LDH hybrid catalyst potentially useful for wide applications in the oxidation of C–H bonds.

Discussion

In summary, we successfully cast 2D Au NSs with thicknesses of single to a few atomic layers through confined-space synthesis using Al-Mg LDH as the host material. These ultra-thin Au NSs are negatively charged and stacked on the non-densely packed {001} family of crystal planes. In comparison with Au NPs supported on various materials, the Au NSs exhibited exceptionally high catalytic activities in the solvent-free oxidation of C–H bonds, which we attribute to the exposure of the low-coordinated edge sites. The LDH host material also stabilizes the Au NSs, making them reusable for multiple reaction runs with nearly unchanged catalytic performance. This synthetic method can be potentially used to prepare other types of 2D metallic nanocatalyst.

Methods

Catalyst preparation. *Synthesis of LDH.* About 30.76 g of Mg(NO$_3$)$_2$·6H$_2$O and 15 g Al(NO$_3$)$_3$·9H$_2$O were dissolved in 400 ml of water, followed by addition of 72 g of urea under stirring. After boiling for 8 h, precipitating at room temperature for 12 h, filtrating and washing with a large amount of water, samples of Mg-Al-LDH (molar ratio of Mg/Al at 3) were obtained.

Synthesis of Au/LDH hybrid. About 3 g of LDH powder was added into a 50 ml aqueous solution of HAuCl$_4$ (4.8×10^{-4} M) and stirred for 8 h at room temperature. After filtrating, washing with a large amount of water and drying under vacuum at 60 °C overnight, 1 g of the obtained solid sample was reduced by 35 mg of NaBH$_4$ to obtain metallic Au in 25 ml of anhydrous toluene and 7 ml of EtOH at room temperature for 6 h. Samples designated as 'Au/LDH hybrid' with Au loading at 0.13 wt% (by ICP) were obtained.

AuNP/SiO$_2$, AuNP/C and AuNP/FeO$_x$ were synthesized by the homogeneous deposition–precipitation method. In a typical run for the synthesis of AuNP/SiO$_2$, 3 g of amorphous SiO$_2$ (mesoporous MCM-41) was added to a 50 ml solution of HAuCl$_4$ (6.9×10^{-4} M) and urea (molar ratio of urea/Au at 100) and stirred at 90 °C for 4 h in a closed reactor kept away from light. Then the solid sample was filtrated, washed with large amount of water, dried at 100 °C for 12 h and calcined at 400 °C for 3 h. After treating the solid sample in anhydrous solvent of toluene and EtOH with NaBH$_4$ at room temperature for 6 h, the sample designated as AuNP/SiO$_2$ was obtained. The AuNP/C and AuNP/FeO$_x$ samples were synthesized in the same way using activated carbon and iron oxide as supports, respectively. The Au loadings on AuNP/SiO$_2$, AuNP/C and AuNP/FeO$_x$ were 0.15 wt%, 0.20 wt% and 0.15 wt%, respectively.

Synthesis of AuNP/LDH. About 700 mg of PVP (molecular weight = 58000) was added into a 50-ml aqueous solution of HAuCl$_4$ ($4.8*10^{-4}$ M). The mixture was further stirred for 30 min under a bath of 0 °C. Then, the aqueous solution of NaBH$_4$ (0.05 M, 15 ml) was added into the mixture under vigorous stirring. After stirring at 0 °C for another 2 h, 3 g of LDH powder was added into the mixture and stirred at room temperature for 3 h. Then, the mixture was stirred at 80 °C overnight to evaporate the water. Finally, the solid powder was calcined at 500 °C for 3 h in pure oxygen and treated in anhydrous solvent of toluene and EtOH with NaBH$_4$ at room temperature for 6 h. AuNP/LDH sample with the Au loading amount of 0.16 wt% was finally obtained.

Characterization. Powder XRDs were obtained with a Rigaku D/MAX 2550 diffractometer with CuKα radiation ($\lambda = 0.154056$ nm). The content of Au was determined from ICP with a Perkin-Elmer plasma 40 emission spectrometer. XPS spectra were performed by a Thermo ESCALAB 250, and the binding energy was calibrated by C1s peak (284.5 eV). The EXAFS of the Au-L$_{III}$ edge in Au foil and other Au samples were measured in a fluorescence mode at room temperature on BL14W1 beam line in the Shanghai Synchrotron Radiation Facility (SSRF). The storage ring was operated at 3.5 GeV with 200 mA as an average storage current. The synchrotron radiation was monochromatized with a Si (111) double crystal monochromator. Data were analysed using Athena and Artemis from the IFef-fit1.2.11 software package. XANES were normalized with edge height and then the first-order derivatives were taken to compare the variation of absorption edge energies. EXAFS oscillation, $\chi(k)$, was extracted using spline smoothing with a Cook-Sayers criterion, and the filtered k^3-weighted $\chi(k)$ was Fourier transformed into R space in the k range of 3–11 Å^{-1} with a Hanning function window. In the curve-fitting step, the possible backscattering amplitude and phase shift were calculated using FEFF8.4 code. HRTEM imaging was carried out on a FEI-Titan ST electron microscope operated at 300 kV with a point resolution of 0.19 nm. TEM image simulation was carried out using the QSTEM code with multi-slice method.

Catalytic tests. The solvent-free aerobic oxidation of ethylbenzene and toluene were performed in a high-pressure autoclave with a magnetic stirrer (1200 r.p.m.). Typically, the substrate, catalyst and initiator were mixed in the reactor by stirring for 1 h at room temperature. Then, the reaction system was heated to a given temperature (the temperature was measured with a thermometer in an oil bath) and oxygen was introduced and kept at the desired pressure. After the reaction, the product was taken out from the reaction system and analysed by gas chromatography (GC-14C, Shimadzu, using a flame ionization detector) with a flexible quartz capillary column coated with FFAP. The TOFs were calculated from the converted substrate per hour over per mol of Au species. The typical reaction conditions were as follows: 47 mmol of substrate, 100 mg of catalyst, 16 h of reaction time, 140 °C, oxygen pressure at 3 MPa, t-butyl hydroperoxide (TBHP in dodecane solution, 3 mol% based on substrate was added as the initiator) and reaction time of 15 min. The recyclability of the catalyst was tested by separating it from the reaction system by successive centrifugation, washing with a large quantity of methanol/water and drying at 80 °C for 3 h.

References

1. Seo, J. W. et al. Two-dimensional nanosheet crystals. Angew. Chem. Int. Ed. 46, 8828–8831 (2007).
2. Kibsgaard, J., Chen, Z. B., Reinecke, B. N. & Jaramillo, T. F. Engineering the surface structure of MoS_2 to preferentially expose active edge sites for electrocatalysis. Nat. Mater. 11, 963–969 (2012).
3. Yuan, W. J. et al. The edge- and basal-plane-specific electrochemistry of a single-layer graphene sheet. Sci. Rep. 3, 2248 (2013).
4. Chen, Z. B. et al. Core-shell MoO_3-MoS_2 nanowires for hydrogen evolution: a functional design for electrocatalytic materials. Nano Lett. 11, 4168–4175 (2011).
5. Merki, D., Fierro, S., Vrubel, H. & Hu, X. L. Amorphous molybdenum sulfide films as catalysts for electrochemical hydrogen production in water. Chem. Sci. 2, 1262–1267 (2011).
6. Li, Y. G. et al. MoS2 nanoparticles grown on graphene: an advanced catalyst for the hydrogen evolution reaction. J. Am. Chem. Soc. 133, 7296–7299 (2011).
7. Jaramillo, T. F. et al. Identification of active edge sites for electrochemical H-2 evolution from MoS_2 nanocatalysts. Science 317, 100–102 (2007).
8. Wang, Q. H., Kalantar-Zadeh, K., Kis, A., Coleman, J. N. & Strano, M. S. Electronics and optoelectronics of two-dimensional transition metal dichalcogenides. Nat. Nanotechnol. 7, 699–712 (2012).
9. Lee, Y. Y. et al. Top laminated graphene electrode in a semitransparent polymer solar cell by simultaneous thermal annealing/releasing method. ACS Nano. 5, 6564–6570 (2011).
10. Lin, T. Q., Huang, F. Q., Liang, J. & Wang, Y. X. A facile preparation route for boron-doped graphene, and its CdTe solar cell application. Energy Environ. Sci. 4, 862–865 (2011).
11. Xiang, Q. J., Yu, J. G. & Jaroniec, M. Synergetic effect of MoS2 and graphene as cocatalysts for enhanced photocatalytic H-2 production activity of TiO_2 nanoparticles. J. Am. Chem. Soc. 134, 6575–6578 (2012).
12. Li, Q. et al. Highly efficient visible-light-driven photocatalytic hydrogen production of CdS-cluster-decorated graphene nanosheets. J. Am. Chem. Soc. 133, 10878–10884 (2011).
13. Hamm, J. M. & Hess, O. Two two-dimensional materials are better than one. Science 340, 1298–1299 (2013).
14. Shi, Y. F. et al. Highly ordered mesoporous crystalline $MoSe_2$ material with efficient visible-light-driven photocatalytic activity and enhanced lithium storage performance. Adv. Funct. Mater. 23, 1832–1838 (2013).
15. Iwase, A., Ng, Y. H., Ishiguro, Y., Kudo, A. & Amal, R. Reduced graphene oxide as a solid-state electron mediator in Z-scheme photocatalytic water splitting under visible light. J. Am. Chem. Soc. 133, 11054–11057 (2011).
16. Meyer, J. C. et al. The structure of suspended graphene sheets. Nature 446, 60–63 (2007).
17. Novoselov, K. S. et al. Electric field effect in atomically thin carbon films. Science 306, 666–669 (2004).
18. Radisavljevic, B., Radenovic, A., Brivio, J., Giacometti, V. & Kis, A. Single-layer MoS_2 transistors. Nat. Nanotechnol. 6, 147–150 (2011).
19. Eda, G. et al. Photoluminescence from chemically exfoliated MoS_2. Nano Lett. 11, 5111–5116 (2011).
20. Oyer, A. J. et al. Stabilization of graphene sheets by a structured benzene/hexafluorobenzene mixed solvent. J. Am. Chem. Soc. 134, 5018–5021 (2012).
21. Wang, H. L., Robinson, J. T., Li, X. L. & Dai, H. J. Solvothermal reduction of chemically exfoliated graphene sheets. J. Am. Chem. Soc. 131, 9910–9911 (2009).
22. Coleman, J. N. et al. Two-dimensional nanosheets produced by liquid exfoliation of layered materials. Science 331, 568–571 (2011).
23. Chen, S. H. & Carroll, D. L. Synthesis and characterization of truncated triangular silver nanoplates. Nano Lett. 2, 1003–1007 (2002).
24. Bradley, J. S., Tesche, B., Busser, W., Masse, M. & Reetz, R. T. Surface spectroscopic study of the stabilization mechanism for shape-selectively synthesized nanostructured transition metal colloids. J. Am. Chem. Soc. 122, 4631–4636 (2000).
25. Jiu, J. T., Suganuma, K. & Nogi, M. Effect of additives on the morphology of single-crystal Au nanosheet synthesized using the polyol process. J. Mater. Sci. 46, 4964–4970 (2011).
26. Wu, Y. W., Hang, T., Wang, N., Yu, Z. Y. & Li, M. Highly durable non-sticky silver film with a microball-nanosheet hierarchical structure prepared by chemical deposition. Chem. Commun. 49, 10391–10393 (2013).
27. Duan, H. H. et al. Ultrathin rhodium nanosheets. Nat. Commun. 5, 3093 (2014).
28. Banu, K. & Shimura, T. Synthesis of large-scale transparent gold nanosheets sandwiched between stabilizers at a solid-liquid interface. N. J. Chem. 36, 2112–2120 (2012).
29. Huang, X. et al. Synthesis of hexagonal close-packed gold nanostructures. Nat. Commun. 2, 292 (2011).
30. Nootchanat, S., Thammacharoen, C., Lohwongwatana, B. & Ekgasit, S. Formation of large H_2O_2-reduced gold nanosheets via starch-induced two-dimensional oriented attachment. RSC Adv. 3, 3707–3716 (2013).
31. Huang, X. et al. Synthesis of gold square-like plates from ultrathin gold square sheets: the evolution of structure phase and shape. Angew. Chem. Int. Ed. 50, 12245–12248 (2011).
32. Parker, S. C. & Campbell, C. T. Reactivity and sintering kinetics of Au/TiO_2(110) model catalysts: particle size effects. Top. Catal. 44, 3–13 (2007).
33. Chen, M. S. & Goodman, D. W. The structure of catalytically active gold on titania. Science 306, 252–255 (2004).
34. Valden, M., Lai, X. & Goodman, D. W. Onset of catalytic activity of gold clusters on titania with the appearance of nonmetallic properties. Science 281, 1647–1650 (1998).
35. Liu, L. et al. The 2-D growth of gold on single-layer graphene/Ru(0001): enhancement of the CO adsorption. Surf. Sci. 605, L47–L50 (2011).
36. Zhou, Z. H., Gao, F. & Goodman, D. W. Deposition of metal clusters on single-layer graphene/Ru(0001): factors that govern cluster growth. Surf. Sci. 604, L31–L38 (2010).
37. Trifiro, F. & Vaccari, A. in Comprehensive Supramolecular Chemistry. Vol. 7 251 (Pergamon Elsevier Science, 1996).
38. Sels, B. et al. Layered double hydroxides exchanged with tungstate as biomimetic catalysts for mild oxidative bromination. Nature 400, 855–857 (1999).
39. Zhao, M. Q. et al. Embedded high density metal nanoparticles with extraordinary thermal stability derived from guest-host mediated layered double hydroxides. J. Am. Chem. Soc. 132, 14739–14741 (2010).
40. Ma, R. Z., Liang, J. B., Takada, K. & Sasaki, T. Topochemical synthesis of Co-Fe layered double hydroxides at varied Fe/Co ratios: unique intercalation of triiodide and its profound effect. J. Am. Chem. Soc. 133, 613–620 (2011).
41. Kwon, T., Tsigdinos, G. A. & Pinnavaia, T. J. Pillaring of layered double hydroxides (LDHs) by polyoxometalate anions. J. Am. Chem. Soc. 110, 3653–3654 (1988).
42. Thyveetil, M. A., Coveney, P. V., Greenwell, H. C. & Suter, J. L. Computer simulation study of the structural stability and materials properties of DNA-intercalated layered double hydroxides. J. Am. Chem. Soc. 130, 4742–4756 (2008).
43. Song, F. Q. et al. Free-standing graphene by scanning transmission electron microscopy. Ultramicroscopy 110, 1460–1464 (2010).
44. Goris, B. et al. Atomic-scale determination of surface facets in gold nanorods. Nat. Mater. 11, 930–935 (2012).
45. Yu, K., Wu, Z. C., Zhao, Q. R., Li, B. X. & Xie, Y. High-temperature-stable Au@SnO_2 sore/shell supported catalyst for CO oxidation. J. Phys. Chem. C 112, 2244–2247 (2008).
46. Kesavan, L. et al. Solvent-free oxidation of primary carbon-hydrogen bonds in toluene using Au-Pd alloy nanoparticles. Science 331, 195–199 (2011).
47. Haruta, M. Catalysis—gold rush. Nature 437, 1098–1099 (2005).
48. Quintanilla, A. et al. Weakly bound capping agents on gold nanoparticles in catalysis: surface poison? J. Catal. 271, 104–114 (2010).
49. Hughes, M. D. et al. Tunable gold catalysts for selective hydrocarbon oxidation under mild conditions. Nature 437, 1132–1135 (2005).
50. Zhao, R. et al. A highly efficient oxidation of cyclohexane over Au/ZSM-5 molecular sieve catalyst with oxygen as oxidant. Chem. Commun. 904–905 (2004).

Acknowledgements

This work was supported by the National Natural Science Foundation of China (21333009, U1462202 and 21403192), the National High-Tech Research and Development program of China (2013AA065301). This work was also supported by the KAUST Office of Competitive Research Funds (OCRF) under Awards No. FCC/1/1972-03-01 and FCC/1/1974-02-01.

Author contributions

L.W. prepared the catalyst, performed the characterizations and catalytic tests. Y.Z. performed the TEM characterization and analysed the catalyst structure. J.-Q. W. performed the EXAFS characterization and analysed the data. F.L., J.H., X.M. and J.-M.B. participated the discussion and analysis of the experimental data. L.W., Y.Z., Y.H. and F.-S.X. designed this study, analysed the data and wrote the paper.

Additional information

4

An allosteric photoredox catalyst inspired by photosynthetic machinery

Alejo M. Lifschitz[1], Ryan M. Young[1,2], Jose Mendez-Arroyo[1], Charlotte L. Stern[1], C Michael McGuirk[1], Michael R. Wasielewski[1,2] & Chad A. Mirkin[1]

Biological photosynthetic machinery allosterically regulate light harvesting via conformational and electronic changes at the antenna protein complexes as a response to specific chemical inputs. Fundamental limitations in current approaches to regulating inorganic light-harvesting mimics prevent their use in catalysis. Here we show that a light-harvesting antenna/reaction centre mimic can be regulated by utilizing a coordination framework incorporating antenna hemilabile ligands and assembled via a high-yielding, modular approach. As in nature, allosteric regulation is afforded by coupling the conformational changes to the disruptions in the electrochemical landscape of the framework upon recognition of specific coordinating analytes. The hemilabile ligands enable switching using remarkably mild and redox-inactive inputs, allowing one to regulate the photoredox catalytic activity of the photosynthetic mimic reversibly and *in situ*. Thus, we demonstrate that bioinspired regulatory mechanisms can be applied to inorganic light-harvesting arrays displaying switchable catalytic properties and with potential uses in solar energy conversion and photonic devices.

[1] Department of Chemistry and International Institute for Nanotechnology, Northwestern University, 2145 Sheridan Road, Evanston, Illinois 60208, USA. [2] Argonne-Northwestern Solar Energy Research (ANSER) Center, Northwestern University, Evanston, Illinois 60208, USA. Correspondence and requests for materials should be addressed to C.A.M. (email: chadnano@northwestern.edu).

Allosteric and recognition-based control of light harvesting in natural photosynthetic systems is critical to sustaining a balanced chemical and redox environment inside cells[1-5]. Achieving this kind of regulation with artificial light-harvesting systems would allow one to control the production of redox and energetic feedstock from light and, subsequently, to apply chemical regulation of photosynthetic activity to systems that extend beyond the reach of biology[6]. Indeed, a general approach for the chemical regulation of inorganic mimics would expand their application in solar energy conversion and photonic devices, as well as enabling their use in signal amplification, catalytic switches and stimuli-responsive materials[7-16]. Furthermore, achieving enzyme-like regulatory properties will become useful as artificial constructs begin to be studied in the context of light-harvesting hybrid systems comprising inorganic and biological units[17]. While there is a rich variety of approaches for controlling light harvesting in natural systems, triggering conformational changes in antenna protein complexes as a response to specific chemical inputs (such as pH fluctuations) represents a common strategy that can form the basis for regulation in inorganic systems[18-20]. Specifically, allosterically triggered conformational changes in antenna protein complexes induce kinetically fast quenching mechanisms (for example, charge transfer and electron transfer) by changing the electronic structure of the embedded dyes or by affecting their electronic coupling with other photoactive moieties[21-23].

The development of a general route for switchable inorganic mimics that can be applied to catalysis and device applications will require a novel approach which, similar to regulatory strategies in nature, allows for inducing large changes in light-harvesting efficiency using mild and selective inputs. Molecular arrays previously studied within this context have enabled remarkable self-regulation and switching of light-harvesting efficiency, yet they cannot be exploited for catalyst control given their intrinsic design limitations. For instance, the activity of light-harvesting antenna/reaction centre mimics has been regulated by appending photochromic switches that can photoisomerize into energy sinks[24-27]. Since the switching inputs cannot address the light-harvesting array and the switching unit independently and photoisomerization quantum yields are low, this approach is best suited to cap the overall yield of light harvesting at high light intensities rather than to create ON/OFF switches. Chemical switching via pH fluctuations has also been explored as a means to regulate light harvesting. Protonation can be used to tune the electronics of the light-harvesting units or to trigger the formation of energy sinks[28,29]. However, the need for large changes in pH results in switching inputs that are far more likely to affect the nature of the redox reactions catalysed by the light-harvesting array than to chemically switch the array itself, thus hampering the application of this approach in catalysis. Furthermore, the regulatory strategies mentioned above rely on quenching via the introduction of a divergent energy transfer step, which does not exhibit intrinsically faster kinetics than the light-harvesting photophysical pathways they set out to out-compete, often limiting switching ratios. Regulation of supramolecular structure in the reaction centre mimics represents a promising alternative since supramolecular switches can be operated using mild coordination inputs[30] or competitive inhibitors[31]. While coordination chemistry has been widely used to regulate molecular logic gates, chemosensors and to manage light-driven charge separation, there are no examples in which allosteric coordination inputs are used to shuttle a photoredox catalyst between active and inactive states reversibly as observed in biological photosynthetic systems.

Against this backdrop, we describe a set of allosterically controlled light-harvesting switches that incorporate photoactive components reminiscent of photosynthetic machinery and that can be used to reversibly regulate photoredox catalytic activity in situ and reversibly ((1-3)-ImC_{60}, Fig. 1). In particular, we employ an established model for a light-harvesting antenna/ reaction centre mimic composed of Bodipy, porphyrin and fullerene units, which are hereby embedded in a Rh(I) coordination complex whose interaction with allosteric

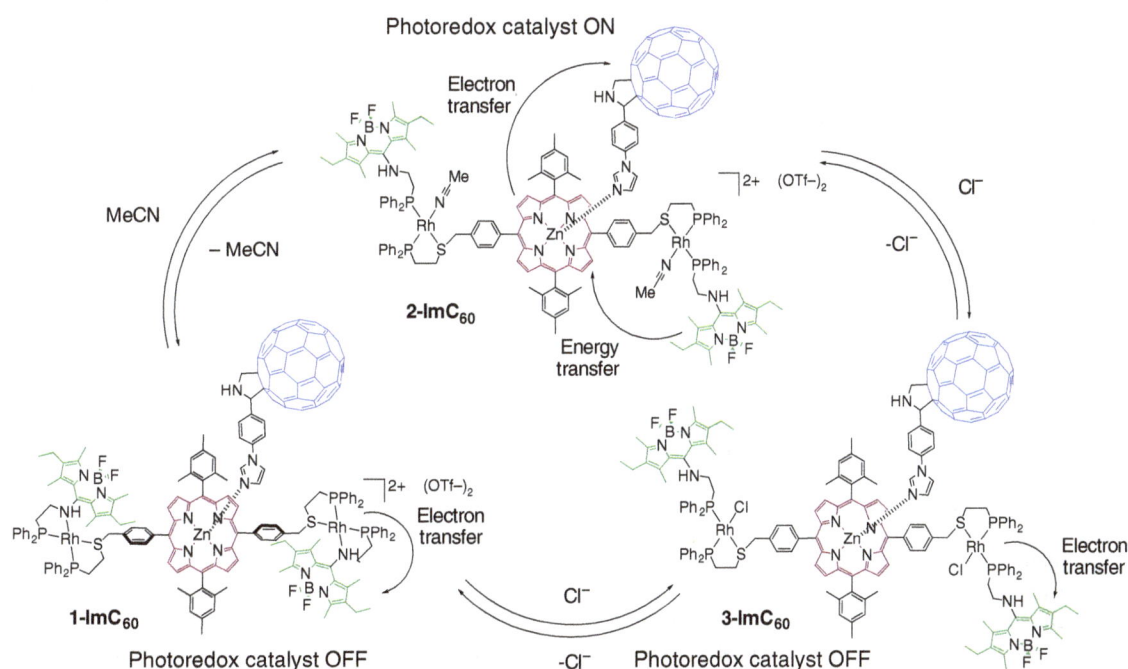

Figure 1 | Allosteric regulation of a light-harvesting antenna/reaction centre mimic. Activation of the catalytically active reaction centre mimic on the excitation of the antenna at 480 nm is controlled via coordination chemistry at a distal, redox-active Rh(I) centre. Light-harvesting antenna is shown in green and the reaction centre is shown in magenta and blue.

coordinating effectors can be used to disrupt the framework's electrochemistry. This regulatory approach is based on reversibly controlling antenna hemilabile ligand coordination to an allosteric Rh(I) receptor, enabling a rare example of chemical switching via coupled electronic and structural changes[12,32–34] typical of photosynthetic systems and proteins[35–37]. The use of a hemilabile coordination chemistry approach enables one to interconvert between active and inactive species in bulk and quantitatively using mild and redox-inactive chemical inputs that render the system compatible with a large array of redox processes. These structures are rapidly and reliably assembled via the weak-link approach (WLA)[38], a synthetic strategy that exploits orthogonal coordination interactions between structural metal centres and functional hemilabile ligands. Analogous to regulation strategies in photosystem II, the switching of catalytic activity depends on regulating energy transfer from the antennae to a reaction centre mimic exploiting an intrinsically faster quenching mechanism. In particular, switch $(1–3)$-ImC_{60} relies on utilizing coordinating effectors that control the redox potential of the antennae relative to the structural metal centres. Thus, photoinduced electron transfer (PeT)[39], which here arises in the sub-picosecond range[40], is used to out-compete picosecond-range energy transfer from the excited antenna to the catalytically active reaction centre. While the quenched state also results in light-driven charge separation, the redox potential of the species it is comprised of does not allow for catalytic turnover. Consequently, we are able to achieve light-harvesting switching ratios of up to 39-fold in a bulk sample. Importantly, the activation of photoredox activity occurs selectively upon binding of allosteric effectors with the proper electron-donating ability. Thus, we demonstrate that coordination chemistry can be exploited to surmount structural, regulatory and selectivity challenges relevant to the design of light-harvesting switches, enabling their potential use in catalytic, photonic and energy conversion application devices[13–15,41–43].

Results

Design and synthesis of inactive coordination framework.
In the past decade, many molecular photoactive ensembles that mimic the interactions between light-harvesting antenna and the reaction centre of photosynthetic systems[44] have been built. This is typically achieved by the incorporating an antenna chromophore whose ground state can be selectively excited at a given wavelength, which triggers excitation energy transfer to an electron transfer pair wherein that energy drives the formation of a charge-separated state. In these constructs, Bodipy has often been employed as the antenna since its HOMO-LUMO (highest occupied molecular orbital and lowest unoccupied molecular orbital) gap and redox properties can be readily tuned via synthetic modifications around its core. Furthermore, Bodipy can be covalently appended to porphyrin energy acceptors, often displaying high energy transfer efficiencies[45]. Incorporation of an electron acceptor into Bodipy–porphyrin ensembles can be achieved via a variety of covalent or supramolecular approaches[30], yet coordination of a nitrogen base-functionalized fullerene to a metallated porphyrin has proven a synthetically simpler approach that allows for PeT from the excited porphyrin to occur with high efficiency[46–47]. Herein, we employed a coordination-driven strategy based on the WLA to assemble a structurally switchable Bodipy–porphyrin-fullerene framework that circumvents the need for complex synthetic procedures. Towards this end, we assembled zinc porphyrin and Bodipy moieties functionalized with hemilabile coordination ligands (**4** and **5**, Supplementary Figs 1 and 2) around Rh(I) structural centres in a one-pot, high-yielding procedure leading to **1**

(Fig. 2a). This is possible due to the WLA and the halide-induced ligand rearrangement reaction[48], which serve to selectively arrange the ligands around Rh(I) yielding a heteroligated complex with porphyrin and Bodipy ligands chelated in a *cis* conformation. Diagnostic $^{31}P[^1H]$ nuclear magnetic resonance (NMR) spectroscopy (δ: 66.8, $J_{P-P} = 38$ Hz, $J_{P-Rh} = 178$ Hz and 50.0, $J_{P-P} = 38$ Hz, $J_{P-Rh} = 166$ Hz), as well as the solid-state structure of a related compound in which the antennae are modified with a thioether spacer (**X1**, Supplementary Fig. 3), support this structural assignment. While the quality of the crystal structure is low ($R1 = 0.14$), the structure provides a good overall picture of the coordination mode of **1** with two bidentate hemilabile ligands coordinated to each Rh(I) centre in a *cis* fashion. Chelation of Bodipy ligand **5** to the charged metal centre in **1** serves to shift the antenna's oxidation potential to higher values by 180 mV (Supplementary Figs 4–6 and Supplementary Table 1), enabling quenching of the excited Bodipy via PeT from Rh(I)[40]. This process can be observed to arise via transient absorption (TA) spectroscopy with a rate of $k_{PeT} = 2.0 \times 10^{12}$ s^{-1} in a model Bodipy-Rh(I) complex **S1**, whereas energy transfer in **1** takes place with a rate of $k_{EnT} = 8.3 \times 10^{10}$ s^{-1} (see Supplementary Figs 7–13). As a result, excitation of the antenna at 480 nm only results in minimal energy transfer to the porphyrin unit, as revealed by the fluorescence emission spectra of **1** (Fig. 2c).

Effects of allosteric effector binding.
By displacing the Bodipy hemilabile ligand with a neutral allosteric effector and expanding the complex, the oxidation potential of the antenna can be shifted to more negative potentials while maintaining the oxidation potential of Rh(I) (Supplementary Fig. 5). Thus, coordination of acetonitrile in **1** to give **2** ($^{31}P[^1H]$ NMR δ: 69.2, $J_{P-P} = 44$ Hz, $J_{P-Rh} = 172$ Hz and 25.8, $J_{P-P} = 44$ Hz, $J_{P-Rh} = 160$ Hz) disables PeT from Rh(I) and an increase in energy transfer efficiency to the porphyrin ensues (Fig. 2c). The change in energy transfer efficiency can be approximated from changes in the lifetime of the excited antenna in **1** and **2**, determined via near-infrared TA (NIR-TA) spectroscopy (Supplementary Figs 14–19), relative to previously reported Rh(I) complexes in which the zinc porphyrin unit is replaced by a benzyl group (complexes **S1** and **S2**, see Supplementary Methods)[40]. For example, comparison of the antenna excited state lifetimes (τ) in **1** and its model complex **S1** allows us to approximate the energy transfer efficiency (E) using the equation $E = 1 - (\tau_1/\tau_{S1})$. When taking into account the change in absorption cross section following acetonitrile coordination (Fig. 2a), our data yield an 11-fold switching ratio (Supplementary Table 2). This value is particularly remarkable given the strong distance dependence of energy transfer[49], suggesting that modulating PeT serves to enhance energy transfer even as the components are spatially further apart.

While energy transfer is enhanced via coordination of a neutral acetonitrile effector to **1**, the incorporation of anionic chloride effectors has the opposite effect. Indeed, the formation of **3** ($^{31}P[^1H]$ NMR δ: 71.4, $J_{P-P} = 40$ Hz, $J_{P-Rh} = 184$ Hz and 30.8, $J_{P-P} = 40$ Hz, $J_{P-Rh} = 166$ Hz) via the addition of two equivalents of N(n-Bu)$_4$Cl to either **1** or **2** results in minimal porphyrin-based emission upon excitation of the antenna (Fig. 2c). The electrochemical basis for the substrate specificity is derived from the fact that chloride lowers the Rh(I) oxidation potential ($\Delta V = -553$ mV) more than it shifts Bodipy's (Supplementary Fig. 6), thus enhancing rather than disabling the thermodynamic driving force for PeT from Rh(I) to Bodipy. By making the same approximations as above, we can estimate that the energy transfer efficiency in **3** is 8 times lower than in **1** and 39 times lower than in **2** (Supplementary Figs 20–25 and Supplementary Table 2).

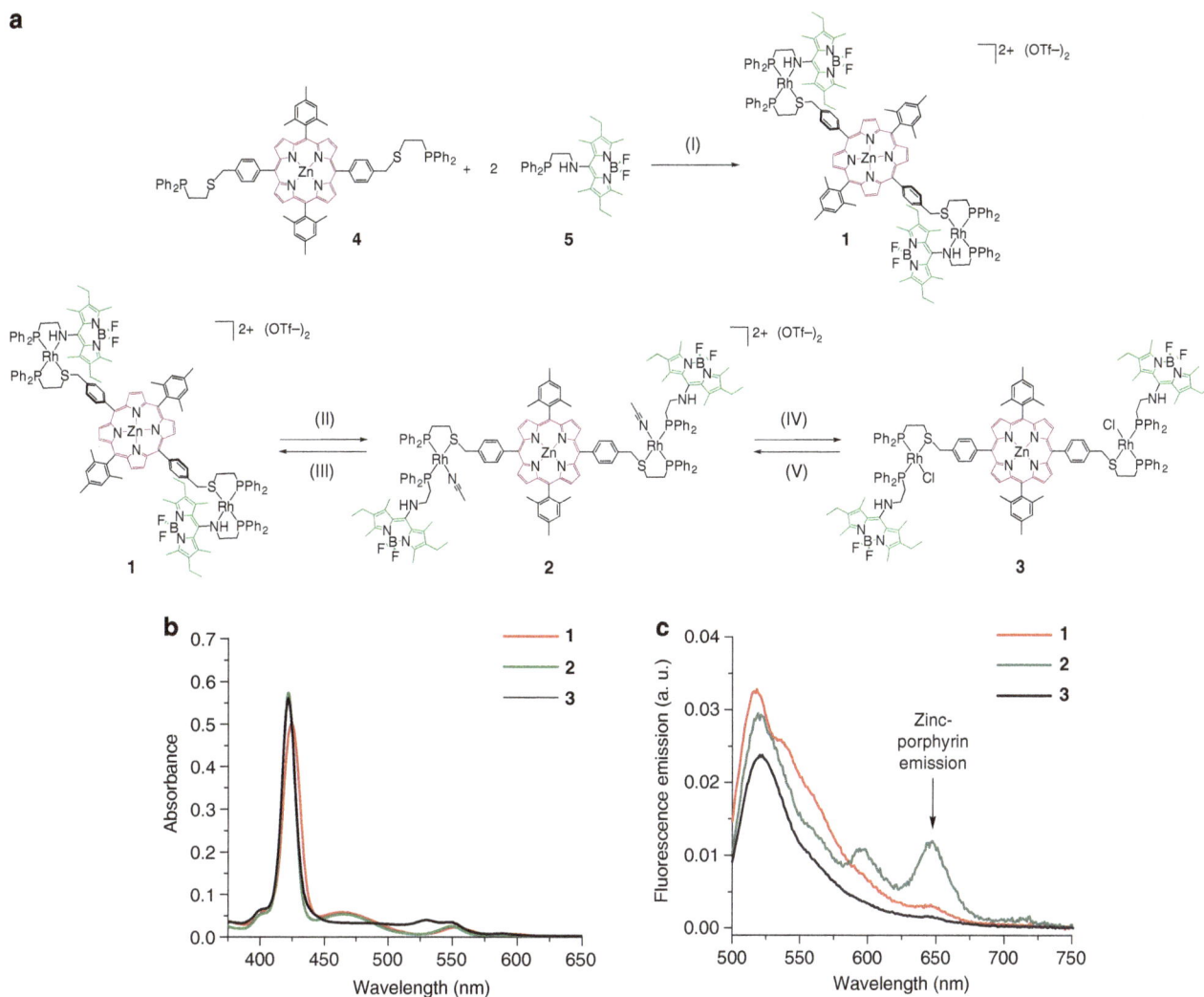

Figure 2 | Allosteric energy transfer switch operated via coordination chemistry. (a) Synthesis of WLA switch and toggling between coordination states with acetonitrile and chloride. Reaction conditions: (I) Rh$_2$Cl$_2$(COE), Tl(OTf), CH$_2$Cl$_2$; (II) 50 μl CH$_3$CN, CH$_2$Cl$_2$; (III) evaporation and redissolution in CH$_2$Cl$_2$; (IV) 2 equiv. N(n-Bu)$_4$Cl, CH$_2$Cl$_2$; (V) 2 equiv. Tl(OTf), CH$_2$Cl$_2$. **(b)** Absorbance spectra and **(c)** fluorescence emission spectra ($\lambda_{ex} = 480$ nm).

Importantly, the benzyl spacer in ligand **4** ensures that through-bond electronic communication between the porphyrin and the rest of the complex is reduced. Therefore, the lifetime of the excited energy acceptor is roughly constant following excitation of the antenna (1,274 ± 207 ps in **1**, 1,126 ± 67 in **2** and 1,121 ± 90 ps in **3**, $\lambda = 1,280$ nm) regardless of the changes in coordination mode. These values suggest that PeT from Rh(I) to Bodipy is indeed the primary contributor to the toggling of energy transfer.

Antenna-reaction centre mimic assembly. Axial coordination to the porphyrin's Zn(II) in **1–3** can be exploited to incorporate an imidazole-modified fullerene moiety (**ImC$_{60}$**, ref. 50), thus giving (**1–3**)-**ImC$_{60}$**. The Zn(II)-imidazole association constant is within the same order of magnitude for the three complexes ($K_a = 1.8 \times 10^5 \pm 5\%$ for **1**, $K_a = 1.1 \times 10^5 \pm 18\%$ for **2** and $K_a = 2.7 \times 10^5 \pm 13\%$ for **3**, see Supplementary Figs 26–28), owing to the fact that the porphyrin ligand exhibits a structural kink introduced by the benzyl bridge. This kink renders the axial coordination site physically accessible regardless of coordination changes at the Rh(I) centre (see solid-state structure of **4** and **X1**, Supplementary Figs 1 and 3). While coordination to Zn(II) requires an excess of **ImC$_{60}$**, the excess does not interfere with

coordination processes at the Rh(I) centre. Indeed, ^{31}P[^{1}H] NMR and optical spectra of (**1–3**)-**ImC$_{60}$** show that the coordination of hemilabile ligands **4** and **5** is not perturbed even after the addition of 40 equiv. of **ImC$_{60}$** and that the framework can be interconverted reversibly and *in situ* between the coordination states (Supplementary Figs 29–32). Thus, a series of orthogonal coordination behaviours particular to each ligand (Bodipy, porphyrin and fullerene) allow for the straightforward assembly of a switchable light-harvesting antenna, reaction centre mimic.

Regulation of charge separation in the reaction centre mimic. The optical properties of (**1–3**)-**ImC$_{60}$** were studied to evaluate their ability to form catalytically active, charge-separated states. In related Bodipy–zinc porphyrin-fullerene molecular assemblies, it has been observed that the rate of electron transfer from the excited porphyrin to the fullerene is significantly faster than the rate of energy transfer from Bodipy to the porphyrin, which results in a large decrease in the porphyrin-based fluorescence quantum yield[45,46]. In the case of (**1–3**)-**ImC$_{60}$**, steady-state fluorescence emission spectra of the three supramolecular structures reveal that the porphyrin excited state is largely depleted following the incorporation of the fullerene moiety and excitation of the antenna (Supplementary Fig. 33). This suggests that, while the formation efficiency of the porphyrin excited state is different in the three coordination modes, the excited porphyrin, however much of it is produced, will undergo electron transfer to the fullerene with high efficiency in all cases. Further, when the antennae are excited in equimolar samples of (1–3)-ImC_{60} using the same excitation intensity, the absorbance resulting from the fullerene radical anion at 1,010 nm, detected via NIR-TA spectroscopy, is noticeably different (Fig. 3). In particular, absorption by the fullerene anion is markedly larger in the case of 2-ImC_{60} (Fig. 3b) relative to both 1-ImC_{60} and 3-ImC_{60} (Fig. 3a,c). This suggests that the PeT switch above, which operates simply by introducing and abstracting allosteric effectors of the proper charge, can also regulate charge separation within the central reaction centre mimic.

Allosteric control of photoredox catalytic activity. To examine whether the modulation of charge-separation efficiency translates into the regulation of photoredox activity, the reduction of methyl viologen by 1-benzyl-1,4-dihydronicotinamide was investigated in the presence of (1–3)-ImC_{60} and an excitation source. This electron transfer process, which is energetically uphill in terms of the two reactants, stores about 1 eV of the energy absorbed by the light-harvesting Bodipy antenna[51,52]. Indeed, in the presence of the active species 2-ImC_{60}, catalytic turnover is observed ($k = 2.8 \times 10^{-7} \pm 0.1\% \, s^{-1}$), as evidenced by absorbance changes in the mixture at 630 nm (Fig. 4a and Supplementary Fig. 34), characteristic of the reduced methyl viologen species[53]. These bioinspired species are thus able to

make use of dihydronicotinamide as a sacrificial reductant, which itself is the basis for the reduced nicotinamide adenine dinucleotide, a common redox feedstock in biology. The catalytic rate for the reduction of methyl viologen with 2-ImC_{60} is similar to that observed when using a covalently appended zinc porphyrin-C_{60} dyad as the photoredox catalyst in which the porphyrin is directly excited[52]. On the other hand, the photoredox catalytic first-order rate constant becomes $k = 5.7 \times 10^{-8} \pm 0.2\% \, s^{-1}$ in 1-ImC_{60}, and catalytic turnover is indistinguishable from the background absorption for 3-ImC_{60} (Fig. 4a, Supplementary Figs 34–37).

The allosteric effectors employed to shuttle through (1–3)-ImC_{60} (one drop of acetonitrile or 2 equiv. of chloride) are remarkably mild, suggesting that they could be introduced in a photoredox catalytic mixture to address the activity of the catalyst without appreciably affecting the intrinsic properties of the reactants themselves. We thus studied the allosteric regulation of the photoredox activity of (1–3)-ImC_{60} in situ. Towards this end, a 10-μM solution of 1-ImC_{60} was prepared in the presence of methyl viologen and 1-benzyl-1,4-dihydronicotinamide mixture and a number of coordination transformations were performed in the presence of an excitation source. The diminished photoredox activity displayed by 1-ImC_{60} is significantly enhanced upon the addition of a single drop of acetonitrile to form 2-ImC_{60}, resulting in a first-order catalytic rate constant of $k = 5.4 \times 10^{-7} \pm 0.1\% \, s^{-1}$ (Fig. 4b, Supplementary Fig. 38). Whereas acetonitrile cannot be abstracted without evacuating the reaction mixture, the catalyst can be deactivated and reactivated in situ via the addition and

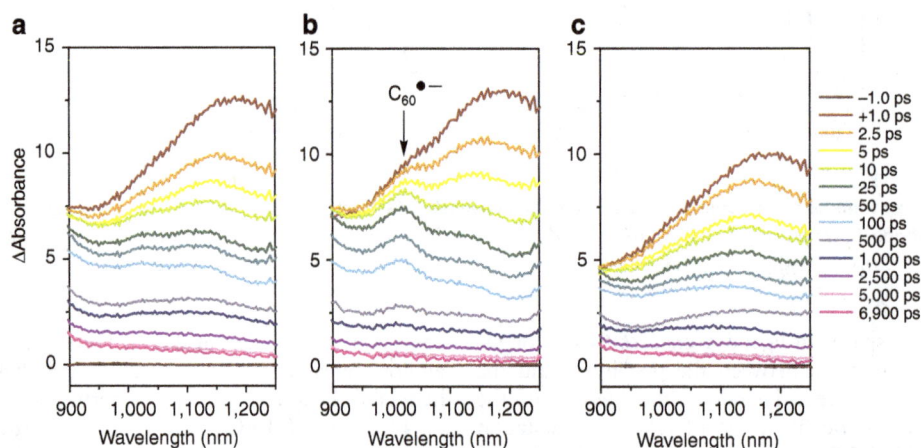

Figure 3 | NIR-TA spectra of (1–3)-ImC_{60}. Excitation of 3 μM solutions of 1 (a), 2 (b) and 3 (c) with 1.1-μJ laser pulses in the presence of 10 equiv. of ImC_{60} in CH_2Cl_2 ($\lambda_{ex} = 477$ nm).

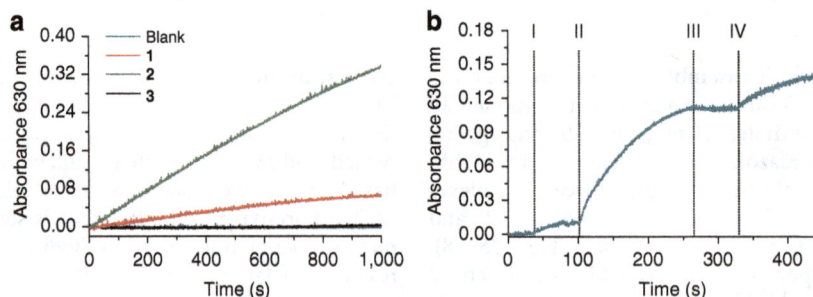

Figure 4 | Catalytic reduction of methyl viologen in the presence of (1–3)-ImC_{60}. (a) Changes in the absorbance at 630 nm in the presence of 1 μM 1, 2, 3 or no complex (blank) with 10 equiv. of ImC_{60} ($\lambda_{ex} = 480$ nm, 0.8 mW) in CH_2Cl_2. (b) Changes in the absorbance at 630 nm in the presence of 5 μM CH_2Cl_2 solution of 1 with 10 equiv. of ImC_{60} on: (I) turning excitation source on ($\lambda_{ex} = 480$ nm, 0.3 mW), (II) addition of acetonitrile (1 drop), (III) addition of chloride (2 equiv.), and (IV) addition of Tl(OTf; 6 equiv.).

abstraction of chloride, and thus two cycles of catalyst activation can be achieved. The reactivated catalyst displays a catalytic rate constant of $k = 5.2 \times 10^{-7} \pm 0.3\% \, \mathrm{s}^{-1}$, suggesting that the light-harvesting framework can be regenerated into its active state reversibly without any significant loss in catalytic activity. $^{31}\mathrm{P}[^1\mathrm{H}]$ NMR measurements of the catalytic sample revealed no degradation of the complex throughout the two cycles. However, degradation via oxidation, detected via the characteristic $^{31}\mathrm{P}[^1\mathrm{H}]$ NMR shift of alkyldiphenylphosphine oxide at δ: 29 ppm, was observed upon undergoing multiple cycles of allosteric catalyst activation and deactivation.

Discussion

In the case of Photosystem II, regulation of energy transfer from light-harvesting antenna is possible since quenching via charge transfer between excited antennae occurs in the picosecond to sub-picosecond time scale[2], kinetically outcompeting other decay pathways that lead to photosynthetic activity. This regulatory strategy provides the basis for switchability in the inorganic system described herein; one can chemically shuttle between two electron transfer processes, one that results in quenching and one that gives rise to catalytic activity, because the former is preceded by an inherently slower energy transfer step. Furthermore, while the charge-separated state that results from quenching the antenna can in principle reduce methyl viologen, the use of a sacrificial reductant whose oxidation potential is significantly higher than that of Rh(I) prevents its catalytic turnover in the inactive state. The multi-state switch $(1–3)$-$\mathbf{ImC_{60}}$ directly borrows the regulatory strategies seen in nature and combines them with allosteric regulation strategies by targeting a fast electron transfer process that is triggered through coupled electronic and conformational changes. As a result, $(1–3)$-$\mathbf{ImC_{60}}$ achieves a bulk switching ratio between the active and inactive states of up to 39-fold in terms of energy transfer efficiency, not observed by any previously developed regulatory strategy. The fact that regulation only occurs at the antenna level in $(1–3)$-$\mathbf{ImC_{60}}$ suggests that this fundamental approach can be readily applied for controlling other light-harvesting and bioinspired photoredox systems.

In addition to demonstrating for the first time that the photoredox catalytic activity of a known light-harvesting, reaction centre mimic can be regulated *in situ* and reversibly, our approach to modulating electron transfer reactions allosterically also shows that coordination-based catalyst switches can be endowed with substrate selectivity. This can become a critical component of WLA catalytic switches, which have been used in the context of PCR-like inorganic substrate amplification, but which are triggered without any input selectivity. In the WLA system described herein, we have demonstrated that the activation of a central catalytic moiety in $(1–3)$-$\mathbf{ImC_{60}}$ only occurs when recognition of a targeted analyte provides the proper Rh(I) oxidation potential for light harvesting to occur. This degree of analyte selectivity will become important when applying the principles outlined herein to develop photoredox catalyst-based signal amplification systems and allosterically controlled photoredox catalysts. Furthermore, $(1–3)$-$\mathbf{ImC_{60}}$ may be coupled to nucleophile-releasing redox reactions[54] to trigger negative feedback loops that chemically limit the efficiency of light harvesting, representing an important step towards self-regulated inorganic systems.

Methods

General methods. All reactions, transformations, characterization and catalytic experiments were performed in strictly oxygen- and water-free conditions inside a nitrogen-atmosphere glove box or in sealed containers under a stream of argon. All glassware, cuvettes and cells were oven-dried for 24 h. Dichloromethane (CH_2Cl_2),

ether, toluene, acetonitrile and tetrahydrofuran (THF) solvents were transferred from an oxygen- and water-free solvent system (J.C. Meyer) into sealed containers and purged with a stream of argon for 20 min before use. All the other solvents were purchased as HPLC grade, similarly purged for 20 min with argon and stored with activated molecular sieves. Deuterated solvents were purchased from Cambridge Isotope Laboratories and were degassed with a stream of argon before use. 5,15-Bis(mesityl)-10,20-bis(4-(chloro-methyl)phenyl)porphyrin]zinc[55], $\mathbf{ImC_{60}}$[50], P,N-Bodipy (5), P,S-benzyl (S4), model complexes S1 and S2[40], and methlyviologen hexafluorophosphate[56] were prepared from literature procedures. 1-benzyl-1,4-dihydronicotinamide was purchased from TCI chemicals and was used as received. All other chemicals were purchased from Aldrich Chemical Co. and used as received. NMR spectra were recorded on a Bruker Avance 400 MHz (Supplementary Figs 39–62). $^1\mathrm{H}$ and $^{13}\mathrm{C}[^1\mathrm{H}]$ NMR spectra were referenced to residual proton and carbon resonances in the deuterated solvents. $^{13}\mathrm{C}[^1\mathrm{H}]$ NMR spectra of complexes $\mathbf{1}$–$\mathbf{3}$ are not provided as WLA complexes give rise to markedly broad and featureless resonances (ref. 15). $^{31}\mathrm{P}[^1\mathrm{H}]$ NMR spectra were referenced to an 85% H_3PO_4 aqueous solution. $^{19}\mathrm{F}$ NMR spectra were referenced to a $CFCl_3$ sample in $CDCl_3$ solution. $^{11}\mathrm{B}[^1\mathrm{H}]$ NMR spectra were referenced to neat $BF_3 \bullet OEt_2$. All the chemical shifts are reported in p.p.m. High-resolution electrospray ionization (ESI) mass spectra measurements were recorded on an Agilent 6120 LC-TOF instrument in positive ion mode.

Ultraviolet–visible absorption measurements were performed in a Varian Cary 50 Bio spectrophotometer utilizing screw-cap 10-mm cell-path quartz cuvettes (VWR). Steady-state fluorescence emission measurements were carried out in a Horiba Jovin-Yvonne Fluorolog fluorimeter and fluorescence quantum yields were derived from the comparative method using fluorescein isothiocyanate in ethanol as standard. In particular, quantum yields were calculated from the least-squares fit of the integrated fluorescence emission versus absorbance values curves comprising seven dilution samples. Cyclic voltammetry measurements were performed with an Epsilon BASi potentiostat in an air-tight cell comprising a glassy carbon working electrode, a Ag wire pseudoreference electrode and a Pt wire auxiliary electrode. Samples were prepared as 5 mM solutions in 0.1 M n-Bu$_4$NPF$_6$ in CH_2Cl_2. Scan rates of 100 mV s^{-1} were used in all measurements. All cyclic voltammetry graphs show voltages versus the ferrocene/ferrocenium couple. Titrations were performed via the addition of aliquots from a 1 mM dichlorobenzene solution into a 1 µM CH_2Cl_2 solution of Rh(I) coordination complex.

Titrations of $\mathbf{ImC_{60}}$ were studied via ultraviolet–visible spectroscopy at room temperature in a screw-cap cuvette with a total volume of 2 ml and a constant hosts $\mathbf{1}$–$\mathbf{3}$ concentration of 1 µM in CH_2Cl_2. Samples were prepared in a glove box. Titrations were performed via the addition of aliquots of 1 mM $\mathbf{ImC_{60}}$ solution in 1,2-dichlorobenzene into the solution of hosts $\mathbf{1}$–$\mathbf{3}$. Binding affinities (K_a) were calculated by monitoring the change in absorbance at 434 nm, corresponding to the axially coordinated porphyrin Soret band. Binding affinity was obtained by non-linear regression analysis of the binding curves utilizing equation (1):

$$\Delta Abs = \varepsilon \times 0.5 \left([G] + [H] + \left(\frac{1}{K} \right) - \sqrt{ \left(x + [H] + \left(\frac{1}{K} \right) \right)^2 - (4 \times [H] \times [G]) } \right)$$

$$(1)$$

Photoredox catalysis. Catalytic experiments were performed in screw-cap 10-mm cell-path quartz cuvettes containing 0.1 mM methylviologen hexafluorophosphate and 0.1 mM 1-benzyl-1,4-dihydronicotinamide in CH_2Cl_2. Methyl viologen was added from a concentrated methanolic solution. $\mathbf{ImC_{60}}$ was added from a concentrated 1,2-dichlorobenzene solution. In a typical experiment, the reagents, catalyst, $\mathbf{ImC_{60}}$ and a stir bar were loaded inside the cuvette in the dark inside a glove box. The cuvette was placed inside a ultraviolet–visible spectrophotometer and the solution was vigorously stirred with a magnetic stirrer. An encasing was built above the cuvette holder to provide a stream of argon while performing the measurements. A laser diode (PicoQuant) was placed inside the spectrophotometer perpendicular to the probe beam and excitation power was monitored regularly using a S130C slim photodiode power sensor connected to PM200 power and energy metre console (Thorlabs). The redox reaction was monitored by tracking the absorbance changes at 630 nm, characteristic of reduced methyl viologen species. To add allosteric effectors to the reaction mixture, excitation was halted and the cuvette was taken into a glove box and then returned to the spectrophotometer.

Experimental catalytic rate constants (k) were obtained by non-linear regression analysis of the kinetic curves produced by plotting the product concentration (derived from Beer's Law) versus time. Equation $[P] = R_0(1-e^{-kt})$ was fit to the experimental value using the program GraphPad using an initial starting material concentration of 0.1 M.

Time-resolved optical spectroscopy. Visible/near-infrared femtosecond TA spectroscopy was performed on an instrument that is described elsewhere[57]. In brief, the 827-nm fundamental output of a commercial Ti:sapphire laser system (Tsunami oscillator / Spitfire amplifier, Spectra-Physics) was frequency doubled to 414 nm (~ 150 µJ per pulse). In our previous study, this was attenuated and used as the pump; here it is used to pump a seeded, two-stage, laboratory-constructed optical parametric amplifier. The signal output of the optical parametric amplifier was tuned to the desired wavelength then attenuated and chopped at 500 Hz. The single-filament continuum probe was generated by focusing into the appropriate non-linear medium for the desired probe range. For visible continuum (360–800 nm), $\sim 1.5\,\mu J$ per pulse was tightly focused into a 5-mm cuvette containing a 1:1 mixture of $H_2O{:}D_2O$, while for the NIR probe (800–1,600 nm), $\sim 5\,\mu J$ per pulse was loosely focused into a proprietary source (Ultrafast Systems, LLC). The residual fundamental beam was filtered using an appropriate edge filter. The polarization of the pump beam was rotated to 54.7° ('magic' angle) relative to the horizontal probe. Both pump and probe were focused to $\sim 100\,\mu m$ at the sample. After interaction, the transmitted probe was then coupled into an optical fibre and detected using the appropriate detector (customized Helios, Ultrafast Systems, LLC).

TA signals were fit to a convolution of a Gaussian instrument response function with the sum of a multi-exponential decay and a step function for signals extending far beyond the experimental window. Uncertainties are reported as the standard error of the fit.

X-ray crystallography. Single crystals were mounted using oil (Infineum V8512) on a glass fibre. All measurements were made on a CCD area detector with MX Optics Cu Kα radiation. Data were collected using Bruker APEXII detector and processed using APEX2 from Bruker. All the structures were solved by direct methods and expanded using Fourier techniques. The non-hydrogen atoms were refined anisotropically. Hydrogen atoms were included in idealized positions, but not refined. Crystallographic information is provided in Supplementary Table 3 and Supplementary Data 1.

Synthesis of P,S-Bz-ZnP(mes)$_2$-Bz-S,P (4). 5,15-Bis(mesityl)-10,20-Bis(4-(chloro-methyl)phenyl)porphyrin]zinc **S5** (1.50 g, 1.75 mmol) and P,S-TIPS **S6** (1.48 g, 3.67 mmol) were dissolved in 100 ml of degassed THF in a Schlenk flask. 1.52 g of CsF were added and the mixture was left to stir in the dark at 60 °C for 24 h under a nitrogen atmosphere. The reaction mixture was filtered and the solvent was evacuated. The resulting solid was redissolved in a small amount of CH_2Cl_2 and poured onto a large silica plug. The P,S-TIPS was washed with excess CH_2Cl_2 and a deep purple band was eluted with a 1:1 mixture of $CH_2Cl_2{:}THF$. The product was isolated, redissolved in hot CH_2Cl_2 with a minimal amount of THF and crashed out solution by adding pentane, affording **4** as deep purple needles (1.92 g, 75% yield). ^1H NMR (400.16 MHz, 25 °C, CD_2Cl_2): δ 8.71 (dd, $J_{H\text{-}H} = 6$ Hz, 8 H), 7.92 (d, $J_{H\text{-}H} = 8$ Hz, 4 H), 7.37 (d, $J_{H\text{-}H} = 8$ Hz, 4 H), 7.30 (s, 4 H), 7.13 (m, 12 H), 6.58 (m, 8 H), 3.77 (s, 4 H), 2.63 (s, 6 H), 1.77 (m, 16 H), 1.60 (m, 4 H). $^{13}C[^1H]$ (100.63 MHz, 25 °C, CD_2Cl_2): δ 150.3 (s), 142.0 (s), 140.3 (s), 139.6 (s), 138.5 (d, $J_{C\text{-}P} = 12$ Hz), 137.9 (s), 136.3 (s), 135.0 (s), 133.1 (s), 132.7 (s), 132.4 (s), 130.9 (s), 129.1 (s), 128.8 (d, $J_{C\text{-}P} = 5$ Hz), 128.1 (s), 127.3 (s), 119.9 (s), 119.1 (s), 36.6 (s), 28.7 (d, $J_{C\text{-}P} = 17$ Hz), 27.2 (d, $J_{C\text{-}P} = 15$ Hz), 22.9 (s), 21.6 (s). $^{31}P[^1H]$ NMR (161.98 MHz, 25 °C, CD_2Cl_2): δ -19.1 (s). HRMS (ESI +) m/z calculated for $[M]^+$: 1276.3809; found: 1276.3802.

Synthesis of [Rh$_2$(κ_2:η_2:κ_2-4)(κ_2-5)$_2$](OTf)$_2$ (1). Rh$_2$Cl$_2$(cyclooctene)$_4$ (3.6 mg, 0.010 mmol) was dissolved in 3 ml of a 1:1 solution of $CH_2Cl_2{:}THF$ and 3 ml of **4** (6.4 mg, 0.005 mmol) in THF were added dropwise over the course of 5 min. Immediately afterwards, **3** (5.3 mg, 0.010 mmol) dissolved in 3 ml of THF was added dropwise. The solution was sonicated for 10 min and was stirred for 3 h. The solution was concentrated to about 1 and 4 ml of CH_2Cl_2 were added. Tl(OTf; 3.6 mg, 0.010 mmol) was added and the mixture was stirred for 1 h and filtrated through a pad of Celite. The product was recrystallized from a CH_2Cl_2/pentane solution twice, yielding **1** as a dark red solid (*in situ* $^{31}P[^1H]$ NMR yields = 90–95%, isolated yield 12.1 mg, 85%). ^1H NMR (400.16 MHz, 25 °C, CD_2Cl_2): δ 7.61 (d, $J_{H\text{-}H} = 4$ Hz, 2 H), 7.45–7.15 (m, 15 H), 7.12–7.00 (m, 8 H), 5.90 (m, 1 H), 4.26 (s, 2 H), 3.91 (m, 2H), 2.53 (m, 2H), 2.38 (m, 10 H), 2.20 (s, 6 H), 2.12–1.95 (m, 4 H), 1.02 (t, $J_{H\text{-}H} = 8$ Hz, 6 H). $^{31}P[^1H]$ NMR (161.98 MHz, 25 °C, CD_2Cl_2): δ 66.8 (dd, $J_{P\text{-}P} = 38$ Hz, $J_{P\text{-}Rh} = 178$ Hz, 2P), 50.0 (dd, $J_{P\text{-}P} = 38$ Hz, $J_{P\text{-}Rh} = 166$ Hz, 2P). ^{19}F NMR (376.49 MHz, 25 °C, CD_2Cl_2): δ −79.4 (s, 6F), −145.6 (m, 4F). $^{11}B[^1H]$ NMR (128.38 MHz, 25 °C, CD_2Cl_2): δ 0.09 (t, $J_{B\text{-}F} = 32$ Hz). HRMS (ESI +) m/z calculated for $[M\text{-}2OTf]^{2+}$: 1272.3746; found: 1272.3762.

Synthesis of [Rh$_2$(CH$_3$CN)$_2$(κ_2:η_2:κ_2-4)(5)$_2$](OTf)$_2$ (2). Complex **1** (5 mg, 1.8×10^{-3} mmol) was dissolved in 1 ml of CH_2Cl_2 and one drop of acetonitrile was added ($^{31}P[^1H]$ NMR yield: quantitative). ^1H NMR (400.16 MHz, 25 °C, CD_2Cl_2): δ 7.61 (d, $J_{H\text{-}H} = 4$ Hz, 2 H), 7.45–7.15 (m, 15 H), 7.12–7.00 (m, 8 H), 5.90 (m, 1 H), 4.26 (s, 2 H), 3.91 (m, 2H), 2.53 (m, 2H), 2.38 (m, 10 H), 2.20 (s, 6 H), 2.12–1.95 (m, 4 H), 1.02 (t, $J_{H\text{-}H} = 8$ Hz, 6 H). $^{31}P[^1H]$ NMR (161.98 MHz, 25 °C, CD_2Cl_2): δ 69.2 (dd, $J_{P\text{-}P} = 44$ Hz, $J_{P\text{-}Rh} = 172$ Hz, 2P), 25.8 (dd, $J_{P\text{-}P} = 44$ Hz, $J_{P\text{-}Rh} = 160$ Hz, 2P). ^{19}F NMR (376.49 MHz, 25 °C, CD_2Cl_2): δ −79.0 (s, 6F), −145.4 (q, $J_{F\text{-}B} = 34$ Hz, 4F). $^{11}B[^1H]$ NMR (128.38 MHz, 25 °C, CD_2Cl_2): δ 0.74 (t, $J_{B\text{-}F} = 32$ Hz). HRMS (ESI +) m/z calculated for $[M\text{-}2OTf]^{2+}$: 1313.3011; found: 1313.3035.

Synthesis of [Rh$_2$Cl$_2$(κ_2:η_2:κ_2-4)(5)$_2$] (3). Complex **1** (5 mg, 1.8×10^{-3} mmol) was dissolved in 1 ml of CH_2Cl_2 and tetrabutylammonium chloride was added. The product was precipitated from solution via the addition of pentane ($^{31}P[^1H]$ NMR yield: quantitative, isolated yield 4.4 mg, 96%). ^1H NMR (400.16 MHz, 25 °C, CD_2Cl_2): δ 7.61 (d, $J_{H\text{-}H} = 4$ Hz, 2 H), 7.45–7.15 (m, 15 H), 7.12–7.00 (m, 8 H), 5.90 (m, 1 H), 4.26 (s, 2 H), 3.91 (m, 2H), 2.53 (m, 2H), 2.38 (m, 10 H), 2.20 (s, 6 H), 2.12–1.95 (m, 4 H), 1.02 (t, $J_{H\text{-}H} = 8$ Hz, 6 H). $^{31}P[^1H]$ NMR (161.98 MHz, 25 °C, CD_2Cl_2): δ 71.4 (dd, $J_{P\text{-}P} = 40$ Hz, $J_{P\text{-}Rh} = 186$ Hz, 2P), 30.8 (dd, $J_{P\text{-}P} = 40$ Hz, $J_{P\text{-}Rh} = 166$ Hz, 2P). ^{19}F NMR (376.49 MHz, 25 °C, CD_2Cl_2): δ − 145.4 (q, $J_{F\text{-}B} = 34$ Hz, 2F). $^{11}B[^1H]$ NMR (128.38 MHz, 25 °C, CD_2Cl_2): δ 0.23 (t, $J_{B\text{-}F} = 33$ Hz). HRMS (ESI +) m/z calculated for $[M\text{-}2Cl]^{2+}$: 1272.3746; found: 1272.3760.

References

1. Horton, P., Ruban, A. V. & Wentworth, M. Allosteric regulation of the light-harvesting system of photosystem II. *Phil. Trans. R. Soc. Lond. B* **355**, 1361–1370 (2000).
2. Ahn, T. K. *et al.* Architecture of a charge-transfer state regulating light harvesting in a plant antenna protein. *Science* **320**, 794–797 (2008).
3. Ruban, A. V., Johnson, M. P. & Duffy, C. D. P. The photoprotective molecular switch in the photosystem II antenna. *Biochim. Biophys. Acta Bioenerg.* **1817**, 167–181 (2012).
4. Pascal, A. A. *et al.* Molecular basis of photoprotection and control of photosynthetic light-harvesting. *Nature* **436**, 134–137 (2005).
5. Allen, J. F. & Forsberg, J. Molecular recognition in thylakoid structure and function. *Trends Plant Sci.* **6**, 317–326 (2001).
6. Scholes, G. D., Fleming, G. R., Olaya-Castro, A. & van Grondelle, R. Lessons from nature about solar light harvesting. *Nat. Chem.* **3**, 763–774 (2011).
7. Kobuke, Y. & Ogawa, K. Porphyrin supramolecules for artificial photosynthesis and molecular photonic/electronic materials. *Bull. Chem. Soc. Jpn* **76**, 689–708 (2003).
8. Rao, K. V., Datta, K. K., Eswaramoorthy, M. & George, S. J. Light-harvesting hybrid assemblies. *Chem. Eur. J.* **18**, 2184–2194 (2012).
9. Frischmann, P. D., Mahata, K. & Wurthner, F. Powering the future of molecular artificial photosynthesis with light-harvesting metallosupramolecular dye assemblies. *Chem. Soc. Rev.* **42**, 1847–1870 (2013).
10. Aguirre-Soto, A., Lim, C. H., Hwang, A. T., Musgrave, C. B. & Stansbury, J. W. Visible-light organic photocatalysis for latent radical-Initiated polymerization via 2e/1H transfers: initiation with parallels to photosynthesis. *J. Am. Chem. Soc.* **136**, 7418–7427 (2014).
11. McQuade, D. T., Hegedus, A. H. & Swager, T. M. Signal amplification of a "turn-on" sensor: harvesting the light captured by a conjugated polymer. *J. Am. Chem. Soc.* **122**, 12389–12390 (2000).
12. Yoon, H. *et al.* A porphyrin-based molecular tweezer: guest-induced switching of forward and backward photoinduced energy transfer. *J. Am. Chem. Soc.* **136**, 1672–1679 (2014).
13. Yoon, H. J. & Mirkin, C. A. PCR-like cascade reactions in the context of an allosteric enzyme mimic. *J. Am. Chem. Soc.* **130**, 11590–11591 (2008).
14. Masar, M. S. *et al.* Allosterically regulated supramolecular catalysis of acyl transfer reactions for signal amplification and detection of small molecules. *J. Am. Chem. Soc.* **129**, 10149–10158 (2007).
15. Yoon, H. J., Kuwabara, J., Kim, J. H. & Mirkin, C. A. Allosteric supramolecular triple-layer catalysts. *Science* **330**, 66–69 (2010).
16. Hecht, D. S. *et al.* Bioinspired detection of light using a porphyrin-sensitized single-wall nanotube field effect transistor. *Nano Lett.* **6**, 2031–2036 (2006).
17. Yadav, R. K. *et al.* A photocatalyst-enzyme coupled artificial photosynthesis system for solar energy in production of formic acid from CO2. *J. Am. Chem. Soc.* **134**, 11455–11461 (2012).
18. Kiss, A. Z., Ruban, A. V. & Horton, P. The PsbS protein controls the organization of the photosystem II antenna in higher plant thylakoid membranes. *J. Biol. Chem.* **283**, 3972–3978 (2008).
19. Horton, P., Ruban, A. V. & Walters, R. G. Regulation of light harvesting in green plants. *Annu. Rev. Plant Phys.* **47**, 655–684 (1996).
20. Barber, J. An Explanation for the relationship between salt-induced thylakoid stacking and the chlorophyll fluorescence changes associated with changes in spillover of energy from Photosystem-II to Photosystem-I. *FEBS Lett.* **118**, 1–10 (1980).
21. Ruban, A. V., Young, A. & Horton, P. Modulation of chlorophyll fluorescence quenching in isolated light-harvesting complex of photosystem-II. *Biochim. Biophys. Acta Bioenerg.* **1186**, 123–127 (1994).
22. Liu, Z. F. *et al.* Crystal structure of spinach major light-harvesting complex at 2.72 angstrom resolution. *Nature* **428**, 287–292 (2004).
23. Holt, N. E. *et al.* Carotenoid cation formation and the regulation of photosynthetic light harvesting. *Science* **307**, 433–436 (2005).
24. Raymo, F. M. & Tomasulo, M. Electron and energy transfer modulation with photochromic switches. *Chem. Soc. Rev.* **34**, 327–336 (2005).
25. Liddell, P. A., Kodis, G., Moore, A. L., Moore, T. A. & Gust, D. Photonic switching of photoinduced electron transfer in a dithienylethene-porphyrin-fullerene triad molecule. *J. Am. Chem. Soc.* **124**, 7668–7669 (2002).

26. Berera, R. *et al.* A simple artificial light-harvesting dyad as a model for excess energy dissipation in oxygenic photosynthesis. *Proc. Natl Acad. Sci.* **103**, 5343–5348 (2006).

27. Straight, S. D. *et al.* Self-regulation of photoinduced electron transfer by a molecular nonlinear transducer. *Nat. Nanotechnol.* **3**, 280–283 (2008).

28. Terazono, Y. *et al.* Mimicking the role of the antenna in photosynthetic photoprotection. *J. Am. Chem. Soc.* **133**, 2916–2922 (2011).

29. Maity, D., Bhaumik, C., Karmakar, S. & Baitalik, S. Photoinduced electron and energy transfer and pH-induced modulation of the photophysical properties in homo- and heterobimetallic complexes of ruthenium(II) and rhodium(III) based on a heteroditopic phenanthroline-terpyridine bridge. *Inorg. Chem.* **52**, 7933–7946 (2013).

30. Sandanayaka, A. S. *et al.* Electron transfer switching in supramolecular porphyrin-fullerene conjugates held by alkylammonium cation-crown ether binding. *Chem. Commun.* 4327–4329 (2006).

31. He, C. *et al.* A photoactive basket-like metal-organic tetragon worked as an enzymatic molecular flask for light driven H_2 production. *Chem. Commun.* **49**, 627–629 (2013).

32. Marchi, E. *et al.* Photoswitchable metal coordinating tweezers operated by light-harvesting dendrimers. *J. Am. Chem. Soc.* **134**, 15277–15280 (2012).

33. Gilbert, M. *et al.* Conformational gating of charge separation in porphyrin oligomer-fullerene systems. *J. Phys. Chem. C* **117**, 26482–26492 (2013).

34. Winters, M. U. *et al.* Control of electron transfer in a conjugated porphyrin dimer by selective excitation of planar and perpendicular conformers. *Chem. Eur. J.* **13**, 7385–7394 (2007).

35. Goodey, N. M. & Benkovic, S. J. Allosteric regulation and catalysis emerge via a common route. *Nat. Chem. Biol.* **4**, 474–482 (2008).

36. Kremer, C. & Lutzen, A. Artificial allosteric receptors. *Chem. Eur. J.* **19**, 6162–6196 (2013).

37. Wiester, M. J., Ulmann, P. A. & Mirkin, C. A. Enzyme mimics based upon supramolecular coordination chemistry. *Angew. Chem. Int. Ed.* **50**, 114–137 (2011).

38. Gianneschi, N. C., Masar, M. S. & Mirkin, C. A. Development of a coordination chemistry-based approach for functional supramolecular structures. *Acc. Chem. Res.* **38**, 825–837 (2005).

39. da Silva, M. F. C. G., Trzeciak, A. M., Ziolkowski, J. J. & Pombeiro, A. J. L. Redox potential, ligand and structural effects in rhodium(I) complexes. *J. Organomet. Chem.* **620**, 174–181 (2001).

40. Lifschitz, A. M. *et al.* Chemically regulating Rh(I)-Bodipy photoredox switches. *Chem. Commun.* **50**, 6850–6852 (2014).

41. Kovbasyuk, L. & Kramer, R. Allosteric supramolecular receptors and catalysts. *Chem. Rev.* **104**, 3161–3187 (2004).

42. Niel, V. *et al.* Crystalline-state reaction with allosteric effect in spin-crossover, interpenetrated networks with magnetic and optical bistability. *Angew. Chem. Int. Ed.* **42**, 3760–3763 (2003).

43. Lifschitz, A. M. *et al.* Boron-dipyrromethene-functionalized hemilabile ligands as "turn-on" fluorescent probes for coordination changes in weak-link approach complexes. *Inorg. Chem.* **52**, 5484–5492 (2013).

44. Roszak, A. W. *et al.* Crystal structure of the RC-LH1 core complex from Rhodopseudomonas palustris. *Science* **302**, 1969–1972 (2003).

45. Maligaspe, E. *et al.* Electronic energy harvesting multi BODIPY-zinc porphyrin dyads accommodating fullerene as photosynthetic composite of antenna-reaction center. *Phys. Chem. Chem. Phys.* **12**, 7434–7444 (2010).

46. D'Souza, F. *et al.* Energy transfer followed by electron transfer in a supramolecular triad composed of boron dipyrrin, zinc porphyrin, and fullerene: a model for the photosynthetic antenna-reaction center complex. *J. Am. Chem. Soc.* **126**, 7898–7907 (2004).

47. Terazono, Y. *et al.* Multiantenna artificial photosynthetic reaction center complex. *J. Phys. Chem. B* **113**, 7147–7155 (2009).

48. Oliveri, C. G., Ulmann, P. A., Wiester, M. J. & Mirkin, C. A. Heteroligated supramolecular coordination complexes formed via the halide-induced ligand rearrangement reaction. *Acc. Chem. Res.* **41**, 1618–1629 (2008).

49. Stryer, L. & Haugland, R. P. Energy transfer: a spectroscopic ruler. *Proc. Natl Acad. Sci* **58**, 719–726 (1967).

50. D'Souza, F. *et al.* Spectroscopic, electrochemical, and photochemical studies of self-assembled via axial coordination zinc porphyrin-fulleropyrrolidine dyads. *J. Phys. Chem. A* **106**, 3243–3252 (2002).

51. Chitta, R. *et al.* Donor-acceptor nanohybrids of zinc naphthalocyanine or zinc porphyrin noncovalently linked to single-wall carbon nanotubes for photoinduced electron transfer. *J. Phys. Chem. C* **111**, 6947–6955 (2007).

52. Fukuzumi, S. *et al.* Uphill photooxidation of NADH analogues by hexyl viologen catalyzed by zinc porphyrin-linked fullerenes. *J. Phys. Chem. A* **106**, 1903–1908 (2002).

53. Watanabe, T. & Honda, K. Measurement of the extinction coefficient of the methyl viologen cation radical and the efficiency of its formation by semiconductor photocatalysis. *J. Phys. Chem.* **86**, 2617–2619 (1982).

54. McNally, A., Prier, C. K. & MacMillan, D. W. Discovery of an α-amino C–H arylation reaction using the strategy of accelerated serendipity. *Science* **334**, 1114–1117 (2011).

55. Jiang, B. W. & Jones, W. E. Synthesis and characterization of a conjugated copolymer of poly(phenylenevinylene) containing a metalloporphyrin incorporated into the polymer backbone. *Macromolecules* **30**, 5575–5581 (1997).

56. Bockman, T. M. & Kochi, J. K. Isolation and oxidation reduction of methylviologen cation radicals. Novel disproportionation in charge-transfer salts by X-ray crystallography. *J. Org. Chem.* **55**, 4127–4135 (1990).

57. Young, R. M. *et al.* Ultrafast Conformational Dynamics of Electron Transfer in ExBox^{4+} ⊂ Perylene. *J. Phys. Chem. A* **117**, 12438–12448 (2013).

Acknowledgements

This material is based on the work supported by the following awards, National Science Foundation CHE-1149314, U.S. Army W911NF-11-1-0229. J.M.-A. acknowledges a fellowship from Consejo Nacional de Ciencia y Tecnología (CONACYT). Time-resolved spectroscopy was supported as part of the ANSER Center, an Energy Frontier Research Center funded by the U.S. Department of Energy, Office of Science, Office of Basic Energy Sciences under Award Number DE-SC0001059 (M.R.W. and R.M.Y.).

Author contributions

A.M.L. and C.A.M. developed the concept, and C.A.M. supervised and guided the research. All experiments were designed and performed by A.M.L., J.M.-A. and C.M.M. A.M.L. and R.M.Y. performed all TA spectroscopy measurements under the supervision of M.R.W. C.L.S. performed all crystallographic studies. A.M.L. and C.A.M. co-wrote the manuscript. All the authors discussed the results and commented on the manuscript during its preparation.

Additional information

Accession codes: The X-ray crystallographic coordinates for structures reported in this Article have been deposited at the Cambridge Crystallographic Data Centre (CCDC), under deposition numbers CCDC 1045099-1045101. These data can be obtained free of charge from The Cambridge Crystallographic Data Centre via www.ccdc.cam.ac.uk/data_request/cif.

5

Nanoscale assembly processes revealed in the nacroprismatic transition zone of *Pinna nobilis* mollusc shells

Robert Hovden[1,*], Stephan E. Wolf[2,3,*], Megan E. Holtz[1], Frédéric Marin[4], David A. Muller[1,5] & Lara A. Estroff[2,5]

Intricate biomineralization processes in molluscs engineer hierarchical structures with meso-, nano- and atomic architectures that give the final composite material exceptional mechanical strength and optical iridescence on the macroscale. This multiscale biological assembly inspires new synthetic routes to complex materials. Our investigation of the prism–nacre interface reveals nanoscale details governing the onset of nacre formation using high-resolution scanning transmission electron microscopy. A wedge-polishing technique provides unprecedented, large-area specimens required to span the entire interface. Within this region, we find a transition from nanofibrillar aggregation to irregular early-nacre layers, to well-ordered mature nacre suggesting the assembly process is driven by aggregation of nanoparticles ($\sim$50–80 nm) within an organic matrix that arrange in fibre-like polycrystalline configurations. The particle number increases successively and, when critical packing is reached, they merge into early-nacre platelets. These results give new insights into nacre formation and particle-accretion mechanisms that may be common to many calcareous biominerals.

[1] School of Applied and Engineering Physics, Cornell University, Ithaca, New York 14853, USA. [2] Department of Materials Science and Engineering, Cornell University, Ithaca, New York 14853, USA. [3] Department of Materials Science and Engineering, Institute of Glass and Ceramics, Friedrich-Alexander-University Erlangen-Nürnberg, 91058 Erlangen, Germany. [4] UMR CNRS 6282 Biogéosciences, Université de Bourgogne Franche-Comté, 6 Boulevard Gabriel, 21000 Dijon, France. [5] Kavli Institute at Cornell for Nanoscale Science, Ithaca, New York 14853, USA. * These authors contributed equally to this work. Correspondence and requests for materials should be addressed to L.A.E. (email: lae37@cornell.edu).

Comprised of calcium carbonate ($CaCO_3$) polymorphs—primarily aragonite and calcite—mollusc shells span a variety of structures[1-3]. Their superior hardness, strength and toughness inspire the artificial synthesis of biomimetic materials[4-8] and motivate understanding the biological mechanisms of formation[9,10]. The most familiar mollusc-derived biomineral is the lustrous nacre (that is, mother of pearl) found in the interior of the animal's shell[11-13]. The 'brick-and-mortar' mesostructure of nacre is a stacked arrangement of polygonal aragonite tablets with a thickness comparable to the wavelength of light (~ 500 nm), giving rise to its iridescence[14-16]. In bivalves, for example, in the order of *Pterioida*, the nacre often grows atop an outer prismatic layer, an assembled array of elongated prismatic calcite crystals aligned perpendicular to the shell's surface.

The transition from the outer shell's prismatic calcite to the nacreous aragonite interior has remained perplexing. In nacroprismatic shells, a thick organic layer (sometimes referred to as the 'green sheet') is present at the prism-to-nacre transition of molluscs and is hypothesized to play a key role in directing the switch from the growth of prismatic calcite to platelet aragonite[10,17]. The nanoscale structure of the $CaCO_3$ assembly occurring throughout this organic layer has, however, not been examined in detail. Previous work has suggested the presence of a disordered aragonite layer at the prismatic–nacre transition zone of both gastropods and bivalves[17-19]. For example, Dauphin *et al.*[18] identified granular material, which they termed 'fibrous aragonite', across the prism–nacre interface of the bivalve *Pinctada margaritifera*. Until now, however, the complete nanoscale morphology of the nacroprismatic transition has not been reported. Employing high-resolution scanning transmission electron microscopy (STEM) and sample

preparation techniques popular in the semiconductor industry (see Methods), we are now able to reveal the nanoscale details of the mineralization process occurring throughout this transition.

The Noble Pen Shell *Pinna nobilis* (*Pterioida*, Linnaeus 1758), endemic to the Mediterranean, is an established model system for bivalve shell formation[20-27] and served as a representative nacroprismatic shell structure for our investigation. As the mollusc shell is deposited by the mantle cells on the periostracum[2,28,29]—a thin organic membrane separating the shell from the surrounding marine environment—the shell is built 'from the outside in'. In addition to mineral components, the epithelial cells secrete extracellular matrix components that provide both the mesoscale structural framework and control of the mineralization process. This organic matrix is composed mainly of various proteins, some of which are highly acidic or glycosylated, and polysaccharides, for example, β-chitin[30]. These components are thought to template and regulate the self-organization process on the micro- to nanoscale during biomineral formation[9,10,31-33].

The switch from prismatically organized calcite to densely stacked aragonite tablets, that is, nacre[11,34-36], is orchestrated by finely tuned secretion of organic and inorganic ingredients into an extracellular compartment—the so-called extrapallial space. The complex interplay between the organic and inorganic constituents then governs the formation of the mineralized shell matrix. It was shown recently that the switch from the prismatic to the nacreous shell layer in *P. margaritifera* and *P. maxima* is accompanied by a marked switch of the secretory repertoires; *in situ* hybridization evidenced a remarkably sharp transition from the prism to the nacre transcript expression in the epithelial cells[31,37,38]. Chemical modifications at the end of prisms, before the formation of the transition organics has also been

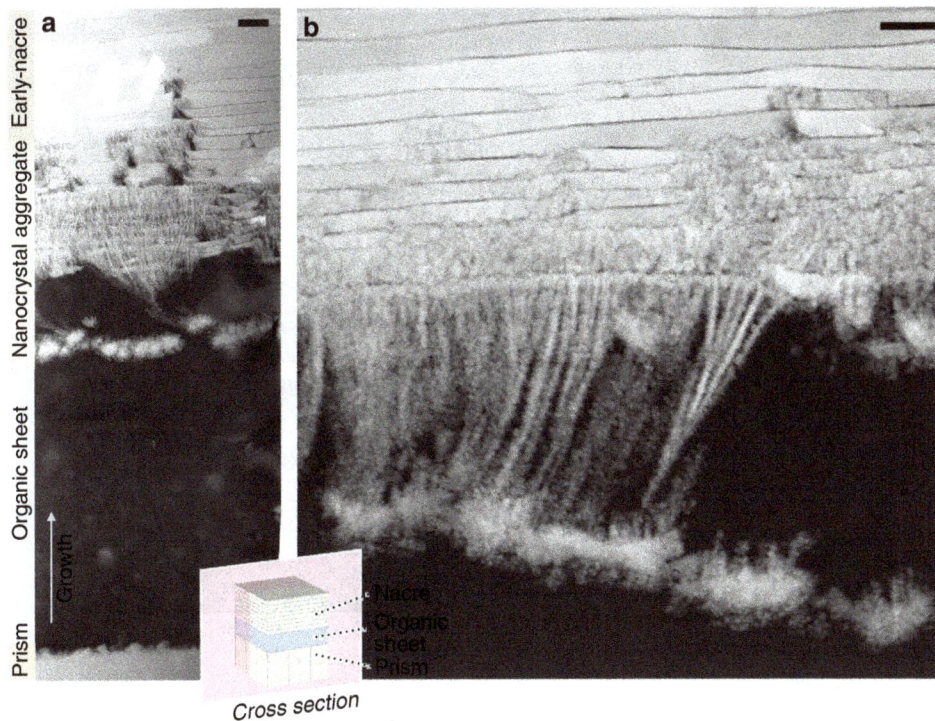

Figure 1 | The start of nacre growth at the nanoscale imaged with ADF-STEM in cross-section across the prismatic to nacreous transition. Prismatic growth is abruptly terminated with a thick organic sheet (**a**, bottom). Several microns later, $CaCO_3$ growth resumes (**a**, middle), first initiated with a layer of larger nanoparticles (~ 700 nm), then followed by deposition and progressive aggregation of small nanoparticles—more clearly seen at higher magnification of another region in **b**. The nanoparticles form a branching fibrillar structure with increasing aggregation of particles. These structures then lead to the onset of disordered nacre layers (early nacre). Scale bars, (**a**) 1 μm; (**b**) 500 nm.

observed[18,36]. It should be noted that both nacreous and prismatic epithelial mantle cells secrete their constituents into the same extrapallial space—there is no membrane separating the two regions.

Previous investigations have proposed that shell growth in adult molluscs proceeds via a nanoparticle-accretion mechanism in which amorphous nanoparticles are deposited, aggregate and crystallize within the extrapallial space[22,39]. Similarly, in bivalve larvae, vibrational spectroscopy has shown amorphous calcium carbonate that later transforms into crystalline aragonite[40] and traces of amorphous calcium carbonate can be found even in adult specimens[22,29,33]. Furthermore, in vitro evidence for a particle-accretion mechanism has been suggested for numerous other calcareous biominerals[41]. Definitive identification of metastable phases, however, is notoriously difficult as it is complicated by both the transient nature of the metastable phase[3,29,33,42–45] and the similarity of the diffraction pattern to nanocrystalline particles[46]. Recently, in situ transmission electron microscopy experiments have observed the nucleation and growth of amorphous $CaCO_3$ in a matrix of polystyrene sulphonate[47]. Direct evidence of similar processes in vivo requires sample preparation and imaging techniques with the capability to reveal nanoscale details.

In this work, we investigate the nacroprismatic transition zone to provide insight into the delicate processes of self-organization by which bivalves control the switch from calcitic prisms to aragonite tablets. Imaging this nacroprismatic transition zone with sub-nanometre resolution, we find a layer of nanocrystallites in fibre-like arrangements within a thick organic layer preceding nacre. This nanofibrillar aragonite branches upward (that is, in the direction of shell growth), with increasing coverage and density in each consecutive growth layer until it finally forms continuous aragonite nacre stacks by the merging and fusion of individual calcium carbonate particles to a dense, space-filling nacre platelet. These early-nacre layers are structurally distinct from mature nacre—with irregular layer thickness, greater layer interface curvature, higher polycrystallinity and frequent stacking faults. The entire transition from prismatic calcite to nacre can occur over tens of microns; however, the fundamental nano-sized building units are only 50 nm and not directly observable by standard optical or X-ray techniques. High-resolution STEM allows us unique, sub-nanometre resolution visualization of the processes of the switch from prismatic to nacre formation. This work provides direct evidence for a particle-assembly pathway occurring during this transition, which is consistent with the model proposed by Gal et al.[41,48,49]

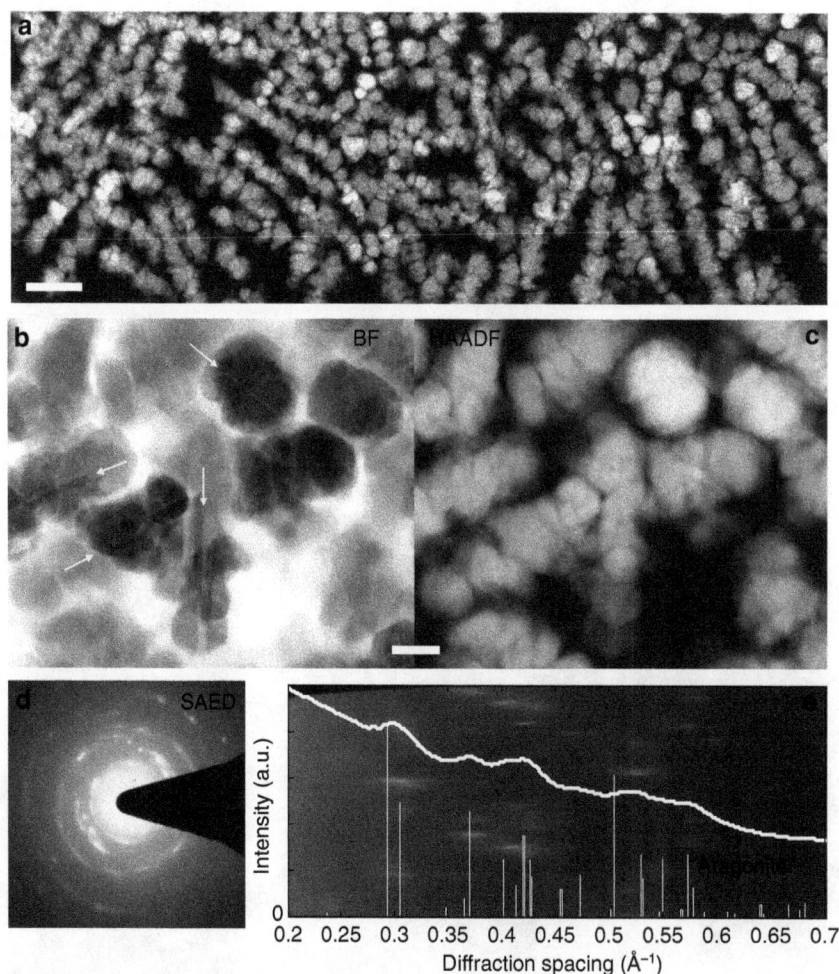

Figure 2 | Structure of the pre-nacre nanocrystals showing a typical diameter of 50–80 nm. (a) Planar view of the packing in pre-nacre aggregation region. (b,c) Pair of annular dark-field (HAADF) and bright-field (BF) STEM images taken in the same region. The BF image of nanocrystals clearly reveals polycrystallinity among the particles and within each particle; twin boundaries running end to end are highlighted with arrows. (d) SAED shows that the nanocrystallites are consistent with the aragonite $CaCO_3$ polymorph—the same polymorph of mature nacre. (e) Plot of radially integrated SAED intensity with known aragonite peaks marked along the x axis; background of plot e is a polar transformation of d over 0–5 radians. Scale bars, (a) 200 nm; (b,c) 50 nm.

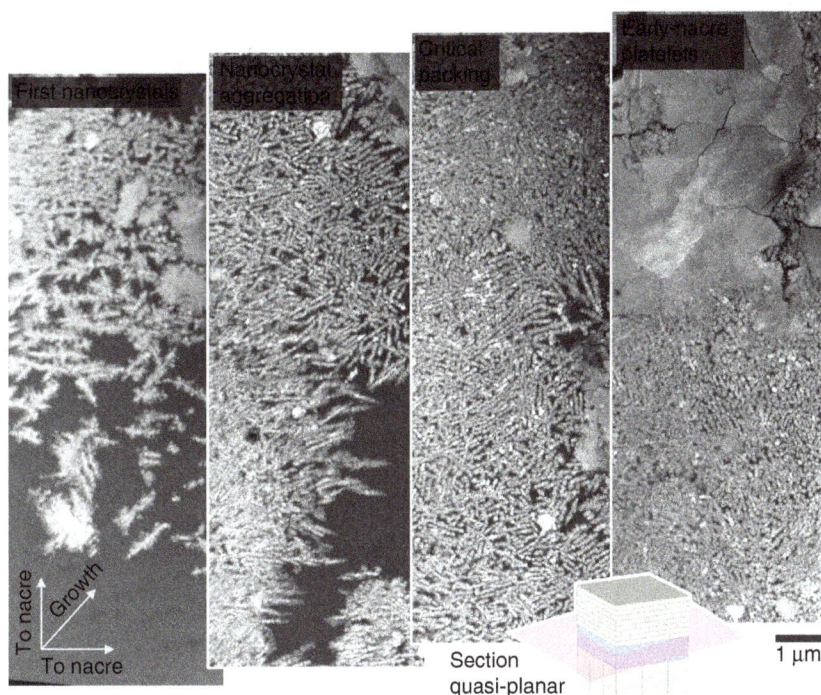

Figure 3 | Quasi-planar nanoscale view at the onset of mollusc nacre growth shown by HAADF-STEM. Nacre growth begins by aggregation of nanocrystals in the organic matrix that separates the prismatic from the nacreous growth (lower left). As growth progresses (towards top right), the nanocrystals aggregate and grow with disordered, often needle-like, geometries. When a critical packing density is reached, continuous early-nacre platelets form—later leading to the familiar large, uniform nacre plates. The quasi-planar section geometry is illustrated in bottom inset.

Results

Observation of the prism-to-nacre transition. Biomineralization at the mollusc's prism-to-nacre transition is governed by nanoscale mechanisms visible to high-resolution STEM. However, the entire prism–nacre interface spans tens of microns, which represents a clear challenge for sample preparation. Pertinent sample preparation methods, such as ion milling and thinning approaches, only provide small areas of a few microns suitable for STEM investigation. Furthermore, biominerals represent composite materials consisting of calcium carbonate and organics that are very sensitive to irradiation damage by electrons or ions. We thus implemented a wedge-polishing technique adapted from the semiconductor industry for sample preparation of remarkably large electron-transparent regions of excellent quality and without the use of ion milling techniques. Several samples from the nacre–prism interface in *P. nobilis* were prepared with this technique (Supplementary Fig. 1). Imaging of these samples demonstrates the power of a wedge-polishing approach: the entire nacroprismatic transition is captured in a single image from the prismatic (Fig. 1a, bottom) to nacreous (top) layer.

Since the bivalve shell is formed from layer-by-layer deposition, the cross-section illustrates sequential nanoscale processes at the onset of early-nacre formation. Early nacre here refers to the first nacre that is formed in the nacroprismatic transition, which actually makes it the oldest nacre in the shell. We can distinguish three stages in the spatial transition from the prismatic to the nacreous layer:

1. Abrupt prism termination: a continuous layer of organic material marks a relatively abrupt termination of prismatic structure. From annular dark-field (ADF)-STEM images, this organic layer has a density indistinguishable from the intra-prismatic organic, appearing as a continuation of the same or similar material (Supplementary Fig. 2). In Fig. 1a, complete coverage of roughly 10-µm-thick organic material was observed before the restart of any significant calcium carbonate deposition.

2. Nanoparticle aggregation: within this organic buffer layer separating the prismatic and the nacreous layers, nacre is preceded by the formation of $CaCO_3$ nanoparticles (Supplementary Fig. 3) that tend to aggregate in a fibre-like geometry. Following initial deposition of these nanoparticles, the particle number increases as the shell growth continues— creating a divergence of densely packed particulate material (Fig. 1) that leads to complete coverage of fibrous aggregates. This region still lacks the mineral density of nacre but often exhibits periodic reduction in density occurring with frequency comparable to the spacings of early nacre. Eventually, the packing density of these aggregates increases until the crystallites merge to form continuous nacre platelets (Fig. 1a,b, top).

3. Formation of early-nacre platelets: the onset of nacre formation is generally incomplete, with early-nacre platelets neighbouring high-density nanoparticle aggregates (Fig. 1, early nacre; and Supplementary Figs 4 and 5). As growth continues, more highly packed aggregate regions become continuous nacre plates. In some instances, nacreous growth returns to aggregate particles. Eventually, the growth forms complete continuous nacre layers (Fig. 1a, top). These early-nacre layers are similar to mature nacre but more disordered at the mesoscale (for example, variation in layer thickness, crystal orientation, and containing stacking faults).

Nanocrystal structure. Closer inspection of the nanoparticles reveals polycrystalline aragonite ∼50 nm in diameter (Fig. 2). The polycrystallinity is most visible in bright-field STEM images (Fig. 2b)—where contrast is sensitive to changes in crystallographic orientation and structural defects. Observation by selected-area electron diffraction (SAED) confirms crystallinity

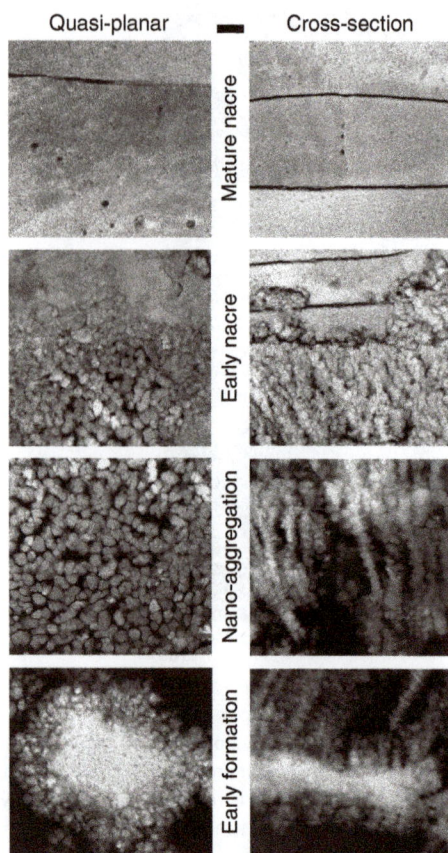

Figure 4 | HAADF-STEM images of nacre formation viewed from orthogonal directions. Early formation begins with the appearance of CaCO$_3$ layers with planar disk geometry (>1 μm diameter, ~300 nm thick), after which nano-aggregation of ~60 nm crystals of aragonite accumulate along the growth direction with increasing density. Once a critical packing density is reached, early-nacre platelets form as a packed continuum with adjacent nanocrystals. Nacre plates are separated by a thin (~8 nm) organic sheet. Eventually, uniform ordered nacre platelets are formed. Organic spheroid inclusions can be seen in nacre. Scale bar, 200 nm.

throughout the nanocrystals and finds them to be consistent with the aragonite polymorph (Fig. 2d,e and Supplementary Fig. 6). Aragonite is also the polymorph found in mature nacre. No amorphous calcium carbonate (ACC) nanoparticles were observed from casual observation via convergent beam electron diffraction nor were they suggested by the bright-field STEM images. As discussed in the Methods section, great care was taken to ensure that no phase transformation occurred during sample preparation and imaging. As such, the observed phase and structure of these particles represents their state at the time the 60-cm mollusc specimen was collected. However, based on this data, there is no way of deducing what the original phase of the particles was at the time of deposition or tracking any phase transformations that may have occurred. Consistent with recent literature regarding the role of ACC in the biomineralization of mollusc shells[3,12,22,29,41] it is likely that the particles were initially ACC and subsequently crystallized, either directly after deposition or during the ageing of the shell.

In-plane observation of the assembly. The structure and packing density of the pre-nacre nanocrystals (nanofibrillar aragonite) is more clearly seen in the plane of the growth. To complement cross-sectional investigation, subsequent planar sections, perpendicular to the growth direction, were prepared. Adding a

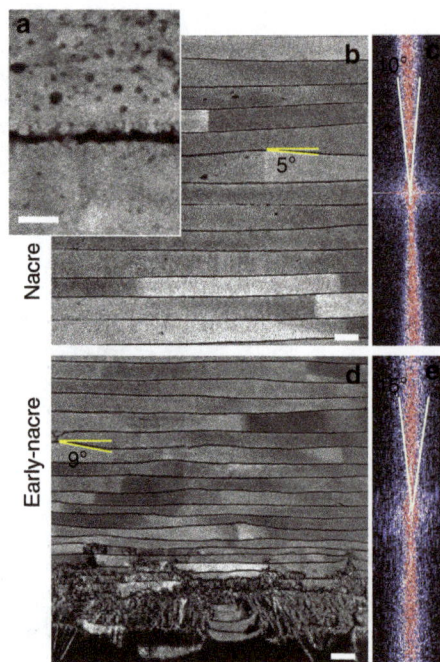

Figure 5 | Comparison of early-nacre and mature nacre growth shown by HAADF-STEM. (**a**) Layers have well-defined interfaces buffered by ~8-nm organic layers. For mature nacre (**b**), the layers are ordered and have interface curvature ranging ±5.0° as clearly shown by the image's Fourier transform (**c**). The first several nacre layers, that is, early nacre (**d**), are more disordered and have an interface curvature ranging roughly ±7.5° as shown by the image's Fourier transform (**e**). As seen in **d**, the interlayer spacing in early nacre increases progressively—starting as small as 120 nm. Mature nacre (**b**) has a spacing of ~450 nm. Scale bars, (**a**) 50 nm; (**b,d**) 500 nm.

small planar tilt (that is, quasi-planar sectioning) allowed in-plane observation throughout the entire prism–nacre interface. The in-plane aggregation process throughout the prism–nacre transition is shown in Fig. 3—beginning in the bottom left-most image and advancing with growth towards the top right-most image (Supplementary Fig. 7). Following the initial appearance of nanofibrillar aragonite within the organic matrix (Fig. 3, first nanocrystals) the growth progresses by increased planar packing of these nanocrystals. The nanocrystals are often connected with a fibrous, or needle-like, geometry. This arrangement may be due to in-plane columnar growth processes or due to a template action of chitin fibres[9,39]; clarification of the underlying processes will be the subject of future research.

The early-nacre platelets appear to be formed by the fusion of individual nanoparticles when a critical particle number density is reached (Fig. 3, top right). The transition from packed nanocrystals to nacre is a continuous process. At the rim of the forming platelets we see nanogranular features and the gradual transitioning from individual nanocrystallites to space-filling nacre platelets (Supplementary Fig. 4). Variation of high-angle annular dark-field (HAADF) image intensity in the early-nacre platelets reflects their polycrystallinity.

Comparison of cross-sectional and quasi-planar views. The three-dimensional structure at the transition can be inferred by comparing the samples that were polished both parallel and perpendicular to the growth direction. Figure 4 compares the two sectioning geometries for high-resolution STEM imaging at the three observed stages in the transition zone. The earliest CaCO$_3$ formations (Fig. 4 early formation) appear with planar disk geometry (>1 μm diameter, ~300 nm thick) and have SAED peaks consistent with aragonite (Supplementary Fig. 8). In some areas, nanofibrilliar aragonite originates from these initial structures leading to the onset of nacre, whereas in other locations, these disk-like structures appear isolated in cross-section (Fig. 1). The nano-aggregation then proceeds: ~ 60 nm aragonite crystals accumulate along the growth direction with increasing density (Fig. 4 nano-aggregation). These nanocrystals are often linearly connected along the growth or in-plane directions. The early formations often—but not always—provide a site of growth for these nanocrystals (as seen in Fig. 1a). Once a critical nanocrystal packing density is reached, continuous early-nacre platelets form adjacent to packed nanocrystals (Fig. 4, early nacre). Eventually, uniform ordered nacre platelets are formed (Fig. 4, mature nacre). Organic spheroid inclusions can be seen in the nacre tablets as previously reported[50].

Structure of early-nacre platelets. Additional structural variation in early-nacre makes it distinguishable from mature nacre, which is characterized by uniform tablets (Fig. 5a–c and Supplementary Fig. 9). Following the first appearance of nacre platelets, subsequent layers contain larger continuous regions, until complete early-nacre layers are formed. However, the early-nacre layers (Fig. 5d) are not structurally equivalent to mature nacre (Fig. 5b). The thin organic layers (~ 10 nm or less) separating all nacre layers (Fig. 5a) provide a well-defined, high-contrast boundary for characterizing the planar variation via real and Fourier space analysis. Mature nacre consists of ordered platelets roughly 450-nm thick with small, slow varying changes in thickness. The Fourier transform reveals a planar variation tightly bound within $\pm 5°$ from the growth plane (Fig. 5c). In comparison, early-nacre layers are thin (as low as ~ 120 nm) and disordered. Rapid variation in layer thickness and interface curvature is clearly seen in the cross-sectional image (Fig. 5d) and reflected in the wide angular intensity in Fourier space (Fig. 5e). We observe a planar variation ca. $\pm 7.5°$ (Fig. 5e) in the early nacre. Changes in HAADF-STEM diffraction contrast, within a single nacre tablet, can be seen in the bright-to-dark variation. This observation corresponds with small variation of the in-plane crystal orientation of nacre plates—crystallographic changes that were previously reported by Gilbert et al.[12,51]. We additionally observe that changes in crystal orientation are more prevalent in the early-nacre when compared with mature nacre. We confirm, with high resolution, the presence of inorganic bridges between nacre layers in the bivalve *P. nobilis*, as, for instance, previously reported by Checa et al.[52]. High-resolution micrographs of the bridges are shown in Fig. 5a and Supplementary Fig. 10.

Discussion

In this work, *P. nobilis* samples were prepared by a wedge-polishing technique for sub-nanometre observation of the nacroprismatic transition zone with high-resolution ADF-STEM and SAED. This contribution illuminates the formation of the first layers of nacre and demonstrates that nacre onset is a complex nanoscale assembly process. The onset of nacre growth occurs from the aggregation of nanoparticles ca. 50–80 nm in size which then pack as polycrystalline fibre-like arrangements (nanofibrillar aragonite), branching outward with increasing density in the growth direction. When the particle number density reaches a critical value, the particles merge and form the first early-nacre platelets. Thus, the highly ordered state of nacre is achieved gradually, transitioning from the densely packed nanofibrillar aragonite, to an early-nacre layer with a higher rate of disorder and eventually, the well-ordered mature nacre layers.

The cross-section of the nacroprismatic transition zone is a representation, frozen in time, of the transition from one mineralization mechanism to another. In other words, using the current techniques, the observed ultrastructure is well preserved and representative of the *in vivo* state (at harvest) and provides insight into the assembly of early nacre. In this way, the observations reported here might be analogous in some respects to the formation process of mature nacre tablets, even in layers further away from the transition zone. We cannot, however, see back in time to the dynamics or material phases that occurred in the tissue during the nanoparticle formation and assembly—including the possible crystallization of ACC. These findings give new insights into the self-organization mechanisms of nacre that may also be common to many calcareous biominerals.

Methods

Specimen. The protected Mediterranean *P. nobilis* specimen was collected near the coast of Villefranche-s-Mer, Département Alpes-Maritimes, France with the authorization of DREAL PACA (Direction Régionale de l'Environnement, de l'Aménagement et du Logement, Provence Alpes-Côte d'Azur).

Wedge-polishing preparation. Small sections were isolated from the 60-cm shell (Supplementary Fig. 1) using a diamond wire saw. Specimen preparation for STEM analysis required undamaged, electron-transparent regions over large areas capable of spanning the entire prism–nacre interface—both the $CaCO_3$ and organic material are easily damaged by electron and ion beams. We successfully implemented a wedge-polishing technique that is particularly popular in the semiconductor industry[53]. The polishing technique is gentle and structural changes have not been observed by this method. Dislocated material or granules did not occur—however would be easily noticed in a STEM image as a vacancy and followed by a damage path caused by the removed material during polishing. The lapping process was conducted with an Allied MultiPrep polishing system and used a waning series of micron-sized diamonds embedded in polishing film (from 30 μm down to 0.1 μm) dictated by the 'trinity of damage': polishing produces a damaging layer, which is approximately three times thick as the size of the grit. Therefore, with each subsequent lapping film, this damage layer has to be removed. To prevent $CaCO_3$ etching during preparation, water-based lubricants and polishing agents were eliminated—instead, Allied alcohol based 'blue' lubricant was used. For the first-side polish, the sample was attached to an aluminum-polishing stub (Allied) with Crystal Bond (Allied) and was polished until a polish free of scratches and imperfections is obtained at the desired location. For the second-side polish, the sample was glued to a Pyrex-stub (Allied) with a very thin layer of superglue. The polishing procedure was essentially repeated, but at a 3° pitch to form a wedge. The orientation, pitch and roll of the sample were additionally adjusted with two micrometer heads so that the sample plane of interest was parallel to the abrasive plane. Finally, the polished wedge samples were mounted to an annular molybdenum transmission electron microscopy grid with MBond 610 epoxy. By optimizing this polishing technique for calcium carbonate biominerals, we could obtain samples with areas up to 3-mm wide, which were accessible to STEM analyses and of high quality.

Electron microscopy. STEM was performed using a 200-keV FEI F20 instrument with a convergence angle ca. 9.6 mrad. A HAADF with 100–300 mm camera lengths was used to provide a Z-contrast image where intensity is sensitive to the atomic number of atoms in the specimen. A bright-field detector with a 200-mm camera length was used to highlight polycrystallinity in the specimen. SAED with an aperture of approximated 8 μm provided a polycrystalline diffraction pattern of nanocrystallites in the prism–nacre transition regions.

References

1. Bøggild, O. B. The shell structure of the mollusks. *Det Kongelige Danske Videnskabernes Selskabs Skrifter, Natruvidenskabelig og Mathematisk, Afdeling* **2**, 231–326 (1930).
2. Kennedy, W. J., Taylor, J. D. & Hall, A. Environmental and biological controls on bivalve shell mineralogy. *Biol. Rev.* **44**, 499–530 (1969).
3. Addadi, L., Joester, D., Nudelman, F. & Weiner, S. Mollusk shell formation: a source of new concepts for understanding biomineralization processes. *Chem. Eur. J.* **12**, 980–987 (2006).
4. Aksay, I. *et al.* Biomimetic pathways for assembling inorganic thin films. *Science* **273**, 892–898 (1996).
5. Sellinger, A. *et al.* Continuous self-assembly of organic-inorganic nanocomposite coatings that mimic nacre. *Nature* **394**, 256–260 (1998).
6. Tang, Z., Kotov, N. A., Magonov, S. & Ozturk, B. Nanostructured artificial nacre. *Nat. Mater.* **2**, 413–418 (2003).
7. Finnemore, A. *et al.* Biomimetic layer-by-layer assembly of artificial nacre. *Nat. Commun.* **3**, 966 (2012).

8. Munch, E. *et al.* Tough, bio-inspired hybrid materials. *Science* **322**, 1516–1520 (2008).
9. Falini, G., Albeck, S., Weiner, S. & Addadi, L. Control of aragonite or calcite polymorphism by mollusk shell macromolecules. *Science* **271**, 67–69 (1996).
10. Belcher, A. M. *et al.* Control of crystal phase switching and orientation by soluble mollusc-shell proteins. *Nature* **381**, 56–58 (1996).
11. Carpenter, W. B. On the microscopic structure of shells. *Rep. Brit. Ass. Adv. Sci.* **14**, 1–24 (1844).
12. Olson, I. C. *et al.* Crystal nucleation and near-epitaxial growth in nacre. *J. Struct. Biol.* **184**, 454–463 (2013).
13. Mutvei, H. The nacreous layer in Mytilus, Nucula, and Unio (Bivalvia). *Calc. Tiss. Res.* **24**, 11–18 (1977).
14. Brewster, D. *A treatise on optics* (Longman, Brown, Green and Longman's, 1853).
15. Liu, Y., Shigley, J. & Hurwit, K. Iridescent color of a shell of the mollusk Pinctada margaritifera caused by diffraction. *Opt. Express* **4**, 177–182 (1999).
16. Choi, S. H. & Kim, Y. L. Hybridized/coupled multiple resonances in nacre. *Phys. Rev. B* **89**, 035115 (2014).
17. Griesshaber, E. *et al.* Homoepitaxial meso- and microscale crystal co-orientation and organic matrix network structure in Mytilus edulis nacre and calcite. *Acta Biomater.* **9**, 9492–9502 (2013).
18. Dauphin, Y. *et al.* Structure and composition of the nacre-prisms transition in the shell of Pinctada margaritifera (Mollusca, Bivalvia). *Anal. Bioanal. Chem.* **390**, 1659–1669 (2008).
19. Gilbert, P. U. P. A. *et al.* Gradual ordering in red abalone nacre. *J. Am. Chem. Soc.* **130**, 17519–17527 (2008).
20. Marin, F., Narayanappa, P. & Motreuil, S. in *Molecular Biomineralization* 52, 353–395 (Springer, 2011).
21. Marin, F. *et al.* Caspartin and calprismin, two proteins of the shell calcitic prisms of the Mediterranean fan mussel Pinna nobilis. *J. Biol. Chem.* **280**, 33895–33908 (2005).
22. Wolf, S. E. *et al.* Merging models of biomineralisation with concepts of nonclassical crystallisation: is a liquid amorphous precursor involved in the formation of the prismatic layer of the Mediterranean Fan Mussel Pinna nobilis? *Farad. Disc.* **159**, 433–448 (2012).
23. Dauphin, Y. Soluble organic matrices of the calcitic prismatic shell layers of two Pteriomorphid bivalves. Pinna nobilis and Pinctada margaritifera. *J. Biol. Chem.* **278**, 15168–15177 (2003).
24. Pokroy, B., Kapon, M., Marin, F., Adir, N. & Zolotoyabko, E. Protein-induced, previously unidentified twin form of calcite. *Proc. Natl Acad. Sci. USA* **104**, 7337–7341 (2007).
25. Gilow, C., Zolotoyabko, E., Paris, O., Fratzl, P. & Aichmayer, B. Nanostructure of biogenic calcite crystals: a view by small-angle X-ray scattering. *Cryst. Grow. Des.* **11**, 2054–2058 (2011).
26. Marin, F., Pokroy, B., Luquet, G., Layrolle, P. & De Groot, K. Protein mapping of calcium carbonate biominerals by immunogold. *Biomaterials* **28**, 2368–2377 (2007).
27. Bayerlein, B. *et al.* Self-similar mesostructure evolution of the growing mollusc shell reminiscent of thermodynamically driven grain growth. *Nat. Mater.* **13**, 1102–1107 (2014).
28. Watabe, N. & Wilbur, K. M. Influence of the organic matrix on crystal type in molluscs. *Nature* **188**, 334–334 (1960).
29. Jacob, D. E., Wirth, R., Soldati, A. L., Wehrmeister, U. & Schreiber, A. Amorphous calcium carbonate in the shells of adult Unionoida. *J. Struct. Biol.* **173**, 241–249 (2011).
30. Evans, J. S. 'Tuning in' to mollusk shell nacre- and prismatic-associated protein terminal sequences. implications for biomineralization and the construction of high performance inorganic-organic composites. *Chem. Rev.* **108**, 4455–4462 (2008).
31. Marie, B. *et al.* Different secretory repertoires control the biomineralization processes of prism and nacre deposition of the pearl oyster shell. *Proc. Natl Acad. Sci. USA* **109**, 20986–20991 (2012).
32. Addadi, L. & Weiner, S. Interactions between acidic proteins and crystals: stereochemical requirements in biomineralization. *Proc. Natl Acad. Sci. USA* **82**, 4110–4114 (1985).
33. Jacob, D. E. *et al.* Nanostructure, composition and mechanisms of bivalve shell growth. *Geochim. Cosmochim. Acta* **72**, 5401–5415 (2008).
34. Olson, I. C. *et al.* Crystal lattice tilting in prismatic calcite. *J. Struct. Biol.* **183**, 180–190 (2013).
35. Freer, A., Greenwood, D., Chung, P., Pannell, C. L. & Cusack, M. Aragonite prism – nacre interface in freshwater mussels Anodonta anatina (Linnaeus, 1758) and Anodonta cygnea (L. 1758). *Cryst. Grow. Des.* **10**, 344–347 (2009).
36. Cuif, J.-P. *et al.* Evidence of a biological control over origin, growth and end of the calcite prisms in the shells of Pinctada margaritifera (Pelecypod, Pterioidea). *Minerals* **4**, 815–834 (2014).
37. Joubert, C. *et al.* Transcriptome and proteome analysis of Pinctada margaritifera calcifying mantle and shell: focus on biomineralization. *BMC Genomics* **11**, 613 (2010).
38. Montagnani, C. *et al.* Pmarg-pearlin is a matrix protein involved in nacre framework formation in the pearl oyster Pinctada margaritifera. *Chembiochem* **12**, 2033–2043 (2011).
39. Nudelman, F., Chen, H. H., Goldberg, H. A., Weiner, S. & Addadi, L. Spiers Memorial Lecture. Lessons from biomineralization: comparing the growth strategies of mollusc shell prismatic and nacreous layers in Atrina rigida. *Farad. Disc.* **136**, 9–25 discussion 107–23 (2007).
40. Weiss, I. M., Tuross, N., Addadi, L. & Weiner, S. Mollusc larval shell formation: amorphous calcium carbonate is a precursor phase for aragonite. *J. Exp. Zool* **293**, 478–491 (2002).
41. Gal, A., Weiner, S. & Addadi, L. A perspective on underlying crystal growth mechanisms in biomineralization: solution mediated growth versus nanosphere particle accretion. *CrystEngComm* **17**, 2606–2615 (2015).
42. Politi, Y. *et al.* Structural characterization of the transient amorphous calcium carbonate precursor phase in sea urchin embryos. *Adv. Funct. Mater.* **16**, 1289–1298 (2006).
43. Günther, C., Becker, A., Wolf, G. & Epple, M. In vitro synthesis and structural characterization of amorphous calcium carbonate. *Z. Anorg. Allg. Chem.* **631**, 2830–2835 (2005).
44. Faatz, M., Gröhn, F. & Wegner, G. Amorphous calcium carbonate: synthesis and potential intermediate in biomineralization. *Adv. Mater.* **16**, 996–1000 (2004).
45. Gebauer, D. *et al.* Proto-calcite and proto-vaterite in amorphous calcium carbonates. *Angew. Chem. Int. Ed.* **122**, 9073–9075 (2010).
46. Rez, P., Sinha, S. & Gal, A. Nanocrystallite model for amorphous calcium carbonate. *J. Appl. Crystallogr.* **47**, 1651–1657 (2014).
47. Smeets, P. J. M., Cho, K. R., Kempen, R. G. E., Sommerdijk, N. A. J. M. & De Yoreo, J. J. Calcium carbonate nucleation driven by ion binding in a biomimetic matrix revealed by in situ electron microscopy. *Nat. Mater.* **14**, 394–399 (2015).
48. Gal, A. *et al.* Particle accretion mechanism underlies biological crystal growth from an amorphous precursor phase. *Adv. Funct. Mater.* **24**, 5420–5426 (2014).
49. Gal, A., Habraken, W., Gur, D. & Fratzl, P. Calcite crystal growth by a solid-state transformation of stabilized amorphous calcium carbonate nanospheres in a hydrogel. *Angew. Chem. Int. Ed.* **52**, 4867–4870 (2013).
50. Gries, K., Kröger, R., Kübel, C., Fritz, M. & Rosenauer, A. Investigations of voids in the aragonite platelets of nacre. *Acta Biomater.* **5**, 3038–3044 (2009).
51. Metzler, R. A. *et al.* Architecture of columnar nacre, and implications for its formation mechanism. *Phys. Rev. Lett.* **98**, 268102 (2007).
52. Checa, A. G., Cartwright, J. H. E. & Willinger, M.-G. Mineral bridges in nacre. *J. Struct. Biol.* **176**, 330–339 (2011).
53. Voyles, P. M., Grazul, J. L. & Muller, D. A. Imaging individual atoms inside crystals with ADF-STEM. *Ultramicroscopy* **96**, 251–273 (2003).

Acknowledgements

L.A.E. acknowledges support from the NSF (DMR-1210304). This work and support for R.H. made use of the Cornell Center for Materials Research Facilities supported by the National Science Foundation under Award Number DMR-1120296. We thank John Grazul and Mick Thomas for experimental assistance, and Sébastien Motreuil (UMR CNRS 6282 Biogéosciences) who sampled *Pinna nobilis*. S.E.W. thanks for partial support by the Cluster of Excellence 315 'Engineering of Advanced Materials—Hierarchical Structure Formation for Functional Devices' funded by the German Research Foundation.

Author contributions

R.H. and S.E.W. contributed equally to this work; sample preparation was developed by R.H., S.E.W. and M.E.H.; electron microscopy was conducted by R.H. and S.E.W.; data analysis and materials interpretation was carried out by R.H., M.E.H. and D.A.M.; biological context was provided by S.E.W. and L.A.E.; protected *Pinna nobilis* specimen and valuable input was provided by F.M.; all authors discussed the results throughout all stages and commented on the manuscript. R.H., S.E.W. and L.A.E. wrote the manuscript.

Additional information

6

A biosynthetic model of cytochrome c oxidase as an electrocatalyst for oxygen reduction

Sohini Mukherjee[1], Arnab Mukherjee[2], Ambika Bhagi-Damodaran[2], Manjistha Mukherjee[1], Yi Lu[2] & Abhishek Dey[1]

Creating an artificial functional mimic of the mitochondrial enzyme cytochrome c oxidase (CcO) has been a long-term goal of the scientific community as such a mimic will not only add to our fundamental understanding of how CcO works but may also pave the way for efficient electrocatalysts for oxygen reduction in hydrogen/oxygen fuel cells. Here we develop an electrocatalyst for reducing oxygen to water under ambient conditions. We use site-directed mutants of myoglobin, where both the distal Cu and the redox-active tyrosine residue present in CcO are modelled. In situ Raman spectroscopy shows that this catalyst features very fast electron transfer rates, facile oxygen binding and O–O bond lysis. An electron transfer shunt from the electrode circumvents the slow dissociation of a ferric hydroxide species, which slows down native CcO (bovine $500\,s^{-1}$), allowing electrocatalytic oxygen reduction rates of $5,000\,s^{-1}$ for these biosynthetic models.

[1] Department of Inorganic Chemistry, Indian Association for the Cultivation of Science, 2A&2B Raja SC Mullick Road, Jadavpur Kolkata 700032, India.
[2] Department of Chemistry, University of Illinois at Urbana-Champaign, Champaign, Illinois 61801, USA. Correspondence and requests for materials should be addressed to A.D. (email: icad@iacs.res.in).

Mimicking the sophistication of naturally occurring enzymes has been a long-term goal of the scientific community. An artificial analogue that can perform equally well as its natural predecessor will not only provide deeper understanding of the native enzymes, but also enable the development of efficient artificial catalysts. For several decades now chemists have embarked on this daunting pursuit of emulating the efficiency and selectivity of naturally occurring enzymes and several important milestones have been achieved. Efforts from synthetic inorganic chemists have resulted in synthetic models of myoglobin (Mb), galactose oxidase, tyrosinase, cytochrome P450 and cytochrome c oxidase (CcO)[1–5]. Alternatively, there has been fervent pursuit of biochemical constructs inspired by natural metalloenzymes. A series of binuclear non-haem iron, cytochrome c, haem oxidases and iron–sulfur enzyme models have resulted from such efforts[4,6–12]. While none of the synthetic or biochemical models reported so far could match the reactivity exhibited by their natural counterparts, fundamental insights regarding the structure–function correlations of several metalloenzymes have been gained in the process[7,13–15]. In addition, key information about the secondary coordination sphere interactions present in the protein-active site, which play a dominating role in determining the electronic structure and reactivity of these metalloenzymes, have been identified[16,17].

In a biosynthetic approach, stable naturally occurring proteins have been used as scaffolds for creating mimics of several metalloenzymes, such as hydrogenases which are involved in the reversible generation of H_2 from water, haem proteins participating in electron transfer and O_2-binding, non-haem iron and copper enzymes active in small molecule activation, and even novel enzymes containing non-native cofactors[18–24]. For example, using this approach, biosynthetic models that structurally and functionally mimic CcO and nitric oxide reductase have been reported[7,25]. Despite decades of focused effort, however, biosynthetic models with catalytic efficiencies approaching those of the naturally occurring metalloenzymes have remained elusive[26–28]. In this report, we communicate a biosynthetic model of CcO bearing the distal Cu_B and a tyrosine residue that is kinetically more competent in reducing O_2 electrochemically than any known synthetic analogue, as well as native CcO itself.

X-ray crystallography of Mb and its mutant have revealed that its two propionate side chains project out of the protein surface into the solvent (Fig. 1)[7]. Taking advantage of this structural feature, we have previously developed an electrocatalytic O_2

reduction system where the native haem cofactor in Mb is replaced by a modified hemin cofactor bearing an alkyne group (Hemin-yne, Fig. 2) so that electrons can be injected directly into the haem from a gold electrode to facilitate O_2 reduction[29]. This method resulted in a Mb-functionalized electrode bearing 2.15×10^{-12} mol per cm^2 of protein, which was characterized using several microscopic and spectroscopic techniques[29]. Over the last few years, a biosynthetic model of CcO has been reported in which two distal residues of Mb (L29 and F43) have been mutated to His, which along with the native His64, form a Cu-binding site, mimicking the distal Cu_B-binding site present in CcO (Cu_BMb)[30]. Furthermore, in an attempt to mimic the conserved Tyr 244 residue in the CcO-active site, a G65Y mutant of Cu_BMb (G65YCu_BMb) containing redox-active tyrosine residue in the distal site and a variant where a tyrosine residue was crosslinked to the active site histidines were also created (Fig. 1)[31].

Herein we report the electrocatalytic properties of the G65YCu_BMb (higher synthetic yields than the tyrosine crosslinked variant) immobilized on an Au electrode using the method developed for WT Mb[29].

Results

Electrode characterization by SERRS. Surface-enhanced resonance Raman spectroscopy (SERRS) data (Fig. 3a) of the electrodes bearing the G65YCu_BMb protein with and without Cu_B show the oxidation and spin state marker v_4, v_3, v_2 and v_{10} bands at 1,375 cm^{-1}, 1,493 cm^{-1}, 1,585 cm^{-1} and 1,641 cm^{-1}, respectively. The v_4, v_3 and v_2 values are consistent with the presence of a five-coordinated high spin haem in the active site on these electrodes bearing the biochemical constructs of CcO[32]. Also the bands at 1,504 cm^{-1} and v_4 at 1,641 cm^{-1} suggest the presence of a mixture of six-coordinate low spin species, which likely has H_2O as the axial ligand. The positions of these bands in the G65YCu_BMb and their relative intensities are different from Hemin-yne (Supplementary Fig. 1)[29].

Electrode characterization by X-ray photoelectron spectroscopy. X-ray photoelectron spectroscopic (XPS) data of a G65YCu_BMb-bound Au electrode clearly indicate the presence of Fe, Cu, C, N and O elements (Supplementary Fig. 2, Supplementary Table 1). The $3p_{3/2}$, $2p_{3/2}$ and $2p_{1/2}$ binding energy peak for the Fe^{III} of haem group appear at 56.5 eV, 709.4 eV and 722.4 eV, respectively[33,34]. The $2p_{3/2}$ and $2p_{1/2}$ binding energy peak for Cu^{II} in the distal site appear at 931.7 eV and 951.8 eV, respectively[35]. The N_{1s}

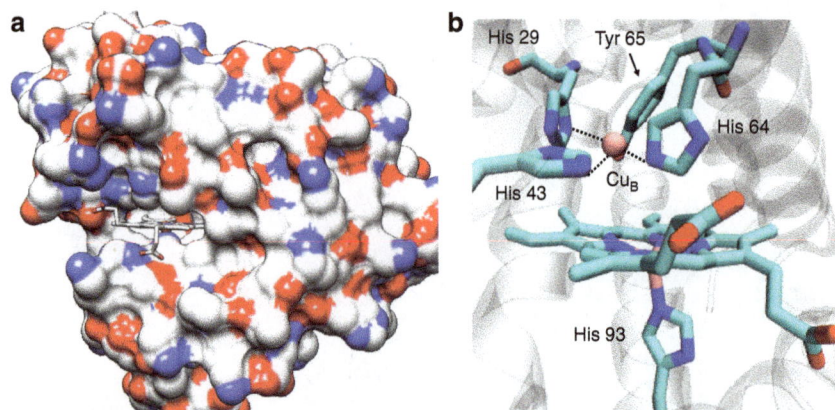

Figure 1 | Crystal structure. Crystal structure of a Mb-based biosynthetic model of CcO, F33Y-Cu_BMb; pdb id: 4FWY. (**a**) The haem cofactor is in a cleft on the molecule protein surface, (colour coded according to the charge of the residues), with the propionate groups exposed to the solvent. (**b**) The computer model of G65YCu_BMb showing its catalytic centre containing the distal Cu_B bound to histidines and a tyrosine 65.

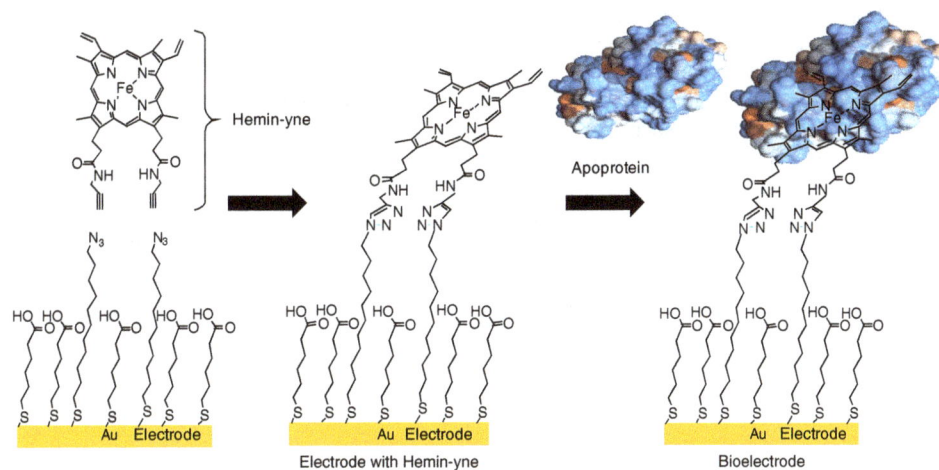

Figure 2 | Construction of the electrode bearing the biosynthetic model. Reconstitution of apoprotein *in situ* with Hemin-yne groups that are covalently attached to mixed self-assembled monolayers of thiols on an Au electrode. The modified hemin is indicated as Hemin-yne.

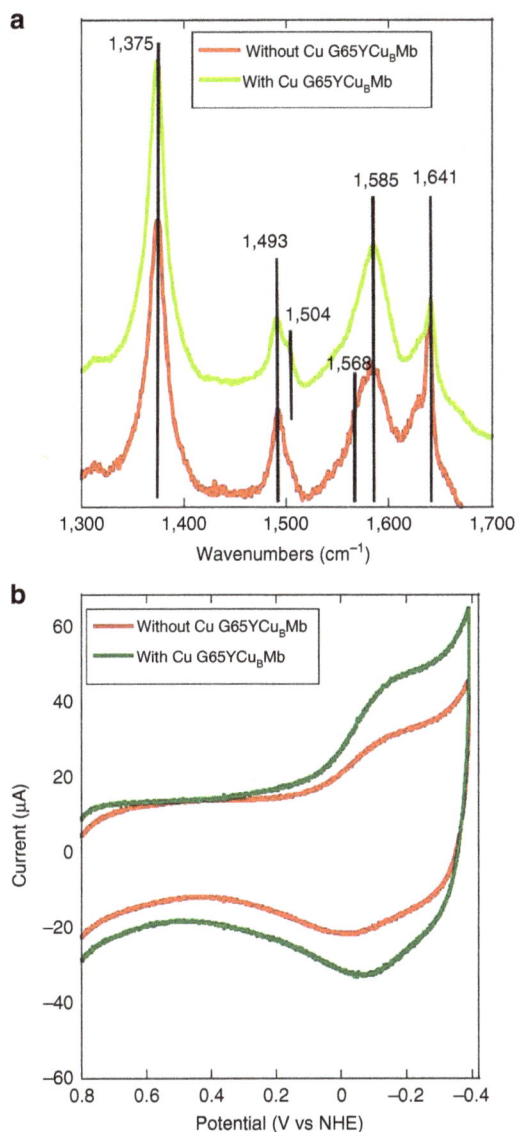

peak is broad (Supplementary Fig. 2), as it contains several components due to the presence of amide, haem pyrroles and the triazole groups (resulting from the covalent attachment of Hemin-yne) on the surface[33]. Similarly, the C_{1s} peak (Supplementary Fig. 2) contains contributions from different types of C atoms (aromatic, aliphatic, haem and so on) on these protein-modified surfaces[36].

Electrode characterization by CV. Cyclic voltammetry (CV) of G65YCu$_B$Mb with and without the distal Cu$_B$ immobilized onto the electrodes in degassed buffer show the haem $Fe^{3+/2+}$ mid-point reduction potential ($E^{1/2}$) at -97 mV and -57.5 mV, respectively (Fig. 3b). The peak separation between the cathodic and the anodic peak for both the cases is ~ 70 mV (ref. 37). Hemin-yne displays the $Fe^{3+/2+}$ reduction potential at -70.0 mV in the absence of a protein and -135.0 mV when bound to wild-type apo Mb[29]. In the case of the G65YCu$_B$Mb protein-bound electrodes, the $Cu^{2+/+}$ process overlaps with the Hemin-yne $Fe^{3+/2+}$ process, resulting in approximately twice the area under these CV peaks relative to the G65YCu$_B$Mb-bearing electrodes prior to Cu$_B$ loading. The integrated area under these CV features in the absence of Cu^{2+} indicates that there are $2.55 \pm 0.05 \times 10^{-12}$ mol of protein per cm^2 of the surface. The ratios of the integrated area under the CV features of G65YCu$_B$Mb functionalized before and after loading the Cu$_B$ is $\sim 1:2$ (Table 1, fourth column) which is consistent with the expected 1:1 stoichiometry (that is, every G65YCu$_B$Mb binds one Hemin-yne and one Cu^{2+} ion). Note that the $E^{1/2}$ values of the hemin and Cu$_B$ measured for these electrodes are slightly different from those estimated from potentiometric titration in solution[7,19]. This is likely due to the interfacial microenvironment of the –COOH-terminated Self Assembled Monolayer (SAM) which is known to shift the apparent formal potentials of redox-active species in its vicinity[38]. Thus the *in situ* reconstitution of the protein with Hemin-yne on the electrode is evident from the SERRS, XPS (Supplementary Fig. 2) and CV data (Supplementary Fig. 3, Supplementary Fig. 4). The presence of the Cu^{2+} at Cu$_B$ site on the electrode is indicated by XPS and CV data. Taken together, these data indicate the assembly of the G65YCu$_B$Mb, biosynthetic model of C*c*O covalently attached to the electrode via the linkage between the Hemin-yne and the azide terminated thiols created using click reaction (Fig. 2).

O_2 reduction reactivity of the electrode. In linear sweep voltammetry experiments performed in aerated buffers, large

Figure 3 | SERRS and CV data for the electrode fabricated with G65YCu$_B$Mb mutant. (a) SERRS spectra of G65YCu$_B$Mb with (green) and without (red) the distal Cu^{2+} in air-saturated 100 mM phosphate buffer (pH 7) solution. (b) Anaerobic CV of G65YCu$_B$Mb without Cu$_B$ (red) and after Cu$_B$ binding (green). 2 V s^{-1} scan rate, in degassed, pH 7, 100 mM phosphate buffer using a Pt counter electrode.

electrocatalytic O_2 reduction currents are observed by the G65YCu$_B$Mb (with and without Cu^{2+}) bearing bio-electrodes at pH 7 at room temperature, as the applied potential is lowered below $+100$ mV versus NHE (Fig. 4a, Supplementary Fig. 5). Thus as the potential of the electrode is lowered such that the iron in these proteins is reduced to Fe^{II}, an electrocatalytic O_2 reduction current is observed. It is important to note that the potential of O_2 reduction reaction (ORR) (E_{ORR}) is -263 mV, which is more negative than the $E^{1/2}(-97$ mV), suggesting that the potential determining step of ORR is not the reduction of resting Fe^{III} to Fe^{II} but the reduction of a different species with -166 mV more negative potential (Fig. 4a inset). A more negative E_{ORR} relative to $E^{1/2}$ is mechanistically significant (*vide infra*).

In these active sites, O_2 may be reduced by $4e^-$ and $4H^+$ to H_2O, or by fewer electrons to produce partially reduced oxygen

species (PROS) like O_2^- and H_2O_2. The extent of $4e^-$ reduction and the second order rate constant (k_{ORR}) of the ORR can be determined using rotating disc electrochemistry (RDE) where the catalytic O_2 reduction current increases with increasing rotation rates (Fig. 4a,b) following the Kouteky–Levich (K–L) equation (equation (1))[39].

$$i_{cat}^{-1} = i_K(E)^{-1} + i_L^{-1} \qquad (1)$$

where, $i_K(E)$ is the potential-dependent kinetic current and i_L is the Levich current. i_L is expressed as

$$i_L = 0.62nFA[O_2](D_{O2})^{2/3}\omega^{1/2}\nu^{-1/6} \qquad (2)$$

where n is the number of electrons transferred to the substrate, A is the macroscopic area of the disc (0.096 cm^2), [O_2] is the concentration of O_2 in an air-saturated buffer (0.26 mM) at 25 °C,

Table 1 | $E^{1/2}$ and coverage.

Protein	Metal-binding sites	$E^{1/2}$ (mV)	Integrated coverage (mol cm^{-2})
Hemin-yne reconstituted myoglobin[29]	Fe	-135.0 ± 5.0	$2.15 \pm 0.05 \times 10^{-12}$
Hemin-yne reconstituted G65YCu$_B$Mb	Fe	-57.5 ± 5.0	$2.55 \pm 0.05 \times 10^{-12}$
	Cu$_B$, Fe	-97.0 ± 5.0	$4.65 \pm 0.05 \times 10^{-12}$

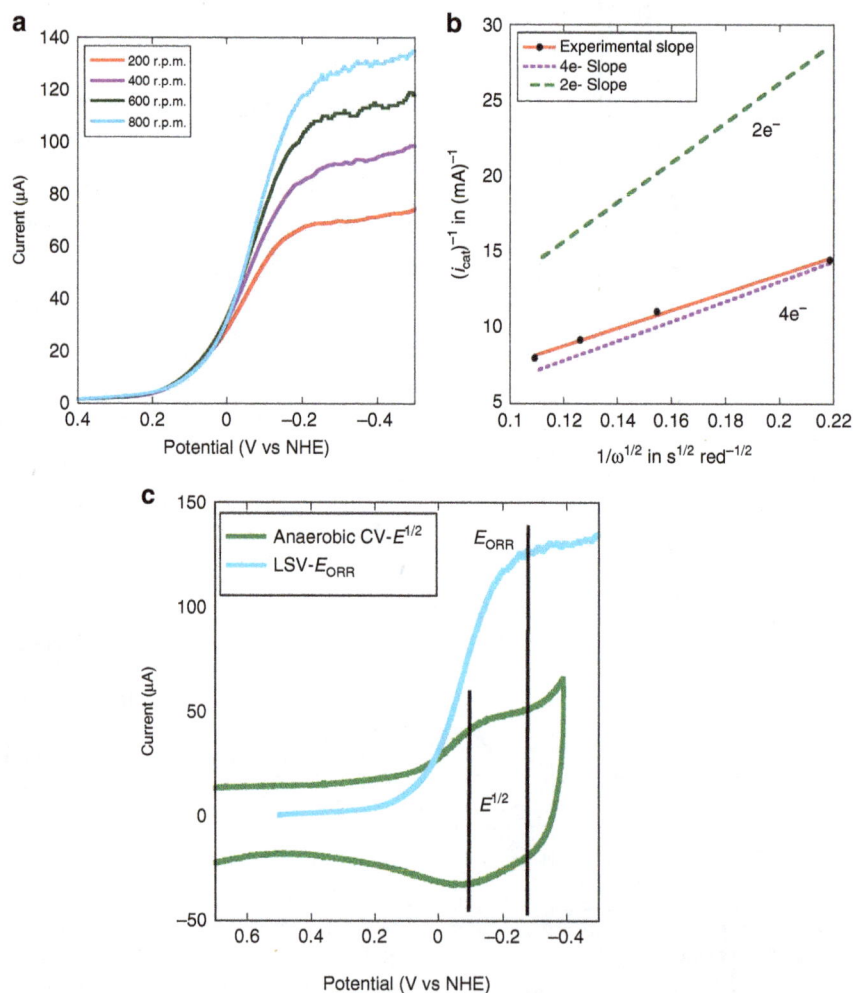

Figure 4 | RDE data for electrode modified with G65YCu$_B$Mb with Cu^{2+}. (**a**) Linear sweep voltammogram of G65YCu$_B$Mb with Cu^{2+} in air-saturated pH 7, 100 mM phosphate buffer solution at 100 mV s^{-1} scan rate, potentials are scaled relative to NHE and a Pt counter electrode is used. Data are collected at different rotation speeds (200 r.p.m.—red, 400 r.p.m.—purple, 600 r.p.m.—green, 800 r.p.m.—blue). (**b**) Plot of i_{cat}^{-1} for G65YCu$_B$Mb-bearing bioelectrode at -300 mV potential and at multiple rotation rates, with the inverse square root of the angular rotation rate ($\omega^{-1/2}$) (**c**) Difference between the potential for O_2 reduction (E_{ORR}) and the midpoint reduction potential of $Fe^{3+/2+}$ redox couple ($E^{1/2}$).

D_{O2} is the diffusion coefficient of O_2 $(1.8 \times 10^{-5} \, cm^2 \, s^{-1})$ at 25 °C, ω is the angular velocity of the disc and v is the kinematic viscosity of the solution $(0.009 \, cm^2 \, s^{-1})$ at 25 °C (ref. 40).

Plot of i_{cat}^{-1} at multiple rotation rates with the inverse square root of the angular rotation rate (i_{cat}^{-1}) for G65YCu$_B$Mb (with Cu^{2+}) (Fig. 4b) is linear. The slope of K–L plot is expressed as $1/[n\{0.62FA(D_{O2})^{2/3}v^{-1/6}\}]$, which can be used to experimentally estimate the value of n where n is the number of electrons donated to the substrate, that is, O_2. The slope obtained from the experimental data for G65YCu$_B$Mb (Fig. 4b) is close to the theoretical slope (Fig. 4b, dotted purple line) expected for a 4e$^-$ process and very different from the slope for a 2e$^-$ process (Fig. 4b, dotted green line). Thus the G65YCu$_B$Mb bioelectrode predominantly catalyses a 4e$^-$/4H$^+$ reduction of O_2 to H_2O at pH 7.

The intercept of the K–L plot is the inverse of the kinetic current $(i_K(E)^{-1})$, where $i_K(E)$ is expressed as[41]

$$i_K(E) = k_{ORR}nFA[O_2]\Gamma_{cat} \quad (3)$$

where, n is the number of electrons, A is the geometric surface area, $[O_2]$ is the bulk concentration of O_2, Γ_{cat} is the surface coverage of the catalyst (obtained from the integration of the anaerobic CV data) and k_{ORR} is the second order rate constant for O_2 reduction estimated at $-300 \, mV$. At this potential, in an oxygenated buffer, the G65YCu$_B$Mb catalyst is involved in substrate diffusion-limited ORR. Using this equation (equation (3)) and the experimentally obtained $i_K(E)$ at $-300 \, mV$, the second order rate constant for O_2 reduction G65YCu$_B$Mb is evaluated to be $1.98 \times 10^7 \, M^{-1} s^{-1}$ (Table 2). The pseudo first order rate can be determined (Table 2) from the second order rate by taking into account the substrate, O_2, concentration under these experimental conditions to be 0.26 mM (Supplementary Fig. 6). The catalytic ORR rate by G65YCu$_B$Mb surpasses those reported for the best artificial synthetic analogues (Table 2).

The G65YCu$_B$Mb biosynthetic Mb scaffold-based bio-electrode for O_2 reduction is remarkably stable. Monolayers bearing covalently attached O_2-reducing electrocatalysts reported so far have never been stable enough to allow these dynamic electrochemical experiments to determine the kinetic parameters (k_{ORR}, number of electrons and so on). Enzymes like laccases, directly attached to chemically modified graphite electrodes, were found to be stable enough to be investigated with these hydrodynamic techniques[27]. The failure to perform these experiments has been attributed to degradation of the catalyst during RDE experiment, that is, very small turnover numbers presumably due to the production of PROS during ORR. Rotating ring disc electrochemistry (RRDE) shows formation of only ~6% PROS by the G65YCu$_B$Mb (Supplementary Fig. 7) during ORR, indicating that it reduces 96% of O_2 to H_2O consistent with the RDE data. During the RDE experiments (Fig. 4a), the G65YCu$_B$Mb-functionalized electrodes bearing 10^{-12} mol of the catalyst reduced $1.8 \pm 0.3 \times 10^{-8}$ mol of O_2 ($7 \pm 1 \times 10^{-3}$ C total charge and 4e$^-$ per O_2 molecule) yielding a turnover

number of at least 10^4. The role of PROS in degrading the catalyst is established by the fact that the electrolytic current (at $-0.3 \, V$) remains stable in the presence of 50 µM catalase in solution (Supplementary Fig. 8).

Discussion

To understand the facile and selective O_2 reduction catalysed by the G65YCu$_B$Mb biochemical model, the recently developed SERRS-RDE technique is employed[42]. In this technique, the rR spectra of the catalyst (that is, G65YCu$_B$Mb) bearing electrode is collected while the system is involved in steady state O_2 reduction and the species accumulated in the steady state can be identified. For any species to accumulate in steady state, its rate of formation has to be greater than its rate of decay. Thus, while the species preceding the rate-determining step (rds) will accumulate at steady state, the accumulation of a species in steady state does not immediately imply its decay as the rds. In the absence of O_2, a high spin ferrous species is formed, characterized by a v_4 and v_3 vibrations at $1,357 \, cm^{-1}$ and $1,473 \, cm^{-1}$ (Fig. 5a, cyan), respectively, when a cathodic potential of $-0.4 \, V$ is applied signifying reduction of the resting ferric state (Fig. 5a, red) to the active ferrous state at these potentials. However, when the same reducing potential is applied in an oxygenated buffer the SERRS-RDE data (Supplementary Fig. 9) clearly show the presence of different species during electrocatalytic ORR, which leads to broadening of the v_4, v_3 and v_2 regions (Fig. 5a, green) relative to the oxidized and reduced states (Fig. 5a, red and cyan). In particular, the v_3 and v_2 vibrations discernibly shift to higher energies as indicated by clear increase in intensities at $1,508 \, cm^{-1}$ and $1,591 \, cm^{-1}$ (Fig. 4b), suggesting the accumulation of Fe^{IV}=O species during steady state ORR[43]. Signals from high spin ferrous, resting high spin ferric, low spin ferric and ferryl species with v_3 at $1,473 \, cm^{-1}$ (Fig. 5a, cyan; and Fig. 5b, brown), $1,493 \, cm^{-1}$ (Fig. 5a, red and Fig. 5b, dashed green), $1,504 \, cm^{-1}$ (Fig. 5a, cyan and Fig. 5b, cyan) and $1,508 \, cm^{-1}$ (Fig. 5a, green and Fig. 5b, green), respectively, could be convoluted by fitting the v_3 region of the spectrum. The lack of significant signal from the high spin ferrous species (weak v_3 at $1,473 \, cm^{-1}$) suggests that O_2 binding to these species is facile in the steady state. These mutants use the basic design of Mb which has a very fast O_2-binding rate ($10^7 \, M^{-1} s^{-1}$) (refs 44,45). This rate is indeed ~10 times faster than O_2 binding to the haem a_3 site of CcO (ref. 46). Similarly, the very weak intensity of the high spin (HS) Fe^{III} species indicates that the ET to Fe^{III} resting state is very facile at these potentials as may be expected due to direct attachment of the Hemin-yne to the electrode. The significant intensity of ferryl species (v_2 at $1,591 \, cm^{-1}$ and v_3 at $1,508 \, cm^{-1}$) entails the O–O bond cleavage leading to its formation to be faster than its decay via reduction under steady state. Thus the reduction of the resting Fe^{III} state, O_2 binding to Fe^{II} are facile in G65YCu$_B$Mb under the reaction conditions. The low spin ferric species accumulated during steady state ORR could be a dioxygen adduct or peroxide adduct similar to those observed in native CcO and its model systems[46-51]. The low frequency region shows $^{18}O_2$-sensitive bands suggestive of the formation of a low spin ferric peroxide and Fe(IV)=O (Supplementary Fig. 10).

If one were to conceive of a Gedanken steady state turnover experiment with CcO where the electron transfer to the active site is very fast (that is, in the hypothetical situation where ET from Cyt c to CcO is not the rds) as the ET from the electrode to the active site is very fast due to direct attachment of the later to the former, the species that would accumulate during turnover, based on the Babkcock–Wikström mechanism (Fig. 6, the parameters of native CcO are indicated in purple and G65YCu$_B$Mb are

Table 2 | k_{ORR} of different ORR catalysts.

ORR catalysts	Metals	k_{ORR}	PROS (%)
G65YCu$_B$Mb	Cu$_B$, Haem	$1.98 \times 10^7 \, M^{-1} s^{-1}$ or $5,148 \, s^{-1}$	~6 ± 1
Synthetic model[60]	Cu$_B$, Haem	$1.2 \times 10^5 \, M^{-1} s^{-1}$	~10 ± 1

ORR, O_2 reduction reaction; PROS, partially reduced oxygen species.
k_{ORR} determined at $-0.4 \, V$ versus NHE; PROS determined at $-0.12 \, V$ versus NHE in both cases.

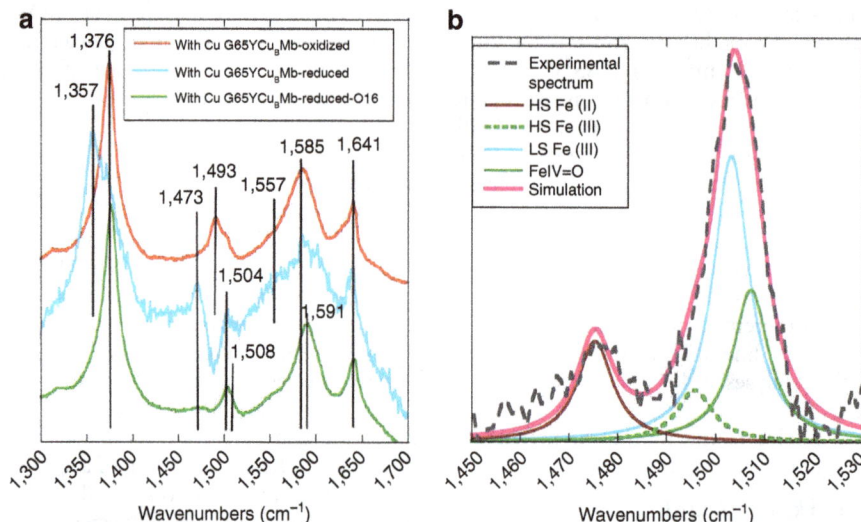

Figure 5 | SERRS-RDE data of G65YCu$_B$Mb-bearing electrode. (**a**) SERRS-RDE data of G65YCu$_B$Mb-bearing electrode at oxidized (applied potential was 0 V with respect to Ag/AgCl reference electrode), reduced (applied potential was − 0.4 V with respect to Ag/AgCl reference electrode) state and in the presence of O$_2$ (O16) saturated 100 mM pH 7 phosphate buffer and (**b**) Components of the rR spectrum determined by simulating the spectra of G65YCu$_B$Mb-bearing electrode in the presence of O$_2$ (O16) saturated 100 mM pH 7 phosphate buffer indicated in green in **a**.

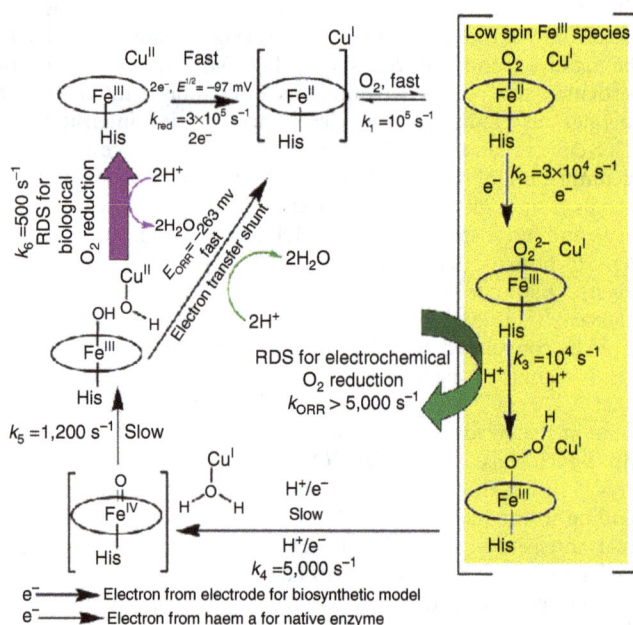

Figure 6 | Mechanisms. Comparison between the mechanism of native CcO (ref. 46) in solution and the biosynthetic G65YCu$_B$Mb model on electrode.

indicated in green[52] and the arrows in black indicate general route for O$_2$ reduction by CcO followed in both the native system, as well as in the G65YCu$_B$Mb-immobilized electrode) are the Fe^{II}, Fe^{II}-O$_2$, Fe^{III}-O$_2^{2-}$, Fe^{III}-OOH, Fe^{IV} = O and Fe^{III}-OH species as the rates of formation of these species are greater than their rates of decay[53,54]. Out of these, the Fe^{II}-O$_2$, Fe^{III}-O$_2^{2-}$ and Fe^{III}-OOH species will have Raman signatures of low spin haem (Fig. 6), Fe^{III}-OH will have rR signature of high spin haem and the Fe^{IV} = O will have signatures unique to haem ferryl species[46,54,55]. The SERRS-RDE data show the presence of species having signatures of low spin Fe^{III} and Fe^{IV} = O. While the later can originate from only a single species, the former can indicate the presence of any of the three species or a combination of them. The lack of significant high spin signal indicates that the biosynthetic model circumvents accumulation of Fe^{III}-OH and

resting Fe^{III} species in the steady state by facile ET. The overall rate-limiting step of native CcO in solution is the dissociation of hydroxide of the Fe^{III}-OH end product of O$_2$ reduction from haem to generate the active ferric resting form and has a first order rate constant of 500 s^{-1} (refs 46,54–56). This dissociation is required during turnover as the potential of this hydroxide-bound form is likely to be more negative that the five-coordinate resting oxidized site (which will be regenerated after hydroxide dissociation) and will not be reduced by haem a. While the $E^{1/2}$ of a haem a$_3$ Fe^{III}-OH species cannot be determined with confidence due to strong interaction potential and co-operativity between the haem a and haem a$_3$ sites, the potential of formate and azide-bound high spin haem a$_3$ site (analogous to hydroxide) is ∼130 to 200 mV more negative than the resting ferric site[57,58]. The potential determining step of ORR (defined as the electron transfer (ET) step in catalysis having lowest potential) by the G65YCu$_B$Mb is 166 mV more negative than the $E^{1/2}$ for the resting high spin ferric state and is likely to be the reduction of the Fe^{III}-OH species. Thus the bioelectrode can circumvent the kinetic barrier associated with the dissociation of the hydroxide by directly reducing it to ferrous at 166 mV lower potential. This direct electron transfer to the ferric hydroxide species, which is an intermediate in the catalytic cycle of CcO, circumventing a slow step in catalysis, is an electron transfer shunt analogous to peroxide shunt in cytochrome P450, which overcomes the rate-determining O$_2$ activation step of the native enzyme[59]. In the mass transfer controlled region of the catalytic current, all ET steps are facile and steps like O$_2$ binding, protonation and O–O bond cleavage can be rds at these potential. The lack of HS Fe^{II} accumulation in the SERRS-RDE indicates that O$_2$ binding to Fe^{II} is very fast and not the rds. The O–O bond cleavage in the Babkock–Wikström mechanism involves ET to the active site and not the rds as well. Thus rds of ORR by these biosynthetic models is likely to be the protonation of the Fe^{III}-O$_2^-$ species with a first order rate of 5,000 s^{-1} (Fig. 6).

In an air-saturated buffer (0.26 mM O$_2$), the pseudo first order rate constant of ORR by G65YCu$_B$Mb is determined to be ∼5–6 × 10^3 s^{-1} (k_{ORR} [O$_2$]). The highest second order O$_2$ reduction rate reported for any synthetic mimic of CcO is 1.2 × 10^5 M^{-1} s^{-1}; that too on a multilayer having 1,000 times more catalyst than the G65YCu$_B$Mb electrodes[60]. The second order rate constant of G65YCu$_B$Mb is $10^7 M^{-1} s^{-1}$ which is, thus, 2 orders of magnitude higher than best synthetic haem/Cu-based O$_2$ reduction electrocatalyst. Thus the selectivity and kinetic rate of the G65YCu$_B$Mb-bearing electrode surpasses those reported for smaller synthetic analogues and illustrates the advantages of using a biochemical scaffold over a synthetic scaffold. Although the pseudo first order rate constant of the G65YCu$_B$Mb is 10 times faster than the rate of native CcO in solution, such a comparison is vulnerable to differences in reaction conditions (for example, G65YCu$_B$Mb is water soluble but CcO exists in membranes). Alternatively, erstwhile efforts resulting in electrodes bearing native CcO in a manner similar to these bio-electrodes show extremely sluggish O$_2$ reduction[60–62]. This is due to improper alignment of this membrane-bound protein on the electrode, which precludes efficient electron transfer to the active site[63–65]. However, the direct attachment of haem to the electrode utilizing its solvent-exposed propionate groups (that is, a short circuit) enables fast electron transfer to the active site[29]. This is further supported by the fact that when ethynylferrocene (Fc) is attached to the same surface ~ 25 mV peak separation is observed even at $5 V s^{-1}$ (Supplementary Fig. 11)[38], suggesting that the ET is indeed fast. As a result when a CcO-functionalized SAM-covered Au electrode produces $<1 \mu A$ electrochemical O$_2$ reduction current at -300 mV, this bio-electrode produce $\sim 100 \mu A$ current at similar potentials[66].

Finally, the G65YCu$_B$Mb mutant has residues in the distal pocket that can help both electron and proton transfer during O$_2$ reduction (Y65 in G65YCu$_B$Mb). In CcO, the involvement of Tyr 244 residues in proton/electron transfer during O$_2$ reduction is now widely accepted[67,68]. Previous biochemical and structural studies on these mutants had indeed indicated the close proximity of this residue to the distal site[7,19,30]. An analogous biochemical model without the Y65 residue, Cu$_B$Mb, is not as stable as the G65YCu$_B$Mb as the former degrades rapidly during the RDE experiments (Supplementary Fig. 12). In summary, a electron transfer shunt which circumvents the rate-determining dissociation of a ferric hydroxide species by directly reducing it at slightly negative potential, fast O$_2$ binding, fast electron transfer to the active site and the presence of a protective Y65 residue in a biochemical model of CcO results in O$_2$ reduction activity 100 times faster than the best synthetic models, order of magnitude faster than CcO immobilized on electrode and follows a mechanism comparable to that of native CcO in solution.

Methods

Materials. 1-Azidoundecane-11-thiol and Hemin-yne were synthesized following the reported procedure[29,69]. 6-Mercaptohexanoic acid was purchased from Sigma Aldrich. Di-sodium hydrogenphosphate dihydrate (Na$_2$HPO$_4$. 2H$_2$O) was purchased from Merck. 2, 6-lutidine was purchased from Avra Synthesis Pvt. Ltd. These chemicals were used without further purification. Au wafers were purchased from Platypus Technologies (1,000 Å of Au on 50 Å of Ti adhesion layer on top of a Si(III) surface). Transparent Au wafers (100 Å of Au on 10 Å of Ti) were purchased from Phasis, Switzerland. Au and Ag discs for the RRDE and SERRS experiments, respectively, were purchased from Pine Instruments, USA. The Mb mutants were prepared as reported in the literature[7,30]. Analysis of the components of the rR spectrum was done by using Lorenztian line shape of peak fit software.

Instrumentation. All electrochemical experiments were performed using a CH Instruments (model CHI710D Electrochemical Analyzer). Bipotentiostat, reference electrode and Teflon plate material evaluating cell (ALS, Japan; http://www.als-japan.com/1398.html) were purchased from CH Instruments. The RRDE set-up from Pine Research Instrumentation (E6 series ChangeDisk tips with AFE6M rotor) was used to obtain the RRDE data. The mutant Mb-functionalized or SAM-covered Au surface (disc of 0.1 cm^2 area for RDE, RRDE and wafer of 0.45 cm^2 area for CV) was always used as the working electrode. The XPS data were collected in a Omicron (model: 1712-62-11) spectrometer using a high-resolution monochromatic Al-Kα source at 1,486.7 eV under 15 kV voltage and 10 mA

current maintaining a base pressure of 5×10^{-10} mbar. The binding energies were calibrated to the Ag 3d$_{5/2}$ peak at 368.2 eV. The resonance Raman experiments were done in the Kr$^+$ Laser (Sabre Innova, Model—SBRC-DBW-K) purchased from Coherent, and the data were collected using the Spectrograph (Model—Trivista 555) from Princeton Instruments.

Formation of mixed SAM and covalent attachment of Hemin-yne to it. Mixed self-assembled monolayer of 1-azidoundecan-11-thiol and 6-mercaptohexanoic acid was formed on immersing the properly cleaned Au wafers or disks into the deposition solution containing 1-azidoundecan-11-thiol and 6-mercaptohexanoic acid in 10 ml of ethanol in a desired ratio (typically 1:49). The total thiol concentration of these deposition solutions were always maintained at 1 mM. On this SAM Hemin-yne was covalently attached using 'Click' reaction[29].

Reconstitution of Apo-G65YCu$_B$Mb mutants to G65YCu$_B$Mb. For all the experiments on heterogeneous SAM surfaces, the Hemin-yne modified –COOH SAM surfaces were incubated with a 20 μM apoprotein (Apo-G65YCu$_B$Mb) solution for 2 h. The supernatant solution was drained and the surface was cleaned with water. The presence of Cu^{2+} in the non-haem-binding site is confirmed by electron paramagnetic resonance (EPR) spectroscopy (Supplementary Fig. 13). The immobilization of the mutant is further confirmed by the absorption spectra and SERRS of the surface fabricated with those mutants (Supplementary Fig. 14, Supplementary Fig. 15).

Cyclic voltammetry. The CV was performed using Au wafers sandwiched between two Teflon blocks of the Plate material evaluating cell. All electrochemical experiments were done in pH 7 phosphate buffer containing potassium hexafluorophosphate. Anaerobic cyclic voltammetric experiments were done by using degassed buffer (three cycles of freeze–pump–thaw). Ag/AgCl reference electrode and Pt counter electrode were used throughout all the electrochemical experiments except the case of anaerobic experiments where only Ag wire was used as the reference electrode.

The peak areas were estimated by integrating the anodic/cathodic peak of the anaerobic CV of the mutant Mb-functionalized SAM-covered Au surface, using the data acquisition software itself. A line collinear with the background is used to subtract the background. The estimated area has been further confirmed by subtracting the background current of a SAM-functionalized electrode (Supplementary Fig. 11) bearing ferrocene. Both these approaches provide the same estimate.

To ensure that the SAM surface is stable during the electrocatalytic investigations, disc bearing just the SAM was subjected to several rotations (200–1,000 r.p.m.) and its capacitive current was found not to change, indicating that the SAM is retained on the electrode during these dynamic electrochemistry experiments (Supplementary Fig. 16). SAM can also be damaged when the protein atop the SAM degrades during ORR due to the reactive oxygen species produced. When an unstable electrocatalyst (Hemin-yne) decayed there was a steady loss of ORR current, indicating degradation of the active site but the capacitive currents of the SAM were unaltered (Supplementary Fig. 17).

SERRS and SERRS-RDE. The excitation wavelength used in the Resonance Raman experiments was 406.7 nm and the power applied to the sample was 10–15 mW. The spectrograph was calibrated against naphthalene. The Ag surfaces were roughened before SERRS experiments following literature protocols[70]. The SERRS-RDE set-up is described in ref. 42. The data for the oxidized state was obtained by holding the potential of the disc at 0 mV versus NHE, and the data during steady state ORR were obtained by holding the disc at -400 mV versus NHE and the disc was rotated at 300 r.p.m. Normally data were acquired over a period of 300 s.

References

1. Collman, J. P. *et al.* A cytochrome c oxidase model catalyzes oxygen to water reduction under rate-limiting electron flux. *Science* **315,** 1565–1568 (2007).
2. Mirica, L. M. *et al.* Tyrosinase reactivity in a model complex: an alternative hydroxylation mechanism. *Science* **308,** 1890–1892 (2005).
3. Wang, Y., DuBois, J. L., Hedman, B., Hodgson, K. O. & Stack, T. D. Catalytic galactose oxidase models: biomimetic Cu(II)-phenoxyl-radical reactivity. *Science* **279,** 537–540 (1998).
4. Beinert, H., Holm, R. H. & Mü̈nck, E. Iron-sulfur clusters: nature's modular, multipurpose structures. *Science* **277,** 653–659 (1997).
5. Cole, A. P., Root, D. E., Mukherjee, P., Solomon, E. I. & Stack, T. D. P. A trinuclear intermediate in the copper-mediated reduction of O2: four electrons from three coppers. *Science (Washington, D. C.)* **273,** 1848–1850 (1996).
6. Robertson, D. E. *et al.* Design and synthesis of multi-heme proteins. *Nature (London)* **368,** 425–432 (1994).
7. Yeung, N. *et al.* Rational design of a structural and functional nitric oxide reductase. *Nature* **462,** 1079–1082 (2009).
8. Tard, C. *et al.* Synthesis of the H-cluster framework of iron-only hydrogenase. *Nature* **433,** 610–613 (2005).

9. Aboelella, N. W. *et al.* Effects of thioether substituents on the O_2 reactivity of β-diketiminate-Cu(I) complexes: probing the role of the methionine ligand in copper monooxygenases. *J. Am. Chem. Soc.* **128**, 3445–3458 (2006).

10. Shin, H., Lee, D.-H., Kang, C. & Karlin, K. D. Electrocatalytic four-electron reductions of O_2 to H_2O with cytochrome c oxidase model compounds. *Electrochim. Acta* **48**, 4077–4082 (2003).

11. Kim, E. *et al.* Superoxo, μ-peroxo, and μ-oxo complexes from heme/O_2 and heme-Cu/O_2 reactivity: Copper ligand influences in cytochrome c oxidase models. *Proc. Natl Acad. Sci. USA* **100**, 3623–3628 (2003).

12. Tolman, W. B. & Que, J. L. Sterically hindered benzoates: a synthetic strategy for modeling dioxygen activation at diiron active sites in proteins. *J. Chem. Soc. Dalton Trans.* **31**, 653–660 (2002).

13. Osyczka, A., Moser, C. C., Daldal, F. & Dutton, P. L. Reversible redox energy coupling in electron transfer chains. *Nature* **427**, 607–612 (2004).

14. Koder, R. L. *et al.* Design and engineering of an O_2 transport protein. *Nature* **458**, 305–309 (2009).

15. Page, C. C., Moser, C. C., Chen, X. & Dutton, P. L. Natural engineering principles of electron tunneling in biological oxidation-reduction. *Nature* **402**, 47–52 (1999).

16. Tani, F., Matsu-ura, M., Nakayama, S. & Naruta, Y. Synthetic models for the active site of cytochrome P450. *Coord. Chem. Rev.* **226**, 219–226 (2002).

17. Dey, A. *et al.* Solvent tuning of electrochemical potentials in the active sites of HiPIP versus ferredoxin. *Science* **318**, 1464–1468 (2007).

18. Astrow, M. L., Peacock, A. F. A., Stuckey, J. A. & Pecoraro, V. L. Hydrolytic catalysis and structural stabilization in a designed metalloprotein. *Nat. Chem.* **4**, 118–123 (2012).

19. Lin, Y.-W. *et al.* Roles of glutamates and metal ions in a rationally designed nitric oxide reductase based on myoglobin. *Proc. Natl Acad. Sci. USA.* **107**, 8581–8586 (2010).

20. Esselborn, J. *et al.* Spontaneous activation of [FeFe]-hydrogenases by an inorganic [2Fe] active site mimic. *Nat. Chem. Biol.* **9**, 607–609 (2013).

21. Fontecave, M. & Artero, V. Bioinspired catalysis at the crossroads between biology and chemistry: A remarkable example of an electrocatalytic material mimicking hydrogenases. *C. R. Chim.* **14**, 362–371 (2011).

22. Lu, Y. & Valentine, J. S. Engineering metal-binding sites in proteins. *Curr. Opin. Struct. Biol.* **7**, 495–500 (1997).

23. Cracknell, J. A. & Blanford, C. F. Developing the mechanism of dioxygen reduction catalyzed by multicopper oxidases using protein film electrochemistry. *Chem. Sci.* **3**, 1567–1581 (2012).

24. dos Santos, L., Climent, V., Blanford, C. F. & Armstrong, F. A. Mechanistic studies of the 'blue' Cu enzyme, bilirubin oxidase, as a highly efficient electrocatalyst for the oxygen reduction reaction. *Phys. Chem. Chem. Phys.* **12**, 13962–13974 (2010).

25. Sigman, J. A., Kim, H. K., Zhao, X., Carey, J. R. & Lu, Y. The role of copper and protons in heme-copper oxidases: kinetic study of an engineered heme-copper center in myoglobin. *Proc. Natl Acad. Sci. USA.* **100**, 3629–3634 (2003).

26. Bhagi-Damodaran, A., Petrik, I. D., Marshall, N. M., Robinson, H. & Lu, Y. Systematic tuning of heme redox potentials and its effects on O2 reduction rates in a designed oxidase in myoglobin. *J. Am. Chem. Soc.* **136**, 11882–11885 (2014).

27. Blanford, C. F., Foster, C. E., Heath, R. S. & Armstrong, F. A. Efficient electrocatalytic oxygen reduction by the 'blue' copper oxidase, laccase, directly attached to chemically modified carbons. *Faraday Discuss.* **140**, 319–335 (2009).

28. Blanford, C. F., Heath, R. S. & Armstrong, F. A. A stable electrode for high-potential, electrocatalytic O_2 reduction based on rational attachment of a blue copper oxidase to a graphite surface. *Chem. Commun.* **43**, 1710–1712 (2007).

29. Mukherjee, S., Sengupta, K., Das, M. R., Jana, S. S. & Dey, A. Site-specific covalent attachment of heme proteins on self-assembled monolayers. *J. Biol. Inorg. Chem.* **17**, 1009–1023 (2012).

30. Sigman, J. A., Kwok, B. C. & Lu, Y. From myoglobin to heme-copper oxidase: design and engineering of a CuB center into sperm whale myoglobin. *J. Am. Chem. Soc.* **122**, 8192–8196 (2000).

31. Miner, K. D. *et al.* A designed functional metalloenzyme that reduces O2 to H2O with over one thousand turnovers. *Angew. Chem. Int. Ed.* **51**, 5589–5592 (2012).

32. Hu, S., Smith, K. M. & Spiro, T. G. Assignment of protoheme resonance Raman spectrum by heme labeling in myoglobin. *J. Am. Chem. Soc.* **118**, 12638–12646 (1996).

33. Bandyopadhyay, S., Mukherjee, S. & Dey, A. Modular synthesis, spectroscopic characterization and in situ functionalization using "click" chemistry of azide terminated amide containing self-assembled monolayers. *RSC Adv.* **3**, 17174–17187 (2013).

34. Guo, L.-H., McLendon, G., Razafitrimo, H. & Gao, Y. Photo-active and electro-active protein films prepared by reconstitution with metalloporphyrins self-assembled on gold. *J. Mater. Chem.* **6**, 369–374 (1996).

35. Ghodselahi, T., Vesaghi, M. A., Shafiekhani, A., Baghizadeh, A. & Lameii, M. XPS study of the Cu@Cu2O core-shell nanoparticles. *Appl. Surf. Sci.* **255**, 2730–2734 (2008).

36. Schmitt, S. K., Murphy, W. L. & Gopalan, P. Crosslinked PEG mats for peptide immobilization and stem cell adhesion. *J. Mater. Chem. B* **1**, 1349–1360 (2013).

37. Boulatov, R., Collman, J. P., Shiryaeva, I. M. & Sunderland, C. J. Functional analogues of the dioxygen reduction site in cytochrome oxidase: mechanistic aspects and possible effects of CuB. *J. Am. Chem. Soc.* **124**, 11923–11935 (2002).

38. Mukherjee, S., Bandyopadhyay, S. & Dey, A. Tuning the apparent formal potential of covalently attached ferrocene using SAM bearing ionizable COOH groups. *Electrochim. Acta* **108**, 624–633 (2013).

39. Bard, A. J. & Faulkner, L. R. *Electrochemical Methods: Fundamentals and Applications* 2nd edn (Wiley, 1980).

40. McCrory, C. C. L. *et al.* Electrocatalytic O2 reduction by covalently immobilized mononuclear copper(I) complexes: evidence for a binuclear Cu2O2 intermediate. *J. Am. Chem. Soc.* **133**, 3696–3699 (2011).

41. Shi, C. & Anson, F. C. Potential-dependence of the reduction of dioxygen as catalysed by tetraruthenated cobalt tetrapyridylporphyrin. *Electrochim. Acta* **39**, 1613–1619 (1994).

42. Sengupta, K., Chatterjee, S., Samanta, S. & Dey, A. Direct observation of intermediates formed during steady-state electrocatalytic O_2 reduction by iron porphyrins. *Proc. Natl Acad. Sci. USA* **110**, 8431–8436 (2013).

43. Oertling, W. A., Kean, R. T., Wever, R. & Babcock, G. T. Factors affecting the iron-oxygen vibrations of ferrous oxy and ferryl oxo heme proteins and model compounds. *Inorg. Chem.* **29**, 2633–2645 (1990).

44. Sono, M., Smith, P. D., McCray, J. A. & Asakura, T. Kinetic and equilibrium studies of the reactions of heme-substituted horse heart myoglobins with oxygen and carbon monoxide. *J. Biol. Chem.* **251**, 1418–1426 (1976).

45. Nagao, S. *et al.* Structural and oxygen binding properties of dimeric horse myoglobin. *Dalton Trans.* **41**, 11378–11385 (2012).

46. Ferguson-Miller, S. & Babcock, G. T. Heme/copper terminal oxidases. *Chem. Rev.* **96**, 2889–2908 (1996).

47. Iwata, S., Ostermeier, C., Ludwig, B. & Michel, H. Structure at 2.8 A resolution of cytochrome c oxidase from Paracoccus denitrificans. *Nature* **376**, 660–669 (1995).

48. Collman, J. P. & Decreau, R. A. Functional biomimetic models for the active site in the respiratory enzyme cytochrome c oxidase. *Chem. Commun.* **44**, 5065–5076 (2008).

49. Garcia-Bosch, I. *et al.* A "naked" Fe^{III}-(O_2^{2-})-Cu^{II} species allows for structural and spectroscopic tuning of low-spin heme-peroxo-Cu complexes. *J. Am. Chem. Soc.* **137**, 1032–1035 (2015).

50. Kim, E., Chufan, E. E., Kamaraj, K. & Karlin, K. D. Synthetic models for heme-copper oxidases. *Chem. Rev.* **104**, 1077–1134 (2004).

51. Kim, E. *et al.* Dioxygen reactivity of copper and heme-copper complexes possessing an imidazole-phenol cross-link. *Inorg. Chem.* **44**, 1238–1247 (2005).

52. Babcock, G. T. & Wikstrom, M. Oxygen activation and the conservation of energy in cell respiration. *Nature* **356**, 301–309 (1992).

53. Varotsis, C., Zhang, Y., Appelman, E. H. & Babcock, G. T. Resolution of the reaction sequence during the reduction of oxygen by cytochrome oxidase. *Proc. Natl Acad. Sci. USA* **90**, 237–241 (1993).

54. Han, S., Takahashi, S. & Rousseau, D. L. Time dependence of the catalytic intermediates in cytochrome c oxidase. *J. Biol. Chem.* **275**, 1910–1919 (2000).

55. Han, S., Ching, Y.-C. & Rousseau, D. L. Ferryl and hydroxy intermediates in the reaction of oxygen with reduced cytochrome c oxidase. *Nature* **348**, 89–90 (1990).

56. Rousseau, D. L. Bioenergetics: two phases of proton translocation. *Nature* **400**, 412–413 (1999).

57. Kojima, N. & Palmer, G. Further characterization of the potentiometric behavior of cytochrome oxidase. Cytochrome a stays low spin during oxidation and reduction. *J. Biol. Chem.* **258**, 14908–14913 (1983).

58. Namslauer, A., Branden, M. & Brzezinski, P. The rate of internal heme-heme electron transfer in cytochrome c oxidase. *Biochemistry* **41**, 10369–10374 (2002).

59. Denisov, I. G., Makris, T. M., Sligar, S. G. & Schlichting, I. Structure and chemistry of cytochrome P450. *Chem. Rev.* **105**, 2253–2278 (2005).

60. Boulatov, R., Collman, J. P., Shiryaeva, I. M. & Sunderland, C. J. Functional analogues of the dioxygen reduction site in cytochrome oxidase:mechanistic aspects and possible effects of CuB. *J. Am. Chem. Soc.* **124**, 11923–11935 (2002).

61. Friedrich, M. G. *et al.* In situ monitoring of the catalytic activity of cytochrome c oxidase in a biomimetic architecture. *Biophys. J.* **95**, 1500–1510 (2008).

62. Haas, A. S. *et al.* Cytochrome c and cytochrome c oxidase: monolayer assemblies and catalysis. *J. Phys. Chem. B* **105**, 11351–11362 (2001).

63. Burgess, J. D., Rhoten, M. C. & Hawkridge, F. M. Cytochrome c oxidase immobilized in stable supported lipid bilayer membranes. *Langmuir* **14**, 2467–2475 (1998).

64. Naumann, R. *et al.* The peptide-tethered lipid membrane as a biomimetic system to incorporate cytochrome c oxidase in a functionally active form. *Biosens. Bioelectron.* **14**, 651–662 (1999).

65. Burgess, J. D., Jones, V. W., Porter, M. D., Rhoten, M. C. & Hawkridge, F. M. Scanning force microscopy images of cytochrome c oxidase immobilized in an electrode-supported lipid bilayer membrane. *Langmuir* **14**, 6628–6631 (1998).

66. Su, L., Kelly, J. & Hawkridge, F. M. Electroreduction of O_2 on cytochrome c oxidase modified electrode for biofuel cell. *ECS Trans.* **2,** 1–6 (2007).

67. Yoshikawa, S., Muramoto, K. & Shinzawa-Itoh, K. Proton-pumping mechanism of cytochrome c oxidase. *Annu. Rev. Biophys.* **40,** 205–223 (2011).

68. Morgan, J. E., Verkhovsky, M. I., Palmer, G. & Wikström, M. Role of the PR intermediate in the reaction of cytochrome c oxidase with O2. *Biochemistry* **40,** 6882–6892 (2001).

69. Collman, J. P., Devaraj, N. K., Eberspacher, T. P. A. & Chidsey, C. E. D. Mixed azide-terminated monolayers: a platform for modifying electrode surfaces. *Langmuir* **22,** 2457–2464 (2006).

70. Arūnas Bulovas, Z. T. *et al.* Double-layered Ag/Au electrode for SERS spectroscopy: preparation and application for adsorption studies of chromophoric compounds. *Chemija* **18,** 9–15 (2007).

Acknowledgements

This research is sponsored by Department of Science and Technology, India and US National Institute of Health (GM062211).

Author contributions

S.M. acquired all the data, analysed the data and wrote the paper. A.M. and A.B. purified and provided the mutant proteins and helped write the paper. M.M. helped S.M. in acquiring some data. Y.L. provided the proteins, helped write the paper and developed the science. A.D. conceived the study, analysed the data, developed the science and wrote the paper

Additional information

Competing financial interests: The authors declare no competing financial interests.

7

Self-assembly of dynamic orthoester cryptates

René-Chris Brachvogel[1], Frank Hampel[1] & Max von Delius[1]

The discovery of coronands and cryptands, organic compounds that can accommodate metal ions in a preorganized two- or three-dimensional environment, was a milestone in supramolecular chemistry, leading to countless applications from organic synthesis to metallurgy and medicine. These compounds are typically prepared via multistep organic synthesis and one of their characteristic features is the high stability of their covalent framework. Here we report the use of a dynamic covalent exchange reaction for the one-pot template synthesis of a new class of coronates and cryptates, in which acid-labile O,O,O-orthoesters serve as bridgeheads. In contrast to their classic analogues, the compounds described herein are constitutionally dynamic in the presence of acid and can be induced to release their guest via irreversible deconstruction of the cage. These properties open up a wide range of application opportunities, from systems chemistry to molecular sensing and drug delivery.

---

[1] Department of Chemistry and Pharmacy, Friedrich-Alexander-University Erlangen-Nürnberg (FAU), Henkestrasse 42, 91054 Erlangen, Germany. Correspondence and requests for materials should be addressed to M.v.D. (email: max.vondelius@fau.de).

Past progress in supramolecular chemistry has been driven chiefly by the development of new macrocyclic molecules[1]. Pedersen's crown ethers (also called coronands)[2] and Lehn's cryptands[3,4] (Fig. 1) are excellent examples of compounds that initially served as platforms for studying non-covalent interactions, but have ultimately found widespread application in industry and medicine[5].

In the last decade, rationally designed three-dimensional cage compounds[6–8] have become larger and larger[9–13], enabling in one extreme case even the accommodation of a small protein[14]. To achieve the self-assembly of such large structures, the method of choice is dynamic constitutional/covalent chemistry (DCC)[15–17], which offers the essential feature of error correction that is needed to avoid significant side-product formation. Besides providing high-yielding syntheses, DCC generally gives rise to target structures that are dynamic and responsive to external stimuli under the conditions of their preparation. In the context of the emerging field of systems chemistry[18,19], dynamic macrocycles and cages have served as a valuable testing ground for the investigation and manipulation of complex molecular networks[20–25]. Constitutionally dynamic cryptates[26,27] would represent the smallest and simplest conceivable three-dimensional platform for studying molecular complexity; however, to the best of our knowledge, there are no reports of monometallic cryptates yet, which can be prepared and manipulated based on a dynamic covalent exchange reaction[28–31].

Here we describe how orthoester exchange, a previously ignored dynamic covalent reaction[32], can be used for the one-pot synthesis of monometallic cryptates (for example, see Fig. 1) from strikingly simple starting materials. We provide comprehensive characterization data (including an X-ray structure) for this new class of compounds and report on their dynamic properties, as well as on the formation of orthoester crown ethers as reaction intermediates and the unexpected finding that 4 Å molecular sieves (MS) can act as a source of sodium guest.

Results

Template synthesis of a dynamic orthoester cryptate. Inspired by reports on dynamic 'scaffolding ligands' (O,N,P-orthoesters)[33,34], we have recently investigated the exchange reaction of carboxylic O,O,O-orthoesters with simple alcohols from a DCC perspective[32]. We realized during the course of these studies that the tripodal architecture and dynamic chemistry of orthoesters[35] could be well suited for establishing two bridgeheads in macrobicyclic compounds (Fig. 1). The synthesis of such orthoester cryptands could be carried out in one step and under thermodynamic control, while a suitable metal ion could serve as a template.

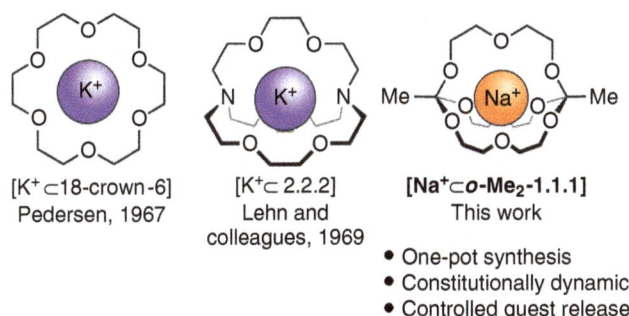

To test these hypotheses, we treated a chloroform solution of two bulk chemicals, trimethyl orthoacetate (**1**) and diethylene glycol (**2**), with catalyst trifluoroacetic acid (TFA) and a stoichiometric metal template (Fig. 2). Analysis of these initial experiments by ^{1}H nuclear magnetic resonance (NMR) spectroscopy and electrospray ionization mass spectrometry revealed that complex mixtures, containing the desired cryptate among other exchange products, had formed. Careful optimization of the reaction conditions (use of MS as a thermodynamic sink for water and methanol; use of the 'non-coordinating' counteranion tetrakis[3,5-bis(trifluoromethyl)phenyl]borate (BArF$^-$))[36–40] eventually led to the formation of cryptate [Na$^+$ ⊂ o-Me$_2$-1.1.1] BArF$^-$ (named in loose analogy to Lehn's classic cryptates; 'o' stands for orthoester) as the predominant reaction product (isolated yields typically 60%–70%).

As shown in the ^{1}H NMR spectra presented in Fig. 2, the reversible reaction between **1** and **2** initially generates a remarkable diversity of different exchange products (various degrees of replacement of MeOH by **2**, as well as formation of macrocyclic and oligomeric products (Fig. 2b)) and it is only on slow removal of MeOH by MS (4 Å) that the system converges to the final reaction product (Fig. 2c).

It should be noted that during the exchange process, water has to be excluded from the reaction mixture, which is not trivial to achieve, because even rigorously dried MS tend to slowly release residual water. As a consequence, hydrolysis of **1** can lead to the slow formation of methyl acetate (MeOAc) as a side product (Fig. 2c). From a preparative standpoint, the formation of MeOAc is not a problem, because it can easily be removed under reduced pressure (Fig. 2d). In addition, its formation as the sole side product provides two valuable pieces of information regarding the dynamic system under study. First, it is remarkable that we find the cryptate as the exclusive reaction product, even though the partial decomposition of orthoester **1** leads to a non-ideal ratio of orthoester to diol (ideal value 2:3). This result indicates that there is a thermodynamic bias for the formation of the final cryptate in the presence of sodium template. Second, the fact that we observe only one type of ester (MeOAc) suggests that exchange products incorporating one or more diethylene glycol chains (generated much more rapidly than MeOAc) are kinetically stabilized against hydrolysis, presumably due to (chelate) binding of sodium.

This kinetic stabilization due to metal binding[41] is most pronounced in pristine cryptate [Na$^+$ ⊂ o-Me$_2$-1.1.1]BArF$^-$. For example, when we mixed [Na$^+$ ⊂ o-Me$_2$-1.1.1]BArF$^-$ with trimethyl orthoacetate (**1**) in water-saturated chloroform, only the simple orthoester **1** was found to hydrolyse, whereas cryptate [Na$^+$ ⊂ o-Me$_2$-1.1.1]BArF$^-$ remained stable for 7 days (Supplementary Fig. 2). In the absence of acid, the cage is in fact stable in dimethyl sulfoxide/water mixtures and can be purified by silica gel chromatography (Fig. 2e). These observations are highly unusual for O,O,O-orthoesters that are not stabilized by the presence of five- or six-membered rings (as in Corey's OBO protecting group)[42]. These properties imply that orthoester-based hosts could have a unique advantage over existing coronands and cryptands: charged guests could be transported across lipophilic membranes[43] and subsequent hydrolysis would trigger the release of the guest ([Na$^+$ ⊂ o-Me$_2$-1.1.1]BArF$^-$ hydrolyses readily in the presence of excess water and acid; Supplementary Fig. 3). The high potential for such a supramolecular approach for drug formulation and delivery is underscored by a recent patent publication, which describes related hydrolysis-prone crown ether compounds (scheduled for phase 1 clinical trials in 2015)[44].

Figure 1 | Coronates and cryptates. Comparison of a classic coronate and cryptate with one of the orthoester-based, constitutionally dynamic cryptates described in this work.

[K$^+$ ⊂ 18-crown-6]
Pedersen, 1967

[K$^+$ ⊂ 2.2.2]
Lehn and colleagues, 1969

[Na$^+$ ⊂ o-Me$_2$-1.1.1]
This work
• One-pot synthesis
• Constitutionally dynamic
• Controlled guest release

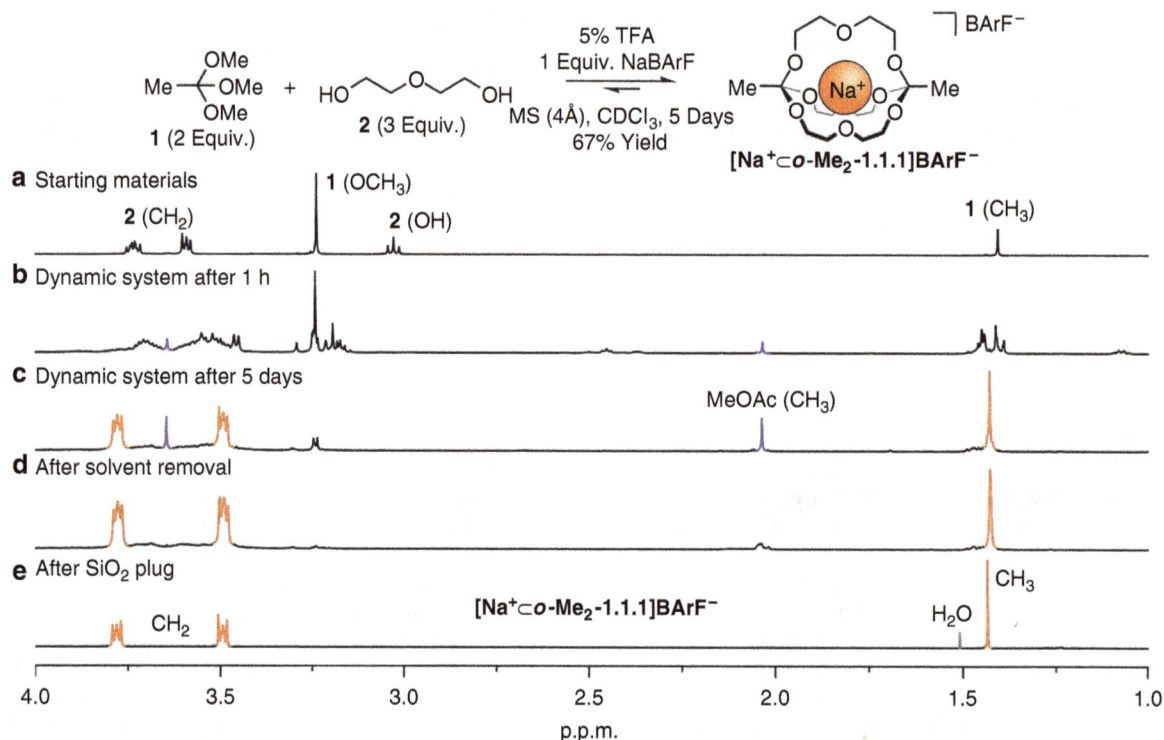

Figure 2 | One-pot self-assembly of dynamic cryptate [Na$^+$ ⊂ o-Me$_2$-1.1.1]BArF$^-$. Partial ^{1}H NMR spectra (400 MHz, CDCl$_3$, 298 K) showing the evolution of the dynamic system over time (see Supplementary Fig. 1 for full spectra). (**a**) Starting materials. (**b**) Complex mixture that forms rapidly after addition of acid catalyst (1 h). (**c**) After 5 days [Na$^+$ ⊂ o-Me$_2$-1.1.1]BArF$^-$ is formed as major product, alongside hydrolysis product MeOAc (singlets at 3.6 and 2.0 p.p.m.). (**d**) The crude product is obtained by removal of MeOAc under reduced pressure. (**e**) Further purification is conveniently achieved by passing the crude product through a short plug of silica gel. Reaction conditions: trimethyl orthoacetate (120 μmol), diethylene glycol (180 μmol), NaBArF (60 μmol), TFA (3.0 μmol; added over 5 days), MS (4 Å, 1 g), CDCl$_3$ (6.0 ml), 5 days, room temperature.

Solid-state structure. Following the comprehensive characterization of [Na$^+$ ⊂ o-Me$_2$-1.1.1]BArF$^-$ by NMR spectroscopy and mass spectrometry (Supplementary Figs 4–13), we turned our attention towards obtaining further structural and dynamic insights on this compound. To our delight, single crystals of [Na$^+$ ⊂ o-Me$_2$-1.1.1]BArF$^-$ suitable for X-ray crystallography could be obtained by slow diffusion of cyclopentane into dilute solutions in chloroform or dichloromethane. The solid-state structure (Fig. 3a, Supplementary Fig. 14, Supplementary Tables 1 and 2, and Supplementary Data 1) shows that the sodium ion is bound to all nine surrounding oxygen atoms with a mean Na–O bond length of 2.56 Å, which is very close to the mean Na–O distance of 2.57 Å found in the solid-state structure of Lehn's classic cryptate [Na$^+$ ⊂ 2.2.2]I$^-$ (ref. 45). An interesting question arises from the relatively small distances between the three diethylene glycol chains (O–O distance between two chains: 4.5 Å; see space-filling model in Fig. 3a) and the relatively rigid architecture of the cage (in contrast to classic cryptates, no inversion is possible at the terminus of the cage): can the metal ion exit from the o-Me$_2$-1.1.1 cage under ambient conditions? Or in other words, is [Na$^+$ ⊂ o-Me$_2$-1.1.1]BArF$^-$ in fact a carceplex, not a cryptate?

Thermodynamics and kinetics of guest exchange. To answer this question, we carried out competition experiments in which complexation agents such as Lehn's cryptate 2.2.1 (Na$^+$ binding constant $K_A = 10^{13}$ M^{-1} in D$_2$O-saturated CDCl$_3$)^{46} were titrated to a freshly deacidified solution of [Na$^+$ ⊂ o-Me$_2$-1.1.1]BArF$^-$ in chloroform. As shown in Fig. 3b (top), addition of classic cryptand 2.2.1 to our orthoester cryptate led to quantitative formation of cryptate [Na$^+$ ⊂ 2.2.1]BArF$^-$ and orthoester

cryptand o-Me$_2$-1.1.1, indicating that 2.2.1 has a significantly higher binding constant under these conditions. The reaction outcome also confirms that sodium can exit from the orthoester cage and the observed ^{1}H NMR spectra demonstrate that in this experiment sodium ion exchange is slow between the two competing cryptands. Following such a titration, we treated a 1:1 mixture of [Na$^+$ ⊂ 2.2.1]BArF$^-$ and o-Me$_2$-1.1.1 with catalytic TFA, resulting in the complete conversion of o-Me$_2$-1.1.1 into orthoester products featuring eight-membered rings (Supplementary Fig. 19). This experiment demonstrates that o-Me$_2$-1.1.1, unlike [Na$^+$ ⊂ o-Me$_2$-1.1.1]BArF$^-$, does not represent a thermodynamic minimum and thus cannot be prepared without template from compounds **1** and **2** via reversible orthoester exchange.

In a second competition experiment, we titrated weaker complexation agent 15-crown-5 (Na$^+$ binding constant $K_A = 10^5$ M^{-1} in acetonitrile)^{46} to cryptate [Na$^+$ ⊂ o-Me$_2$-1.1.1]BArF$^-$. As evident from the series of ^{1}H NMR spectra (Fig. 3b, bottom), at equimolar addition of 15-crown-5 the equilibrium lies on the side of the orthoester cryptate, although broadening of the peaks indicates that exchange of sodium is fast in this case. Titration with up to 20 equivalents of 15-crown-5 gave rise to binding isotherms, from which we could deduce that the binding constant of o-Me$_2$-1.1.1 is about one order of magnitude higher than that of 15-crown-5 (see Supplementary Figs 20–22 for further thermodynamic data). In a pristine mixture of cryptate [Na$^+$ ⊂ o-Me$_2$-1.1.1]BArF$^-$ and cryptand o-Me$_2$-1.1.1 (*vide infra* for preparation method), we were able to determine an exchange rate of 0.6 s^{-1} for sodium exchange between degenerate orthoester cryptands (NMR exchange spectroscopy (EXSY); Supplementary Fig. 23 and Supplementary Note 1).

a

X-ray structure:

Front view

Side view

Space filling

b

Competition for binding of Na⁺:

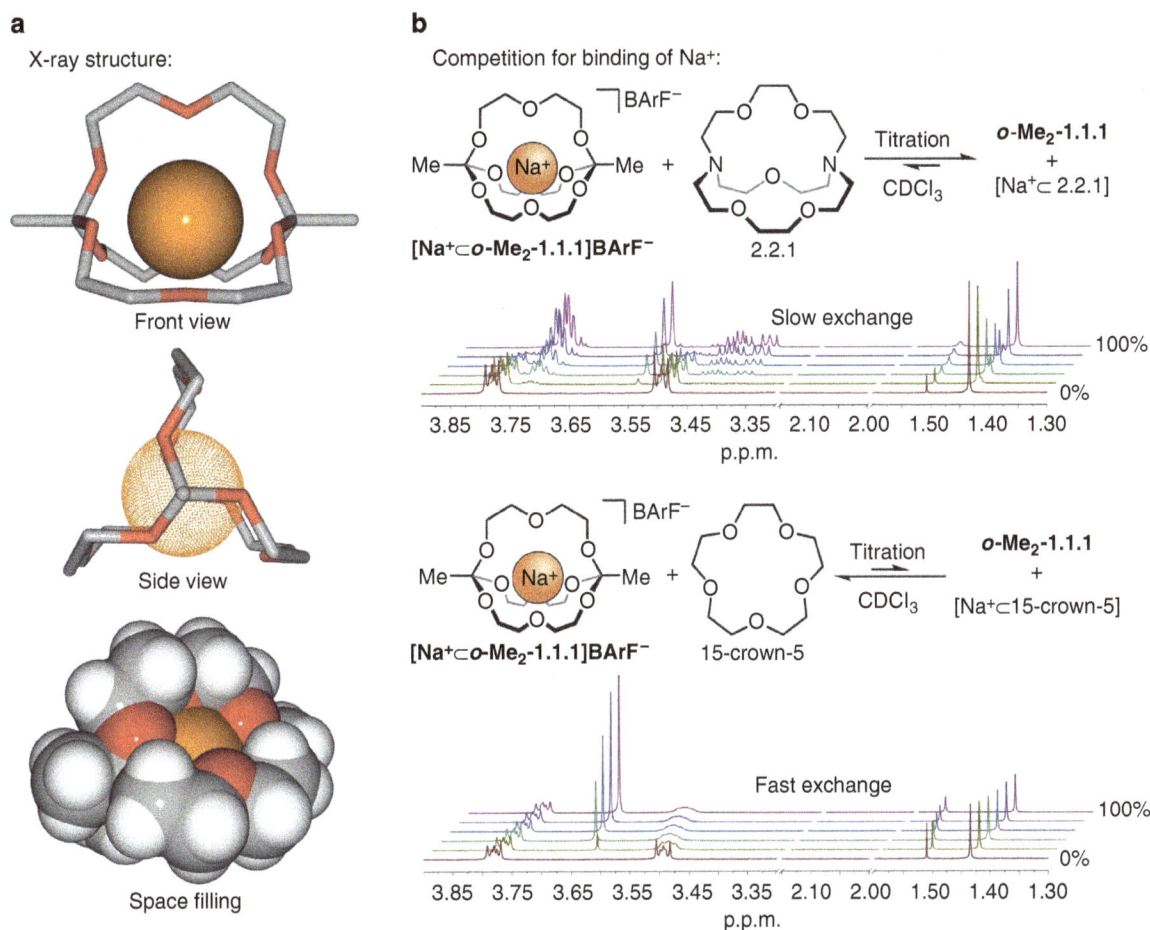

Figure 3 | Solid-state structure of [Na⁺ ⊂ o-Me₂-1.1.1]BArF⁻ and competition experiments with classic hosts for Na⁺. (a) Solid-state structure determined by single-crystal X-ray diffraction. Crystals were obtained by slow diffusion of cyclopentane into dichloromethane. Counteranion, hydrogen atoms and disorder (along the CH₂-CH₂-O-CH₂-CH₂ chains, disorder with a population of 1:1 was observed; see Supplementary Fig. 14) are omitted for clarity. Oxygen atoms are shown in red, carbon atoms in grey. Sodium (orange) is shown at 65% of the van der Waals radius in stick model representations. Na–O bond lengths: 2.51–2.58 Å (six orthoester oxygens), 2.55–2.62 Å (three chain oxygens). (b) ¹H NMR titration experiments (400 MHz, CDCl₃, 298 K) using classic complexation agents 2.2.1 (top) and 15-crown-5 (bottom). Pristine orthoester cryptate is shown at the front, 100% addition of competing host is shown at the back. The region around 2.1 p.p.m. is included to demonstrate that orthoester hydrolysis is not occurring during these experiments (despite the presence of water; singlet at 1.6 p.p.m.). For details, see Supplementary Figs 15–18.

Other metal templates and orthoester crown ethers. To confirm that a metal template effect is responsible for the remarkably clean formation of [Na⁺ ⊂ o-Me₂-1.1.1]BArF⁻ (Fig. 1), we studied the exchange reaction between 1 and 2 in the absence of template and in the presence of different metal templates under otherwise identical reaction conditions. As shown in Fig. 4 (top), the exchange reaction without template mainly gave rise to exchange product 3, a monomeric orthoester featuring an eight-membered ring that results from one molecule of diethyleneglycol (2) having displaced two molecules of methanol (Supplementary Fig. 24). Theoretically, it should be possible to remove the last equivalent of methanol by increasing the time during which the mixture is exposed to MS, but we found that the system has a strong tendency to remain at this particular state (that is, product 3). However, when such a dynamic mixture was treated with one equivalent of Sodium tetrakis[3,5-bis(trifluoromethyl)phenyl]borate (NaBArF), the dynamic system responded by forming cryptate [Na⁺ ⊂ o-Me₂-1.1.1]BArF⁻ quantitatively and in a relatively short time (Fig. 4, right-hand side, and Supplementary Fig. 25).

When we used metal salts LiTPFPB, NaBArF or KBArF for the self-assembly reaction, we observed that after 2–3 days reaction time two distinct reaction products had formed in a 1:1 ratio.

Careful analysis of one- and two-dimensional NMR spectra, as well as high-resolution mass spectrometry, indicated that unprecedented orthoester crown ethers[1] o-Me₂-(OMe)₂-16-crown-6 had formed as a mixture of *syn* and *anti* diastereomers (Fig. 4, centre). These crown ethers are the main products at a stage of the reaction where two equivalents of methanol have been removed through the effect of MS (2–3 days reaction time). The dynamic chemical system is thus not only responsive to the described metal template effect, but also to the precise quantity of available methanol.

Molecular sieves (4 Å) as an unexpected sodium source. To our initial surprise, the crown ethers originating from the lithium and potassium salts eventually transformed into the corresponding sodium cryptates [Na⁺ ⊂ o-Me₂-1.1.1]X⁻ (Fig. 4, X⁻ = tetraarylborate anion). Using mass spectrometry, ²³Na and ⁷Li NMR spectroscopy and, most notably, atom absorption and emission spectroscopy, we were able to confirm that these cage compounds were indeed the sodium cryptates [Na⁺ ⊂ o-Me₂-1.1.1]X⁻ (Supplementary Figs 26 and 27), and that the sodium source is type A zeolite (4 Å MS)[47], a porous framework material that contains accessible sodium ions. A search

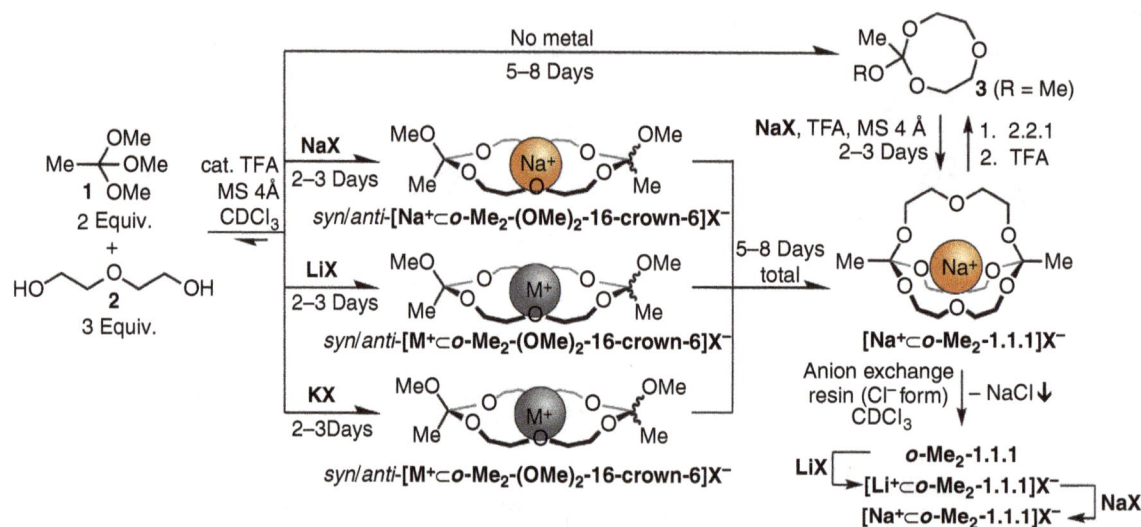

Figure 4 | Complex behaviour of a dynamic orthoester system. Structures of predominant products as a function of time, template and other chemical stimuli. Reaction conditions: trimethyl orthoacetate (120 μmol), diethylene glycol (180 μmol), metal salt (60 μmol), TFA (0.6 μmol added per day), MS (4 Å, 1g), CDCl$_3$ (6.0 ml), room temperature. Eight-membered ring products (**3** for R = Me) could be quantitatively converted into cryptate [**Na$^+$ ⊂ o-Me$_2$-1.1.1**]X$^-$ and vice versa. Pristine cryptand **o-Me$_2$-1.1.1** could be prepared using anion exchange resin Lewatit MP64 (chloride form). Cryptand **o-Me$_2$-1.1.1** could be transformed into lithium cryptate [**Li$^+$ ⊂ o-Me$_2$-1.1.1**]X$^-$ and from there back into [**Na$^+$ ⊂ o-Me$_2$-1.1.1**]X$^-$. NaX: NaBArF; LiX: LiTPFPB (lithium tetrakis(pentafluorophenyl)borate); KX: KBArF; M$^+$: mixture of Na$^+$ with residual Li$^+$ or K$^+$ due to cation exchange with MS.

of the literature revealed several reports on this type of ion exchange in aqueous[47,48] and one report in organic medium[49]. When we used metal salt KBArF in conjunction with MS 3 Å, which contain potassium instead of sodium ions, we did not observe self-assembly of a potassium cryptate from starting materials **1** and **2**. Collectively, our experiments with different metal ions point towards a pronounced preference for sodium cryptate [**Na$^+$ ⊂ o-Me$_2$-1.1.1**] over the potential lithium or potassium cryptates, which could be explained by the differences in effective ionic radii between lithium (0.9 Å), sodium (1.2 Å) and potassium (1.6 Å)[50]. Only sodium appears to have the right size for forming nine efficient metal–oxygen bonds, while not inducing energetically costly conformations within the organic host.

Preliminary experiments on scope and limitations. We conducted preliminary studies on the scope of the described orthoester cryptates and coronates. A cryptate derived from orthoester trimethyl orthopropanoate, [**Na$^+$ ⊂ o-Et$_2$-1.1.1**]BArF$^-$ (terminal substituent: Et), could be prepared without difficulty, following a similar procedure to that used for cryptate [**Na$^+$ ⊂ o-Me$_2$-1.1.1**]BArF$^-$. Using trimethyl orthoformate as the starting material (terminal substituent: H) led to the formation of remarkably stable and pure crown ether complexes [**Na$^+$ ⊂ o-H$_2$-(OMe)$_2$-16-crown-6**], which did not further react to the corresponding cryptates under our standard conditions. A notable limitation of the self-assembly reaction concerns the counteranion of the sodium template. Thus far, we were able to prepare cryptate [**Na$^+$ ⊂ o-Me$_2$-1.1.1**]X$^-$ with three different tetraarylborate anions (X$^-$ = BArF$^-$, TPFPB$^-$ and tetrakis(4-chlorophenyl)borate), while simpler anions such as PF$_6^-$ or BF$_4^-$ have failed, suggesting that a truly 'non-coordinating'[37] anion needs to be present during self-assembly.

In an attempt to exchange the counteranion in pristine [**Na$^+$ ⊂ o-Me$_2$-1.1.1**]BArF$^-$ to chloride, we discovered a convenient method for preparing cryptand **o-Me$_2$-1.1.1** (Fig. 4, bottom right). When a solution of [**Na$^+$ ⊂ o-Me$_2$-1.1.1**]BArF$^-$ in CDCl$_3$ was treated with anion exchange resin Lewatit MP-64 (Cl form), we observed the clean formation of **o-Me$_2$-1.1.1**, for

which the precipitation of NaCl is presumably the driving force. With pristine **o-Me$_2$-1.1.1** in our hands, we were able to prepare cryptate [**Li$^+$ ⊂ o-Me$_2$-1.1.1**]TPFPB$^-$ by exposing the cryptand to a solution of the lithium salt (Fig. 4, bottom right; structure confirmed by ^{1}H/^{7}Li hetero nuclear overhauser effect NMR spectroscopy; Supplementary Fig. 28). The lithium cryptate could be transformed back into [**Na$^+$ ⊂ o-Me$_2$-1.1.1**]BArF$^-$ by addition of one equivalent of NaBArF, confirming the preference of this cryptand for Na$^+$. Based on preliminary NMR data, K$^+$ (salt: KBArF) appears to not enter the cage, but presumably 'nests' on the crown-ether-type faces of the cryptand. Further experiments to such ends are ongoing in our laboratory.

Discussion

We have shown that a dynamic system based on two strikingly simple organic starting materials converges to three distinct types of exchange products under the influence of dry MS: (i) in the absence of a template, a simple exchange product featuring an eight-membered ring is formed; (ii) in the presence of a sodium template, an unprecedented dynamic orthoester cryptate is formed, in which nine oxygen donors are bound to the metal ion; (iii) *en route* to the sodium cryptates, novel orthoester coronates can be observed and, in some cases, isolated as a mixture of *syn* and *anti* isomers. Sodium cryptate [**Na$^+$ ⊂ o-Me$_2$-1.1.1**]BArF$^-$ was found to be surprisingly stable against water in neutral solution, but susceptible to hydrolysis in the presence of water and acid. We believe that this property will make orthoester cages useful for the traceless delivery of cations into biological systems[44]. Competition experiments in solution suggest that the encapsulated metal ion is in slow exchange with the bulk ($k_{obs} = 0.6$ s^{-1}) and the binding constant for Na$^+$ lies between the classic complexation agents 15-crown-5 and 2.2.1. Besides their interesting structural and dynamic properties, orthoester cryptates offer preparative advantages over their classic analogues: their synthesis relies on a one-pot, dynamic covalent ring-closing reaction, and substituted cages **o-R$_2$-1.1.1** are accessible simply by using different orthoesters as starting materials. We are currently working towards further diversifying the target structures, as well as increasing the dynamic system's complexity (for example, self-sorting and response to external stimuli).

Methods

Preparation of stock solutions. NaBArF (0.14 mmol, 127.8 mg) and diethylene glycol (0.42 mmol, 39.9 μl) were dissolved in 14 ml CDCl$_3$ and the mixture was dried over MS (4 Å, 1 g) for 3 days. TFA (0.24 mmol, 18.4 μl) was dissolved in CDCl$_3$ (total volume 2.00 ml).

Self-assembly of cryptate [Na$^+$ ⊂ o-Me$_2$-1.1.1]BArF$^-$. To 6 ml of stock solution were added fresh MS (4 Å, 1 g) and the reaction mixture was left to stand at room temperature. After 16 h, 1 mol% TFA (10 μl) was added from stock solution, the mixture was shaken and trimethyl orthoacetate (0.12 mmol, 15.4 μl) was added. Every 24 h, 1 mol% TFA was added to keep the exchange reaction active (MS slowly transform the acid catalyst into inactive anhydride and/or esters). The reaction progress was monitored regularly by ^{1}H NMR spectroscopy. After 5 days, the solvent was removed under reduced pressure and [Na$^+$ ⊂ o-Me$_2$-1.1.1]BArF$^-$ was obtained as a colourless solid (67% yield). Characterization data: M.p. 124 °C – 128 °C. ^{1}H NMR (400 MHz, CDCl$_3$, 298 K): δ = 7.68 (t, J = 2.8 Hz, 8H), 7.51 (s, 4H), 3.79–3.77 (m, 12H), 3.50–3.48 (m, 12H), 1.43 p.p.m. (s, 6H). ^{13}C NMR (100 MHz, CDCl$_3$, 298 K): δ = 162.8, 162.3, 161.8, 161.3, 135.1, 129.7, 129.4, 129.1, 128.9, 128.8, 126.2, 123.5, 120.8, 117.7, 113.0, 69.2, 62.0, 17.7 p.p.m. ^{11}B NMR (128 MHz, CDCl$_3$, 298 K): δ = − 6.7 p.p.m. ^{19}F NMR (282 MHz, CDCl$_3$, 298 K): δ = − 62.1 p.p.m. ^{23}Na NMR (132 MHz, CDCl$_3$, 298 K): δ = − 5.9 p.p.m. HRMS (ESI$^+$): m/z = 389.1794 [M + Na]$^+$ (calcd. 389.1782 for C$_{16}$H$_{30}$O$_9$Na). For further experimental details and characterization data, see Supplementary Methods and Supplementary Figs 29–47.

Exclusion of moisture. Molecular sieves were dried by heating for 3 days at 150 °C under reduced pressure (10^{-2} mbar). All solvents were dried over MS for at least 24 h. All orthoester exchange reactions (catalysed by TFA) were carried out under nitrogen. After the acid was quenched (for example, by addition of triethylamine or basic aluminum oxide), most orthoester complexes described herein were found to be unusually stable against water and could be handled on the benchtop without further precautions (Supplementary Fig. 2).

References

1. Steed, J. W. & Gale, P. A. (eds) *Supramolecular Chemistry* (Wiley-VCH, 2012).
2. Pedersen, C. J. Cyclic polyethers and their complexes with metal salts. *J. Am. Chem. Soc.* **89**, 7017–7036 (1967).
3. Dietrich, B., Lehn, J. M. & Sauvage, J. P. Diaza-polyoxa-macrocycles et macrobicycles. *Tetrahedron Lett.* **10**, 2885–2888 (1969).
4. Lehn, J. M. Cryptates: the chemistry of macropolycyclic inclusion complexes. *Acc. Chem. Res.* **11**, 49–57 (1978).
5. Schneider, H.-J. (eds) *Applications of Supramolecular Chemistry* (CRC Press, 2012).
6. Harris, K., Fujita, D. & Fujita, M. Giant hollow M$_n$L$_{2n}$ spherical complexes: structure, functionalisation and applications. *Chem. Commun.* **49**, 6703–6712 (2013).
7. Han, M., Engelhard, D. M. & Clever, G. H. Self-assembled coordination cages based on banana-shaped ligands. *Chem. Soc. Rev.* **43**, 1848–1860 (2014).
8. Zhang, G. & Mastalerz, M. Organic cage compounds - from shape-persistency to function. *Chem. Soc. Rev.* **43**, 1934–1947 (2014).
9. Tozawa, T. *et al.* Porous organic cages. *Nat. Mater.* **8**, 973–978 (2009).
10. Sun, Q. -F. *et al.* Self-assembled M$_{24}$L$_{48}$ polyhedra and their sharp structural switch upon subtle ligand variation. *Science* **328**, 1144–1147 (2010).
11. Granzhan, A., Schouwey, C., Riis-Johannessen, T., Scopelliti, R. & Severin, K. Connection of metallamacrocycles via dynamic covalent chemistry: a versatile method for the synthesis of molecular cages. *J. Am. Chem. Soc.* **133**, 7106–7115 (2011).
12. Zhang, G., Presly, O., White, F., Oppel, I. M. & Mastalerz, M. A permanent mesoporous organic cage with an exceptionally high surface area. *Angew. Chem. Int. Ed.* **53**, 1516–1520 (2014).
13. Kim, J. *et al.* Reversible morphological transformation between polymer nanocapsules and thin films through dynamic covalent self-assembly. *Angew. Chem. Int. Ed.* **54**, 2693–2697 (2015).
14. Fujita, D. *et al.* Protein encapsulation within synthetic molecular hosts. *Nat. Commun.* **3**, 1093 (2012).
15. Corbett, P. T. *et al.* Dynamic combinatorial chemistry. *Chem. Rev.* **106**, 3652–3711 (2006).
16. Jin, Y., Yu, C., Denman, R. J. & Zhang, W. Recent advances in dynamic covalent chemistry. *Chem. Soc. Rev.* **42**, 6634–6654 (2013).
17. Herrmann, A. Dynamic combinatorial/covalent chemistry: a tool to read, generate and modulate the bioactivity of compounds and compound mixtures. *Chem. Soc. Rev.* **43**, 1899–1933 (2014).
18. Ludlow, R. F. & Otto, S. Systems chemistry. *Chem. Soc. Rev.* **37**, 101–108 (2008).
19. Li, J., Nowak, P. & Otto, S. Dynamic combinatorial libraries: from exploring molecular recognition to systems chemistry. *J. Am. Chem. Soc.* **135**, 9222–9239 (2013).
20. Hiraoka, S., Harano, K., Shiro, M. & Shionoya, M. Quantitative dynamic interconversion between AgI-mediated capsule and cage complexes accompanying guest encapsulation/release. *Angew. Chem. Int. Ed.* **44**, 2727–2731 (2005).
21. Carnall, J. M. A. *et al.* Mechanosensitive self-replication driven by self-organization. *Science* **327**, 1502–1506 (2010).
22. Riddell, I. A. *et al.* Anion-induced reconstitution of a self-assembling system to express a chloride-binding Co$_{10}$L$_{15}$ pentagonal prism. *Nat. Chem.* **4**, 751–756 (2012).
23. Zarra, S., Wood, D. M., Roberts, D. A. & Nitschke, J. R. Molecular containers in complex chemical systems. *Chem. Soc. Rev.* **44**, 419–432 (2015).
24. Stefankiewicz, A. R., Sambrook, M. R. & Sanders, J. K. M. Template-directed synthesis of multi-component organic cages in water. *Chem. Sci.* **3**, 2326–2329 (2012).
25. Ayme, J. -F., Beves, J. E., Campbell, C. J. & Leigh, D. A. The self-sorting behavior of circular helicates and molecular knots and links. *Angew. Chem. Int. Ed.* **53**, 7823–7827 (2014).
26. Saalfrank, R. W. *et al.* Topologic equivalents of coronands, cryptands and their inclusion complexes: synthesis, structure and properties of {2}-metallacryptands and {2}-metallacryptates. *Chem. Eur. J.* **3**, 2058–2062 (1997).
27. Saalfrank, R. W., Maid, H. & Scheurer, A. Supramolecular coordination chemistry: the synergistic effect of serendipity and rational design. *Angew. Chem. Int. Ed.* **47**, 8794–8824 (2008).
28. Jazwinski, J. *et al.* Polyaza macrobicyclic cryptands: synthesis, crystal structures of a cyclophane type macrobicyclic cryptand and of its dinuclear copper(I) cryptate, and anion binding features. *J. Chem. Soc., Chem. Commun.* 1691–1694 (1987).
29. MacDowell, D. & Nelson, J. Facile synthesis of a new family of cage molecules. *Tetrahedron Lett.* **29**, 385–386 (1988).
30. Voloshin, Y. Z. *et al.* Template synthesis, structure and unusual series of phase transitions in clathrochelate iron(II) α-dioximates and oximehydrazonates formed by capping with functionalized boron-containing agents. *Polyhedron* **20**, 2721–2733 (2001).
31. Wise, M. D. *et al.* Large, heterometallic coordination cages based on ditopic metallo-ligands with 3-pyridyl donor groups. *Chem. Sci.* **6**, 1004–1010 (2015).
32. Brachvogel, R. -C. & von Delius, M. Orthoester exchange: a tripodal tool for dynamic covalent and systems chemistry. *Chem. Sci.* **6**, 1399–1403 (2015).
33. Lightburn, T. E., Dombrowski, M. T. & Tan, K. L. Catalytic scaffolding ligands: an efficient strategy for directing reactions. *J. Am. Chem. Soc.* **130**, 9210–9211 (2008).
34. Tan, K. L. Induced intramolecularity: an effective strategy in catalysis. *ACS Catal.* **1**, 877–886 (2011).
35. DeWolfe, R. H. Synthesis of carboxylic and carbonic ortho esters. *Synthesis* 153–172 (1974).
36. Rosenthal, M. R. The myth of the non-coordinating anion. *J. Chem. Educ.* **50**, 331 (1973).
37. Krossing, I. & Raabe, I. Noncoordinating anions—fact or fiction? a survey of likely candidates. *Angew. Chem. Int. Ed.* **43**, 2066–2090 (2004).
38. Lee, S., Chen, C. -H. & Flood, A. H. A pentagonal cyanostar macrocycle with cyanostilbene CH donors binds anions and forms dialkylphosphate [3]rotaxanes. *Nat. Chem.* **5**, 704–710 (2013).
39. Lin, Y. -H., Lai, C. -C., Liu, Y. -H., Peng, S. -M. & Chiu, S. -H. Sodium Ions template the formation of rotaxanes from BPX26C6 and nonconjugated amide and urea functionalities. *Angew. Chem. Int. Ed.* **52**, 10231–10236 (2013).
40. Wu, K. -D., Lin, Y. -H., Lai, C. -C. & Chiu, S. -H. Na$^+$ ion templated threading of oligo(ethylene glycol) chains through BPX26C6 allows synthesis of [2]Rotaxanes under solvent-free conditions. *Org. Lett.* **16**, 1068–1071 (2014).
41. Gold, V. & Sghibartz, C. M. Crown ether acetals: detection of cation binding by kinetic measurements. *J. Chem. Soc., Chem. Commun.* 507–508 (1978).
42. Corey, E. J. & Raju, N. A new general synthetic route to bridged carboxylic ortho esters. *Tetrahedron Lett.* **24**, 5571–5574 (1983).
43. Kirch, M. & Lehn, J.-M. Selective transport of alkali metal cations through a liquid membrane by macrobicyclic carriers. *Angew. Chem. Int. Ed.* **14**, 555–556 (1975).
44. Botti, P., Tchertchian, S. & Theurillat, D. Orthoester derivatives of crown ethers as carriers for pharmaceutical and diagnostic compositions. *Eur. Pat. Appl.* EP 2 332 929 A1 (2009).
45. Moras, P. D. & Weiss, R. Etude structurale des cryptates. III. Structure cristalline et moléculaire du cryptate de sodium C$_{18}$H$_{36}$N$_2$O$_6$.NaI. *Acta Cryst.* **B29**, 396–399 (1973).
46. Izatt, R. M., Pawlak, K., Bradshaw, J. S. & Bruening, R. L. Thermodynamic and kinetic data for macrocycle interactions with cations and anions. *Chem. Rev.* **91**, 1721–2085 (1991).
47. Breck, D. W., Eversole, W. G., Milton, R. M., Reed, T. B. & Thomas, T. L. Crystalline zeolites. I. The properties of a new synthetic zeolite, type A. *J. Am. Chem. Soc.* **78**, 5963–5972 (1956).

48. Breck, D. W. Crystalline molecular sieves. *J. Chem. Educ.* **41,** 678 (1964).
49. Golden, J. H., Mutolo, P. F., Lobkovsky, E. B. & DiSalvo, F. J. Lithium-mediated organofluorine hydrogen bonding: structure of lithium tetrakis(3,5-bis(trifluoromethyl)phenyl)borate tetrahydrate. *Inorg. Chem.* **33,** 5374–5375 (1994).
50. Shannon, R. D. Revised effective ionic radii and systematic studies of interatomic distances in halides and chalcogenides. *Acta Cryst.* **A32,** 751–767 (1976).

Acknowledgements

This paper is dedicated to Professor Jean-Marie Lehn on the occasion of his 75th birthday. We gratefully acknowledged funding from the Fonds der Chemischen Industrie (FCI, Liebig Fellowship Li 191/01 for M.v.D., doctoral fellowship for R.-C.B.) and the Deutsche Forschungsgemeinschaft (DFG, Emmy-Noether Programme DE 1830/2-1). We thank Dr Harald Maid for help with visualizing the solid-state structures, Professor Walter Bauer for measuring ^{23}Na and ^{7}Li NMR spectra, and Dr Ralph Puchta for preliminary DFT calculations. Dr Jörg Sutter, Stefanie Zürl and Franziska Popp are acknowledged for recording atom absorption and atom emission spectra.

Author contributions

R.-C.B. and M.v.D. performed the experiments, contributed to the design of the experiments and the analysis of the data. F.H. solved the crystal structure. M.v.D. conceived the project and wrote the manuscript.

Additional information

Accession codes: The X-ray crystallographic coordinates for structures reported in this study have been deposited at the Cambridge Crystallographic Data Centre (CCDC), under CCDC number 1038394. These data can be obtained free of charge from The Cambridge Crystallographic Data Centre via www.ccdc.cam.ac.uk/data_request/cif.

Competing financial interests: M.v.D. and R.-C.B. are co-inventors of a European patent application ('Novel crown ether complexes and methods for producing the same', EP14186189). F.H. declares no competing financial interests.

8

Quantification of thickness and wrinkling of exfoliated two-dimensional zeolite nanosheets

Prashant Kumar[1], Kumar Varoon Agrawal[1], Michael Tsapatsis[1] & K Andre Mkhoyan[1]

Some two-dimensional (2D) exfoliated zeolites are single- or near single-unit cell thick silicates that can function as molecular sieves. Although they have already found uses as catalysts, adsorbents and membranes precise determination of their thickness and wrinkling is critical as these properties influence their functionality. Here we demonstrate a method to accurately determine the thickness and wrinkles of a 2D zeolite nanosheet by comprehensive 3D mapping of its reciprocal lattice. Since the intensity modulation of a diffraction spot on tilting is a fingerprint of the thickness, and changes in the spot shape are a measure of wrinkling, this mapping is achieved using a large-angle tilt-series of electron diffraction patterns. Application of the method to a 2D zeolite with MFI structure reveals that the exfoliated MFI nanosheet is 1.5 unit cells (3.0 nm) thick and wrinkled anisotropically with up to 0.8 nm average surface roughness.

[1]Department of Chemical Engineering and Materials Science, University of Minnesota, Minneapolis, Minnesota 55455, USA. Correspondence and requests for materials should be addressed to M.T. (email: tsapatsis@umn.edu) or to K.A.M. (email: mkhoyan@umn.edu).

Zeolites are three-dimensional (3D) framework silicates with precisely sized pores of molecular dimensions[1]. Two-dimensional (2D) zeolites and zeolite nanosheets[2–4] with single, double or fractional unit cell (UC) thickness are particularly desirable for separation[5,6] and catalysis of bulky molecules[7–10]. They may also emerge as candidates for device fabrication requiring low-k dielectric materials[11].

Since their structure is crucial in order to predict or interpret their adsorption, transport and catalytic properties[10,12,13], and can vary depending on the synthesis procedure[14,15], its quantification is highly desirable. Unlike many 2D materials, such as graphene, BN, phosphorene, MoS_2 and other transition metal dichalcogenides, where the UC is 1-, 2- or 3-atom-layers thick, the UC of zeolites can be >10-atom-layers thick. Moreover, zeolite nanosheets can be synthesized with thicknesses that are non-integer multiples of the UC thickness[5]. Therefore, methods developed to characterize other 2D materials are not necessarily applicable to 2D zeolites.

Thickness measurements of 2D materials are often performed by atomic force microscopy (AFM)[5,16,17], X-ray reflectivity experiments[18–20] or imaging cross-sectional samples with conventional transmission electron microscopy (TEM)[8]. However, AFM data cannot provide crystallographic information over the entire nanosheet thickness, and X-ray reflectivity measurements often require fabrication of a periodic multilayer, which is not feasible for nanosheets with sub-micron lateral dimensions such as 2D zeolites. Preparation of cross-sections for TEM imaging can be challenging, and conventional TEM images require in-depth analysis as contrast is strongly sensitive to imaging conditions and specimen thickness[21]. Moreover, zeolites are electron beam-sensitive materials that suffer from knock-on damage and radiolysis at high and low accelerating voltages[22]. Therefore, a method for unambiguously determining the thickness and structure of 2D zeolites remains elusive.

In addition to thickness, it is also important to determine deviations from the nominal (ideal, wrinkle-free) structure of 2D zeolites. These deviations can affect their internal and external surface structure and pore openings, which, in turn, can also affect their adsorption, transport and catalytic properties. Atomic-scale wrinkles have been observed in 2D materials such as graphene[23] through lattice imaging using scanning TEM (STEM)[24–27]. However, using this approach for zeolites is challenging due to their rapid amorphization under the required high-dose electron irradiation.

Here we demonstrate a method that, based on a set of TEM experiments, provides complete quantitative characterization of the atomic structure of zeolite nanosheets, that is, crystal structure and uniformity through high-resolution TEM (HR-TEM) and annular dark-field (ADF)-STEM imaging, and thickness and wrinkling through electron diffraction. Although applicable for all 2D materials, it is particularly well suited for 2D zeolites. The all-silica 2D zeolite with the MFI structure type (for a list of structure types see http://www.iza-online.org) is used as the prototype material[5,8] and its anisotropic wrinkling is quantified.

Results

In-plane structure determination. MFI belongs to the pentasil family of zeolites, where the periodic building unit is composed of 12 interconnected SiO_4 tetrahedral (T12) units[28] (Fig. 1a). Rotation of T12 units about the c axis (right- or left rotation), along with translation by half of a UC in the c-direction, forms right-handed or left-handed pentasil chains consisting of five membered rings. Alternating left- and right-handed chains, when connected along the a-direction through inversion symmetry, form the MFI zeolite[29] structure with an orthorhombic UC (Fig. 1b,c and Supplementary Fig. 1).

The crystal structure of MFI nanosheets was assessed by a combination of HR-TEM imaging, electron diffraction pattern analysis and high-angle annular dark-field STEM (HAADF-STEM) imaging. The HR-TEM image viewed along the [010] crystallographic direction (Fig. 1d) and the [010] zone axis electron diffraction pattern (Fig. 1e) are in agreement with the standard bulk MFI structure, confirming preservation of bulk crystal structure in nanosheets. Thickness-sensitive

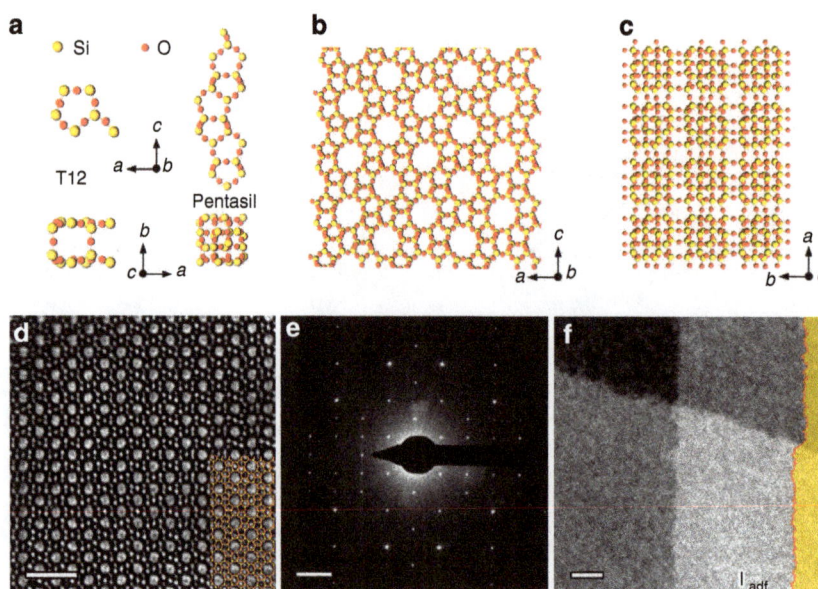

Figure 1 | MFI nanosheet crystal structure. (**a**) Construction of MFI pentasil chain from the smallest T12 building unit, which contains 12 silicon atoms. (**b**) Projection of MFI crystal structure along b-direction formed by linking pentasil chains. MFI nanosheets are typically ~100–200 nm wide along a- and c-directions. (**c**) Projection of MFI nanosheet along c-direction showing three complete pentasil chains in b-direction (1.5 UCs along b-direction). (**d**) Bragg-filtered HR-TEM image of MFI nanosheet with the overlaid crystal structure along [010] direction. (**e**) Diffraction pattern of MFI nanosheet along [010] zone axis. (**f**) HAADF-STEM image of two overlaid MFI nanosheets with intensity-scan taken from the overlaid area. It shows homogenous thickness across nanosheets and doubling of intensity (that is, doubling of thickness) in the overlaid region. Scale bars, 3 nm (**d**), 1 nm^{-1} (**e**) and 10 nm (**f**).

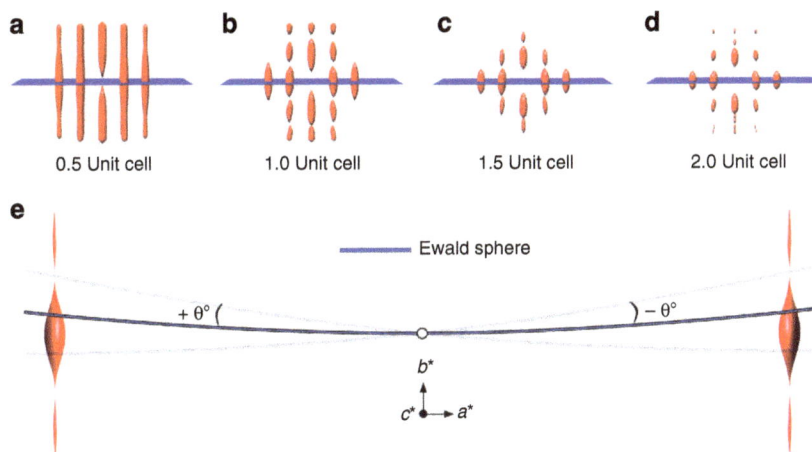

Figure 2 | Thickness dependence of rel-rods. Isosurfaces of reciprocal lattices at 5% of maximum intensity simulated for four crystal models of MFI nanosheets with thickness of (**a**) 0.5, (**b**) 1.0, (**c**) 1.5 and (**d**) 2.0 UC in [010] or *b*-direction. The blue plane represents the (010) diffraction plane. (**e**) Schematic description of the method of rel-rod mapping by tilting the Ewald sphere in positive and negative directions (+ θ and -θ).

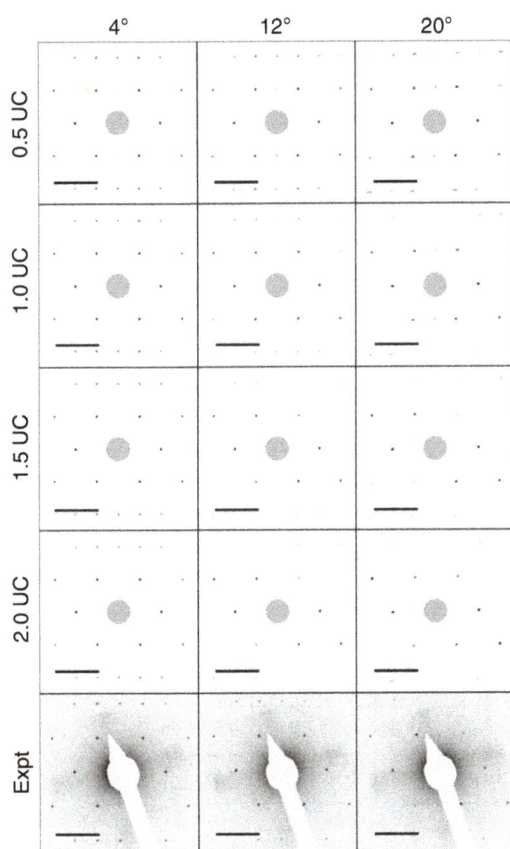

Figure 3 | Comparison of experimental and simulated diffraction pattern tilt-series. Simulated diffraction pattern tilt-series (with tilt-axis: $[\bar{1}\bar{1}05]$) for MFI zeolite nanosheets of thickness 0.5, 1.0, 1.5 and 2.0 UC. Corresponding experimental (Expt) diffraction patterns of MFI nanosheets are presented in the bottom row. All results are presented in reverse grey-scale colour map for better visibility. Scale bars, 1 nm^{-1}. All data sets show clear changes in diffraction pattern contrast as a function of sample tilt.

HAADF-STEM images (Fig. 1f) reveal that the synthesized nanosheets used in this study are uniform and have the same thickness. As can be seen from Fig. 1f, the intensity of ADF image doubles between the areas containing single nanosheets to that with two overlapped nanosheets. While these experiments

provide unambiguous determination of crystal structure, they are not useful for thickness evaluation.

Quantification of thickness. To determine the thickness of a 2D zeolite, we took advantage of the fact that the reciprocal lattice of the nanosheet, which is uniquely defined by the sample thickness, can be easily mapped by electron diffraction. As the thickness of the nanosheets (that is, the number of UCs spanning top and bottom surfaces) decreases, the reciprocal lattice points elongate in reciprocal space to form rod-like structures, which are often referred to as 'rel-rods'[30]. Figure 2a–d shows reciprocal lattices modelled for MFI nanosheets with thicknesses 0.5, 1.0, 1.5 and 2.0 UC, respectively, (the details of calculations are explained in the Methods section). Since AFM data show that these nanosheets are ~ 3 nm thick[5], we limited our analysis to a maximum thickness of 3.98 nm or 2.0 UC. As can be seen, the rel-rods are sensitive even to fractional UC increments of thickness.

Using the kinematical theory of electron diffraction[31], the patterns recorded in the TEM can be described as the intersections of the Ewald sphere with the reciprocal lattice (Fig. 2e). The rel-rods of nanosheets can then be discretely mapped by tilting the sheets and acquiring diffraction patterns at each tilt angle (θ). Here, to simplify the analysis, the initial orientation ($\theta = 0°$) of the nanosheets was selected to be along the [010] zone axis, which is normal to the nanosheet surface. A tilt-series of diffraction patterns from a MFI nanosheet was acquired from $\theta = -60°$ to $\theta = 60°$ with tilt-axis perpendicular to the [010] direction. While any tilt-axis perpendicular to [010] allows rel-rod mapping, knowledge of the axis is critical for accurate analysis. In this experiment, it was determined to be $[\bar{1}\bar{1}05]$ (see tilt-axis determination in the Methods section and Supplementary Fig. 2).

To correlate the shape of the experimentally mapped rel-rod to the thickness of the nanosheet, we simulated diffraction pattern tilt-series and mapped rel-rods of nanosheets with four different thicknesses: 0.5, 1.0, 1.5 and 2.0 UC. Simulations were performed using the multislice method[32] and the TEMSIM simulation package developed by Kirkland[21], using parameters closely matching the experimental conditions (see 'Multislice simulations' in the Methods section for details). The experimentally obtained tilt-series of diffraction patterns from a MFI nanosheet, along with a set of simulated patterns, are presented in Fig. 3 (additional data are provided in Supplementary Fig. 3).

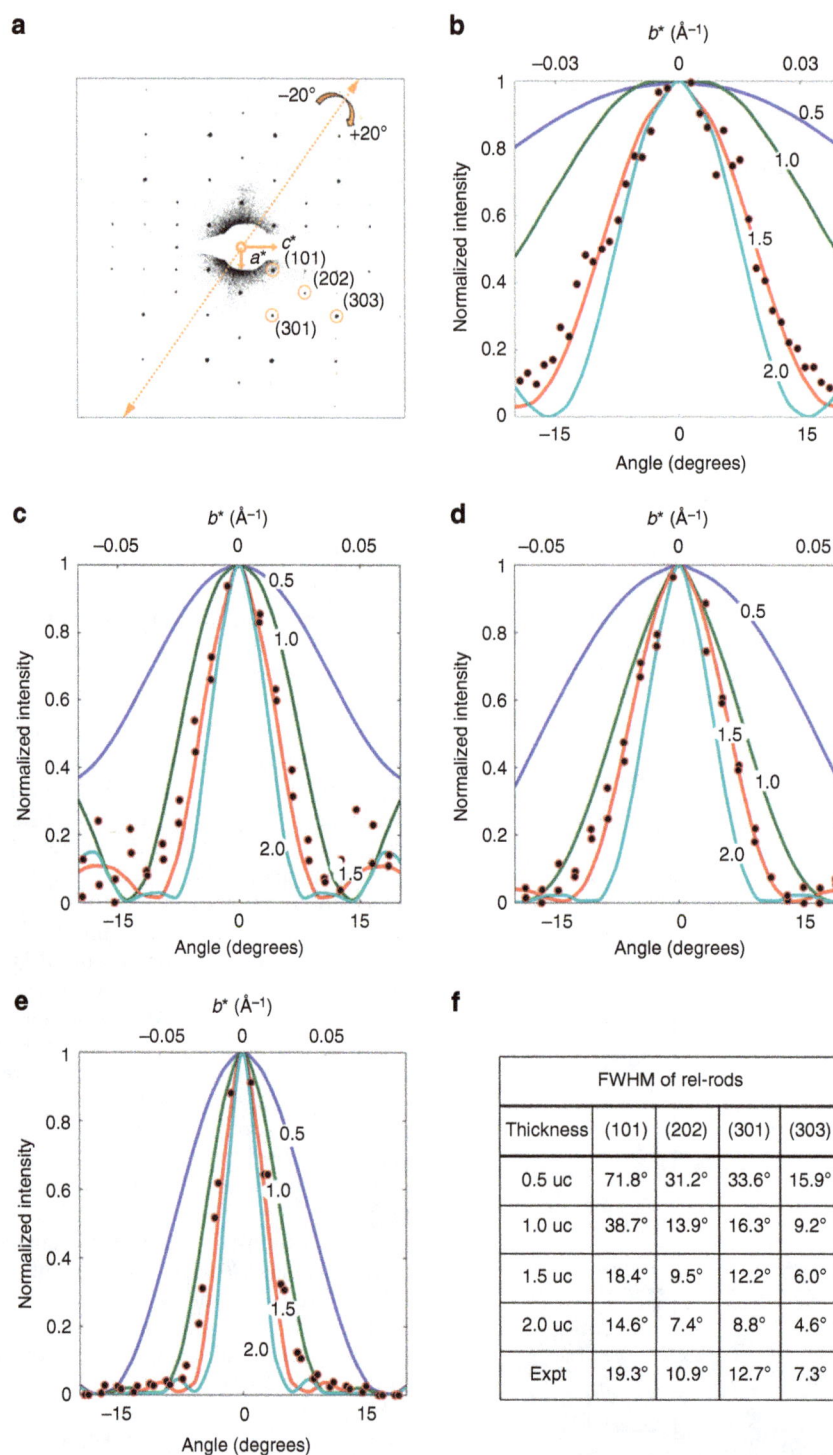

Figure 4 | Diffraction spot intensity modulation with tilt. (**a**) TEM electron diffraction pattern taken along [010] zone axis. It is presented in reverse grey-scale colour map for better visibility. The $[\bar{1}\bar{1}05]$ tilt-axis is indicated with an orange arrow. Diffraction spots used for analysis are circled. The variation in intensity with tilt angle (rel-rod map) of (**b**) (101), (**c**) (202), (**d**) (301) and (**e**) (303) spots are plotted for simulated (lines) and experimental (circles) data, respectively. (**f**) Calculated FWHMs of simulated and experimentally mapped rel-rods show best agreement for a 1.5-UC-thick nanosheet. FWHMs were calculated by fitting a 1D Gaussian function.

	FWHM of rel-rods			
Thickness	(101)	(202)	(301)	(303)
0.5 uc	71.8°	31.2°	33.6°	15.9°
1.0 uc	38.7°	13.9°	16.3°	9.2°
1.5 uc	18.4°	9.5°	12.2°	6.0°
2.0 uc	14.6°	7.4°	8.8°	4.6°
Expt	19.3°	10.9°	12.7°	7.3°

To quantify diffraction intensities, the intensity (I_θ) of any diffraction spot in the tilt-series was calculated as the volume under a 2D Gaussian function fitted to that spot (see 'Gaussian fitting' in the Methods section and Supplementary Fig. 4). The same procedure was applied to the simulated tilt-series as well. Moreover, for quantitative analysis of the experimental data, the intensities obtained were corrected for beam damage (see 'Compensation for beam damage' in the Methods section and

Supplementary Fig. 5). For direct comparison of experimentally obtained and simulated rel-rods, they are both normalized to 1 at $\theta = 0°$. Figure 4b–e shows the resulting experimental and simulated rel-rod maps corresponding to (101), (202), (301) and (303) diffraction spots for the MFI nanosheet. Comparison of full-width at half-maxima (FWHM) of experimentally mapped rel-rods (see Supplementary Fig. 6 for evaluation of FWHM of rel-rods) with the simulated ones is given in Fig. 4f. All the experimental data agree with 1.5-UC-thick nanosheet simulations, regardless of the diffraction spot chosen for analysis. This finding confirms the tentative conclusion made earlier by inspecting X-ray diffraction (XRD) data, comparing them with simulations and combining them with AFM data[5]. In future studies, it would be interesting to compare the single particle data obtained here, with quantitative analysis of bulk XRD measurements (for example, powder diffraction and/or grazing incidence on monolayers).

Quantification of wrinkling. In addition to observing intensity modulations of diffraction spots with tilt, we also tracked changes in the shapes of the spots, as they are a measure of wrinkling of the MFI nanosheet. The wrinkling of nanosheets in real space corresponds to precession of the rel-rods into 'cones' in reciprocal space; therefore, the diffraction spot shape changes are particularly pronounced at larger tilt angles (for more details, see Supplementary Fig. 7). It should be noted here that to ensure accurate analysis, it is essential to exclude the possibilities of shape changes due to uncompensated astigmatism of the lenses or slight convergence of the electron beam. Our TEM alignments confirmed that the probe had minimal astigmatism during acquisition. Moreover, we found from simulations (shown in Supplementary Fig. 8) that changes in convergence angle of the electron beam cause changes in the size of diffraction spots at all tilt angles but do not lead to detectable shape asymmetry. Therefore, the changes in diffraction spots' shape observed here can be confidently attributed to wrinkling of nanosheets.

Wrinkling was also detected by HR-TEM images of the MFI nanosheets, obtained along the [010] zone axis, showing that the nanosheet structure is not uniformly in focus. It exhibits in- and out-of-focus domains with an average characteristic dimension of $l_w \approx 20$ nm (an example of one such HR-TEM image is shown in Supplementary Fig. 9a).

In order to quantify the wrinkles in the MFI nanosheet, the FWHM of the (011) diffraction spot was measured in reciprocal a^*- (FWHM$_{a^*}$) and c^*- (FWHM$_{c^*}$) directions. This spot, which is favourably close to the tilt-axis, is quantifiable within the tilt angle range 18°–50°, even though it was not observed in the diffraction pattern recorded at [010] zone axis with no tilt ($\theta = 0°$). It should be noted here that selecting the diffraction spot in which to analyse broadening of the corresponding rel-rod is critical, as some of the spots will show splitting instead of broadening with tilt. While the method is still valid, more rigorous analysis of modified rel-rods will be required to determine the level of wrinkling from those diffracted spots.

The values of FWHM$_{a^*}$ and FWHM$_{c^*}$ were determined by fitting a 2D Gaussian function to the (011) diffraction spot in each diffraction pattern of the tilt-series as described previously. Broadened due to wrinkling, the (011) rel-rod is then reconstructed in reciprocal space by calculating the line segments in a^*- and c^*-directions corresponding to intersection of the Ewald sphere and the rel-rod (details are provided in Supplementary Fig. 9b,c). The results are presented in Fig. 5a,b. Projections of this reconstructed rel-rod show that the rel-rod is broadened more in the a^*-direction than in the c^*-direction, with maximum tilt angles of $\theta_{a^*} = 2.66°$ and $\theta_{c^*} = 1.18°$, correspondingly. This difference in broadening of rel-rod in a^*- versus c^*-directions indicates that MFI nanosheets wrinkle differently in a- and c-directions. The greater resistance of the MFI nanosheets to bending in the c-direction likely results from the stiff pentasil periodic bond chains that extend along the c-direction.

To accurately translate the measured broadening of rel-rods into a measure of MFI nanosheet wrinkling, measured FWHMs of (011) rel-rods are compared with those generated theoretically

using multislice simulations. The experimentally determined values of $\theta_{a^*} = 2.66°$, $\theta_{c^*} = 1.18°$ and domain size $l_w = 20$ nm were used to set up an initial model of a wrinkled MFI nanosheet. The structural model is created by shifting all atoms of an ideally flat nanosheet in the b-direction (y_i) using the superposition of two perpendicular sine waves with a wavelength $2l_w$. The new position of atoms in the wrinkled nanosheet can be expressed as:

$$y_i^{\text{new}} = y_i + A_a \sin(k_w x_i) + A_c \sin(k_w z_i),$$

where (x_i, y_i, z_i) are the coordinates of atom i, $A_a = (l_w/2)$ $\tan(\theta_{a^*})$ and $A_c = (l_w/2)\tan(\theta_{c^*})$ are amplitudes of sine waves in a- and c- direction, respectively, and, $k_w = \pi/l_w$ (see Supplementary Fig. 9d for details of creating wrinkled nanosheet models). The values of A_a and A_c are then determined from fitting of FWHM$_{a^*}$ and FWHM$_{c^*}$ of (011) spot in simulated diffraction pattern tilt-series to the corresponding experimental data as shown in Fig. 5c. The best fit resulted in the following values: $A_a = 5.0 \pm 0.5$ Å, $A_c = 2.0 \pm 0.5$ Å and $k_w = 0.013$ Å^{-1} (Fig. 5d). The total deviation from flatness for this wrinkled nanosheet is estimated to be in the range of -8 Å to $+8$ Å (up to 0.04 Å shift in b-direction for every 1.0 Å traversed laterally) with wrinkling along the a-direction being more pronounced (Fig. 5e). To our knowledge, this is the first time that wrinkling of highly crystalline zeolite nanosheets is quantified.

With MFI nanosheets as an example, we have shown that by using TEM and quantitatively mapping the reciprocal lattice, it is possible to fully characterize the atomic structure of a 2D zeolite, including determination of crystal structure, sheet thickness and the level of wrinkling. We showed that the method is sufficiently simple and robust to be applicable for 2D zeolites, which are fractional multiples of UCs in thickness. This analytical technique of mapping the reciprocal lattice is based on monitoring the modulations in intensity and changes in shape of diffraction spots with tilt, allowing the thickness and wrinkling of 2D zeolites to be determined from a tilt-series of diffraction patterns. This technique does not even require the use of double tilt TEM holders. The issue of beam damage that typically limits any TEM study of zeolites is accounted for and quantitatively incorporated into this method. The method should be applicable to zeolites including aluminosilicates and to other porous materials including metal–organic frameworks. Application of this method on a 2D zeolite with the MFI structure reveals that the exfoliated MFI nanosheets are 1.5 UCs or 3 nm thick, and while they have the same nominal crystal structure as bulk MFI, they are notably wrinkled. The wrinkling is non-uniform, with up to 0.8 nm deviations from flatness. It is possible that the anisotropic flexibility of 2D MFI zeolite nanosheets will have considerable effects on their application as adsorbents, membranes and catalysts, and should be taken into account in future studies.

Methods

Materials. MFI nanosheet suspensions in octanol were prepared from multi-lamellar MFI as reported by Choi et al.[8] followed by exfoliation by melt blending and purification by density gradient centrifugation as described by Varoon et al.[5,6]. TEM samples were prepared by drop-casting an octanol suspension onto standard 400-mesh TEM copper grids coated with holey carbon film supported on an ultrathin carbon layer (from Ted Pella Inc.).

Instrumentation. FEI Tecnai G2 F30 (S)TEM equipped with TWIN pole piece (Cs = 2 mm) and a Schottky field emission electron gun operating at 300 kV with extraction voltage of 4,000 V was used for conventional bright-field TEM imaging. Low-dose setup of the microscope was used to minimize beam exposure of the sample between tilts. The emission current during the experiment was 90 μA. Selected area electron diffraction patterns were acquired with an integration time of 8 s at a camera length of 3.7 m on a Gatan 4k × 4k Ultrascan CCD at a 4 × 4 binning to yield a final 1k × 1k pixel[2] image. HAADF-STEM imaging was performed on FEI Tecnai G2 F30 (S)TEM equipped with S-TWIN pole piece (Cs = 1.3 mm) and a Schottky field emission electron gun operating at 300 kV. HAADF detector collection inner and outer angles were 50 and 200 mrad, respectively.

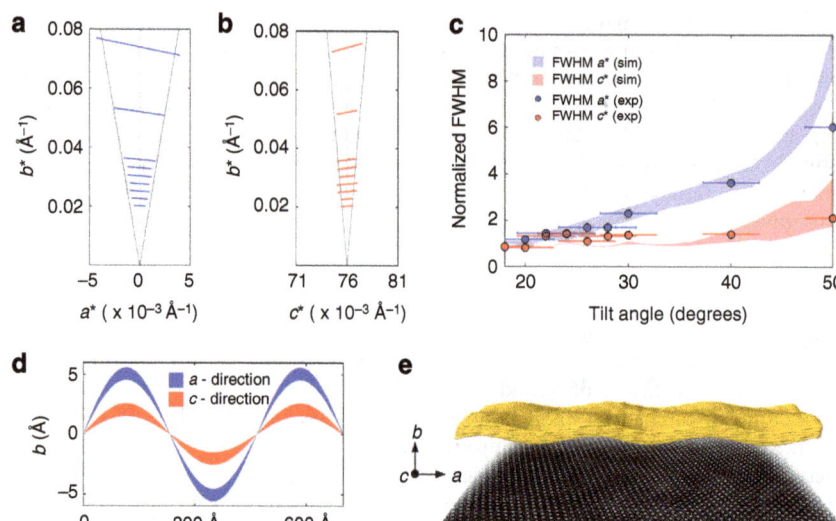

Figure 5 | Quantification of diffraction spot shape change with tilt (**a**) a^*b^* and (**b**) b^*c^* projections of (011) rel-rods from experimental diffraction patterns of MFI nanosheet. Ewald sphere intersections with rel-rods are plotted as solid blue and red lines. The cones determine the broadening of the rel-rods due to nanosheet wrinkling. (**c**) Comparison of experimental and simulated $FWHM_{a^*}$ and $FWHM_{c^*}$ of (011) diffraction spot for a 1.5-UC-thick MFI nanosheet as a function of tilt (normalization of FWHM is done with FWHM at $\theta = 18°$). The error bars represent mechanical accuracy of TEM and holder in tilt angle determination (see 'Error analysis' in Methods section). (**d**) Two sine waves in a- and c-directions, superposition of which provides the best description of the wrinkles in these MFI nanosheets. (**e**) Bragg-filtered HR-TEM image of a MFI nanosheet showing variations in intensity across the sheet overlaid with the estimated wrinkled nanosheet model.

Reciprocal lattice modelling. A 3D reciprocal lattice of MFI nanosheet was constructed by plotting the square of the structure factor $F(\mathbf{q})$, which is defined as $F(\mathbf{q}) = \sum_{k=1}^{n} f_k \cdot \exp[-2\pi i \mathbf{q} \cdot \mathbf{r_k}] \cdot \sum_{N_x} \sum_{N_y} \sum_{N_z} \exp[-2\pi i \mathbf{q} \cdot \mathbf{r_g}]$, where $\mathbf{q}$ is a lattice vector in reciprocal space, f_k is the atomic scattering factor, and $\mathbf{r_k}$ is the position vector for atom k in the UC. The lattice amplitude, which can be simplified to $\sum_{N_x} \sum_{N_y} \sum_{N_z} \exp[-2\pi i \mathbf{q} \cdot \mathbf{r_g}] = \frac{\sin(N_x \pi a_x q_x)}{\sin(\pi a_x q_x)} \frac{\sin(N_y \pi b_y q_y)}{\sin(\pi b_y q_y)} \frac{\sin(N_z \pi c_z q_z)}{\sin(\pi c_z q_z)}$, depends on the number of UCs (N_x, N_y and N_z) in the crystal in the x-, y- and z-direction, respectively, and on real space position vector $\mathbf{r_g}$ of each UC. The orthorhombic UC of MFI ($a = 20.09$ Å, $b = 19.74$ Å, $c = 13.14$ Å, $\alpha = \beta = \gamma = 90°$) was used to create models of MFI nanosheets ($N_x = 10$ UC; $N_y = 0.5$, 1.0, 1.5 and 2.0 UC and $N_z = 15$ UC) and the respective reciprocal lattices.

Tilt-axis determination. Diffraction spots move upon tilting. We track the motion paths of higher-order diffraction spots moving parallel to each other throughout the tilt (Supplementary Fig. 2a,b). Crystallographically equivalent diffraction spots on either side of the tilt-axis move in opposite directions. The common direction in the diffraction pattern that satisfies symmetric movements of equivalent diffraction spots is therefore the tilt-axis. The estimated tilt-axis for the experimental data analysed here is the $[\bar{1}105]$ crystallographic direction, which passes through the plane of the nanosheet.

Multislice simulations. Supercells of MFI nanosheets with dimensions 1,100 Å × 1,100 Å in a- and c-directions, respectively, and thickness of 9.94 Å (0.5 UC), 19.89 Å (1.0 UC), 29.83 Å (1.5 UC) and 39.78 Å (2.0 UC) in the b-direction were created from an orthorhombic MFI UC. Nanosheet models were tilted using rotation matrices from $\theta = -60°$ to $60°$ in steps of $2°$ to generate the atomic positions for tilted nanosheet structures. Diffraction pattern tilt-series were simulated from these models using the TEMSIM multislice simulation package. A $4k \times 4k$ pixel^2 grid was used in these simulations, which provide pixel sizes of 0.27 Å per pixel in a- and c-direction. The electron probe was generated using a spherical aberration coefficient $C_s = 2$ mm, convergence angle $\alpha = 0.02$ mrad, defocus $\Delta f = 800$ Å and beam energy $E_0 = 300$ keV.

Gaussian fitting. 32 × 32 Pixel sections from the peaks of interest (Fig. 4a) were extracted from the experimental and simulated diffraction patterns using a custom script in Matlab. Using the curve fitting algorithm in Matlab, a 2D Gaussian function, defined as $f(x, y) = a_1 \exp\left(-\left(\frac{(x-x_0)^2}{2\sigma_x^2} + \frac{(y-y_0)^2}{2\sigma_y^2}\right)\right) + a_2$ was fitted to the entire 32 × 32 pixel region, using σ_x, σ_y, a_1, a_2, x_0 and y_0 as fitting parameters (shown in Supplementary Fig. 4). This fitting also estimates the background level in diffraction patterns due to the central beam with the parameter a_2. Intensity of diffraction spot at each tilt angle is the integrated intensity of the fitted 2D Gaussian function ($I_\theta = 2\pi a_1 \sigma_x \sigma_y$). The broadening of the diffraction spot is measured in two directions as: $FWHM_x = 2\sqrt{2\ln 2}\sigma_x$ and $FWHM_y = 2\sqrt{2\ln 2}\sigma_y$.

Compensation for beam damage. MFI nanosheets, such as all zeolites, are electron beam sensitive. Therefore, diffraction patterns were acquired under a low-dose condition. Loss of diffraction intensity was compensated by separately measuring the reduction of diffraction spot intensities at $\theta = 0°$ as a function of dose within the first 100 min of exposure (typical tilt experiments take ~ 100 min). As shown in Supplementary Fig. 5, in experiments with emission current of ~ 90 μA, loss of diffraction intensity with time due to amorphization was found to be linear with slopes: $k_{\{101\}} = 2.8 \times 10^{-5}$ min^{-1}, $k_{\{202\}} = 8.7 \times 10^{-4}$ min^{-1}, $k_{\{301\}} = 7.2 \times 10^{-4}$ min^{-1} and $k_{\{303\}} = 4.0 \times 10^{-3}$ min^{-1}. In order to compensate for the effects of amorphization, the diffraction spot intensities obtained in each tilt experiment are adjusted to damage-free intensity by multiplying them by a factor of $(1 - kt)^{-1}$, where t is the corresponding time at which they were acquired.

Error analysis. (101) and $(\bar{1}0\bar{1})$ being crystallographically equivalent spots are expected to have same intensity values at all time points for tilt angle $\theta = 0°$. However, a mismatch in experimental intensity values was observed, which is attributed to the limited mechanical precision of the TEM goniometer in tilting the nanosheets to a specific tilt angle (θ). The relative error in intensity measurement i.e., $\sigma_I / \bar{I}$, was calculated from the recorded measurements to be 0.1, where $\sigma_I = \sqrt{\frac{1}{8} \sum_{n=1}^{8} n^2}$, $d_n = \bar{I} - I_n$ and $\bar{I} = \frac{\sum_{n=1}^{8} I_n}{8}$. The damage-free intensities of (101) and $(\bar{1}0\bar{1})$ spots from experimental diffraction patterns are listed in Supplementary Table 1. To correlate the error in intensity to error in θ, the expected values for intensity modulation of crystallographically equivalent (101) and $(\bar{1}0\bar{1})$ for a 1.5-UC-thick nanosheet (Supplementary Fig. 10a) was determined. This known intensity modulation was then used to estimate the maximum possible error in tilt angle ($|\Delta \theta|$) to be $3.4°$ (Supplementary Fig. 10b), as indicated in Fig. 5c with the horizontal error bars.

References

1. Davis, M. E. Ordered porous materials for emerging applications. *Nature* **417**, 813–821 (2002).
2. Tsapatsis, M. 2-dimensional zeolites. *AIChE J.* **60**, 2374–2381 (2014).
3. Roth, W. J., Nachtigall, P., Morris, R. E. & Čejka, J. Two-dimensional zeolites: current status and perspectives. *Chem. Rev.* **114**, 4807–4837 (2014).
4. Čejka, J., Morris, R. E., Nachtigall, P. & Roth, W. J. Layered inorganic solids. *Dalton Trans.* **43**, 10274 (2014).
5. Varoon, K. *et al.* Dispersible exfoliated zeolite nanosheets and their application as a selective membrane. *Science* **334**, 72–75 (2011).
6. Agrawal, K. V. *et al.* Solution-processable exfoliated zeolite nanosheets purified by density gradient centrifugation. *AIChE J.* **59**, 3458–3467 (2013).
7. Corma, A., Fornes, V., Pergher, S. B., Maesen, T. L. M. & Buglass, J. G. Delaminated zeolite precursors as selective acidic catalysts. *Nature* **396**, 353–356 (1998).

8. Choi, M. *et al.* Stable single-unit-cell nanosheets of zeolite MFI as active and long-lived catalysts. *Nature* **461,** 246–249 (2009).

9. Na, K. *et al.* Pillared MFI zeolite nanosheets of a single-unit-cell thickness. *J. Am. Chem. Soc.* **132,** 4169–4177 (2010).

10. Zhang, X. *et al.* Synthesis of self-pillared zeolite nanosheets by repetitive branching. *Science* **336,** 1684–1687 (2012).

11. Lew, C. M., Cai, R. & Yan, Y. Zeolite thin films: from computer chips to space stations. *Acc. Chem. Res.* **43,** 210–219 (2010).

12. Maheshwari, S. *et al.* Influence of layer structure preservation on the catalytic properties of the pillared zeolite MCM-36. *J. Catal.* **272,** 298–308 (2010).

13. Bai, P., Olson, D. H., Tsapatsis, M. & Siepmann, J. I. Understanding the unusual adsorption behavior in hierarchical zeolite nanosheets. *ChemPhysChem* **15,** 2225–2229 (2014).

14. Park, W. *et al.* Hierarchically structure-directing effect of multi-ammonium surfactants for the generation of MFI zeolite nanosheets. *Chem. Mater.* **23,** 5131–5137 (2011).

15. Maheshwari, S. *et al.* Layer structure preservation during swelling, pillaring, and exfoliation of a zeolite precursor. *J. Am. Chem. Soc.* **130,** 1507–1516 (2008).

16. Haselberg, R., Flesch, F. M., Boerke, A. & Somsen, G. W. Thickness and morphology of polyelectrolyte coatings on silica surfaces before and after protein exposure studied by atomic force microscopy. *Anal. Chim. Acta* **779,** 90–95 (2013).

17. Balkose, D., Oguz, K., Ozyuzer, L., Tari, S. & Arkis, E. Morphology, order, light transmittance, and water vapor permeability of aluminum-coated polypropylene zeolite composite films. *J. Appl. Polym. Sci.* **120,** 1671–1678 (2010).

18. Häussler, D. *et al.* Aperiodic W/B4C multilayer systems for X-ray optics: quantitative determination of layer thickness by HAADF-STEM and X-ray reflectivity. *Surf. Coat. Technol.* **204,** 1929–1932 (2010).

19. Jiang, H. *et al.* Determination of layer-thickness variation in periodic multilayer by x-ray reflectivity. *J. Appl. Phys.* **107,** 103523 (2010).

20. Arac, E., Burn, D. M., Eastwood, D. S., Hase, T. P. A. & Atkinson, D. Study of focused-ion-beam–induced structural and compositional modifications in nanoscale bilayer systems by combined grazing incidence X ray reflectivity and fluorescence. *J. Appl. Phys.* **111,** 044324 (2012).

21. Kirkland, E. J. *Advanced Computing in Electron Microscopy.* 40 (Springer, 2009).

22. Ugurlu, O. *et al.* Radiolysis to knock-on damage transition in zeolites under electron beam irradiation. *Phys. Rev. B* **83,** 113408 (2011).

23. Meyer, J. C. *et al.* The structure of suspended graphene sheets. *Nature* **446,** 60–63 (2007).

24. Fasolino, A., Los, J. H. & Katsnelson, M. I. Intrinsic ripples in graphene. *Nat. Mater.* **6,** 858–861 (2007).

25. Bangert, U., Gass, M. H., Bleloch, A. L., Nair, R. R. & Geim, A. K. Manifestation of ripples in free-standing graphene in lattice images obtained in an aberration-corrected scanning transmission electron microscope. *Phys. Status Solidi* **206,** 1117–1122 (2009).

26. Wang, W. L. *et al.* Direct imaging of atomic-scale ripples in few-layer graphene. *Nano Lett.* **12,** 2278–2282 (2012).

27. Miranda, R. & Vázquez de Parga, A. L. Graphene: Surfing ripples towards new devices. *Nat. Nanotechnol.* **4,** 549–550 (2009).

28. Kokotailo, G., Lawton, S. & Olson, D. Structure of synthetic zeolite ZSM-5. *Nature* **272,** 437–438 (1978).

29. First, E. L., Gounaris, C. E., Wei, J. & Floudas, C. A. Computational characterization of zeolite porous networks: an automated approach. *Phys. Chem. Chem. Phys.* **13,** 17339–17358 (2011).

30. Ludwig, R. & Helmut, K. *Transmission Electron Microscopy: Physics of Image Formation.* 587 (Springer Science & Business Media, 2008).

31. Hirsch, P., Howie, A., Nicholson, R., Pashley, D. W. & Whelan, M. J. *Electron Microscopy of Thin Crystals.* 563 (Krieger Publishing Company, 1977).

32. Cowley, J. M. & Moodie, A. F. The scattering of electrons by atoms and crystals. I. A new theoretical approach. *Acta Crystallogr.* **10,** 609–619 (1957).

Acknowledgements

This work was supported as part of the Catalysis Center for Energy Innovation, an Energy Frontier Research Center funded by the US Department of Energy, Office of Science, Basic Energy Sciences under Award DE-SC0001004. We thank M. Odlyzko and Professor J. Jeong for their helpful discussions.

Author contributions

P.K. conceived, executed and analysed TEM experiments. K.V.A. synthesized the zeolite samples. P.K., M.T. and K.A.M. wrote the manuscript. M.T. and K.A.M. conceived and directed the project. All the authors participated in the discussion and interpretation of data, read the manuscript and provided input.

Additional information

Competing financial interests: The authors declare no competing financial interests.

9

Homochiral D_4-symmetric metal–organic cages from stereogenic Ru(II) metalloligands for effective enantioseparation of atropisomeric molecules

Kai Wu[1,*], Kang Li[1,*], Ya-Jun Hou[1], Mei Pan[1], Lu-Yin Zhang[1], Ling Chen[1] & Cheng-Yong Su[1,2]

Absolute chiral environments are rare in regular polyhedral and prismatic architectures, but are achievable from self-assembly of metal–organic cages/containers (MOCs), which endow us with a promising ability to imitate natural organization systems to accomplish stereo-chemical recognition, catalysis and separation. Here we report a general assembly approach to homochiral MOCs with robust chemical viability suitable for various practical applications. A stepwise process for assembly of enantiopure $\Delta\Delta\Delta\Delta\Delta\Delta\Delta\Delta$- and $\Lambda\Lambda\Lambda\Lambda\Lambda\Lambda\Lambda\Lambda$-$Pd_6(RuL_3)_8$ MOCs is accomplished by pre-resolution of the Δ/Λ-Ru-metalloligand precursors. The obtained Pd-Ru bimetallic MOCs feature in large D_4-symmetric chiral space imposed by the predetermined Ru(II)-octahedral stereoconfigurations, which are substitutionally inert, stable, water-soluble and are capable of encapsulating a dozen guests per cage. Chiral resolution tests reveal diverse host–guest stereoselectivity towards different chiral molecules, which demonstrate enantioseparation ability for atropisomeric compounds with C_2 symmetry. NMR studies indicate a distinctive resolution process depending on guest exchange dynamics, which is differentiable between host–guest diastereomers.

[1] MOE Laboratory of Bioinorganic and Synthetic Chemistry, State Key Laboratory of Optoelectronic Materials and Technologies, Lehn Institute of Functional Materials, School of Chemistry and Chemical Engineering, Sun Yat-Sen University, Guangzhou 510275, China. [2] State Key Laboratory of Applied Organic Chemistry, Lanzhou University, Lanzhou 730000, China. * These authors contributed equally to this work. Correspondence and requests for materials should be addressed to C.-Y.S. (email: cesscy@mail.sysu.edu.cn) or to M.P. (email: panm@mail.sysu.edu.cn).

The design and synthesis of discrete nanoscale metal–organic cages/containers (MOCs) with specific configurations and cavities applying directional bridging ligands and geometrically prefixed metals is emerging as an appealing topic in recent supramolecular coordination chemistry[1–3]. Among this, the controlled assembly of enantiopure chiral cages is of special importance because of their potential applications in stereoselective recognition, catalysis and enzyme mimics[4–11]. Since the chiral space in regular polyhedra only rarely presents in snub dodecahedron and snub cube (all other Platonic, Archimedean, prismatic and antiprismatic solids are achiral)[12,13], assembly of chiral polyhedral MOCs is usually achieved by introducing stereogenic centres into the faces, edges or vertices of a polyhedron to remove inversion and mirror symmetries. In this way, a number of homochiral MOCs of T-symmetry[14–19] have been constructed, whereas the chiral MOCs of O-symmetry or higher were proved to be more formidable because of more possible stereoisomers and the demand to transmit single chirality from more subcomponents[20,21]. In principle, the chirality of an MOC can be generated either by the organic stereocentres (such as chiral tetrahedral C*) or the metal stereogenic centres. The latter strategy provides a versatile platform for stereochemistry of MOCs because the plentiful metal coordination geometries can afford innumerable stereogenic metal centres for assembly of chiral structures even from achiral components in a supramolecular sense[22–24]. The overall MOC symmetry can be restricted or reduced by the stereochemical coupling between metal centres. For example, transfer of stereoconfiguration information between vertices of a tetrahedron enables absolute assembly[24] of homoconfigurational $\Delta\Delta\Delta\Delta$- or $\Lambda\Lambda\Lambda\Lambda$-cages based on the stereogenic tris-chelate metal centres[15–17]. However, the lability of metal–ligand exchange often causes enantiomerization between opposite enantiomers[25], and racemic mixture cannot be prevented during the assembly process. Resolution of the enantiopure product usually has to be accomplished with the aid of chiral auxiliaries to form diastereomers, and stabilization of the dynamic metal centre often needs synergistic effect[15–17].

An alternative way to construct stable and robust homochiral MOCs based on the stereogenic metal centres is to design a metalloligand[26] containing a stereoconfigurationally inert metal centre in lieu of the C* stereocentre in organic ligand. Formation of MOCs by virtue of various metalloligands has been achieved in many excellent lines of works[27–32], in which spontaneous resolution and geometric isomerism were observed[33–35], yet construction of enantiopure MOCs from predetermined chiral metalloligands remains unexplored. On the basis of the well-known stereochemistry of D_3-symmetric [Ru(bpy)$_3$]$^{2+}$- or [Ru(phen)$_3$]$^{2+}$-type compounds, which are widely explored in DNA interactions, asymmetric catalysis and supramolecular chiral assemblies[36–40], we initiated the design of [Ru(phen)$_3$]$^{2+}$-type metalloligand for homochiral MOC self-assembly[26]. Since the stereoconfiguration of such tri-chelate Ru-octahedral centres is substitutionally inert and stable in solution assembly and crystallization process, we expect that the predetermined chirality of the Ru metalloligands can direct the assembly of homochiral MOCs with sufficient stability in practical applications. Although stereoselective recognition and catalysis using chiral hosts has been well established[14–19,41,42], enantioseparation of racemic guest molecules by means of homochiral coordination cages remains a challenge. Only a few examples are known to achieve moderate to good diastereoselectivity[43–46], thus urging an extensive study to solve the common problems in this field; for example, (a) efficient resolution of enantiopure cages, (b) effective stabilization of cage stereochemistry and (c) high guest inclusion capacity (more than three guests per host). Herein we report a general approach to assemble homochiral MOCs without post resolution based on the pre-resolved stereogenic Ru-octahedral centres, offering huge cages capable of large amounts of guest encapsulation (>10 guests per host). Specifically, stereoselective separation of atropisomeric molecules rather than C*-based chiral compounds is achieved, and a dynamic resolution process based on differentiable guest exchange by formation of diastereomers is proposed.

Results

Assembly of enantiopure MOCs. We have previously assembled heteronuclear Δ/Λ-Pd$_6$(RuL$_3$)$_8$ MOCs racemate (hereafter assigned as **rac-Δ/Λ-MOCs-16**, Fig. 1) from the racemic RuL$_3$ metalloligands (**rac-Δ/Λ-3**), which show the shape of an octahedron (defined by Pd$_6$ centres) or a rhombic dodecahedron (defined by Pd$_6$Ru$_8$ centres)[26]. It was noted that the cage assembly proceeded in a homochiral manner, with each individual **MOC-16** integrating the same handed Δ- or Λ-3 enantiomers to display either $\Delta\Delta\Delta\Delta\Delta\Delta\Delta\Delta$ or $\Lambda\Lambda\Lambda\Lambda\Lambda\Lambda\Lambda\Lambda$

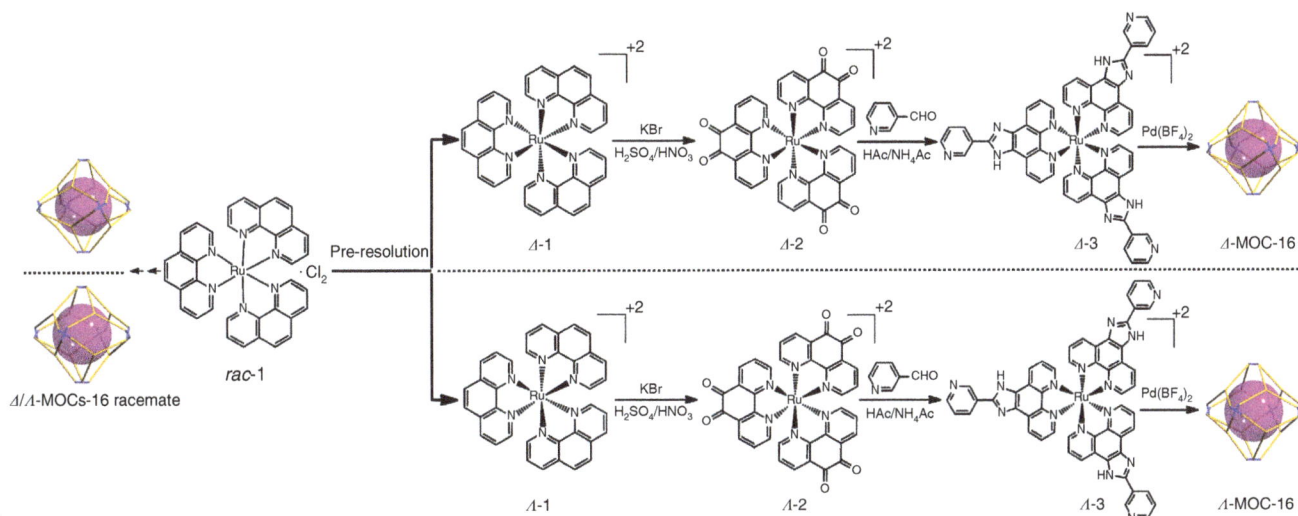

Figure 1 | Assembly procedures. Formation of racemic Δ/Λ-Pd$_6$(RuL$_3$)$_8$ cages (rac-Δ/Λ-MOCs-16) from mixed precursors, and stepwise syntheses of enantiopure Δ- and Λ-Pd$_6$(RuL$_3$)$_8$ cages (Δ-/Λ-MOCs-16) from pre-resolved Δ-3 and Λ-3 metalloligands.

homoconfigurations, indicative of strong cooperative stereochemical coupling between the metal centres[14–19,22–24] to direct the absolute self-organization[24] and exclusive formation of single homochiral Δ- or Λ-MOC-16. However, thus assembled chiral Δ- and Λ-MOCs co-crystallize simultaneously to give racemic products that are not ready for practical applications.

To make use of these homochiral cages, we started chiral resolution from well-established pre-resolution of Δ/Λ-[Ru(phen)$_3$]$^{2+}$ precursors and developed a pair of enantiomeric triangular metalloligands incorporating fixed chiral octahedral Ru(II) centres and pyridyl (Py) terminals ready for assembly of enantiopure Δ- and Λ-Pd$_6$(RuL$_3$)$_8$ MOCs separately. As shown in Fig. 1 and described in detail in Supplementary Figs 1–10, racemic Δ/Λ-[Ru(phen)$_3$]$^{2+}$ was first resolved into a pair of enantiomers (Δ- and Λ-1) in good yields using K$_2$[Sb$_2${(+)-tartrate}$_2$]·3H$_2$O as chiral induction agent, and then oxidized into Δ- and Λ-[Ru(Phendione)$_3$]$^{2+}$ (Δ- and Λ-2). With the aid of chiral shift reagent Eu((+)tfc)$_3$, the enantiopurity was tested to be 94.8% for Δ-1 and 95.3% for Λ-1 (ref. 47). The absolute configurations of the two pairs of Δ-/Λ-1 and Δ-/Λ-2 enantiomers have been well established by the single-crystal structural analyses (Supplementary Figs 1 and 2), which are in excellent agreement with the experimental resolution and syntheses. The phase purity of the bulk products of Δ-/Λ-1 and Δ-/Λ-2 enantiomers has also been verified using the powder X-ray diffraction measurements (Supplementary Figs 1 and 2). Further reaction of Δ- and Λ-2 with 3-pyridinecarboxaldehyde afforded a pair of stereogenic bulky Δ- and Λ-RuL$_3$ metalloligands (Δ- and Λ-3), and, finally, the coordination assembly of Δ- and Λ-3 enantiomers with Pd^{2+} ions unambiguously resulted in a pair of homochiral Δ- and Λ-Pd$_6$(RuL$_3$)$_8$ cages, namely Δ-MOC-16 and Λ-MOC-16, respectively. ^{1}H NMR spectra of two optically pure Δ-/Λ-MOCs-16 enantiomers give well-resolved proton patterns basically identical to previously reported racemic rac-Δ/Λ-MOCs-16 (Supplementary Fig. 8), showing distinguishable H resonance between the protons inside and outside cage (Supplementary Fig. 9). The ^{1}H-^{1}H-COSY and high-resolution mass spectrometry (HR-ESI-TOF-MS) have also been performed to verify formation of Pd$_6$(RuL$_3$)$_8$ cage structures (Supplementary Fig. 10).

The absolute configurational arrangement of the Δ- or Λ-3 metalloligands in Δ-MOC-16 or Λ-MOC-16, respectively, has been undoubtedly established by the single-crystal analyses (Supplementary Fig. 3). The single crystals of Δ-MOC-16 and Λ-MOC-16 were grown from their MeCN solutions in the presence of S-BINOL and R-BINOL, respectively, as absolute structural reference compounds for further authentication of the crystal chirality. Both Δ-MOC-16 and Λ-MOC-16 crystallize in the chiral space groups I422. In Δ-MOC-16, eight Δ-3 metalloligands are assembled by six square-coordinative Pd^{2+} ions to form Pd$_6$(RuL$_3$)$_8$ cage with the $\Delta\Delta\Delta\Delta\Delta\Delta\Delta\Delta$ homoconfigurations (Fig. 2a). The crystal is packed by the identical Δ-MOC-16 cages in together with S-BINOL molecules, giving rise to enantiopure product with the absolute chirality exactly according to the chiral Δ-3 metalloligands and reference S-BINOL used in syntheses and crystal growth. In contrast, Λ-MOC-16 integrates eight Λ-3 metalloligands and six Pd^{2+} ions to form Pd$_6$(RuL$_3$)$_8$ cage with the $\Lambda\Lambda\Lambda\Lambda\Lambda\Lambda\Lambda\Lambda$ homoconfigurations (Fig. 2a), and co-crystallizes with R-BINOLs to result in enantiopure crystals. For both Δ-MOC-16 and Λ-MOC-16, the powder X-ray diffraction patterns of the bulk samples well match those of the single-crystal simulations, indicating satisfactory phase purity (Supplementary Fig. 3).

Careful examination of the crystal structures of Δ-/Λ-MOCs-16 enantiomers reveals that the cage molecule possesses crystallographically imposed D_4 symmetry (Fig. 2 and Supplementary Fig. 3). If regarding the cage as a pseudo-octahedron, the C_4 axis passes two vertices occupied by Pd1 ions, while two pairs of Pd2 ions are located on the C_2 axes. Therefore, the cage symmetry may be considered to degrade from chiral O-symmetry owing to disposition of the same handed Ru-stereocentres on eight faces of octahedron, or, on eight C_3 vertices of rhombic dodecahedron to impose $\Delta\Delta\Delta\Delta\Delta\Delta\Delta\Delta$ or $\Lambda\Lambda\Lambda\Lambda\Lambda\Lambda\Lambda\Lambda$ homoconfigurations in Δ-MOC-16 and Λ-MOC-16, respectively. In another word, the assembly of the homochiral Δ-/Λ-MOCs-16 enantiomers proceeds in a way of octahedral face-control or rhombic dodecahedral vertex-control, thus removing inversion i and mirror σ symmetries to turn an achiral O_h group into a chiral D_4 group (Fig. 2c). Furthermore, the stereoconfigurations around six Pd^{2+} vertices are also induced by the fixed Ru-stereocentres. In Δ-MOC-16, six Pd-Py$_4$ subcomponents are all in Λ-configurations, with the four Py rings showing anticlock fan-like arrangement and vice versa in Λ-MOC-16. In contrast to other completely labile coordination cages[14–24], in the present cases, the stereoconfiguration around the Pd^{2+} corner is fixed by inserting Ru-stereocentres and cage integrity; therefore, enantiomerization through labile Pd-ligand exchange is inhibited for the whole cage. It is worth mentioning that eight S-BINOLs are captured by a Δ-MOC-16, or reversely, eight R-BINOLs by a Λ-MOC-16, on its window pockets but not completely into its cavity (Fig. 2b and Supplementary Figs 3a,b) probably because the crystallization takes place in the MeCN solution where hydrophobic effect is absent and the host–guest inclusion behaviour is different from that in aqueous medium discussed below (vide infra).

The circular dichroism (CD) spectra were also employed to monitor the whole synthetic and assembly processes to confirm that the absolute chirality of the starting Δ-/Λ-[Ru(phen)$_3$]$^{2+}$ precursors were well preserved all the way down (Fig. 3 and Supplementary Fig. 11). In MeCN solution, Δ-/Λ-[Ru(phen)$_3$]$^{2+}$ mainly presents three absorption peaks at 225, 265 and 450 nm (Fig. 3a), with the first two corresponding to the n–π^* and π–π^* transitions of phen groups, while the last one originating from the metal-to-ligand charge transfer (MLCT) transition between Ru^{2+} and phen ligands. All three absorption bands are reflected in the corresponding CD spectra of the resolved Δ-1 and Λ-1 with the middle peak at 265 nm, giving the most prominent CD signal. Taking Δ-1 as an example, the same tendency of first negative and second positive Cotton effect from longer to shorter wavelength in the three CD bands is in accordance with the Δ-type octahedral chirality established for the Ru^{2+} coordination centre[48] and vice versa for the Λ-1 compound. For the rest three pairs of enantiomers, because of the cutoff effect of the solvents (dimethylsulphoxide (DMSO) or H$_2$O), the CD signals corresponding to the absorption at 225 nm were not fully presented; however, the other bands were clearly detected in the whole synthetic process, preserving the same chirality attributes for the same series of enantiomers (Δ-1, 2, 3, MOC-16 versus Λ-1, 2, 3, MOC-16, respectively). Furthermore, because of the accumulation effect (eightfold in Δ- or Λ-MOC-16 compared with Δ- or Λ-3) of multiple chiral Ru centres in one entity in the final enantiopure Δ-/Λ-MOCs-16, a remarkable increase in CD signal intensities was observed for Δ- or Λ-MOC-16 ($\Delta\varepsilon = \sim 720$ M^{-1}cm^{-1}) in comparison with Δ- or Λ-3 ($\Delta\varepsilon = \sim 120$ M^{-1}cm^{-1}). Optical rotation tests also manifested the absolute configurations in Δ- and Λ-MOCs-16 (Δ, $[\alpha]^{20}_D = -266°$; Λ, $[\alpha]^{20}D = 272°$, $c = 0.5$, H$_2$O). From these CD studies we see that the stereochemistry of octahedral Ru centres is robust enough to survive all reaction conditions, exactly in agreement with the observations of chirality preservation in crystallographic study. The stereochemical stability of Δ- and Λ-MOCs-16 has also been testified against heating and longtime stay in solution (Fig. 3d), confirming that the absolute

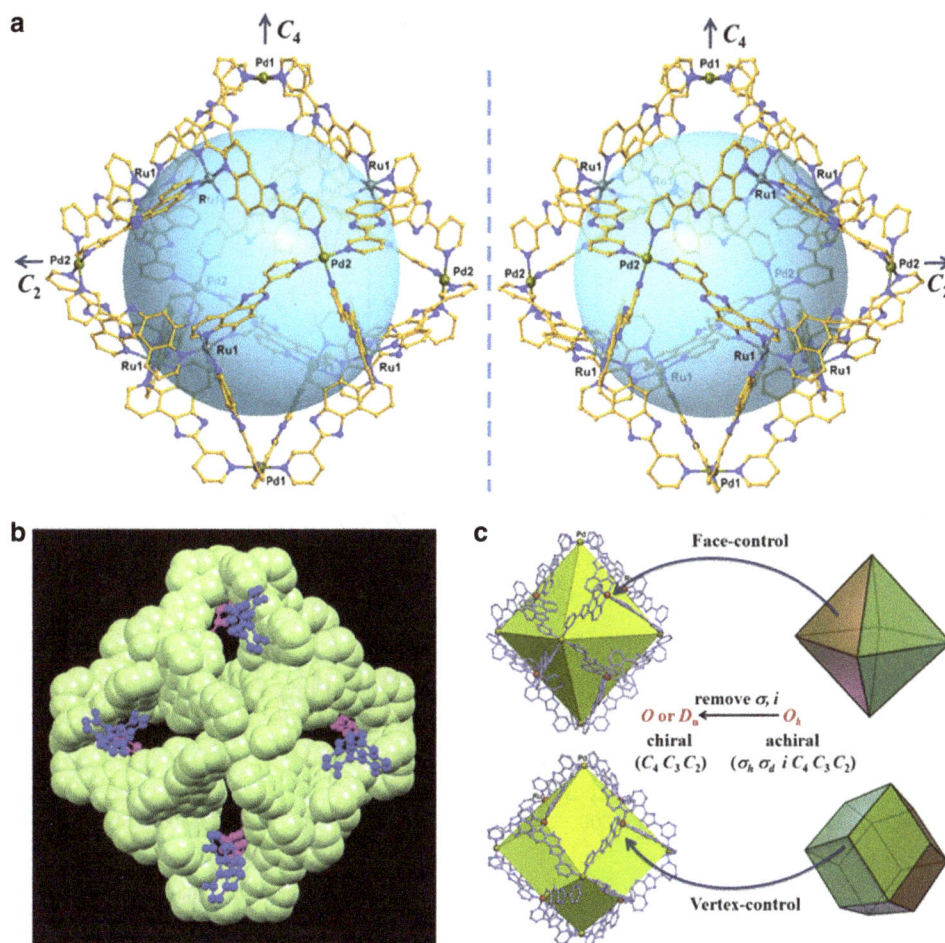

Figure 2 | Crystal structures. (**a**) A pair of D_4-symmetric homochiral Δ- and Λ-MOCs-16 showing $\Delta\Delta\Delta\Delta\Delta\Delta\Delta\Delta$ and $\Lambda\Lambda\Lambda\Lambda\Lambda\Lambda\Lambda\Lambda$ configurations of eight RuL$_3$ metalloligands whereas $\Lambda\Lambda\Lambda\Lambda\Lambda\Lambda$- and $\Delta\Delta\Delta\Delta\Delta\Delta$-configurations of six Pd-Py$_4$ subcomponents. (**b**) A Δ-MOC-16 cage (in space-filling mode) capturing eight S-BINOL guests (in ball-and-stick mode) on the windows pockets. (**c**) The demonstration how to form D_4 Λ-MOC-16 from O_h regular polyhedra by introducing eight Λ-3 metalloligands on the faces of an octahedron, or, on the C_3 vertices of a rhombic dodecahedron, to reduce molecular symmetry, and further direct Δ-arrangement of four pyridyl rings around six vertices of Pd centres.

chirality of each enantiomeric Δ- and Λ-MOC-16 is well retained on heating to 373 K and staying in solution for 50 days. Such a stable and substitutionally inert nature of stereogenic Ru centres plays a key role in fixing absolute chirality of Δ- and Λ-MOCs-16, despite intrinsic dynamics of Pd^{2+} centres subject to metal–ligand exchange, thereof paving the way for utilization of these enantiopure cages in, for example, stereoselective catalysis and separation.

Stereoselective separation of racemic guests. In an attempt to test enantioseparation ability of Δ-/Λ-MOC-16 cages, we selected two types of racemic organic molecules, one carrying a chiral C* centre and the other characteristic of C_2-symmetric chirality (Table 1). The host–guest inclusion examined by ^{1}H NMR in the D$_2$O system revealed that all chiral molecules can be well encapsulated by the **MOC-16** host owing to hydrophobic effect (Supplementary Figs 12–16), showing typical upfield shift of guest protons and further splitting of cage protons[26]. Moreover, the host–guest stereochemical relationship between enantiomeric Δ-/Λ-**MOCs-16** and R-/S-BINOLs has been examined using ^{1}H NMR enantiodifferentiation experiments, where two pairs of host–guest diastereomers, namely S-BINOL $\subset$ Δ-**MOC-16**, R-BINOL $\subset$ Δ-**MOC-16** and S-BINOL $\subset$ Λ-**MOC-16**, R-BINOL $\subset$ Λ-**MOC-16**, and two pairs of host–guest enantiomers, namely S-BINOL $\subset$ Δ-**MOC-16**, R-BINOL $\subset$ Λ-**MOC-16** and

R-BINOL $\subset$ Δ-**MOC-16**, S-BINOL $\subset$ Λ-**MOC-16**, are formed. As shown in Fig. 4, the solution dynamics is obviously distinguishable between the diastereomeric pairs, while that between the enantiomeric pairs is similar[41]. This means the homochiral Δ- and Λ-**MOC-16** cages are able to recognize and differentiate R- and S-BINOL enantiomeric guests in solution because of their diastereomeric host–guest relationship. As a consequence, the chiral resolution of racemic molecules was carried out by Δ- and Λ-**MOCs-16** separately in pure D$_2$O solution based on either a homogeneous or a heterogeneous method (see details in Methods or Supplementary Methods). The resolved guests were determined using high-performance liquid chromatography (HPLC) with enantiomeric excess (ee) averaged from three parallel experiments (Table 1 and Supplementary Figs 17–24).

The resolution results unveil that the homochiral Δ- or Λ-**MOCs-16** have rather poor stereoselectivity towards chiral compounds containing C* stereocentres. As seen in Table 1, no obvious resolution effect can be detected for naproxen, 1-(1-naphthyl)ethanol and benzoin, despite ^{1}H NMR-proved inclusion of these guests by the host **MOC-16** (Supplementary Figs 14–16). However, through the same separation process, a pair of R-/S-BINOL atropisomers was successfully resolved, with the ee values reaching 34% or more by Δ-/Λ-**MOCs-16**. Relatively low enantioseparation results were obtained for R-/S-3-Br-BINOL

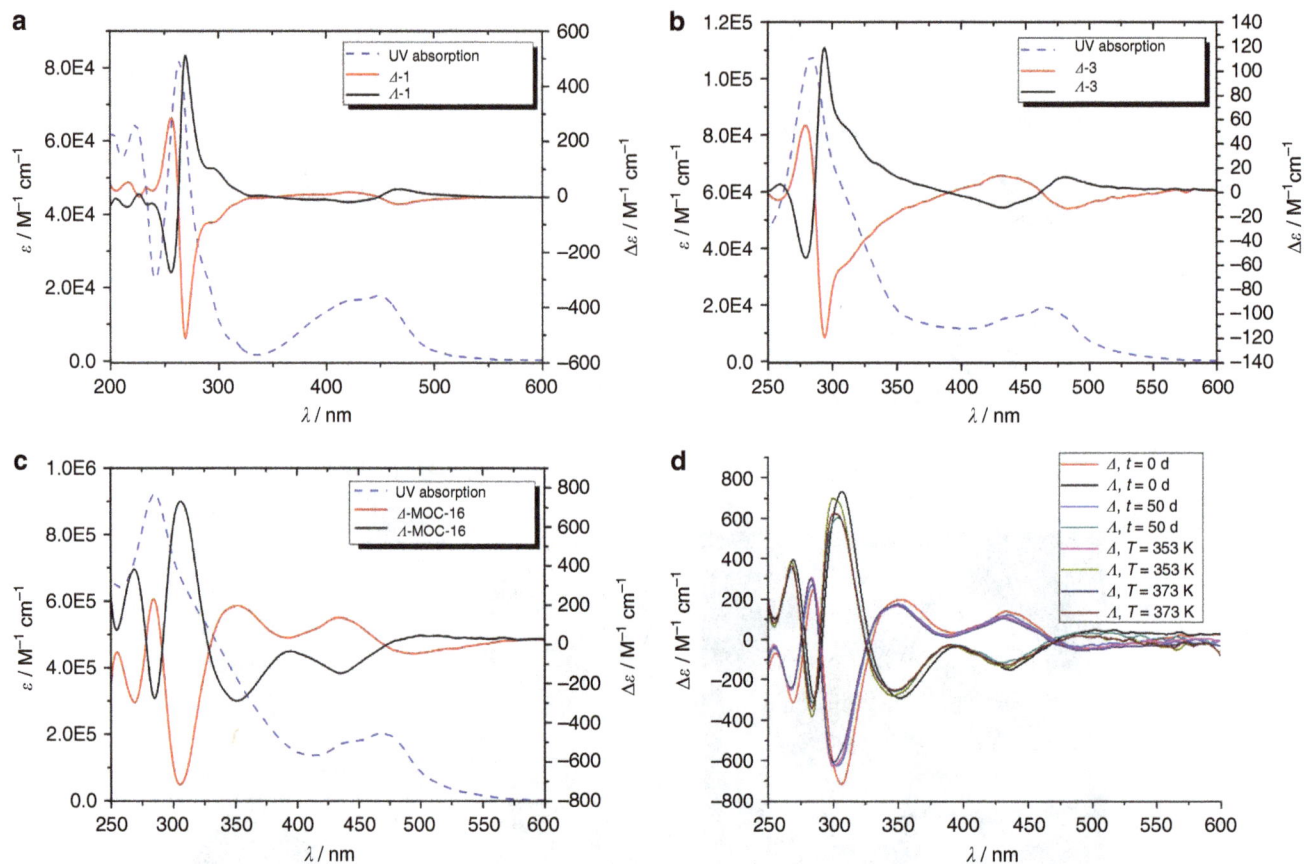

Figure 3 | CD and UV spectra. CD (solid lines) and ultraviolet (dotted lines). (**a**) Δ- and Λ-1 in MeCN, (**b**) Δ- and Λ-3 in DMSO, (**c**) Δ- and Λ-MOCs-16 in H₂O and (**d**) the stereochemical stability depending on time and temperature.

Table 1 | Enantioselective resolution of chiral organic molecules*.

Atropisomers of C_2 symmetry: Enantiomers based on C* centre:

R-BINOL   S-BINOL   R-3-Br-BINOL   S-3-Br-BINOL   R-1-(1-Naphthyl)ethanol   S-1-(1-Naphthyl)ethanol

R-6-Br-BINOL   S-6-Br-BINOL   R-Spirodiol   S-Spirodiol   R-Benzoin   S-Benzoin   R-Naproxen   S-Naproxen

Guests	Δ-MOC-16		Λ-MOC-16	
	R,S ratio	ee (%)	R,S ratio	ee (%)
BINOL	67:33	34	32:68	36
3-Br-BINOL	54:46	8	46:54	8
6-Br-BINOL	77:23	54	19:81	62
1,1'-spirobiindane-7,7'-diol	67:33	34	28:72	44
Naproxen	50:50	0	50:50	0
1-(1-Naphthyl)ethanol	51:50	1	50:50	0
Benzoin	50:50	0	50:50	0

ee, enantiomeric excess; MOC, metal–organic cage/container.
*The estimated uncertainty is about ± 2% as calculated for the averaged values from the results of three repeating resolution processes.

racemate; however, the resolution effect was greatly improved for the chiral discrimination of R-/S-6-Br-BINOL enantiomers. By applying Δ-MOC-16, the resolved product contains 77% of R-isomer and 23% of S-isomer, giving an *ee* value of 54%. Surprisingly, an *ee* value up to 62% was obtained by Λ-MOC-16 with the product dominant in S-isomer. Similar

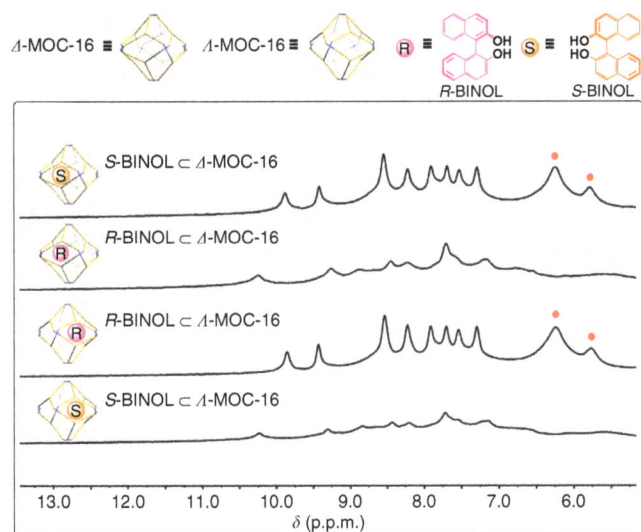

Figure 4 | ^{1}H NMR enantiodifferentiation experiments. Sequestration of enantiomeric R- or S-BINOL guests by homochiral Δ- or Λ-MOCs-16 (d_6-DMSO/D$_2$O = 1/5, 298 K). Red circles denote signals of encapsulated guests.

enantioseparation ability of Δ-/Λ-MOCs-16 was able to extend to another kind of atropisomeric compound R-/S-spirodiol, exhibiting the same host–guest stereoselectivity. The Δ-MOC-16 got 34% predominance of R-isomer, while a higher ee value of 44% was obtained for S-isomer by Λ-MOC-16. In general, Δ-MOC-16 shows a preferable stereoselectivity towards R-isomer, while Λ-MOC-16 prefers S-isomer for all chiral guests of C_2 symmetry. The higher-resolution effect from S-isomer ⊂ Λ-MOC-16 inclusion than from R-isomer ⊂ Δ-MOC-16 inclusion is unexpected, probably owing to the slight difference in optical attribute based on their optical rotation tests. To the best of our knowledge, such a preferable enantiorecognition of chiral guests with C_2 symmetry has not been observed before for cage compounds, and the enantioseparation ability of Δ-/Λ-MOCs-16 reaches high level within the known chiral organic and coordination cages[43–46,49]. In addition, the adequate solubility of MOCs-16 in water (2.6 g per 100 ml) makes it convenient to implement enantioseparation either in a homogeneous two-phase way (*Method I*: organic-water transfer as shown in Supplementary Fig. 25) or simply in a heterogeneous suspension way (*Method II*: solid-solution transfer as described in Methods). In comparison with the normally insoluble metal–organic frameworks for chiral separation, the water solubility of MOCs-16 offers advantages by using the hydrophobic effect to transfer water-insoluble organic guests into the aqueous phase, and the guest transformation between the organic-water phases is easy to accomplish. Extraction of the resolved chiral guests from the water phase of Δ-/Λ-MOCs-16 readily leads to recovery of the empty cages, which can be reused for the next runs of chiral resolution without a significant loss of the enantioseparation ability as tested by four cycles of R/S-6-Br-BINOL resolution with Δ-MOCs-16 (ee 51–57%, Supplementary Table 1). On the other hand, chiral resolution test of R/S-6-Br-BINOL racemate within Δ-MOC-16 and Λ-MOC-16 using 10 times the amount of host and guest indicates that the enantioselectivity is retained almost the same for the scaling up separation (ee 55 and 60%).

Resolution process study. To further understand the host–guest interactions for insight into the resolution mechanism, ^{1}H NMR titration was performed in an attempt to acquire association constants[42] for the pairs of host–guest diastereomers. However,

the experimental results obviously reveal intricate host–guest solution dynamics (Fig. 5 and Supplementary Figs 26–28). Since the MOC-16 cage has a huge molecular size (3.3 × 3.3 nm) and cavity (2895 Å^3 based on VOIDOO calculations) where a large amount of guests could be hosted (for example, 18 Phen guests per host)[26], the proton signals of both cage and guests are generally broadened and poorly resolved because of slow rotational diffusion and dynamics typical of large molecules, thereof preventing us from quantitative study with regard to thermodynamic or kinetic details by using the known methods for relatively simple host–guest systems (usually more than three guests per host)[50–53]. Nevertheless, it is evident that titration of enantiopure Λ-MOC-16 (or Δ-MOC-16) with R- and S-BINOL guests of C_2 symmetry undergoes remarkably different host–guest interaction processes, showing distinguishable guest inclusion behaviours for R- and S-BINOL atropisomers as demonstrated in Fig. 5a,b. This is in agreement with the observation from above-mentioned ^{1}H NMR enantiodifferentiation experiments because of the formation of a pair of host–guest diastereomers R-BINOL ⊂ Λ-MOC-16 and S-BINOL ⊂ Λ-MOC-16. In contrast, titration of Δ- and Λ-MOC-16 cages with the same C*-chiral S-1-(1-naphthyl)ethanol guest, which should also give a pair of diastereomers S-1-(1-naphthyl)ethanol ⊂ Δ-MOC-16 and S-1-(1-naphthyl)ethanol ⊂ Λ-MOC-16, just results in rather similar ^{1}H NMR chemical shift patterns (Fig. 5d,e), indicating that the homochiral Λ-MOC-16 (or Δ-MOC-16) cage exhibits the same guest inclusion behaviours for R- and S-enantiomeric guests carrying C* stereocentres.

As demonstrated in Fig. 5, stepwise inclusion of R-BINOLs by Λ-MOC-16 at 298 K causes inverse chemical shifts of cage protons, with those inside the cage moving upfield while those outside the cage moving downfield (Fig. 5a). The guest protons appear as severely broadened doublet and remain almost unmoved up to 12 guest inclusion. Addition of more than 12 equivalent R-BINOLs shows little influence on cage protons, but leads to downfield shift and further broadening of guest protons. These results suggest that at least 12 R-BINOLs are encapsulated inside the cage, and further guest uptake may speed up dynamic exchange. Inclusion of guests inside the cage is also supported with ^{1}H-^{1}H-COSY and NOESY measurements (Supplementary Fig. 26). For comparison, inclusion of S-BINOLs at 298 K does not lead to distinct bidirectional shifts of cage protons, while the resonance of guests is even broadened and becomes poorly visible together with the host protons on inclusion of more than 10 S-BINOLs (Fig. 5b). To observe guest signals more clearly, titration at a higher temperature 353 K was performed (Fig. 5c), which presents better resolved guest resonances but basically same overall chemical shift patterns as observed at 298. Therefore, similar host–guest interacting manners may be expected at these temperatures (*vide infra*). It is notable that the guest signals display a continuously downfield shift, characteristic of fast guest exchange. These NMR observations imply more dynamic host–guest interactions for S-BINOL ⊂ Λ-MOC-16 inclusion compared with R-BINOL ⊂ Λ-MOC-16 inclusion at the room temperature. Broadening of H resonance is indicative of slow and restricted molecular rotation and tumbling[54,55] as well as of a comparable guest exchange rate with the NMR timescale. For the S-BINOL ⊂ Λ-MOC-16 system, faster guest exchange dynamics may present, showing averaged influence on host protons either inside or outside. When the cage cavity is getting fulfilled (∼12 guest per cage), the overall host–guest dynamics is slowed down to make NMR unable to discriminate resonating frequency. In contrast, guest exchange in the R-BINOL ⊂ Λ-MOC-16 system is slow enough at room temperature, thus showing distinguishable impact on host protons inside and outside. Such a guest dynamic difference between two host–guest diastereomers

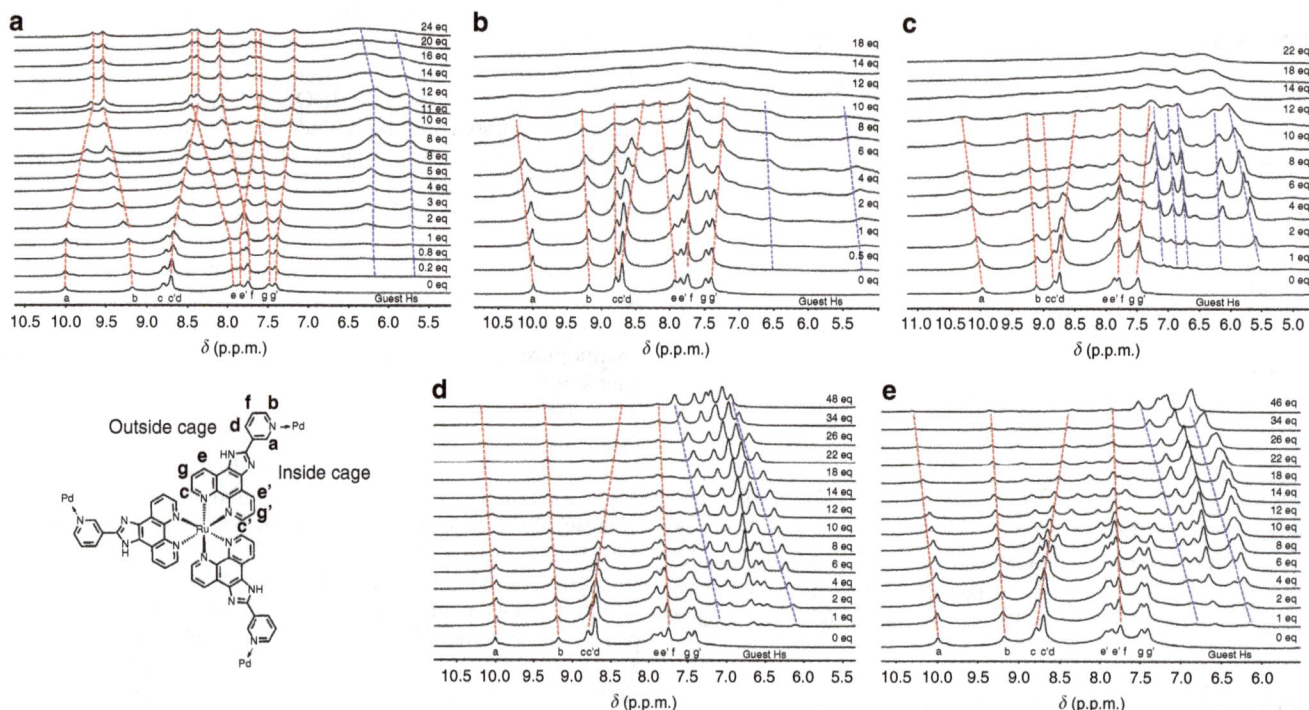

Figure 5 | ^{1}H NMR titration in DMSO-d_6/D$_2$O = 1/5. (**a**) R-BINOL ⊂ Λ-MOC-16 at 298 K, (**b**) S-BINOL ⊂ Λ-MOC-16 at 298 K, (**c**) S-BINOL ⊂ Λ-MOC-16 at 353 K, (**d**) S-1-(1-naphthyl)ethanol ⊂ Λ-MOC-16 at 298 K and (**e**) S-1-(1-naphthyl)ethanol ⊂ Δ-MOC-16 at 298 K.

may account for the intrinsic factor that determines the enantioseparation ability of homochiral Δ- or Λ-MOCs-16 towards racemic R/S-BINOLs. On the contrary, titration of Δ- and Λ-MOCs-16 with S-1-(1-naphthyl)ethanol guests shows a similar guest exchange dynamics (Fig. 5d,e), where fast guest exchange is evident for both host–guest diastereomers. This may explain why the homochiral Δ-/Λ-MOCs-16s are unable to discriminate R/S-stereomers carrying opposite C* stereocentres. Although a confinement effect of a cage is usually expected to enhance the intrinsic chirality of the C* guests, discriminable stereoselectivity was not observed for the present C* molecules because of fast guest exchange.

Variable-temperature ^{1}H NMR study has been carried out to testify the above-proposed resolution process (Fig. 6 and Supplementary Fig. 29). It is clear that, for both R-BINOL ⊂ Λ-MOC-16 and S-BINOL ⊂ Λ-MOC-16 diastereomers, heating boosts guest dynamics and accelerates guest exchange, with two broadened signals getting better resolved and moving constantly downfield to approach free guests. This kind of host–guest solution dynamics might be comparable to the NMR-observable molecular dynamics[53–55]. However, the accelation of the guest exchange dynamics from the sufficiently slow state to the fast state may not undergo a normal peak coalescence, but display a turning point where guest H resonances start to resolve apparently owing to NMR-observable freedom of guests from cage restriction. If taking the resonance frequency separation between two slowly restricted guest signals and the turning point of guest signal shifts for Eyring analysis, the guest exchange rates and energy barriers might be estimated at 1,021, 488 s^{-1} and 55.7, 60.5 kJ mol^{-1}, for S-BINOL ⊂ Λ-MOC-16 and R-BINOL ⊂ Λ-MOC-16 diastereomers, respectively. This means that Λ-MOC-16 can capture S-BINOL faster at lower energy cost than R-BINOL to accomplish a host–guest inclusion process.

Discussion

The host–guest dynamics and guest exchange mechanism have been vigorously explored for the insight of fundamental host–guest interactions and more complex encapsulation system design, in which both thermodynamics and kinetics play important roles in guest binding[50–55]. On the basis of above enantioseparation and NMR studies, we may speculate that the resolution process of homochiral Δ- or Λ-MOCs-16 towards chiral molecules of C_2 symmetry is mainly controlled by guest exchange dynamics, in comparison with the more popular thermodynamic resolution of racemic guests by chiral cages[43]. As demonstrated in Fig. 7, encapsulation of racemic R/S stereomers by, for example, Λ-MOC may proceed in a dynamic way depending on host–guest interactions and R/S-guest competition. If inclusion of S-stereomers is faster than R-stereomers via a lower guest exchange energy barrier, preferable resolution of S-stereomers over R-stereomers is achievable. It should be noted that such a dynamic resolution based on guest exchange dynamics might be comparable but inherently different from the well-known 'kinetic resolution' based on different reaction rates between a chiral catalyst and enantiomers[41,42,56]. Guest exchange and displacement process in a host–guest system is often sensitive to the synergistic effect from both thermodynamic and kinetic contributions[50–55]. Elongating resolution time may not influence ee results so much as by the catalytic kinetic resolution. We have tested the time-dependent chiral resolution of R/S-BINOL racemate by Λ-MOC-16. The results indicate an increase in the ee value within first 2 h, but remaining nearly unchanged afterwards (Supplementary Tables 2 and 3). We believe that the distinctive host–guest dynamics between R-BINOL ⊂ Λ-MOC-16 and S-BINOL ⊂ Λ-MOC-16 diastereomers should originate from the stereoconfigurations of the octahedral Ru centres. The twisted arrangement of three Phen motifs around Ru centres in helical sense may not be able to differentiate inclusion behaviour of configurationally free racemic guests carrying C* stereocentres, but significantly affect the interactions between Δ-/Λ-MOCs-16 and atropisomeric guests bearing C_2 symmetry owing to their intrinsic helical configurations. Formation of adaptive or mismatched host–guest diastereomers through dynamic guest exchange may be more

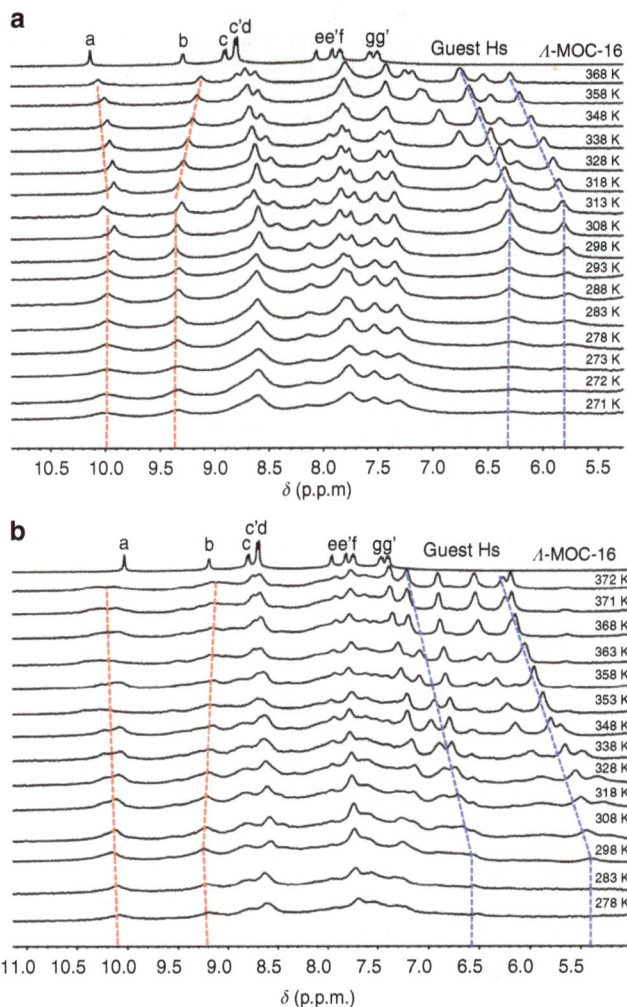

Figure 6 | VT ^{1}H NMR study of guest inclusion dynamics in solvent DMSO-d_6/D$_2$O = 1/5. (**a**) **R-BINOL** $\subset$ Λ-MOC-16. (**b**) **S-BINOL** $\subset$ Λ-MOC-16. Red lines show host protons, while blue lines show guest protons.

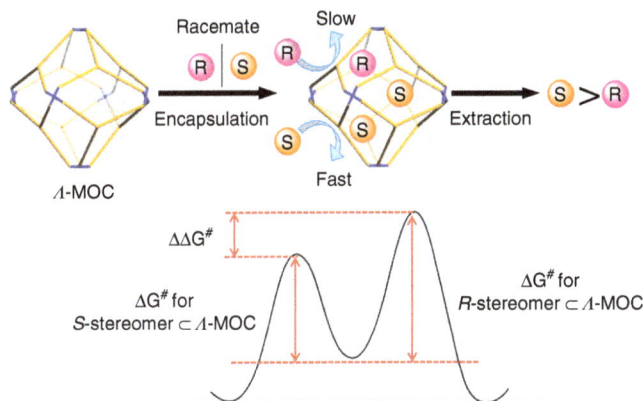

Figure 7 | Enantioseparation mechanism. A possible resolution process relying on guest exchange dynamics of atropisomers with homochiral Λ-MOC-16. $\Delta G^{\#}$ represents guest exchange energy barrier.

dominated by stereochemical compatibility than by binding constant. Therefore, such stereoconfigurationally predetermined **MOCs** could afford better adaptive inclusion of one atropisomer over the opposite one, thus resulting in stereoselective separation.

In conclusion, pre-resolution of a pair of enantiomeric Δ-/Λ-Ru metalloligands has been successfully implemented based on the

stereogenic octahedral Ru centres in Δ-/Λ-[Ru(phen)$_3$]$^{2+}$ precursors, giving rise to the assembly of enantiopure D_4-symmetric Δ- and Λ-MOC-16 cages separately, which feature in high guest inclusion capacity and substantial stereochemical stability. The single-crystal diffraction analyses of the individual Δ- and Λ-MOC-16 cages verified the formation of absolute $\Delta\Delta\Delta\Delta\Delta\Delta\Delta\Delta$- and $\Lambda\Lambda\Lambda\Lambda\Lambda\Lambda\Lambda\Lambda$ homoconfigurations, respectively, in corresponding Pd$_6$(RuL$_3$)$_8$ cages, and the crystallization of optically pure cage products. The stereoselective inclusion of chiral molecules has been tested for two kinds of organic racemates, that is, classic chiral compounds having C* centres and atropisomeric compounds characteristic of C_2 symmetry, with the phase transformation resolution processes. Successful enantioseparation of atropisomers has been accomplished by the use of these homochiral Δ- and Λ-MOCs-16, manifesting an unprecedented dynamic resolution process based on the kinetically driven guest exchange. The possible resolution mechanism has been investigated by the means of ^{1}H NMR titration, ^{1}H NMR enantiodifferentiation experiments as well as variable-temperature ^{1}H NMR study. In general, this kind of assembly process may provide a new platform to study the stereochemical transmission of optically stable metal centres to versatile homochiral entities in coordination chemistry, and the dynamic resolution behaviour imposed by stereoconfiguration of metal centres might be useful in various chiral resolution of synthetic and industrial significance.

Methods

Materials and measurements. Unless otherwise stated, all commercial reagents and solvents were used as commercially purchased without additional purification. The NMR spectra were recorded on Bruker AVANCE III 400 (400 MHz). Circular dichroism spectra and ultraviolet–visible absorption spectra were measured with a JASCO J-810 spectropolarimeter. Specific rotations were recorded on ADP440 + B + S. HR-ESI-TOF mass spectra were tested on Bruker Maxis 4G, and data analyses were processed with the Bruker Data Analysis software. HPLC spectra were measured on Agilent-2000. Diffraction data for the single crystals were collected on an Agilent SuperNova X-ray diffractometer using micro-focus dual X-ray sources (Supplementary Data 1). Syntheses and characterization details for all compounds are given in Supplementary Methods. Selected bond lengths (Å) and bond angles (°) are listed in Supplementary Tables 4–9.

Crystal data for {Δ-[Pd$_6$(RuL$_3$)$_8$](S-BINOL)$_4$}·anion·solvent (Δ-MOC-16).
I422 space group, $a = 32.2284(4)$ Å, $c = 38.2801(7)$ Å, $V = 39,760.5(13)$ Å^3, $Mr = 9,727.82$, $Dx = 0.813$ g cm^{-3}, $Z = 2$, $\mu = 2.599$ mm^{-1}, 16,396 independent reflections, of which 9,021 observed ($I > 2\sigma(I)$), $R_1 = 0.0711$, $wR_2 = 0.2375$, $S = 1.008$, Flack parameter = 0.131(13).

Crystal data for {Λ-[Pd$_6$(RuL$_3$)$_8$](R-BINOL)$_4$}·anion·solvent (Λ-MOC-16).
I422 space group, $a = 32.5722(6)$ Å, $c = 38.7258(9)$ Å, $V = 41,086.1(19)$ Å^3, $Mr = 9,727.82$, $Dx = 0.786$ g cm^{-3}, $Z = 2$, $\mu = 2.515$ mm^{-1}, 17,111 independent reflections, of which 7,718 observed ($I > 2\sigma(I)$), $R_1 = 0.0913$, $wR_2 = 0.2680$, $S = 1.038$, Flack parameter = 0.139(16).

Crystal data for {Δ-[Ru(Phen)$_3$](PF$_6$)$_2$}·(C$_6$H$_5$CH$_3$)·(CH$_3$CN)$_2$ (Δ-1).
P4$_1$ space group, $a = 25.5619(2)$ Å, $c = 12.5769(2)$ Å, $V = 8,217.88(18)$ Å^3, $Mr = 2,037.48$, $Dx = 1.647$ g cm^{-3}, $Z = 4$, $\mu = 0.556$ mm^{-1}, 19,736 independent reflections, of which 17,221 observed ($I > 2\sigma(I)$), $R_1 = 0.0678$, $wR_2 = 0.1759$, $S = 1.050$, Flack parameter = 0.00(4).

Crystal data for {Λ-[Ru(Phen)$_3$](PF$_6$)$_2$}·(C$_6$H$_5$CH$_3$)·(CH$_3$CN)$_2$ (Λ-1).
P4$_3$ space group, $a = 25.5802(1)$ Å, $c = 12.5709(1)$ Å, $V = 8,225.73(9)$ Å^3, $Mr = 2,037.48$, $Dx = 1.645$ g cm^{-3}, $Z = 4$, $\mu = 0.556$ mm^{-1}, 20,142 independent reflections, of which 17,755 observed ($I > 2\sigma(I)$), $R_1 = 0.0408$, $wR_2 = 0.1071$, $S = 1.025$, Flack parameter = $-0.033(8)$.

Crystal data for Δ-[Ru(Phendione)$_3$](ClO$_4$)$_2$·(H$_2$O)·(CH$_3$CN)$_2$ (Δ-2).
P2$_1$2$_1$2$_1$ space group, $a = 13.8114(2)$ Å, $b = 14.0525(2)$ Å, $c = 20.7957(3)$ Å, $V = 4,036.13(10)$ Å^3, $Mr = 1,028.64$, $Dx = 1.693$ g cm^{-3}, $Z = 4$, $\mu = 5.107$ mm^{-1}, 7,890 independent reflections, of which 7,536 observed ($I > 2\sigma(I)$), $R_1 = 0.0438$, $wR_2 = 0.1195$, $S = 1.025$, Flack parameter = $-0.015(4)$.

Crystal data for Λ-[Ru(Phendione)$_3$](ClO$_4$)$_2$ · (H$_2$O) · (CH$_3$CN)$_2$ (Λ-2).
$P2_12_12_1$ space group, $a = 13.7734(2)$ Å, $b = 14.0148(2)$ Å, $c = 20.7100(3)$ Å, $V = 3,997.68(10)$ Å^3, $Mr = 1,022.59$, $Dx = 1.699\,\mathrm{g\,cm}^{-3}$, $Z = 4$, $\mu = 5.156\,\mathrm{mm}^{-1}$, 7,980 independent reflections, of which 7,737 observed ($I > 2\sigma(I)$), $R_1 = 0.0409$, $wR_2 = 0.1084$, $S = 1.032$, Flack parameter = $-0.011(3)$.

General chiral resolution of racemic guests by enantiopure Δ/Λ-MOCs-16.
Two kinds of methods were used to resolve racemic guests depending on whether the guest inclusion leads to precipitation. For racemic R/S-BINOL, R/S-3-Br-BINOL, R/S-6-Br-BINOL and R/S-naproxen molecules, *Method I* based on a solution–solution transfer was applied to avoid host–guest precipitation (Supplementary Fig. 25). An aqueous solution of Δ- or Λ-MOC-16 and an ethereal solution of racemic guest were mixed and stirred vigorously at room temperature, and then the bottom layer was taken out and extracted with CHCl$_3$. The extractants were combined and the solvent was removed using rotary evaporator to afford white solid as resolved guests by the homochiral MOC host. The solid was redissolved in isopropanol and the *ee* of guest molecules was determined using HPLC. For racemic R/S-spirodiol, R/S-1-(1-naphthyl)ethanol and R/S-benzoin molecules, Method II based on a solid-solution transfer was applied directly. The powder of guest racemate was suspended in the aqueous solution of Δ-or Λ-MOC-16. The mixture was stirred vigorously at room temperature. After centrifugation, the filtrate was collected and extracted with CHCl$_3$. The extractants were combined and the solvent was removed by rotary evaporator to afford white solid as resolved guest. The *ee* analysis is the same as in Method I.

References

1. Amouri, H., Desmarets, C. & Moussa, J. Confined nanospaces in metallocages: guest molecules, weakly encapsulated anions, and catalyst sequestration. *Chem. Rev.* **112**, 2015–2041 (2012).

2. Cook, T. R., Zheng, Y.-R. & Stang, P. J. Metal–organic frameworks and self-assembled supramolecular coordination complexes: comparing and contrasting the design, synthesis, and functionality of metal–organic materials. *Chem. Rev.* **113**, 734–777 (2013).

3. Meng, W. *et al.* A self-assembled M$_8$L$_6$ cubic cage that selectively encapsulates large aromatic guests. *Angew. Chem. Int. Ed.* **50**, 3479–3483 (2011).

4. Smulders, M. M. J., Riddell, I. A., Browne, C. & Nitschke, J. R. Building on architectural principles for three-dimensional metallosupramolecular construction. *Chem. Soc. Rev.* **42**, 1728–1754 (2013).

5. Pluth, M. D., Bergman, R. G. & Raymond, K. N. Proton-mediated chemistry and catalysis in a self-assembled supramolecular host. *Acc. Chem. Res.* **42**, 1650–1659 (2009).

6. Nishioka, Y., Yamaguchi, T., Kawano, M. & Fujita, M. Asymmetric [2 + 2] olefin cross photoaddition in a self-assembled host with remote chiral auxiliaries. *J. Am. Chem. Soc.* **130**, 8160–8161 (2008).

7. Argent, S. P., Riis-Johannessen, T., Jeffery, J. C., Harding, L. P. & Ward, M. Diastereoselective formation and optical activity of an M$_4$L$_6$ cage complex. *Chem. Commun.* 4647–4649 (2005).

8. Ousaka, N. *et al.* Efficient long-range stereochemical communication and cooperative effects in self-assembled Fe$_4$L$_6$ cages. *J. Am. Chem. Soc.* **134**, 15528–15537 (2012).

9. Dong, J., Zhou, Y., Zhang, F. & Cui, Y. A highly fluorescent metallosalalen-based chiral cage for enantioselective recognition and sensing. *Chem. Eur. J.* **20**, 6455–6461 (2014).

10. Wang, J., He, C., Wu, P. Y., Wang, J. & Duan, C. Y. An amide-containing metal–organic tetrahedron responding to a spin-trapping reaction in a fluorescent enhancement manner for biological imaging of NO in living cells. *J. Am. Chem. Soc.* **133**, 12402–12405 (2011).

11. Samanta, D. & Mukherjee, S. P. Component selection in the self-assembly of palladium(II) nanocages and cage-to-cage transformations. *Chem. Eur. J.* **20**, 12483–12492 (2014).

12. MacGillivray, L. R. & Atwood, J. L. Structural classification and general principles for the design of spherical molecular hosts. *Angew. Chem. Int. Ed.* **38**, 1018–1033 (1999).

13. Hamilton, T. D. & MacGillivray, L. R. Enclosed chiral environments from self-assembled metal-organic polyhedral. *Cryst. Growth Des.* **4**, 419–430 (2004).

14. Stang, P. J. & Olenyuk, B. Transition-metal-mediated rational design and self-assembly of chiral, nanoscale supramolecular polyhedra with unique *T* symmetry. *Organometallics* **16**, 3094–3096 (1997).

15. Ousaka, N., Clegg, J. K. & Nitschke, J. R. Nonlinear enhancement of chiroptical response through subcomponent substitution in M$_4$L$_6$ cages. *Angew. Chem. Int. Ed.* **51**, 1464–1468 (2012).

16. Davis, A. V., Fiedler, D., Ziegler, M., Terpin, A. & Raymond, K. N. Resolution of chiral, tetrahedral M$_4$L$_6$ metal-ligand hosts. *J. Am. Chem. Soc.* **129**, 15354–15363 (2007).

17. Terpin, A. J., Ziegler, M., Johnson, D. W. & Raymond, K. N. Resolution and kinetic stability of a chiral supramolecular assembly made of labile components. *Angew. Chem. Int. Ed.* **40**, 157–160 (2001).

18. Meng, W., Clegg, J. K., Thoburn, J. D. & Nitschke, J. R. Controlling the transmission of stereochemical information through space in terphenyl-edged Fe$_4$L$_6$ cages. *J. Am. Chem. Soc.* **133**, 13652–13660 (2011).

19. Schweiger, M., Seidel, S. R., Schmitz, M. & Stang, P. J. Rational design of chiral nanoscale adamantanoids. *Org. Lett.* **2**, 1255–1257 (2000).

20. Gütz, C. *et al.* Enantiomerically pure [M$_6$L$_{12}$] or [M$_{12}$L$_{24}$] polyhedra from flexible bis(pyridine) ligands. *Angew. Chem. Int. Ed.* **53**, 1693–1698 (2014).

21. Yang, Y. *et al.* Diastereoselective synthesis of O symmetric heterometallic cubic cages. *Chem. Commun.* **51**, 3804–3807 (2015).

22. Knof, U. & von Zelewsky, A. Predetermined chirality at metal centers. *Angew. Chem. Int. Ed.* **38**, 302–322 (1999).

23. Castilla, A. M., Ramsay, W. J. & Nitschke, J. R. Stereochemistry in subcomponent self-assembly. *Acc. Chem. Res.* **47**, 2063–2073 (2014).

24. Northrop, B. H., Zheng, Y.-R., Chi, K.-W. & Stang, P. J. Self-organization in coordination-driven self-assembly. *Acc. Chem. Res.* **42**, 1554–1563 (2009).

25. Saalfrank, R. W. *et al.* Enantiomerisation of tetrahedral homochiral [M$_4$L$_6$] clusters: synchronised four bailar twists and six atropenantiomerisation processes monitored by temperature-dependent dynamic ^{1}H NMR spectroscopy. *Chem. Eur. J.* **8**, 2679–2683 (2002).

26. Li, K. *et al.* Stepwise assembly of Pd$_6$(RuL$_3$)$_8$ nanoscale rhombododecahedral metal–organic cages via metalloligand strategy for guest trapping and protection. *J. Am. Chem. Soc.* **136**, 4456–4459 (2014).

27. Garrison, J. C. *et al.* Synthesis and characterization of a trigonal bipyramidal supramolecular cage based upon rhodium and platinum metal centers. *Chem. Commun.* 4644–4646 (2006).

28. Smulders, M. M. J., Jiménez, A. & Nitschke, J. R. Integrative self-sorting synthesis of a Fe$_8$Pt$_6$L$_{24}$ cubic cage. *Angew. Chem. Int. Ed.* **51**, 6681–6685 (2012).

29. Hiraoka, S., Sakata, Y. & Shionoya, M. Ti(IV)-centered dynamic interconversion between Pd(II), Ti(IV)-containing ring and cage molecules. *J. Am. Chem. Soc.* **130**, 10058–10059 (2008).

30. Sakata, Y., Hiraoka, S. & Shionoya, M. Site-selective ligand exchange on a heteroleptic Ti^{IV} complex towards stepwise multicomponent self-assembly. *Chem. Eur. J.* **16**, 3318–3325 (2010).

31. Kryschenko, Y. K., Seidel, S. R., Arif, A. M. & Stang, P. J. Coordination-driven self-assembly of predesigned supramolecular triangles. *J. Am. Chem. Soc.* **125**, 5193–5198 (2003).

32. Li, H., Yao, Z.-J., Liu, D. & Jin, G.-X. Multi-component coordination-driven self-assembly toward heterometallic macrocycles and cages. *Coord. Chem. Rev.* **293**, 139–157 (2015).

33. Wu, H.-B. & Wang, Q.-M. Construction of heterometallic cages with tripodal metalloligands. *Angew. Chem. Int. Ed.* **48**, 7343–7345 (2009).

34. Metherell, A. J. & Ward, M. D. Stepwise assembly of an adamantoid Ru$_4$Ag$_6$ cage by control of metal coordination geometry at specific sites. *Chem. Commun.* **50**, 10979–10982 (2014).

35. Zhou, X.-P. *et al.* A high-symmetry coordination cage from 38- or 62-component self-assembly. *J. Am. Chem. Soc.* **134**, 8042–8045 (2012).

36. Carlucci, L., Ciani, G., Maggini, S., Proserpio, D. M. & Visconti, M. Heterometallic modular metal-organic 3D frameworks assembled via new tris-beta-diketonate metalloligands: nanoporous materials for anion exchange and scaffolding of selected anionic guests. *Chem. Eur. J.* **16**, 12328–12341 (2010).

37. Hamelin, O., Rimboud, M., Pecaut, J. & Fontecave, M. Chiral-at-metal ruthenium complex as a metalloligand for asymmetric catalysis. *Inorg. Chem.* **46**, 5354–5360 (2007).

38. Hiort, C., Lincoln, P. & Norden, B. DNA binding of Δ- and Λ-[Ru(phen)$_2$DPPZ]$^{2+}$. *J. Am. Chem. Soc.* **115**, 3448–3454 (1993).

39. Smith, J. A., Collins, J. G., Patterson, B. T. & Keene, F. R. Total enantioselectivity in the DNA binding of the dinuclear ruthenium (II) complex [{Ru(Me$_2$bpy)$_2$}$_2$(μ-bpm)]$^{4+}$ {bpm = 2,2'-bipyrimidine. Me$_2$bpy = 4,4'-dimethyl-2,2'-bipyridine}. *Dalton Trans.* **9**, 1277–1283 (2004).

40. MacDonnell, F. M., Kim, M. J. & Bodige, S. Substitutionally inert complexes as chiral synthons for stereospecific supramolecular syntheses. *Coord. Chem. Rev.* **185**, 535–549 (1999).

41. Zhao, C. *et al.* Chiral amide directed assembly of a diastereo- and enantiopure supramolecular host and its application to enantioselective catalysis of neutral substrates. *J. Am. Chem. Soc.* **135**, 18802–18805 (2013).

42. Kida, T., Iwamoto, T., Asahara, H., Hinoue, T. & Akashi, M. Chiral recognition and kinetic resolution of aromatic amines via supramolecular chiral nanocapsules in nonpolar solvent. *J. Am. Chem. Soc.* **135**, 3371–3374 (2013).

43. Fiedler, D., Leung, D. H., Bergman, R. G. & Raymond, K. N. Enantioselective guest binding and dynamic resolution of cationic ruthenium complexes by a chiral metal-ligand assembly. *J. Am. Chem. Soc.* **126**, 3674–3675 (2004).

44. Hastings, C. J., Pluth, M. D., Biros, S. M., Bergman, R. G. & Raymond, K. N. Simultaneously bound guests and chiral recognition: a chiral self-assembled supramolecular host encapsulates hydrophobic guests. *Tetrahedron* **64**, 8362–8367 (2008).

45. Sawada, T., Matsumoto, A. & Fujita, M. Coordination-driven folding and assembly of a short peptide into a protein-like two-nanometer-sized channel. *Angew. Chem. Int. Ed.* **53**, 7228–7232 (2014).

46. Liu, T. F., Liu, Y., Xuan, W. M. & Cui, Y. Chiral nanoscale metal–organic tetrahedral cages: diastereoselective self-assembly and enantioselective separation. *Angew. Chem. Int. Ed.* **49**, 4121–4124 (2010).

47. Torres, A. S., Maloney, D. J., Tate, D., Saad, Y. & MacDonnell, F. M. Retention of optical activity during conversion of *Λ*-[Ru(1,10-phenanthroline)₃]²⁺ to *Λ*-[Ru(1,10-phenanthroline-5,6-dione)₃]²⁺ and *Λ*-[Ru(dipyrido[a:3,2-h:2′,3′-c-]-phenazine)₃]²⁺. *Inorg. Chim. Acta* **293**, 37–43 (1999).

48. McGee, K. A. & Mann, K. R. Inefficient crystal packing in chiral [Ru(phen)₃](PF₆)₂ enables oxygen molecule quenching of the solid-state MLCT emission. *J. Am. Chem. Soc.* **131**, 1896–1902 (2009).

49. Chen, L. *et al.* Separation of rare gases and chiral molecules by selective binding in porous organic cages. *Nat. Mater.* **13**, 954–960 (2014).

50. Smulders, M. M. J., Zarra, S. & Nitschke, J. R. Quantitative understanding of guest binding enables the design of complex host-guest behavior. *J. Am. Chem. Soc.* **135**, 7039–7046 (2013).

51. Zarra, S., Smulders, M. M. J., Lefebvre, Q., Clegg, J. K. & Nitschke, J. R. Guanidinium binding modulates guest exchange within an [M₄L₆] capsule. *Angew. Chem. Int. Ed.* **51**, 6882–6885 (2012).

52. Jiang, W., Ajami, D. & Rebek, Jr J. Alkane lengths determine encapsulation rates and equilibria. *J. Am. Chem. Soc.* **134**, 8070–8073 (2012).

53. Davis, A. V. *et al.* Guest exchange dynamics in an M₄L₆ tetrahedral host. *J. Am. Chem. Soc.* **128**, 1324–1333 (2006).

54. Mugridge, J. S., Szigethy, G., Bergman, R. G. & Raymond, K. N. Encapsulated guest-host dynamics: guest rotational barriers and tumbling as a probe of host interior cavity space. *J. Am. Chem. Soc.* **132**, 16256–16264 (2010).

55. Wang, B.-Y. *et al.* The entrapment of chiral guests with gated baskets: can a kinetic discrimination of enantiomers be governed through gating? *Chem. Eur. J.* **19**, 4767–4775 (2013).

56. Ma, G., Deng, J. & Sibi, M. P. Fluxionally chiral DMAP catalysts: kinetic resolution of axially chiral biaryl compounds. *Angew. Chem. Int. Ed.* **53**, 11818–11821 (2014).

Acknowledgements

We gratefully acknowledge the financial support from the 973 Program (2012CB821701), the NSFC Projects (Grants 91222201, 21373276, 21573291) of China, NSFGP (S2013030013474) and the RFDP of Higher Education (20120171130006).

Author contributions

C.-Y.S. and M.P. designed the research and wrote the paper. K.W. and K.L. carried out most of the syntheses and measurements. Y.-J.H., L.-Y.Z. and L.C. helped in experiments and data analyses. All authors discussed the results and commented on the manuscript.

Additional information

Accession codes: The X-ray crystallographic coordinates for structures reported in this Article have been deposited at the Cambridge Crystallographic Data Centre (CCDC), under deposition number CCDC 1432349–1432354. These data can be obtained free of charge from The Cambridge Crystallographic Data Centre via www.ccdc.cam.ac.uk/data_request/cif.

Competing financial interests: The authors declare no competing financial interests.

10

Active porous transition towards spatiotemporal control of molecular flow in a crystal membrane

Yuichi Takasaki[1] & Satoshi Takamizawa[1]

Fluidic control is an essential technology widely found in processes such as flood control in land irrigation and cell metabolism in biological tissues. In any fluidic control system, valve function is the key mechanism used to actively regulate flow and miniaturization of fluidic regulation with precise workability will be particularly vital in the development of microfluidic control. The concept of crystal engineering is alternative to processing technology in microstructure construction, as the ultimate microfluidic devices must provide molecular level control. Consequently, microporous crystals can instantly be converted to microfluidic devices if introduced in an active transformability of porous structure and geometry. Here we show that the introduction of a stress-induced martensitic transition mechanism converts a microporous molecular crystal into an active fluidic device with spatiotemporal molecular flow controllability through mechanical reorientation of subnanometre channels.

[1] Graduate School of Nanobioscience, Yokohama City University, 22-2 Seto, Kanazawa-ku, Yokohama, Kanagawa 236-0027, Japan. Correspondence and requests for materials should be addressed to S.T. (email: staka@yokohama-cu.ac.jp).

Orderly fluid transportation can be seen in a wide range of processes, from irrigation to biological tissues such as flood control or cell metabolism, through programmed pathways with adequate motive force to generate flow. Although the scale and means used in fluidic control depend on purpose, the valve function that controls the directivity and mass rate of flow is a key mechanism to actively regulate flow in any fluidic transportation operation. In microfluidic control systems[1,2], precision in the 'valve' mechanism driven by an active procedure such as magnetic[3], electrostatic[4] or piezoelectric[5] actuation is especially important. Currently, gas fluidic research of man-made microfluidic-oriented devices are on a submillimetre scale[6-8] and that of passages are on submicrometre scale by present-day progressing technology[9]. Although processing technologies will continue to advance, the technological limits will be continuously addressed as sizes and architectural styles change and novel methods will be explored in future micro- or nanofluidic devices. Considering the ultimate miniaturization of microfluidic systems, passage widths will reach the sizes of the fluid components and the fluid should behave as a diffusion of the included component particles. Thus, towards the theoretically finest flow control, microporous solids seem to have the fundamental requirements for such the ultimately miniature devices. They could immediately convert into attractive 'nanofluidic' devices if some sort of molecular flow controllability is embedded in them. Therefore, we have focused on microporous molecular crystals with flexible subnanometre channels and have reported on the anisotropic gas permeation through channels[10,11]. In the past, the several interchangeabilities of a channel geometry by gas adsorption-induced phase transitions have been reported[12-14]. The remaining requirement was how to instill active controllability into such microporous crystals. We recently found a superelastic property in organic molecular crystals (organosuperelasticity)[15,16] that can actively rearrange the porous structure by mechanically induced transition if the organosuperelastic crystal was porous. Herein, we report a microporous crystal with superelasticity that actively controls accurate molecular flow by means of channel rearrangement. Active transformability in a porous crystal will offer a novel theory for the spatiotemporal control of molecular flow.

Results

Martensitic transition involving reorientation of channels. We accidentally found stress-induced martensitic transition behaviour on a microporous single-crystal host, $[Cu(II)_2(bza)_4 (pyz)]_n$ (bza: benzoate; pyz: pyrazine) (**1**), which can be a microfluidic crystal for gaseous fluid (Fig. 1). By pushing the crystal surface (00-1) of **1** with a glass needle at room temperature, while one edge of the crystal is fixed to a base, a stress-induced daughter phase began to grow out from {1-1-1} with an unchanging bandwidth of about 5 μm sandwiched by two parallel planar interfaces. The edge of the band ran 133 μm ms^{-1} along [010] direction (Fig. 2a). After going across the crystal, the thin band broadened in 0.5 μm ms^{-1} to separate the interfaces (Supplementary Fig. 1). The daughter domain spontaneously contracted and disappeared by removal of the stress through the reverse transition with organosuperelasticity[15,16], which simplifies the reverse operation. This is the first example of a superelastic crystal consisting of metal complexes. Crystal phase indexing under the coexisting state of the mother (α) and daughter (α') phases revealed a rotation twin in which the crystal lattice was maintained but was rotated accompanied by the rotation of channel direction in rearranging 0.8-nm-width pore units during the structural phase transition on the boundary (Fig. 2b, Supplementary Figs 2–4 and Supplementary Table 1). The

channels in the α and α' phases run in a skewed position in the twinned crystal, which is slightly bent at the phase boundary by 14.6° along the projected direction of $[010]_\alpha$ and $[010]_{\alpha'}$, as seen in Fig. 2b. Thus, the generation of the α' phase can change the direction of gas permeation by mechanical twinning and the width or number of channels in the generated α' domains are precisely regulated by the shear range of the mother α crystal, as shown in Fig. 2c–f and Supplementary Fig. 5.

Gas permeability under twinning state. We confirmed the interchange of gas flow direction and the regulation of flow rate by crystal twinning of **1** by a gas permeation technique on a single-crystal membrane (Fig. 3a). In the α phase crystal where nothing was operated, gases permeate the crystal surfaces of $\{100\}_\alpha$ (indicated as the open surface), whereas they were effectively blocked on $\{001\}_\alpha$ (indicated as the closed surface), agreeing with the previous report[10] (white bars in Fig. 3b,c). In the mechanically generated α' phase within the α phase crystal, the permeation/barrier directions were interchanged by the reorientation of channel directions estimated in crystallography discussed in Fig. 2b (black bars in Fig. 3b,c). In addition, the extension of the open surface area by widening the α' domain linearly increased the flow rate of H_2 gas (inset of Fig. 3b), which showed that the number of the channels precisely determines the flow rate due to the identical microflux through the uniform subnanometre channels. Therefore, a finer flux control in molecular scale can be realized by minimizing the α' domain to a nanometre scale or by using additive slits or masks on a crystal

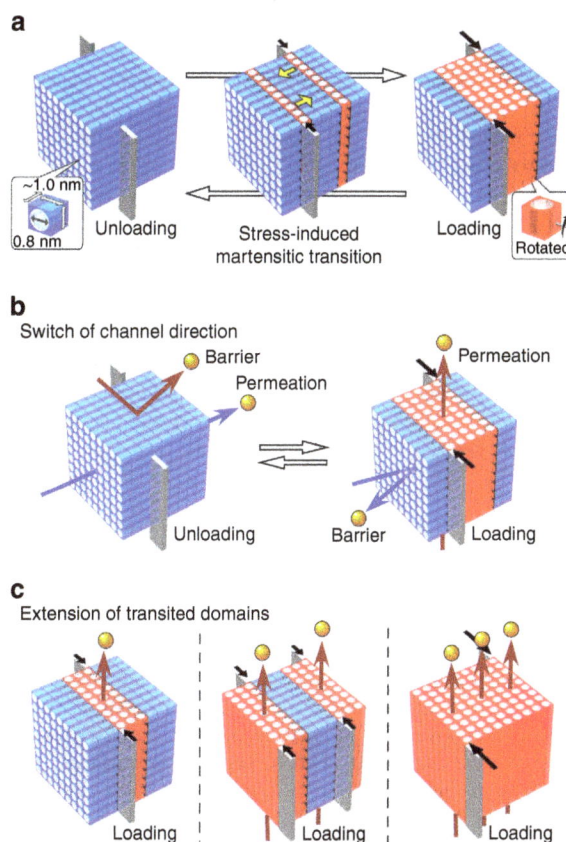

Figure 1 | Pore rearrangement in stress-induced transition. (a) Rotation of pore directions by generating twinned daughter crystal phase (α': red) within mother crystal phase (α: blue) during mechanical loading. Expected gas flow switch in direction (**b**) and in positions (**c**) where the area determines the rate of flow mass through subnanometre uniform channels.

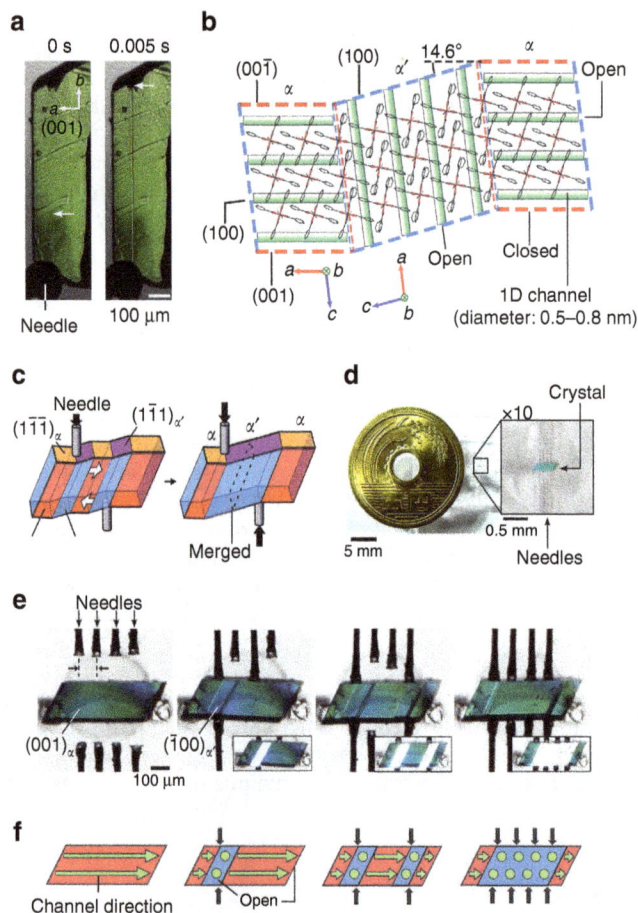

Figuree 2 | Active positional controllability of martensitic transition.
(**a**) Growth of 5-μm-wide band from the pushed edge of the (1-1-1) crystal surface at room temperature. (**b**) Connection of mother (α phase) and daughter (α' phase) crystals at 298 K accompanied by the rotation of channel direction (green bands) under the twinned state with a bending angle of 14.6° along the projected direction of $[010]_\alpha$ and $[010]_{\alpha'}$ based on crystallography. (**c**) Schematic explanation for the regulation of the α' crystal domains sandwiched by the pushing positions as a shear on $\{1\text{-}1\text{-}1\}_\alpha$. (**d**) Picture of an experimental system. (**e**) Active generation/degeneration of daughter crystal domains by shearing the microcrystal of **1** (0.48 × 0.17 × 0.10 mm) with movable needles (**f**) and figures indicating the directions of the penetrating channels. (Inset pictures in **e**: highlighted α' domains by reflecting light due to the parallel crystal surfaces in each crystal phase.; see Supplementary Movie 1 for **e**).

if considering the practical limit of the minimum size of a well-controlled domain (5 μm in width in the current crystal membrane).

Active gas flow control. To demonstrate the dynamic switch of the permeated gas flow at the designed time and positions on the crystal **1**, we measured spatiotemporal gas permeation, which was traced by the movement of silicone oil inside capillaries guided through the capillaries attached to the crystal surfaces (Fig. 4a, Supplementary Figs 8 and 9 and Supplementary Movies 2 and 3). CO_2 was selected in this experiment because of the advantages offered by its permeability among the gases, as summarized in Fig. 3. The directions of gas flow through the crystal specimen were defined as horizontal (H) and vertical (V) derived from crystal orientation (see Fig. 4b). In the α phase crystal, gas flowed out from the open surface in the H direction, whereas it was blocked on the closed surface in the V direction (blue region in

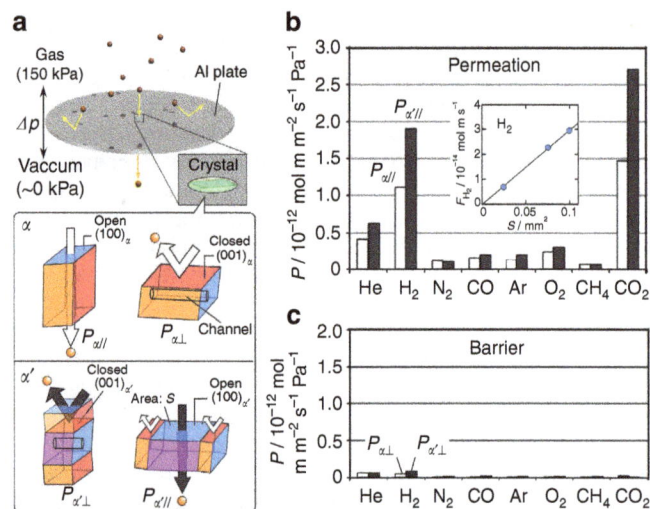

Figure 3 | Gas permeation on a single-crystal membrane. (**a**) Schematic explanation for single-crystal membrane and orientations of the embedded crystals in a hole of an Al plate. Gas permeability (P) in α (white bars) and in α' phase (black bars) through open surface (**b**) and closed surface (**c**) of the crystal at 293 K and Δp of 150 kPa. (Inset figure in **b**: correlation between open surface area in α' phase (S) and flow rate of H_2 gas normalized by crystal thicknesses (F_{H_2}).) The permeability of H_2 and CO_2 in the α' phase become higher than those in the α phase due to the slight change in channel structure caused by molecular distortion (see Supplementary Fig. 7).

Fig. 4d). By applying shear stress on the crystal, the permeation/barrier directions were dynamically interchanged in generating α' crystal domain through the rotation of the channel direction (red region in Fig. 4d). The gas permeation starts immediately after α' phase appears (Supplementary Figs 11–16 and Supplementary Tables 6–10). The interchange was spontaneously switched back by removal of the shear stress and could be reproduced repeatedly (Fig. 4d). Applying shear stress at multiple positions dynamically generated the gas flow positions in the crystal (Fig. 4c), which was demonstrated by alternate gas permeations at positions V_1 and V_2, depending on the switch of the sheared positions (black and red regions in Fig. 4e). Therefore, a single crystal of **1** provides spatiotemporal controllability of molecular flow by mechanical twinning.

Discussion

With respect to the controllability of molecular flow, a single-crystal host of **1** can be regarded as an assemblage of pore units, which can be called 'porons (pore + on (s))' as they are quasi-particles, with transformability assisted by flexible host skeletons. Dynamic rearrangement of 'porons' can alter the pore connections, which produces the spatiotemporal controllability of flow directivity within a single solid. In this study, organosuperelasticity is an optimal property for the rearrangement of 'porons', although other techniques to have an effect on assembly may be available. This theory raises a novel strategy for constructing microfluidic devices.

We have demonstrated the active switchability of gas flow in directions and positions through a microporous molecular crystal of **1** at a designed time caused by stress-induced martensitic transition. The generated daughter domain in mechanical twining involved rotation of the channel directions. The domain spontaneously contracted and disappeared by removal of the stress as a superelasticity, which is the first example in metal complexes. In fact, a single crystal of **1** provides spatiotemporal

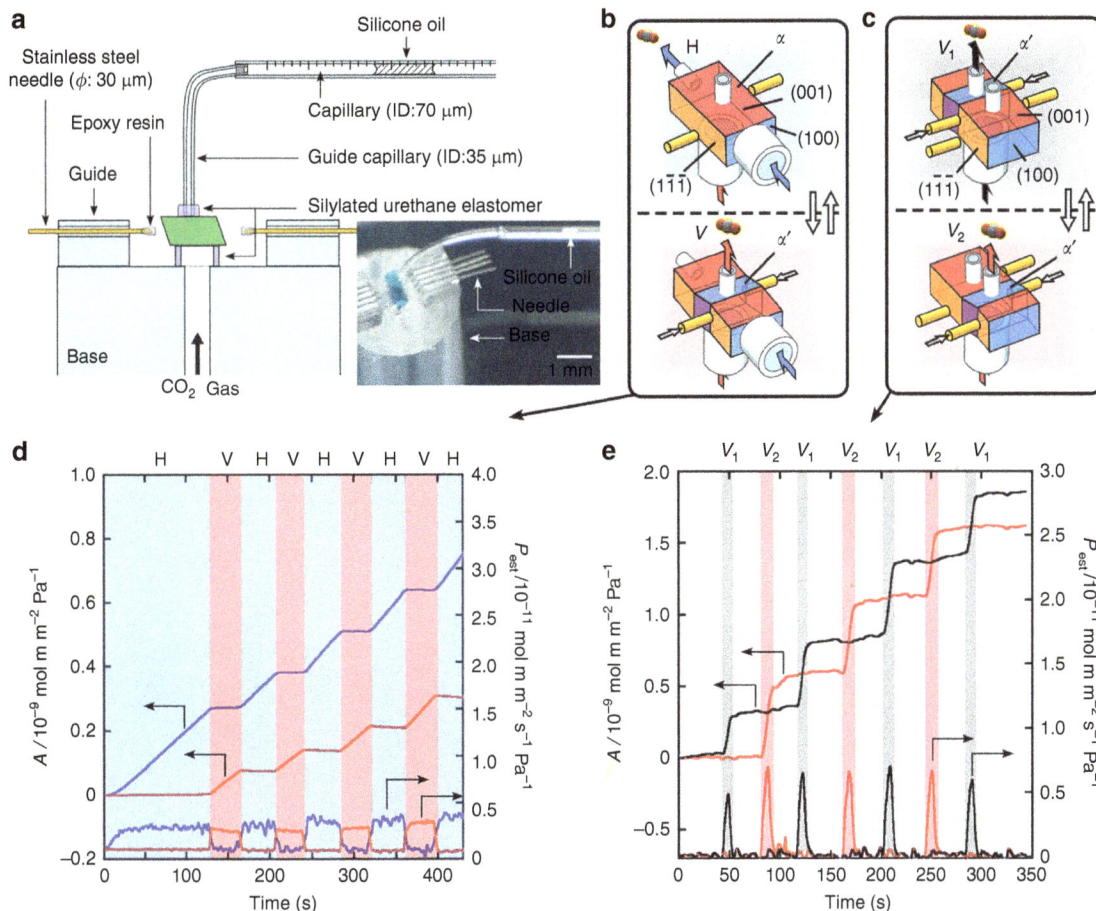

Figure 4 | Dynamic gas flow switch. (**a**) Experimental system of spatiotemporal gas permeation measurement with a shearing procedure at room temperature in continuous introduction of CO_2 gas from the base at Δp_{CO2} (difference of partial pressure) of 100 kPa, as shown in Fig. 2e. Schematic explanation for gas flow switch in directions (**b**) and positions of the crystal (**c**). Permeated amount of CO_2 gas (A) and estimated permeability (P_{est}) through the crystal in H (blue line) and V (red line) directions (**d**) and through two positions (V_1 and V_2) in the V direction (black and red lines) (**e**). (P_{est} were estimated as time-derivative values of A; Supplementary Movies 2 and 3).

controllability of molecular flow by mechanical twinning. This microporous single crystal would be used as a device with precise flux and/or directional controllability of a molecular flow manipulated by mechanical stress. Furthermore, by using a stress-induced mechanism in a microdevice, flammable fluid such as high-pressure hydrogen gas would be safely controlled due to the lack of need for electric actuation. Consequently, the introduction of stress-induced transition phenomena into channel solids enables dynamic control of molecular flow in the solids.

Methods

X-ray single crystal diffraction analysis. Single-crystal X-ray structural analysis of **1** in α/α' coexisting state was performed at 298 K on a Bruker Smart APEX CCD (charge-coupled device) area diffractometer (Bruker AXS K.K.) with a nitrogen-flow temperature controller using graphite-monochromated Mo $K\alpha$ radiation ($\lambda = 0.71073$ Å). Empirical absorption corrections were applied using the SADABS programme. The structure was solved by direct methods (SHELXS-97) and refined by full-matrix least-squares calculations on F^2 (SHELXL-97) using the SHELX TL programme package. Non-hydrogen atoms were refined anisotropically; hydrogen atoms were fixed at calculated positions by riding model approximation.

Gas permeation measurement. Single crystals of **1** in the α' phase were cut to suitable sizes with an area of the crystal surface such as $1.00 \times 10^5\ \mu m^2$ for (100) and thickness of 80 μm along [100], and $4.78 \times 10^4\ \mu m^2$ for (001) and thickness of 185 μm along [001], which were embedded in a hole of each aluminum plate. These crystal membranes were used in gas permeation measurements along [100] and [001] directions in the α' phase (Supplementary Fig. 6 and Supplementary Tables 2–4). The results are displayed as black bars in Fig. 3b,c, respectively, by using a GTR-20XAYU

Analyzer (GTR Tech Corporation) at a differential pressure of 150 kPa and 293 K. Gas permeation was monitored by gas chromatography with a thermal conductivity detector on a GC-2014 Gas Chromatograph (Shimadzu Corporation).

Active gas flow control. Single crystals of **1** ($0.33 \times 0.19 \times 0.15\ mm^3$ for the test of Fig. 4d and $0.36 \times 0.10 \times 0.10\ mm^3$ for the test of Fig. 4e) were placed on a cylindrical glass base (diameter: 2 mm; inner diameter: 0.4 mm), which was equipped with a pair of four stainless-steel needles (diameter: 30 μm; spacing of the pushing positions: 100 μm) for applying force to the crystal membrane (Supplementary Table 5). Glass capillaries (inner diameter: 35 μm) were attached to the crystal by using silylated urethane elastomers and were also connected to gauge capillaries (inner diameter: 65 or 28 μm) including silicone oil (Shin-Etsu Chemical Co., Ltd) as a probe medium for CO_2 gas permeation. Permeability was calculated with a differential partial pressure of 100 kPa in the CO_2/membrane/air system. The movement of silicone oil and martensitic transition were observed separately under two microscopes, while making videos at 60 fps.

References

1. Laser, D. J. & Santiago, J. G. A review of micropumps. *J. Micromech. Microeng.* **14,** R35–R64 (2004).
2. Oh, K. W. & Ahn, C. H. A review of microvalves. *J. Micromech. Microeng.* **16,** R13–R39 (2006).
3. Terry, S. C., Jerman, J. H. & Angell, J. B. A gas chromatographic air analyzer fabricated on a silicon wafer. *IEEE Trans. Electron Devices* **26,** 1880–1886 (1979).
4. Sato, K. & Shikida, M. An electrostatically actuated gas valve with an S-shaped film element. *J. Micromech. Microeng.* **4,** 205–209 (1994).
5. Marinescu, D. Low dead volume piezoelectric valve. *Rev. Sci. Instrum.* **61,** 1749–1750 (1990).
6. Garstecki, P., Gañán-Calvo, A. M. & Whitesides, G. M. Formation of bubbles and droplets in microfluidic systems. *Bull. Pol. Ac.: Tech.* **53,** 361–372 (2005).

7. Unger, M. A., Chou, H.-P., Thorsen, T., Scherer, A. & Quake, S. R. Monolithic microfabricated valves and pumps by multilayer soft lithography. *Science* **288**, 113–116 (2000).

8. Sackmann, E. K., Fulton, A. L. & Beebe, D. J. The present and future role of microfluidics in biomedical research. *Nature* **507**, 181–189 (2014).

9. Liang, X. & Chou, S. Y. Nanogap detector inside nanofluidic channel for fast real-time label-free DNA analysis. *Nano Lett.* **8**, 1472–1476 (2008).

10. Takamizawa, S., Takasaki, Y. & Miyake, R. Single crystal membrane for anisotropic and efficient gas permeation. *J. Am. Chem. Soc.* **132**, 2862–2863 (2010).

11. Takasaki, Y. & Takamizawa, S. Gas permeation in a molecular crystal and space expansion. *J. Am. Chem. Soc.* **136**, 6806–6809 (2014).

12. Takamizawa, S., Kojima, K. & Akatsuka, T. Channel-switching crystal with guest stress drive. *Inorg. Chem.* **45**, 4580–4582 (2006).

13. Takasaki, Y. & Takamizawa, S. Reversible crystal deformation of a single-crystal host of copper(II)1-naphthoate—pyrazine through crystal phase transition induced by methanol vapor sorption. *Chem. Commun.* **51**, 5024–5027 (2015).

14. Jones, J. T. A. *et al.* On–off porosity switching in a molecular organic solid. *Angew. Chem. Int. Ed.* **50**, 749–753 (2011).

15. Takamizawa, S. & Miyamoto, Y. Superelastic organic crystals. *Angew. Chem. Int. Ed.* **53**, 6970–6973 (2014).

16. Takamizawa, S. & Takasaki, Y. Superelastic shape recovery of mechanically twinned 3,5-difluorobenzoic acid crystals. *Angew. Chem. Int. Ed.* **54**, 4815–4817 (2015).

Acknowledgements

This work was partially supported by the Grant for Strategic Research Project of YCU, IZUMI Science and Technology Foundation, SUZUKI Foundation and IKEYA Science and Technology Foundation.

Author contributions

Y.T. performed experiments, analysed data and wrote the paper. S.T. designed the study and wrote the paper.

Additional information

Accession code: The X-ray crystallographic data reported in this paper have been deposited at the Cambridge Crystallographic Data Centre (CCDC), under deposition number CCDC 1421863 and 1421864. These data can be obtained free of charge from The Cambridge Crystallographic Data Centre via www.ccdc.cam.ac.uk/data_request/cif.

Competing financial interests: The authors declare no competing financial interests.

11

Emergence of californium as the second transitional element in the actinide series

Samantha K. Cary[1], Monica Vasiliu[2], Ryan E. Baumbach[3], Jared T. Stritzinger[1], Thomas D. Green[1], Kariem Diefenbach[1], Justin N. Cross[1], Kenneth L. Knappenberger[1], Guokui Liu[4], Mark A. Silver[1], A. Eugene DePrince[1], Matthew J. Polinski[1], Shelley M. Van Cleve[5], Jane H. House[1], Naoki Kikugawa[6], Andrew Gallagher[3], Alexandra A. Arico[1], David A. Dixon[2] & Thomas E. Albrecht-Schmitt[1]

A break in periodicity occurs in the actinide series between plutonium and americium as the result of the localization of 5f electrons. The subsequent chemistry of later actinides is thought to closely parallel lanthanides in that bonding is expected to be ionic and complexation should not substantially alter the electronic structure of the metal ions. Here we demonstrate that ligation of californium(III) by a pyridine derivative results in significant deviations in the properties of the resultant complex with respect to that predicted for the free ion. We expand on this by characterizing the americium and curium analogues for comparison, and show that these pronounced effects result from a second transition in periodicity in the actinide series that occurs, in part, because of the stabilization of the divalent oxidation state. The metastability of californium(II) is responsible for many of the unusual properties of californium including the green photoluminescence.

[1] Department of Chemistry and Biochemistry, Florida State University, Tallahassee, Florida 32306, USA. [2] Department of Chemistry, The University of Alabama, Tuscaloosa, Alabama 35487, USA. [3] National High Magnetic Field Laboratory, Tallahassee, Florida 32310, USA. [4] Chemical Sciences and Engineering Division, Argonne National Laboratory, Argonne, Illinois 60439, USA. [5] Nuclear Materials Processing Group, Oak Ridge National Laboratory, Oak Ridge, Tennessee 37830, USA. [6] National Institute for Materials Science, Tsukuba, Ibaraki 305-0047, Japan. Correspondence and requests for materials should be addressed to T.E.A.-S. (email: albrecht-schmitt@chem.fsu.edu).

Advances in theory coupled with sophisticated spectroscopic and structural analyses of actinide complexes and materials have transformed the way in which we view these elements from what was once considered mundane to utter fascination[1–4]. While tantalizing evidence that 5f elements might utilize their valence orbitals in bonding was uncovered shortly after the Manhattan Project in the 1950s (ref. 5), it was not until more compelling techniques were applied to this problem that the broader chemical community started to become aware that a simplistic description of the 5f series as being ostensibly the same as that of lanthanides is indefensible. Among the more convincing probes of the nature of bonding in these compounds is ligand K-edge XANES[6], which when coupled with advanced electronic structure modelling has revealed that the metal–ligand interactions can be quite similar to those of d orbital interactions in transition metal coordination complexes[2,7]. However, the majority of this progress is restricted to early actinides for quite practical reasons that include a lack of structural and spectroscopic data from mid- to late actinides, and increasing difficulties in quantum mechanical calculations that come with the larger number of 5f electrons and variety of available acceptor orbitals that include the 6d, 7s and 7p.

Californium is the last element in the periodic table where it is possible to measure the properties of a bulk sample; albeit these measurements come with considerable experimental challenges[8]. Nevertheless, one of the hallmarks of the nuclear era was the maturation of ultramicrochemical techniques that had been perfected by the time californium became available in appreciable quantities. This enabled characterization of binary compounds such as halides and oxides[9], and even in these systems peculiarities were observed that include abnormally broad f-f transitions in electronic absorption spectra, and consistently reduced magnetic moments with respect to that calculated for the free ion[10,11]. A testament to the unexpected features of californium compounds is that CfCp₃ (Cp = cyclopentadienyl) has a deep red colouration instead of the expected bright green[12]. These deviations from isoelectronic Dy(III) compounds point to changing chemical behaviour and alterations in electronic characteristics as the result of complexation that are not paralleled by lanthanides or, more importantly, by lighter actinides.

There are several lines of reasoning that shed some light on the departure of electronic behaviour late in the actinide series. The first of these arguments is electrochemical (that is, thermodynamic). The increasing stability of the 3 + oxidation state among heavier 5f elements is typically ascribed to the contraction, localization and lowering in energy of the 5f orbitals, such that by americium oxidation states beyond 3 + are difficult to achieve. What is seldom recognized is that the divalent oxidation state is also becoming increasingly thermodynamically accessible late in the actinide series[13]. In fact, the solution chemistry of nobelium could not be explained until it was determined that unlike earlier 5f elements, its most stable oxidation state in aqueous media is 2 + because ostensibly this provides a closed-shell 5f¹⁴ configuration[14]. Californium is the first element in the actinide series where the 2 + oxidation state is chemically accessible at reasonable potentials, and, in fact, Cf(II) compounds, such as CfCl₂, have been successfully prepared and characterized[15].

Further support for the stability of Cf(II) comes from ambiguity in the valency of metallic californium. While the connection of the behaviour of elements in their metallic state with that of their ions in solution may seem tenuous, in this case many parallels can be discerned. Early reports indicated that californium acted as a divalent metal much like europium and ytterbium[16,17]. However, it is now understood that thin films of californium trap a metastable divalent state, whereas the bulk material is trivalent as expected[18]. Similar behaviour is found with samarium, which possesses a reduction potential similar to that of californium. The coupling of these two lines of evidence indicates that californium is located at an electronic tipping point in the actinide series, and it plays a role much like that of plutonium where an earlier departure in 5f character takes place. In the case of plutonium, its unique electronic properties that stem from the changing roles of the 5f orbitals allow it to undergo six phase transitions before melting, and it can simultaneously equilibrate four oxidation states in solution; both of these features are unmatched anywhere else in the periodic table.

We recently reported on the preparation, structure, properties and results of quantum chemical calculations of Cf[B₆O₈(OH)₅] (ref. 8). The structure of this compound is not paralleled by the lighter actinides and, more importantly, the electronic properties are rather unexpected from an f element. These unusual features include broad f-f transitions, strongly vibronically coupled photoluminescence and a massive reduction in the magnetic moment with respect to that calculated for the free ion. All of these features point to complexation perturbing the 5f orbitals in a way that is more typical of ligand-field effects on transition metal ions. In fact, density functional theory (DFT) and multi-reference molecular orbital calculations support the donation of electron density from the borate ligands into the 5f orbitals, as well as the 6d, 7s and 7p, and the latter method reveals strong ligand-field splitting that is among the largest observed for an f element.

However, this prior study[8] opened up as many questions as it answered. First, are these perturbations of the ground and excited states of Cf(III) unique to the highly electron-rich environment that borate provides, or are these effects achievable with much simpler and better understood ligands? Second, is it possible that indicators of these effects have been observed since the earliest developments of californium chemistry (vide supra), but simply not recognized for their significance? In the present study, we answer these questions and provide a hypothesis to explain why this chemical behaviour is not observed earlier in the actinide series, and why we predict that these effects will only become more pronounced later in the heaviest 5f elements.

Results

Synthesis and characterization. The reactions of hydrous ²⁴³AmCl₃, ²⁴⁸CmCl₃ and ²⁴⁹CfCl₃ with excess 2,6-pyridinedicarboxylic acid (dipicolinic acid, DPA) at 150 °C in a 1:1 ethanol/H₂O mixture results in the formation of crystals of An(HDPA)₃·H₂O (An = Am, Cm, Cf). These compounds are isomorphous and consist of nine-coordinate, tricapped trigonal prismatic An(III) ions bound by three, tridentate, monoprotonated DPA ligands. These tris-chelate complexes are necessarily chiral, and the structure, as determined by single crystal X-ray diffraction, reveals a racemic mixture of the Δ and Λ enantiomers, as expected. A view of both enantiomers is shown in Fig. 1. Crystallographic details are provided in Supplementary Table 1. Crystallographic information files (CIFs) are available in Supplementary Data 1–3.

Examination of the volume changes of the unit cells across the actinide series show the expected reduction ascribed to the actinide contraction (Supplementary Table 1). In fact, we collected high-angle diffraction data to reduce the s.d. values in the bond distances as much as possible. Inspection of the average Am–O, Cm–O, and Cf–O bond distances, as well as the An–N distances, reveals a decrease of ∼0.05 Å from Am(III) to Cf(III) as provided in Tables 1 and 2 (see also Supplementary Table 2).

The expected contraction between neighbouring f elements is on the order of 0.01 Å. Therefore, a decrease in average bond lengths of ~ 0.03 Å between Am(III) and Cf(III) is expected if a fully ionic model is imposed. The Cf–O and Cf–N bonds are slightly shorter than expected, and this might be an indicator of increased effects of covalency across the series[19,20]. However, this system presents a unique opportunity in that both enantiomers are present in the asymmetric unit. A comparison of the bond distance variations between enantiomers of the same element reveals that the average differences between bond lengths between the Δ and Λ enantiomers of the Cf(III) complex are on the same order as the differences observed in the bond lengths between the Am(III) and Cf(III) molecules. However, the entire structure is a part of a hydrogen bonding network, and the hydrogen bond contacts for the Δ and Λ enantiomers are not the same. Hence, these interactions cause minor distortions of the different enantiomers. If one instead compares the same enantiomers between the Am(III), Cm(III) and Cf(III) complexes, the Cf–N

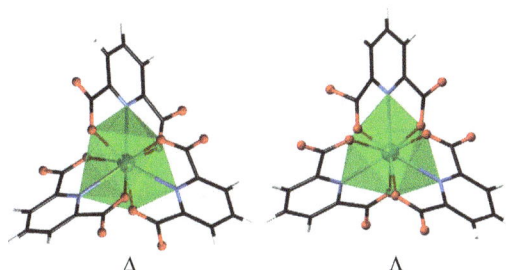

Figure 1 | Views of the Δ and Λ enantiomers of Cf(HDPA)$_3 \cdot$ H$_2$O. Views of the Δ and Λ enantiomers of Cf(HDPA)$_3 \cdot$ H$_2$O showing the nine-coordinate, tricapped trigonal prismatic coordination environments of the Cf(III) ions created via chelation by three, monoprotonated, 2,6-dipicolinate ligands, HDPA$^-$. The coordination environment of the Cf(III) centre in the Δ enantiomer is more distorted than in Λ enantiomer because of differences in hydrogen bonding with co-crystallized water molecules.

and Cf–O bond distances are slightly shorter than anticipated; albeit the statistical significance of these differences depends on how one treats the errors in the bond distances. It must also be kept in mind that berkelium lies between curium and californium, but preparation of the Bk(III) complex is not currently possible owing to the half-life of ^{249}Bk being only 320 days.

There are more convincing indicators that the electronic structure of the Cf(III) complex deviates significantly from expectations. For example, magnetic susceptibility data were collected from 300 to 1.8 K for Cf(HDPA)$_3 \cdot$ H$_2$O under an applied field of 0.1 T as shown in Fig. 2. The large temperature-independent paramagnetic effects that are indicative of low-lying excited states observed for Cf[B$_6$O$_8$(OH)$_5$] are not found in Cf(HDPA)$_3 \cdot$ H$_2$O. Instead the data are essentially Curie-Weiss like with a measured μ_{eff} of 9.3(1) μ_B. This value is significantly lower than that calculated for the free-ion of 10.65 μ_B (ref. 21). Examination of previously reported magnetic moments for Cf(III) compounds shows that values of ~ 9.3 μ_B are, in fact, the most commonly reported. However, most investigators have attributed these deviations from the calculated free-ion moment to experimental artefacts that result from the very small sample sizes employed in the measurements (~ 1 µg of ^{249}Cf), and hence large errors in the masses of ^{249}Cf (refs 10,11). It should be pointed out, however, that the quantity of ^{249}Cf used in these prior studies was not determined by weighing, but rather by far more accurate radiation counting methods, and later studies conducted with tens of micrograms of ^{249}Cf yielded similar values as those with much smaller quantities[10,11].

To demonstrate that the reduced moment measured for Cf(HDPA)$_3 \cdot$ H$_2$O is not an artefact of the small sample size, our measurements were performed on the largest quantity of ^{249}Cf ever used in a magnetic susceptibility study (2.2 mg of ^{249}Cf), and the data were collected using a VSM-SQUID, which provides significantly more sensitivity than a traditional SQUID. We conclude that the earlier reports are probably correct, and that Cf(III) commonly displays reduced magnetic moments unlike isoelectronic Dy(III), whose measured μ_{eff} from a variety of compounds are typically close to the calculated free-ion

Table 1 | Selected Bond lengths (Å) for An(HDPA)$_3$ (An = Am, Cm, Cf) complexes

	Am Δ	Cm Δ	Cf Δ		Am Λ	Cm Λ	Cf Λ
O13	2.472 (3)	2.462 (3)	2.455 (4)	O1	2.507 (3)	2.496 (3)	2.476 (4)
O14	2.491 (3)	2.483 (3)	2.443 (4)	O2	2.441 (3)	2.433 (3)	2.413 (4)
O15	2.430 (3)	2.417 (3)	2.387 (4)	O3	2.405 (3)	2.389 (3)	2.363 (3)
O16	2.468 (3)	2.459 (3)	2.422 (3)	O4	2.512 (3)	2.501 (3)	2.476 (3)
O17	2.516 (3)	2.508 (3)	2.494 (4)	O5	2.494 (3)	2.480 (3)	2.441 (3)
O18	2.499 (3)	2.481 (3)	2.411 (4)	O6	2.457 (3)	2.448 (3)	2.417 (4)
N1	2.550 (4)	2.533 (4)	2.512 (4)	N2	2.556 (3)	2.520 (4)	2.518 (4)
N3	2.551 (4)	2.536 (4)	2.508 (4)	N4	2.531 (4)	2.535 (4)	2.506 (4)
N5	2.591 (3)	2.581 (4)	2.545 (4)	N6	2.573 (4)	2.569 (4)	2.526 (4)

Table 2 | Comparison of bond lengths between the Δ and Λ enantiomers of An(HDPA)$_3$.

	Am Δ (Å)	Cm Δ (Å)	Cf Δ (Å)	Am Λ (Å)	Cm Λ (Å)	Cf Λ (Å)
Longest An-O	2.515 (4)	2.509 (4)	2.494 (4)	2.519 (4)	2.501 (4)	2.477 (4)
Shortest An-O	2.431 (4)	2.416 (4)	2.386 (4)	2.401 (4)	2.388 (4)	2.363 (4)
Average An-O	2.482 (4)	2.468 (4)	2.436 (4)	2.468 (4)	2.457 (4)	2.431 (4)
Longest An-N	2.589 (4)	2.582 (4)	2.545 (4)	2.573 (4)	2.566 (4)	2.526 (4)
Shortest An-N	2.551 (4)	2.532 (4)	2.508 (4)	2.532 (4)	2.519 (4)	2.506 (4)
Average An-N	2.564 (4)	2.550 (4)	2.522 (4)	2.554 (4)	2.540 (4)	2.517 (4)

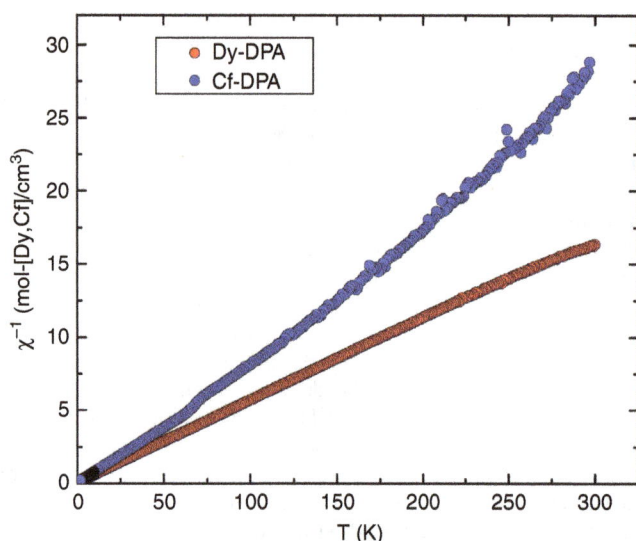

Figure 2 | Inverse magnetic susceptibility as a function of temperature.
Inverse magnetic susceptibility of polycrystalline samples of
$Dy(HDPA)_3 \cdot H_2O$ and $Cf(HDPA)_3 \cdot H_2O$ as a function of temperature.
The measured μ_{eff} of the $Dy(HDPA)_3 \cdot H_2O$ sample reaches the theoretical
free-ion moment of 10.65 μ_B expected for an f^9 system, whereas the Cf(III)
compound shows a reduced μ_{eff} of 9.3(1) μ_B.

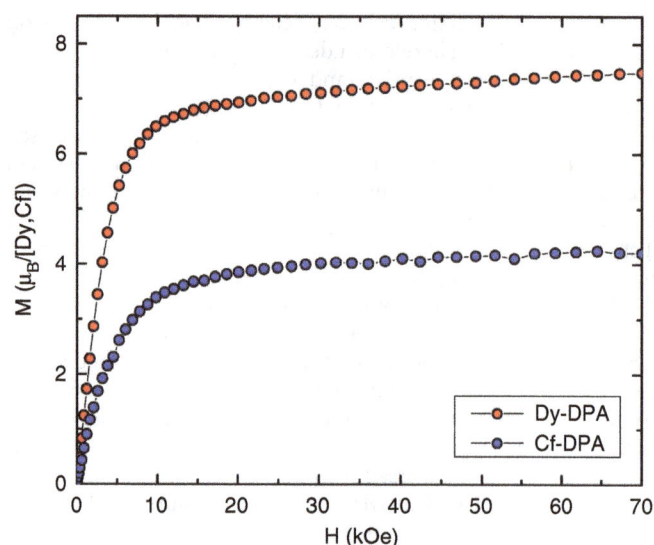

Figure 3 | Magnetization as a function of magnetic field. Magnetization of
polycrystalline samples of $Dy(HDPA)_3 \cdot H_2O$ ($4f^9$) and $Cf(HDPA)_3 \cdot H_2O$
($5f^9$) as a function of magnetic field. The Cf(III) sample has a μ_{sat} value
approximately half of that of the Dy(III) sample. The Dy(III) sample displays
typical hard magnetism of lanthanides, whereas the Cf(III) complex shows
much softer behaviour.

moment[21]. In fact, we prepared $Dy(HDPA)_3 \cdot H_2O$, and measured
its magnetic susceptibility in the same way, to find its exhibited
moment reaches the calculated free-ion moment value, as
expected. The inverse magnetic susceptibility data are compared
in Fig. 2, and a large deviation between the two metals ions is
clearly apparent. In addition, much like what is found in
$Cf[B_6O_8(OH)_5]$, $Cf(HDPA)_3 \cdot H_2O$ behaves as a much softer
magnetic system than expected for an f-element, and
magnetization versus field measurements reveals saturation of
the Cf sample at approximately half of the value of the Dy
complex as provided in Fig. 3.

A deeper understanding of the electronic structure of the
Cf(III) complex was achieved by measuring electronic absorption
and photoluminescence spectra from single crystals. The former
data are provided in Supplementary Fig. 3, and the latter in Fig. 4.
We were fortunate that fairly large crystals of $Cf(HDPA)_3 \cdot H_2O$
can be prepared, whereas $Cf[B_6O_8(OH)_5]$ is microcrystalline.
A comparison of these spectra with An^{3+} ions in crystals with
weak ligand interactions reveals that the $5f$-$5f$ transitions are
abnormally broad even at 79 K. In addition, there is a very broad
absorption band in the short wavelength region that cannot be
attributed to $5f$-$5f$ transitions. Again, we examined earlier
published data on the absorption spectra of Cf(III) compounds,
and these features are typically at least an order of magnitude
broader than expected[22]. Photoluminescence data, which were
collected at variable-temperature using either 365 or 420 nm
excitation wavelengths, show green photoluminescence centred at
~525 nm. Historically, this photoluminescence has been assigned
to the $J = 5/2$ excited state transition to the $J = 15/2$ ground
state[8,23,24]. However, as we will delineate below, this assignment
is most likely incorrect, and the actual origin is far more
intriguing. Much like in $Cf[B_6O_8(OH)_5]$, strong vibronic coupling
is observed, and the photoluminescence peak width at half-height
is massive at ~126 nm. In $Cf(HDPA)_3 \cdot H_2O$, the vibrational
progression is more clearly resolved than in $Cf[B_6O_8(OH)_5]$,
perhaps because the former is a single crystal sample.

In both $Cf[B_6O_8(OH)_5]$ and $Cf(HDPA)_3 \cdot H_2O$, photo
luminescence of the daughter of ^{249}Cf α decay, ^{245}Cm, is
also observed. In the $Cf[B_6O_8(OH)_5]$ sample, the Cm(III)

photoluminescence peak is nearly as broad as that of Cf(III),
leaving unanswered whether the highly electron-rich coordination environment that borate provides is solely responsible for the
changes in electronic structure, or whether these features only
occur because the Cm(III) is effectively being doped into the
Cf(III) sample, or whether it is some combination of the two.
In $Cf(HDPA)_3 \cdot H_2O$, the photoluminescence of $^{245}Cm(III)$
at 611 nm is a sharp transition as expected for a Cm(III)
complex[25-27]. We also prepared the pure $Cm(HDPA)_3 \cdot H_2O$
complex using a $^{248}Cm(III)$ starting material, and fully
characterized this sample, including its variable-temperature
photoluminescence spectra as illustrated in Fig. 5. As
anticipated, the $Cm(HDPA)_3 \cdot H_2O$ compound produces a
single sharp emission line at 611 nm and vibronic coupling is
absent. This supports the postulate that Cf(III) represents a
transition point in the actinide series where emergent phenomena
are apparent. Buttressing of this argument is also provided by the
measured μ_{eff} of the Cm(III) complex being 8.0(1) μ_B, which
agrees well the calculated moment of 7.94 μ_B (refer to
Supplementary Fig. 6), and with magnetic moments measured
from other Cm(III) compounds[28]. The absorption spectrum is
also typical for a Cm(III) material (refer to Supplementary Fig. 4).

Crystal-field analysis. Crystal-field analysis allows for the electronic energy levels of Cf(III) in $Cf(HDPA)_3$ to be calculated, and
the resultant states are compared with the absorption spectrum as
shown in Fig. 6 (ref. 29). The free-ion parameters were taken
from previously obtained values for An(III) doped into $LaCl_3$
crystals. Assuming C_{3v} site symmetry for Cf(III), the crystal-field
parameters were first calculated using a superposition model from
ligand-field theory[30]. The values of the calculated crystal-field
parameters were varied proportionally from the calculated ones
together with the spin-orbit coupling constant ζ_{5f} to obtain the
best fit with the experimental data. Detailed analysis of the
electronic energy levels and crystal-field calculations for Cf(III)
and Am(III) in $An(HDPA)_3$ will be reported separately. The
calculated energy levels are indicated with the leading free-ion
states by the vertical lines on the top of the spectrum shown in
Fig. 6. These calculated levels match the experimental data quite

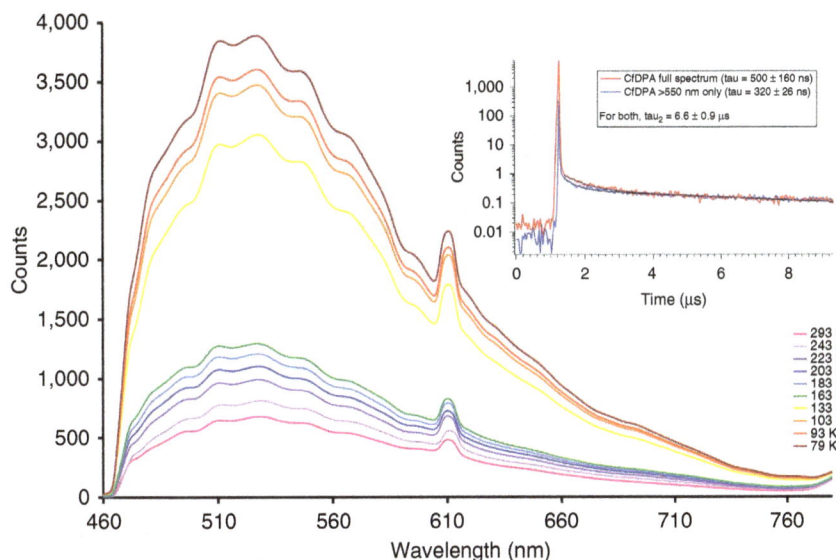

Figure 4 | Photoluminescence spectra of Cf(HDPA)$_3 \cdot$ H$_2$O. Photoluminescence spectra from a single crystal of Cf(HDPA)$_3 \cdot$ H$_2$O on excitation with 420 nm light as a function of temperature. The emission from Cf(III) is centred at 525 nm, whereas the emission from the ^{245}Cm(III) daughter occurs at 611 nm. Strong vibronic coupling is found for Cf(HDPA)$_3 \cdot$ H$_2$O, but not with Cm(HDPA)$_3 \cdot$ H$_2$O. Inset shows the decay lifetimes of 500 ± 160 ns for Cf(III) and 320 ± 26 ns for the Cm(III) daughter.

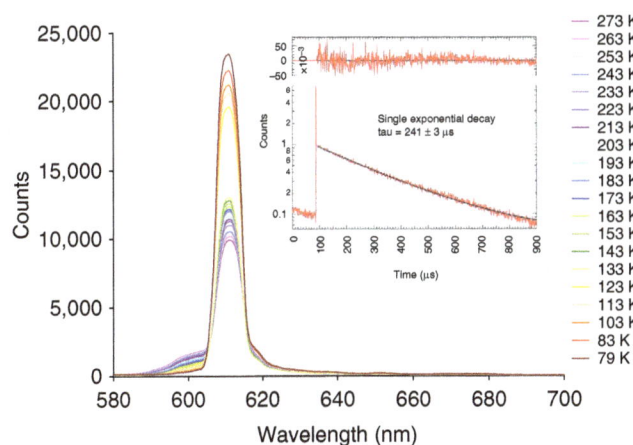

Figure 5 | Photoluminescence spectra of ^{248}Cm(HDPA)$_3 \cdot$ H$_2$O.
Photoluminescence spectra from a single crystal of ^{248}Cm(HDPA)$_3 \cdot$ H$_2$O on excitation with 420 nm light as a function of temperature. The photoluminescence from the Cm(III) complex is centred at 611 nm as found when it is doped into the Cf(III) compound in the form of the ^{245}Cm daughter. The inset shows the decay lifetime of 241 ± 160 μs, which is much longer than that found in the Cf(III) sample. This lifetime is typical of Cm(III) compounds, and is substantially shortened by the rapid creation of colour centres in the Cf(III) sample because of radiation damage.

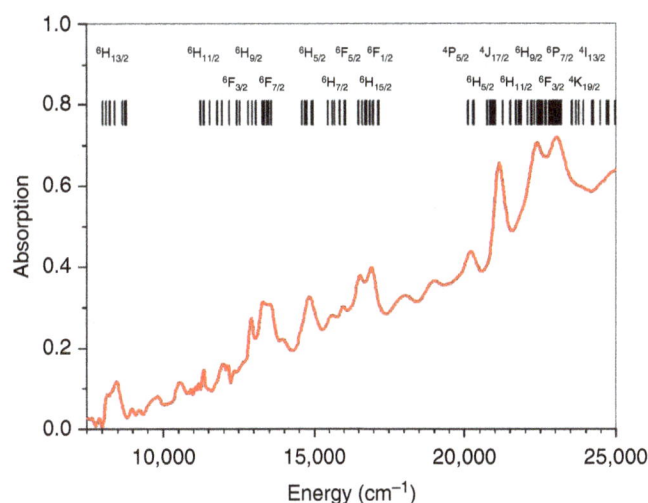

Figure 6 | Absorption spectrum. Absorption spectrum of a single crystal of Cf(HDPA)$_3 \cdot$ H$_2$O at 83 K. The narrower bands are due to 5f^9-5f^9 transitions. Calculated using an effective-operator Hamiltonian, the crystal-field splittings of the 5f^9 states between 8,000 and 25,000 cm^{-1} are marked by the vertical bars along with the leading SLJ multiplets. Several weaker bands with energy levels at 9,700, 10,570, 17,900 and 18,960 cm^{-1} are not predicted from the crystal-field calculation, and are thus attributed to vibronic structures coupled to charge-transfer transition.

well. The resultant crystal-field parameters are reduced by 20% from the calculated values. They are $B_0^2 = 588$ cm^{-1}, $B_0^4 = 2,820$ cm^{-1}, $B_3^4 = -1,932$ cm^{-1}, $B_0^6 = 3,387$ cm^{-1}, $B_3^6 = 580$ cm^{-1} and $B_6^6 = 2,154$ cm^{-1}. The spin-orbit coupling constant is $\zeta_{5f} = 3536$ cm^{-1}, which is only 1% less than that for Cf(III):LaCl$_3$. The ligand-field strength (that is, N'_v)[31] in Cf(HDPA)$_3 \cdot$ H$_2$O calculated from the crystal-field parameters is 1,632 cm^{-1}, which is much larger than the strength of 610 cm^{-1} found for Cf(III) in a chloride environment (that is, doped into a LaCl$_3$ lattice)[31]. The crystal-field splitting of the ground multiplet ^{6}H$_{15/2}$ is 824 cm^{-1}. For comparison, the total splitting of ^{6}H$_{15/2}$ for the 4f^9 ion Dy(III) in LaCl$_3$ is only 141 cm^{-1} (ref. 32) The increased ground-state splitting would significantly reduce the magnetic susceptibility (μ_{eff}).

A compilation of all of the electronic spectroscopy and magnetism data gathered for this work, as well as that measured from Cf[B$_6$O$_8$(OH)$_5$][8], allows us to reach the following conclusions. (1) Under 365 nm excitation, the origin, or zero-phonon line of the photoluminescence of Cf(HDPA)$_3 \cdot$ H$_2$O is above 25,000 cm^{-1} (400 nm). However, as shown in Fig. 6, this is far above the expected energy level of the $J = 5/2$ emitting state (dominated by the ^{4}P$_{5/2}$ level with an energy near 20,000 cm^{-1})[31]. A large hypsochromic shift of the 5f-5f transitions is not observed in the absorption spectrum. (2) There is a substantial Stokes shift and photoluminescence band broadening with strong vibronic features. In both Cf[B$_6$O$_8$(OH)$_5$] and Cf(HDPA)$_3 \cdot$ H$_2$O, the californium photoluminescence band stretches more than $10,000 \, \text{cm}^{-1}$ across the narrow emission from the Cm(III) daughter centred at 611 nm. If the photoluminescence is due to a 5f-5f transition, it should have the same bandwidth as that of the 5f-5f absorption bands. In addition, no photoluminescence is observed from the corresponding Am(III) complex. This indicates that radiative 5f-5f relaxation is quenched even from the metastable $5f^6$ state ($^5D_1$) of Am(III). This state of Am(III) has an energy gap between $^5D_1$ and $^7F_6$ that is much larger than the difference between the Cf(III) $^4P_{5/2}$ state and the next low-lying state $^6F_{1/2}$. According to the energy-gap law[33], if photoluminescence occurs from the $^4P_{5/2}$ state of Cf(III), one should also see red $^5D_1$ emission from the Am(III) complex.

The crystal-field strength is often used as an index of covalency in actinide compounds. Earlier work conducted by Edelstein and co-workers demonstrated that N'_v strongly depends on the oxidation state of the actinide ion and the valence orbitals of its ligands[34–36]. Summarized in Supplementary Fig. 9, the value of N'_v varies from $600 \, \text{cm}^{-1}$ for An(III) in chlorides and bromides to above $7,000 \, \text{cm}^{-1}$ for An(V) in fluorides. The value of N'_v is insensitive to the electronic configuration of ions in the same oxidation states as was realized in the An(III):LaCl$_3$ and An(IV):CeF$_4$ series[37,38], with values of N'_v close to 600 and $2,000 \, \text{cm}^{-1}$, respectively, for actinides across the series. Systematically, once N'_v is above $2,000 \, \text{cm}^{-1}$, covalency is thought to be dominant in ion-ligand bonding. For instance, the value of N'_v for U(V) in (NEt$_4$)UCl$_6$ and (NEt$_4$)UBr$_6$ is between 3,000 and $4,000 \, \text{cm}^{-1}$, and these uranium complexes are considered to have predominantly covalent bonding[39]. For Cf(HDPA)$_3 \cdot$H$_2$O, N'_v is $1,623 \, \text{cm}^{-1}$, which is much larger than that for trivalent actinides in many compounds, and suggests the presence of some covalency in the bonding. This result supports the ligand hyperpolarizability arguments that we have previously put forth as well as the change in periodicity at californium[3]. The analysis of the crystal-field strength and $5f^9$ energy levels supports our assignment of the broadband photoluminescence to a charge-transfer transition instead of an intra 5f transition.

Given the above observations and analysis, there is no basis to assign the photoluminescence to a Cf(III) 5f-5f transition. This calls into question whether the assignment of the green photoluminescence has ever been correct, and we are unable to find a photoluminescence spectrum in the literature that is clearly indicative of a 5f-5f transition. The assignment most likely stems from an extension of the self-luminescence of short-lived ^{242}Cm and ^{244}Cm compounds, whose orange luminescence is exactly the same as that found from exciting long-lived ^{248}Cm(III) compounds, where the photoluminescence is clearly 5f in origin. A mechanism that provides a satisfying interpretation of the spectroscopic data and is also consistent with the magnetic and thermodynamic studies is that the photoluminescence is from a ligand-to-metal charge-transfer transition that is best described as a Cf(III) to Cf(II) + h$^+$ (h$^+$ = hole in the valence band) occurring in the energy region at $\sim 25,000 \, \text{cm}^{-1}$ (400 nm). Transitions that correspond to this charge-transfer are clearly apparent in the absorption spectrum (see Supplementary Fig. 1). A photon is emitted when the hole recombines with Cf(II) returning it to the Cf(III) ground state. Charge-transfer transitions are often strongly coupled to vibronic interactions. These vibronic progressions lead to significantly broader photoluminescence bands than observed for 5f-5f transitions[40]. As previously discussed, starting at californium, the divalent state becomes metastable. The mechanism of charge-transfer photoluminescence is the recombination of Cf(II) with the hole created in the ligand valence band. This occurs partially through a radiative process to the ground state of Cf(III) without complete

non-radiative phonon cross relaxation into the low-lying Cf(III) excited states. To further support this mechanism of photoluminescence, we also prepared the californium sulfate fluoride, KCf$_2$(SO$_4$)$_2$F$_3$, where it would be expected that creating a hole in the ligands would be energetically challenging. This compound is not luminescent, even at low temperatures.

Discussion

All of the data disclosed in this report support Cf(III) complexes displaying emergent phenomena that cannot be predicted from either extrapolating from its isoelectronic lanthanide analogue, Dy(III), or from other trivalent actinides[3,8,41]. A juxtaposition of factors that include the relative ease of reducing Cf(III) to Cf(II) and presence of available 5f, 6d, 7s and 7p acceptor orbitals creates an unusually large ligand-field strength that significantly alters the electronic properties anticipated for an f-element[8]. The atypical features include a reduced magnetic moment and vibronically coupled, charge-transfer-based photoluminescence. Reduction in the magnetic moment of transition metal and f-element complexes can occur via a variety of mechanisms that include the size of the crystal-field splitting of the ground state, spin-orbit coupling, and delocalization of electrons from the metal ions onto the ligands (that is, covalent bonding and ligand-field effects)[42,43]. Crystal-field theory is based purely on electrostatics, but is adequate for lanthanides because the 4f orbitals lie within the xenon core and are effectively nonbonding. We have shown here that the moment reductions can arise from several factors. In the case of the Cm(III) complex, the magnetic susceptibility is normal for a $5f^7$ system[28]. However, for the Cf(III) complex the reduced magnetic moment is caused by ligand-field effects on the 5f electrons that are quite pronounced as indicated by a large ligand-field strength and notable splitting of the ground state.

We offer the following hypothesis that both explains these observations and predicts the outcome of future studies with late actinides: the alterations in californium's physical and chemical properties are caused by ligand-field effects that manifest because of the relative ease with which formally Cf(III) can be reduced to Cf(II)[13]. Fajan's Rules also support that the smaller size of the Cf(III) ion relative to earlier actinides should create greater polarization of the electron density from the ligands to the metal centre[44,45]. Therefore, these perturbations will become more pronounced with later actinides because the 2 + oxidation state becomes progressively more stable and the metal ions continue to diminish in size. This combination of thermodynamics and ion size overrides the minor contraction of the 5f orbitals in the actinide series. In fact, even in biphasic extraction studies used for separating 5f elements from one another, a marked break is observed between Cm(III) and Cf(III) that supports californium's chemistry deviating from earlier actinides regardless of whether it is in the solid-state or a coordination complex in solution[46]. While the data are still scant, californium consistently shows larger binding constants than would be anticipated[47]. Gathering the data needed to bolster this hypothesis will require renewed production of einsteinium and heavier elements. We predict that complexes of einsteinium, fermium and mendelevium will show greater perturbations from their isoelectronic lanthanide analogues than even californium because of the increasing stability of the divalent state and the involvement of valence orbitals in bonding.

Methods

Experimental. ^{243}Am ($t_{1/2} = 7.38 \times 10^3$ years) and ^{248}Cm ($t_{1/2} = 3.48 \times 10^5$ years) represent potential health risks owing to their α and γ emission, and the emission of their daughters. ^{243}Am decays to ^{239}Np ($t_{1/2} = 2.35$ days) which is a β- and γ-emitter; ^{248}Cm decays to ^{244}Pu ($t_{1/2} = 8.08 \times 10^7$ years) as well as undergoing spontaneous fission (which accounts for 8.3% of its decay) releasing a large flux of neutrons that can have a specific activity of $\sim 100\,\mathrm{mRem\,h^{-1}}$ for the sample size used. ^{249}Cf ($t_{1/2} = 351$ years; specific activity $= 4.1\,\mathrm{Ci/g}$) represent a serious external hazard because of its γ (0.388 MeV) emission. ^{249}Cf decays to ^{245}Cm ($t_{1/2} = 8{,}500$ years), which also has a high specific activity. All studies with transuranium elements were conducted in a laboratory dedicated to these studies. This laboratory is equipped with HEPA-filtered hoods and negative pressure glove boxes that are ported directly into the hoods. A series of counters continually monitor radiation levels in the laboratory. The laboratory is licensed by the State of Florida (an NRC-compliant state). All experiments were carried out with approved safety operating procedures. All free-flowing solids are worked with in glove boxes, and products are only examined when coated with either water or Krytox oil and water. The ^{249}Cf sample used produces $1.7\,\mathrm{R\,h^{-1}}$ at 40 mm, and $\sim 10\,\mathrm{R\,h^{-1}}$ at contact, and therefore represents a serious external hazard that required the experiments to be carefully choreographed to minimize exposure times. Thick lead sheets and long lead vests were used as much as possible to shield researchers from the γ- emission.

Syntheses. All Ln(HDPA)$_3 \cdot$ H$_2$O and An(HDPA)$_3 \cdot$ H$_2$O compounds were prepared using 5 mg of the appropriate f element, which was then combined with a fivefold excess of DPA in 100 µl of a 1:1 mixture of ethanol and water. The resultant reaction mixture was heated in a PTFE-lined Parr 4749 autoclave with a 10 ml internal volume for 4 h at 150 °C, and then slowly cooled to 23 °C over a 12-h period. While reactions with the lanthanides were conducted in a standard muffle furnace in a hood, the furnaces for heating the ^{243}Am and ^{249}Cf autoclaves were inside a negative-pressure glovebox and were surrounded by thick lead sheets. The reactions result in the formation of crystals of the appropriate colours for the f elements with block and columnar habits (see Supplementary Figs 2,5 and 8).

Crystallographic studies. Single crystals of the Ln$_2$(HDPA)$_6 \cdot$ 2H$_2$O (see Supplementary Table 3) and An$_2$(HDPA)$_6 \cdot$ 2H$_2$O compounds were glued to Mitogen mounts with epoxy and optically aligned on a Bruker D8 Quest X-ray diffractometer using a digital camera. Initial intensity measurements were performed using a IµS X-ray source (MoKα, $\lambda = 0.71073\,\text{Å}$) with high-brilliance and high-performance focusing multilayered optics. Standard software was used for determination of the unit cells and data collection control. The intensities of reflections of a sphere were collected by a combination of multiple sets of exposures (frames). Each set had a different φ angle for the crystal and each exposure covered a range of 0.5° in ω. A variety of data collection strategies were employed including standard hemispheres, and more complex data sets with higher angles and greater degrees of redundancy. The SAINT software was used for data integration including Lorentz and polarization corrections. The structure was solved by direct methods and refined on F^2 by full-matrix least squares techniques using the program suite SHELX (Supplementary Figs 1 and 7). Parameters for Am, Cm and Cf are not present in the SHELX software and have to be input manually. Solutions were checked for missed symmetry using PLATON[48].

UV–vis–NIR and photoluminescence spectroscopy. UV–vis–NIR and photoluminescence data were acquired from single crystals using a Craic Technologies microspectrophotometer. Crystals were placed on quartz slides under Krytox oil, and the data were collected from 200 to 1,700 nm. The exposure time was auto optimized by the Craic software. Photoluminescence data were acquired using the same microspectrophotometer with an excitation wavelength of 365 or 420 nm (Figs 4 and 5; see also Supplementary Figs 2,3 and 5). Temperature control was achieved by using a Linkam temperature control stage. Raman measurements were also attempted, but these were impeded by the self-luminescence and rapid decomposition of the sample in laser beam.

Life-time measurements. Time-correlated single-photon counting (TCSPC) photoluminescence measurements were carried out using a femtosecond laser system. A regeneratively amplified titanium:sapphire laser system (Spectra Physics Tsunami coupled to Spitfire amplifier) produced laser pulses centred at 800 nm with a duration of 100 fs. The fundamental output was frequency doubled and subsequently attenuated to produce a 10 µJ per pulse, 400 nm excitation source for the TCSPC experiment. The photoluminescence was isolated from residual excitation pulses using long-pass filters, and detected using an avalanche photodiode (Quantique) coupled to a single photon-counting unit (Becker Hickl). The low-energy portion of the photoluminescence was isolated using a 550-nm long-pass filter. The resulting lifetime data were fit to a biexponential decay function (Figs 4 and 5).

Magnetic measurements. Magnetic measurements were performed on polycrystalline samples that were encapsulated in tightly closed PTFE sample holders with a Quantum Design SQUID magnetometer MPMS-XL or a VSM-SQUID MPMS. DC magnetic susceptibility measurements were carried out in an applied field of 0.100 T in the 1.8–300 K temperature range. Field-dependent magnetization was recorded at 1.8 K under an applied magnetic field that was varied from 0 to 7T.

The data were corrected for the diamagnetic contribution from the sample holder and constituent elements (Figs 2 and 3).

References

1. Kaltsoyannis, N. Does covalency increase or decrease across the actinide series? Implications for minor actinide partitioning. *Inorg. Chem.* **52**, 3407–3413 (2013).
2. Neidig, M. L., Clark, D. L. & Martin, R. L. Covalency in f-element complexes. *Coord. Chem. Rev.* **257**, 394–406 (2013).
3. Polinski, M. J. *et al.* Differentiating between trivalent lanthanides and actinides. *J. Am. Chem. Soc.* **134**, 10682–10692 (2012).
4. Galbis, E. *et al.* Solving the hydration structure of the heaviest actinide aqua ion known: The californium(III) case. *Angew. Chem. Int. Ed.* **49**, 3811–3815 (2010).
5. Diamond, R. M., Street, K. & Seaborg, G. T. An ion-exchange study of possible hybridized 5 f-bonding in the actinides. *J. Am. Chem. Soc.* **76**, 1461–1469 (1954).
6. Solomon, E. I., Hedman, B., Hodgson, K. O., Dey, A. & Szilagyi, R. K. Ligand K-edge X-ray absorption spectroscopy: covalency of ligand-metal bonds. *Coord. Chem. Rev.* **249**, 97–129 (2005).
7. Minasian, S. G. *et al.* New evidence for 5 f covalency in actinocenes determined from carbon K-edge XAS and electronic structure theory. *Chem. Sci.* **5**, 351–359 (2014).
8. Polinski, M. J. *et al.* Unusual structure, bonding and properties in a californium borate. *Nat. Chem.* **6**, 387–392 (2014).
9. Copeland, J. C. & Cunningham, B. B. Crystallography of the compounds of californium—II crystal structure and lattice parameters of californium oxychloride and californium sesquioxide. *J. Inorg. Nucl. Chem.* **31**, 733 (1969).
10. Haire, R. G. & Gibson, J. K. Selected systematic properties and some recent investigations of actinide metals and alloys. *J. Radioanal. Nucl. Chem.* **143**, 35–51 (1989).
11. Moore, J. R., Nave, S. E., Haire, R. G. & Huray, P. G. Magnetic susceptibility of californium oxides. *J. Less Common Met.* **121**, 187–192 (1986).
12. Laubereau, P. G. & Burns, J. H. Microchemical preparation of tricyclopentadienyl compounds of berkelium, californium, and some lanthanide elements. *Inorg. Chem.* **9**, 1091–1095 (1970).
13. Edelstein, N. M., Fuger, J., Katz, J. J. & Morss, L. R. *The Chemistry of the Actinide and Tranactinide Elements* Ch. 15, 1779 (Springer, 2006).
14. Maly, J., Sikkeland, T., Silva, R. J. & Ghiorso, A. Nobelium: tracer chemistry of the divalent and trivalent Ions. *Science* **160**, 1114–1115 (1968).
15. Peterson, J. R., Fellows, R. L., Young, J. P. & Haire, R. G. Stabilization of californium(II) in the solid-state – californium dichloride, 249CfCl2. *Radiochem. Radioanal. Lett.* **31**, 277–282 (1977).
16. Haire, R. G. & Baybarz, R. D. Crystal structure and melting point of californium metal. *J. Inorg. Nucl. Chem.* **36**, 1295–1302 (1974).
17. Haire, R. G. & Asprey, L. B. Studies on californium metal system. *J. Inorg. Nucl. Chem. Lett.* **12**, 73–84 (1976).
18. Heathman, S., Le Bihan, T., Yagoubi, S., Johansson, B. & Ahuja, R. Structural investigation of californium under pressure. *Phys. Rev. B* **87**, 214111–214118 (2013).
19. Jones, M. B. & Gaunt, A. J. Recent developments in synthesis and structural chemistry of nonaqueous actinide complexes. *Chem. Rev.* **113**, 1137–1198 (2013).
20. Jones, M. B. *et al.* Unovering f-element bonding differences and electronic structure in a series of 1:3 and 1:4 complexes with a diselenophosphinate ligand. *Chem. Sci.* **4**, 1189–1203 (2013).
21. Skanthakumar, S., Soderholm, L. & Movshovich, R. Magnetic properties of Dy in Pb$_2$Sr$_2$DyCu$_3$O$_8$. *J. Alloys Compd.* **303**, 298–302 (2000).
22. Haire, R. G. *The Chemistry of the Actinide and Transactinide Elements.* Ch. 11, 1544 (Springer, 2006).
23. Fields, P. R., Wybourne, B. G. & Carnall, W. T. The electronic energy levels of the heavy actinides Bk^{3+} (5f8), Cf^{3+} (5f9), Es^{3+} (5f10), and Fm^{3+} (5f11). Argonne National Laboratory AEC Research and Development Report (U.S. Atomic Energy Commission), ANL-6911 (1964).
24. Sykora, R. E., Assefa, Z., Haire, R. G. & Albrecht-Schmitt, T. E. The first structural determination of a trivalent californium compound with oxygen coordination. *Inorg. Chem.* **45**, 475–477 (2006).
25. Skanthakumar, S., Antonio, M. R., Wilson, R. E. & Soderholm, L. The curium aqua ion. *Inorg. Chem.* **46**, 3485–3491 (2007).
26. Lindqvist-Reis, P. *et al.* The structures and optical spectra of hydrated transplutonium ions in the solid state and solution. *Angew. Chem. Int. Ed.* **46**, 919–922 (2007).
27. Apostolidis, C. *et al.* [An(H$_2$O)$_9$][CF$_3$SO$_3$]$_3$ (An = U–Cm, Cf): Exploring their stability, structural chemistry, and magnetic behavior by experiment and theory. *Angew. Chem. Int. Ed.* **49**, 6343–6347 (2010).
28. Soderholm, L., Skanthakumar, S. & Williams, C. W. Structure and magnetic properties of the high Tc related phase Cm$_2$CuO$_4$. *Phys Rev. B* **60**, 4302–4308 (1999).

29. Liu, G. & Beitz, J. V. *The Chemistry of the Actinide and Transactinide Elements* Ch. 18, 2013–2111 (Springer, 2006).

30. Newman, D. J. & Ng, B. *Crystal Field Handbook* Ch. 5, 83–139 (Cambridge University Press, 2000).

31. Carnall, W. T. A systematic analysis of the spectra of trivalent actinide chlorides in D_{3h} site symmetry. Argonne National Laboratory Report, ANL-89/39 (1989).

32. Dieke, G. H. *Spectra and energy levels of rare earth ions in crystals* (Interscience Publishers, 1968).

33. Riseberg, L. A. & Moos, H. W. Multiphonon Orbit-Lattice Relaxation in $LaBr_3$, $LaCl_3$, and LaF_3. *Phys. Rev. Lett.* **25**, 1423–1426 (1967).

34. Hubert, S., Thouvenot, P. & Edelstein, N. Spectroscopic studies and crystal-field analyses of Am^{3+} and Eu^{3+} in the cubic-symmetry site of ThO_2. *Phys. Rev. B* **48**, 5751–5790 (1993).

35. Edelstein, N., Brown, D. & Whittaker, B. Covalency effects on ligand-field splittings of octahedral 5f1 compounds. *Inorg. Chem.* **13**, 563–567 (1974).

36. Sytama, J., Murdoch, K. M., Edelstein, N. M., Boatner, L. A. & Abraham, M. M. Spectroscopic studies and crystal-field analysis of Cm^{3+} and Gd^{3+} in $LuPO_4$. *Phys. Rev. B* **52**, 12668–12676 (1995).

37. Carnall, W. T. A systematic analysis of the spectra of trivalent actinide chlorides in D_{3h} site symmetry. *J. Chem. Phys.* **96**, 8713–8726 (1992).

38. Liu, G. K., Carnall, W. T., Jursich, G. & Williams, C. W. Analysis of the crystal-field spectra of the actinide tetrafluorides. II. AmF_4, CmF_4, $Cm^{4+}:CeF_4$ and $Bk^{4+}:Cef_4$. *J. Chem. Phys.* **101**, 8277–8289 (1994).

39. Edelstein, N., Brown, D. & Whittaker, B. Covalency effects on the ligand field splittings of octahedral 5f1 compounds. *Inorg. Chem.* **13**, 563–567 (1974).

40. Liu, G., Deifel, N. P., Cahill, C. L., Zhurov, V. V. & Pinkerton, A. A. Charge transfer vibronic transitions in uranyl tetrachloride compounds. *J. Phys. Chem. A* **116**, 855–864 (2012).

41. Kot, W. K., Edelstein, N. M., Abraham, M. M. & Boatner, L. A. Electron paramagnetic resonance of Pu^{3+} and Cf^{3+} in single crystals of LuPO4. *Phys. Rev. B* **47**, 3412–3414 (1993).

42. Castro-Rodriguez, I. *et al.* Uranium tri-aryloxide derivatives supported by triazacyclononane: Engendering a reactive uranium(III) center with a single pocket for reactivity. *J. Am. Chem. Soc.* **125**, 4565–4571 (2003).

43. Bray, T. H., Nelson, A.-G. D., Jin, G. B., Haire, R. G. & Albrecht-Schmitt, T. E. *In Situ* hydrothermal reduction of Np(VI) as a route to Np(IV) phosphonates. *Inorg. Chem.* **46**, 10959–10961 (2007).

44. Fajans, K. Struktur und Deformation der Elektronenhullen in ihrer bedeutung fur die chemischen und optischen Eigenschaften anorganischer Verbindungen. *Naturwiss* **11**, 165–172 (1923).

45. Fajans, K. & Joos, G. Molrefraktion von Ionen und Molekulen im Lichte der Atomstruktur. *Z. Phys.* **23**, 1–46 (1924).

46. Weaver, B. & Kappelman, F. A. Preferential extraction of lanthanides over trivalent actinides by monoacidic organophosphates from carboxylic acids and from mixtures of carboxylic and aminopolyacetic acids. *J. Inorg. Nucl. Chem.* **30**, 263–272 (1968).

47. Choppin, G. R. & Jensen, M. P. *Actinides in Solution: Complexation and Kinetics. The Chemistry of the Actinide and Transactinide Elements* Vol. 4 Ch. 23, 2574–2575 (Springer, 2006).

48. Spek, A. L. Single-crystal structure validation with the program PLATON. *J. Appl. Cryst.* **36**, 7–13 (2003).

Acknowledgements

This material is based on work supported by the U.S. Department of Energy, Office of Science, Office of Basic Energy Sciences, Heavy Elements Chemistry Program, under Award Number DE-FG02-13ER16414 (TEA-S) and DE-AC02-06CH11357 (GL and DAD). We are especially grateful for the assistance and supervision by the Office of Environmental Health and Safety at FSU; specifically Jason A. Johnson and Ashley L. Gray of the Office of Radiation Safety for their facilitation of these studies. D.A.D. thanks the Robert Ramsay Chair Fund of The University of Alabama for partial support. The isotopes used in this research were supplied by the U.S. Department of Energy, Office of Science, by the Isotope Program in the Office of Nuclear Physics. The ^{243}Am, ^{248}Cm and ^{249}Cf were provided to Florida State University via the Isotope Development and Production for Research and Applications Program through the Radiochemical Engineering and Development Center at Oak Ridge National Laboratory. The ^{249}Cf was purchased via the Gregory R. Choppin Chair Endowment. Magnetization measurements using the VSM SQUID MPMS were performed at the National High Magnetic Field Laboratory, which is supported by National Science Foundation Cooperative Agreement No. DMR-1157490, the State of Florida, and the U.S. Department of Energy.

Author contributions

S.K.C., J.N.C., J.T.S., M.J.P. and T.E.A.-S. conceived, designed and carried out the synthetic and crystallographic experiments. M.V., D.A.D., M.A.S. and A.E.D. aided in the development of the bonding concepts. S.K.C., J.T.S. and J.H.H. carried out variable-temperature absorption and photoluminescence experiments. G.L. analysed all electronic spectroscopy experiments. S.K.C., T.E.A.-S., N.K., K.D., A.G. and R.E.B. designed and carried out the magnetic susceptibility experiments. T.B.G. and K.L.K. carried out the photoluminescence life-time measurements. A.A.A. carried out PXRD measurements. S.M.V.C. prepared and manipulated the original stock of ^{249}Cf at ORNL. All authors discussed and co-wrote the manuscript.

Additional information

Accession codes: The X-ray crystallographic coordinates for structures reported in this Article have been deposited at the Cambridge Crystallographic Data Centre (CCDC), under deposition numbers CCDC 1028642, 1028643 and 1028646. These data can be obtained free of charge from The Cambridge Crystallographic Data Centre via www.ccdc.cam.ac.uk/data_request/cif.

Competing financial interests: The authors declare no competing financial interests.

12

Efficient hydrogen evolution in transition metal dichalcogenides via a simple one-step hydrazine reaction

Dustin R. Cummins[1,2], Ulises Martinez[1], Andriy Sherehiy[2], Rajesh Kappera[1,3], Alejandro Martinez-Garcia[2], Roland K. Schulze[4], Jacek Jasinski[2], Jing Zhang[5], Ram K. Gupta[6], Jun Lou[5], Manish Chhowalla[3], Gamini Sumanasekera[2], Aditya D. Mohite[1], Mahendra K. Sunkara[2] & Gautam Gupta[1]

Hydrogen evolution reaction is catalysed efficiently with precious metals, such as platinum; however, transition metal dichalcogenides have recently emerged as a promising class of materials for electrocatalysis, but these materials still have low activity and durability when compared with precious metals. Here we report a simple one-step scalable approach, where MoO_x/MoS_2 core-shell nanowires and molybdenum disulfide sheets are exposed to dilute aqueous hydrazine at room temperature, which results in marked improvement in electrocatalytic performance. The nanowires exhibit $\sim 100\,mV$ improvement in overpotential following exposure to dilute hydrazine, while also showing a 10-fold increase in current density and a significant change in Tafel slope. *In situ* electrical, gate-dependent measurements and spectroscopic investigations reveal that hydrazine acts as an electron dopant in molybdenum disulfide, increasing its conductivity, while also reducing the MoO_x core in the core-shell nanowires, which leads to improved electrocatalytic performance.

[1] Materials Physics and Applications (MPA-11), Los Alamos National Laboratory, Los Alamos, New Mexico 87545, USA. [2] Chemical Engineering and Conn Center for Renewable Energy Research, University of Louisville, Louisville, Kentucky 40292, USA. [3] Materials Science and Engineering, Rutgers University, Piscataway, New Jersey 08854, USA. [4] Materials Science and Technology (MST-6), Los Alamos National Laboratory, Los Alamos, New Mexico 87545, USA. [5] Materials Science and NanoEngineering, Rice University, Houston, Texas 77005, USA. [6] Chemistry, Pittsburg State University, Pittsburg, Kansas 66762, USA. Correspondence and requests for materials should be addressed to M.K.S. (email: mahendra@louisville.edu) or to G.G. (email: gautam@lanl.gov).

Hydrogen has the potential to be a zero-emission, renewable fuel; however, today it is primarily obtained from thermal steam reforming of natural gas[1–4]. It can be produced via water splitting, but the high cost of precious metal catalysts and rare earth materials that are currently used present a challenge to large scale implementation[5–8]. Recently, layered transition metal dichalcogenides (TMDs), such as WS_2, MoS_2 and so on, have been explored as a viable alternative to precious metal catalysts[9,10]. Bulk MoS_2 powders have limited catalytic activity due to an inert crystal basal plane and low in-plane conductivity[11]. High electrocatalytic activity can be achieved by either increasing the exposure of the active edge planes[12–17], increasing the conductivity of MoS_2 (refs 18–20), phase transformation[21–26], use of a co-catalyst[27] or a combination of these approaches[28–30]. Of these techniques, phase transformation from semiconducting (bulk hexagonal 2H-MoS_2) to a metastable trigonal crystal structure (1T-MoS_2), which has metallic properties and does not suffer from anisotropy, has recently shown the most promise for electrochemical and opto-electronic applications[28,31–33]. Phase-transformed TMDs, obtained via lithium intercalation, result in very efficient hydrogen evolution reaction (HER) catalytic characteristics; however, the processing conditions are expensive, time consuming (days), involve use of inert glove box atmosphere and often require elevated temperatures (100 °C). Lithium processing has also been shown to result in the formation of Li_2S nanoparticle contaminants[9]. High-aspect ratio structures, such as one-dimensional core-shell nanowires[29], also have the potential to achieve high HER activity. Although these structures possess properties such as high surface area and a conductive reduced oxide core, the primary drawback is that synthesis methods lead to the relatively inert basal plane of the MoS_2 shell growing parallel to the length of the nanowire. This reduces the exposure of available active edge sites, resulting in lower catalytic activity for HER than theoretically achievable.

Hydrazine has been well researched as a reducing agent in two-dimensional (2D) reduced graphene oxide[34] and demonstrated as an n-type dopant in graphene[35,36], single-walled carbon nanotubes, both semiconducting and metallic[37,38], as well as observed in inorganic nanocrystalline systems[39,40], but its effects on layered TMDs and electrocatalytic properties have not been previously investigated to date. Chemical modification of the MoS_2 inert basal plane to increase its charge carrier concentration, that is, electron doping, could lead to marked improvement in electrocatalytic activity.

Here we report a simple process, in which exposure of the MoO_x/MoS_2 core-shell nanowire arrays, as well as pure MoS_2 particles and 2D sheets, to dilute hydrazine (N_2H_4) results in a marked improvement in catalytic activity towards HER, that is, both a significant improvement in overpotential (~ 100 mV versus the reversible hydrogen electrode (RHE)), which is among the lowest reported HER overpotentials for any MoS_2 architecture, and a 10-fold increase in current density (~ 2 to 22 mA cm^{-2}). Detailed characterization and conductivity measurements of MoO_x/MoS_2 core-shell nanowires, as well as pure MoS_2 particles and sheets, show that hydrazine is acting as an electron dopant, donating electrons to increase the conductivity of MoS_2, which leads to improved electrocatalytic activity. In the case of the core-shell nanowires, hydrazine further reduces the oxide core, which enhances conductivity and facilitates the charge transfer kinetics in the system, synergistically improving the HER performance of core-shell nanowires after exposure to hydrazine. This 'activation' of the normally inert TMD basal plane by electron doping from hydrazine presents a unique opportunity to serves as a novel direction for efficient catalytic development and the use of simple processing techniques that can rival state-of-art platinum catalysts.

Results

Synthesis and characterization of MoO_x/MoS_2 architectures.

Figure 1a shows a schematic representation of a simple hydrazine treatment; the as-grown MoO_x/MoS_2 core-shell nanowires are exposed to a dilute hydrazine (1% in water) at room temperature for 10 min. Figure 1b shows the scanning electron microscopic (SEM) image of an as-grown, vertically oriented MoO_x/MoS_2 core-shell nanowire array on SiO_2 substrates. Nanowires are typically 1–2 µm in length, with diameters of 20–50 nm. A thin (2–5 nm) highly oriented crystalline MoS_2 shell is grown epitaxially on a single-crystal reduced MoO_x core, as can be seen using high-resolution transmission electron microscopy (Fig. 1d). MoO_3 nanowire arrays are deposited on SiO_2 substrates using chemical vapor deposition (CVD), followed by the sulfurization reaction at 300 °C under low pressures of 99% H_2S (100 mTorr) for 2 h, leading to a thin (2–5 nm) single crystalline MoS_2 on a single crystal MoO_x core. The synthesis methods of the MoO_x/MoS_2 core-shell nanowires are described in more detail

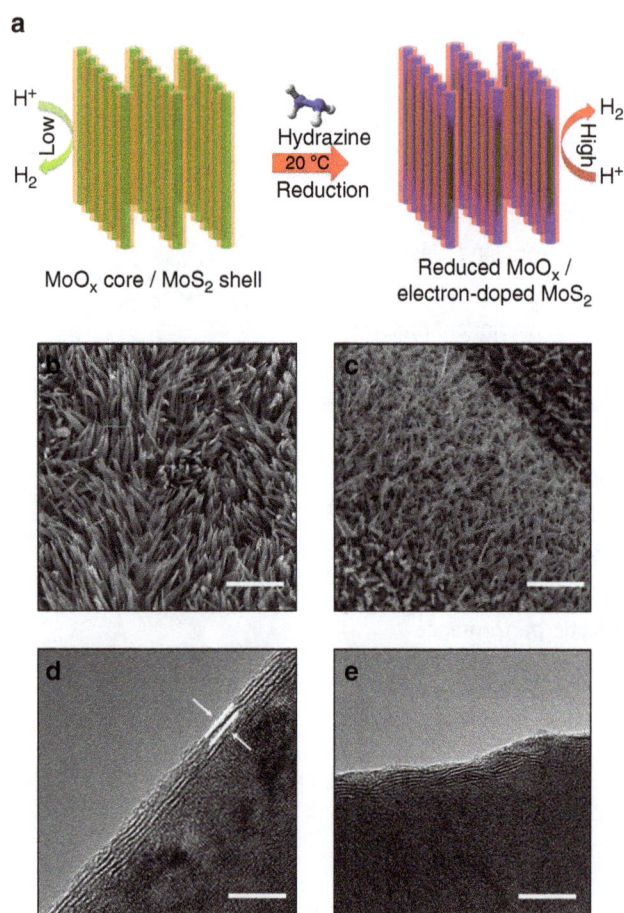

Figure 1 | Schematic representation of hydrazine treatment and microscopic analysis of MoO_x/MoS_2 core-shell nanowires. (a) Schematic representation of exposure of MoO_x/MoS_2 core-shell nanowire array to dilute hydrazine under ambient conditions. (b) SEM image of as-grown MoO_x/MoS_2 core-shell nanowire array. Scale bar, 2 µm. (c) SEM imaging MoO_x/MoS_2 core-shell nanowires following dilute hydrazine treatment, showing that the overall nanowire morphology is maintained. Scale bar, 2 µm. (d) High-resolution TEM (HRTEM) of as-grown nanowire, showing thin (~3-5 nm) MoS_2 shell on a single crystal MoO_x core. The MoS_2 has the typical interlayer spacing of 6.2 Å, denoted in image by two parallel lines. Scale bar, 5 nm. (e) HRTEM of the MoO_x/MoS_2 core-shell nanowire after exposure to hydrazine. Scale bar, 5 nm.

elsewhere[29,30,41]. After exposure to hydrazine, SEM (Fig. 1c) shows minimal disruption of the nanowire morphology and there is no evidence of crystallographic disruption of the MoS_2 shell, as shown in Fig. 1e.

Electrochemical performance of materials. The nanowires are dispersed in distilled water, removing any agglomerated hydrazine on the surface and transferred ($\sim 70\,\mu g\,cm^{-2}$) to a glassy carbon electrode for electrochemical characterization. Figure 2a shows a linear sweep voltammogram of the as-grown and hydrazine-treated nanowires. All of these electrocatalytic

measurements are corrected for ohmic potential (iR) losses in the system (~ 12 ohms); this resistance calculation is shown in Supplementary Fig. 1. The measured HER onset potential for the as-grown nanowires is approximately $-200\,mV$ versus RHE and a current density of $\sim 2\,mA\,cm^{-2}$ at $-0.35\,V$ versus RHE is obtained. After exposure to 1% hydrazine, the onset potential improves to approximately $-100\,mV$ versus RHE and the current density increases to $\sim 22\,mA\,cm^{-2}$ at $-0.35\,V$. Hydrazine treatment on multiple nanowire samples indicate that this electrochemical performance is highly reproducible (Supplementary Fig. 2).

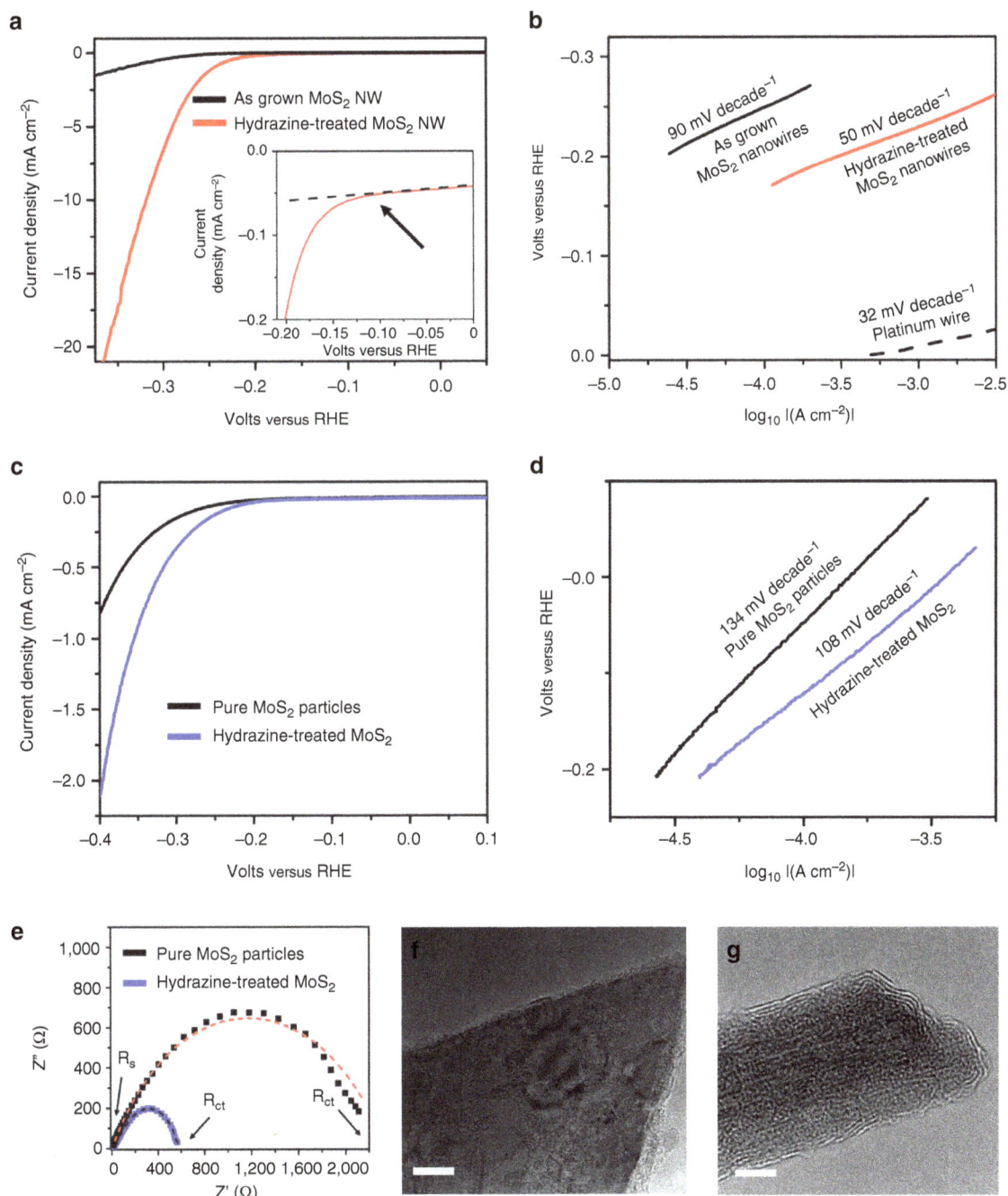

Figure 2 | Electrochemical analysis of MoO_x/MoS_2 core-shell nanowires and MoS_2 particles. (a) Linear sweep voltammetry for as-grown MoO_x/MoS_2 core-shell nanowire (black curve) and after exposure to 1% Hydrazine (red curve). **(b)** Tafel slopes for as-grown MoO_x/MoS_2 core-shell nanowires (black curve), 1% hydrazine-treated nanowires (red curve), and a platinum wire (dotted black curve). **(c)** Linear voltammograms and corresponding **(d)** Tafel slope analysis of MoS_2 particles before (black curve) and following exposure to dilute hydrazine (blue curve). **(e)** Electrochemical impedance spectroscopy (EIS) Nyquist plots of the MoS_2 particles following exposure to hydrazine, with line fits shown by dotted lines. **(f)** High-resolution TEM (HRTEM) of 2H-MoS_2 particle. Scale bar, 5 nm. **(g)** HRTEM of as-grown MoO_x/MoS_2 core-shell nanowire. Scale bar, 10 nm.

The rate of hydrogen evolution is limited by either proton adsorption onto an active site or evolution of the formed hydrogen from the surface. A high Tafel slope (120 mV per decade) is indicative of proton adsorption (Volmer step) as the rate-limiting step, while a lower Tafel slope (30 or 40 mV per decade) indicates that the evolution of molecular hydrogen from the catalyst is rate limiting (Heyrovsky or Tafel step, respectively)[7,42,43]. Figure 2b shows the Tafel plots for these voltammograms. The as-grown MoO_x/MoS_2 core-shell nanowires show a Tafel slope of 90 mV per decade, suggesting that adsorption of protons from the electrolyte is the rate-limiting step. Since the relatively inactive basal plane of the $2H\text{-}MoS_2$ shell is grown parallel to the MoO_x core nanowire, there are fewer catalytically active sites available for proton adsorption. After exposure to 1% hydrazine, the Tafel slope decreases to 50 mV per decade, indicating that the evolution of hydrogen via the combination of two adsorbed protons becomes the rate-limiting step. These results show that the hydrazine treatment significantly facilitates the adsorption of protons onto the catalyst surface. For comparison, the Tafel slope of a platinum wire is shown in Fig. 2b (dashed black curve); platinum has a Tafel slope of ~ 30 mV per decade, which indicates proton adsorption is favourable and hydrogen evolution is the rate-limiting step[42,43].

To corroborate the effects of hydrazine on TMDs without a core-shell nanowire architecture, pure MoS_2 particles are exposed to dilute hydrazine. Figure 2c shows the linear voltammetry plots of bulk MoS_2 powder with and without hydrazine. Bulk powder shows poor catalytic activity for the HER; following the hydrazine treatment, an improvement in both the current density and HER overpotential is observed. Tafel analysis (Fig. 2d) shows a decrease in the Tafel slope of the pure $2H\text{-}MoS_2$ particles, indicative of increased favourability for proton absorption onto the catalysis surface. Furthermore, electrochemical impedance spectroscopy of the MoS_2 particles before and after hydrazine treatment (Fig. 2e) show a significant decrease in the charge transfer resistance. A $R_s\text{-}(CPE\text{-}R_{ct})$ circuit diagram is used to fit the experimental electrochemical impedance spectroscopy data. The solution resistance (R_s) remains nearly constant ($\sim 12\,\Omega$), whereas the charge transfer resistance (R_{ct}) decreases from $\sim 2,340$ to $\sim 625\,\Omega$, indicative of enhanced conductivity after the hydrazine treatment.

To quantify the improvement in catalytic activity following the hydrazine treatment, turnover frequency (TOF) is calculated for the MoO_x/MoS_2 core-shell nanowires and MoS_2 particles using the following equation:

$$\text{TOF}\left(s^{-1}\right) = \frac{i_0\left(\frac{A}{cm^2}\right)}{\#\,\frac{\text{sites}}{cm^2}*1.602\times 10^{-19}\left(\frac{C}{e^-}\right)*2\left(\frac{e^-}{H_2}\right)} \quad (1)$$

Exchange current densities (i_0) are calculated from the Tafel equation, while the number of active sites is calculated from cyclic voltammograms of MoS_2 particles and MoO_x/MoS_2 core-shell nanowires, both as-grown and treated with hydrazine. Following

HER measurements, the potential applied to the working electrode was driven to high oxidation potentials (~ 1.4 V versus RHE) to convert MoS_2 particles to MoO_3 (ref. 44). Calculation of active sites was obtained from the reduction charge transfer ($MoO_3 \rightarrow Mo^0$) occurring at ~ -0.3 to 0.0 V (ref. 29), assuming that each Mo^{3+} reduced to Mo^0 corresponded to one MoS_2 site. The results of these electrochemical decompositions to calculate surface area are shown in Supplementary Figs 3 and 4. The number of active sites for MoS_2 bulk particles are similar before and after hydrazine treatment, 9.0×10^{14} and 9.8×10^{14} MoS_2 sites per cm^2, while Tafel analysis indicates an increase in exchange current density from 8.4×10^{-6} to 1.7×10^{-5} A cm^{-2}. Using these values, the calculated TOF values increase for the MoS_2 particles following the hydrazine treatment, from 0.03–$0.05\,s^{-1}$, while the number of active sites remain relatively unchanged. Summary of these obtained values, shown in Table 1, are consistent with reports by other researchers[31,45,46].

A similar analysis process is performed to calculate TOF values for the MoO_x/MoS_2 core-shell nanowires. The -0.3 to 0.0 V region used in the particle calculations is more convoluted in the nanowires' case with the combination of core and shell oxidation–reduction peaks (Supplementary Fig. 4). Thus, the oxidation of MoS_2 to MoO_3 involving 11 e^- was used as the region for determining the number of MoS_2 sites (as explained by Chen et al.)[29]. Interestingly, the series of oxidation–reduction decomposition peaks observed in the as-grown nanowires, resulting from the MoO_x core, are not observed following hydrazine treatment. This suggests that the hydrazine, in addition to doping of MoS_2 shell, also reduces the oxide core, which increases its conductivity. The calculated number of MoS_2 sites per surface area for the as-grown MoO_x/MoS_2 core-shell nanowires is $\sim 5.9\times 10^{14}$ sites per cm^2. There is almost no change in the concentration of active sites in the nanowires following exposure to hydrazine, 6.0×10^{14} sites per cm^2; this is confirmed by no observed physical change from SEM and TEM analysis (Fig. 1). Exchange current densities calculated from the obtained Tafel equations are 7.5×10^{-6} and 4.5×10^{-5} A cm^{-2} for the as-grown and hydrazine-treated MoO_x/MoS_2 core-shell nanowires, respectively. This results in an increase of TOF from 0.04 to $0.2\,s^{-1}$. This fivefold improvement in TOF quantitatively shows the effect of hydrazine treatment on the electrocatalytic properties of MoS_2 shelled nanowires.

Although, there is clear improvement in HER performance characteristics, the magnitude of the change in the bulk MoS_2 is lower when compared with nanowires. This can be attributed to the difference in structure; the MoS_2 powder particles are composed of several tens of molecular MoS_2 layers with random orientations, as seen in Fig. 2f. This is in contrast to the core-shell nanowires, which have a few molecular layer thick MoS_2 shell on an oxide core, as shown in Fig. 2g.

Hydrazine is a reducing agent, as well as an electron dopant; we hypothesize that it interacts with the oxide core, further reducing it and increasing the intra-particle conductivity of the nanowire,

Table 1 | Electrochemical parameters with hydrazine treatment of MoS_2.

	HER onset (Volts versus RHE)	Current density at -0.4 V (mA cm^{-2})	Tafel slope (mV per decade)	Exchange current density i_0 (A cm^{-2})	No of sites per cm^2	Turnover frequency (s^{-1}) $\eta=150\,mV,200\,mV$
As-grown MoO_x/MoS_2 Nanowires	-0.200	2.0	90	7.5×10^{-6}	5.9×10^{14}	0.04
Hydrazine-treated MoO_x/MoS_2 Nanowires	-0.100	22.0	50	4.5×10^{-5}	6.0×10^{14}	0.2
Bulk MoS_2 particles	-0.250	0.75	134	8.4×10^{-6}	9.0×10^{14}	0.03
Hydrazine-treated MoS_2 particles	-0.200	2.0	108	1.7×10^{-5}	9.8×10^{14}	0.05

while hydrazine also electron dopes the MoS_2 surface. To test the hypothesis of reduction of oxide core, pure MoO_3 nanowires are treated with dilute 1 % hydrazine, reported in the Supplementary Fig. 5. Pure MoO_3 is catalytically inactive and decomposes quickly in acid solutions[47] As expected, the MoO_3 nanowires show no catalytic activity and decompose in the 0.5 M H_2SO_4. When exposed to hydrazine before testing, the MoO_3 reduction and oxidation peaks are no longer observed, suggesting that the oxide has been reduced. This experiment supports a synergistic mechanism; hydrazine improves electron conductivity of the nanowire core, that is, intra-particle conductivity, while also electron doping the MoS_2 surface (shown in pure MoS_2 powder). The MoO_x/MoS_2 core-shell nanowire morphology, combined with the hydrazine treatment, uniquely provides architecture for an optimized electrocatalyst. The reduction of the MoO_x core by hydrazine is also confirmed by X-ray photoelectron spectroscopy (XPS) of a MoO_x/MoS_2 core-shell nanowire array in Fig. 3.

Spectroscopic investigation of MoO_x/MoS_2 architectures. The core level binding energies of molybdenum and sulfur in as-grown MoO_x/MoS_2 core-shell nanowires are analysed using XPS, depicted in Fig. 3. The as-grown MoO_x/MoS_2 core-shell nanowires show strong doublet peaks at 229 eV, which are consistent with literature values for Mo^{4+} $3d_{5/2}$ and Mo^{4+} $3d_{3/2}$ of 2H-MoS_2, shown by the red curves[21]. The MoO_x core results in a convolution of two oxidation states for Mo^{6+} in the reduced oxide core, shown by the dark and light blue signals at ~ 232.6 and 230.6 eV; these binding energies suggest MoO_3 and reduced MoO_{3-x}. This is a result of incomplete reaction between the original MoO_3 nanowires and the H_2S during synthesis[29]. Following exposure to hydrazine, the partially reduced oxide core continues to reduce, indicated by the shift in Mo 3d signal towards lower binding energies of the further reduced oxide core (MoO_x), ~ 230.6 and 229.6 eV, respectively. It is established that reduced molybdenum oxide has almost metallic conductivity[48]; hydrazine, as a reducing agent, is increasing the intra-particle conductivity of the nanowire core. Despite the shift in Mo corresponding to the molybdenum oxide core, there is no shift in the MoS_2 Mo 3d binding energies. When observing the S 2p signal for the MoS_2 nanowire shell, the as-grown sample shows the clear doublet signal at ~ 162 eV, typical for 2H-MoS_2. After

exposure to hydrazine, there is no detectable shift in the peak positions of the Mo 3d binding energies for 2H-MoS_2 and only a slight broadening of the S 2p binding energy, but no noticeable shift. For comparison, the XPS spectra for chemically exfoliated 1T-MoS_2 and 2H-MoS_2 sheets are shown, modified from Cummins et al.[30] It is well established that the phase transition from semiconducting 2H-MoS_2 to metallic 1T-MoS_2 has a corresponding large XPS binding energy shift of ~ 0.9 eV (ref. 21), which is clearly evident in the chemically exfoliated control sample, but is not observed during the hydrazine treatment; it appears that hydrazine does not induce phase transition of MoS_2 from 2H to 1T phase. This lack of phase transformation from 2H to 1T is also corroborated using Raman spectroscopy. The Raman analysis of MoS_2 is discussed in Supplementary Note 1, with the Raman spectra shown in Supplementary Fig. 6. It has been shown that the 1T-MoS_2 crystal phase results in unique Raman excitations, which can distinguish the metastable phase from the stable 2H-MoS_2 structure[21,30]; these excitations are not detected in the hydrazine-treated samples.

Electrical measurements on nanowires and 2D materials. Spectroscopic studies indicate that the oxide core is reduced, but do not reveal the effects of hydrazine on the MoS_2 shell or pure MoS_2 particles; therefore, in situ four-probe resistance and conductivity measurements, as well as gate-dependent measurements, are performed to elucidate the effects of hydrazine on both core-shell nanowires and CVD grown flakes. Initially, the MoO_x/MoS_2 core-shell nanowire array is grown directly on a non-conducting glass substrate and mounted on a ceramic holder; two thermocouples and two copper wires (as current leads) are attached to the sample with silver epoxy to measure four point probe resistance prior and following hydrazine vapour exposure. The probe is loaded into a quartz reactor, which was placed inside a tube furnace. The experimental setup is described schematically in the Supplementary Fig. 7. Initial measurements of the resistance at atmospheric pressure and room temperature result in a value of 1.588 kΩ. The system is evacuated to $\sim 10^{-5}$ Torr and annealed at 150 °C to remove surface adsorbed moisture; an initial resistance of $\sim 706\,\Omega$ for the as-grown MoO_x/MoS_2 core-shell nanowire array is observed. To compare the effect of air and moisture on the resistivity of the core-shell nanowires, the sample is first exposed to ambient air, to a maximum pressure of 350 Torr, which results in a minimal increase in the sample resistance ($\sim 708\,\Omega$), as seen in Fig. 4a. The system is then evacuated again and the resistance stabilized. Then, hydrazine vapours (15 Torr) are introduced to the system. Almost immediately (< 30 s), a drop in the sample's resistance is observed, decreasing from ~ 710 to $\sim 495\,\Omega$, which stabilizes after ~ 30 min, as shown in Fig. 4b. On evacuation of the system and, therefore, the removal of the hydrazine, the sample's resistance does not significantly increase, maintaining an average resistance of $\sim 500\,\Omega$. This experimentally confirms that the hydrazine markedly lowers the sample's resistance (increases the conductivity), and thereby, improves the charge transfer characteristics during HER.

For further confirmation, conductivity measurements are performed on a core-shell nanowire back-gated device. The nanostructures are transferred onto Si^{++}/SiO_2 substrates and the electrical contacts (Au) are defined using e-beam lithography (inset of Fig. 4c). Figure 4c shows the change in resistance of a MoO_x/MoS_2 core-shell nanowire cluster, before and after hydrazine treatment. It can be clearly observed that after hydrazine treatment (red curve) the current increases sharply in comparison with the untreated device (black curve) for the

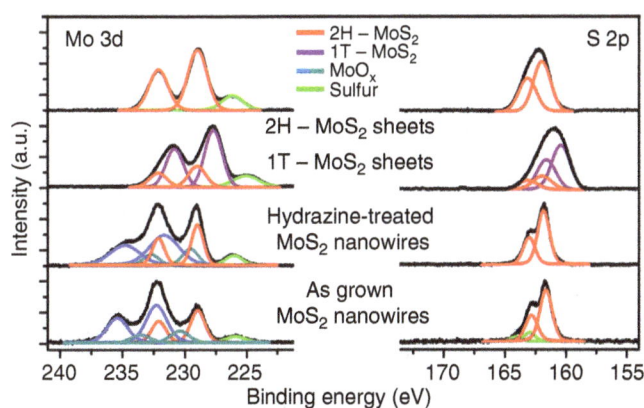

Figure 3 | XPS Spectroscopy following hydrazine treatment. Spectra for MoO_x/MoS_2 core-shell nanowire array before and after hydrazine treatment, showing the Mo 3d, S 2s, and S 2p core level binding energies. The red curves denote the Mo 3d and S 2p signals corresponding to 2H-MoS_2, with the purple curves showing the shift resulting from the phase transformation to 1T-MoS_2. The MoO_x core seems to be a mixed phase valence, the Mo 3d oxidation states are shown by the dark and light blue curve, and sulfur signals are denoted by green.

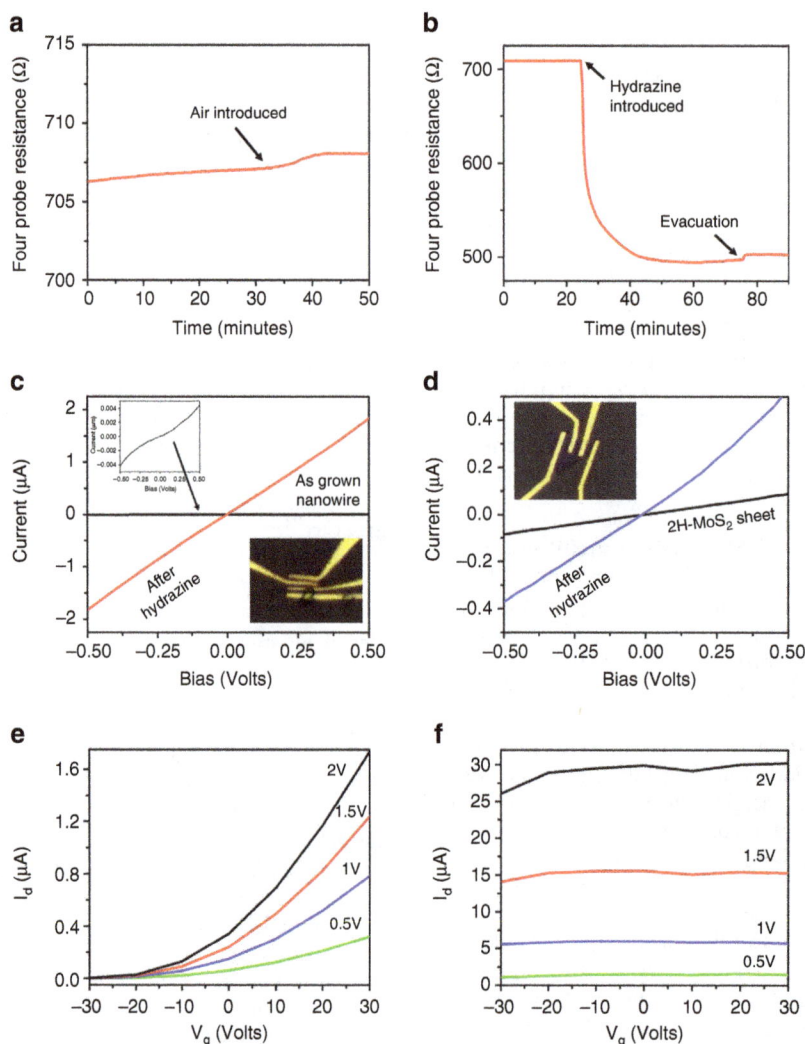

Figure 4 | Effects of hydrazine treatment on conductivity of MoS₂ architectures. (**a**) Four-probe resistance measurement of MoO_x/MoS_2 core-shell nanowire array, grown on a non-conductive glass substrate, with exposure to ambient air (shown by arrow) to ~ 350 Torr. (**b**) The effects on the four-probe resistance of the MoO_x/MoS_2 core-shell nanowire array following the introduction of a small amount (~ 15 Torr) of hydrazine (N_2H_4) vapour, leading to an almost instantaneous, and irreversible drop in the system resistance. (**c**) Resistance measurement of device fabricated on a small cluster of MoO_x/MoS_2 core-shell nanowires (optical micrograph shown in inset). The resistance of the nanowires decreases from ~ 133 to 0.3 MΩ following exposure to hydrazine. (**d**) Resistance measurement of device fabricated on single CVD grown 2H-MoS₂ sheet (optical micrograph shown in inset). The resistance decreases from ~ 5.8 to 1.2 MΩ following exposure to hydrazine. (**e**) Drain current-gate voltage analysis of single CVD grown 2H-MoS₂ sheet and (**f**) following exposure to dilute hydrazine.

nanowires. The MoO_x/MoS_2 core-shell nanowires show a resistance of ~ 133 MΩ before hydrazine treatment, decreasing markedly to ~ 0.3 MΩ following the hydrazine treatment.

To isolate the MoS₂ system from contributions of the reduced oxide core, the resistance measurements are then performed on a single layered CVD grown 2H-MoS₂ flake[49]. Figure 4d shows the resistance change in the single flake before and after the hydrazine treatment. The MoS₂ flake shows a resistance of ~ 5.8 MΩ before hydrazine treatment; following exposure to hydrazine, the resistance decreases to ~ 1.2 MΩ. These results directly support the hypothesis that the hydrazine treatment leads to a decrease in the resistance (or increased conductivity) of MoS₂. In addition, field-effect gating experiments are performed on a single CVD-grown 2H-MoS₂ sheet. The effect on drain current (I_d) with changing gate bias (V_g) held at constant drain-source voltages (V_{ds}) on the untreated MoS₂ sheet is shown in Fig. 4e. The as-grown MoS₂ shows a ON/OFF ratio of $\sim 10^3$ (for $V_{ds} = 2$ V) and exhibits n-type behaviour, consistent with literature reports for CVD grown MoS₂ (ref. 50). Following these

measurements, this MoS₂ sheet is treated with hydrazine, thoroughly rinsed with distilled water, and then the field effect gating is measured again (Fig. 4f). In contrast to the untreated device, there is no observable modulation with changing gate voltage and the drain current increases by an order of magnitude. This lack of modulation and increase in current shows the emergence of metallic behaviour following exposure to hydrazine, most likely due to the new states and increased carrier density at the Fermi energy. These experiments support the conclusion that hydrazine electron dopes MoS₂, in pure sheets as well as the nanowire shell, improving conductivity and electrocatalytic properties.

Gate-dependent electrochemical HER measurements. The increase in the conductivity and its correlation to HER catalysis is further corroborated with back gate-dependent electrochemical HER measurements. A single layer 2H-MoS₂ flake is patterned with gold contacts using e-beam lithography; a schematic

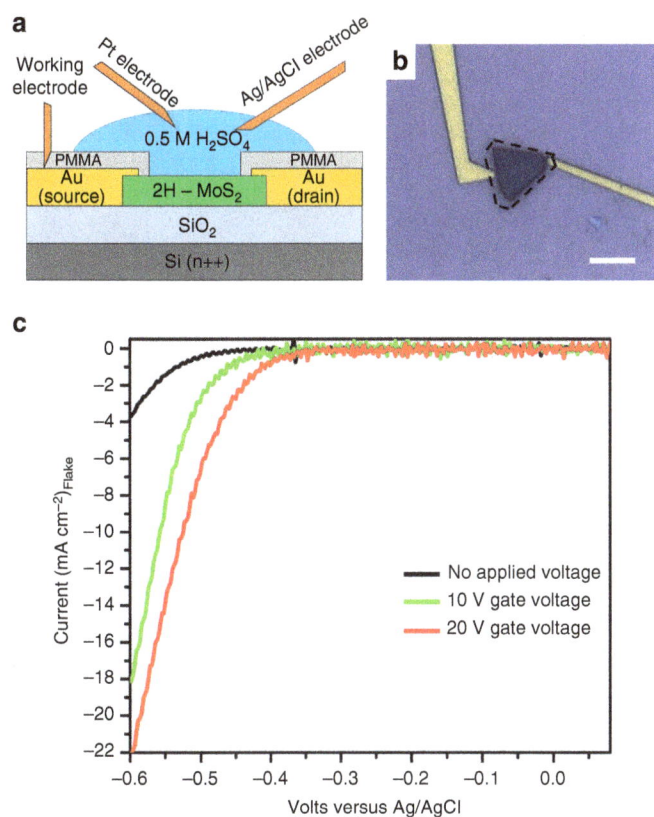

Discussion

It is essential to understand the interactions between hydrazine and MoS$_2$, as well as which mechanism leads to increased conductivity and enhancement in electrocatalysis. Hydrazine has been shown to be an intercalating compound,[51] can act as a pseudo-reducing agent in TMD systems[52] and also is a strong reducing agent, which can repair oxidized sulfur sites and inhibit further oxidation of the TMD surface[39,53]. Finally, hydrazine has also been shown to be an effective electron dopant in graphene[35,36], carbon nanotubes[37,38], and experiments suggest it can improve the conductivity in PbSe quantum dots[39,53–56]. Due to these numerous possible pathways, it is somewhat challenging to pinpoint the exact mechanism; we investigate these possible mechanisms by identifying key experiments. First, no change in the d-spacing of the MoS$_2$ nanowire shell or MoS$_2$ sheets is observed following exposure to hydrazine (Supplementary Fig. 8); therefore, hydrazine acting as an intercalating agent is ruled out. Second, hydrazine has been proposed as pseudo-reducing agent[52] in alkaline conditions. The proposed mechanism of hydrazine forming OH$^-$ to act as a pseudo-reduction agent is outlined in Supplementary Note 2. To test this hypothesis, MoS$_2$ particles are treated with 0.1 M KOH, but show no improvement in the electrocatalytic characteristics (Supplementary Fig. 9), therefore, the pseudo-reduction of the MoS$_2$ surface purely by OH$^-$ groups does not adequately explain the mechanism of hydrazine interaction. Third, in the case of MoO$_x$/MoS$_2$ core-shell nanowires, hydrazine can interact with the oxide core, further reducing the MoO$_x$ to make it more conductive[48]. Hydrazine acting as a strong reducing agent is confirmed by CV measurements (Supplementary Fig. 5) as well as XPS studies (Fig. 3).

Finally, electron doping of MoS$_2$ by hydrazine should facilitate the adsorption of protons; slight broadening of the sulfur 2p binding energies in the XPS spectra (Fig. 3) is consistent with this modification of the surface energies. Moreover, ultraviolet photoelectron spectroscopy analysis of the hydrazine-treated MoO$_x$/MoS$_2$ core-shell nanowire system supports this mechanism, providing evidence of electron donation from hydrazine to the conduction band of the MoS$_2$ (Supplementary Fig. 10). To characterize the chemical form of hydrazine in the MoO$_x$/MoS$_2$ core-shell nanowire system, the nitrogen XPS signal (N 1s) is analysed, shown in Supplementary Fig. 11. Following exposure to hydrazine, a N 1s signal arises at ∼400.3 eV, which is not observed in the as-grown sample; the nitrogen signal is in the vicinity of Mo 3p binding energies (analogue to Mo 3d). This nitrogen binding energy corresponds to a surface absorbed amine phase, such as NH$_3$ or a sub-amine (NH, NH$_2$ and so on)[57]. While XPS makes it difficult to exactly identify the dissociated hydrazine (N$_2$H$_4$) species, it is clear that amine groups have absorbed on the surface. It has been shown that ammonia, NH$_3$, can potentially act as an n-type dopant in metal oxides[58,59]. However, it is more likely that a dissociated radical of hydrazine is acting as the electron donor. Thermal decomposition studies show that N$_2$H$_4$ readily decomposes to form 2 NH$_2^-$ (ref. 60), a reactive radical species that could contribute electrons to the MoS$_2$ surface. Recent theoretical work by Zhang et al.[61] shows that at room temperatures, hydrazine hydrate readily dissociates at a catalyst surface to form radicals, which can donate electrons to the semiconductor surface. The XPS signal of an absorbed amine group supports this theorized mechanism, where dissociated hydrazine radicals (NH$_2$* and NH$_3$) are present at the MoS$_2$ surface and capable of donating electrons. While there are amine groups on the surface, there is no evidence that there is a chemical bonding between the nitrogen and the MoS$_2$ shell; the amine group acts as the electron donor. To conclude, in case of core-shell nanowires, hydrazine reduces the oxide core and also

Figure 5 | Gate-Dependent HER. (a) Schematic of the gate-dependent electrochemical device, with the SiO$_2$ layer acting as the gate. **(b)** Optical micrograph of gold contacts and 2H-MoS$_2$ single layer flake. The edges of the MoS$_2$ flake masked with PMMA are outlined by the black dotted line, the exposed window can be seen as the darker region. Scale bar, 10 µm **(c)** Linear voltammograms from the gate-dependent HER measurements on the single MoS$_2$ flake device. The black curve is the activity of the flake with no applied voltage. The green and red curves show the improvement in electrocatalytic activity after applying a positive gate voltage of 10 and 20 V, respectively.

representation of the experimental setup is shown in Fig. 5a. The flake and contacts are covered completely with polymethylmethacrylate (PMMA) polymer and a window (∼140 µm^2) is opened over to the surface of the flake to allow for electrocatalysis, taking special care to ensure the gold contacts are still covered by the polymer. An optical microscopic image of the device can be seen in Fig. 5b. To test the HER activity, a small drop of 0.5 M H$_2$SO$_4$ is placed on the flake and linear voltammograms are taken, with a thin platinum wire acting as the counter electrode and AgCl-coated Ag wire acting as the pseudo reference electrode. The results of the gate-dependent HER catalysis measurements are shown in Fig. 5c. At 0 V bias, there is a small, but detectable, HER obtained for the 2H-MoS$_2$ flake (black curve). A 10 V positive bias is applied to the back gate, inducing a negative charge at the MoS$_2$ surface (green curve). The overpotential to drive the HER is reduced (∼500 to ∼400 mV versus the Ag/AgCl wire) and the current density increases by over four times. When the gate voltage is increased to 20 V, the overpotential to drive the HER continues to decrease by an additional 50–100 mV and the current increases by five times, when compared with generated current density with no gate bias. This *in situ* observation of the effect of surface electron concentration on HER catalysis directly shows that increasing charge concentration at the MoS$_2$ surface can enhance the electrocatalytic activity.

acts as an electron dopant for the MoS_2 shell; in the case of pure MoS_2 sheets, the change in conductivity is entirely due to electron doping.

Finally, for commercialization and technologically viable use of TMDs for hydrogen production, thermal stability and long-term durability are required. Chemically exfoliated 1T-MoS_2 sheets are sensitive to temperature as the metastable phase transformation is reversed on annealing (loss of catalytic activity)[21,31]. Electrocatalytic activity of hydrazine-treated nanowires before and after annealing at 150 °C for 1 h under argon are shown in Fig. 6a. There is no significant change in HER onset potential or current density, which demonstrates that the effect of the hydrazine treatment is not merely due to physisorption. The electrochemical durability at room temperature of the hydrazine-treated nanowires is also investigated, shown in Fig. 6b. The initial current density at 0.35 V versus RHE is $\sim 24\,\mathrm{mA\,cm^{-2}}$ following the hydrazine treatment, which is set as 100%. After 10 scans, the current density at 0.35 V actually increases, which has been seen in other reports[29], but slowly decreases to $\sim 60\%$ of its initial activity, stabilizing after around 400 cycles to an average of $\sim 13\,\mathrm{mA\,cm^{-2}}$. This degradation may be due, in part, to decomposition of the oxide core, since the MoS_2 shell is thinner than in other reports[29].

In conclusion, exposure of MoO_x/MoS_2 core-shell nanowires to an aqueous hydrazine solution leads to a marked improvement in electrocatalytic activity. We observe a 100 mV improvement in the HER onset potential (from approximately $-200\,\mathrm{mV}$ to approximately $-100\,\mathrm{mV}$ versus RHE) and also an exponential increase in generated current density (from 2 to $22\,\mathrm{mA\,cm^{-2}}$ at $-0.35\,\mathrm{V}$ versus RHE). Furthermore, the TOF for the core-shell nanowires increases five-fold following the hydrazine treatment, from 0.04 to $0.2\,\mathrm{s^{-1}}$, due to the synergistic reduction of the oxide core and the electron doping of the MoS_2 shell. In the case of MoS_2 bulk powder, the TOF increases by nearly twofold, from 0.03 to $0.05\,\mathrm{s^{-1}}$, since electron doping is the only contributing factor. Surface characterization studies reveal that the change in catalytic properties does not result from a phase transformation, but is due to enhanced conductivity. The increased conductivity, as result of hydrazine treatment, is shown in MoO_x/MoS_2 core-shell nanowires, 2H-MoS_2 particles, and single layer sheets, by utilizing field effect gating experiments, conductivity, and spectroscopic techniques. Hydrazine is shown to act as an electron dopant; dissociated amine radicals donate electrons to the MoS_2 surface, facilitating electrocatalysis. The reported hydrazine modifications can be performed in ambient conditions and on the order of minutes, when compared to conventional techniques. This is one of the first known investigations into the effect of hydrazine exposure in 2D layered chalcogenides for electrocatalysis application. Understanding the effects of hydrazine on TMDs in catalysis can lead to a fundamental breakthrough in the areas of electrochemistry and material science.

Methods

Electrochemical testing measurements. To examine the electronic properties of the nanowires, 1% by volume hydrazine (N_2H_4) and water solution (20 µl) was dropped directly on the nanowire array and allowed to dry in air ($\sim 10\,\mathrm{min}$). Electrodes were prepared by gently dispersing the hydrazine-treated nanowire array in distilled water, then transferred to 3 mm diameter glassy carbon tip electrodes ($\sim 70\,\mathrm{\mu g\,cm^{-2}}$). A thin layer of 5% Nafion ($\sim 2.7\,\mathrm{\mu l\,cm^{-2}}$) was further applied on the electrodes to prevent delamination of the active material into solution, but still allowing for proton transport. The electrochemical analysis was performed by cyclic voltammetry in 0.5 M H_2SO_4 solution (pH = 0) using a graphite rod as counter electrode and Ag/AgCl as a reference electrode ($+0.210\,\mathrm{V}$ versus RHE). Nitrogen was bubbled vigorously through the electrolyte to remove any oxygen from solution. The sample was cycled multiple times to remove surface contamination and ensure steady state conditions.

For electrochemical measurements of bulk MoS_2 powder, the powder was dispersed by sonication in distilled water (100 mg MoS_2 per 10 ml H_2O) and drop-cast onto glassy carbon electrodes ($\sim 1\,\mathrm{mg\,cm^{-2}}$). For the hydrazine treatment, the dried MoS_2 dispersion on glassy carbon was drop-cast with 1% hydrazine solution and allowed to dry, then coated with Nafion, following the exact methods as with the core-shell nanowires.

Gate device fabrication and conductivity measurements. For fabrication of the MoS_2 sheet devices, MoS_2 monolayer sheets were grown using standard CVD procedure[49] and transferred onto 300 nm SiO_2/Si^{++} substrates by the PMMA assisted transfer method. PMMA layers were then washed by acetone under 50 °C followed by IPA rinse. In the case of the MoO_x/MoS_2 core-shell nanowire device, the nanowires are first dispersed by sonication in water, then directly drop-casted onto the 300 nm SiO_2/Si^{++} substrates. PMMA was spin-coated at 3,000 r.p.m. for 60 s, followed by a soft bake at 180 °C. The electrode patterns were defined by using standard e-beam lithography method (5/40 nm Ti/Au contacts). This was followed by a lift-off process to achieve the final device configuration. Initial conductivity measurements and gate-dependent measurements were performed on these devices in vacuum. To test the effect of hydrazine, another e-beam lithography step was performed to selectively expose a region within the device channel to isolate the contacts. Hydrazine treatment was performed by placing a 1% dilute hydrazine solution for 5 min to a few hours. The devices are then thoroughly rinsed with distilled water and electrical measurements performed again in vacuum.

Characterization instrumentation. FEI Tecnai F20 TEM was used for high-resolution TEM. SEM imaging utilizes a FEI Quanta FEG 400 Scanning Electron Microscope. XPS analysis is performed using a Physical Electronics 5600ci XPS system with an Al Kα radiation source. All XPS spectra are calibrated by the position of the C 1s peak. The carbon signal used for calibration results from surface absorbed hydrocarbons, which have a characteristic peak location of 284.5 eV. The Raman spectroscopic analysis of the nanowire arrays is performed

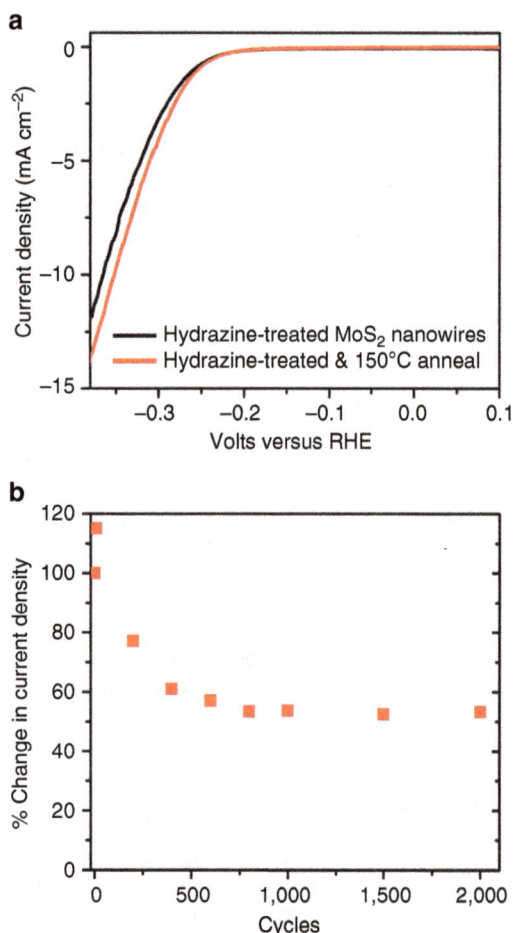

Figure 6 | Stability studies of hydrazine-treated MoO_x/MoS_2 core-shell nanowire. (**a**) Linear voltammetry for MoO_x/MoS_2 core-shell nanowires treated with Hydrazine (red curve), then annealed at 150 °C for 1 h (red dashed line). (**b**) Stability of hydrazine-treated MoO_x/MoS_2 core-shell nanowires over 2,000 cycles. Current densities are normalized to $\sim 24\,\mathrm{mA\,cm^{-2}}$ as 100%.

using a Renishaw Invia Micro Raman system with a 633 nm HeNe laser. Raman system is calibrated using single crystal Si wafer, with characteristic peak at $520.0\,cm^{-1}$.

References

1. James, B. D., Baum, G. N., Perez, J. & Baum, K. N. *Technoeconomic Analysis of Photoelectrochemical (PEC) Hydrogen Production*. Report No. GS-10F-009J, published on U.S. DOE EERE website.www1.eere.doe.gov/hydrogenandfuelcells/pdfs/pec_technoeconomic_analysis.pdf (US Department of Energy, 2009).
2. Ashcroft, A. T., Cheetham, A. K., Green, M. L. H. & Vernon, P. D. F. Partial oxidation of methane to synthesis gas-using carbon-dioxide. *Nature* **352**, 225–226 (1991).
3. Cortright, R. D., Davda, R. R. & Dumesic, J. A. Hydrogen from catalytic reforming of biomass-derived hydrocarbons in liquid water. *Nature* **418**, 964–967 (2002).
4. Rostrup-Nielsen, J. R., Sehested, J. & Norskov, J. K. in *Advances in Catalysis* **Vol. 47** (eds Gates, B. C. & Knozinger, H.) 65–139 (2002).
5. Joo, S. H. *et al.* Ordered nanoporous arrays of carbon supporting high dispersions of platinum nanoparticles. *Nature* **412**, 169–172 (2001).
6. Si, Y. C. & Samulski, E. T. Exfoliated graphene separated by platinum nanoparticles. *Chem. Mater.* **20**, 6792–6797 (2008).
7. Sheng, W. C., Gasteiger, H. A. & Shao-Horn, Y. Hydrogen oxidation and evolution reaction kinetics on platinum: acid vs alkaline electrolytes. *J. Electrochem. Soc.* **157**, B1529–B1536 (2010).
8. Khaselev, O. & Turner, J. A. A monolithic photovoltaic-photoelectrochemical device for hydrogen production via water splitting. *Science* **280**, 425–427 (1998).
9. Chhowalla, M. *et al.* The chemistry of two-dimensional layered transition metal dichalcogenide nanosheets. *Nat. Chem.* **5**, 263–275 (2013).
10. Benck, J. D. *et al.* Catalyzing the hydrogen evolution reaction (HER) with molybdenum sulfide nanomaterials. *ACS Catal.* **4**, 3957–3971 (2014).
11. Chianelli, R. R. *et al.* The reactivity of MoS₂ single-crystal edge planes. *J. Catal.* **92**, 56–63 (1985).
12. Kibsgaard, J., Chen, Z. B., Reinecke, B. N. & Jaramillo, T. F. Engineering the surface structure of MoS₂ to preferentially expose active edge sites for electrocatalysis. *Nat. Mater.* **11**, 963–969 (2012).
13. Lauritsen, J. V. *et al.* Size-dependent structure of MoS2 nanocrystals. *Nat. Nanotechnol.* **2**, 53–58 (2007).
14. Kong, D. S. *et al.* Synthesis of MoS₂ and MoSe₂ films with vertically aligned layers. *Nano Lett.* **13**, 1341–1347 (2013).
15. Wu, Z. Z. *et al.* MoS₂ nanosheets: a designed structure with high active site density for the hydrogen evolution reaction. *ACS Catal.* **3**, 2101–2107 (2013).
16. Tsai, C., Abild-Pedersen, F. & Norskov, J. K. Tuning the MoS₂ edge-site activity for hydrogen evolution via support interactions. *Nano Lett.* **14**, 1381–1387 (2014).
17. Chen, Z. B., Forman, A. J. & Jaramillo, T. F. Bridging the gap between bulk and nanostructured photoelectrodes: the impact of surface states on the electrocatalytic and photoelectrochemical properties of MoS₂. *J. Phys. Chem. C* **117**, 9713–9722 (2013).
18. Ho, W. K. *et al.* Preparation and photocatalytic behavior of MoS₂ and WS₂ nanocluster sensitized TiO₂. *Langmuir* **20**, 5865–5869 (2004).
19. Hinnemann, B. *et al.* Biornimetic hydrogen evolution: MoS₂ nanoparticles as catalyst for hydrogen evolution. *J. Am. Chem. Soc.* **127**, 5308–5309 (2005).
20. Li, Y. *et al.* MoS₂ nanoparticles grown on graphene: an advanced catalyst for the hydrogen evolution reaction. *J. Am. Chem. Soc.* **133**, 7296–7299 (2011).
21. Eda, G. *et al.* Photoluminescence from chemically exfoliated MoS₂. *Nano Lett.* **11**, 5111–5116 (2011).
22. Py, M. A. & Haering, R. R. Structural destabilization induced by lithium iintercalation in MoS₂ and related-compounds. *Can. J. Phys.* **61**, 76–84 (1983).
23. Joensen, P., Frindt, R. F. & Morrison, S. R. Single-layer MoS₂. *Mater. Res. Bull.* **21**, 457–461 (1986).
24. Matte, H. *et al.* MoS₂ and WS₂ analogues of graphene. *Angew. Chem. Int. Ed.* **49**, 4059–4062 (2010).
25. Coleman, J. N. *et al.* Two-dimensional nanosheets produced by liquid exfoliation of layered materials. *Science* **331**, 568–571 (2011).
26. Voiry, D. *et al.* Enhanced catalytic activity in strained chemically exfoliated WS₂ nanosheets for hydrogen evolution. *Nat. Mater.* **12**, 850–855 (2013).
27. Gao, M. R. *et al.* An efficient molybdenum disulfide/ cobalt diselenide hybrid catalyst for electrochemical hydrogen generation. *Nat. Commun.* **6**, 5982 (2015).
28. Wang, H. T. *et al.* Electrochemical tuning of vertically aligned MoS₂ nanofilms and its application in improving hydrogen evolution reaction. *Proc. Natl Acad. Sci. USA* **110**, 19701–19706 (2013).
29. Chen, Z. *et al.* Core-shell MoO₃-MoS₂ nanowires for hydrogen evolution: a functional design for electrocatalytic materials. *Nano Lett.* **11**, 4168–4175 (2011).
30. Cummins, D. R. *et al.* Catalytic activity in lithium-treated core–shell MoOₓ/MoS₂ nanowires. *J. Phys. Chem. C* **119**, 22908–22914 (2015).
31. Voiry, D. *et al.* Conducting MoS₂ nanosheets as catalysts for hydrogen evolution reaction. *Nano Lett.* **13**, 6222–6227 (2013).
32. Lukowski, M. A. *et al.* Enhanced hydrogen evolution catalysis from chemically exfoliated metallic MoS₂ nanosheets. *J. Am. Chem. Soc.* **135**, 10274–10277 (2013).
33. Wang, H. T. *et al.* Electrochemical tuning of MoS₂ nanoparticles on three-dimensional substrate for efficient hydrogen evolution. *ACS Nano* **8**, 4940–4947 (2014).
34. Eda, G., Fanchini, G. & Chhowalla, M. Large-area ultrathin films of reduced graphene oxide as a transparent and flexible electronic material. *Nat. Nanotechnol.* **3**, 270–274 (2008).
35. Lee, I.-Y. *et al.* Hydrazine-based n-type doping process to modulate Dirac point of graphene and its application to complementary inverter. *Org. Electron.* **14**, 1586–1590 (2013).
36. Feng, T. *et al.* Electron-doping of graphene-based devices by hydrazine. *J. Appl. Phys.* **116**, 224511-1–224511-6 (2014).
37. Mistry, K. S. *et al.* n-type transparent conducting films of small molecule and polymer amine doped single-walled carbon nanotubes. *ACS Nano* **5**, 3714–3723 (2011).
38. Klinke, C., Chen, J., Afzali, A. & Avouris, P. Charge transfer induced polarity switching in carbon nanotube transistors. *Nano Lett.* **5**, 555–558 (2005).
39. Talapin, D. V. & Murray, C. B. PbSe nanocrystal solids for n- and p-channel thin film field-effect transistors. *Science* **310**, 86–89 (2005).
40. Lee, I. *et al.* Non-degenerate n-type doping by hydrazine treatment in metal work function engineered WSe₂ field-effect transistor. *Nanotechnology* **26**, 455203 (2015).
41. Cummins, D. R. *Synthesis of Molybdenum Oxide Nanowires and Their Facile Conversion to Molybdenum Sulfide*. Master of Engineering thesis, Univ. Louisville (2009).
42. Pentland, N., Bockris, J. O. & Sheldon, E. Hydrogen evolution reaction on copper, gold, molybdenum, palladium, rhodium, and iron. *J. Electrochem. Soc.* **104**, 182–194 (1957).
43. Conway, B. E. & Tilak, B. V. Interfacial processes involving electrocatalytic evolution and oxidation of H₂, and the role of chemisorbed H. *Electrochim. Acta* **47**, 3571–3594 (2002).
44. Bonde, J. *et al.* Hydrogen evolution on nano-particulate transition metal sulfides. *Faraday Discuss.* **140**, 219–231 (2008).
45. Merki, D. & Hu, X. Recent developments of molybdenum and tungsten sulfides as hydrogen evolution catalysts. *Energy Environ. Sci.* **4**, 3878–3888 (2011).
46. Jaramillo, T. F. *et al.* Identification of active edge sites for electrochemical H-2 evolution from MoS₂ nanocatalysts. *Science* **317**, 100–102 (2007).
47. Pourbaix, M. *Atlas of Electrochemical Equilibria in Aqueous Solutions* 2nd edn (NACE International, 1974).
48. Hu, B., Mai, L. Q., Chen, W. & Yang, F. From MoO₃ nanobelts to MoO₂ nanorods: structure transformation and electrical transport. *ACS Nano* **3**, 478–482 (2009).
49. Bilgin, I. *et al.* Chemical vapor deposition synthesized atomically thin molybdenum disulfide with optoelectronic-grade crystalline quality. *ACS Nano* **9**, 8822–8832 (2015).
50. Kappera, R. *et al.* Phase-engineered low-resistance contacts for ultrathin MoS₂ transistors. *Nat. Mater.* **13**, 1128–1134 (2014).
51. Subba Rao, G. V. & Shafer, M. W. *Intercalation in Layered Transition Metal Dichalcogenides* 99–199 (D. Reidel Publishing Company, 1979).
52. Schollhorn, R., Sick, E. & Lerf, A. Reversible topotacti redox reactions of layered dichalcogenides. *Mater. Res. Bull.* **10**, 1005–1012 (1975).
53. Steckel, J. S., Coe-Sullivan, S., Bulovic, V. & Bawendi, M. G. 1.3 μm to 1.55 μm tunable electroluminescence from PbSe quantum dots embedded within an organic device. *Adv. Mater.* **15**, 1862–1866 (2003).
54. Williams, K. J. *et al.* Strong electronic coupling in two-dimensional assemblies of colloidal PbSe quantum dots. *ACS Nano* **3**, 1532–1538 (2009).
55. Law, M. *et al.* Structural, optical, and electrical properties of PbSe nanocrystal solids treated thermally or with simple amines. *J. Am. Chem. Soc.* **130**, 5974–5985 (2008).
56. Talapin, D. V. *et al.* Alignment, electronic properties, doping, and on-chip growth of colloidal PbSe nanowires. *J. Phys. Chem. C* **111**, 13244–13249 (2007).
57. Bischoff, J. L., Lutz, F., Bolmont, D. & Kubler, L. Use of multilayer techniques for xps indentification of various nitrogen environments in the Si/NH₃ system. *Surf. Sci.* **251**, 170–174 (1991).
58. Bang, J. *et al.* Molecular doping of ZnO by ammonia: a possible shallow acceptor. *J. Mater. Chem. C* **3**, 339–344 (2015).
59. Huang, J. Y. *et al.* Growth of N-doped p-type ZnO films using ammonia as dopant source gas. *J. Mater. Sci. Lett.* **22**, 249–251 (2002).
60. Szwarc, M. The dissociation energy of the N-N bond in hydrazine. *Proc. R Soc. Lond. Ser. A* **198**, 267–284 (1949).

61. Zhang, C. *et al.* Transfer hydrogenation of nitroarenes with hydrazine at near-room temperature catalysed by a MoO$_2$ catalyst. *Green Chem.* **18**, 2435–2442 (2016).

Acknowledgements

This work was funded primarily by Los Alamos Directed Research Grant. This work was performed, in part, at the Center for Integrated Nanotechnologies, an Office of Science User Facility operated for the U.S. Department of Energy (DOE) Office of Science. Los Alamos National Laboratory, an affirmative action equal opportunity employer, is operated by Los Alamos National Security, LLC, for the National Nuclear Security Administration of the US Department of Energy under contract DE-AC52-06NA25396. The authors would also like to acknowledge the Conn Center for Renewable Energy Research at the University of Louisville for facilities and access to characterization equipment. Development of samples and characterization was supported partially by DOE EPSCoR (DE-FG02-07ER46375) and by a graduate fellowship funded by NASA Kentucky under NASA award No: NNX10AL96H. We thank Dan Kelly and Joseph Dumont at Los Alamos National Laboratory for assistance with XPS analysis. We thank National Science Foundation NSF EPSCoR Grant 1355438 for supporting this work.

Author contributions

G.G., A.D.M. and M.K.S. designed the experiments and organization of the manuscript. D.R.C. developed the synthesis of the core/shell nanowires, applied the dilute hydrazine treatments, electrochemical measurements, and characterization (SEM, Raman and XPS). D.R.C. and G.G. co-wrote the manuscript. U.M. along with D.R.C. performed the electrochemical analysis. A.S. and G.S. designed and conducted *in situ* conductivity experiments. R.K. developed the gate-dependent HER experimental setup and assisted with measurements. A.M.-G. synthesized MoS$_2$ core-shell nanowires. R.K.S. performed XPS on core-shell nanowires and assisted with interpretation. J.J. conducted TEM analysis and assisted with XPS and UPS interpretation. R.K.G. performed electrochemical tests on MoS$_2$ bulk powders. J.Z. and J.L. fabricated nanowire and CVD flake field effect devices and performed conductivity measurements. M.C. provided valuable input and understanding of 2D materials and overall organization of the manuscript.

Additional information

Competing financial interests: The authors declare no competing financial interests.

13

Dynamic protein coronas revealed as a modulator of silver nanoparticle sulphidation *in vitro*

Teodora Miclăuş[1], Christiane Beer[2], Jacques Chevallier[3], Carsten Scavenius[4], Vladimir E. Bochenkov[1,5], Jan J. Enghild[4] & Duncan S. Sutherland[1]

Proteins adsorbing at nanoparticles have been proposed as critical toxicity mediators and are included in ongoing efforts to develop predictive tools for safety assessment. Strongly attached proteins can be isolated, identified and correlated to changes in nanoparticle state, cellular association or toxicity. Weakly attached, rapidly exchanging proteins are also present at nanoparticles, but are difficult to isolate and have hardly been examined. Here we study rapidly exchanging proteins and show for the first time that they have a strong modulatory effect on the biotransformation of silver nanoparticles. Released silver ions, known for their role in particle toxicity, are found to be trapped as silver sulphide nanocrystals within the protein corona at silver nanoparticles in serum-containing cell culture media. The strongly attached corona acts as a site for sulphidation, while the weakly attached proteins reduce nanocrystal formation in a serum-concentration-dependent manner. Sulphidation results in decreased toxicity of Ag NPs.

[1] Interdisciplinary Nanoscience Center (iNANO), Aarhus University, Gustav Wieds Vej 14, 8000 Aarhus, Denmark. [2] Department of Public Health, Aarhus University, Bartholins Alle 2, 8000 Aarhus, Denmark. [3] Department of Physics, Aarhus University, Ny Munkegade 120, 8000 Aarhus, Denmark. [4] Department of Molecular Biology and Genetics, Aarhus University, Gustav Wieds Vej 10, 8000 Aarhus, Denmark. [5] Department of Chemistry, Lomonosov Moscow State University, Leninskie gory 1/3, 119991 Moscow, Russia. Correspondence and requests for materials should be addressed to D.S.S. (email: duncan@inano.au.dk).

The biological effects of engineered nanomaterials as drug delivery vehicles or as unintentionally released nanoparticles (NPs) are of strong current interest. Biomolecules—mainly proteins—adsorbing at NPs modify their surface properties and are proposed as important modulators of particle–cell interactions[1–8]. A pragmatic distinction has been made between the relatively easily studied, strongly attached proteins as long-lived, hard coronas and the weakly attached, rapidly exchanging proteins as soft coronas[9–13]. The former are under focus with residence at the particles on timescales relevant for cellular binding and uptake[4,6,14], whereas the role of the latter in modulating NP behaviour has yet to be established. Specific and different profiles of molecules concentrated within the hard corona at particles in biological media have been observed for different surface coatings[15], charges[16,17], sizes[15,18] and shapes[19]. The concept of a biological identity imprinted within the protein corona and which determines NP–cellular interactions[1–6] has been proposed[20]. Although the long-lived layer has been linked to particle aggregation[19] and cell association[6,14,21], the correlation of protein composition to cellular uptake/toxicity is still relatively weak[4,22,23]. The involvement of soft corona in physical and/or chemical transformations of particle with potential implications for toxicity is so far unstudied, despite it forming a dense second layer around the strongly attached biomolecules[24].

In addition to protein corona formation, ion release is central to the toxicity of silver NPs and is an important parameter studied in vitro[25–27] and in vivo[28]. Oxidation contributes to ion release through the formation of Ag_2O on the particles[29,30], which is then dissolved in aqueous media[31–33]. Oxidative dissolution is an important step in Ag_2S formation from/at Ag NPs[34]. Silver NP sulphidation has been receiving increasing attention, as the resulting sulphide is insoluble in water, decreasing the availability of Ag^+ and impacting antibacterial[35] and toxicological[36–38] effects. After identification of silver sulphide in sewage sludge[39], interest in studying Ag_2S was focused on wastewater plants[40,41] and aquatic environments[42]. Although most toxicity experiments are conducted in vitro, much less is known about such transformations of Ag NPs under these conditions. Thiols (for example, cysteine) have been proven to bind Ag^+ in biological environments[43,44]. Tracking the oxidation state of intracellular silver showed an evolution from Ag^0 to

oxygen-bound and sulphur-bound Ag ions[45,46]. Formation of Ag_2S in alveolar cells was proposed to explain decreased toxicity of silver nanowires[47] and their sulphidation in protein-free culture medium was recently studied[48]. A further step involves exploring chemical changes that occur in full culture media, in the presence of protein coronas, before cellular uptake.

Here we demonstrate one clear role for the soft corona in modulating silver NP sulphidation in vitro, and highlight the interplay between strongly and weakly attached proteins for the chemical transformation of Ag NPs. We also suggest some potential implications for toxicity, without, however, establishing a clear direct connection between the soft corona and observed toxicological effects. We show for the first time a functional effect of rapidly exchanging proteins, which decreased the amount of nano-Ag_2S formed at polyvinylpyrrolidone (PVP)-coated Ag NPs incubated in serum-supplemented cell culture media. We propose and study a mechanism for soft corona protein-assisted Ag^+ transport explaining reduced sulphide formation. Striking differences when going from in vitro to in vivo relevant protein concentrations are observed and discussed. As it is known that sulphidation decreases silver toxicity[36–38,47,49–51], it is not surprising that under conditions where Ag NPs were partially or completely transformed into Ag_2S in cell culture media, much less toxicity to J774 macrophages and different cytokine secretion profiles are seen compared with silver NPs.

Results

Protein coronas modulate nano-Ag_2S formation at Ag NPs. Upon incubation of PVP-coated, cubic or quasi-spherical Ag NPs in RPMI-1640 cell culture medium supplemented with fetal bovine serum (FBS), new NPs were observed to form close to the surface of the silver. Details regarding incubation are available in the Methods section, Particle incubation in cell culture media subsection. Figure 1a shows a typical transmission electron microscopy (TEM) image of nanocubes after 7 days in 1% serum, with the NPs forming a dispersed layer around the silver core (highlighted by arrows). X-rays elemental mapping (Fig. 1b) and energy-dispersive X-ray spectroscopy (EDS, Fig. 1c) revealed the presence of sulphur. Co-localization of Ag and S matches the small NPs in the proximity of the silver surface (Fig. 1b). The

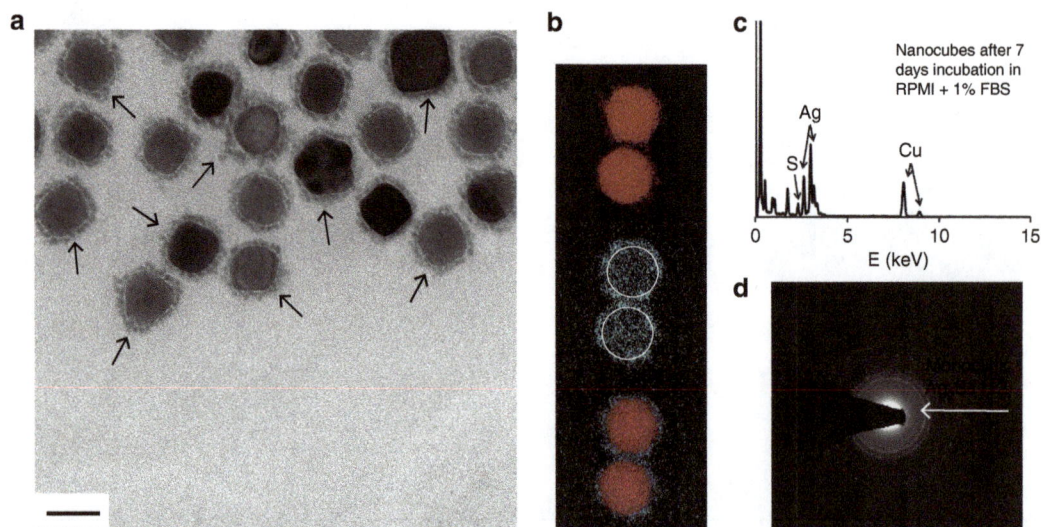

Figure 1 | Silver sulphide forms close to the surface of Ag NPs. TEM image with arrows highlighting nano-Ag_2S (**a**, scale bar 50 nm), X-rays elemental mapping of Ag (red), S (blue, with white rings marking the approximate contour of the Ag NPs) and overlaid Ag and S (**b**), EDS spectrum—with arrows pointing at the peaks corresponding to each element—(**c**) and diffraction pattern—arrow pointing at the diffraction line corresponding to monoclinic Ag_2S—(**d**) of silver nanocubes after 7 days incubation in RPMI-1640 supplemented with 1% FBS and formation of Ag_2S at the surface of the Ag NPs.

diffraction line at 2.80 (Fig. 1d) corresponds to monoclinic Ag$_2$S (ref. 52).

When in contact with biological media, NPs get covered with biomolecules[1–4]. Hard and soft protein coronas around silver nanocubes have previously been quantified and it has been shown that the polymer coating is replaced during the first hour in 1% serum[24]. We observe no sulphide within 1 h, before PVP replacement (Supplementary Fig. 1). The later appearance of Ag$_2$S close to the silver surface suggests its formation is related to the layers of adsorbed biomolecules. We hypothesize a mechanism where protein coronas and ion release govern the formation of Ag$_2$S at silver NPs (Fig. 2a). We propose that released Ag$^+$ can get trapped in the long-lived protein corona where, if enough reduced sulphur and Ag$^+$ are available, Ag$_2$S may form. In contrast, the rapidly exchanging soft corona proteins prevent sulphide formation by transporting Ag$^+$ away

from the particle, thus decreasing local ion concentration. To test this hypothesis, one must account for both hard and soft corona proteins, as well as bulk biomolecules.

Silver nanocubes (Supplementary Fig. 2) were incubated in RPMI-1640 with 1% FBS for 24 h (Fig. 2b) to provide a stable hard corona (Supplementary Discussion). The particles were washed, removing unbound and loosely bound proteins while retaining the hard corona (Fig. 2c, cartoon), and re-suspended in RPMI-1640 without serum for 6 days. Similarly, particles were incubated for 7 days in 1 or 10% FBS to ensure the presence of hard and soft coronas, and bulk proteins (Fig. 2d,e cartoon). The same behaviour was seen for quasi-spherical NPs (Supplementary Fig. 3). Ag$_2$S is observed under all incubation conditions where serum was present for at least the 24 h needed to fully form the hard corona[24]. Figure 2c–e shows typical TEM images of the different conditions–6 days at 0, 1 and 10% FBS after 24 h in

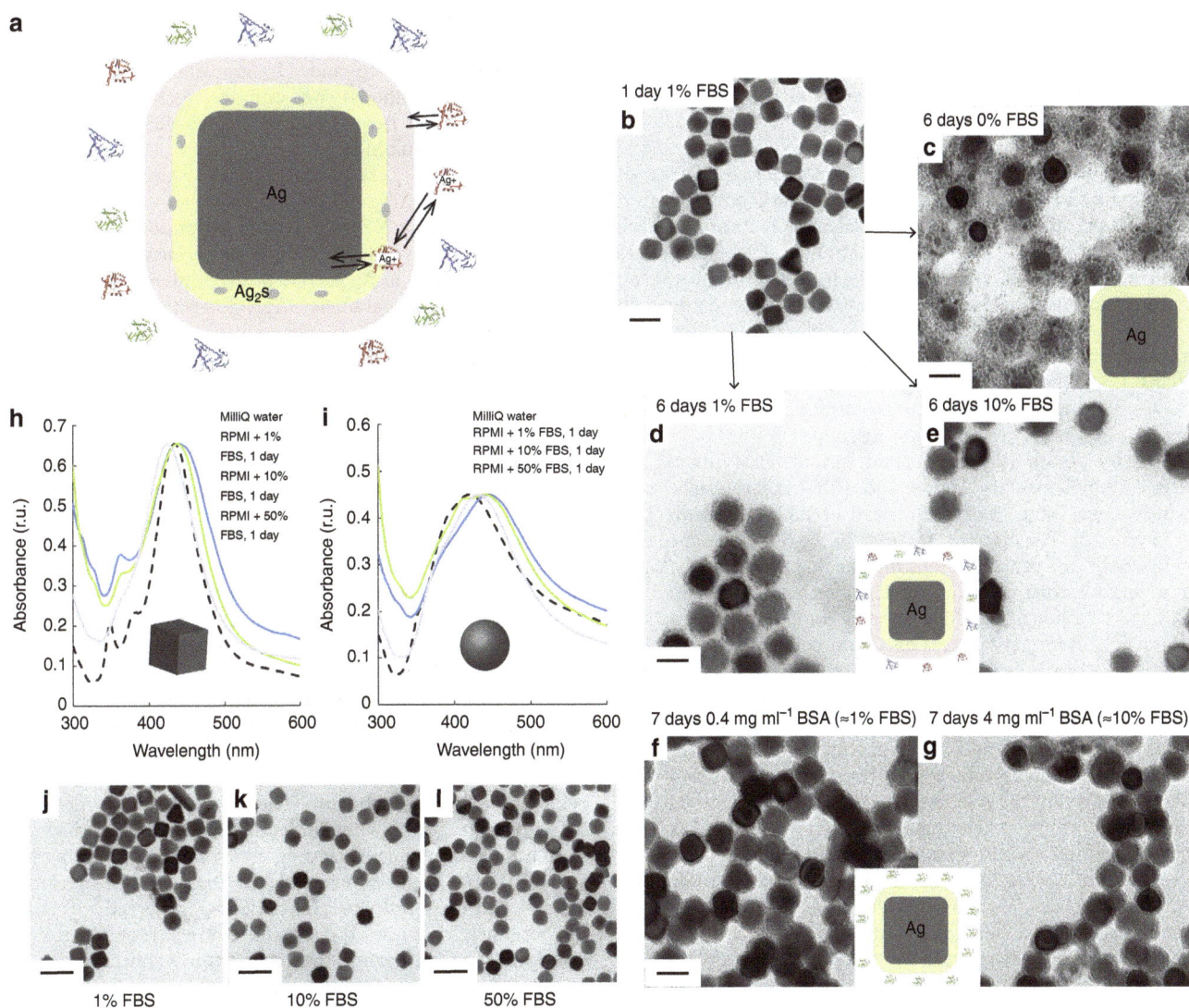

Figure 2 | Protein coronas modulate sulphide formation. Proposed mechanism of protein corona-modulated nano-Ag$_2$S formation at Ag NPs, with hard corona proteins trapping Ag$^+$ released from the nanoparticle surface and soft corona proteins transporting said ions away from the sulphide-formation centres in the long-lived corona (**a**); TEM images of silver nanocubes after 24 h in RPMI-1640 cell culture medium supplemented with 1% FBS (**b**), followed by 6 days incubation in RPMI-1640 with 0% FBS (inset cartoon showing only hard corona around Ag NPs) (**c**), 1% FBS (**d**) or 10% FBS (**e**; common inset cartoon showing hard and soft coronas, as well as free bulk proteins around Ag NPs); TEM images of silver nanocubes after 7 days incubation in RPMI-1640 with 0.4 mg ml^{-1} BSA (**f**) or 4 mg ml^{-1} BSA (**g**) (common inset cartoon showing hard corona and free bulk proteins around Ag NPs); Ultraviolet-visible spectra of cubic (**h**) and quasi-spherical (**i**) Ag NPs after 24 h incubation in RPMI-1640 cell culture medium supplemented with 1, 10 or 50% FBS; TEM images of silver nanocubes after 24 h in RPMI-1640 supplemented with 1% FBS (**j**), 10% FBS (**k**) and 50% FBS (**l**). Scale bars are 100 nm (**b**, **j**–**l**) or 50 nm (**c**–**g**).

serum-containing media. EDS confirmed the presence of sulphur (Supplementary Fig. 4). Although the total incubation time is the same (7 days) for all samples, the differences in Ag_2S amounts are striking, with significantly more sulphide observed at the particle surfaces when soft corona and free bulk proteins are absent after the first day. Incubating Ag NPs in serum-containing media for the initial 24 h establishes a long-lived corona, stable after the transition to 0% FBS. It has previously been shown that the hard corona is similar after 24 h in 1 and 10% FBS for different silver nanocubes[24], and in the present study, we observed the same protein profiles for NPs of various sizes (Supplementary Fig. 5). Here there is a similar long-lived layer on all Ag NPs, as confirmed by mass spectrometry (Supplementary Tables 1–3); what varies (0, 1 and 10% FBS) are the soft corona and the bulk protein concentrations. Nano-Ag_2S formation at Ag NPs decreased with soft corona presence and the increase in free proteins, with significantly more sulphide at 0% than 1% and 10% FBS. The variation does not appear linear: differences between 1 and 10% FBS are not as striking as between 0 and 1%. Furthermore, although in the presence of serum the particles are dispersed even after 7 days (Supplementary Fig. 6), at 0% FBS the nano-Ag_2S forms bridges between Ag NPs, which result in particle agglomeration (Fig. 2c).

Both soft corona and bulk proteins could bind Ag^+; distinguishing between the two categories requires absence of one of them. To test the role of soft coronas, bovine serum albumin (BSA), at concentrations equivalent to 1, 10 or 50% FBS, was used to form a hard corona around silver nanocubes (Supplementary Fig. 7). When only BSA is present in the system, the protein–protein interactions needed to form soft coronas do not occur (Supplementary Fig. 8 and Supplementary Methods); incubation in albumin provides only hard coronas and bulk proteins (Fig. 2f,g, cartoon). Figure 2f,g shows slightly more nano-Ag_2S formation when BSA concentration is increased from equivalent of 1% FBS to 10% FBS; the opposite phenomenon is observed for serum (Fig. 2d,e). Full sets of TEM images are available (Supplementary Figs 9 and 10); similar results were obtained when using lysozyme instead of BSA (Supplementary Figs 11 and 12). The observed increase in Ag_2S content at higher albumin/lysozyme concentrations may be due to a thicker, faster-formed hard corona, allowing for trapping and sulphidation of more Ag^+. Although sulphide appears for incubation in both FBS and BSA/lysozyme, increasing serum concentration—unlike increasing albumin/lysozyme concentration—decreases Ag_2S formation. This is attributed to the lack of a soft corona when a single type of protein exists in the system[11]—as shown in Supplementary Fig. 8 and Supplementary Methods—as hard coronas and free bulk proteins are present in both cases.

To further study the soft corona influence, experiments were performed at serum concentrations from 1 to 50%, over 24 h. Ultraviolet–visible spectra were collected and plasmon peak shifts were observed. Owing to localized surface plasmon resonances, variations in peak position indicate refractive index changes around particles. Red shifts of 2–4 nm upon binding of proteins to Ag NPs have been seen previously[24]. Here, we observe changes of up to 45 nm indicating the presence of a material with high refractive index—like Ag_2S—close to the Ag NPs. Comparison to TEM images (Fig. 2j,l and Supplementary Figs 13 and 14) shows the more sulphide present at the surface, the larger the red-shift in the plasmon peaks is. Increased protein concentration and, therefore, soft corona protein content[24], results in a visible decrease of nano-Ag_2S formation not just at prolonged incubation, but also after 24 h. Analysis of multiple TEM images revealed almost no sulphide at the NP surface in 50% serum. Furthermore, TEM and ultraviolet–visible data indicate a lower sulphide content after 24 h in 10% compared with 1% FBS.

Figure 2h,i shows spectra for cubic and quasi-spherical Ag NPs after 24 h in 1, 10 or 50% serum. Decreasing FBS concentration and, implicitly, the amount of soft corona[24], results in the formation of more Ag_2S at Ag NPs, as suggested by the plasmon peak shifted towards higher wavelengths when going from 50% to 10% and 1% serum. NP dissolution (Supplementary Fig. 15) results in reshaping and resizing on a similar timescale for these nanocubes, with concomitant changes of the optical spectra. This leads to the decreased intensity of the signal around 350 nm, where a more prominent peak is characteristic of larger silver nanocubes, with sharper edges[53]. Reduction in particle diameter, coupled with shape changes from cubes to sphere-like, also explains the blue-shift around 435 nm (Fig. 2h) after incubation in 50% serum. No reshaping is observed for the quasi-spherical NPs, while slight diameter decrease may occur.

It has been shown that shifts in plasmon peak position, coupled with finite-difference time-domain simulations of plasmonic behaviour may be used to quantify refractive index changes caused by protein adsorption at Ag NPs[24]. Via a similar approach (Supplementary Figs 16 and 17, Supplementary Table 4 and Supplementary Methods) we estimate that, at the Ag NP concentration used here ($10 \, \mu g \, ml^{-1}$), 15–40% of the silver is transformed into sulphide.

Ion release from Ag NPs is necessary for nano-Ag_2S formation. Sulphidation of silver NPs involves the presence of ionic silver therefore variations in Ag^+ content may influence the formation of Ag_2S. To test this parameter, $AgNO_3$—with Ag^+ representing 10% by weight of the particulate silver—was added to the NP suspension at the beginning of 1 or 7 days incubation in RPMI-1640 with 1% FBS (Supplementary Figs 18–21 for 10% serum). ultraviolet–visible spectra of quasi-spherical particles with or without extra ions were collected (Fig. 3a); a less pronounced red-shift was seen when free Ag^+ were added, especially at short incubation, suggesting a slower NP dissolution in the presence of extra ions.

Nanocubes have several characteristic plasmon peaks, of which a quadrupole (≈ 350 nm) and a dipole (≈ 435 nm; Supplementary Fig. 18). The quadrupole indicates the degree to which a NP is cubic, as well as its size[53], and, as such, it is used to track the synthesis of silver nanocubes[54] (Supplementary Fig. 22). Dissolution reduces particle size and blunts cube edges, resulting in flattening of the quadrupole peak (Fig. 3b). After 7 days in 1% FBS, the signal disappears almost completely.. At 24 h, both samples exhibit a pronounced quadrupole peak, with a visibly sharper signal when 10% free Ag^+ were added at the beginning of the incubation. The results suggest Ag NPs in samples with added ions undergo slower and, perhaps, less dissolution, observable in resizing and reshaping of the particles. This behaviour could be explained by the existence of a plateau ion concentration in the bulk, as previously observed[32,45]. Adding free ions may decreases the amount of Ag^+ required from NP dissolution for achieving this equilibrium concentration, but further investigations would be necessary to elucidate the mechanism.

Inspection of multiple TEM images suggests there is no difference in the amount of Ag_2S at silver NPs with or without added 10% Ag^+ at 7 days, but some decrease in sulphide incidence occurs at 24 h (Supplementary Fig. 19). This indicates the formation of nano-Ag_2S requires the release of Ag^+ from NPs and not just the existence of free ions in the bulk. The observation is strengthened by control experiments with silica particles in RPMI-1640 with FBS and free silver ions; no nano-Ag_2S was detected after 7 days (Supplementary Fig. 23).

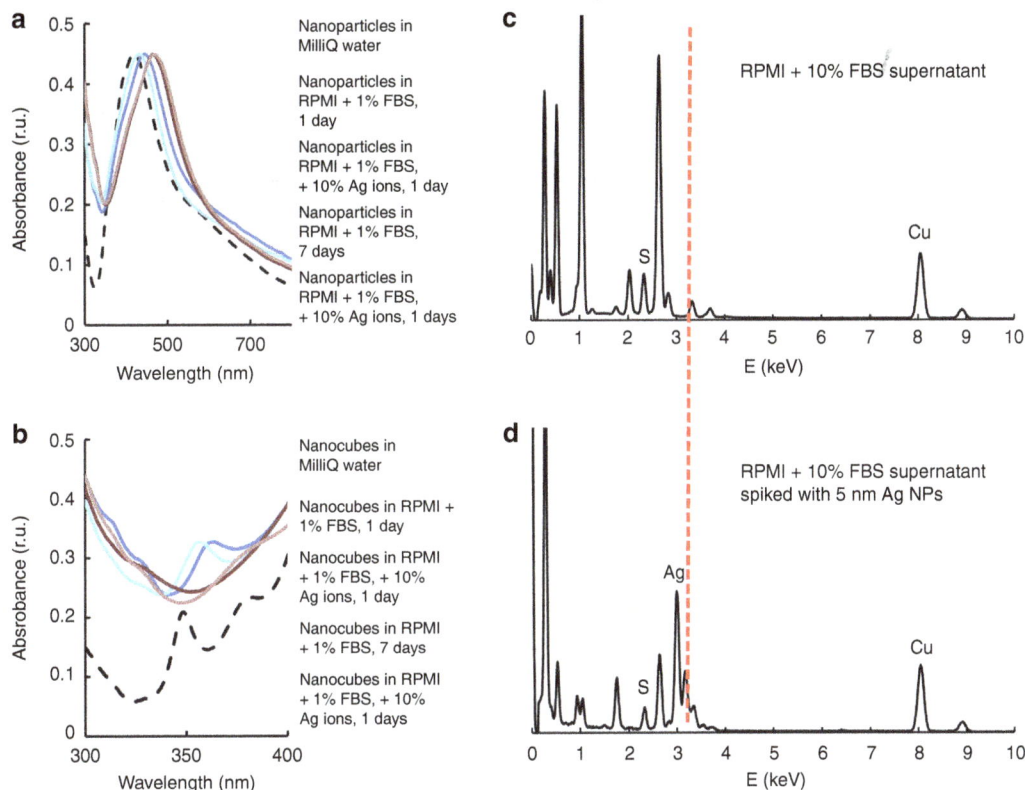

Figure 3 | Silver nanoparticle dissolution is involved in nano-Ag₂S formation. Ultraviolet–visible full spectra of quasi-spherical Ag NPs (**a**) and quadrupole peak detail of nanocubes (**b**) incubated (1 day: blue or 7 days: pink) in RPMI-1640 cell culture medium supplemented with 1% FBS, with or without added extra 10% (by mass) Ag⁺ ions from AgNO₃; EDS spectra of the supernatant obtained after centrifugation of Ag NPs incubated for 7 days in RPMI-1640 with 1% FBS, before (**c**) and after (**d**) spiking with 5 nm PVP-coated Ag NPs, with dotted red line highlighting the presence of a silver signal only in the spiked sample.

Atomic absorption spectroscopy (Supplementary Fig. 15) confirmed the presence of silver in the bulk, strengthening the soft corona ion transport mechanism; it did not, however, indicate the chemical form of the silver. Although the timescale of soft corona exchange is on the order of seconds and minutes and nano-Ag₂S formation occurs over hours, we verified that at prolonged incubation the rapidly exchanging proteins do not transport Ag₂S nano-crystals away from the Ag NP surface. We measured EDS spectra and mapped supernatants collected after 7 days incubation in 1% FBS, before and after spiking in PVP-coated Ag NPs (Fig. 3c,d and Supplementary Fig. 24) of similar size (5 nm) and amount to that of nano-Ag₂S. If Ag₂S nanocrystals were transported to the bulk, a peak would appear in the EDS spectrum around 3 keV. Absence of this signal in the un-spiked supernatant indicates nano-Ag₂S is not transported by the soft corona.

Sulphur sources and Ag/S ratio influence nano-Ag₂S formation. Sulphur-containing gases can contribute to the formation of Ag₂S at Ag NPs exposed to air, but this is a lengthy and slow process, extended over 24 weeks[55]. In the liquid phase, RPMI-1640 provides several sulphur sources (Supplementary Tables 5 and 6), with L-cysteine and L-methionine accounting for most of the reduced S; furthermore, many serum proteins contain cysteine residues. To pin-point the sulphur responsible for Ag₂S formation in our case, we incubated Ag NPs in phosphate-buffered saline (PBS) supplemented with either 1% FBS or L-cysteine and L-methionine at the same concentrations as in RPMI-1640. After 7 days, no sulphide was seen in PBS or in buffer with serum (Fig. 4a,b), but nano-Ag₂S was present in the

amino-acid-supplemented buffer (Fig. 4c and Supplementary Fig. 25). EDS spectra (Fig. 4d) confirmed the observations from TEM images. These experiments show reduced sulphur from small molecules and not from proteins is the one forming Ag₂S.

We further tested the effect of Ag/S ratios by varying the amount of nanocube stock suspension added to a given volume of serum-supplemented RPMI-1640. After 7 days, increasing the initial silver concentration from 2 to 10 and then to 100 μg ml⁻¹ (Fig. 4e–g and Supplementary Fig. 26) drastically decreases Ag₂S formation by limiting the L-cysteine and L-methionine available per Ag⁺. *In vivo* cysteine concentration is more than double that in RPMI, making more S available for Ag₂S formation[56,57]. At the lowest Ag/S ratio, the particles are almost entirely transformed into Ag₂S, forming 'pockets' of sulphide that conserve the shape of the initial particle, with the metal core still visible in some cases (Fig. 4e). These observations suggest sulphidation is confined to the protein hard corona, in agreement with EDS showing the absence of nano-Ag₂S from the suspension supernatant (Fig. 3c,d).

Protein corona-mediated sulphidation impacts cell toxicity. Previous research has already shown that trapping Ag⁺ in the form of insoluble Ag₂S decreases the toxicity of silver[50,51] and Ag NPs[35,49], mostly in the case of aquatic environments[36,37,58] and in soil[38], which are settings with low protein contents, but higher concentrations of other components that are not prevalent under *in vitro* cell study conditions. Although the effects under *in vitro* parameters have not been studied to the same extent, some information is also available on the decreased toxicity of sulphidated Ag NPs to cultured cells[47]. Our results confirm

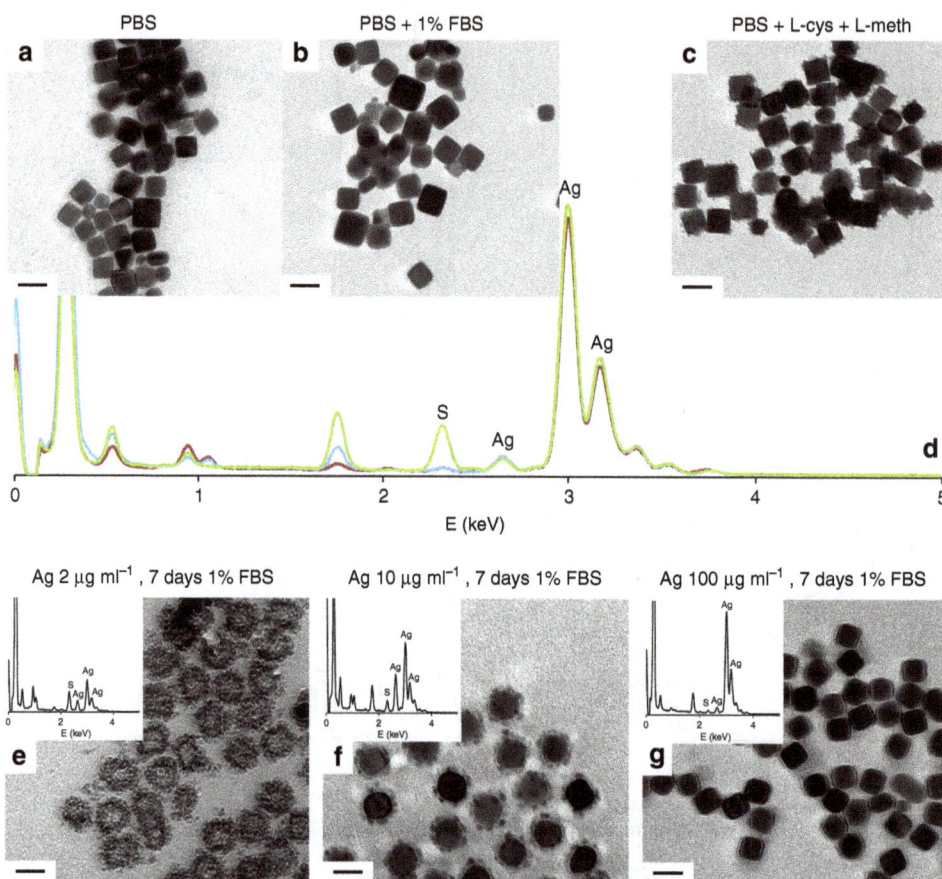

Figure 4 | Sulphur sources and the Ag:S ratio influence Ag$_2$S formation. TEM images of Ag NPs after 7 days incubation in PBS (**a**), PBS supplemented with 1% FBS (**b**) and PBS supplemented with L-cysteine and L-methionine at the same concentrations of amino acids as those found in RPMI-1640 (**c**) and corresponding EDS spectra (**d**); TEM images and corresponding EDS spectra (insets) of Ag NPs after 7 days incubation in RPMI-1640 supplemented with 1% FBS, with initial silver concentrations of 2 μg ml^{-1} (f), 10 μg ml^{-1} (f) and 100 μg ml^{-1} (**g**), with elemental mapping images provided in Supplementary Fig. 26. Scale bars are 100 nm (**a-c**) or 50 nm (**e-g**).

such already published findings and extend the observations to a cell line where the consequences of sulphidation have not previously been investigated. We indirectly tested the effects of corona-mediated sulphidation on the toxicity of Ag NPs to J774 macrophages by exposing cells to partially and completely sulphidated NPs (obtained by pre-incubation in 10 and 1% FBS respectively, Fig. 5a,b) to mimic *in vitro* conditions, as well as NPs with no sulphide, to mimic *in vivo* settings (Supplementary Fig. 27). Detailed experimental procedures are available in the Methods section and in Supplementary Methods. Ag NPs disrupted mitochondrial activity, as measured using a 3-(4,5-dimethylthiazol-2-yl)-2,5-diphenyltetrazolium bromide (MTT) assay, and caused cell death at concentrations above 25 μg ml^{-1}, whereas partially and completely transformed particles showed no effect at concentrations as high as 100 μg ml^{-1} (Fig. 5c). Silver ions were lethal to the cells even at the lowest concentration (2 μg ml^{-1}), but pristine, partially and completely sulphidated Ag NPs were suitable for analysis of potential effects at sub-lethal doses. Of all the molecules measured in the cell supernatants, responses above the detection limit were seen for interleukin-1beta, interleukin-6, interleukin-18, tumour necrosis factor alpha (TNFα) and macrophage inflammatory protein 2 (MIP-2; Supplementary Fig. 28 and Fig. 5). Granulocyte–macrophage colony-stimulating factor was also measured at the highest particle doses in the Ag NPs samples, but for most systems, granulocyte–macrophage colony-

stimulating factor values were below the detection limit and, therefore, they are not included here. The most pronounced impact is observed on TNFα and MIP-2. TNFα is a cytokine released by macrophages at early inflammatory stages together with interleukin 6 (ref. 59), which we also observed (Supplementary Fig. 28). A threefold increase in its production is caused by pristine and partially sulphidated Ag NPs, but not by completely transformed ones (Fig. 5d). This result indicates upregulation of TNFα production by Ag requires the presence of nano-particulate or ionic silver, as previously seen[60,61]. A similar trend (Fig. 5e) was observed for MIP-2, a cytokine involved in cell recruitment to the site of infection following the initiation of the inflammatory response[62]. Together, these results confirm that sulphidation, which we show is mediated by the dynamic protein coronas and is more likely to occur at the serum concentrations used *in vitro* than *in vivo*, decreases the toxicity of silver NPs both at lethal and sub-lethal doses. However, although we show, for the first time, a link between both soft and hard protein coronas and Ag NPs sulphidation and confirm previous findings connecting sulphidation to decreased toxicity, we cannot, at this time, provide a direct link between the soft corona and the diminished toxicological effects of sulphidated Ag NPs. Furthermore, technical limitations in conducting *in vitro* studies at different FBS concentrations do not allow us to account for the Ag$^+$ transported by soft corona proteins in the bulk of the pre-incubation system. As silver ions are known for their toxicity

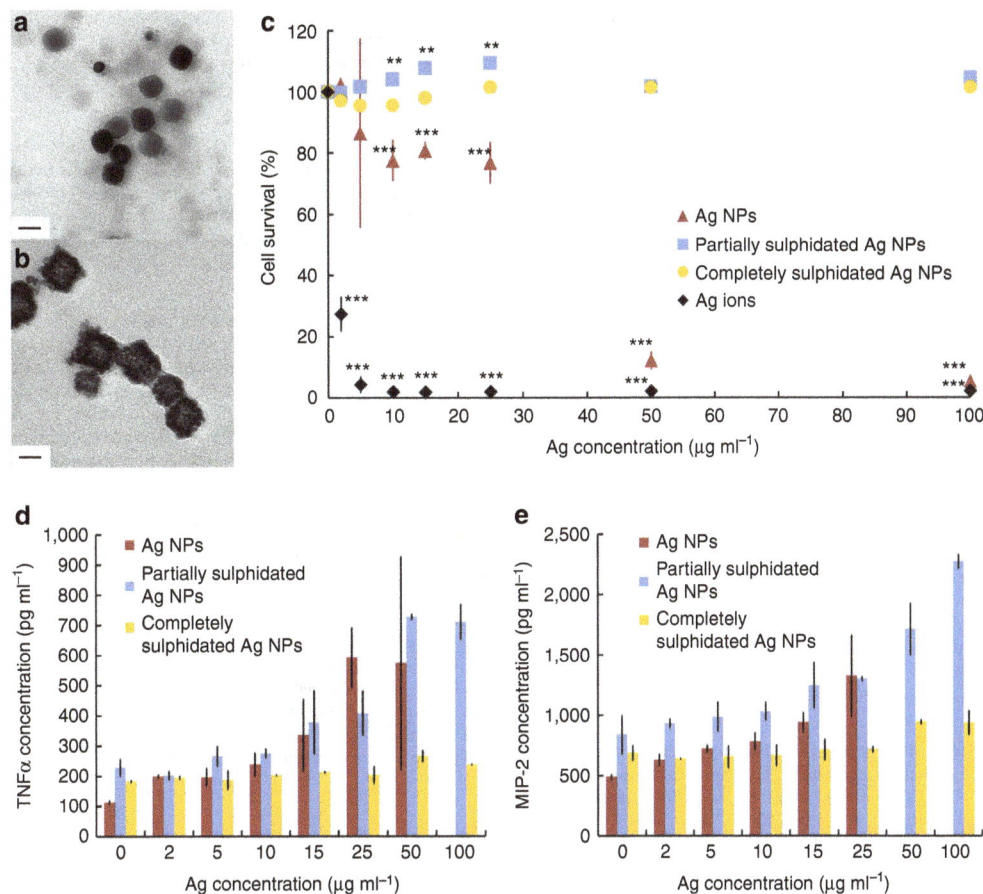

Figure 5 | Corona-mediated sulphidation of Ag NPs impacts particle toxicity. TEM images of partially sulphidated Ag NPs after pre-incubation in RPMI-1640 with 10% FBS (**a**) and completely sulphidated Ag NPs after pre-incubation in RPMI-1640 with 1% FBS (**b**); scale bars are 50 nm; Viability of J774 murine macrophages (as measured with MTT assays) after 24 h exposure to various concentrations (2, 5, 10, 15, 25, 50 and 100 µg ml^{-1}) of Ag$^+$ ions (black diamonds), pristine Ag NPs (red triangles), partially sulphidated Ag NPs (blue squares) and completely sulphidated Ag NPs (orange circles); error bars are provided as standard deviation; statistically significant differences (two-tailed t-test, with all data sets showing normal distribution and similar variance values) as compared with the control are marked with **$P < 0.005$ or ***$P < 0.0005$ ($n = 6$), with all the P values available in Supplementary Table 7 (**c**); release profiles of TNFα (**d**) and MIP-2 (**e**) after 24 h exposure of J774 macrophages to various concentrations (2, 5, 10, 15, 25, 50 and 100 µg ml^{-1}) of pristine (red), partially sulphidated (blue) and completely sulphidated (orange) Ag NPs; the missing concentrations of TNFα and MIP-2 after exposure to pristine Ag NPs are above the measuring limit (see calibration curves in Supplementary Fig. 29).

and our experimental setting does not permit investigation of their interaction with cells when bound to rapidly exchanging proteins, we cannot make a general claim about the overall toxicological impact that the soft corona-mediated biotransformation of Ag NPs has under various *in vitro* or *in vivo* scenarios.

Discussion

We have shown for the first time a situation where the weakly attached protein layer forming the soft corona has a visible and measurable effect on the transformation of Ag NPs in a complex biological environment. We demonstrated the presence of crystalline nano-Ag$_2$S at the surface of silver NPs upon incubation in cell culture medium. Reduced sulphur, an organic layer at the Ag NPs and release of ions from the metal core are necessary for sulphide formation. Protein concentration greatly impacted the amount of nano-Ag$_2$S observed at the particles through a soft corona protein-assisted mechanism of Ag$^+$ removal. In the absence of a rapidly exchanging corona, the decrease in sulphide formation upon increased bulk protein concentration was no longer observed, in agreement with the proposed mechanism. Greater free silver ion concentrations

introduced in the media did not increase sulphide formation. We have studied well-defined Ag nanocubes and quasi-spherical particles formed via PVP stabilization. Although we expect the proposed mechanism to apply to Ag NPs having other sizes, shapes and coatings, the particulars of the experimental situation—such as incubation time, media, Ag$_2$S formation rate and crystal size, protein concentration—are likely to influence the specific outcomes.

The low water solubility of Ag$_2$S decreases Ag$^+$ ions bioavailability and, although this phenomenon has been studied extensively in ecotoxicology settings[32,37,39–42], little is known about the effect of NP sulphidation in cell culture media[48]. Ions are an important component in silver toxicity to cells; transformations impacting their availability should be taken into account when analysing the stability of Ag NPs in protein-containing media relevant for *in vitro* experiments and interpreting subsequent toxicity studies. We show that even partial sulphidation of Ag NPs prevents cell death, whereas complete sulphidation also prevents the increased pro-inflammatory cytokines production seen with pristine particles. The observed decrease in Ag$_2$S formation at increased protein contents may raise a question regarding using *in vitro* results to predict *in vivo* scenarios, as the bulk biomolecule concentration in the latter settings is much higher than in the former. However, for this to become an issue, a direct link between the dynamic protein coronas and the toxic effects of Ag NPs should first be established in future research. Further studies into the bioavailability and effects of soft corona-bound Ag^+ from particle dissolution in protein-containing media are also necessary to obtain a clear and full picture of how biomolecules-modulated biotransformations may change NPs' toxicity effects *in vitro* and *in vivo*.

Methods

Particle synthesis and characterization. Silver NPs, both cubic and quasi-spherical, were prepared using the polyol method[63], where particle shape is controlled by the ratio between the capping agent (PVP) and the silver precursor[64]. Briefly, a specific amount of silver trifluoroacetate dissolved in anhydrous ethylene glycol is reduced by ethylene glycol at high temperature (145–155 °C) in the presence of PVP (Mw $\approx$ 55,000 Da). The ratio of PVP to CF_3COOAg dictates the outcome regarding particle shape. For the cube synthesis, trace amounts of HCl and $NaSH \cdot xH_2O$ are added to the reaction mixture, as described elsewhere[63]. The particles were purified by repeated washing with acetone, ethanol and MilliQ water[24]. The synthesis was tracked by collecting ultraviolet–visible spectra (Shimadzu UV-visible-NIR spectrophotometer, UV-3600) of the reaction mixture (a few drops in MilliQ water) at various times (Supplementary Fig. 22). The resulting silver particles were quantified using flame atomic absorption spectroscopy (F-AAS, PerkinElmer Analyst 300 atomic absorption spectrometer mounted with a silver lumina hollow cathode lamp), after digestion in 65% HNO_3. NP size was obtained using the SPIP scanning probe image software (Image Metrology) to analyse TEM images of at least 500 particles.

Particle incubation in cell culture media. RPMI-1640 (Invitrogen) is a medium widely used for cell cultures, including for the J774 murine macrophages employed here. We incubated Ag NPs (cubic or quasi-spherical) in RPMI-1640 with or without added supplements of heat-inactivated FBS (HyClone; 1–50% by volume), for 1 or 7 days. As we are studying a model system from the perspective of a mechanism of chemical transformation, our choice of serum concentrations and incubation times is only partially based on real toxicology settings. As such, 10% FBS is used as the typical serum supplement for *in vitro* toxicity studies[65-67], with 24 h being the go-to time for acute toxicity experiments[26,45,65-69]. However, *in vitro* Ag NPs exposure studies have been performed at lower serum concentrations[45,68,69], down to 1% (ref. 25), which is the concentration selected in our work. Furthermore, the 1% provides a better model system to study and understand mechanisms, as the low protein concentration makes the protein exchange-dependant process at the NP surface take longer, with fewer biomolecules available to participate. Although FBS contents higher than 10% are not common for *in vitro* studies, they are closer to an *in vivo* situation, so 50% serum was selected to better investigate the behaviour of Ag NPs in a realistic setting. However, we perform a mechanistic study on model systems, therefore the incubation times were chosen to have a fully formed hard corona on Ag NPs[24] (24 h) and to observe an extensive chemical transformation of the NPs (7 days), even though the onset of the sulphidation occurs, as we show, much earlier and clear changes are visible at 24 h. To study the influence of free ions on nano-Ag_2S formation, $AgNO_3$ (Sigma-Aldrich) aqueous solutions (10% Ag^+ by weight of the total Ag NP mass) were added to the media at the beginning of the incubation process for some of the samples.

Sample analysis post-incubation. Ultraviolet–visible spectra were collected against a MilliQ water background, over a wavelength domain of 300–800 nm. The peak positions of the resulting spectra were assessed using the Savitzky–Golay method to derive the spectral plots in the SpecManager software (ACD Labs). After incubation, free proteins were eliminated by centrifugation of particles in a Heraeus Multifuge X1R table top centrifuge (Thermo Scientific) and removal of the supernatant. The silver ions in the supernatants were quantified through F-AAS. Ag NPs were washed with MilliQ water three times through re-dispersion of the pellets and TEM/STEM samples were prepared by dropping 5–10 μl of suspension on a formvar/carbon-supported copper grid (Ted Pella) and leaving it to dry overnight. Imaging and diffraction studies were performed using a Phillips CM20 transmission electron microscope working at 200 keV. EDS and X-ray elemental mapping experiments were performed on a Talos 200X STEM instrument.

EDS and elemental mapping of supernatants. Silver nanocubes were incubated (10 μg ml^{-1}) for 7 days in RPMI-1640 cell culture medium supplemented with 1% FBS. The resulting sample was centrifuged to pellet the particles. A drop of the supernatant was deposited on a formvar/carbon-supported TEM copper grid and left to dry overnight. 150 μl of the remaining supernatant was separated from the pelleted particles, moved to a clean Eppendorf tube and spiked with 40 μl of PVP-coated NanoXact 5 nm Ag NPs (nanoComposixm, stock concentration 20 μg ml^{-1}). The NanoXact Ag NPs used for spiking are comparable in size to the observed Ag_2S nanocrystals and the amount added provides a mass of silver

comparable to that in the Ag_2S nanocrystals formed at the Ag NPs. A drop of the spiked supernatant was deposited on a TEM grid and left to dry overnight. EDS spectra and elemental maps from the resulting samples were collected using a Talos 200X STEM machine.

Cell line. The J774A.1 murine macrophage cell line (referred to as J774) was obtained from Cell Line Service (#400220). The cells were cultured in RPMI-1640 cell culture medium supplemented with penicillin (100 μg ml^{-1}), streptomycin (100 U ml^{-1}), GlutaMAX (1 $\times$) and 10% heat-inactivated FBS. The cells were kept in at 37 °C humidified atmosphere with 5% CO_2. The cell culture medium and all the supplements were purchased from Gibco.

Particle pre-incubation and toxicity studies. Pristine Ag nanocubes, as well as partially and completely sulphidated Ag nanocubes were used for toxicity testing. Silver NPs (2 μg ml^{-1}) were pre-incubated in RPMI-1640 supplemented with 10 or 1% FBS, resulting in either their partial or their complete transformation into Ag_2S (Supplementary Fig. 27). After pre-incubation, the samples were centrifuged, the supernatants were removed and the pelleted particles were re-suspended in RPMI-1640 with 10% FBS. Re-suspension was done in such a way as to up-concentrate the samples to the desired concentrations, namely 2, 5, 10, 15, 25, 50 and 100 μg ml^{-1} silver. For each of the final concentrations, pre-incubation was done in separate tubes. For the partially and completely transformed Ag NPs, up-concentration was done correcting for the amount of silver lost through ion release during pre-incubation. The resulting samples were fed to the J774 cells. Dosing of the cells was performed in a blinded experimental setting with the researcher performing the toxicity studies not being informed of the specific content of each tube provided for cell treatment. Cytokine production was quantified using a multiplex assay and cell mitochondrial activity was measured using an MTT assay (Supplementary Methods) in six separate replicates. Sample sizes down to three are routinely used in MTT assays. We believe larger sample numbers (six in this case) only serve to strengthen the significance of the results. Values are calculated as mean ± standard deviation.

References

1. Monopoli, M. P., Åberg, C., Salvati, A. & Dawson, K. A. Biomolecular coronas provide the biological identity of nanosized materials. *Nat. Nanotechnol* **7**, 779–786 (2012).
2. Walczyk, D., Baldelli Bombelli, F., Monopoli, M. P., Lynch, I. & Dawson, K. A. What the cell "sees" in bionanoscience. *J. Am. Chem. Soc.* **132**, 5761–5768 (2010).
3. Tenzer, S. *et al.* Rapid formation of plasma protein corona critically affects nanoparticle pathophysiology. *Nat. Nanotechnol* **8**, 772–781 (2013).
4. Walkey, C. D. *et al.* Protein corona fingerprinting predicts the cellular interaction of gold and silver nanoparticles. *ACS Nano* **8**, 2439–2455 (2014).
5. Albanese, A. *et al.* Secreted biomolecules alter the biological identity and cellular interactions of nanoparticles. *ACS Nano* **8**, 5515–5526 (2014).
6. Walkey, C. D., Olsen, J. B., Guo, H., Emili, A. & Chan, W. C. W. Nanoparticle size and surface chemistry determine serum protein adsorption and macrophage uptake. *J. Am. Chem. Soc.* **134**, 2139–2147 (2012).
7. Pelaz, B. *et al.* Interfacing engineered nanoparticles with biological systems: anticipating adverse nano-bio interactions. *Small* **9**, 1573–1584 (2013).
8. Docter, D. *et al.* The nanoparticles biomolecule corona: lessons learned – challenge accepted? *Chem. Soc. Rev.* **44**, 6094–6121 (2015).
9. Del Pino, P. *et al.* Protein corona formation around nanoparticles – from the past to the future. *Mater. Horiz* **1**, 301–313 (2014).
10. Casals, E., Pfaller, T., Duschl, A., Oostingh, G. J. & Puntes, V. Time evolution of the nanoparticle protein corona. *ACS Nano* **4**, 3623–3632 (2010).
11. Röcker, C., Pötzl, M., Zhang, F., Parak, W. J. & Nienhaus, G. U. A quantitative fluorescence study of protein monolayer formation on colloidal nanoparticles. *Nat. Nanotechnol* **4**, 577–580 (2009).
12. Milani, S., Baldelli Bombelli, F., Pitek, A. S., Dawson, K. A. & Rädler, J. Reversible versus irreversible binding of transferrin to polystyrene nanoparticles: soft and hard corona. *ACS Nano* **6**, 2533–2541 (2012).
13. Monopoli, M. P. *et al.* Physical-chemical aspects of protein corona: relevance to *in vitro* and *in vivo* biological impacts of nanomaterials. *J. Am. Chem. Soc.* **133**, 2525–2534 (2011).
14. Fedeli, C. *et al.* The functional dissection of the plasma corona of SiO_2-NPs spots histidine rich glycoprotein as a major player able to hamper nanoparticle capture by macrophages. *Nanoscale* **7**, 17710–17728 (2015).
15. Lundqvist, M. *et al.* Nanoparticle size and surface properties determine the protein corona with possible implications for biological impacts. *Proc. Natl Acad. Sci. USA* **105**, 14265–14270 (2008).
16. Eigenheer, R. *et al.* Silver nanoparticle protein corona composition compared across engineered particle properties and environmentally relevant reaction conditions. *Environ. Sci. Nano* **1**, 238–247 (2014).

17. Nel, A. E. *et al.* Understanding the biophysicochemical interactions at the nano-bio interface. *Nat. Mater.* **8**, 543–557 (2009).

18. Tenzer, S. *et al.* Nanoparticle size is a critical physic-chemical determinant of the human blood plasma corona: a comprehensive quantitative proteomic analysis. *ACS Nano* **5**, 7155–7167 (2011).

19. Deng, Z. J. *et al.* Differential plasma protein binding to metal oxide nanoparticles. *Nanotechnology* **20**, 455101–455109 (2009).

20. Walkey, C. D. & Chan, W. C. W. Understanding and controlling the interaction of nanomaterials with proteins in physiological environment. *Chem. Soc. Rev.* **41**, 2780–2799 (2012).

21. Lesniak, A. *et al.* Effects of the presence or absence of a protein corona on silica nanoparticle uptake and impact on cells. *ACS Nano* **6**, 5845–5857 (2012).

22. Liu, R., Jiang, W., Walkey, C. D., Chan, W. C. W. & Cohen, Y. Prediction of nanoparticles-cell association based on corona proteins and physicochemical properties. *Nanoscale* **7**, 9664–9675 (2015).

23. Setyawati, M. I., Tay, C. Y., Docter, D., Stauber, R. H. & Leong, D. T. Understanding and exploiting nanoparticles' intimacy with the blood vessel and blood. *Chem. Soc. Rev.* **44**, 8174–8199 (2015).

24. Miclăuş, T., Bochenkov, V. E., Ogaki, R., Howard, K. A. & Sutherland, D. S. Spatial mapping and quantification of soft and hard protein coronas at silver nanocubes. *Nano Lett.* **14**, 2086–2093 (2014).

25. Beer, C., Foldbjerg, R., Hayashi, Y., Sutherland, D. S. & Autrup, H. Toxicity of silver nanoparticles – nanoparticle or silver ion? *Toxicol. Lett.* **208**, 286–292 (2012).

26. Foldbjerg, R. *et al.* Silver nanoparticles – wolves in sheep's clothing? *Toxicol. Res* **4**, 563–575 (2015).

27. Gliga, A. R., Skoglund, S., Wallinder, I. O., Fadeel, B. & Karlsson, H. L. Size-dependent cytotoxicity of silver nanoparticles in human lung cells: the role of cellular uptake, agglomeration and Ag release. *Part. Fibre. Toxicol* **11**, 11–27 (2014).

28. Johnston, H. J. *et al.* A review of the in vivo and in vitro toxicity of silver and gold particulates: particle attributes and biological mechanism responsible for the observed toxicity. *Crit. Rev. Toxicol.* **40**, 328–346 (2010).

29. Henglein, A. Colloidal silver nanoparticles: photochemical preparation and interaction with O_2, CCl_4, and some metal ions. *Chem. Mater.* **10**, 444–450 (1998).

30. Lok, C.-N. *et al.* Silver nanoparticles: partial oxidation and antibacterial activities. *J. Biol. Inorg. Chem.* **12**, 527–534 (2007).

31. Sotiriou, G. A., Meyer, A., Knijnenburg, J. T. N., Panke, S. & Pratsinis, S. E. Quantifying the origin of released Ag^+ ions from nanosilver. *Langmuir* **28**, 15929–15936 (2012).

32. Liu, J. & Hurt, R. H. Ion release kinetics and particle persistence in aqueous nano-silver colloids. *Environ. Sci. Technol.* **44**, 2169–2175 (2010).

33. Ma, R. *et al.* Size-controlled dissolution of organic-coated silver nanoparticles. *Environ. Sci. Technol.* **46**, 752–759 (2012).

34. Liu, J., Pennell, K. G. & Hurt, R. H. Kinetics and mechanism of nanosilver oxysulfidation. *Environ. Sci. Technol.* **45**, 7345–7353 (2011).

35. Reinsch, B. C. *et al.* Sulfidation of silver nanoparticles decreases *Escherichia coli* growth inhibition. *Environ. Sci. Technol.* **46**, 6992–7000 (2012).

36. Levard, C. *et al.* Sulfidation of silver nanoparticles: natural antidote to their toxicity. *Environ. Sci. Technol.* **47**, 13440–13448 (2013).

37. Devi, G. P. *et al.* Sulfidation of silver nanoparticles reduces its toxicity in zebrafish. *Aquat. Toxicol.* **158**, 149–156 (2015).

38. Starnes, D. L. *et al.* Impact of sulfidation on the bioavailability and toxicity of silver nanoparticles to *Caenorhabditis elegans*. *Environ. Pollut.* **196**, 239–246 (2015).

39. Kim, B., Park, C.-S., Murayama, S. & Hochella, Jr. M. F. Discovery and characterization of silver sulfide nanoparticles in final sewage sludge products. *Environ. Sci. Technol.* **44**, 7509–7514 (2010).

40. Kent, R. D., Oser, J. G. & Vikesland, P. J. Controlled evaluation of silver nanoparticle sulfidation in full-scale wastewater treatment plant. *Environ. Sci. Technol.* **48**, 8564–8572 (2014).

41. Thalmann, B., Voegelin, A., Sinnet, B., Morgenroth, E. & Kaegi, R. Sulfidation kinetics of silver nanoparticles reacted with metal sulfides. *Environ. Sci. Technol.* **48**, 4885–4892 (2014).

42. Yu, S.-J., Yin, Y.-G., Chao, J.-B., Shen, M.-H. & Liu, J.-F. Highly dynamic PVP-coated silver nanoparticles in aquatic environments: chemical and morphology change induced by oxidation of Ag^0 and reduction of Ag^+. *Environ. Sci. Technol.* **48**, 403–411 (2014).

43. Liu, J., Sonshine, D. A., Shervani, S. & Hurt, R. H. Controlled release of biologically active silver from nanosilver surfaces. *ACS Nano* **4**, 6903–6913 (2010).

44. Gondikas, A. P. *et al.* Cysteine-induced modifications of zero-valent silver nanomaterials: implications for particles surface chemistry, aggregation, dissolution, and silver speciation. *Environ. Sci. Technol.* **46**, 7037–7045 (2012).

45. Jiang, X. *et al.* Fast intracellular dissolution and persistent cellular uptake of silver nanoparticles in CHO-K1 cells: implication for cytotoxicity. *Nanotoxicology* **9**, 181–189 (2015).

46. Wang, L. *et al.* Use of synchrotron radiation-analytical techniques to reveal chemical origin of silver-nanoparticle cytotoxicity. *ACS Nano* **9**, 6532–6547 (2015).

47. Chen, S. *et al.* Sulfidation of silver nanowires inside human alveolar epithelial cells: a potential detoxification mechanism. *Nanoscale* **5**, 9839–9847 (2013).

48. Chen, S. *et al.* High-resolution analytical electron microscopy reveals cell culture media-induced changes to the chemistry of silver nanowires. *Environ. Sci. Technol.* **47**, 13813–13821 (2013).

49. Choi, O. *et al.* Role of sulfide and ligand strength in controlling nanosilver toxicity. *Water. Res* **43**, 1879–1886 (2009).

50. Bianchini, A. *et al.* Evaluation of the effect of reactive sulfide on the acute toxicity of silver (I) to Daphnia magna. Part 2: toxicity results. *Environ. Toxicol. Chem.* **21**, 1294–1300 (2002).

51. Mann, R. M., Ernste, M. J., Bell, R. A., Kramer, J. R. & Wood, C. M. Evaluation of the protective effects of reactive sulfide on the acute toxicity of silver to rainbow trout (Oncorhynchus Mykiss). *Enviro. Toxicol. Chem* **23**, 1204–1210 (2004).

52. *Selected Powder Diffraction Data for Metals & Alloys Data Book. 1st edn.* vol. 1JCPDS International Center for Diffraction Data, 1982).

53. Cia, X., Zeng, J., Zhang, Q., Moran, C. H. & Xia, Y. Recent developments in shape-controlled synthesis of silver nanocrystals. *J. Phys. Chem. C* **116**, 21647–21656 (2012).

54. Wang, Y., Zheng, Y., Huang, C. Z. & Xia, Y. Synthesis of Ag nanocubes 18-32 nm in edge length: the effects of polyol on reduction kinetics, size control and reproducibility. *J. Am. Chem. Soc.* **135**, 1941–1951 (2013).

55. Elechiguerra, J. L. *et al.* Corrosion at the nanoscale: the case of silver nanowires and nanoparticles. *Chem. Mater.* **17**, 6042–6052 (2005).

56. El-Khairy, L., Ueland, P. M., Refsum, H., Graham, I. M. & Vollset, S. E. Plasma total cysteine as a risk factor for vascular disease. *Circulation* **103**, 2544–2549 (2001).

57. Salemi, G. *et al.* Blood levels of homocysteine, cysteine, glutathione, folic acid, and vitamin B_{12} in the acute phase of atherothrombosis stroke. *Neural. Sci* **30**, 361–364 (2009).

58. Levard, C., Hotze, E. M., Lowry, G. V. & Brown, Jr. G. E. Environmental transformations of silver nanoparticles: Impact on stability and toxicity. *Environ. Sci. Technol.* **46**, 6900–6914 (2012).

59. Bopst, M., Haas, C., Car, B. & Engster, H. P. The combined inactivation of tumor necrosis factor and interleukin-6 prevents induction of the major acute phase proteins by endotoxin. *Eur. J. Immunol.* **28**, 4130–4137 (1998).

60. Martinez-Gutierrez, F. *et al.* Antibacterial activity, inflammatory response, coagulation and cytotoxicity effects of silver nanoparticles. *Nanomed. Nanotechnol* **8**, 328–336 (2012).

61. Park, M.V.D.Z. *et al.* The effect of particle size on the cytotoxicity, inflammation, developmental toxicity and genotoxicity of silver nanoparticles. *Biomaterials* **32**, 9810–9817 (2011).

62. Driscoll, K. E. TNFα and MIP-2 production: role in particle-induced inflammation and regulation by oxidative stress. *Toxicol. Lett.* **112-113**, 177–184 (2000).

63. Zhang, Q., Li, W., Wen, L.-P., Chen, J. & Xia, Y. Facile synthesis of Ag nanocubes of 30 to 70 nm in edge length with CF_3COOAg as a precursor. *Chem. Eur. J* **16**, 10234–10239 (2010).

64. Wiley, B., Sun, Y., Mayers, B. & Xia, Y. Shape-controlled synthesis of metal nanoparticles: the case of silver. *Chem. Eur. J* **11**, 454–463 (2005).

65. Foldbjerg, R., Dang, D. A. & Autrup, H. Cytotoxicity and genotoxicity of silver nanoparticles in the human lung cancer cell line, A549. *Arch. Toxicol.* **85**, 743–750 (2011).

66. Reidy, B., Haase, A., Luch, A., Dawson, K. A. & Lynch, I. Mechanism of silver nanoparticle release, transformation and toxicity: a critical review of current knowledge and recommendations for future studies and applications. *Materials* **6**, 2295–2350 (2013).

67. Helmlinger, J. *et al.* Silver nanoparticles with different size and shape: equal cytotoxicity, but different antibacterial effects. *RSC Adv* **6**, 18490–18501 (2016).

68. Foldbjerg, R. *et al.* Global gene expression profiling of human lung epithelial cells after exposure to nanosilver. *Toxicol. Sci.* **130**, 145–157 (2012).

69. Jiang, X. *et al.* Multi-platform genotoxicity analysis of silver nanoparticles in the model cell line CHO-K1. *Toxicol. Lett.* **222**, 55–63 (2013).

Acknowledgements

This work was funded through a Danish strategic research council grant (SIDANA 09-067185). The research leading to these results has received funding from the European Community's Seventh Framework Programme under grant agreement no 602699 (DIREKT). V.E.B. acknowledges the support from RFBR grant 15-03-99582. We thank Mauro Porcu from FEI for his help with the preliminary STEM study.

Author contributions

T.M. and D.S.S. designed the experiments, except for the cell studies. T.M., C.B. and D.S.S. designed the toxicology experiments and C.B. performed them. T.M. performed all other experiments except FDTD simulations (V.E.B.), mass spectrometry (C.S. and J.E.) and EDS, elemental mapping and TEM diffraction (T.M and J.C). T.M and D.S.S. analysed the data, except for EDS and diffraction (T.M., J.C. and D.S.S), mass spectrometry (T.M., C.S. and J.E.), toxicology (T.M., C.B. and D.S.S.) and simulations data (T.M, V.E.B.). T.M and D.S.S. wrote the manuscript and the Supplementary Information.

Additional information

Competing financial interests: The authors declare no competing financial interests.

14

Polarized three-photon-pumped laser in a single MOF microcrystal

Huajun He[1,*], En Ma[2,*], Yuanjing Cui[1,*], Jiancan Yu[1], Yu Yang[1], Tao Song[1], Chuan-De Wu[3], Xueyuan Chen[2], Banglin Chen[1,4] & Guodong Qian[1]

Higher order multiphoton-pumped polarized lasers have fundamental technological importance. Although they can be used to *in vivo* imaging, their application has yet to be realized. Here we show the first polarized three-photon-pumped (3PP) microcavity laser in a single host–guest composite metal–organic framework (MOF) crystal, via a controllable *in situ* self-assembly strategy. The highly oriented assembly of dye molecules within the MOF provides an opportunity to achieve 3PP lasing with a low lasing threshold and a very high-quality factor on excitation. Furthermore, the 3PP lasing generated from composite MOF is perfectly polarized. These findings may eventually open up a new route to the exploitation of multi-photon-pumped solid-state laser in single MOF microcrystal (or nanocrystal) for future optoelectronic and biomedical applications.

[1] State Key Laboratory of Silicon Materials, Cyrus Tang Center for Sensor Materials and Applications, School of Materials Science and Engineering, Zhejiang University, Hangzhou 310027, China. [2] Key Laboratory of Optoelectronic Materials Chemistry and Physics, Fujian Institute of Research on the Structure of Matter, Chinese Academy of Sciences, Fuzhou, Fujian 350002, China. [3] Department of Chemistry, Zhejiang University, Hangzhou 310027, China. [4] Department of Chemistry, University of Texas at San Antonio, San Antonio, Texas 78249-0698, USA. * These authors contributed equally to this work. Correspondence and requests for materials should be addressed to X.C. (email: xchen@fjirsm.ac.cn) or to B.C. (email: banglin.chen@utsa.edu) or to G.Q. (email: gdqian@zju.edu.cn).

Polarization has been used in various fields, particularly in the field of biophotonics due to its ability to reduce multiple scattering, while to enhance the contrast and to improve tissue imaging resolution[1-3]. Through the measurement of the polarization state of the scattered light, a wealth of structural information of scatters (for example, lesions information in the tissue) can be collected given the fact that the microscopic structure of a scattering media is closely related to changes in the polarization state of the photon during a scattering process[1,3]. On the other hand, high-order multiphoton excitation can offer stronger spatial confinement, deeper tissue penetration and less Rayleigh scattering, which are significantly beneficial to the biological imaging[4-8]. To make use of the uniqueness of both polarization and high-order multiphoton excitation, the polarized three-photon and/or higher order pumped laser in single solid-state microcrystal is potentially useful for a new kind of biological imaging, so called multiphoton pumped (MPP) polarized emission biological imaging (Supplementary Fig. 1), but has never been realized. In order to produce such a unique laser, the gain medium not only needs to have a high multiphoton absorption (MPA) cross-section and lasing efficiency[4,5], but more importantly needs to be assembled into a suitable microcavity of high concentration and orientation (especially in the case that the absorption transition moment of gain medium is anisotropic) without significant luminescent quenching to enforce the high optical gain and to generate controllable and directional laser. This is really a daunting challenge. In fact, although extensive research endeavours have been pursued to target such a goal, progress has been very slow. The initial effort to diminish the significant quenching effects on the solid state was to homogeneously disperse the gain medium such as the dye molecules with high multiphoton absorption cross-section into its solution[6,9]. By employing such a strategy, it still remains extremely difficult to provide with a sufficiently high quenching concentration, which limits the realization of necessary optical gain for compensating the losses. The quenching concentration means that the aggregation-caused quenching (ACQ) gradually becomes dominant when the concentration of gain medium is higher than the quenching concentration. Furthermore, the molecules in the solution are randomly oriented, which would limit their capacities to maximize the optical gain. So far, this dispersed solution methodology can only lead to the amplified spontaneous emission instead of generating three-photon or higher order pumped laser[4,5]. Although the luminescent properties of quantum dots are intriguing, they only have generated three-photon-pumped (3PP) random lasing in which the emission direction, position and numbers of mode frequency, and the uniformity of light-emitting region of such lasers are very difficult to control[7,10]. Recently, the 3PP lasing from colloidal nanoplatelets in solution has been demonstrated by Li et al.[11]; however, no polarization property of the 3PP lasing has been realized. Furthermore, its liquid nature has limited practical applications. To take advantage of the pore confinement of porous materials, zeolites and nanoporous silica have been explored to incorporate dye molecules and semiconducting polymers into the corresponding crystals and thin films to develop solid-state lasing[12,13]. However, zeolite/dye composites can only generate single-photon pumped lasing, mainly due to the incompatibility between the inorganic framework and organic guest, leading to the low loading concentration (0.005 ~ 0.0005 M), uneven distribution of dye molecules and poor crystal morphology; while nanoporous silica/semiconducting polymer matrix basically leads to the single-photon pumped polarized amplified spontaneous emission.

Previously, we have used a porous metal–organic framework (MOF) for its pore confinement of a dye molecule bearing moderately high two-photon absorption cross-section, and realized the two-photon pumped lasing from a composite crystal bio-MOF-1 $\supset$ DMASM (DMASM = 4-[p-(dimethylamino)styryl] -1-methylpyridinium) at room temperature[14]. However, the pores (two types of channels along the c-axis of about 7.0 and 10.0 Å, respectively) within bio-MOF-1 are still too large to exactly match the dye molecules of DMASM, thus the orientation of the dye molecules inside the pore cavities is not of a high order, particularly when the high concentration of the dye molecules are applied. Such a moderate pore confinement of bio-MOF-1 apparently has limited us to further realize the higher order multiphoton pumped laser in this solid-state crystal. In order to enhance the pore confinement efficiency of a porous MOF crystal, the pore sizes within a porous MOF need to be tuned to match the size of the dye molecule better. But the dilemma is that when the pore sizes of a porous MOF can exactly match the size of the dye molecule, the dye molecules cannot diffuse into the pore channels through the simple post-synthetic exchange process. To overcome this problem, we have developed an in situ self-assembly strategy[15,16]: the components for building a MOF crystal (metal ion and organic linker) and the organic dye molecule are simultaneously assembled together to form the MOF/dye single crystals. Such a methodology has enabled us to tightly incorporate the dye molecules into the porous MOF crystals, and thus the dye molecules are highly ordered and oriented. We have also managed to immobilize high concentration of the dye molecules into the MOF crystal ZJU-68 $\supset$ DMASM ((DMASM)$_{0.33}$H$_{1.67}$[Zn$_3$O(CPQC)$_3$], CPQC, 7-(4-carboxyphenyl)quinoline-3-carboxylate) (the average pore size of the one-dimensional channel along the c-axis is 6.0 Å) with the dye content over 0.4 M. Furthermore, the suitable refraction index and well-faceted MOF composite crystals of certain morphology symmetries can be naturally and efficiently utilized as the laser resonant cavities without any other fabrications. The powerful in situ self-assembly strategy, highly efficient pore confinement of ZJU-68 for DMASM dye molecule, and suitable refraction index as well as perfect crystal morphology have enabled us to target the first example of polarized three-photon-pumped laser in single solid-state microcrystal.

Results

Synthesis and characterization. Reaction of a new organic linker 7-(4-carboxyphenyl)quinoline-3-carboxylic acid (H$_2$CPQC) containing quinolone group and Zn(NO$_3$)$_2$·6H$_2$O in N,N-dimethylformamide/acetonitrile/H$_2$O/HBF$_3$ at 100 °C affords colourless hexagonal prism crystals of H$_2$[Zn$_3$O(C$_{17}$H$_9$NO$_4$)$_3$] · 2.5H$_2$O · 0.5DMF · MeCN (ZJU-68, Fig. 1a). Single crystal X-ray diffraction studies reveal that ZJU-68 crystallizes in the P$\bar{3}$ space group (see Supplementary Table 1 for detailed crystallographic data). As shown in Fig. 2a, trinuclear secondary building units (SBUs) of [Zn$_3$O]$^{4+}$ are linked by the ligands CPQC^{2-} to form an anionic framework of [Zn$_3$O(C$_{17}$H$_9$NO$_4$)$_3$]$^{2-}$. In this structure, nine coordination sites of [Zn$_3$O]$^{4+}$ are completely occupied by six carboxylates and three of nitrogen atoms from the quinoline moieties, which are different from most of metal–organic frameworks with [M$_3$O]$^{3n-2}$ (n = 3 for M = Cr^{3+}, Fe^{3+} and so on or n = 2 for M = Zn^{2+}, Cu^{2+}) SBUs in which three sites are occupied by small capping ligands such as water and hydroxide[17,18].

The crystal has one-dimensional (1D) sub-nano channels along the c-axis with a hexagonal cross-section (Fig. 2b; Supplementary Fig. 2). The edge of the hexagon is about 3.0 Å. For the synthesis of laser dye functionalized crystals, we tried to introduce linear-shaped laser dye cations DMASM via an ion-exchange process, as described in our previous work[14], but failed. This is because the DMASM molecule (about 6.3 Å in the width, Supplementary

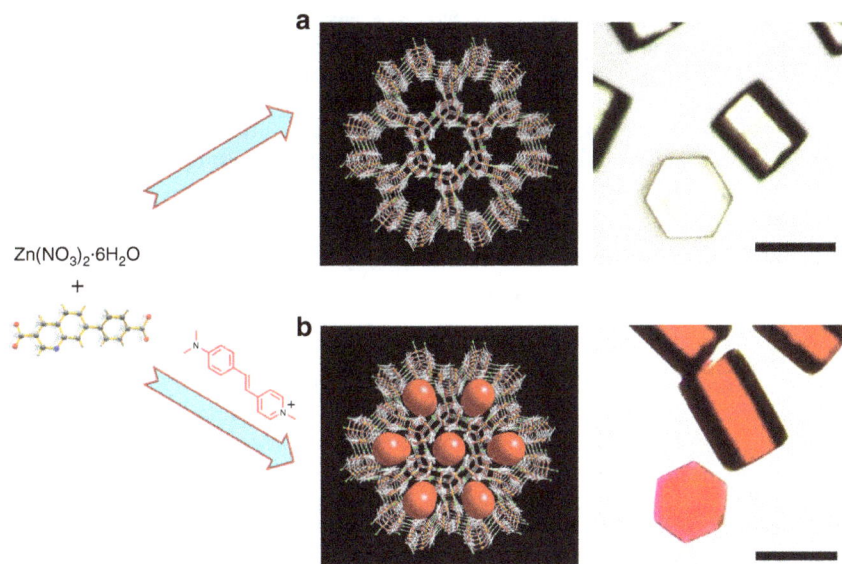

Figure 1 | Schematic synthesis of ZJU-68 and ZJU-68 ⊃ DMASM. (**a**) The synthesis and micrograph of a novel metal–organic framework ZJU-68. (**b**) *In situ* synthesis of laser dye incorporated metal–organic framework crystals ZJU-68 ⊃ DMASM. The inclusion of the red dye DMASM molecules leads to the color change from the original colourless ZJU-68 to red ZJU-68 ⊃ DMASM. Scale bar, 50 µm.

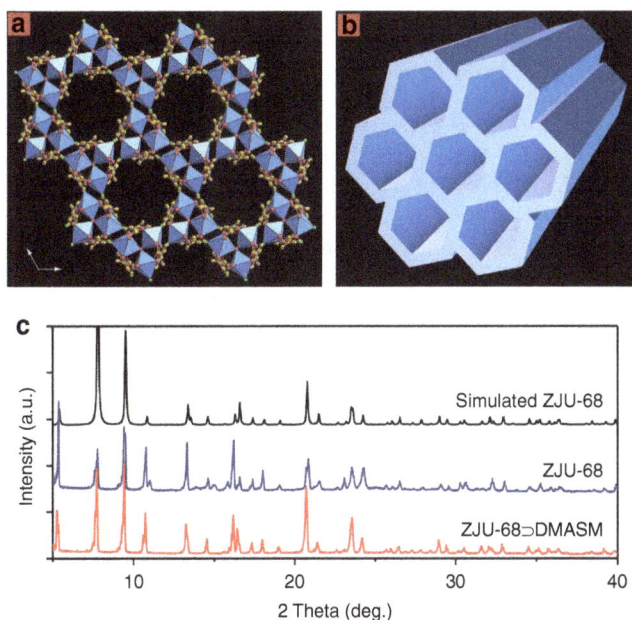

Figure 2 | The structure of a novel metal–organic framework crystal ZJU-68. (**a**) Crystal structure of ZJU-68 viewed along the crystallographic c direction (C, orange; N, green; O, red; Zn, blue polyhedra). H atoms and solvent molecules are omitted for clarity. In this structure, nine coordination sites of a trinuclear SBU $[Zn_3O]^{4+}$ are completely occupied by six carboxylates and three of nitrogen atoms from the quinoline moieties, which may play a crucial role in the stabilization of the resulting MOF, ZJU-68. (**b**) The simplified network structure of ZJU-68, displaying 1D channels along the c-axis. Different objects are not drawn to scale. (**c**) PXRD patterns of ZJU-68 and ZJU-68 ⊃ DMASM, which indicate that the ZJU-68 ⊃ DMASM has the identical framework structure with ZJU-68.

Fig. 2e) is too large to diffuse into the channels of ZJU-68 (ref. 19). We thus developed the *in situ* self-assembly synthetic approach in which the dye DMASM molecules were simultaneously incorporated into framework during the solvothermal synthesis of ZJU-68 by simply adding the dye molecules into the reaction

solution (Fig. 1b). The resulting dye DMASM included ZJU-68 ⊃ DMASM has the same hexagonal prism crystal morphology. The inclusion of the red dye DMASM molecules leads to the colour change from the original colourless ZJU-68 to red ZJU-68 ⊃ DMASM. Both single crystal and the powder X-ray diffraction studies (Fig. 2c) confirmed that the ZJU-68 ⊃ DMASM has the identical framework structure with ZJU-68. Furthermore, both ZJU-68 and ZJU-68 ⊃ DMASM demonstrate excellent stability in the air and in the common solvents such as water, ethanol and dimethylformamide (Supplementary Fig. 3). Of course, most of the 1D hexagonal channel spaces have been occupied by DMASM molecules in ZJU-68 ⊃ DMASM. Supplementary Fig. 4 shows the fluorescence micrographs of ZJU-68 ⊃ DMASM, taken by confocal laser scanning microscope. The flat and uniform intensity profiles suggest that the DMASM dyes are homogeneously distributed inside the ZJU-68 ⊃ DMASM composite crystals. The dye contents in this composite can be finely tuned by the addition of different amount of DMASM dyes during the *in situ* self-assembly solvothermal synthesis. Generally speaking, the relatively weak MPA responses require high dye content for MPA lasing measurements[5,6], it is thus necessary to encapsulate as much dye molecules as possible into the pore space of ZJU-68. However, the high dye contents (ingredient mole ratio of $n_{DMASM}/n_{H_2CPQC} \geq 70\%$) in the reaction mixtures not only affect the *in situ* self-assembly process (formation of other MOF phases) but also lead to the formation of poor crystalline ZJU-68 ⊃ DMASM. As such, we adjusted the dye concentration in the reaction solution, which produced the optimized ZJU-68 ⊃ DMASM crystals when the ingredient mole ratio of n_{DMASM}/n_{H_2CPQC} is 35%. Accordingly, per gram of resulting ZJU-68 ⊃ DMASM crystals contain 67.7 mg dye molecules corresponding to the concentration of 0.46 M (the molar amount of dye in per unit volume of solid composite; Supplementary Fig. 5). The optimized ingredient mole ratio (35%) is determined by the measurement of fluorescence quantum yield. Among the ZJU-68 ⊃ DMASM samples with different dye loading concentrations, the ZJU-68 ⊃ DMASM composite crystals with ingredient mole ratio of 35% exhibit the strongest emission at around 635 nm with the highest quantum yield φ of $24.28 \pm 5\%$ on excitation at 450 nm (Supplementary Fig. 6). This is much higher than the quantum yield of 0.45% in dye solutions and of 1.48% solid powder[14]. These results demonstrate that the good confinement of the DMASM molecules within the size-matched channels of ZJU-68 can effectively restrain the intramolecular torsional motion and increase the conformational rigidity of the dye, thus diminishing the ACQ and populating its radiative decay pathway[20].

Multiphoton-excited fluorescence in ZJU-68 ⊃ DMASM. Figure 3a compares the single-photon-, two-photon- and three-photon-excited fluorescence spectra of a single ZJU-68 ⊃ DMASM crystal with the dye concentration of 0.46 M under the excitation of a femtosecond laser at different wavelengths. The ZJU-68 ⊃ DMASM shows a strong emission peaked at 627 nm on excitation at 532 nm, whereas the emission peak is red-shifted by 11 to 638 nm when excited at 1,064 nm. The full-width at half-maximum (FWHM) is 53.6 and 42.5 nm, respectively, in the single-photon-, two-photon-excited fluorescence spectra of ZJU-68 ⊃ DMASM. The emission spectrum on excitation at 1,380 nm is basically similar to that excited at 1,064 nm except one additional small peak at around 690 nm attributed to the second harmonic generation response. The red shift of 11 nm under multiphoton excitation can be ascribed to the reabsorption effect[14]. The diffuse reflectance ultraviolet-visible (vis) spectrum of ZJU-68 ⊃ DMASM was shown in Supplementary Fig. 7. There exists overlap between the long wavelength side of the absorption band and the short wavelength side of the fluorescence band (Fig. 3a). Furthermore, all MPP fluorescence bands in Fig. 3a are asymmetric with their left part seeming to be cut off[21]. In addition, the emission peaked at 627 nm from a single ZJU-68 ⊃ DMASM crystal on excitation at 532 nm is blue shifted relative to the spontaneous emission (maximum at 635 nm, see Supplementary Fig. 14) from multiple ZJU-68 ⊃ DMASM crystals, which also suggests the presence of reabsorption effect in ZJU-68 ⊃ DMASM (Supplementary Fig. 8). Fig. 3b shows the fluorescence intensity of the crystal with respect to the pump polarization direction when excited at 1,380 nm. The ZJU-68 ⊃ DMASM exhibits a strong emission when the pump polarization direction is parallel to the crystal channels (along the c-axis, denoted as 0°), but hardly emits any light when the pump polarization direction is perpendicular to the excitation direction (90°). Such significant directional fluorescence (dichroic ratio ~ 365 (ref. 15)) behaviours indicate that the absorption transition moments (approximately along the dye molecule axis[22]) are highly oriented along the crystal channels.

3PP lasing in ZJU-68 ⊃ DMASM. 3PP lasing properties were investigated on an isolated single crystal of ZJU-68 ⊃ DMASM with the dye concentration of 0.46 M under a microscope. A femtosecond laser at 1,380 nm was used to pump the crystal at room temperature. This laser beam was directed from a femtosecond optical parametric amplifier (OPA), and then was coupled to the microscope. The emission beam from the crystal was focused and collected with a fibre optic spectrometer. Representative emission spectra near the lasing threshold are shown in Fig. 4a. Under the low-pump energy (E) of 113 nJ, the emission spectrum shows a broad peak centred at ~ 649 nm with a FWHM of 57.6 nm, which corresponds to the spontaneous emission. The pump energy is defined as the laser energy directly received by the MOF crystal (after going through the objective lens and before being incident on the MOF crystal). As the pump energy increases to ≥ 230 nJ, a highly progressional emission pattern centred at 642.7 nm appears and grows rapidly with increasing pump energy, while the intensity of the broad spontaneous emission remains almost constant. The visible stimulated emission spectrum centred at 642.7 nm is between one half and one-third of the pumped wavelength of 1,380 nm, which means that the sum energy of two photons at 1,380 nm is not large enough to overcome the bandgap between the ground state (S_0) and excited state (S_1) of ZJU-68 ⊃ DMASM. The stimulated emission of ZJU-68 ⊃ DMASM is therefore induced by the simultaneous absorption of more than two near-infrared photons. To unravel how many photons involved in such a simultaneous absorption process, we further measured the dependence of the stimulated emission intensity on the pump energy. The right inset in Fig. 4a illustrates the pump energy dependence of the fluorescence intensity and FWHM plot as a function of pump energy, giving rise to a linear relationship with cubic pump energy and a low lasing threshold of $E_{th} \sim 224$ nJ as compared with other 3PP stimulated emission[4,6,7]. The FWHM plot shows a constant value below E_{th} and a sudden drop by more than two orders of magnitude when above E_{th}. The presence of a significant spectral narrowing and a threshold energy coupled with the linearly rapid increase in intensity with cubic pump energy suggest that the 3PP lasing has occurred in the ZJU-68 ⊃ DMASM crystal. The quality factor (Q) is given by $Q = \lambda/\delta\lambda$, where λ and $\delta\lambda$ are the peak wavelength and its FWHM, respectively. At pump energy of 369 nJ, the FWHM of lasing peaked at 642.7 nm is ~ 0.38 nm. This records a high-quality factor $Q \sim 1,691$ for 3PP lasing, which indicates the high crystal quality supported by our simple chemical approach without etching and coating.

For hexagonal ZJU-68 ⊃ DMASM crystal, the opposing facets can act as the mirrors of a Fabry–Pérot (F–P) cavity, or the six

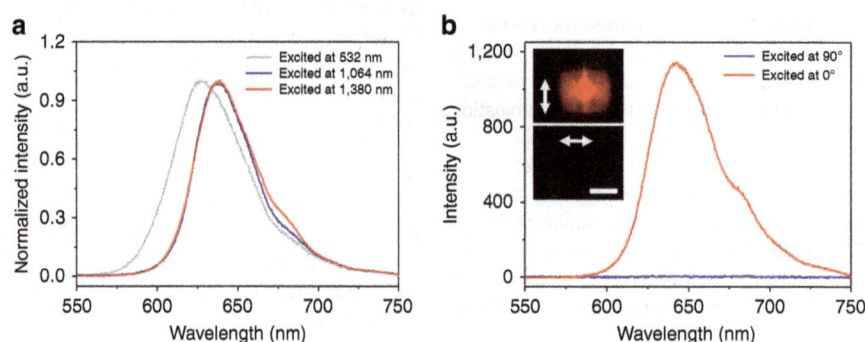

Figure 3 | Multiphoton-pumped fluorescence performance of a ZJU-68 ⊃ DMASM single crystal. (a) Single-photon-(532 nm), two-photon-(1,064 nm) and three-photon-(1,380 nm) excited emission spectra of ZJU-68 ⊃ DMASM. **(b)** The emission intensity versus pump polarization at two angles $\theta = 0°$ (parallel to the crystal channels) and $\theta = 90°$ (perpendicular to the crystal channels), excited at 1,380 nm. Insets: micrographs of a ZJU-68 ⊃ DMASM single crystal ($R = 36.5 \mu m$) with different pump polarizations excited at 1,380 nm, Scale bar, 50 μm. The high intensity ratio between the two angles indicates the high orientation of dye molecules within the channels of ZJU-68.

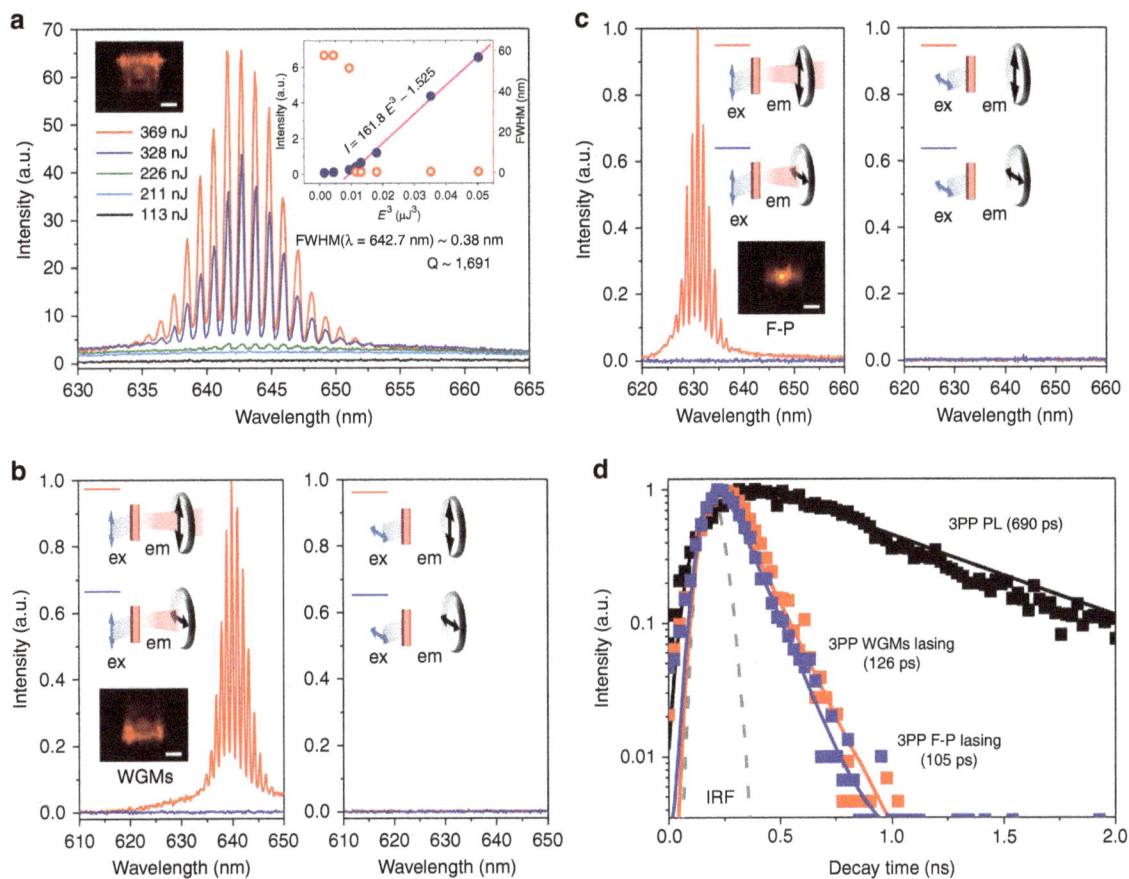

Figure 4 | Three-photon-pumped lasing performance of ZJU-68 ⊃ DMASM. (**a**) 1,380 nm pumped emission spectra of ZJU-68 ⊃ DMASM around the lasing threshold. Insets: the micrograph of a ZJU-68 ⊃ DMASM single crystal ($R = 26.2\,\mu m$) excited at 1,380 nm (left) and emission intensity and FWHM as a function of pump energy showing the lasing threshold at ~ 224 nJ (right). At pump energy of 369 nJ, the FWHM of lasing peaked at 642.7 nm is ~ 0.38 nm, corresponding to a Q factor ~ 1,700. (**b,c**) Intensity-dependent emission spectra of 3PP WGMs (**b**, pump energy at 433 nJ) and F–P lasing (**c**, pump energy at 837 nJ) from two isolated crystals ($R = 26.1$ and $27.6\,\mu m$, respectively) with pump/emission-detected polarization combinations at two angles $\theta = 0°$ (parallel to the crystal channels) and $\theta = 90°$ (perpendicular to the crystal channels), excited at 1,380 nm. Insets: schematic diagrams of the measurement geometry for an individual crystal and micrographs of two kinds of lasing spot patterns due to the F–P (one-spot) and WGMs (two-spot) mechanisms, Scale bar, 20 μm. Both 3PP WGMs and F–P lasing exhibit perfectly polarized emission with DOP > 99.9%, which are attributed to the highly oriented assembly of dye molecules within the host–guest composite ZJU-68 ⊃ DMASM microcrystal. (**d**) TRPL decay kinetics measurements of ZJU-68 ⊃ DMASM under photoluminescence (PL), F–P and WGMs lasing excited at 1,380 nm. TRPL, time-resolved photoluminescence

facets can form a whispering gallery modes (WGMs) or other quasi-WGMs cavities[23] (Supplementary Fig. 9). We observed two kinds of lasing spot pattern on isolated ZJU-68 ⊃ DMASM crystals when excited at 1,380 nm: (1) the strong emission with spatial interference from two-side facets of a hexagonal prism crystal (two-spot pattern)[24], as shown in the insets of microscopy image in Figs. 4a and 4b; the strong emission from a round bright spot at the central facet of the crystal (one-spot pattern)[14], as shown in the inset of Fig. 4c. Such two lasing patterns can be attributed to the WGMs and F–P feedback mechanisms, respectively, as confirmed later. Fig. 4b,c show the anisotropic study of 3PP WGMs and F–P lasing from two crystals with side lengths of 26.1 and 27.9 μm, respectively. The red emission light from the crystal passed through a polarizer first and then was focused and collected with a fibre optic spectrometer. The schematic diagrams of the measurement geometry for an individual crystal are shown in Fig. 4b,c, where the polarization directions of the pump light and the polarizer are parallel (0°)/perpendicular (90°) to the crystal channels (along *c*-axis). We can see that both 3PP WGMs and F–P lasing with highly structured spectra occur when excited at 0° and emission polarization detected at 0°, while hardly any emission intensity

can be detected in all other configurations. The corresponding pump energy is 433 nJ for WGMs lasing and 837 nJ for F–P lasing. It should be noted that the pump energy at almost 3.2 E_{th} of 3PP WGMs lasing can realize the 3PP F–P lasing in our experiments, indicating that the WGMs mechanism is more conducive to the realization of 3PP lasing due to the total internal reflection for less loss of light in such size of the crystal. The degree of polarization can be defined as DOP = $(I_{max} - I_{min})/(I_{max} + I_{min})$ in our experiments[25], and we calculated that both 3PP WGMs and F–P lasing exhibited DOP > 99.9% (limited by the spectral intensity sensitivity of our measurement system) when the excitation polarization is fixed parallel to the crystal channels, indicating a perfectly polarized 3PP lasing operation. Compared with the WGMs lasing, the parallel mirrors in a F–P cavity cannot be utilized as Brewster windows for the polarization selectivity. Therefore, the perfectly polarized 3PP F–P lasing is attributed to the highly oriented assembly of dye molecules within the host–guest composite ZJU-68 ⊃ DMASM microcrystal given the fact that the angle between the absorption transition moment and emission transition moment is close to zero in the dye molecule DMASM[26]. These anisotropic results indicate that ZJU-68 ⊃ DMASM can only be excited at the polarization direction parallel to the crystal channels, and can produce lasing perfectly polarized along the crystal channels, which exhibit a great potential for bioimaging, optical sensing and future optoelectronic integration.

To further confirm the optical-feedback mechanisms for 3PP lasing in these hexagonal ZJU-68 ⊃ DMASM crystals, several single crystals with different side lengths R were chosen for the further measurements (Supplementary Fig. 10a and b). For both feedback mechanisms, the spectra of lasing exhibit an increased mode spacing with the decrease of side length of the MOF crystals. For possible resonant modes, the mode spacing $\Delta\lambda_s$ is defined as[27]

$$\Delta\lambda_s = \frac{\lambda^2}{Ln_g} \qquad (1)$$

where λ is the resonant wavelength, L is the cavity path length ($3\sqrt{3}R$ for WGMs and $2\sqrt{3}R$ for F–P cavity), and n_g is the group index of refraction. The measured mode spacing, $\Delta\lambda_s$, around 635 nm, demonstrates a linear relationship with $1/R$ for each feedback mechanism, which agrees well with Equation (1) (see Supplementary Fig. 10c,d). This result indicates that the lasing with the similar output spot pattern (one-spot or two-spot) can be attributed to the same feedback mechanism. According to the fitting formula, we calculated the ratio of slopes ($S_{\text{one-spot}}/S_{\text{two-spot}}$) to be 1.47, which is very close to the ratio of the cavity path lengths ($L_{\text{WGMs}}/L_{\text{F–P}} = 1.5$), verifying that the lasing with these two kinds of output pattern (one-spot and two-spot) should be attributed to the F–P cavity and WGMs, respectively. On the basis of Equation (1), we also derived $n_g \sim 3.27$ for F–P cavity, and $n_g \sim 3.21$ for WGMs at the wavelength of 635 nm. The relatively high group index n_g value may result from the unusual dispersion relation near the absorption band or the strong exciton–photon coupling in organic materials[28]. Further insight into the 3PP emission performance of the single crystal ZJU-68 ⊃ DMASM arises from time-resolved photoluminescence measurements (Fig. 4d). The pulse durations of the 3PP F–P lasing and WGMs lasing were determined to be 105 and 126 ps, respectively, which are much shorter than the corresponding 3PP fluorescence (below the E_{th}) decay time of 690 ps. Such temporal narrowing can be ascribed to the depletion in the population inversion of the gain medium with photon-stimulated amplification[6]. The lasing pulse durations from WGMs and F–P are almost in the same order of magnitude, indicating that the optical-feedback mechanism may have little effect on the pulse duration. Subtle differences in our measured decay times of lasing may be ascribed to the proportion of stimulated emission and spontaneous emission in the emitted light, which depends on multiple factors, for example, pump energy, crystal size and crystalline quality[29,30].

Discussion

In summary, we have achieved an unprecedented solid-state polarized frequency-upconversion lasing in a novel composite single microcrystal ZJU-68 ⊃ DMASM by simultaneous three-photon absorption in the near-infrared region. The tightly confined and highly oriented cationic DMASM dye molecules in anionic ZJU-68 nano-channels through an *in situ* assembly process efficiently increase the loaded concentration, minimize the aggregation and optimize the orientation of dye molecules within the framework, which fulfilled the high-gain lasing with highly polarized excitation response and perfectly polarized emission in a micro-sized laser cavity. Particularly, the 3PP lasing, with a low lasing threshold of ∼ 224 nJ centred at 642.7 nm on excitation at 1,380 nm, has been successfully achieved with a record high-quality factor of ∼ 1,700. Both F–P

and WGMs optical-feedback mechanisms have been confirmed to be responsible for 3PP lasing in ZJU-68 ⊃ DMASM microcrystals. Owing to the highly oriented assembly of dye molecules within ZJU-68 ⊃ DMASM, the 3PP WGMs and especially F–P lasing show a perfect emission polarization with DOP > 99.9%. The observed solid-state frequency-upconversion polarized lasing induced by 3PP may find great potentials in practical applications such as photonics, information storage and biomedicine, to name a few. For instance, the wavelength of 1,380 nm belongs to the near-infrared-IIa window (1,300-1,400 nm), which is very promising in biological applications (especially for *in vivo* imaging) because such wavelength region not only can reach deeper penetration depths and minimize the scattering/auto-fluorescence of biological tissues, but also avoid an increased light absorption from water above 1,400 nm (ref. 31). Because the MOF strategy and design can provide us with rich structures of the systematically tuned pore/channel sizes to encapsulate various chromophores with controlled concentration and orientation[32–34], we anticipate that higher order multiphoton-pumped lasing in solid state can also be realized given that the chromophores (or other nano-sized materials) with great multiphoton absorption properties are well incorporated into the structurally matched MOFs. These findings may eventually open up a new route to the exploitation of multiphoton-pumped solid-state laser in single MOF microcrystal (or nanocrystal) for future optoelectronic and biomedical applications.

Methods

Synthesis of ZJU-68 ⊃ DMASM. A mixture of $Zn(NO_3)_2 \cdot 6H_2O$ (0.34 mmol, 149 mg), H_2CPQC (0.17 mmol, 50 mg), DMF (10 ml), MeCN (2 ml), H_2O (0.05 ml), HBF_3 (0.05 ml) and DMASM iodide (0.03 mmol, 11 mg) were sealed in a 15 ml Teflon-lined stainless-steel bomb at 100 °C for 24 h, which was then slowly cooled to room temperature. After decanting the mother liquor, the fine red hexagonal crystalline product was rinsed three times with fresh DMF (5 ml × 3) and dried in air. The synthesis of the new organic linker H_2CPQC can be found in Supplementary Fig. 11 and Supplementary Methods.

Measurements. For MPP, an optical parametric amplifier (TOPAS-F-UV2, Spectra-Physics) pumped by a regeneratively amplified femtosecond Ti:sapphire laser system (800 nm, 1 kHz, pulse energy of 4 mJ, pulse width < 120 fs, Spitfire Pro-FIKXP, Spectra-Physics), which was seeded by a femtosecond Ti-sapphire oscillator (80 MHz, pulse width < 70 fs, 710-920 nm, Mai Tai XF-1, Spectra-Physics) was used for generating the excitation pulse (1 kHz, 240–2,600 nm, pulse width < 120 fs). The incident laser was coupled to the microscope (Ti-U, Nikon), focusing on crystals through an objective lens (CFI TU Plan Epi ELWD 50 ×, numerical aperture = 0.60, work distance = 11.0 mm) with an exposure region of diameter around 30 μm (supplementary Fig. 12). The excited red light was then focused and collected by the fibre optic spectrometer (QE65Pro, Ocean Optics).

The decay curves of multiphoton-pumped emissions were measured by a picosecond lifetime spectrometer (Lifespec-ps, Edinburgh Instruments). For the lifetime measurement of upconverted fluorescence, the pump energy was under the lasing threshold to ensure that no stimulated emission was generated. To measure the decay of the multiphoton-pumped lasing, the pump energy was enhanced over the threshold so that the ultra-strong lasing could be achieved.

Contents of well-dried dye-included ZJU-68 ⊃ DMASM crystals were determined by ^1H NMR. As shown in Supplementary Figs 5 and 13c, we calibrated and obtained peak area values of peaks that belong to H_2CPQC and DMASM, respectively. The ratio (R_a) of their peak area values represents the ratio of their contents in the crystal. The dye concentration of the ZJU-68 ⊃ DMASM composite is calculated from $c = 2R_a/N_AV$, where $V = 2403.91$ Å^3 and $N_A = 6.02 \times 10^{23}$ mol^{-1} is Avogadro's constant.

References

1. Jameson, D. M. & Ross, J. A. Fluorescence polarization/anisotropy in diagnostics and imaging. *Chem. Rev.* **110**, 2685–2708 (2010).
2. Ghosh, N. & Vitkin, I. A. Tissue polarimetry: concepts, challenges, applications, and outlook. *J. Biomed. Opt.* **16**, 110801 (2011).
3. Gurjar, R. S. *et al.* Imaging human epithelial properties with polarized light-scattering spectroscopy. *Nat. Med.* **7**, 1245–1248 (2001).
4. He, G. S., Tan, L. S., Zheng, Q. & Prasad, P. N. Multiphoton absorbing materials: molecular designs, characterizations, and applications. *Chem. Rev.* **108**, 1245–1330 (2008).

5. Guo, L. & Wong, M. S. Multiphoton excited fluorescent materials for frequency upconversion emission and fluorescent probes. *Adv. Mater.* **26,** 5400–5428 (2014).
6. Zheng, Q. D. *et al.* Frequency-upconverted stimulated emission by simultaneous five-photon absorption. *Nat. Photon.* **7,** 234–239 (2013).
7. Wang, Y. *et al.* Stimulated emission and lasing from CdSe/CdS/ZnS core-multi-shell quantum dots by simultaneous three-photon absorption. *Adv. Mater.* **26,** 2954–2961 (2014).
8. Hoover, E. E. & Squier, J. A. Advances in multiphoton microscopy technology. *Nat. Photon.* **7,** 93–101 (2013).
9. He, G. S., Markowicz, P. P., Lin, T. C. & Prasad, P. N. Observation of stimulated emission by direct three-photon excitation. *Nature* **415,** 767–770 (2002).
10. Gomes, A. S., Carvalho, M. T., Dominguez, C. T., de Araujo, C. B. & Prasad, P. N. Direct three-photon excitation of upconversion random laser emission in a weakly scattering organic colloidal system. *Opt. Express* **22,** 14305–14310 (2014).
11. Li, M. *et al.* Ultralow-threshold multiphoton-pumped lasing from colloidal nanoplatelets in solution. *Nat. Commun.* **6,** 8513 (2015).
12. Vietze, U. *et al.* Zeolite-dye microlasers. *Phys. Rev. Lett.* **81,** 4628–4631 (1998).
13. Martini, I. B. *et al.* Controlling optical gain in semiconducting polymers with nanoscale chain positioning and alignment. *Nat. Nanotechnol.* **2,** 647–652 (2007).
14. Yu, J. *et al.* Confinement of pyridinium hemicyanine dye within an anionic metal-organic framework for two-photon-pumped lasing. *Nat. Commun.* **4,** 2719 (2013).
15. Martinez-Martinez, V., Garcia, R., Gomez-Hortiguela, L., Perez-Pariente, J. & Lopez-Arbeloa, I. Modulating dye aggregation by incorporation into 1D-MgAPO nanochannels. *Chemistry* **19,** 9859–9865 (2013).
16. Martínez-Martínez, V. *et al.* Highly luminescent and optically switchable hybrid material by one-pot encapsulation of dyes into MgAPO-11 unidirectional nanopores. *ACS Photon.* **1,** 205–211 (2014).
17. Mao, C. *et al.* Anion stripping as a general method to create cationic porous framework with mobile anions. *J. Am. Chem. Soc.* **136,** 7579–7582 (2014).
18. Ferey, G. *et al.* A chromium terephthalate-based solid with unusually large pore volumes and surface area. *Science* **309,** 2040–2042 (2005).
19. Zhao, C. F., He, G. S., Bhawalkar, J. D., Park, C. K. & Prasad, P. N. Newly synthesized dyes and their polymer/glass composites for one-photon and 2-photon pumped solid-state cavity lasing. *Chem. Mater.* **7,** 1979–1983 (1995).
20. Cui, Y. J. *et al.* Dye encapsulated metal-organic framework for warm-white LED with high color-rendering index. *Adv. Funct. Mater.* **25,** 4796–4802 (2015).
21. Ren, Y. *et al.* Synthesis, structures and two-photon pumped up-conversion lasing properties of two new organic salts. *J. Mater. Chem.* **10,** 2025–2030 (2000).
22. Weiß, Ö. *et al.* in *Host-Guest-Systems Based on Nanoporous Crystals* 544–557 (Wiley-VCH Verlag GmbH & Co., 2005).
23. Wang, X. *et al.* Whispering-gallery-mode microlaser based on self-assembled organic single-crystalline hexagonal microdisks. *Angew. Chem. Int. Ed.* **53,** 5863–5867 (2014).
24. Braun, I. *et al.* Hexagonal microlasers based on organic dyes in nanoporous crystals. *Appl. Phys. B* **70,** 335–343 (2000).
25. Zhu, H. *et al.* Lead halide perovskite nanowire lasers with low lasing thresholds and high quality factors. *Nat. Mater.* **14,** 636–642 (2015).
26. Gozhyk, I. *et al.* Polarization properties of solid-state organic lasers. *Phys. Rev. A* **86,** 043817 (2012).
27. Choi, S., Ton-That, C., Phillips, M. R. & Aharonovich, I. Observation of whispering gallery modes from hexagonal ZnO microdisks using cathodoluminescence spectroscopy. *Appl. Phys. Lett.* **103,** 171102 (2013).
28. Takazawa, K., Inoue, J., Mitsuishi, K. & Takamasu, T. Fraction of a millimeter propagation of exciton polaritons in photoexcited nanofibers of organic dye. *Phys. Rev. Lett.* **105,** 067401 (2010).
29. Zhang, C. *et al.* Two-photon pumped lasing in single-crystal organic nanowire exciton polariton resonators. *J. Am. Chem. Soc.* **133,** 7276–7279 (2011).
30. Liu, X. *et al.* Whispering gallery mode lasing from hexagonal shaped layered lead iodide crystals. *ACS Nano* **9,** 687–695 (2015).
31. Hong, G. S. *et al.* Through-skull fluorescence imaging of the brain in a new near-infrared window. *Nat. Photon.* **8,** 723–730 (2014).
32. Furukawa, H., Cordova, K. E., O'Keeffe, M. & Yaghi, O. M. The chemistry and applications of metal-organic frameworks. *Science* **341,** 1230444 (2013).
33. Kitagawa, S., Kitaura, R. & Noro, S. Functional porous coordination polymers. *Angew. Chem. Int. Ed.* **43,** 2334–2375 (2004).
34. Chen, B., Xiang, S. & Qian, G. Metal-organic frameworks with functional pores for recognition of small molecules. *Acc. Chem. Res.* **43,** 1115–1124 (2010).

Acknowledgements

We acknowledge the financial support from the National Natural Science Foundation of China (Nos. 51229201, 51272229, 51272231, 51402259, 51472217, 51432001, U1305244 and 21325104) and Zhejiang Provincial Natural Science Foundation of China (Nos. LR13E020001 and LZ15E020001). This work is also partially supported by Welch Foundation (AX-1730) and National Science Foundation of United States (ECCS-1407443). X.C. and E.M. acknowledge the support from Special Project of National Major Scientific Equipment Development of China (No. 2012YQ120060) and the CAS/SAFEA International Partnership Program for Creative Research Teams. We also thank Dr Ghezai Musie for proof-reading the manuscript.

Author contributions

H.H., E.M., Y.C., J.Y. and G.Q. conceived and designed the experiments. H.H. synthesized the materials. J.Y. and C.W. Analysed the crystal structure. H.H. and E.M. performed the multiphoton experiments. Y.C., J.Y., Y.Y. and T.S. assisted with the linear optical property measurements and characterization of the material. H.H., E.M., Y.C., X.C., B.C. and G.Q. analysed the data and co-wrote the manuscript. All authors discussed the results and commented on the manuscript.

Additional information

Accession codes: The X-ray crystallographic coordinates for structure reported in this study has been deposited at the Cambridge Crystallographic Data Centre (CCDC), under deposition number 1046524. These data can be obtained free of charge from the Cambridge Crystallographic Data Centre via www.ccdc.cam.ac.uk/data_request/cif.

15

A zwitterionic gel electrolyte for efficient solid-state supercapacitors

Xu Peng[1], Huili Liu[1], Qin Yin[1], Junchi Wu[1], Pengzuo Chen[1], Guangzhao Zhang[1], Guangming Liu[1], Changzheng Wu[1] & Yi Xie[1]

Gel electrolytes have attracted increasing attention for solid-state supercapacitors. An ideal gel electrolyte usually requires a combination of advantages of high ion migration rate, reasonable mechanical strength and robust water retention ability at the solid state for ensuring excellent work durability. Here we report a zwitterionic gel electrolyte that successfully brings the synergic advantages of robust water retention ability and ion migration channels, manifesting in superior electrochemical performance. When applying the zwitterionic gel electrolyte, our graphene-based solid-state supercapacitor reaches a volume capacitance of $300.8\,F\,cm^{-3}$ at $0.8\,A\,cm^{-3}$ with a rate capacity of only 14.9% capacitance loss as the current density increases from 0.8 to $20\,A\,cm^{-3}$, representing the best value among the previously reported graphene-based solid-state supercapacitors, to the best of our knowledge. We anticipate that zwitterionic gel electrolyte may be developed as a gel electrolyte in solid-state supercapacitors.

[1] Hefei National Laboratory for Physical Sciences at the Microscale, iChEM (Collaborative Innovation Centre of Chemistry for Energy Materials), Hefei Science Center (CAS), and CAS Key Laboratory of Mechanical Behavior and Design of Materials, University of Science and Technology of China, Hefei, Anhui 230026, China. Correspondence and requests for materials should be addressed to C.W. (email: czwu@ustc.edu.cn) or to G.L. (email: gml@ustc.edu.cn).

The development of highly efficient energy storage devices has greatly satisfied growing energy demands for our daily life, of which supercapacitors have emerged as high-performance energy storage devices for long operating lifetimes and high power densities[1–3]. Recently, gel electrolytes have attracted increasing attention in solid-state supercapacitors, due to their capability to fulfil multiple roles of electrolyte, separator and binder in solid-state supercapacitors[4–6]. Developing gel electrolytes accelerates the evolution of solid-state supercapacitors from traditional sandwich-type supercapacitors, to flexible, transparent and planar super-capacitors (micro-supercapacitors)[7,8], and thus offering power support to flexible and even transparent electronics[9,10]. Generally, gel electrolytes are made of a polymeric material as matrix and an electrolyte salt to provide mobile ions[4]. Nowadays, non-aqueous gel electrolytes, such as block-copolymer-based gel electrolyte and silica-based gel electrolyte, have been developed dissolved in ionic liquid to solve the evaporation problem, achieving the enhancement in the ion migration rate and good mechanical strength, and making a great advance in electrochemical performance[11–14]. Aqueous gel electrolytes are dominantly based on polyvinyl alcohol (PVA) matrix, such as PVA/H_2SO_4 (refs 15,16), PVA/KOH (refs 17,18) and PVA/LiCl (refs 19,20). The good abilities of PVA gel electrolyte, with a wide range of pH values like the aqueous electrolyte solution and serving as an elastic coating with a certain mechanical strength to avoid structure degradation of electrode materials, render excellent performance of PVA gel electrolytes for solid-state supercapacitors[21]. Despite the fact that PVA gel electrolytes offer convenience to fabricate solid-state supercapacitors, the development of aqueous polymeric gel electrolytes is still at a primary stage and the inner electrochemical mechanism remains to be explored. This leaves plenty of room to improve electrochemical performance of solid-state supercapacitors via chemical design of gel electrolytes.

Polyzwitterions are a type of charged polymer with robust water retention ability coming from the presence of a zwitterionic group in a repeat unit[22], serving as a potential gel electrolyte catering for highly efficient solid-state supercapacitors. Polyzwitterions can be highly hydrated due to the strong interactions between the charged groups and water molecules, which make them suitable as a kind of superabsorbent polymeric material[23]. Also, their zwitterionic nature makes the cationic and anionic counterions of polyzwitterions easily separated during ion migration, ensuring a high ionic conductivity[24,25]. Moreover, polyzwitterions can form physical gels through the dipole–dipole interactions between the zwitterionic groups, thereby rendering the polyzwitterionic gel a certain mechanical strength[26]. In contrast to conventional polyelectrolytes, polyzwitterions usually exhibit the so-called 'anti-polyelectrolyte' effect, thus it has a good solubility in aqueous solutions with a high salt concentration[27,28]. Besides, the charged and polar groups associated with polyzwitterions can strengthen the adhesion between the gel electrolyte and the electrodes, so that the polyzwitterionic gel can be employed as a polymer binder to hold the electrodes together. More importantly, zwitterions have been used as electrolytes for high ionic conductivity and high lithium ion transfer number[29–32]. For example, zwitterionic gel electrolyte with $66.1\,mS\,cm^{-1}$ with the help of high ion mobility of H^+ was reported recently[33]. But there still presents great demand for introducing them to energy storage applications, especially for enhancing electrochemical performance via the electrolyte. In short, the combined advantages of water retention ability, high ion conductivity, reasonable mechanical strength and anti-polyelectrolyte effect renders polyzwitterions as a promising gel electrolyte for energy storage devices.

In this work, we develop a class of zwitterionic gel electrolyte, realising superior electrochemical performance in solid-state supercapacitors. In our case, the zwitterionic nature of poly (propylsulfonate dimethylammonium propylmethacrylamide) (PPDP) not only offers this gel electrolyte robust water retention ability at the solid-state through a combination of about eight water molecules around the charged groups but also brings ion migration channels to the electrolyte ions, leading to better electrochemical performance. When applying PPDP as a gel electrolyte, our graphene-based solid-state supercapacitor reaches a volume capacitance of $300.8\,F\,cm^{-3}$ at $0.8\,A\,cm^{-3}$, with a rate capacity of only 14.9% capacitance loss as current density increases from 0.8 to $20\,A\,cm^{-3}$, representing the best value among the previously reported graphene-based solid-state supercapacitors, to the best of our knowledge.

Results

Chemical structures of PPDP/LiCl gel electrolyte. PPDP is a kind of polyzwitterion bearing both positively charged quaternary ammonium group and negatively charged sulfonate group on the same monomeric unit. In this regard, PPDP is virtually neutral in a whole due to its zwitterionic character, as illustrated in Fig. 1a. Owing to the strong electrostatic interactions between the charged groups and water molecules, PPDP is highly hydrated by the surrounding water molecules with a robust water retention ability, showing a potential for superior electrochemical performance when supercapacitors operate at solid state. Meanwhile, the ion migration channel can be developed within

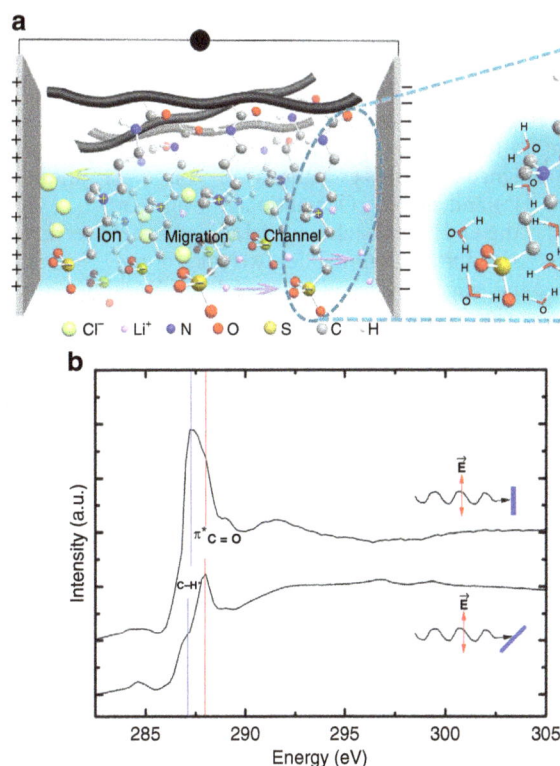

Figure 1 | Zwitterionic PPDP under external electric field. (a) Schematic illustration of the PPDP gel electrolyte applied on electrodes. The ion migration channel is formed by applying external electric field. PPDP is strongly hydrated by water molecules with robust water retention ability due to the electrostatic interactions between the zwitterionic groups and water molecules. **(b)** Angular-dependent C K-edge X-ray absorption near-edge spectroscopy (XANES) of zwitterionic PPDP sample after applying external field, with linearly polarized soft X-ray beam.

the hydration layer along the polyzwitterion chains between two electrodes by applying external electric field due to the robust water retention ability of PPDP gel electrolyte. Such a kind of ion migration channel strongly boosts the efficiency of ion transport in the gel electrolyte. Therefore, the zwitterionic nature of PPDP makes the electrolyte ions of cationic Li$^+$ and anionic Cl$^-$ in gel electrolyte easily separated without overcoming the strong electrostatic attractions between the charged groups and the counterions during the ion migration and readily transferred onto the surface of graphene electrodes to realize the equivalent electric double-layer capacitance (EDLC), thereby enhancing the capacitance of solid-state supercapacitors.

Although the PPDP main chains are randomly dispersed in the gel electrolyte, the ion migration channel could be formed along the aligned zwitterionic side groups within the PPDP gel electrolyte induced by the external electric field. To verify the aligned zwitterionic groups in gel electrolyte under the external electric field, the angular-dependent X-ray absorption near-edge spectroscopy (XANES) is employed to probe the orientation of the molecular bonds, taking advantage of the polarized nature of the synchrotron radiation sources[34]. Figure 1b shows the total electron yield XANES spectra measured at the normal director ($\theta = 90°$) and $\theta = 45°$ while the PPDP sample was applied a potential. The spectra were normalized to the incident photon flux and further processed by levelling the pre- and post-edges[35]. The sample contains a lot of the carbon bonds, such as C–C, C–N, C$=$O, C–H and so on. So, the C K-edge XANES spectra present rich absorption features between 282 and 295 eV. Therein, the two main features of the XANES spectra at ~287.3 and 287.9 eV present the significant angular dependence. For E-vector of the incident soft X-ray parallel to the surface of the substrate ($\theta = 90°$), the XANES presents a strong feature at 287.3 eV and a weak one at 287.9 eV. While in the case of oblique incidence of the X-ray beam ($\theta = 45°$), the former feature becomes weaker and the later becomes stronger. According to the reported literatures[36], these two main features can be assigned to the excitations of the C–H antibond orbital (C–H*) and the C$=$O π antibond orbital ($\pi^*_{C=O}$). It is known that the C–H* molecular orbital is along the direction of the C–H axis and $\pi^*_{C=O}$ orbital lies two sides of the C$=$O axis. Therefore, the angular dependences of both XANES features indicate that the C–H and C$=$O bonds should be parallel to the substrate plane when a potential was applied. Since the main chains of the sample are fixedly attached on the surface of the substrate, these significant orientations of the C–H and C$=$O bonds point out that the zwitterionic side groups of the sample become ordered (vertical to the external field) in the electrochemical process. This fact indicates that the zwitterionic group can be aligned in some extent by the external electric field, accompanied by the formation of an ion migration channel in the gel electrolyte at solid state.

Electrochemical performances of solid-state supercapacitor. The PPDP gel electrolyte with the ion migration channel and robust water retention ability significantly enhances the electrochemical performances of the as-fabricated graphene-based supercapacitor, as shown in Figs 2 and 3. To verify the superior electrochemical performances of the PPDP gel electrolyte, we conducted graphene-based symmetric super-capacitors and tested in a two-electrode configuration. Cyclic voltammetry (CV) was first studied with a potential range of 0–1.0 V at the scan rate of 10, 50, 100 and 400 mV s^{-1}. As shown in Fig. 2a–c and Supplementary Fig. 1, the nearly rectangular CV curves indicated nearly ideal capacitive behaviours of graphene electrodes. Note that the resulting rectangle areas from the PPDP

gel electrolyte are substantially larger than those for the PVA gel electrolyte, revealing that specific capacitance values of graphene-based supercapacitors applying PPDP gel electrolyte are much higher than those of PVA gel electrolyte. Considering that there is no pseudocapacitive material in electrode, the capacitance enhancement is mainly due to the better EDLC behaviour. Moreover, galvanostatic charge–discharge (CD) measurements were performed at the current density of 0.8, 1, 2, 4, 8 and 20 A cm^{-3} (Fig. 2e,f; Supplementary Fig. 2). The discharging times of PPDP sample were 384.0 s for liquid state and 376.0 s for solid state at the current density of 0.8 A cm^{-3}, longer than those of PVA sample 250.4 s at liquid state and 242.0 s at solid state. The CD time of PPDP gel electrolyte sample continues to provide higher capacitance than that of PVA gel electrolyte sample. The as-fabricated graphene-based solid-state supercapacitor applying PPDP gel electrolyte reaches a volume capacitance of 300.8 F cm^{-3} at 0.8 A cm^{-3}, recording the best value among the previously reported graphene-based solid-state supercapacitors to the best of our knowledge.

Accordingly, the as-fabricated supercapacitors applying PPDP gel electrolyte yielded specific capacitances of 300.8, 298.2, 292.4, 279.6, 270.4 and 256.0 F cm^{-3} at current densities of 0.8, 1, 2, 4, 8 and 20 A cm^{-3} at solid state, respectively, obviously larger than those of the as-fabricated supercapacitors applying PVA gel electrolyte, as summarized in Fig. 3a. Meanwhile, the as-fabricated supercapacitor applying PPDP gel electrolyte exhibited better rate capacity, with only 14.9% capacitance loss when the current density increases by a factor of 25 (from 0.8 to 20 A cm^{-3}) compared with 23.6% loss for the as-fabricated supercapacitor applying PVA gel electrolyte. The PPDP gel electrolyte shows more advantages than the PVA gel electrolyte: (i) The capacitance loss of the PPDP gel electrolyte (1.4%) is much less than that of the PVA gel electrolyte (7.6%) after the electrolyte is transformed from liquid to solid state. (ii) The capacitance of as-fabricated supercapacitor applying PPDP gel electrolyte is 55.4% higher than the PVA gel electrolyte, as shown in Fig. 3b. The enhanced capacitance and rate capacity can be attributed to the unique property of PPDP gel electrolyte, which synergizes the effects of robust water retention ability and the ion migration channel with higher ion conductivity, and thus enhancing the capacitance, rate capacity and durability of the as-fabricated supercapacitor. In a word, applying the zwitterionic gel electrolyte, the graphene-based solid-state supercapacitor reaches a volume capacitance of 300.8 F cm^{-3} at 0.8 A cm^{-3} with a rate capacity of only 14.9% capacitance loss as the current density increases from 0.8 to 20 A cm^{-3}. Cycling performance is a critically important characteristic for evaluating the stability of solid-state supercapacitors. Owing to the same electrode material in our as-fabricated symmetric supercapacitors, cycling performance is largely attributed to the long-term stability of gel electrolytes (Supplementary Figs 3 and 4; Supplementary Note 1). We tested the 10,000 times of charging and discharging cycles at 4 A cm^{-3}. The PPDP gel electrolyte-based supercapacitor delivers a remarkable cycling stability of 103% retention up to 10,000 cycles, better than 91.8% retention of PVA gel electrolyte-based supercapacitor, as shown in Fig. 3c. These splendid electrochemical performances make such a kind of zwitterionic gel electrolyte attractive for fabricating the next generation of solid-state supercapacitors.

Mechanism of PPDP gel electrolyte. To explore the physical mechanism of PPDP gel electrolyte, static contact angle and viscoelastic properties were tested to verify its favourable wetting behaviour and reasonable mechanical strength that is vital for construction of graphene-based solid-state supercapacitors.

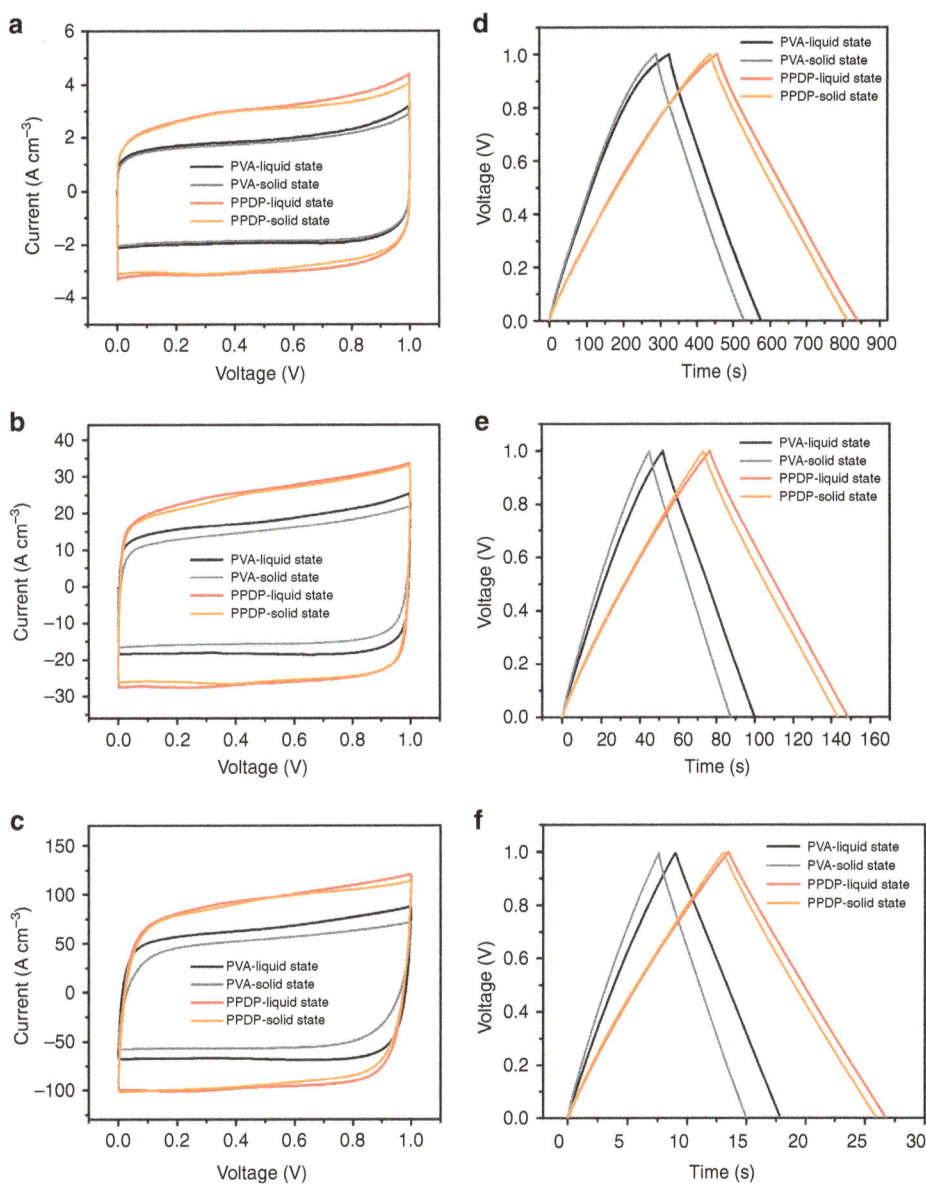

Figure 2 | Electrochemical performance of graphene-based supercapacitors. (**a–c**) CV curves of graphene-based solid-state supercapacitors applying PPDP and PVA gel electrolytes at liquid state and solid state, the scan rates were 10, 100 and 400 mV s^{-1}. (**d–f**) Galvanostatic charge–discharge curves of graphene-based solid-state supercapacitors applying PPDP and PVA electrolytes at liquid state and solid state at the current density of 0.8, 4 and 20 A cm^{-3}.

The affinity between the electrode and the gel electrolyte is crucial for supercapacitors because EDLC is strongly dependent on the surface contact areas between the electrode and the gel electrolyte. The wetting behaviour of the gel electrolytes on the graphene electrode is shown in Fig. 4a. For the initially prepared liquid-state electrolytes, the static contact angles are 72° and 61° for the PVA and PPDP electrolytes, respectively. The contact angles, respectively, decrease to 37° and 22° for the PVA and PPDP gel electrolytes after 24 h when the liquid state is transformed into the solid state due to evaporation. These facts suggest that the PPDP gel electrolyte can more effectively penetrate into the multilayer graphene electrodes and enhance the contact areas between the gel electrolyte and the working electrode compared with the PVA gel electrolyte as shown in the inset of Fig. 4a, due to the relatively low viscosity and high fluidity of the PPDP gel electrolyte than that of the PVA gel electrolyte (Supplementary Table 1). This characteristic of PPDP gel electrolyte would bring superior capacitance for solid-state

supercapacitors. An ideal gel electrolyte not only can provide a superior electrochemical performance but also can serve as a binder with a certain mechanical strength to hold two electrodes together. Figure 4b shows the viscoelastic properties of the PPDP gel electrolyte. At the liquid state of PPDP electrolyte, the storage modulus (G') is smaller than the loss modulus (G''). In contrast, the G' dominates over G'' at the solid state, indicating that PPDP actually forms physically crosslinked gel due to the dipole–dipole interaction between the zwitterionic groups (Supplementary Fig. 5). The inter-chain hydrogen bonding between the amide parts of PPDP may also contribute to the formation of physical gel. Therefore, such a kind of gel with a storage modulus of 140 Pa provides enough mechanical strength to hold the two electrodes together, which can be supported from the formation of freestanding solid thin film by the PPDP gel electrolyte at a low water content as shown in the inset of Fig. 4b. Likewise, PVA gel electrolyte also exhibits similar viscoelastic properties at liquid and solid states (Supplementary Fig. 6; Supplementary Note 2).

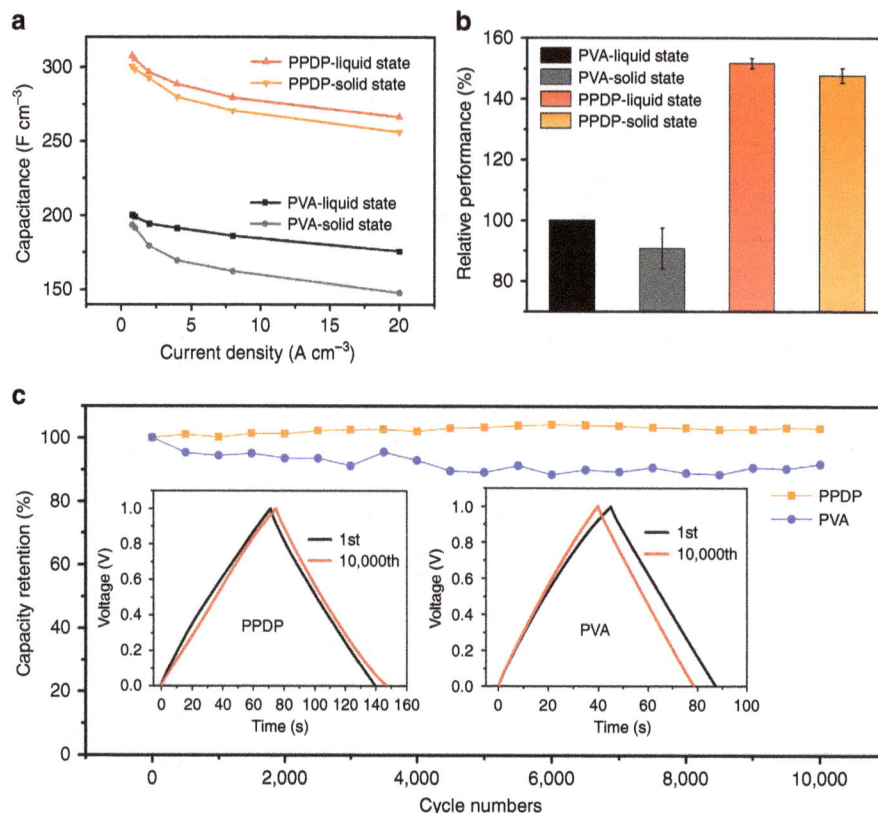

Figure 3 | Capacitance and cycling performance of graphene-based supercapacitors. (**a**) Comparison of specific capacitance values for graphene-based supercapacitors between PPDP and PVA gel electrolytes at different current densities. (**b**) Relative performance of PVA and PPDP electrolytes in liquid and solid states, where the error bars are obtained based on the capacitances at different current densities. (**c**) Cycling performance of graphene electrodes applying PPDP or PVA gel electrolyte at solid state. Inset: comparison of CD curves between the 1st CD cycle and the 10,000th cycle for graphene-based solid-state supercapacitors applying PPDP (left of panel) and PVA (right of panel) gel electrolyte at $4\,A\,cm^{-3}$.

To explore the chemical mechanism of PPDP gel electrolyte, differential scanning calorimetry (DSC) and electrochemical impedance spectroscopy were performed to verify the unique characteristics of robust water retention ability and efficient ion migration channel. The hydration capacity obtained from the DSC measurements demonstrates that PPDP has high water retention ability (Fig. 4c). Specifically, no endothermic peak can be observed in the thermogram during the heating of PPDP without water from -35 to $60\,°C$, suggesting that the polyzwitterion itself does not contribute to the thermal transition behaviour. Similarly, no thermal transition is observed during the heating process when the mole ratio of H_2O to PDP equals 6:1 and 7:1, indicating that all the water molecules around one zwitterionic group (about eight water molecules) in the hydration shell are tightly bound to the polyzwitterion through the electrostatic interactions with the positively charged $N^+(CH_3)_3$ and negatively charged SO_3^- and no freezable water exists in the system. However, an obvious endothermic peak is observed as the mole ratio of H_2O to PDP increases to 8:1, which means that the freezable water can be detected in the system when all binding sites of the polyzwitterion are saturated by water molecules[37]. Therefore, one PDP unit is strongly hydrated by seven to eight water molecules, as highlighted in Fig. 1. By contrast, Supplementary Fig. 7 and Supplementary Note 3 show that only one to two water molecules are tightly bound to one PVA monomeric unit via hydrogen bonding between the hydroxide group and the water molecules. Similar results are also observed in the low-field nuclear magnetic resonance (NMR) measurements (Supplementary Fig. 8; Supplementary Note 4). Consequently, the PPDP gel electrolyte gives rise to a higher

water retention ability than PVA gel electrolyte at solid state (Supplementary Fig. 9; Supplementary Table 2; Supplementary Note 5). To confirm the efficient ion migration channel of the PPDP gel electrolyte, we performed electrochemical impedance spectroscopy tests using a two-electrode configuration from $100\,mHz$ to $100\,kHz$. Usually, the intersection of the curve at the real part reflects the equivalent series resistance at the high frequency ($100\,kHz$), which is contributed from the resistance of both electrolyte and electrode, while the linear region corresponds to the Warburg diffusion process (W), reflecting the ion diffusion into the electrode materials. As the Nyquist plots shown in Fig. 4d, the equivalent series resistance of PPDP-solid-state sample was $40.54\,\Omega$, much smaller than that of PVA-solid-state sample of $122.6\,\Omega$. Meanwhile, the slopes of PPDP gel electrolyte at liquid state and solid state are apparently larger than those of PVA samples at the low frequency, revealing that the transport and diffusion of ions into electrode materials of PPDP gel electrolyte are much better than those of PVA gel electrolyte, which answers for the superior electrochemical performance of the zwitterionic gel electrolyte.

The ion migration channel along with the robust water retention ability of PPDP gel electrolyte greatly facilitates the ion transport during charging and discharging process. To further evaluate the energy efficiency of the as-fabricated graphene-based solid-state supercapacitors applying PPDP gel electrolyte, energy density and power density were calculated from CD curves. The Ragone plot of the as-fabricated graphene-based solid-state supercapacitors was shown in Fig. 5a. The as-fabricated solid-state supercapacitor applying PPDP gel electrolyte delivers a high energy density of $41.78\,Wh\,l^{-1}$ at a power density of

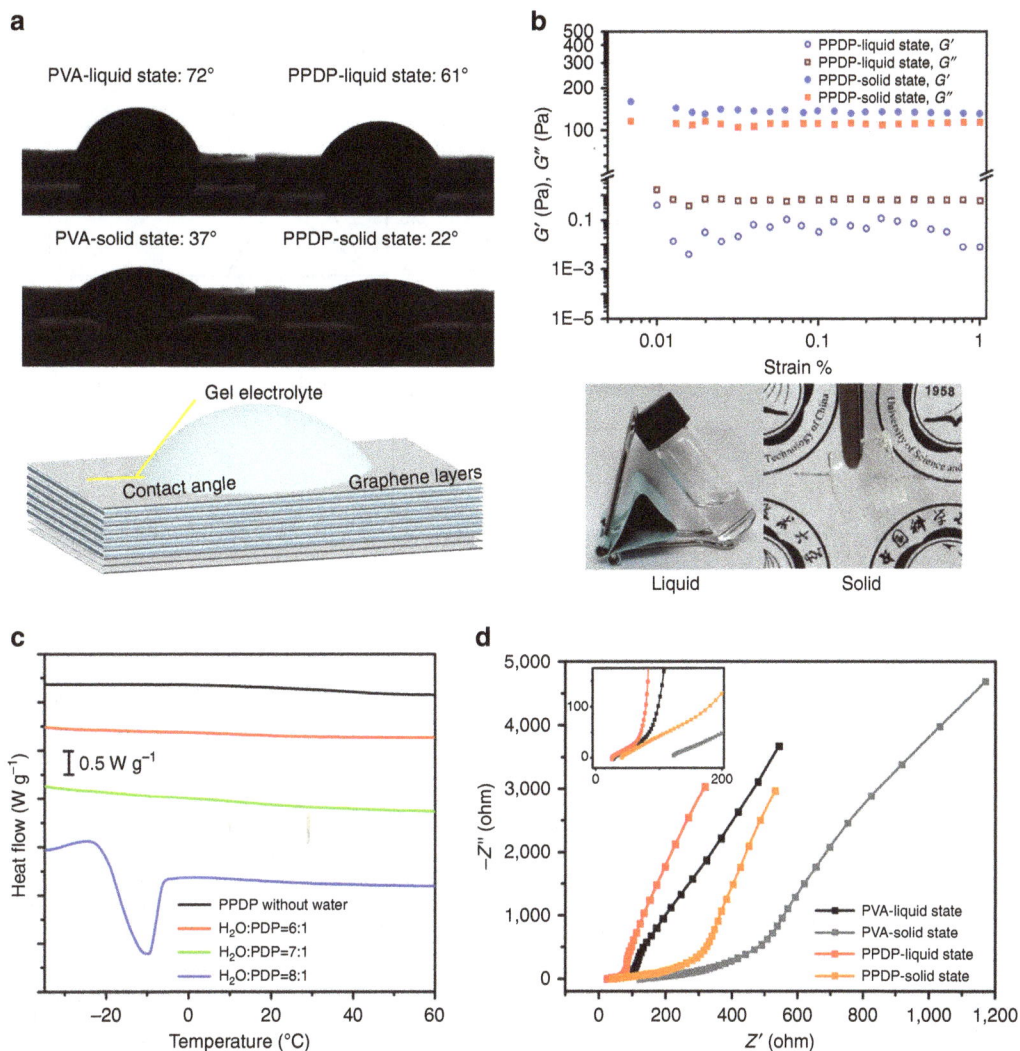

Figure 4 | Physical and chemical mechanisms of PPDP gel electrolyte. (**a**) Static contact angles of PVA and PPDP gel electrolytes on the graphene electrode. Inset is the schematic illustration of the penetration of gel electrolyte into the multilayer graphene electrode. (**b**) Viscoelastic properties of the PPDP gel electrolyte at liquid and solid states. PPDP gel electrolyte is demonstrated at liquid state in a vial and at solid state with a freestanding solid thin film. (**c**) DSC thermograms of PPDP at different water contents. (**d**) Electrochemical impedance spectroscopy (EIS) of graphene-based solid-state supercapacitors applying PPDP and PVA gel electrolytes at liquid state and solid state.

$400 \ W l^{-1}$, superior than $26.89 \ Wh l^{-1}$ of the as-fabricated solid-state supercapacitor applying PVA gel electrolyte, which is also excellent among other previously reported state-of-the-art graphene-based solid-state supercapacitors[8,17,38], revealing that polyzwitterions especially PPDP would be a promising candidate to serve as gel electrolyte in solid-state supercapacitors. Besides, the feasibility of the as-fabricated solid-state supercapacitors is demonstrated in Fig. 5b. Generally, an effective way to enhance the potential or energy density can be achieved using serial and parallel assemblies. Compared with single supercapacitor, two as-fabricated supercapacitors in series show twice potential of 2.0 V, and two as-fabricated supercapacitors in parallel show almost twice CD time, indicating that the as-fabricated solid-state supercapacitor has good practical value. Meanwhile, there is an only slight difference in CV curves of the as-fabricated solid-state supercapacitor between normal and under bent status (90°), indicating that PPDP gel electrolyte has a certain mechanical strength with good flexibility and stability (Fig. 5c), which is suitable for establishing ultraflexible and even transparent energy storage devices and other electronics. The similar test about ionic conductivity of the PPDP gel electrolyte under different bending states is shown in Supplementary Fig. 10 and

Supplementary Note 6. In addition, our proposed zwitterionic PPDP gel electrolyte also has general applicability for solid-state supercapacitors under acidic and basic conditions (Supplementary Fig. 11; Supplementary Note 7). Furthermore, the PPDP gel electrolyte has better self-discharging performance than that of the PVA gel electrolyte (Supplementary Fig. 12; Supplementary Note 8).

Discussion

Zwitterionic gel electrolyte has been proven to be a promising gel electrolyte catering for solid-state supercapacitors. The zwitterionic nature of PPDP as gel electrolyte not only offers robust water retention ability but also brings ion migration channels to the electrolyte ions, leading to superior electrochemical performances. When applying PPDP as a gel electrolyte, the as-fabricated graphene-based solid-state supercapacitor reaches a volume capacitance of $300.8 \ F \ cm^{-3}$ at $0.8 \ A \ cm^{-3}$, with a rate capacity of only 14.9% capacitance loss when the current density increases from 0.8 to $20 \ A \ cm^{-3}$, recording the best value among the previously reported graphene-based solid-state supercapacitors, to the best of our knowledge. Moreover, the high water retention ability of the zwitterionic

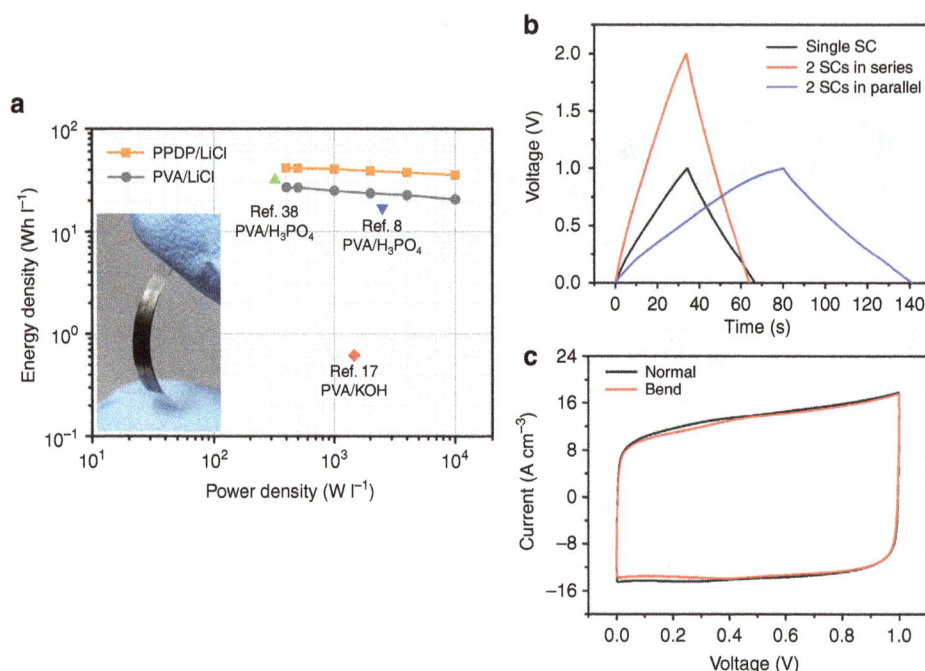

Figure 5 | Applications of PPDP gel electrolyte. (a) A typical Ragone plot of the as-fabricated graphene-based solid-state supercapacitors. Inset is a prototype of graphene-based solid-state supercapacitor applying PPDP gel electrolyte with a planar configuration. **(b)** Galvanostatic charge–discharge curves of single supercapacitor (SC) and two SCs connected in series and in parallel at $8\,A\,cm^{-3}$. **(c)** Comparison of CV curves between normal and bent status (90°), the scan rate is $50\,mV\,s^{-1}$.

groups brings a robust cyclability, the capacitance remaining 103% retention after undergoing 10,000 CD cycles. We anticipate that zwitterionic gel electrolyte will be a promising gel electrolyte for the next generation of solid-state supercapacitors.

Methods

Preparation of gel electrolytes. The PVA/LiCl gel electrolyte was synthesized according to the procedure reported in previous literature[20]. An amount of 4.0 g PVA powder was put into 40 ml distilled water with stirring at 95 °C. After the PVA powder was completely dissolved, 4.0 g LiCl•H_2O was added into the solution under vigorous stirring. When PVA/LiCl turned to be transparent and clear gel, it was cooled down to room temperature and the gel electrolyte was successfully prepared. The PPDP/LiCl gel electrolyte was prepared as follows. An amount of 1.0 g propylsulfonate dimethylammonium propylmethacrylamide (PDP), 1.0 mg 4,4′-azobis(4-cyanovaleric acid) (ACVA) and 4.0 ml distilled water were added into a round-bottom reactor. After three times of frozen–degassing–thawing cycles, the free-radical polymerization was conducted at 70 °C for 10 h. The polymerization is completed and cooled down to room temperature. Afterwards, the as-prepared PPDP was mixed homogeneously with 6.0 ml aqueous solution including 1.0 g LiCl•H_2O under vigorous stirring to prepare the zwitterionic gel electrolyte, PPDP physical gel was successfully prepared.

To compare the electrochemical performances between PVA and PPDP gel electrolytes, the optimal initial inventory mass ratio of H_2O:polymer:LiCl•H_2O is fixed at 10:1:1 for the both two gel electrolytes as the concentration of LiCl•H_2O has important influences on the physical state, ionic conductivity and specific capacitance of the gel electrolytes (Supplementary Figs 13 and 14; Supplementary Notes 9 and 10). And the link between ionic conductivity of gel electrolyte and the capacitance of solid-state supercapacitors is discussed in Supplementary Fig. 15 and Supplementary Note 11. The role of LiCl in enhancing the water retention of gel electrolytes is clarified in Supplementary Fig. 16 and Supplementary Note 12.

Fabrication of graphene electrodes and solid-state supercapacitors. Graphene was obtained from the chemical reduction of graphene oxides. Graphene oxide was synthesized in a modified Hummer's method[39]. Then, graphene was reduced using hydrazine according to a reported method[40]. The graphene thin film on the surface of cellulose acetate membrane was obtained by a vacuum filtration process, followed by tightly pressing thin film onto polyethylene terephthalate (PET) substrate and slowly peeling off the cellulose acetate membrane, leading to freestanding graphene thin film on PET. Then, the graphene thin film was scraped along the channel to fabricate working electrodes, while two columns of Au current collectors were thermally evaporated on each side of the working electrodes. Finally, the solid-state supercapacitor was finally established after coating PPDP/LiCl gel electrolyte on the parallel region filling the channel between two working

electrodes. The working electrodes were prepared with the same thickness and the surface areas of graphene as well as the gel electrolyte were applied onto the graphene with the same amounts for PVA and PPDP gel electrolytes, so that the electrochemical behaviour between PVA and PPDP gel electrolytes can be compared fairly. After the gel electrolyte became solid state, the graphene-based solid-state supercapacitor was fabricated. The volume of the graphene electrodes was calculated through multiplying the thickness by areas. The exact thickness of the graphene electrodes was ~250 nm determined by the field-emission scanning electron microscopy on a FEI Sirion-200 SEM instrument (Supplementary Fig. 17). The distance between the two electrodes is ~150 μm (Supplementary Fig. 18) and the exact surface area of each graphene electrode is 10×1 mm (Supplementary Fig. 19; Supplementary Note 13).

Electrochemical tests. Electrochemical performances of the graphene-based supercapacitors were measured in a two-electrode system by CV and galvanostatic CD at an electrochemical station (CHI 760E). Potential range set for CV and CD tests was from 0 to 1.0 V. The as-fabricated graphene electrodes were directly used as the working electrodes for the electrochemical tests. Electrochemical impedance measurements of the supercapacitors were performed in the same configuration from 100 mHz to 100 kHz with a Zahner IM6 electrochemical workstation. The cycling stability of the as-fabricated supercapacitors was performed at room temperature with a sweep charge and discharge rate at the current density of $4\,A\,cm^{-3}$ for 10,000 cycles.

Other characterizations. Proton NMR (^{1}H NMR) spectra were recorded on a Bruker AV400 spectrometer, using deuterium water as the solvent and tetramethylsilane as an internal standard. The ^{1}H NMR spectra of PPDP and PVA were shown in Supplementary Fig. 20 and Supplementary Note 14. The number-average molar mass ($M_n = 1.87 \times 10^5\,g\,mol^{-1}$, $M_w/M_n \sim 1.6$) of PPDP was determined by size-exclusion chromatography (Waters 1,515), using monodisperse poly(ethylene glycol) as the standard and a 1.0 M NH_4NO_3 solution as the eluent with a flow rate of $1.0\,ml\,min^{-1}$. The viscoelastic property of PPDP gel electrolyte was determined by a rheological measurement on a TA AR-G2 rheometer. The storage modulus (G') and loss modulus (G'') were measured as a function of strain with 1 Hz frequency at 25 °C. The DSC scans were conducted on a TA Instruments Q2000 scanning calorimeter at a heating rate of $5\,°C\,min^{-1}$. The angular-dependent C K-edge XANES spectra, using the linearly polarized X-ray beam, were measured at the beamline BL12B of National Synchrotron Radiation Laboratory (NSRL, Hefei) in the total electron yield mode by collecting the sample drain current under a vacuum better than 10^{-7} Pa. The beam from the bending magnet was monochromatized utilizing a varied line-spacing plane grating and refocused by a toroidal mirror. The energy range is 100–1,000 eV with an energy resolution of ca. 0.2 eV. The PPDP sample was applied CV from 0 to 1.0 V for 1 h at the scan rate of $10\,mV\,s^{-1}$ and underwent freeze-drying process before the measurement of XANES spectra. Low-field NMR measurements were performed by Bruker Minispec MQ-20 NMR analyser operating at a resonance frequency of 20 MHz (0.47 T). Sample was inserted in the NMR probe. Spin–Spin relaxation time, T_2, was measured using the Carr-Purcell-Meiboom-Gill (CPMG) sequence.

References

1. Simon, P., Gogotsi, Y. & Dunn, B. Where do batteries end and supercapacitors begin? *Science* **343**, 1210–1211 (2014).
2. Simon, P. & Gogotsi, Y. Materials for electrochemical capacitors. *Nat. Mater.* **7**, 845–854 (2008).
3. Peng, X., Peng, L., Wu, C. & Xie, Y. Two dimensional nanomaterials for flexible supercapacitors. *Chem. Soc. Rev.* **43**, 3303–3323 (2014).
4. Lu, X., Yu, M., Wang, G., Tong, Y. & Li, Y. Flexible solid-state supercapacitors: design, fabrication and applications. *Energy Environ. Sci.* **7**, 2160–2181 (2014).
5. Sun, G. *et al.* Hybrid fibers made of molybdenum disulfide, reduced graphene oxide, and multi-walled carbon nanotubes for solid-state, flexible, asymmetric supercapacitors. *Angew. Chem. Int. Ed.* **127**, 4734–4739 (2015).
6. Li, L. *et al.* A flexible quasi-solid-state asymmetric electrochemical capacitor based on hierarchical porous V_2O_5 nanosheets on carbon nanofibers. *Adv. Energy Mater.* **5**, 1500753 (2015).
7. Feng, J. *et al.* Metallic few-layered VS_2 ultrathin nanosheets: high two-dimensional conductivity for in-plane supercapacitors. *J. Am. Chem. Soc.* **133**, 17832–17838 (2011).
8. Peng, L. *et al.* Ultrathin two-dimensional MnO_2/graphene hybrid nanostructures for high-performance, flexible planar supercapacitors. *Nano Lett.* **13**, 2151–2157 (2013).
9. Liu, Z., Xu, J., Chen, D. & Shen, G. Flexible electronics based on inorganic nanowires. *Chem. Soc. Rev.* **44**, 161–192 (2015).
10. Wang, X. *et al.* Flexible energy-storage devices: design consideration and recent progress. *Adv. Mater.* **26**, 4763–4782 (2014).
11. Kang, Y. *et al.* All-solid-state flexible supercapacitors fabricated with bacterial nanocellslose papers, carbon nanotubes, and triblock-copolymer ion gels. *ACS Nano* **6**, 6400–6406 (2012).
12. Kang, Y., Chung, H., Han, C. & Kim, W. All-solid-state flexible supercapacitors based on papers coated with carbon nanotubes and ionic-based gel electrolytes. *Nanotechnology* **23**, 065401 (2012).
13. Lee, J., Kim, W. & Kim, W. Stretchable carbon nanotube/ion-gel supercapacitors with high durability realized through interfacial microroughness. *ACS Appl. Mater. Interface* **6**, 13578–13586 (2014).
14. Kim, D., Shin, G., Kang, Y., Kim, W. & Ha, J. Fabrication of a stretchable solid-state micro-supercapacitor array. *ACS Nano* **7**, 7975–7982 (2013).
15. Cao, X. *et al.* Reduced graphene oxide-wrapped MoO_3 composites prepared by using metal–organic frameworks as precursor for all-solid-state flexible supercapacitors. *Adv. Mater.* **27**, 4695–4701 (2015).
16. Wang, G. *et al.* Solid-state supercapacitor based on activated carbon cloths exhibits excellent rate capability. *Adv. Mater.* **26**, 2676–2682 (2014).
17. Wang, X. *et al.* Fiber-based flexible all-solid-state asymmetric supercapacitors for integrated photodetecting system. *Angew. Chem. Int. Ed.* **53**, 1849–1853 (2014).
18. Xie, J. *et al.* Layer-by-layer β-Ni(OH)$_2$/graphene nanohybrids for ultraflexible all-solid-state thin-film supercapacitors with high electrochemical performance. *Nano Energy* **2**, 65–74 (2013).
19. Zhu, C. *et al.* All metal nitrides solid-state asymmetric supercapacitors. *Adv. Mater.* **27**, 4566–4571 (2015).
20. Wu, C. *et al.* Two-dimensional vanadyl phosphate ultrathin nanosheets for high energy density and flexible pseudocapacitors. *Nat. Commun.* **4**, 2431 (2013).
21. Wang, G. *et al.* LiCl/PVA gel electrolyte stabilizes vanadium oxide nanowire electrodes for pseudocapacitors. *ACS Nano* **6**, 10296–10302 (2012).
22. Kudaibergenov, S., Jaeger, W. & Laschewsky, A. in *Supramolecular Polymers Polymeric Betains Oligomers* **201**, 157–224 (Springer (2006).
23. Lalani, R. & Liu, L. Electrospun zwitterionic poly(sulfobetaine methacrylate) for nonadherent, superabsorbent, and antimicrobial wound dressing applications. *Biomacromolecules* **13**, 1853–1863 (2012).
24. Tiyapiboonchaiya, C. *et al.* The zwitterion effect in high-conductivity polyelectrolyte materials. *Nat. Mater.* **3**, 29–32 (2004).
25. Cardoso, J., Huanosta, A. & Manero, O. Ionic conductivity studies on salt-polyzwitterion systems. *Macromolecules* **24**, 2890–2895 (1991).
26. Vasantha, V. A., Jana, S., Parthiban, A. & Vancso, J. G. Water swelling, brine soluble imidazole based zwitterionic polymers - synthesis and study of reversible UCST behaviour and gel-sol transitions. *Chem. Commun.* **50**, 46–48 (2014).
27. Lowe, A. B. & McCormick, C. L. Synthesis and solution properties of zwitterionic polymers. *Chem. Rev.* **102**, 4177–4190 (2002).
28. Wang, T., Wang, X., Long, Y., Liu, G. & Zhang, G. Ion-specific conformational behavior of polyzwitterionic brushes: exploiting it for protein adsorption/desorption control. *Langmuir* **29**, 6588–6596 (2013).
29. Narita, A. *et al.* Lithium ion conduction in an organoborate zwitterion-LiTFSI mixture. *Chem. Commun.* 1926–1928 (2006).
30. Matsumi, N., Sugai, K., Miyake, M. & Ohno, H. Polymerized ionic liquid via hydroboration polymerization as single ion conductive polymer electrolytes. *Macromolecules* **39**, 6924–6927 (2006).
31. Brown, R. H. *et al.* Effect of ionic liquid on mechanical properties and morphology of zwitterionic copolymer menbranes. *Macromolecules* **43**, 790–796 (2010).
32. Yoshizawa, M., Hirao, M., Ito-Akita, K. & Ohno, H. Ion conduction in zwitterionic-type molten salts and their polymers. *J. Mater. Chem.* **11**, 1057–1062 (2001).
33. Zhou, T., Gao, X., Dong, B., Sun, N. & Zheng, L. Poly(ionic liquid) hydrogels exhibiting superior mechanical and electrochemical properties as flexible electrolytes. *J. Mater. Chem. A* **4**, 1112–1118 (2016).
34. Stöhr, J. NEXAFS spectroscopy. *Springer Ser. Surf. Sci.* **392**, 276–290 (1992).
35. Zhong, J. *et al.* Bio-nano interaction of proteins adsorbed on single-walled carbon nanotubes. *Carbon* **47**, 967–973 (2009).
36. Li, Z., Zhang, L., Resasco, D., Mun, B. & Requejo, F. Angle-resolved X-ray absorption near edge structure study of vertically aligned single-walled carbon nanotubes. *Appl. Phys. Lett.* **90**, 103115 (2007).
37. Wu, J., Lin, W., Wang, Z., Chen, S. & Chang, Y. Investigation of the hydration of nonfouling material poly(sulfobetaine methacrylate) by low-field nuclear magnetic resonance. *Langmuir* **28**, 7436–7441 (2012).
38. Niu, Z. *et al.* All-solid-state flexible ultrathin micro-supercapacitors based on graphene. *Adv. Mater.* **25**, 4035–4042 (2013).
39. Marcano, D. *et al.* Improved synthesis of graphene oxide. *ACS Nano* **4**, 4806–4814 (2010).
40. Park, S. *et al.* Hydrazine-reduction of graphite- and graphene oxide. *Carbon* **49**, 3019–3023 (2011).

Acknowledgements

This work was financially supported by the National Basic Research Program of China (2015CB932302 and 2012CB933802), the National Natural Science Foundation of China (21222101, U1432133, U1532265, 21331005, 11321503, J1030412, 21374110, 21574121, 21234003 and 91127042), National Program for Support of Top-notch Young Professionals the Chinese Academy of Sciences (XDB01020300), the Fok Ying-Tong Education Foundation, China (grant no.141042), Users with Excellence of Hefei Science Center (2015HSC-UE008) and the Fundamental Research Funds for the Central Universities (WK2060190027 and WK2340000066). We thank Professor Wangsheng Chu and Professor Wensheng Yan in National Synchrotron Radiation Laboratory, University of Science and Technology of China for assistance and discussion with the sample measurement by angular-dependent XANES test.

Author contributions

C.W. conceived the idea and experimentally realized the study, co-wrote the paper and supervised the project and is responsible for the infrastructure and project direction. X.P. and H.L. experimentally realized the idea and wrote the paper; J.W. and P.C. carried out the examples synthesis, characterization and data analysis. G.L. and G.Z. are responsible for the design of zwitterionic gel electrolyte and co-wrote the paper. Y.X. supervised the whole experimental procedure and co-wrote the paper. All the authors discussed the results, commented on and revised the manuscript. X.P. and H.L. contributed equally to this research.

Additional information

16

A large family of filled skutterudites stabilized by electron count

Huixia Luo[1], Jason W. Krizan[1], Lukas Muechler[1,2], Neel Haldolaarachchige[1], Tomasz Klimczuk[3], Weiwei Xie[1], Michael K. Fuccillo[1], Claudia Felser[2] & Robert J. Cava[1]

The Zintl concept is important in solid-state chemistry to explain how some compounds that combine electropositive and main group elements can be stable at formulas that at their simplest level do not make any sense. The electronegative elements in such compounds form a polyatomic electron-accepting molecule inside the solid, a 'polyanion', that fills its available energy states with electrons from the electropositive elements to obey fundamental electron-counting rules. Here we use this concept to discover a large family of filled skutterudites based on the group 9 transition metals Co, Rh, and Ir, the alkali, alkaline-earth, and rare-earth elements, and Sb_4 polyanions. Forty-three new filled skutterudites are reported, with 63 compositional variations—results that can be extended to the synthesis of hundreds of additional new compounds. Many interesting electronic and magnetic properties can be expected in future studies of these new compounds.

[1] Department of Chemistry, Princeton University, Princeton, New Jersey 08544, USA. [2] Max-Planck-Institut für Chemische Physik Fester Stoffe, Dresden 01187, Germany. [3] Faculty of Applied Physics and Mathematics, Gdansk University of Technology, Narutowicza 11/12, Gdansk 80-233, Poland. Correspondence and requests for materials should be addressed to H.L. (email: huixial@princeton.edu) or to R.J.C. (email: rcava@princeton.edu).

The chemical stability of simple ionic compounds can be explained by fundamental concepts through which an electropositive ion donates electrons to an electronegative ion and both achieve closed shell electron configurations. NaCl, with Na^{1+}, configuration $2s^2 2p^6$, and Cl^{1-}, configuration $3s^2 3p^6$, is the canonical example. For many compounds, however, the electron-counting and stability rules are not as straightforward. The Zintl phases are a family of solid compounds typically made from electropositive metallic elements combined with non-metals or metalloids capable of forming polyanions[1]. The more electropositive elements donate their valence electrons to the more electronegative elements, and the electronegative elements form polyanions to accept the donated electrons. The bonding and architecture of the polyanions is understood in terms of the octet rule[2–4]. Due to the filled valence states of the polyanion array and the empty valence states of the electropositive electron donors, semiconducting behaviour is favoured for the resulting compounds.

The binary skutterudites are one well known class of Zintl compounds, deriving from the archetypal mineral skutterudite, $CoAs_3$ (refs 4,5). They have the general formula BX_3, where B is a transition metal element such as Fe, Co, Rh or Ir and X is a pnictogen such as P, As or Sb (see Fig. 1a)[5–7]; they typically crystallize in cubic space group $Im\bar{3}$ (#204) with two B_4X_{12} formula units and two large empty cages per unit cell. The B ions are in the 8c (1/4, 1/4, 1/4) site and the X ions are in the 24g (0, y, z) site with $y \sim 0.15$ and $z \sim 0.35$. Featured in the structure are distorted square X_4 Zintl polyanions that have a formal charge of $4-$. Thus, the simple binary compounds have a semiconducting electron count for B elements from group 9; the B ions, Co, Rh and Ir, formally $3+$, have the electron configuration nd^6. In octahedral coordination and low spin, they thus have a filled t_{2g} band and an empty e_g band, with an energy gap between them. $CoAs_3$, for example, can be understood as $Co^{3+}_4 (As_4)^{4-}_3$ (refs 8–10).

The empty cages within the binary skutterudite framework can be filled by up to one ion (A) per B_4X_{12} formula unit, leading to the 'filled skutterudites' of formula $A_y B_4 X_{12}$, where A = alkali, alkaline-earth, rare-earth, actinide or Tl (see Fig. 1b)[11–13]. These are also in space group #204, with the A ions (in the ideal structures) in the 2a (0,0,0) site and the B and X ions as in the unfilled case. The known filled skutterudites are virtually all based on the group 8 metals Fe, Ru and Os. The A ion donates its charge to the B_4X_{12} framework. By judicious choice of constituents, complete filling of the cages ($y=1$), a Zintl (electron precise) electron count, and thus semiconducting behaviour can be achieved. In addition to those with an electron-precise formula, some compounds such as $LaRu_4Sb_{12}$ (with one electron in deficit of the Zintl count) are also known. A very small number of filled skutterudites based on Co, Ni and Pt have been reported (see Fig. 1c)[14–46].

In addition to their chemical interest, skutterudites are of interest to a broad community of materials scientists and physicists due to their interesting physical properties. Because they make very good thermoelectrics[21,22,47], host unusual metal–insulator transitions[23–25], can be rare-earth magnets, heavy-fermion compounds[26–30], non-Fermi liquids[31–35], itinerant ferromagnets[36–38] and superconductors[27,39–46], and have been proposed as topological insulators[48], filled skutterudites are a very important class of non-molecular solids. Because the electronic and magnetic properties of solids depend critically on the actual atoms making up the compounds in addition to the compound's electron count, expanding the filled skutterudites from primarily group 8 to group 9 metal-based compounds, as we have done in this study, affords the opportunity to observe many new physical properties.

Here we report that an extremely large family of filled skutterudites based on the group 9 metals Co, Rh and Ir can be chemically stabilized if the electron count is stabilized by partial substitution on the X ion site to yield electron-precise formulas. These new filled skutterudites were designed by filling or partial filling the cages in the binary $CoSb_3$, $RhSb_3$ and $IrSb_3$ frameworks with alkali (Li, Na and K), alkaline-earth metal (Ca, Sr and Ba) and rare-earth atoms (La, Ce, Pr, Nd, Gd and Yb), by compensating for the extra-positive charge of the filling ion by partial substitution of Sn on the Sb site or Si on the P site to yield electron-precise formulas, guided by the predictions of first principles electronic structure calculations. The pure antimonides and pure phosphides (with no Sb/Ns or P/Si mixing) are not stable, but we find that electron-precise formulas are not strictly required for compound stability in some cases. In addition to their synthesis, crystal structures and electronic structures, we briefly survey the magnetic properties of selected members of this large new family of compounds.

Results

Electronic structure and prediction. The prototypical binary BX_3 skutterudites with $B =$ Co, Rh and Ir are non-magnetic semiconductors with a Zintl electron count of 24 valence electrons per formula unit. (The 24 electron count rule per BX_3 unit for semiconducting behaviour also holds for a recently proposed non-Zintl view of the formal electron configuration of the constituent elements in skutterudites[49].) Filled AB_4X_{12} skutterudites that have B in a d^6 low spin configuration and a p^6 X atom configuration are expected to be at electron precise, semiconducting, non-magnetic compounds when they have a valence electron count of 96 (4×24) per formula unit. The power of the Zintl concept is that these simple electron-counting rules enable the prediction of new thermodynamically stable compounds. A cobalt-based filled skutterudite compound can be stabilized, for example, by using Ba as an A^{2+} ion and compensating for the added electrons by removing two electrons from the X ion site, and thus $BaCo_4Sb_{10}Sn_2$ should be stable, and an electron precise, non-magnetic semiconductor. In the same manner, the Rh and Ir compounds can be obtained as can the filled skutterudite semiconductors for any A ion.

This simple electron count principle is confirmed by our *ab initio* calculations for several examples in this family, which are shown in Fig. 2a,c,e. The valence band consists of strongly hybridized metal d and pnictogen p states and is separated by a gap from the conduction band showing the expected semiconducting character. Our results for the current compounds are similar to those of *ab initio* calculations performed previously on other skutterudites[50–52]. Surprisingly, the electronic structure of the Rh variant calculated with the generalized gradient approximation (GGA) is semimetallic. Use of the Tran-Blaha Modified Becke-Johnson (MBJ) functional in the calculation[49], which accounts for electronic correlations more effectively, on the other hand predicts the Rh variant to be a semiconductor. The results of the MBJ functional calculation are shown in Supplementary Fig. 1. The calculated band structure of skutterudites around the Fermi level is very sensitive to changes in the lattice constants and internal parameters, as well as the chosen density functional theory (DFT)[48,53]. Since we are interested in the stability of the compounds, the qualitatively correct behaviour shown here is sufficient for our discussions[54].

The chemical concept behind the stability of these compounds is, furthermore, supported by crystal orbital Hamilton population (COHP) calculations using the tight-binding linear-muffin-tin-orbital (TB-LMTO) method, with the corresponding calculations shown in Fig. 2b,d,f. The COHP curves are used to indicate

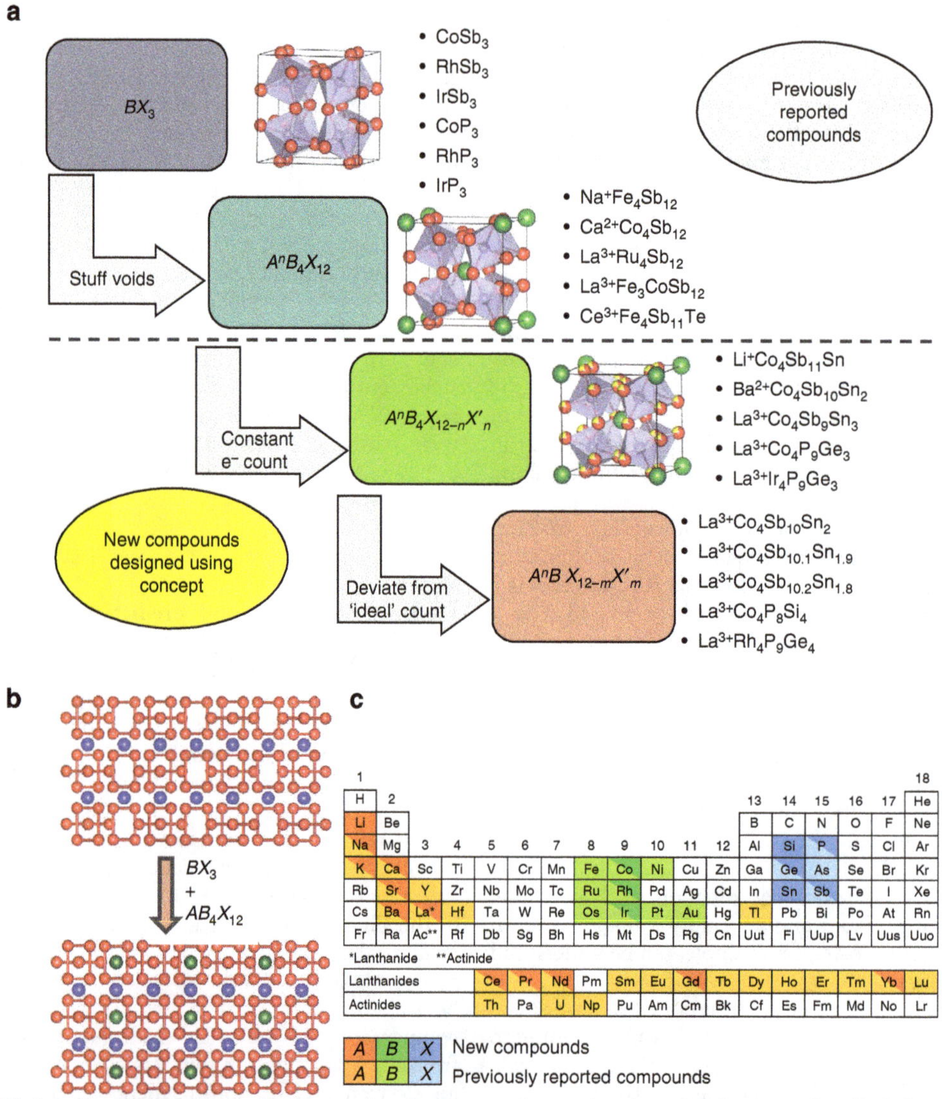

Figure 1 | Design of filled skutterudites using the Zintl concept and summary of current and previously reported stuffed skutterudites. (**a**) Flowchart for designing new skutterudites using the Zintl concept. (**b**) Structures of BX_3 and AB_4X_{12} showing the covalently bonding regions (red atoms) separate from the ionic regions (blue and green atoms). The atoms in the two regions can be partially substituted for tuning of the electron concentration to change the electronic structure. (**c**) Periodic table of the elements highlighted to show the new stuffed skutterudites found in this work and those reported previously.

bonding and anti-bonding interactions. Here they show the presence of bonding interactions below the Fermi level in La[Co/Rh/Ir]$_4$Sb$_9$Sn$_3$, while anti-bonding contributions are found above the Fermi level. This indicates that the filled skutterudite structure maximizes its bonding interactions at the electron count of 96 valence electrons per formula unit—the principle used here to synthesize the large new family reported.

Structural characterization. The compositions of the filled skutterudites was determined by standard solid-state phase equilibria methods, which, using the characterization of reaction products by powder X-ray diffraction (XRD), especially for the heavy elements in compounds such as these, is quite sensitive to the presence of impurities. In addition, the crystal structures of all 63 variants reported here were determined by quantitative fits to the powder diffraction patterns. Figure 3 shows the powder XRD patterns for the quantitative Rietveld structure refinements for 12 examples of compounds the Co-series, Rh-series, Ir-series and P-series filled skutterudites. The agreement between the observed

and calculated patterns in all cases is excellent; a selection of the structural refinements is shown in Table 1 with the remainder given in the Supplementary Information in Supplementary Table 1. The cubic crystallographic cell parameters for all the compositions synthesized, determined by least squares fitting to 20 or more observed reflections in the powder XRD patterns using profile fits, are found in Table 2.

Supplementary Figures 2–4 show the Rietveld refinements of the laboratory powder XRD data for the filled skutterudites made in this work. The ideal filled skutterudite structure is primarily found, with random Sb/Sn and P/Si mixing on the X ion site. For all the filled skutterudites based on P, and for the Sb-based variants based on Co, the A ions are found in the centres of the skutterudite cages (the 2a, (0,0,0) site), their ideal positons, as is frequently assumed for filled skutterudites in other chemical families. It is of interest, however, that for the compounds with the larger B ions, and consequently larger unit cells and X cages, we find that the A ions are sometime displaced from the centres of the cages, that is, off the ideal 2a site. This is found for the Ir compounds in particular. The qualitative evidence for this

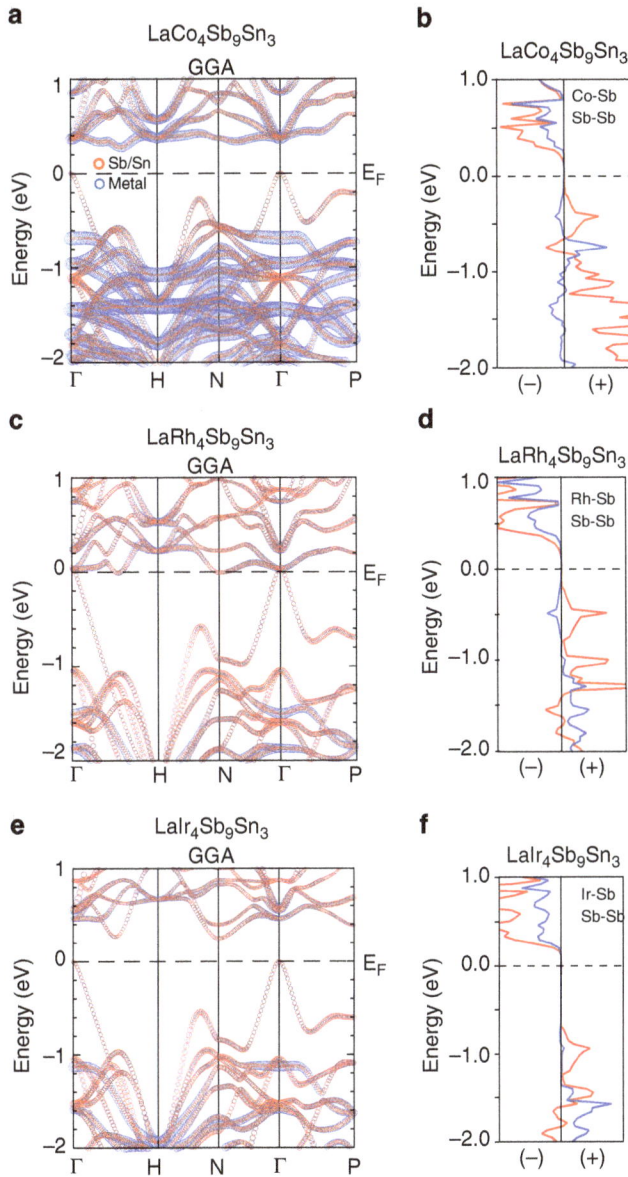

Figure 2 | Eelectronic structures of representative group 9 filled skutterudites. (**a,c,e**) Calculated electronic structures of the filled skutterudites $LaCo_4Sb_9Sn_3$, $LaRh_4Sb_9Sn_3$ and $LaIr_4Sn_9Sn_3$ using the GGA functional. (**b,d,f**) Results of TB-LMTO-ASA (local density approximation)[66,69] electronic structure calculations on La[Co/Rh/Ir]$_4$Sb$_{12}$ model compounds with La on 2a sites, Co/Rh/Ir on 8c sites and Sb/Sn on 24g sites.COHP curves between Co/Rh/Ir-Sb and Sb-Sb.

positions. The alternative model (c), where the A atom is randomly displaced to partially occupy sites along a <111> type direction to minimize the A-X distances in the cage, seems like the most physically realistic scenario and was the model used here to describe the rare-earth ion disorder. In a handful of cases, $BaIr_4Sb_{10}Sn_2$ and $La_{0.9}Ir_4Sb_{10.2}Sn_{1.8}$, for example, realistic off-centre A atom positions cannot sufficiently fully account for the observed intensities of the low-angle peaks, and thus it may be that a small fraction of the A ions are substituted on the B atom sites. Although the existence of 5d Os and Pt rare-earth filled skutterudites been reported in the past[9,10,54,55], few quantitative structure determinations have been reported. A model like (b), with anomalously large atomic displacement parameters for the lanthanides, has been used to describe the Ln site disorder for Os-based pyrochlores[56]. The very low thermal conductivity of the filled skutterudites based on group 8 elements has been attributed to the anharmonic thermal vibration of the A site ion 'rattling' in the skutterudite cage[47,57], implying a tendency towards off-centre positioning of the A site ion in the cages in this family, in agreement with our structural refinements and those where the A site ion disorder is modelled by large thermal parameters. It is not yet known whether the A site disordered position model we observe here is a general feature of all larger filled skutterudites. Whether the disorder is static or dynamic or a combination of both, and whether there is any complex defect chemistry present in some of these compounds, would be of interest to study further by other structural characterization methods.

The important structural features of the filled skutterudites studied here are presented in Fig. 4. This figure shows the geometry of the A site and group 9 site polyhedra and the manner in which they share faces (leading to the strong anti-bonding interactions described above), the displacements of the A site ions from the cage centres and the X_4 squares. The lower part of the figure compares the X–X bond lengths within the slightly distorted X_4 antimonide and phosphide squares and the B–X bond lengths (all six are equivalent) for several representative members of the group 9 filled skutterudites. Neither the X_4 'squares' nor the BX_6 octahedra are constrained by symmetry to be regular, and they are not, although they are all close to regular and the distortions are relatively small. Because the same electron count is nearly perfectly maintained for all the compounds studied quantitatively here, large bond-length differences in these critical structural components are not expected to occur.

displacement is seen in comparing Fig. 3a to Fig. 3b,c; it can be seen that the intensities of the low-angle diffraction peaks are suppressed by when the Co-skutterudites are filled, but that there is residual intensity seen in the Rh and Ir skutterudites. The three models evaluated to best model the observed diffraction peak intensities, rather than employing the ideal filled skutterudite structure, were: (a) refine the 2a site occupancy, (b) refine the 2a site atomic displacement parameter and (c) move the A atom off the ideal 2a site towards the surrounding X cage. Model (a) is not appropriate because the samples are single phase at the stoichiometry determined by phase equilibria studies; the formulas were further confirmed by energy dispersive X-ray spectroscopy. Model (b) provides an adequate fit, yielding excessively large atomic displacement parameters, implying that the A site ions are dynamically displaced from their ideal

Magnetic properties. Filled skutterudites based on rare-earth metals and group 8 transition elements show diverse and interesting rare-earth magnetism, and they have been widely studied (see, for example, refs 28,29,31). To generally survey the magnetic properties for our new filled group 9-based skutterudites, the magnetizations of all the rare-earth-based filled skutterudites were measured from 2 to 200 K with an applied magnetic field of $\mu_0 H = 1$ Tesla. The compounds are strongly paramagnetic as expected due to the rare-earth magnetism. We consider the analysis of several set of compounds in more detail as examples. Figure 5a–c shows in the main panel the inverse magnetic susceptibilities (χ is defined as $M/\mu_0 H$) for the $Ce_{0.9}[Co/Rh/Ir]_4Sb_{10.2}Sn_{1.8}$, $Yb_{0.9}[Co/Rh/Ir]_4Sb_{10.2}Sn_{1.8}$ and $Pr_{0.9}[Co/Rh/Ir]_4Sb_{10.2}Sn_{1.8}$ samples. The temperature dependences of the inverse magnetic susceptibilities, corrected for χ_0, are linear at high temperatures for all three compounds, indicating Curie Weiss behaviour. Fits to the susceptibilities were performed in the temperature range of 100–200 K, according to $\chi - \chi_0 = C/(T - \theta_{cw})$, where C is the Curie constant, θ_{cw} is the Weiss temperature and χ_0 is the temperature independent contribution to the susceptibility. From these fits, the effective

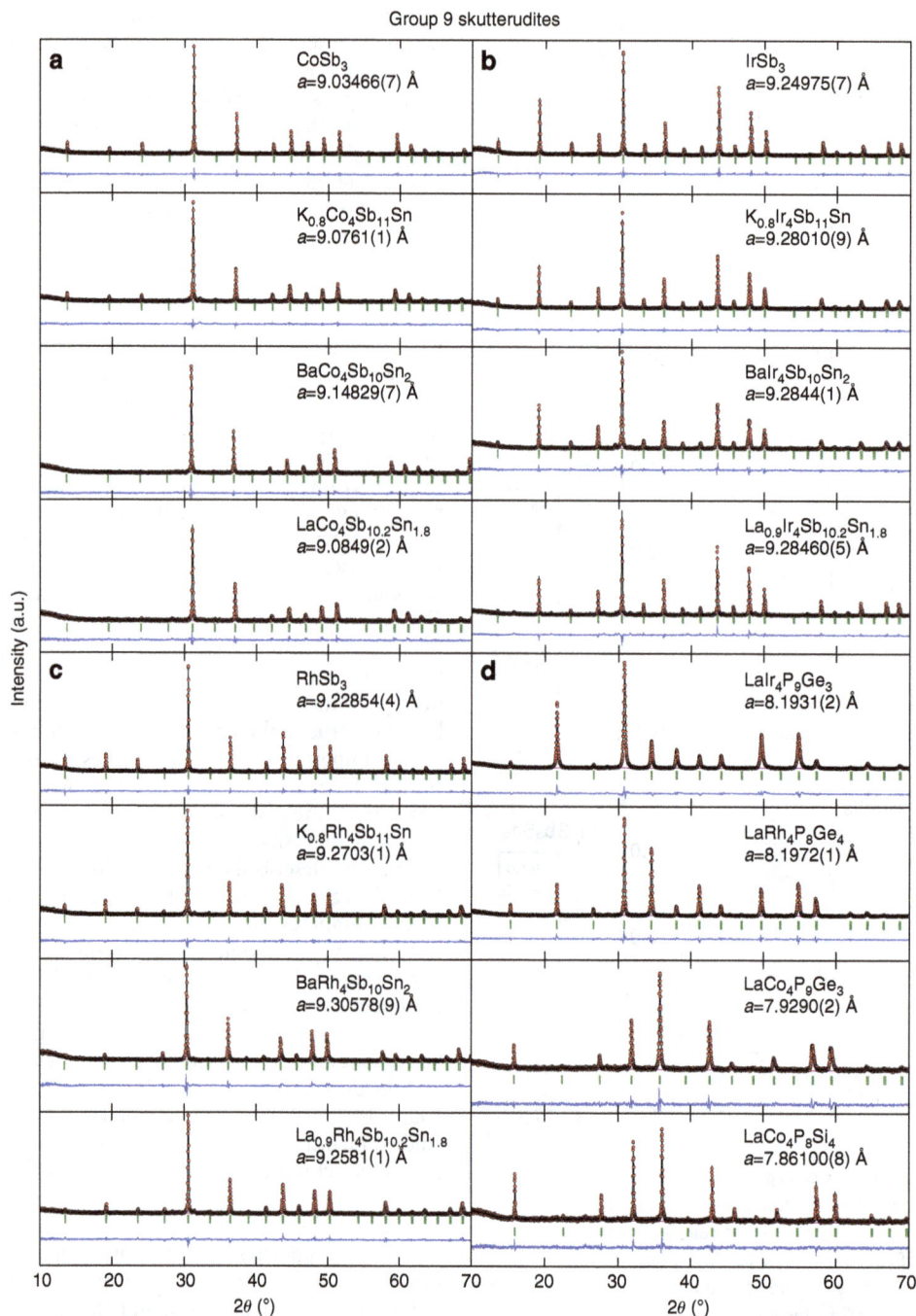

Figure 3 | Rietveld refinements for representative group 9 filled skutterudites. (**a**) $CoSb_3$, $K_{0.8}Co_4Sb_{11}Sn$, $BaCo_4Sb_{10}Sn_2$ and $LaCo_4Sb_{10.2}Sn_{1.8}$. (**b**) $RhSb_3$, $K_{0.8}Rh_4Sb_{11}Sn$, $BaRh_4Sb_{10}Sn_2$ and $La_{0.9}Rh_4Sb_{10.2}Sn_{1.8}$. (**c**) $IrSb_3$, $K_{0.8}Ir_4Sb_{11}Sn$, $BaIr_4Sb_{10}Sn_2$ and $LaIr_4Sb_{10.2}Sn_{1.8}$. (**c**) $LaCo_4P_9Ge_3$, $LaRh_4P_8Ge_4$, $LaIr_4P_9Ge_3$ and $LaCo_4P_9Si_3$. Red points are observed data; black curves are the calculated pattern; green tics are expected peak positions; the lower blue curve is the difference between observed and calculated diffraction pattern. Refined structural parameters are found in Table 1.

magnetic moment (P_{eff}) per Ln (Ln = La, Ce, Pr, Nd, Gd and Yb) ion was obtained by using $P_{eff} = (8C)^{1/2}$. The thus derived basic magnetic characteristics (that is, θ_{cw} and P_{eff}) for all our new compounds containing magnetic rare earths are summarized in Table 2. Fitting the magnetic susceptibility using the Curie Weiss law in the range 100–200 K, we obtain the effective moments for $Pr_{0.9}Co_4Sb_{10.2}Sn_{1.8}$, $Pr_{0.9}Rh_4Sb_{10.2}Sn_{1.8}$ and $Pr_{0.9}Ir_4Sb_{10.2}Sn_{1.8}$ of $P_{eff} = 3.43$, 3.66 and 3.66 μ_B/Pr, respectively, close to the effective moment value expected for the free Pr^{3+}-free ion ($P_{eff} = 3.58$ μ_B/Pr). In addition, the magnetic susceptibility data show a broad peak at around 3.5 K for $Pr_{0.9}Rh_4Sb_{10.2}Sn_{1.8}$ (Fig. 5d,e), which implies the onset of

antiferromagnetic ordering. To better estimate the Neel temperature, we follow standard procedure[58,59] and plot $d(\chi T)/dT$ (Fig. 5f). The maximum of $d(\chi T)/dT$ is observed at 3.5 K, which can be taken as T_N. Similar fits were performed for all the rare-earth compounds synthesized.

The effective moments for the $Ce_{0.9}[Co/Rh/Ir]_4Sb_{10.2}Sn_{1.8}$ samples are in the range $P_{eff} = 2.1$–2.5 μ_B/Ce, which are close to the expected free-ion Hund's rule value of 2.54 μ_B/Ce^{3+}. These values are similar to the effective moment ($P_{eff} = 2.1$–2.26 μ_B/Ce) reported for the group 8 filled skutterudites $CeRu_4Sb_{12}$ (ref. 32), but are different from those reported for $CeFe_4Sb_{12}$ ($P_{eff} = 3.5$–3.8 μ_B/Ce, which is larger than the value expected for a

Table 1 | Structural parameters for refined crystal structures of selected filled skutterudites and comparison to binary skutterudites.

Compound	a_0 (Å)	B-site(y)	B-site(z)	A-site coordinate in (x,x,x)	% A on B sites	χ^2
$CoSb_3$	9.0347 (1)	0.1577 (3)	0.3344 (3)	—		1.24
$K_{0.8}Co_4Sb_{11}Sn$	9.0761 (1)	0.1590 (4)	0.3368 (4)	0.026 (4)		1.31
$BaCo_4Sb_{10}Sn_2$	9.1483 (1)	0.1626 (6)	0.3417 (5)	0.012 (2)		1.25
$LaCo_4Sb_{10.2}Sn_{1.8}$	9.0849 (2)	0.1589 (6)	0.3363 (6)	0.027 (2)		1.34
$LaCo_4P_9Si_3$	7.8610 (1)	0.1505 (9)	0.3525 (8)	0		1.38
$LaCo_4P_9Ge_3$	7.9290 (2)	0.1507 (8)	0.3538 (7)	0		1.5
$RhSb_3$	9.2285 (0)	0.1538 (3)	0.3395 (3)	—		1.36
$K_{0.8}Rh_4Sb_{11}Sn$	9.2703 (1)	0.1546 (4)	0.3419 (4)	0.014 (5)		1.42
$BaRh_4Sb_{10}Sn_2$	9.30578 (9)	0.1569 (6)	0.3442 (6)	0	2%	1.56
$La_{0.9}Rh_4Sb_{10.2}Sn_{1.8}$	9.2581 (1)	0.1542 (5)	0.3389 (5)	0.039 (2)	2%	1.56
$LaRh_4P_8Ge_4$	8.1973 (1)	0.1484 (7)	0.3569 (6)	0		1.87
$IrSb_3$	9.2498 (1)	0.1524 (3)	0.3396 (3)	—		1.28
$K_{0.8}Ir_4Sb_{11}Sn$	9.2801 (1)	0.1536 (3)	0.3414 (3)	0.019 (4)		1.28
$BaIr_4Sb_{10}Sn_2$	9.2844 (1)	0.1542 (5)	0.3420 (5)	0.035 (3)	13%	1.55
$LaIr_4P_9Ge_3$	8.1931 (2)	0.147 (2)	0.357 (2)	0		2.69
$La_{0.9}Ir_4Sb_{10.2}Sn_{1.8}$	9.28460 (5)	0.1530 (7)	0.3407 (7)	0.037 (3)	7%	1.86

Table 2 | Cubic cell parameters and selected physical properties of group 9-based filled skutterudites.

Compound	a_0 (Å)	χ at 150 K ($\times 10^{-5}$ emu f.u. mol^{-1})	P_{eff} μB/Ln, θ_{cw}	ρ at 100 K ($\times 10^{-4}$ Ohm cm)	Compound	a_0 (Å)	χ at 150 K ($\times 10^{-5}$ emu f.u. mol^{-1})	P_{eff} μB/Ln, θ_{cw}	ρ at 100 K ($\times 10^{-4}$ Ohm cm)
$CoSb_3$	9.0347 (1)	−140 (4)		60 (1)	$LaRh_4Sb_9Sn_3$	9.2720 (1)	−601 (3)		34 (4)
$Li_{0.8}Co_4Sb_{11}Sn$	9.0591 (1)	−330 (4)		10 (4)	$La_{0.9}Rh_4Sb_{10.2}Sn_{1.8}$	9.2581 (1)	−490 (1)		20 (3)
$Na_{0.8}Co_4Sb_{11}Sn$	9.0660 (1)	−300 (5)		20 (6)	$LaRh_4Sb_{11}Sn$	9.2342 (1)	—		40 (5)
$K_{0.8}Co_4Sb_{11}Sn$	9.0761 (1)	−421 (1)		10 (0)	$CeRh_4P_8Ge_4$	8.1919 (1)		2.13, −4.57	50 (4)
$CaCo_4Sb_{11}Sn$	9.0634 (1)	−400 (3)		110 (9)	$CeRh_4Sb_8Sn_4$	9.2681 (1)		2.53, −10.9	20 (3)
$BaCo_4Sb_{10}Sn_2$	9.1483 (1)	−280 (5)		330 (9)	$CeRh_4Sb_9Sn_3$	9.2682 (1)		2.70, −13.0	20 (3)
$LaCo_4P_8Si_4$	7.8610 (1)	−210 (8)		1,420 (0)	$Ce_{0.9}Rh_4Sb_{10.2}Sn_{1.8}$	9.2646 (1)		1.53, −0.0044	800 (1)
$LaCo_4P_9Ge_3$	7.9290 (2)	−850 (1)		90 (1)	$PrRh_4P_8Ge_4$	8.1713 (1)		3.43, −11.95	50 (5)
$LaCo_4Sb_9Sn_3$	9.1020 (1)	−460 (0)		40 (3)	$PrRh_4Sb_8Sn_4$	9.2674 (1)		3.86, −28.5	11 (5)
$LaCo_4Sb_{10}Sn_2$	9.0907 (1)	−520 (4)		70 (5)	$PrRh_4Sb_9Sn_3$	9.2668 (1)		3.78, −19.3	30 (4)
$LaCo_4Sb_{10.1}Sn_{1.9}$	9.0814 (2)	−240 (9)		650 (2)	$Pr_{0.9}Rh_4Sb_{10.2}Sn_{1.8}$	9.2542 (1)		3.66, 3.6	10 (6)
$LaCo_4Sb_{10.2}Sn_{1.8}$	9.0849 (2)	−90 (9)		130 (1)	$NdRh_4P_8Ge_4$	8.1835 (1)		2.56, 0.64	60 (3)
$LaCo_4Sb_{10.3}Sn_{1.7}$	9.0842 (1)	−440 (9)		50 (2)	$Gd_{0.9}Rh_4Sb_{10.2}Sn_{1.8}$	9.2705 (1)		8.5, −16.5	100 (3)
$CeCo_4P_9Ge_3$	7.9306 (1)		2.16, −5.05	90 (4)	$Yb_{0.9}Rh_4Sb_{10.2}Sn_{1.8}$	9.2647 (1)		2.76, −16.0	50 (5)
$Ce_{0.9}Co_4Sb_{10.2}Sn_{1.8}$	9.0740 (1)		2.28, −4.3	80 (2)	$IrSb_3$	9.2498 (1)	−173 (3)		6,160 (1)
$PrCo_4P_9Ge_3$	7.9170 (1)		3.59, −24.8	40 (3)	$Li_{0.8}Ir_4Sb_{11}Sn$	9.2672 (0)	−280 (2)		20 (1)
$Pr_{0.9}Co_4Sb_{10.2}Sn_{1.8}$	9.0727 (1)		3.43, −9.7	50 (9)	$Na_{0.8}Ir_4Sb_{11}Sn$	9.2708 (1)	−424 (7)		11 (0)
$NdCo_4P_8Ge_4$	7.9082 (2)		2.59, 0.38	60 (3)	$K_{0.8}Ir_4Sb_{11}Sn$	9.2801 (1)	−361 (0)		320 (3)
$Yb_{0.9}Co_4Sb_{10.2}Sn_{1.8}$	9.0814 (1)		1.56, 4.7	10 (1)	$CaIr_4Sb_{10.2}Sn_{1.8}$	9.2871 (1)	−501 (1)		1,100 (2)
$RhSb_3$	9.2285 (0)	−180 (9)		50 (2)	$CaIr_4Sb_{11}Sn$	9.2694 (2)	−780 (5)		4,650 (4)
$Li_{0.8}Rh_4Sb_{11}Sn$	9.2484 (1)	−390 (5)		90 (0)	$SrIr_4Sb_{11}Sn$	9.2522 (1)	−470 (3)		921 (0)
$Na_{0.8}Rh_4Sb_{11}Sn$	9.2575 (1)	−410 (5)		10 (2)	$BaIr_4Sb_{10}Sn_2$	9.2844 (1)	−380 (4)		113 (4)
$K_{0.8}Rh_4Sb_{11}Sn$	9.2703 (1)	−290 (4)		10 (3)	$LaIr_4P_9Ge_3$	8.1931 (2)	−320 (8)		310 (4)
$KRh_4Sb_{10}Si_2$	9.2484 (1)	−500 (2)		220 (2)	$LaIr_4Sb_8Sn_4$	9.2887 (1)	310 (2)		0.34 (5)
$KRh_4Sb_{10}Ge_2$	9.2171 (1)	−500 (1)		40 (2)	$La_{0.9}Ir_4Sb_{10.2}Sn_{1.8}$	9.2846 (1)	−1,060 (3)		53 (3)
$KRh_4Sb_{10}Sn_2$	9.2824 (1)	−540 (7)		11 (4)	$CeIr_4P_9Ge_3$	8.1860 (1)		2.17, −6.29	273 (4)
$CaRh_4Sb_{10.2}Sn_{1.8}$	9.2714 (1)	−490 (5)		2,880 (1)	$Ce_{0.9}Ir_4Sb_{10.2}Sn_{1.8}$	9.2695 (1)		2.55, −2.2	88 (1)
$CaRh_4Sb_{11}Sn$	9.2493 (1)	−521 (1)		1,510 (2)	$PrIr_4P_9Ge_3$	8.1998 (1)		3.23, −0.19	170 (4)
$SrRh_4Sb_{11}Sn$	9.2364 (1)	−820 (6)		206 (7)	$PrIr_4Sb_9Ge_3$	9.2933 (1)	−480 (6)		21 (5)
$BaRh_4Sb_9Ge_3$	9.1969 (1)	−220 (0)		53 (4)	$Pr_{0.9}Ir_4Sb_{10.2}Sn_{1.8}$	9.2773 (1)		3.66, −5.4	54 (5)
$BaRh_4Sb_{10}Sn_2$	9.3058 (1)	−450 (5)		100 (8)	$NdIr_4P_9Ge_3$	8.1961 (1)		2.54, −2.28	93 (4)
$LaRh_4P_8Ge_4$	8.1973 (1)	−330 (8)		20 (3)	$Gd_{0.9}Ir_4Sb_{10.2}Sn_{1.8}$	9.2888 (1)		8.48, −19.6	41 (6)
$LaRh_4Sb_8Sn_4$	9.2756 (1)	−571 (2)		20 (5)	$Yb_{0.9}Ir_4Sb_{10.2}Sn_{1.8}$	9.2782 (1)		2.71, −6.9	44 (0)

Ce^{3+} (refs 59,60)). For the Yb-filled phases, P_{eff} = 1.56, 2.76 and 2.71 μB/Yb for $Yb_{0.9}Co_4Sb_{10.2}Sn_{1.8}$, $Yb_{0.9}Rh_4Sb_{10.2}Sn_{1.8}$ and $Yb_{0.9}Ir_4Sb_{10.2}Sn_{1.8}$, respectively. For all three samples, the fits to Curie Weiss laws in the low-temperature regions yield effective moments that appear to be intermediate between those for Yb^{3+} (P_{eff} = 4.54 μB/Yb for the free ion) and non-magnetic Yb^{2+} (P_{eff} = 0), however, due to the small energy differences between ground state and excited state electron configurations in Yb compounds, fits to the Curie Weiss law often do not reflect the true magnetic state of the system. Depressed values of effective moment for Yb have also been inferred from such fits for $YbFe_4Sb_{12}$ (P_{eff} = 3.09 μB/Yb), for example ref. 61.

The new filled group 9 skutterudites that do not contain rare earths display temperature independent susceptibilities, with the exception of weak 'Curie tails' at low temperatures due to the presence of impurity spins. Almost all of the intrinsic susceptibilities are in fact diamagnetic for these materials, indicating that they are dominated by core diamagnetism, as expected from the electron-precise formulas. The magnetic susceptibilities at 150 K for the non-magnetic samples, also presented in Table 2, are between − 0.001 and − 0.01 emu (mol formula unit)$^{-1}$. This indicates that the compounds do not display local moment behaviour, but rather band behaviour, even for the Co variants. It is of interest that the Co variants all display susceptibilities where no local moments are observed, as Co can be magnetic in many of its compounds. The implication is that the Co^{3+} is in a low spin state, consistent with the chemical picture for these compounds, which has the t_{2g} levels filled and the e_g levels empty. More detailed study of the magnetic properties of the group 9 filled skutterudites will be of future interest.

a

Geometry comparisons

$La_{0.9}Ir_4Sb_{10.2}Sn_{1.8}$:
Sb cage, polyanion
$IrSb_6$ and displaced La

Legend
- ● Co
- ◎ Rh
- ○ Ir
- ◉ Sb/Sn
- ○ P/Ge

b
a
c

b

$LaCo_4Sb_{10.2}Sn_{1.8}$	$La_{0.9}Rh_4Sb_{10.2}Sn_{1.8}$	$La_{0.9}Ir_4Sb_{10.2}Sn_{1.8}$
2.89 Å	2.86 Å	2.84 Å
2.97 Å	2.98 Å	2.96 Å
2.54 Å	2.61 Å	2.63 Å

c

$LaCo_4P_9Ge_3$	$LaRh_4P_8Ge_4$	$LaIr_4P_9Ge_3$
2.39 Å	2.43 Å	2.40 Å
2.32 Å	2.34 Å	2.34 Å
2.28 Å	2.38 Å	2.39 Å

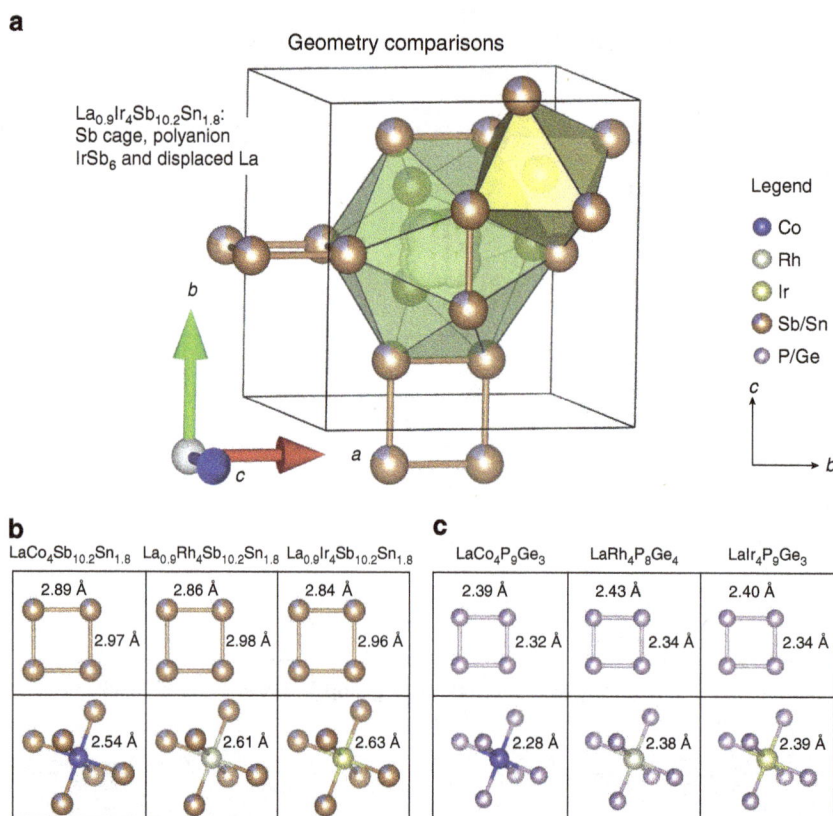

Figure 4 | Fragment of the filled skutterudite structure and comparisons of the X_4 square and BX_6 octahedron geometries for selected group 9 filled skutterudites. (a) Details of the AX_{12} polyhedron, the BX_6 octahedron, the A site displacement and the X_4 squares, and the manner in which they are related in the $La_{0.9}Ir_4Sb_{10.2}Sn_{1.8}$ filled skutterudite. **(b,c)**: Comparisons of the X_4 square and BX_6 octahedra geometries for selected filled group 9-based skutterudite compounds, from the current structure refinements.

Discussion

A 96 valence electron rule for stability of filled skutterudites is strongly supported by the current results and is analogous to the powerful 18 electron rule for half Heusler compounds. Using this counting rule, we have shown the existence of a very large new family of filled skutterudites based on the group 9 skutterudites $CoSb_3$, $RhSb_3$ and $IrSb_3$, stabilized by X site substitution to yield compounds with electron-precise formulas. The compounds were found through a combination of experimental studies and theoretical first principle calculations. In this work, 63 new filled skutterudite compositions based on the group 9 metals are reported, but by simple extension, for example, to the full 14 member rare-earth family, and to phosphides and arsenides in addition to the antimonides, we expect that several hundred new group 9-based filled skutterudite compounds can be found and characterized based on the concept presented here. This greatly expands the family of known filled skutterudites, an important solid structural family. The properties of the rare-earth-based filled skutterudites in other chemical families have proven to be very interesting especially for compounds based on the beginning (that is, Ce and Pr) and end (that is, Yb) of the magnetic rare-earth series. Those properties are strongly dependent on the spacing between rare earths, modified by varying the size of the X atoms, and also on the hybridization of the rare-earth orbitals with those of the transition metals present. The same effects are very likely to be exposed in the new skutterudite family described here through detailed study; their different transition metals in particular add a new degree of freedom by which the magnetic properties can be manipulated. Further, filled skutterudites with electron counts different from the electron-precise values have been made based on group 8 elements. In these cases, surprising

properties like superconductivity can sometimes occur, for example, in $LaRu_4P_{12}$ (ref. 62). The same kind of interesting electronic properties are also likely in our group 9 series with the correct combination of electron count and elemental constituents. Finally, the general approach employed here—to search for stable solid compounds of interesting transition element ions, even complex ones, by aiming at electron-precise (that is, Zintl) formulas—is likely to be a fruitful approach for expanding other families of solid compounds in the future.

Methods

Calculation. The electronic band structure calculations were performed in the framework of DFT using the WIEN2k[54] code with a full-potential linearized augmented plane-wave and local orbitals basis together with the Perdew Burke Ernzerhof parameterization[55] of the GGA as the exchange–correlation functional. In one case (see text), the MBJ[63] functional was also used (see Supplementary Fig. 1) The plane wave cutoff parameter RK_{MAX} was set to 7 and the Brillouin zone was sampled by 500 k-points. To simulate the substitution of Sb by Sn, the virtual crystal approximation[64,65] was employed. The COHP[66] were generated by TB-LMTO atomic-sphere approximation (ASA) calculations using the Stuttgart code[65]. Exchange and correlation were treated by the local density approximation[67]. In the ASA method, space is filled with overlapping Wigner–Seitz (WS) spheres. The empty spheres were necessary in the calculation, and the WS sphere overlap was limited to no >16%. The basis set for the calculations included La ($6s$, $6p$, $5d$ and $4f$), Co ($4s$, $4p$ and $3d$)/Rh ($5s$, $5p$ and $4d$)/Ir ($6s$, $6p$ and $5d$) and Sb ($5s$ and $5p$) wavefunctions. The convergence criterion was set to 10^- eV. A mesh of $6 \times 6 \times 6$ k-points in the irreducible wedge of the first Brillouin zone was used to obtain all values. Experimental lattice constants were used and the free internal parameters were optimized by minimizing the forces.

Synthesis and experiment. Single-phase polycrystalline samples were synthesized from starting compositions $A[Co/Rh/Ir]_4Sb_{11}Sn$ for A^{1+} ions (A = Li, Na and K), $A[Co/Rh/Ir]_4Sb_{10}Sn_2$ for A^{2+} ions (A = Ca, Sr and Ba) and $A_{0.9}[Co/Rh/Ir]_4Sb_{10.2}Sn_{1.8}$ for A^{3+} ions (A = Ln = La, Ce, Pr, Gd and Yb). (Due to the large

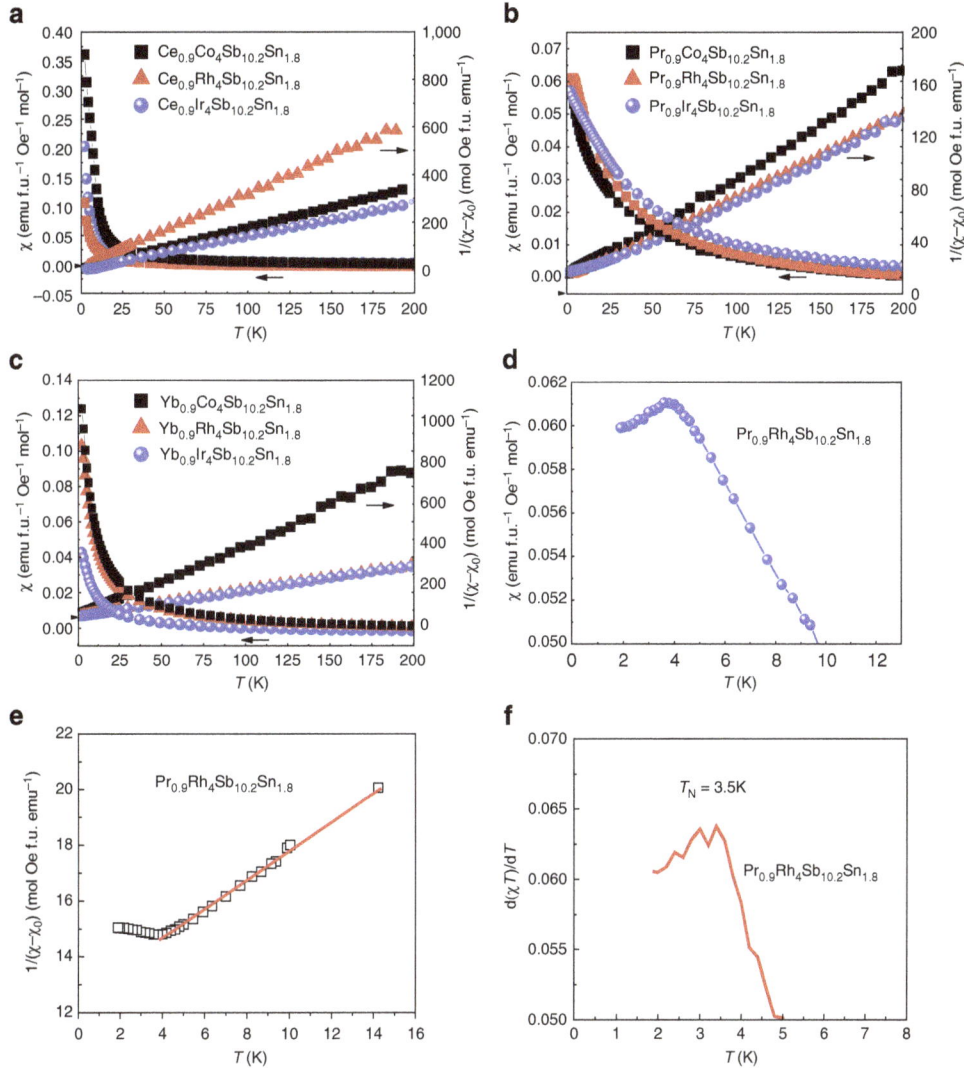

Figure 5 | Magnetic properties of selected group 9 filled skutterudites. (**a**) $Ce_{0.9}[Co/Rh/Ir]_4Sb_{10.2}Sn_{1.8}$; (**b**) $Yb_{0.9}[Co/Rh/Ir]_4Sb_{10.2}Sn_{1.8}$; and (**c**) $Pr_{0.9}[Co/Rh/Ir]_4Sb_{10.2}Sn_{1.8}$. (**d**) Enlarged low-temperature region of the magnetic susceptibility of $Pr_{0.9}Rh_4Sb_{10.2}Sn_{1.8}$. (**e**) Enlarged low-temperature region of the inverse susceptibility. (**f**) Enlarged low-temperature region of $d(\chi T)/dT$.

number of compounds described here, the formulas for the groups are abbreviated by the use of the square brackets for the group 9 element—in each case, pure Co, pure Rh and pure Ir variants were made.) The single-phase compositions reported here were determined in all cases by performing a series of systematic syntheses in extensive phase equilibria studies. The use of phase equilibria studies facilitated by powder XRD is a powerful, common method for determining the compositions of solids. The single-phase compositions reported here are those that yielded filled skutterudite compounds with y closest to 1 in $A_yB_4X_{12}$.

Samples were prepared by a solid-state reaction method. The alkali- and alkaline-earth-based groups are single phase for starting compositions at the ideal Zintl electron count, but for the lanthanide-antimony-based compounds, a more complex formula, with electrons in excess of the electron-precise count, was needed to make single-phase materials under our conditions. For $A^{1+}[Co/Rh/Ir]_4Sb_{11}Sn$ and $A^{2+}[Co/Rh/Ir]_4Sb_{10}Sn_2$, stoichiometric quantities of high-purity elemental Li/Na/K (99.999%), Ca/Sr/Ba (99.9%), Co (99.8%), Rh (99.99%) or Ir (99.99%), Sb (99.999%) and Sn (99.8%) were mixed, pressed into a pellet, put into an alumina crucible and then sealed in clean evacuated silica ampoules. The ampoules were slowly heated up to 600 °C and held for 10 h, then slowly heated up to 700 °C and held for 10 h. Then, they were removed from the furnace, thoroughly ground into powder, repressed into pellets and reheated at 700 °C for 10 h. This process was repeated two to four times until single-phase material was obtained. All the alkali- and alkaline-earth metals samples were made in two steps. For the first step, we mixed and ground the Co/Rh/Ir and Sb and Sn powders, and then put half the mixed powder in the bottom of the pellet die (in a glove box). The small pieces of alkali or alkaline-earth metals were then added into the middle, and the other half of the mixed power was put on the top, sandwiching the alkali/alkaline earth. The pellet was then pressed and put into a silica glass tube and heated to 600 °C at a rate of 1 °C min^{-1} for 10 h. Then, we opened the tube in a glove box and reground and

repelletized the sample and reheated at 700 °C for 10 h. Critically, especially for the case of alkali filling of the skutterudite cavities, where a naive view might speculate that volatility or tube reaction might disrupt the synthesis, the internal surfaces of the silica ampoules were completely clean after the syntheses, showing no sign of attack or mass loss during the synthetic procedure. All the mixing and grinding processes were performed in a glove box (P_{O2} and $P_{H2O} < 1$ p.p.m.).

For comparison purposes, the binary skutterudites $CoSb_3$, $RhSb_3$ and $IrSb_3$ were prepared by the same method. For $Ln_{0.9}[Co/Rh/Ir]_4Sb_{10.2}Sn_{1.8}$ (Ln = La, Ce, Pr, Gd and Yb), precursor LnSn binaries or Ln[Co/Rh/Ir]Sn ternaries were first made by arc melting the elements. The as-prepared binaries and ternaries were then ground into powder and mixed with the appropriate stoichiometric quantities of high-purity elemental Co (99.8%), Rh (99.99%), Ir (99.99%), Sb (99.999%) and Sn (99.8%) powder, and heated as described above. Compounds based on phosphorous as the most electronegative element were also made, but for the rare-earth family only; Si or Ge were partially substituted for P to yield the Zintl electron count. For $Ln[Co/Rh/Ir]_4P_{12-x}[Ge/Si]_x$, (Ln = La, Ce, Pr and Nd), the precursor binaries or $Ln[Co/Rh/Ir]_4[Ge/Si]_x$ ternaries were first made by arc melting the elements. Weight loss in this process was <1%. The as-prepared binaries or ternaries were then ground into powder and mixed with the appropriate stoichiometric quantities of elemental red P (99.9%), pressed into pellets and then sealed in clean evacuated silica ampoules. These ampoules were slowly heated to 400 °C, held for 10 h, and then slowly heated to 700 °C and held for 10 h; they were then slowly heated to 900 °C and held overnight. Finally, they were thoroughly ground into a powder, repressed into pellets, and reheated at 900 °C and held there overnight.

Structural determination and physical properties. The single-phase compositions determined in the phase equilibria studies were specified by powder

XRD using a Bruker D8 Focus diffractometre with Cu Kα radiation and a graphite diffracted beam monochromator. The initial structural model for the structures that were quantitatively refined from the powder diffraction data was taken from that of $LaRu_4Sb_{12}$ (ref. 68). The FullProf software suite was used for the Rietveld refinements. Peak shapes were modelled with the Thompson–Cox–Hastings pseudo–Voight profile convoluted with axial divergence asymmetry. The background was modelled with a Chebychev polynomial. The structures were refined in space group Im QUOTE with Co, Rh or Ir in the 8c sites and Sb/Sn in the 24g sites. The position of the A site ion was dependent on the specific compound, as described. In the final structural models, the structural parameters refined were the two positional coordinates (x and y) of the X atoms in position 24g, and when necessary, the position of the atom on the A atom site along the <111> direction and a small per cent of site mixing of A from the 2a onto the 8c site; all other structural parameters are fixed by symmetry. The formulas observed in the refinements were all consistent with the compositions determined from the phase equilibria studies. Similarly, the compositions of representative compounds tested by EDX were consistent with those compositions. Measurements of the temperature dependence of the electrical resistivity and magnetization were performed in a Quantum Design Physical Property Measurement System (PPMS) from 2 to 300 K.

References

1. Sevov, S. C. in *Intermetallic Compounds - Principles and Practice: Progress, Volume 3* (eds Westbrook, J. H. & Freisher, R. L.) 113–132 (John Wiley & Sons. Ltd., 2002).
2. Kauzlarich, S. M. *Chemistry, Structure, and Bonding of Zintl Phases and Ions* (VCH, 1996).
3. Fässler, T. F. *Zintl Phases-Principles and Recent Developents* (Springer, 2010).
4. Sales, B. C., Jin, R. Y. & Mandrus, D. Zintl compounds: from power generation to the anomalous hall effect. *J. Phys. Soc. Jpn* **77**, 48–53 (2008).
5. Kjekshus, A. & Rakke, T. Compounds with the skutterudite type crystal structure. III. structural data for arsenides and antimonides. *Acta Chem. Scand. A* **28**, 99–103 (1974).
6. Schmidt, T., Kliche, G. & Lutz, H. D. Structure refinement of skutterudite-type cobalt triantimonide, $CoSb_3$. *Acta Cryst. C* **43**, 1678–1679 (1987).
7. Kjekshus, A. & Pedersen, G. The crystal structures of $IrAs_3$ and $IrSb_3$. *Acta Crystallogr.* **14**, 1065–1070 (1961).
8. Jung, D., Whangbo, M. H. & Alvarez, S. Importance of the X_4 ring orbitals for the semiconducting, metallic, or superconducting properties of skutterudites MX_3 and RM_4X_{12}. *Inorg. Chem.* **29**, 2252–2255 (1990).
9. Papoian, G. A. & Hoffmann, R. Hypervalent bonding in one, two, and three dimensions: extending the Zintl–Klemm concept to nonclassical electron-rich networks. *Angew. Chem. Intl Ed.* **39**, 2408–2448 (2000).
10. Uher, C. in *Thermoelectrics Handbook: Macro to Nano* 34.1–34.17 (Taylor and Francis, 2006).
11. Aleksandrov, K. S. & Beznosikov, B. V. Crystal chemistry and prediction of compounds with a structure of skutterudite type. *Crystallogr. Rep.* **52**, 28–36 (2007).
12. Gumeniuk, R. et al. Filled platinum germanium skutterudites MPt_4Ge_{12} (M = Sr, Ba, La-Nd, Sm, Eu): crystal structure and chemical bonding. *Z. Kristallogr.—Cryst. Mater.* **225**, 531–543 (2010).
13. Gumeniuk, R. et al. Superconductivity in the platinum germanides MPt_4Ge_{12} M = rare-earth or alkaline-earth metal) with filled skutterudite structure. *Phys. Rev. Lett* **100**, 017002 (2008).
14. Kihou, K. et al. Magnetic properties of $TbRu_4P_{12}$ studied by neutron diffraction. *Phy. B Condens. Matter* **359–361**, 859–861 (2005).
15. Nakanishi, Y. et al. Elastic property of $TbRu_4P_{12}$ under pressure. *Phy. B Condens. Matter* **404**, 3271–3274 (2009).
16. Shirotani, I. et al. Electrical and magnetic properties of new filled skutterudites $LnFe_4P_{12}$ (Ln = Ho, Er, Tm and Yb) and YRu_4P_{12} with heavy lanthanide (including Y) prepared at high pressure. *J. Phys. Condens. Matter* **17**, 4383–4391 (2005).
17. Danebrock, M. E., Evers, C. B. H. & Jeitschko, W. Magnetic properties of alkaline earth and lanthanoid iron antimonides AFe_4Sb_{12} (A = Ca, Sr, Ba, La, Nd, Sm, Eu) with the $LaFe_4P_{12}$ structure. *J. Phys. Chem. Solids* **57**, 381–387 (1996).
18. Galván, D. H. et al. Extended Huckel tight-binding calculations of the electronic structure of $YbFe_4Sb_{12}$, UFe_4P_{12}, and $ThFe_4P_{12}$. *Phys. Rev. B* **68**, 115110 (2003).
19. Aoki, D. et al. First single crystal growth of the transuranium filled-skutterudite compound $NpFe_4P_{12}$ and its magnetic and electrical properties. *J. Phys. Soc. Jpn* **75**, 073703 (2006).
20. Leithe-Jasper, A. et al. $TlFe_4Sb_{12}$: weak itinerant ferromagnetic analogue to alkali-metal iron-antimony skutterudites. *Phys. Rev. B* **77**, 064412 (2008).
21. Nolas, G. S., Yoon, G., Sellinschegg, H., Smalley, A. & Johnson, D. C. Synthesis and transport properties of $HfFe_4Sb_{12}$. *Appl. Phys. Lett.* **86**, 042111 (2005).
22. Shi, X. et al. Multiple-filled skutterudites: high thermoelectric figure of merit through separately optimizing electrical and thermal transports. *J. Am. Chem. Soc.* **133**, 7837–7846 (2011).

23. Sekine, C., Uchiumi, T., Shirotani, I. & Yagi, T. Metal-insulator transition in $PrRu_4P_{12}$ with skutterudite structure. *Phys. Rev. Lett.* **79**, 3218–3221 (1997).
24. Lee, C. H. et al. Structural phase transition accompanied by metal-insulator transition in $PrRu_4P_{12}$. *J. Phys. Condens. Matter* **13**, L45 (2001).
25. Matsuhira, K. et al. Specific heat study on Sm-based filled skutterudite phosphides SmT_4P_{12} (T = Fe, Ru and Os). *J. Phys. Soc. Jpn* **74**, 1030–1035 (2005).
26. Kikuchi, J., Takigawa, M., Hitoshi, S. & Sato, H. Quadrupole order and field-induced heavy-fermion state in the filled skutterudite $PrFe_4P_{12}$ via 31 P NMR. *Phy. B Condens. Matter* **359–361**, 877–879 (2005).
27. Torikachvili, M. S. et al. Low-temperature properties of rare-earth and actinide iron phosphide compounds MFe_4P_{12} (M = La, Pr, Nd, and Th). *Phys. Rev. B* **36**, 8660–8664 (1987).
28. Butch, N. P. et al. Ordered magnetic state in $PrFe_4Sb_{12}$ single crystals. *Phys. Rev. B* **71**, 214417 (2005).
29. Nicklas, M. et al. Magnetic order in the filled skutterudites RPt_4Ge_{12} (R = Nd, Eu). *J. Phys. Conf. Ser.* **273**, 012118 (2011).
30. Gumeniuk, R. et al. High-pressure synthesis and exotic heavy-fermion behaviour of the filled skutterudite $SmPt_4Ge_{12}$. *New J. Phys.* **12**, 103035 (2010).
31. Nicklas, M. et al. Charge-doping-driven evolution of magnetism and non-fermi-liquid behavior in the filled skutterudite $CePt_4Ge_{12-x}Sb_x$. *Phys. Rev. Lett.* **109**, 236405 (2012).
32. Baumbach, R. E. et al. Non-Fermi liquid behavior in the filled skutterudite compound $CeRu_4As_{12}$. *J. Phys. Condens. Matter* **20**, 075110 (2008).
33. Dordevic, S. V. et al. Heavy fermion fluid in high magnetic fields: an infrared study of $CeRu_4Sb_{12}$. *Phys. Rev. Lett.* **96**, 017403 (2006).
34. Adroja, D. T. et al. Spin gap formation in the heavy fermion skutterudite compound $CeRu_4Sb_{12}$. *Phys. Rev. B* **68**, 094425 (2003).
35. Bauer, E. D. et al. Electronic and magnetic investigation of the filled skutterudite compound $CeRu_4Sb_{12}$. *J. Phys. Condens. Matter* **13**, 5183–5193 (2001).
36. Leithe-Jasper, A. et al. Weak itinerant ferromagnetism and electronic and crystal structures of alkali-metal iron antimonides: $NaFe_4Sb_{12}$ and KFe_4Sb_{12}. *Phys. Rev. B* **70**, 214418 (2004).
37. Gippius, A. et al. Crossover between itinerant ferromagnetism and antiferromagnetic fluctuations in filled skutterudites MFe_4Sb_{12} (M = Na, Ba, La) as determined by NMR. *J. Magn. Magn. Mater.* **300**, e403–e406 (2006).
38. Yoshizawa, M. et al. Elastic properties of $HoFe_4P_{12}$ polycrystal. *J. Magn. Magn. Mater.* **310**, 1786–1788 (2007).
39. Meisner, G. P. Superconductivity and magnetic order in ternary rare earth transition metal phosphides. *Phys. B + C* **108**, 763–764 (1981).
40. Goshchitskii, B., Naumov, S., Kostromitina, N. & Karkin, A. Superconductivity and transport properties in $LaRu_4Sb_{12}$ single crystals probed by radiation-induced disordering. *Phys. C Supercond.* **460–462**, 691–693 (2007).
41. Shirotani, I. et al. Electrical conductivity and superconductivity of metal phosphides with skutterudite-type structure prepared at high pressure. *J. Phys. Chem. Solids* **57**, 211–216 (1996).
42. Shirotani, I. et al. Superconductivity of filled skutterudites $LaRu_4As_{12}$ and $PrRu_4As_{12}$. *Phys. Rev. B* **56**, 7866–7869 (1997).
43. B. Maple, M. et al. Heavy fermion superconductivity in the filled skutterudite compound $PrOs_4Sb_{12}$. *J. Phys. Soc. Jpn* **71**, 23–28 (2002).
44. Bauer, E. D., Frederick, N. A., Ho, P. C., Zapf, V. S. & Maple, M. B. Superconductivity and heavy fermion behavior in $PrOs_4Sb_{12}$. *Phys. Rev. B* **65**, 100506 (2002).
45. Bauer, E. et al. Superconductivity in novel Ge-based skutterudites: {Sr,Ba}Pt_4Ge_{12}. *Phys. Rev. Lett.* **99**, 217001 (2007).
46. Bauer, E. et al. $BaPt_4Ge_{12}$: a skutterudite based entirely on a Ge framework. *Adv. Mater.* **20**, 1325–1328 (2008).
47. Sales, B. C., Mandrus, D. & Williams, R. K. Filled skutterudite antimonides: a new class of thermoelectric materials. *Science* **272**, 1325–1328 (1996).
48. Pardo, V., Smith, J. C. & Pickett, W. E. Linear bands, zero-momentum Weyl semimetal, and topological transition in skutterudite-structure pnictides. *Phys. Rev. B* **85**, 214531 (2012).
49. David, J. Singh, electronic structure calculations with the Tran-Blaha modified Becke-Johnson density functional. *Phys Rev. B* **82**, 205102 (2010).
50. Bocquet, A. et al. Electronic structure of early 3d-transition-metal oxides by analysis of the 2p core-level photoemission spectra. *Phys. Rev. B* **53**, 1161–1170 (1996).
51. Singh, D. & Mazin, I. Calculated thermoelectric properties of La-filled skutterudites. *Phys. Rev. B* **56**, R1650–R1653 (1997).
52. Singh, D. & Pickett, W. Skutterudite antimonides: quasilinear bands and unusual transport. *Phys. Rev. B* **50**, 11235–11238 (1994).
53. Smith, J. C., Banerjee, S., Pardo, V. & Pickett, W. E. Dirac point degenerate with massive bands at a topological quantum critical point. *Phys Rev. Lett.* **106**, 056401 (2011).
54. Blaha, P., Schwarz, K., Madsen, G., Kvasnicka, D. & Luitz, J. *WIEN2K, An Augmented Plane Wave + Local Orbitals Program for Calculating Crystal Properties* (Technische Universitat Wien, 2001).

55. Perdew, J. P., Burke, K. & Ernzerhof, M. Generalized gradient approximation made simple. *Phys. Rev. Lett.* **77,** 3865–3868 (1996).

56. Yasumoto, Y. *et al.* Off-center rattling and tunneling in filled skutterudite $LaOs_4Sb_{12}$. *J. Phys. Soc. Jpn* **77,** 242–244 (2008).

57. Yamaura, J. i. & Hiroi, Z. Rattling vibrations observed by means of single-crystal X-ray diffraction in the filled skutterudite ROs_4Sb_{12} (R = La, Ce, Pr, Nd, Sm). *J. Phys. Soc. Jpn* **80,** 054601 (2011).

58. Klimczuk, T. *et al.* Negative thermal expansion and antiferromagnetism in the actinide oxypnictide NpFeAsO. *Phys. Rev. B* **85,** 174506 (2012).

59. Morelli, D. T. & Meisner, G. P. Low temperature properties of the filled skutterudite $CeFe_4Sb_{12}$. *J Appl. Phys.* **77,** 3777–3781 (1995).

60. Gajewski, D. A. *et al.* Heavy fermion behaviour of the cerium-filled skutterudites $CeFe_4Sb_{12}$ and $Ce_{0.9}Fe_3CoSb_{12}$. *J. Phys. Condens. Matter* **10,** 6973–6985 (1998).

61. Dilley, N. R., Freeman, E. J., Bauer, E. D. & Maple, M. B. Intermediate valence in the filled skutterudite compound $YbFe_4Sb_{12}$. *Phys. Rev. B* **58,** 6287–6290 (1998).

62. Uchiumi, T. *et al.* Superconductivity of $LaRu_4X_{12}$ (X = P, As and Sb) with skutterudite structure. *J. Phys. Chem. Solids* **60,** 689–695 (1999).

63. Tran, F. & Blaha, P. Accurate band gaps of semiconductors and insulators with a semilocal exchange-correlation potential. *Phys. Rev. Lett.* **102,** 226401 (2009).

64. Schoen, J. M. Augmented-plane-wave virtual-crystal approximation. *Phys. Rev* **184,** 858–863 (1969).

65. Bellaiche, L. & Vanderbilt, D. Virtual crystal approximation revisited: Application to dielectric and piezoelectric properties of perovskites. *Phys. Rev. B* **61,** 7877–7882 (2000).

66. Dronskowski, R. & Bloechl, P. E. Crystal orbital Hamilton populations (COHP): energy-resolved visualization of chemical bonding in solids based on density-functional calculations. *J. Phys. Chem.* **97,** 8617–8624 (1993).

67. Kotani, T. Exact exchange potential band-structure calculations by the linear muffin-tin orbital–atomic-sphere approximation method for Si, Ge, C, and MnO. *Phys. Rev. Lett.* **74,** 2989–2992 (1995).

68. Braun, D. J. & Jeitschko, W. Preparation and structural investigations of antimonides with the $LaFe_4P_{12}$ structure. *J. Less Common Met* **72,** 147–156 (1980).

69. Jepsen, O., Burkhardt, A. & Andersen, O. The STUTTGART TB-LMTO-ASA program. http://www2.fkf.mpg.de/andersen/LMTODOC/LMTODOC.html (2000).

Acknowledgements

This work was supported by the AFOSR MURIs in thermoelectric and superconducting materials, grants FA9550-10-1-0553 and FA9550-09-1-0953. We also acknowledge Brendan F. Phelan and Quinn Gibson for stimulating discussions.

Author contributions

R.J.C., H.L., L.M., N.H. and T.K. conceived and designed the experiments, and H.L., T.K. and N.H. performed the synthetic experiments. H.L. and R.J.C. analysed and interpreted the data. T.K. performed the magnetic characterization. J.W.K. performed and analysed the XRD refinement data. C.F., L.M. and W.X. performed and analysed the calculations. M.K.F. supported the Seebeck measurement and designed the TOC. H.L. and R.J.C. wrote the paper. All authors approved the content of the manuscript.

Additional information

17

In situ X-ray diffraction monitoring of a mechanochemical reaction reveals a unique topology metal-organic framework

Athanassios D. Katsenis[1], Andreas Puškarić[2], Vjekoslav Štrukil[1,2], Cristina Mottillo[1], Patrick A. Julien[1], Krunoslav Užarević[2], Minh-Hao Pham[3], Trong-On Do[3], Simon A.J. Kimber[4], Predrag Lazić[2], Oxana Magdysyuk[5], Robert E. Dinnebier[5], Ivan Halasz[2] & Tomislav Friščić[1]

Chemical and physical transformations by milling are attracting enormous interest for their ability to access new materials and clean reactivity, and are central to a number of core industries, from mineral processing to pharmaceutical manufacturing. While continuous mechanical stress during milling is thought to create an environment supporting non-conventional reactivity and exotic intermediates, such speculations have remained without proof. Here we use *in situ*, real-time powder X-ray diffraction monitoring to discover and capture a metastable, novel-topology intermediate of a mechanochemical transformation. Monitoring the mechanochemical synthesis of an archetypal metal-organic framework ZIF-8 by *in situ* powder X-ray diffraction reveals unexpected amorphization, and on further milling recrystallization into a non-porous material via a metastable intermediate based on a previously unreported topology, herein named *katsenite* (**kat**). The discovery of this phase and topology provides direct evidence that milling transformations can involve short-lived, structurally unusual phases not yet accessed by conventional chemistry.

[1] Department of Chemistry, McGill University, 801 Sherbrooke Street West, Montreal, Québec, Canada H3A 0B8. [2] Ruđer Bošković Institute, Bijenička cesta 54, Zagreb, HR-10000, Croatia. [3] Department of Chemical Engineering, Université Laval, Quebec City, Québec, Canada G1V 0A6. [4] ESRF—The European Synchrotron, CS 40220, Grenoble 38043, France. [5] Scientific Service Group X-ray Diffraction, Max Planck Institute for Solid State Research, Heisenbergstrasse 1, 70569 Stuttgart, Germany. Correspondence and requests for materials should be addressed to I.H. (email: ihalasz@irb.hr) or to T.F. (email: tomislav.friscic@mcgill.ca).

Mechanochemical reactions by milling or grinding have emerged as excellent, rapid and cleaner alternatives to conventional chemical synthesis in a number of areas, from nanomaterials and alloys to pharmaceuticals and metal-organic frameworks[1–4]. However, despite significant and rapid advances in synthetic scope and methodology[5–7], the mechanistic understanding of mechanochemical reactions has remained elusive[8–10], creating a major obstacle for the development and optimization of attractive proof-of-principle laboratory experiments into viable large scale processes[11]. It has been proposed that continuous mechanical stress and impact in a mechanochemical reaction environment can give rise to unusual reactivity or intermediate phases different than those normally accessible using conventional chemistry[12]. So far, there has been very little evidence[13] in support of or against such expectations due to a lack of methods that would permit direct, real-time insight into the rapidly agitated environment of an operating ball mill. Recently, however, we presented the first methodology for time-resolved in situ powder X-ray diffraction (PXRD) monitoring of mechanochemical reactions[14] as they take place[15,16]. In principle, this synchrotron radiation-based methodology now offers a unique opportunity to monitor mechanochemical reaction pathways and potentially observe unusual intermediate crystalline or amorphous phases whose existence might otherwise be only proposed or deduced by non-direct methods.

The initial target of our study, the popular sodalite topology (SOD) framework ZIF-8 is known to form readily from ZnO and 2-methylimidazole (**HMeIm**) by liquid-assisted grinding (LAG), that is, milling using a small amount of a liquid additive that facilitates reactivity (Fig. 1a,b)[5–7].

Due to its high porosity and excellent resistance to pressure, temperature and steam, ZIF-8 has been extensively studied for applications in carbon sequestration and is also one the four metal-organic frameworks being manufactured commercially (Basolite Z1200)[17–25]. Crystallization of ZIF-8 from solution has been explored by several groups[26,27], and a non-porous polymorph of ZIF-8 with a diamondoid (dia) topology was recently reported[28,29]. With the initial aim to simplify the mechanosynthesis of ZIF-8, which requires the presence of weakly acidic salts and an organic liquid (for example, N,N-dimethylformamide), we decided to explore LAG synthesis using only aqueous acetic acid.

Here we report that in situ and real-time X-ray diffraction monitoring of mechanochemical ZIF-8 synthesis reveals an unexpected amorphization–crystallization process involving the intermediate formation of a previously not known metal-organic framework. The new framework is a metastable polymorph of ZIF-8, based on a hitherto unknown net topology, which we name katsenite (**kat**), demonstrating that monitoring of mechanochemical transformations in real time can reveal materials and structures that are difficult or even impossible to access by other means.

Results

Initial experiments. The reaction mixture consisted of 0.8 mmol ZnO, 1.6 mmol **HMeIm**, giving a total of ≈ 200 mg of solid reactants and a designated volume of an aqueous solution of acetic acid with a concentration of either 2.50 or 1.25 mol dm^{-3}. The reaction mixture, along with two stainless-steel balls of 7-mm diameter (weight 1.3 g), was placed into an in-house designed 14 ml volume poly(methyl)methacrylate reaction jar and milled using a modified Retsch MM200 mill operating at 30 Hz (Supplementary Fig. 1). The reaction was monitored in situ and in real time by diffraction of high-energy X-rays (87.4 keV, $\lambda = 0.14202$ Å) at the European Synchrotron Radiation Facility (ESRF) Beamline ID15B[14]. For the initial experiment, conducted using 32 µl of 2.5 M aqueous acetic acid (~ 10 mol% acetic acid with respect to ZnO), the time-resolved diffractogram (Fig. 2a) revealed an almost immediate formation of ZIF-8, rapid disappearance of X-ray reflections of **HMeIm** and a much slower consumption of ZnO (see Methods, Supplementary Figs 2–7). Formation of ZIF-8 was recognized by its distinct PXRD pattern. Surprisingly, however, the intensity of X-ray reflections of ZIF-8 started to decrease after several minutes of milling. Within 40 min, the PXRD pattern exhibited only weak features of ZnO, indicating complete amorphization of the ZIF-8 formed initially. Such amorphization contrasts all previously reported mechanochemical LAG syntheses, where ZIF-8 was found to persist during milling[8,15]. Indeed, amorphization of ZIF-8 was previously reported only on dry milling of nano-sized crystals[30,31], or on exposure to pressures > 10 GPa (ref. 21). To elucidate the relative importance of the amount of acetic acid and the total volume of the liquid for the unexpected amorphization, the experiment was repeated, but using either half the amount of acetic acid in the same volume of liquid (32 µl of 1.25 M solution, Fig. 2b) or using the same amount of acid, but in a larger volume of liquid (64 µl of 1.25 M solution, Fig. 2c).

The results (Fig. 2a–d) reveal that the volume of liquid affects the amorphization of ZIF-8 more than the acid content: with 32 µl of aqueous acetic acid, complete amorphization of ZIF-8 was observed after ~ 30 min regardless of acetic acid concentration. With 64 µl of the 1.25 M solution, ZIF-8 was still clearly detectable after 55 min. Moreover, slow amorphization also took place with pure water as the grinding liquid, indicating that the weak acid facilitates, but is not mandatory for, the mechanically induced collapse of the ZIF-8 structure (Supplementary Fig. 8).

Properties of amorph-Zn(MeIm)₂. The amorphous material, designated amorph-Zn(**MeIm**)₂, could be dried in vacuum at room temperature or 40 °C and stored in a vial without recrystallization. Samples of amorph-Zn(**MeIm**)₂ gave Brunauer-Emmett-Teller (BET) surface areas of 49–65 m² g^{-1}, indicating that the collapse of the initially formed open ZIF-8. The lack of porosity is consistent with the structure previously proposed for amorphous zinc bis(2-methylimidazole) on the basis of pair distribution function analysis[31]. Magic-angle spinning solid-state

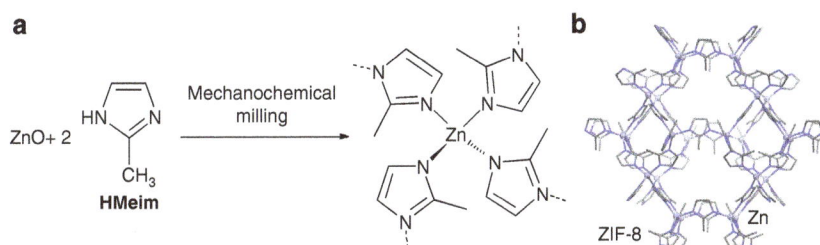

Figure 1 | Mechanosynthesis of ZIF-8. (**a**) Mechanochemical synthesis of ZIF-8; (**b**) Fragment of the crystal structure of ZIF-8.

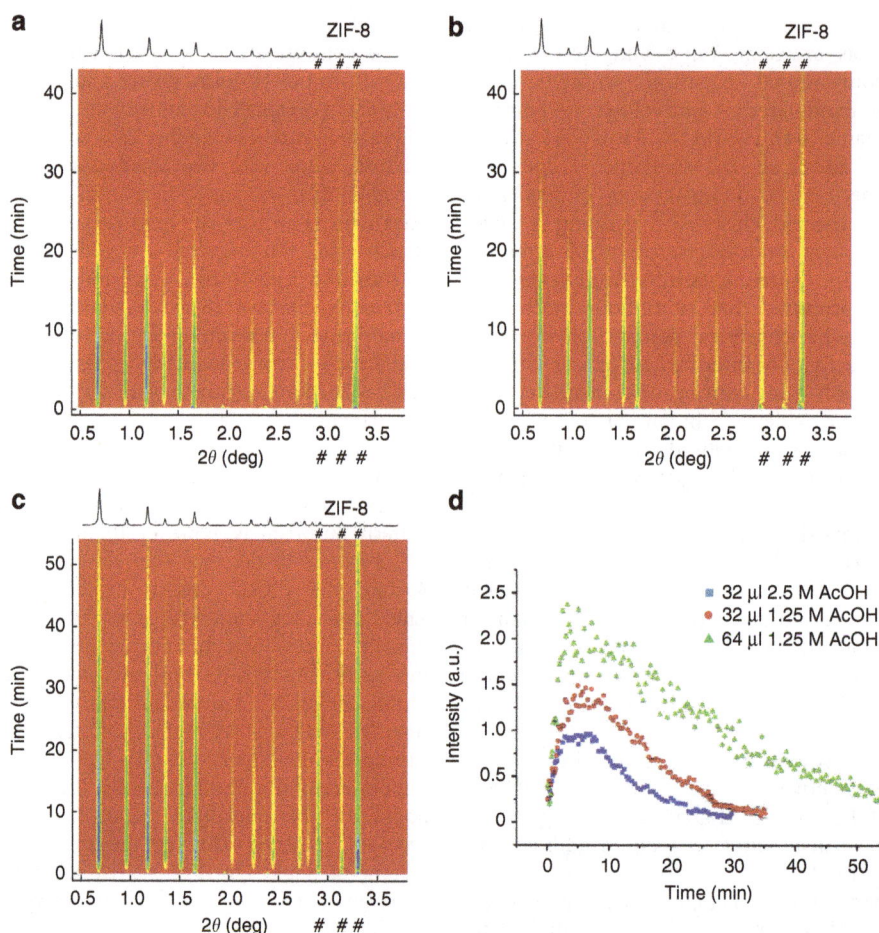

Figure 2 | Time-resolved diffractograms for the mechanochemical formation and amorphization of ZIF-8. Reactions were performed by milling ZnO and **HMeIm** in the presence of aqueous acetic acid. The calculated PXRD pattern of ZIF-8 (CSD code VELVOY) is shown above each time-resolved diffractogram: (**a**) LAG with 32 μl of 2.5 M acetic acid ($\eta = 0.16\,\mu l\,mg^{-1}$); (**b**) LAG with 32 μl of 1.25 M acetic acid ($\eta = 0.16\,\mu l\,mg^{-1}$) and (**c**) LAG with 64 μl of 1.25 M acetic acid ($\eta = 0.32\,\mu l\,mg^{-1}$). (**d**) The formation and amorphization of ZIF-8 illustrated by the change in intensity of the (011) reflection of ZIF-8, with error bars corresponding to s.d. In the time-resolved diffractograms, the reflections of ZnO starting material are highlighted by '#'. Selected examples and details of Rietveld analysis for experiments (**a–c**) are given in the Supplementary Figs 2–7 and Supplementary Note 1.

nuclear magnetic resonance measurements (Fig. 3) on *amorph*-Zn(**MeIm**)$_2$ revealed broad signals from three chemically distinct types of carbon atoms. The observed chemical shifts resemble those of ZIF-8, confirming the retention of a zinc imidazolate network, while the broadening of the NMR signals is consistent with the lack of long-range order.

Discovery of a new phase. Monitoring of mechanochemical reactions is affected by a changing and non-uniform distribution of the sample in the reaction vessel during milling, which results in scattering of the values of the measured diffraction intensity (Fig. 2d). To account for this effect, crystalline silicon ($\sim$20% by weight) was added into the reaction mixture as an internal scattering standard[32]. However, milling with silicon unexpectedly led to *in situ* recrystallization of the initial *amorph*-Zn(**MeIm**)$_2$ after $\approx$30 min of milling, as shown by the appearance of reflections in the time-resolved diffractogram (Fig. 4a,b). The reflections did not correspond to either ZIF-8 or *dia*-Zn(**MeIm**)$_2$. Further milling led to the disappearance of these novel reflections and the appearance of reflections of the close-packed, non-porous *dia*-Zn(**MeIm**)$_2$ (Cambridge Structural Database (CSD) OFERUN01, Supplementary Figs 9–13)[23].

We subsequently established that the same unknown intermediate also appears in the absence of silicon, but only after

longer milling ($>$50 min, Supplementary Figs 14–16). While it is tempting to conclude that crystallization of the new phase was facilitated by silicon acting as a heterogeneous nucleating agent, we found that the exact time of appearance of the new phase was difficult to reproduce, and that the crystallization of amorphous material sometimes directly gave *dia*-Zn(**MeIm**)$_2$ (Fig. 4c,d). Such behaviour is characteristic of stochastic processes of nucleation from an amorphous matrix, and highlights the importance of *in situ* monitoring for the discovery of intermediate phases[33]. With the reaction course known, we were also able to reproduce the preparation of the new phase using conventional steel milling equipment in slightly over 50% of deliberate synthesis attempts, that is, the new *kat* phase was observed after milling in 9 out of 17 reactions (see Methods, Supplementary Fig. 17). Crystallization of *amorph*-Zn(**MeIm**)$_2$ indicates a high degree of mobility imparted by wet milling, which contrasts its stability under storage and the earlier reports of stability of amorphized ZIF-8 under high pressures or dry milling[30,31]. As further evidence of the dynamic nature of the amorphous phase under milling conditions, we milled a dried sample of *amorph*-Zn(**MeIm**)$_2$ with 100 μl of *N,N*-dimethylformamide, a solvent normally used for preparation of ZIF-8. Indeed, ZIF-8 was produced, indicating that *amorph*-Zn(**MeIm**)$_2$ can be re-arranged back into the open SOD topology by wet milling in a suitable environment (Supplementary Fig. 18).

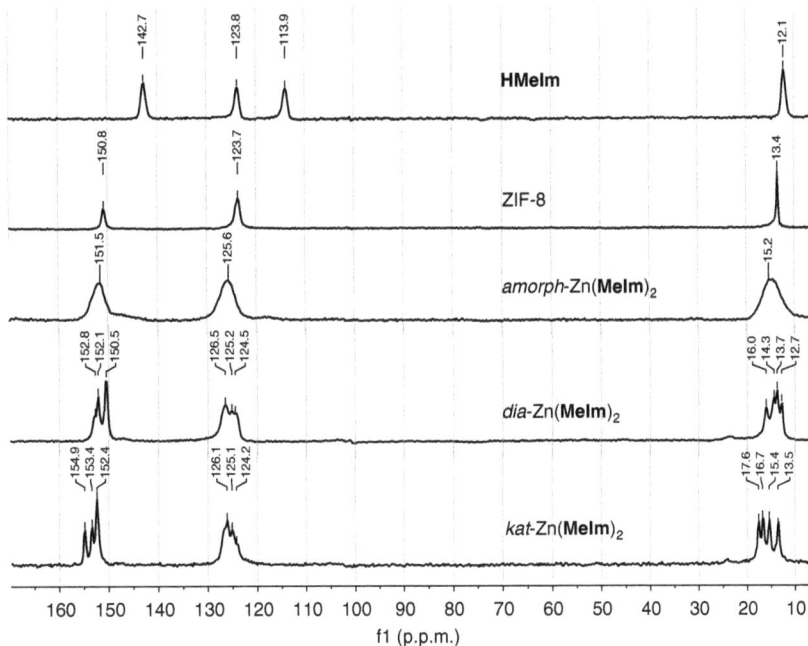

Figure 3 | CP-MAS ^{13}C SSNMR spectra. CP-MAS ^{13}C SSNMR spectra for **HMeIm**, ZIF-8, *amorph*-Zn(**MeIm**)$_2$, *dia*-Zn(**MeIm**)$_2$ and *kat*-Zn(**MeIm**)$_2$. The weak signal ~25 p.p.m. is assigned to the acetic acid additive in the reaction mixture.

The *katsenite* topology. The crystal structure of the new phase was solved from PXRD data (Supplementary Fig. 19, Supplementary Note 2) and revealed a new polymorph of Zn(**MeIm**)$_2$ with a hitherto unknown framework topology which we named *katsenite* (three letter symbol: *kat*). The new phase *kat*-Zn(**MeIm**)$_2$ crystallizes in the tetragonal space group $P\bar{4}2c$ ($a = 16.139(1)$ Å, $b = 16.321(1)$ Å) with four crystallographically independent Zn(II) sites. Each Zn(II) ion is in a tetrahedral environment defined by four 2-methylimidazolate ligands. Each ligand bridges two Zn(II) ions to form a three-dimensional (3D) framework (Fig. 5a). Although this framework comprises only four-coordinated tetrahedral nodes, the underlying net is unique due to four non-equivalent 4-c nodes, each adopting a different coordination sequence and point symbol. Thus, we identified a quadrinodal network based on tetrahedral nodes with transitivity 4463 (see Methods, Supplementary Figs 20 and 21, Supplementary Table 1). An augmented view of the *kat*-net, in which the nodes of the original net are replaced by their vertex figures, is shown in Fig. 5b.

The new phase *kat*-Zn(**MeIm**)$_2$ could be dried at room temperature in vacuum and stored for at least 3 months, which allowed a more extensive characterization and comparison to other zinc bis(2-methylimidazole) forms using thermogravimetric analysis and Fourier-transform infrared-attenuated total reflectance spectroscopy (Supplementary Figs 22 and 23). Although stable at room temperature, *kat*-Zn(**MeIm**)$_2$ readily rearranges into the *dia* phase by mild heating, for example, on attempted drying at 40 °C or on exposure to organic liquids(for example, on attempted washing with methanol). Consequently, samples of *kat*-Zn(**MeIm**)$_2$ were investigated with minimum exposure to heat or washing. Examination of samples of *amorph*-, *kat*- and *dia*-Zn(**MeIm**)$_2$ by scanning electron microscopy (Supplementary Figs 24–26) did not reveal any significant differences in particle morphology. Attempts to use previously made *kat*-Zn(**MeIm**)$_2$ to seed the recrystallization of *amorph*-Zn(**MeIm**)$_2$ during milling were only partially successful, always yielding impure samples containing small amounts of ZIF-8. Tentatively, we rationalize the appearance of ZIF-8 impurity by epitaxial nucleation, due to the

similarity of unit cell parameters between *kat*-Zn(**MeIm**)$_2$ ($a = 16.1$ Å, $c = 16.3$ Å) and ZIF-8 ($a = 16.9$ Å). The solid-state ^{13}C NMR spectrum of *kat*-Zn(**MeIm**)$_2$ indicates four symmetrically non-equivalent **MeIm**$^-$ ligands, in agreement with the crystal structure obtained from X-ray powder diffraction data. Although the structure of *dia*-Zn(**MeIm**)$_2$ also features four non-equivalent ligands, these two phases can be clearly distinguished by solid-state ^{13}C NMR chemical shifts of the ligand methyl groups (Fig. 3). The structure of *kat*-Zn(**MeIm**)$_2$ contains pores consisting of tight channels and pockets, which amount to 8% of the unit cell volume. Consistent with a low porosity structure, nitrogen gas sorption experiments revealed a low BET surface area of 37 m^2 g^{-1} for freshly prepared samples of *kat*-Zn(**MeIm**)$_2$ dried in vacuum at room temperature. This is much less than that expected for a microporous material like ZIF-8 (without washing and activation, ZIF-8 prepared from ZnO in aqueous environment exhibits a BET surface area of ~1,300 m^2 g^{-1})[29]. Samples of *dia*-Zn(**MeIm**)$_2$ prepared by mechanochemical collapse of ZIF-8 are non-porous, exhibiting an almost completely flat BET nitrogen sorption curve. The comparison of BET surface areas is consistent with the comparison of calculated densities for these materials: *dia*-Zn(**MeIm**)$_2$ (1.578 g cm^{-3}), *kat*-Zn(**MeIm**)$_2$ (1.423 g cm^{-3}) and ZIF-8 (0.922 g cm^{-3}).

Discussion

The herein described crystalline → amorphous → crystalline transformation of ZIF-8 into its polymorphs *kat*-Zn(**MeIm**)$_2$ or *dia*-Zn(**MeIm**)$_2$ by milling in a water-containing environment is remarkable for several reasons. While amorphization on milling is a well-known phenomenon, spontaneous recrystallization of the mechanochemically obtained amorphous phase by continued milling has, to the best of our knowledge, not yet been described. Furthermore, the synthesis of crystalline ZIF-8 in aqueous solution has been described by several reports[34,35], making the herein observed structural instability in the presence of water highly unexpected. Finally, previous studies[36] have consistently established that amorphous ZIF-8, which can be prepared either

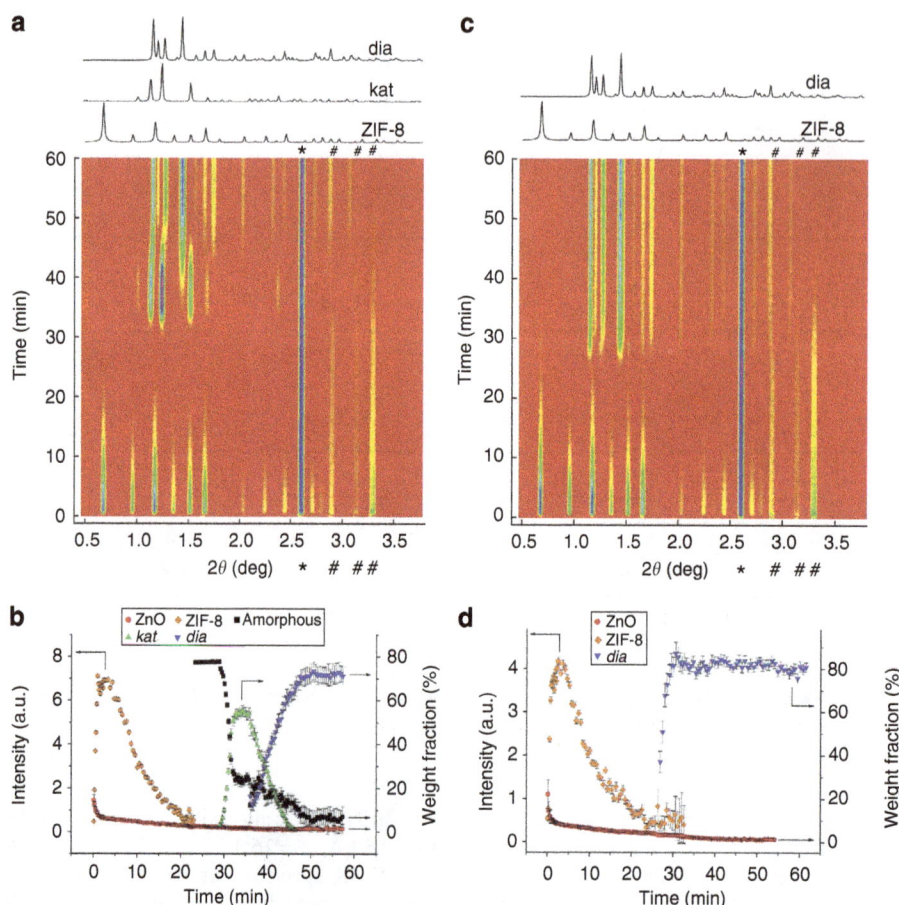

Figure 4 | Time-resolved diffractograms of mechanochemical amorphization and recrystallization. Mechanochemical LAG conversion of ZnO and **HMeIm** in the presence of 2.5 M acetic acid with 20 wt% crystalline silicon as an internal scattering standard. The diffractogram (**a**) and the corresponding quantitative plot of the evolution of each phase (**b**) reveal the formation of the new (*kat*) phase at ∼30 min milling. The diffractogram (**c**) and the corresponding quantitative plot for the evolution of each phase (**d**) demonstrate an experiment in which the *kat*-Zn(**MeIm**)$_2$ did not form and *dia*-Zn(**MeIm**)$_2$ crystallized directly from the amorphous phase. Final weight fraction of ∼80% of *dia*-Zn(**MeIm**)$_2$ for (**d**) indicates complete crystallization of the reaction mixture (remaining 20% being the silicon standard). Calculated PXRD patterns for crystalline phases are given on top of the time-resolved diffractograms. Due to the inability to establish the nature and arrangement of included guests in nascent ZIF-8, this phase was not included in Rietveld analysis. Instead, changes in the amount of ZIF-8 are represented with variations in the intensity of its (011) reflection, while changes in the amounts of other, non-porous phases are represented with variations of their weight fractions calculated by Rietveld analysis. Error bars correspond to s.d. obtained from the refinement procedure. To keep the liquid-to-solid ratio (η) after addition of silicon consistent with initial experiments, the volume of aqueous acid was 40 µl. Three characteristic ZnO reflections are marked with '#', while the reflection of Si is marked with '*'. Examples of Rietveld analysis plots are given in the Supplementary Figs 9–13 and Supplementary Note 1.

by neat milling of ZIF-8 or by exposing it to high pressures, does not rearrange into a crystalline material. Consequently, the observed transformation of *amorph*-Zn(**MeIm**)$_2$ into different porous and non-porous crystalline frameworks is very much in contrast to established knowledge. It is, however, known that ZIFs are generally sensitive to acidic environments[37] and we believe that the observed transformations of ZIF-8 and of its amorphous form are the result of structural lability induced by a mildly acidic aqueous environment.

The course of structural transformations observed taking place during milling is consistent with calculated order of stabilities with the *dia*-Zn(**MeIm**)$_2$ being the most stable (see Supplementary Note 3). If *dia*-Zn(**MeIm**)$_2$ is set as a reference point, *kat*-Zn(**MeIm**)$_2$ is less stable by 0.16 kcal mol^{-1} (7 meV per atom) and ZIF-8 by 0.29 kcal mol^{-1} (12.7 meV per atom). The reaction course can also be more qualitatively rationalized by considering the number of tetrahedra (T) per volume (V) for each involved framework[38]. The T/V value of 3.8 nm^{-3} places *kat*-Zn(**MeIm**)$_2$ between ZIF-8 ($T/V = 2.4$ nm^{-3}) and *dia*-Zn(**MeIm**)$_2$ ($T/V = 4.2$ nm^{-3}). This reveals that the

stepwise transformation of ZIF-8 into *dia*-Zn(**MeIm**)$_2$ follows the Ostwald's rule of stages[33] and is driven by the formation of increasingly stable solid phases (Fig. 5c). Although the stepwise interconversion of ZIF-8 into *kat*-Zn(**MeIm**)$_2$ and *dia*-Zn(**MeIm**)$_2$ does not preserve any long-range topological features, a comparison of the secondary coordination sphere around crystallographically non-equivalent zinc nodes in *kat*-Zn(**MeIm**)$_2$ reveals structural elements that are found in ZIF-8 and in *dia*-Zn(**MeIm**)$_2$, consistent with its role as an intermediate phase in the sequence SOD → **kat** → **dia**.

In conclusion, the presented results demonstrate the potential of *in situ* and real-time monitoring not only for understanding the processes taking place during milling, but also for the discovery of new phases and unexpected transformations in systems that are considered to be well-established. These results show that mechanical agitation under wet conditions provides means to access unconventional behaviour and structures different from those observed at static conditions or even under dry milling. In the context of understanding mechanochemical reactivity, the described results are an important advance as the

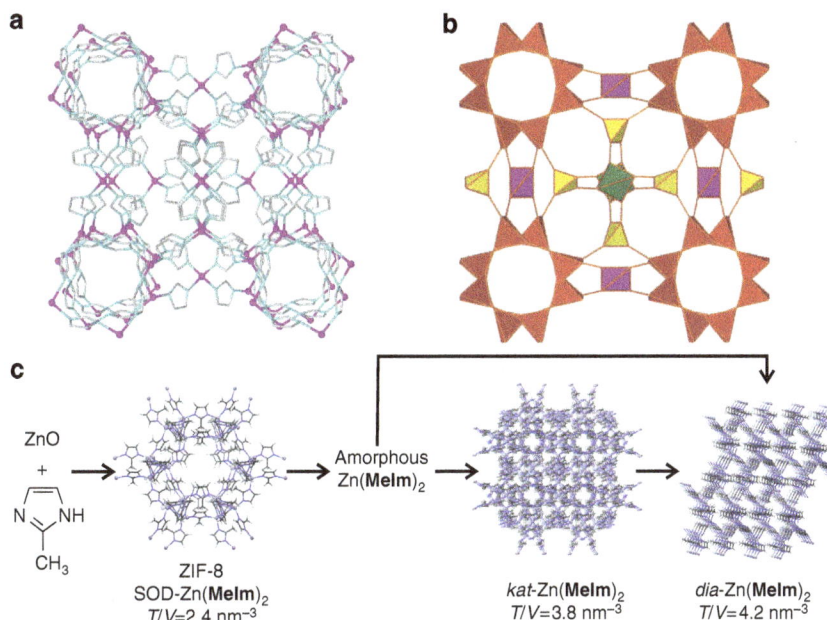

Figure 5 | Crystal structure and topology of katsenite and framework transformations. Structure of *kat*-Zn(**MeIm**)$_2$ viewed along the crystallographic *c*-direction: (**a**) ball-and-stick representation and (**b**) the *kat* framework with different colouring for each type of vertex, represented by its vertex figure. (**c**) The sequence of solid-state transformations in the LAG reaction of ZnO and **HMeIm**: the transformations of ZIFs resemble the Ostwald's rule of stages[33] as they follow the order of increasing *T/V* values, and hence the expected increase in the thermodynamic stability.

first direct observation of an amorphization–recrystallization mechanism in mechanosynthesis and the first evidence of a metastable, structurally unusual intermediate in a mechanochemical process. The ability to use mechanochemical reaction monitoring to discover a new net topology in a system that has so far been subject to hundreds of studies is of outstanding importance in solid-state chemistry, and especially in the area of metal-organic frameworks[39]. Thus, our findings highlight the role of *in situ* and real-time PXRD monitoring as a complement to other experimental and computational studies of the structural landscape of metal-organic frameworks[40,41]. We note that the use of *in situ* monitoring to detect new phases appearing during milling may not be limited to metal-organic frameworks, but could very likely impact the understanding of structural transformations in related inorganic materials, such as zeolites and silicates, during mechanical processing or geological transformations[42].

Methods

Experimental set-up. Real-time, *in situ* experiments were conducted at the ESRF beamline ID15B in a modified MM200 Retsch mill operating at 30 Hz[15]. The mill used was a Retsch MM200 ball mill with the sample holder custom modified to enable the access of the incident beam. Each reaction was conducted in a jar of 14 ml volume using two stainless-steel balls of 7-mm diameter (weight 1.33 g). Milling jars were fabricated so that the two complementary parts easily snapped together and did not leak liquid during the experiments[14]. The procedure of snapping the jar together and mounting it onto the mill was performed within 20 s or less, after which the milling was initiated by remote control and reaction course monitored in real time and *in situ*.

The synthesis of *kat*-Zn(**MeIm**)$_2$, *dia*-Zn(**MeIm**)$_2$ and *amorph*-Zn(**MeIm**)$_2$ was reproduced outside the synchrotron facility using conventional equipment, that is, using a commercial Retsch MM400 mill operating at 30 Hz, with 200 mg of a mixture of ZnO and **HMeIm** in a stoichiometric ratio of 1:2 placed in a 10 ml stainless-steel jar along with a liquid additive and stainless-steel milling media. As milling media, we used either two 7 mm (each weighing 1.33 g) balls or a single 10-mm diameter ball (weight 4 g), with similar results. The formation of *kat*-Zn(**MeIm**)$_2$, *dia*-Zn(**MeIm**)$_2$ and *amorph*-Zn(**MeIm**)$_2$ was confirmed by PXRD analysis immediately after milling (Supplementary Fig. 17). Reproducible synthesis of pure *kat*-Zn(**MeIm**)$_2$ was challenging and the samples often contained a variable amount of *dia*-Zn(**MeIm**)$_2$. In contrast, the *amorph*-Zn(**MeIm**)$_2$ was readily obtained by milling ZnO and **HMeIm** with either pure water or aqueous acetic acid

(*c* = 2.5 M) for 30–50 min. Reproducible mechanosynthesis of *dia*-Zn(**MeIm**)$_2$ was achieved by milling ZnO and **HMeIm** with1.25 M aqueous acetic acid for 90 min.

Powder X-ray diffraction. For *in situ* reaction monitoring, incident X-rays were selected using a bent Laue Si crystal, and the beam size on the sample was $300 \times 300\,\mu m^2$. Diffracted X-rays were detected with a flat panel Perkin-Elmer detector. Each diffractogram was typically obtained by summing 10 frames (or less), collected each with exposure time of 0.4 s, thus giving time resolution between successive diffractograms in seconds. The incident energy and detector distance (1,225.76 mm) were calibrated using a NIST CeO$_2$ standard sample and the Fit2D software package (ESRF Internal Report, ESRF98HA01T, FIT2D V9.129 Reference Manual V3.1, 1998). Raw data frames were integrated using Fit2D. For time-resolved 2D diffractograms the background for each pattern was subtracted using the Sonneveld–Visser[43] algorithm implemented in Powder3D[44]. Pawley[45] and Rietveld[46] refinements have been carried out on raw integrated diffraction patterns using Topas[47].

Laboratory PXRD experiments were conducted on a Bruker D2 Phaser equipped with a CuK_α X-ray source and a Ni filter. Reaction products with known crystal structures were identified by comparing the measured PXRD patterns to the ones simulated for known structures in the CSD (version 5.35, November 2013 update) or determined in this work.

The structure of *kat*-Zn(MeIm)$_2$. The structure of *kat*-Zn(MeIm)$_2$ was solved by global optimization in direct space using the program Topas[47], from PXRD data collected on a Bruker D8 Discovery X-ray diffractometer in the 2θ range 5° to 65° using a Cu-K_α ($\lambda = 1.54\,Å$) source, equipped with a Våntec area detector and a nickel filter. The X-ray tube was operating at the power setting of 40 kV and 40 mA. Topological analysis was performed using the program TOPOS[48]. The structure is based on a 3D framework with an unprecedented topology and point symbol $\{4.7^3.8^2\}_4\{4.7^4.8\}_2\{4^2.7^4\}\{7^4.8^2\}$. In the asymmetric unit, there are four crystallographically independent zinc ions (Zn(1), Zn(2), Zn(3) and Zn(4)), each in a tetrahedral environment defined by four nitrogen atoms of four different ligands. Although this new framework comprises only four-coordinated distorted tetrahedral nodes, the underlining network is unique due to the presence of four non-equivalent 4-c nodes, each adopting a different coordination sequence and point symbol (Supplementary Table 1, Supplementary Fig. 20). There are six kinds of non-equivalent essential rings: two 4-rings designated **4a** and **4b** (Supplementary Fig. 21a), two 7-rings and two 8-rings that are constructed by regarding the ligands as spacers between metal nodes. Vertices of the **4a**-rings are alternating Zn(1) and Zn(3) nodes, while the **4b**-rings are defined by Zn(2) nodes. The **4a**-rings form infinite chains of corner-sharing rectangles by sharing Zn(3) nodes. The tetrahedral environment around each Zn(1) node in a **4a**-ring is completed by two Zn(2) nodes. In that way, each **4a**-ring in the *kat* structure is linked to four different **4b**-rings. Finally, four **4b**-rings connect to a Zn(4) node in a distorted tetrahedral arrangement. The above-mentioned arrangement of four-membered rings yields three different kinds of cages (tiles) with tiling $[4^2.8^2] + 2[4.7^4] + 2[7^4.8^2]$ (Supplementary Fig. 21b). Before the final Rietveld refinement the structure model was optimized at 0 K in the space group $P1$ using *ab initio* calculations (Supplementary Note 2,3).

Ab initio calculations. *Ab initio* calculations of structural and electronic properties of all participating crystalline phases were performed using the VASP[49] code with the projector augmented wave method (Supplementary Note 2,3)[50]. Consideration of van der Waals interactions was necessary to correctly evaluate relative stabilities of the investigated structures, namely, the density functional theory (DFT) calculations with self-consistently implemented van der Waals functional (vdW-DF)[51-53] give the order of stability in agreement with chemical behaviour and expectations based on differences in the density of tetrahedral centres. However, DFT calculations using general gradient approximation (GGA) type of the exchange correlation functional yield results that are clearly not consistent with the experiment, with relative energies of zinc(2-methylimidazolate) structures increasing in the order, that is, reverse from that expected: ZIF-8 [0] $< kat$-Zn(**MeIm**)$_2$ (0.11 kcal mol^{-1}, 4.7 eV per atom) $< dia$-Zn(**MeIm**)$_2$ (0.168 kcal mol^{-1}, 7.3 meV per atom). This demonstrates that in such situations[54] where total energies of investigated systems are close in value, the difference between results obtained with different functionals may not only be quantitative but also qualitative. Our results demonstrate that vdW-DF, as a nonlocal functional, is a better choice than the semi-local ones of the GGA type[55].

References

1. James, S. L. *et al.* Mechanochemistry: opportunities for new and cleaner synthesis. *Chem. Soc. Rev.* **41**, 413–447 (2012).
2. Boldyreva, E. Mechanochemistry of inorganic and organic systems: what is similar, what is different? *Chem. Soc. Rev.* **42**, 7719–7738 (2013).
3. Stolle, A., Szuppa, T., Leonhardt, S. E. S. & Ondruschka, B. Ball milling in organic synthesis: solutions and challenges. *Chem. Soc. Rev.* **40**, 2317–2329 (2011).
4. Baláž, P. *et al.* Hallmarks of mechanochemistry: from nanoparticles to technology. *Chem. Soc. Rev.* **42**, 7571–7637 (2013).
5. Bowmaker, G. A. Solvent-assisted mechanochemistry. *Chem. Commun.* **49**, 334–348 (2013).
6. Braga, D., Maini, L. & Grepioni, F. Mechanochemical preparation of co-crystals. *Chem. Soc. Rev.* **42**, 7638–7648 (2013).
7. Beldon, P. J. *et al.* Rapid room-temperature synthesis of zeolitic imidazolate frameworks by using mechanochemistry. *Angew. Chem. Int. Ed.* **49**, 9640–9643 (2010).
8. Rothenberg, G., Downie, A. P., Raston, C. L. & Scott, J. L. Understanding solid/solid reactions. *J. Am. Chem. Soc.* **123**, 8701–8708 (2001).
9. Michalchuk, A. A. L., Tumanov, I. A. & Boldyreva, E. V. Complexities of mechanochemistry: elucidation of processes occurring in mechanical activators via implementation of a simple organic system. *CrystEngComm* **15**, 6403–6412 (2013).
10. McKissic, K. S., Caruso, J. T., Blair, R. G. & Mack, J. Comparison of shaking versus baking: further understanding the energetics of a mechanochemical reaction. *Green Chem.* **16**, 1628–1632 (2014).
11. Burmeister, C. F. & Kwade, A. Process engineering with planetary ball mills. *Chem. Soc. Rev.* **42**, 7660–7667 (2013).
12. Gilman, J. J. Mechanochemistry. *Science* **274**, 65 (1996).
13. Wang, G.-W., Komatsu, K., Murata, Y. & Shiro, M. Synthesis and X-ray structure of dumb-bell-shaped C120. *Nature* **387**, 583–586 (1997).
14. Halasz, I. *et al.* In situ and real-time monitoring of mechanochemical milling reactions using synchrotron X-ray diffraction. *Nat. Protoc.* **9**, 1718–1729 (2013).
15. Friščić, T. *et al.* Real-time and in situ monitoring of mechanochemical milling reactions. *Nat. Chem.* **5**, 66–73 (2013).
16. Halasz, I. *et al.* Real-time *in situ* powder X-ray diffraction monitoring of mechanochemical synthesis of pharmaceutical cocrystals. *Angew. Chem. Int. Ed.* **52**, 11538–11541 (2013).
17. Phan, A. *et al.* Synthesis, structure, and carbon dioxide capture properties of zeolitic imidazolate frameworks. *Acc. Chem. Res.* **43**, 58–67 (2010).
18. Huang, X.-C., Lin, Y.-Y., Zhang, J.-P. & Chen, X.-M. Ligand-directed strategy for zeolite-type metal–organic frameworks: zinc(II) imidazolates with unusual zeolitic topologies. *Angew. Chem. Int. Ed.* **45**, 1557–1559 (2006).
19. Zhang, J.-P., Zhang, Y.-B., Lin, J.-B. & Chen, X.-M. Metal azolate frameworks: from crystal engineering to functional materials. *Chem. Rev.* **112**, 1001–1033 (2012).
20. Czaja, A. U., Trukhan, N. & Müller, U. Industrial applications of metal–organic frameworks. *Chem. Soc. Rev.* **38**, 1284–1293 (2009).
21. Moggach, S. A., Bennett, T. D. & Cheetham, A. K. The effect of pressure on ZIF-8: increasing pore size with pressure and the formation of a high-pressure phase at 1.47 GPa. *Angew. Chem. Int. Ed.* **48**, 7087–7089 (2009).
22. Park, K. S. *et al.* Exceptional chemical and thermal stability of zeolitic imidazolate frameworks. *Proc. Natl Acad. Sci. USA* **103**, 10186–10191 (2006).
23. Banerjee, R. *et al.* High-throughput synthesis of zeolitic imidazolate frameworks and application to CO2 capture. *Science* **319**, 939–943 (2008).
24. Lu, G. *et al.* Imparting functionality to a metal–organic framework material by controlled nanoparticle encapsulation. *Nat. Chem.* **4**, 310–316 (2012).
25. Morris, W. *et al.* NMR and X-ray study revealing the rigidity of zeolitic imidazolate frameworks. *J. Phys. Chem. C* **116**, 13307–13312 (2012).
26. Venna, S. R., Jasinski, J. B. & Carreon, M. A. Structural evolution of zeolitic imidazolate framework-8. *J. Am. Chem. Soc.* **132**, 18030–18033 (2010).
27. Cravillon, J. *et al.* Fast nucleation and growth of ZIF-8 nanocrystals monitored by time-resolved *in situ* small-angle and wide-angle X-ray scattering. *Angew. Chem. Int. Ed.* **50**, 8067–8071 (2011).
28. Shi, Q., Chen, Z., Song, Z., Li, J. & Dong, J. Synthesis of ZIF-8 and ZIF-67 by steam-assisted conversion and an investigation of their tribological behaviors. *Angew. Chem. Int. Ed.* **50**, 672–675 (2011).
29. Mottillo, C. *et al.* Mineral neogenesis as an inspiration for mild, solvent-free synthesis of bulk microporous metal–organic frameworks from metal (Zn, Co) oxides. *Green Chem.* **15**, 2121–2131 (2013).
30. Bennett, T. D. *et al.* Facile mechanosynthesis of amorphous zeolitic imidazolate frameworks. *J. Am. Chem. Soc.* **133**, 14546–14549 (2011).
31. Cao, S. *et al.* Amorphization of the prototypical zeolitic imidazolate framework ZIF-8 by ball-milling. *Chem. Commun.* **48**, 7805–7807 (2012).
32. Halasz, I. *et al.* Quantitative *in situ* and real-time monitoring of mechanochemical reactions. *Faraday Discuss.* **170**, 203–221 (2014).
33. Burley, J. C., Duer, M. J., Stein, R. S. & Vrcelj, R. M. Enforcing Ostwald's rule of stages: isolation of paracetamol forms III and II. *Eur. J. Pharm. Sci.* **31**, 271–276 (2007).
34. Kida, K., Okita, M., Fujita, K., Tanaka, S. & Miyake, Y. Formation of high crystalline ZIF-8 in an aqueous solution. *CrystEngComm* **15**, 1794–1801 (2013).
35. Gross, A. F., Sherman, E. & Vajo, J. J. Aqueous room temperature synthesis of cobalt and zinc sodalite zeolitic imidizolate frameworks. *Dalton Trans.* **41**, 5458–5460 (2012).
36. Bennett, T. D. & Cheetham, A. K. Amorphous metal-organic frameworks. *Acc. Chem. Res.* **47**, 1555–1562 (2014).
37. Mottillo, C. & Friščić, T. Carbon dioxide sensitivity of zeolitic imidazolate frameworks. *Angew. Chem. Int. Ed.* **53**, 7471–7474 (2014).
38. Lewis, D. W. *et al.* Zeolitic imidazole frameworks: structural and energetics trends compared with their zeolite analogues. *CrystEngComm* **11**, 2272–2276 (2009).
39. Guillerm, V. *et al.* Discovery and introduction of a (3,18)-connected net as an ideal blueprint for the design of metal–organic frameworks. *Nat. Chem.* **6**, 673–680 (2014).
40. Vaidhyanathan, R. *et al.* Direct observation and quantification of CO2 binding within an amine-functionalized nanoporous solid. *Science* **330**, 650–653 (2010).
41. Schröder, C. A., Baburin, I. A., van Müllen, L., Wiebcke, M. & Leoni, S. Subtle polymorphism of zinc imidazolate frameworks: temperature-dependent ground states in the energy landscape revealed by experiment and theory. *CrystEngComm* **15**, 4036–4040 (2013).
42. Haines, J. Topologically ordered amorphous silica obtained from the collapsed siliceous zeolite, Silicalite F-1: a step towards "perfect" glasses. *J. Am. Chem. Soc.* **131**, 12333–12338 (2009).
43. Sonneveld, E. J. & Visser, J. W. Automatic collection of powder data from photographs. *J. Appl. Cryst.* **8**, 1–7 (1975).
44. Hinrichsen, B., Dinnebier, R. E. & Jansen, M. Powder3D: an easy to use program for data reduction and graphical presentation of large numbers of powder diffraction patterns. *Z. Kristallogr. Suppl.* **23**, 231–236 (2006).
45. Pawley, G. S. Unit-cell refinement from powder diffraction scans. *J. Appl. Cryst.* **14**, 357–361 (1981).
46. Rietveld, H. M. A profile refinement for nuclear and magnetic structures. *J. Appl. Cryst.* **2**, 65–71 (1969).
47. Topas, version 4.2, Bruker-AXS, Karlsruhe, Germany. www.bruker.com.
48. Blatov, V. A., Shevchenko, A. P. & Proserpio, D. M. Applied topological analysis of crystal structures with the program package ToposPro Cryst. *Growth Des.* **14**, 3576–3586 (2014).
49. Kresse, G. & Furthmüller, J. Efficiency of ab-initio total energy calculations for metals and semiconductors using a plane-wave basis set. *Comput. Mater. Sci.* **6**, 15–50 (1996).
50. Kresse, G. & Joubert, D. From ultrasoft pseudopotentials to the projector augmented-wave method. *Phys. Rev. B* **59**, 1758–1775 (1999).
51. Dion, M., Rydberg, H., Schröder, E., Langreth, D. C. & Lundqvist, B. I. Van der Waals density functional for general geometries. *Phys. Rev. Lett.* **92**, 246401 (2004).
52. Dion, M., Rydberg, H., Schröder, E., Langreth, D. C. & Lundqvist, B. I. Van der waals density functional for general geometries. *Phys. Rev. Lett.* **95**, 109902E (2005).
53. Langreth, C. *et al.* Van der Waals density functional theory with applications. *Int. J. Quantum Chem.* **101**, 599 (2005).

54. Lazić, P. *et al.* Density functional theory with nonlocal correlation: a key to the solution of the CO adsorption puzzle. *Phys. Rev. B* **81,** 045401 (2010).

55. Lazić, P. *et al.* Rationale for switching to nonlocal functionals in density functional theory. *J. Phys. Condens. Matter* **24,** 424215 (2012).

Acknowledgements

We acknowledge the financial support of the NSERC Discovery Grant, McGill University, FRQNT Doctoral Scholarship (C.M.), FRQNT Nouveaux Chercheurs Grant, FRQNT Centre for Green Chemistry and Catalysis (CCVC/CGCC), Ministry of Science, Education and Sport of the Republic of Croatia (Grant Nos. 098-0982915-2950, 098-0982904-2953). Mr Vitomir Stanišić (the Ruđer Bošković Institute) and Mr Jean-Philippe Guay (the McGill University) are acknowledged for the manufacture of grinding jars and Mr Hrvoje Dagelić (the Ruđer Bošković Institute) for programming. Professor Vladislav A. Blatov (the Samara State University) is acknowledged for his kind advice in characterization of the *katsenite* topology.

Author contributions

In situ experiments were performed by A.P., V.Š., C.M., P.A.J., K.U., S.A.J.K., O.M., I.H. and T.F. Raw data were analysed by A.P. and I.H. Crystal structure of *katsenite* was solved by I.H. P.L. performed crystal structure optimization. A.D.K. performed laboratory syntheses and topological analysis; C.M. performed solid-state NMR studies. M.-H.P. and T.-O.D. performed BET and SEM analyses. Figures were prepared by A.D.K., A.P., I.H. and T.F. All authors discussed the results. The manuscript was written by T.F. and I.H. with input from all authors.

Additional information

Accession codes: The X-ray crystallographic coordinates for the structure reported in this Article has been deposited at the Cambridge Crystallographic Data Centre (CCDC) under deposition number CCDC 989593. These data can be obtained free of charge from the CCDC via www.ccdc.cam.ac.uk/data_request/cif.

18

Multinuclear metal-binding ability of a carotene

Shinnosuke Horiuchi[1], Yuki Tachibana[1,2], Mitsuki Yamashita[1,3], Koji Yamamoto[1,3], Kohei Masai[1,3], Kohei Takase[2], Teruo Matsutani[2], Shiori Kawamata[1], Yuki Kurashige[4,5], Takeshi Yanai[4,5] & Tetsuro Murahashi[1,3,6]

Carotenes are naturally abundant unsaturated hydrocarbon pigments, and their fascinating physical and chemical properties have been studied intensively not only for better understanding of the roles in biological processes but also for the use in artificial chemical systems. However, their metal-binding ability has been virtually unexplored. Here we report that β-carotene has the ability to assemble and align ten metal atoms to afford decanuclear homo- and heterometal chain complexes. The metallo–carotenoid framework shows reversible metalation–demetalation reactivity with multiple metals, which allows us to control the size of metal chains as well as the heterobimetallic composition and arrangement of the carotene-supported metal chains.

[1] Research Center of Integrative Molecular Systems (CIMoS), Institute for Molecular Science, National Institutes of Natural Sciences, Myodaiji, Okazaki, Aichi 444-8787, Japan. [2] Department of Applied Chemistry, Graduate School of Engineering, Osaka University, Suita, Osaka 565-0871, Japan. [3] Department of Structural Molecular Science, The Graduate University for Advanced Studies, Myodaiji, Okazaki, Aichi 444-8787, Japan. [4] Department of Theoretical and Computational Molecular Science, Institute for Molecular Science, National Institutes of Natural Sciences, Myodaiji, Okazaki, Aichi 444-8585, Japan. [5] Department of Functional Molecular Science, The Graduate University for Advanced Studies, Myodaiji, Okazaki, Aichi 444-8585, Japan. [6] PRESTO, Japan Science and Technology Agency (JST), Myodaiji, Okazaki, Aichi 444-8787, Japan. Correspondence and requests for materials should be addressed to T.M. (email: mura@ims.ac.jp).

arotenoids are naturally abundant pigments containing extended π-conjugated C=C double-bond arrays. The fascinating physical and chemical properties of carotenoids, such as light-harvesting, photo-protective, antioxidative and conductive properties have been explored not only for better understanding of their roles in biological processes but also for the use in artificial chemical systems[1-4]. An attractive, although yet undeveloped function of carotenoids is their metal-binding ability. While several early reports showed that carotenoids bind mononuclear metal moieties through conventional 1,3-butadiene-type tetrahapto–π-coordination[5-9], the extended π-conjugated polyene moieties in their backbones may impart the ability to bind a large metal–metal (M-M)-bonded array through continuous multiple π-coordination bonds (Fig. 1). Such potential multidentate bridging π-coordination ability of carotenoids is highly attractive because of its usefulness in inorganic synthesis. Indeed, chemists have sought to use poly-nucleating multidentate ligands as the scaffolds or templates to control the metal assembly that provides methodology to synthesize molecularly well-defined metal clusters for catalysis and materials science[10-14]. However, the scaffold strategy has been applied mostly to the construction of metal clusters with small size owing to the difficulty to design practically useful large scaffold ligands that can assemble and then hold many metal atoms. In fact, only few synthetic scaffold ligands that can bind 10 or more metal atoms through multidentate bridging coordination have been developed. Recently, Peng and colleagues[15] developed synthetic multidentate N-donor ligands that lead to isolation of Ni[11] chain complexes. Shionoya and co-workers developed artificial metallo–DNA motifs that brought a method to construct a heterometal array of 10 metal ions, although direct M–M bonds are absent in the metal array[16]. While these previous studies used the σ-type scaffolds bearing multiple hetero-atom σ-binding sites, π-type extended π-conjugated unsaturated hydrocarbon scaffolds are also promising in light of their preferable rows of C=C π-binding sites at regular intervals comparable to M–M bond lengths[17-20]. Furthermore, relatively weak C=C π-coordination bond may cause dynamic metal binding at each C=C site, which would be a key for assembling many metal atoms in a convergent manner. Carotenes are the rational choice as the extended π-conjugated scaffold for many metal atoms, because carotenes are one of the most readily available extended unsaturated hydrocarbons containing more than ten π-conjugated C=C double bonds.

Herein, we report remarkable multinuclear metal-binding ability of β-carotene through synthesis and characterization of bis-(β-carotene) decanuclear metal chain complexes. The metallo–carotene framework shows interesting multinuclear metalation–demetalation reactivity, allowing us to construct heterobimetallic decanuclear chain. The bis-(β-carotene) decanuclear metal chain complexes are stable rod-like sandwich molecules, exhibiting parallel π–π stacking self-assembly in the crystalline state.

Results

Synthesis and structure of bis-(β-carotene) decanuclear Pd complexes. We examined the homoleptic carotene–metal systems in which all auxiliary ligands contained in the starting metal complexes are replaced by carotene to afford a sandwich-type multimetal-binding motif. Pd was used because of its feasibility to undergo convergent metal assembly with the aids of relatively weak metal–ligand and metal–metal bonds[21,22]. At first, we investigated the full metalation of the bis-carotene π-framework. The redox–condensation reaction of $[Pd_2(CH_3CN)_6][BF_4]_2$ (ref. 23) and excess $Pd_2(dba)_3 \cdot C_6H_6$ (ref. 24) in the presence of β-carotene at 60 °C gave the decanuclear palladium complex $[Pd_{10}(\mu_{10}\text{-}\beta\text{-carotene})_2][B(Ar^F)_4]_2$ (**1**, $B(Ar^F)_4 = B(3,5\text{-}(CF_3)_2 C_6H_3)_4$) in 33% yield as a mixture of two isomers (Fig. 2a, 1-**meso:1-rac** = 7:3). It is noted that use of $Pd_2(dba)_3 \cdot CHCl_3$ instead of $Pd_2(dba)_3 \cdot C_6H_6$ gave a complicated mixture from which **1** was not obtained. The major isomer (**1-meso**) is poorly soluble in CH_3CN, whereas another isomer (**1-rac**) is soluble, enabling us to separate the two isomers. The molecular structures of thus obtained yellow complexes **1-meso** and **1-rac** were determined by X-ray crystallographic analysis (Fig. 3). Remarkably, two β-carotene molecules flank an array of 10 Pd atoms through unprecedentedly large μ_{10}-bridging π-coordination. The two β-carotene ligands are stacking in an eclipsed form in **1-meso** and in a staggered one in **1-rac**. The nine Pd–Pd bond lengths (2.5827(7)–2.7172(6) Å in **1-meso**; 2.6010(11)–2.7111(11) Å in **1-rac**) are shorter than that of bulk Pd (2.76 Å), indicating that 10 Pd atoms in **1-meso** or **1-rac** are connected through Pd–Pd bonds (Supplementary Table 1). The calculated indices of Mayer bond order for each Pd–Pd are ranged from 0.23 to 0.14 for **1-meso**; cf., for a typical Pd–Pd-bonded complex, $Pd_2Cl_2(PH_3)_4$, the index of Meyer bond order for Pd–Pd was calculated to be 0.60. The β-carotene ligands in **1-meso** showed reduced and inversed C=C/C–C bond length alternation, that is, the long/short alternation (1.46/1.41 Å) was found for the inner nonaene substructure in the β-carotene ligands in **1-meso**, being in between the C–C bond lengths in ethane (1.54 Å) and ethylene (1.34 Å; cf. 1.33/1.47 Å for the bond length alternation of free β-carotene; Fig. 3e). It is noted that the bis-β-carotene Pd_{10} chain dication of **1-meso** formed infinite intermolecular π–π stacking columns (the shortest intermolecular C · · · C distance is 3.51 Å) in the crystalline state (Fig. 3f,g). Such intermolecular backbone π–π stacking represents the typical property of the planar π-conjugated system. The dication of **1-rac** formed the π–π stacking dimer instead of the infinite column in the crystalline state. The ^1H and $^{13}C\{^1H\}$ nuclear magnetic resonance (NMR) spectra of **1-meso** and **1-rac** in CD_2Cl_2 showed that all olefinic proton and carbon resonances of the β-carotene ligands appeared at the high-field region (olefinic moieties appeared at $\delta = 3.5$–2.6 ppm for ^1H; $\delta = 111$–69 ppm for ^{13}C), being consistent with the solid state structures determined by X-ray crystallography. The decanuclear sandwich complexes **1-meso** and **1-rac** are stable in solution even in the aerobic condition. Thus, it has been proven that bis-β-carotene

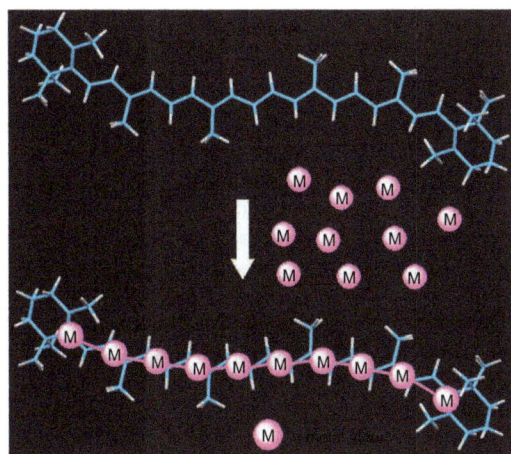

Figure 1 | Schematic representation of multinuclear metal binding by β-carotene. Many metal atoms are assembled on the π-conjugated plane of β-carotene.

a

b

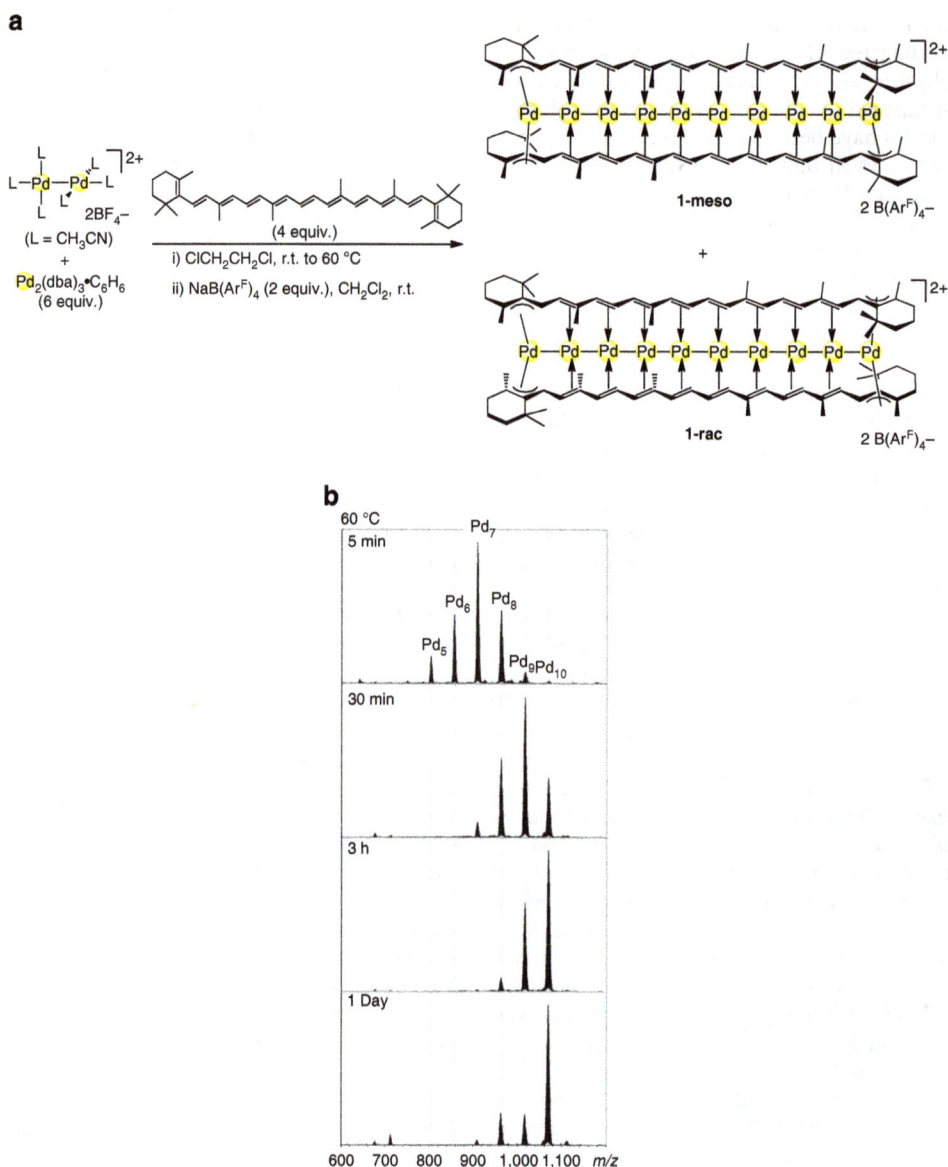

Figure 2 | Synthesis of bis-(β-carotene) Pd₁₀ chain complexes. (a) Synthesis of $[Pd_{10}(\mu_{10}\text{-}\beta\text{-carotene})_2][B(Ar^F)_4]_2$ (**1**), (**b**) ESI-MS monitoring of the formation of **1** at 60 °C showing the formation of metal-deficient intermediates $[Pd_n(\beta\text{-carotene})_2]^{2+}$ ($n = 5, 6, 7, 8$ and 9).

π-framework can accommodate 10 Pd atoms array through remarkable multidentate bridging π-coordination. The decanuclear complexes **1-meso** and **1-rac** are the soluble and isolable organometallic clusters having a long metal chain. Existence of long inorganic palladium wires in solution was recently reported, where di- or tetranuclear palladium units are self-assembled through Pd–Pd interactions[25,26]. The extended (π-carbon framework)–(metal clusters) contact found in **1-meso** and **1-rac** may be related with the interface structure of the sp²-carbon material and metal clusters that is of current interests in materials science and catalysis[27,28].

Synthesis and structure of bis-(β-carotene) decanuclear PdPt complexes. To further explore the metal-binding ability of β-carotene, we next examined the binding of bimetallic chains by bis-β-carotene π-framework. It was difficult to obtain a single bimetallic chain product simply by using Pt₂(dba)₃ together with Pd₂(dba)₃ in the synthetic reaction. We then thought a stepwise synthesis, that is, if metal-deficient bis-β-carotene complexes $[Pd_n(\beta\text{-carotene})_2]^{2+}$ ($n \leq 9$) can be selectively constructed with

Pd, subsequent incorporation of Pt may give bimetallic chains. The metal-deficient dications $[Pd_n(\beta\text{-carotene})_2]^{2+}$ ($5 \leq n \leq 9$) were indeed observed when the formation of **1** (Fig. 2a) was monitored by electrospray ionization mass spectroscopy (ESI-MS; Fig. 2b). Upon mixing the starting materials at ambient temperature, Pd₆ and Pd₇ complexes of β-carotene were detected as the major MS-detectable species after 3 h. Relatively small MS signals for Pd₅, Pd₈ and Pd₉ complexes were also detected. Further incorporation of Pd into the bis-β-carotene framework proceeded gradually but was incomplete at ambient temperature after 2 days, resulting in that Pd₇ and Pd₈ complexes were the major MS-detectable species. Heating at 60 °C resulted in shift of the distribution of products to higher nuclearity species, eventually affording the Pd₁₀ chain complexes as the major MS-detectable product. However, it has been difficult to isolate and characterize each of metal-deficient products from the reaction mixtures of the build-up reaction. We confirmed the existence of regioisomers for a short chain model, that is, $[Pd_2(1,10\text{-diphenylpentaene})_2][B(Ar^F)_4]_2$, which was obtained by the reaction of $[Pd_2(1,4\text{-diphenyl-1,3-butadiene})_2][BF_4]_2$ (ref. 29) with 1,10-diphenyl-1,3,5,7,9-decapentaene at room temperature

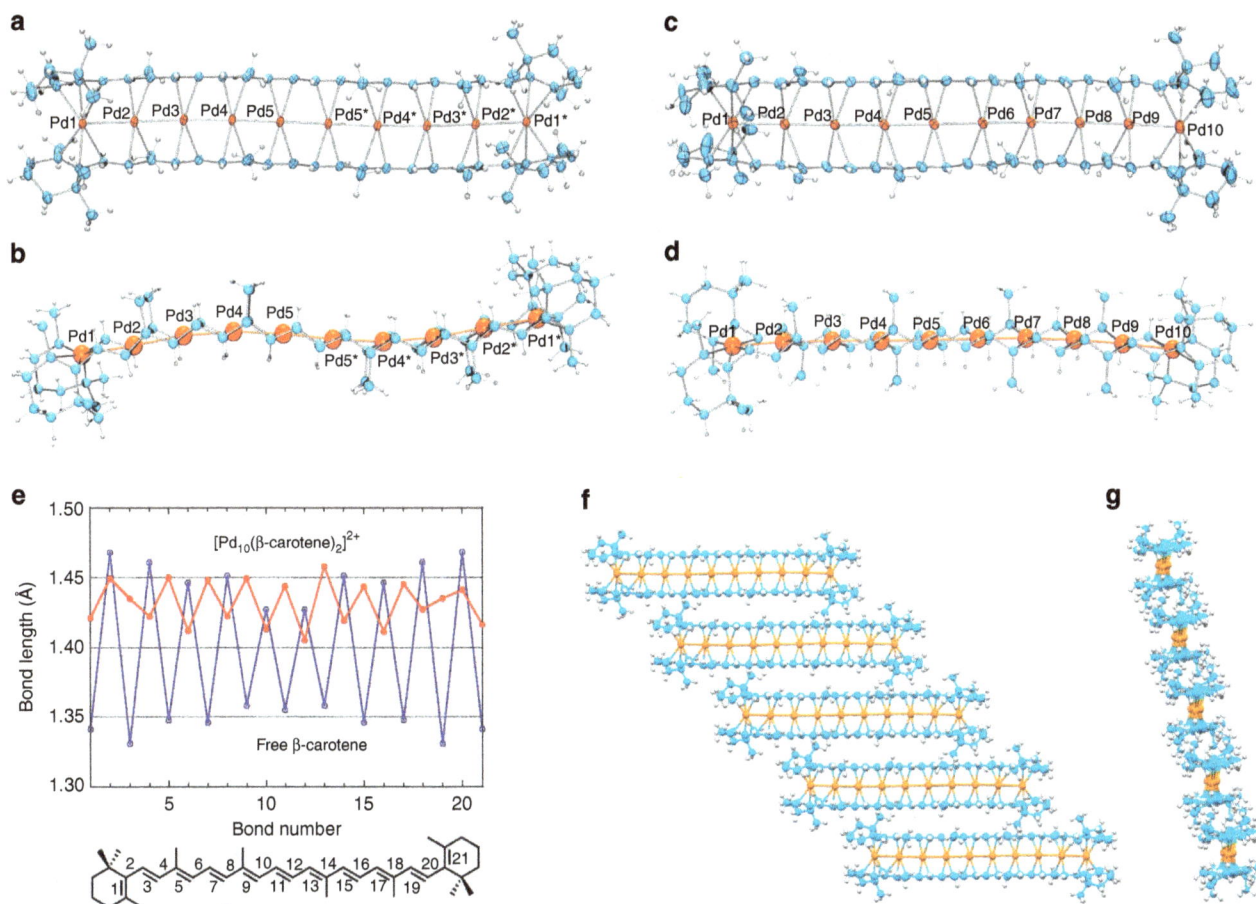

Figure 3 | Structures of bis-(β-carotene) Pd₁₀ chain complexes. (**a**) Thermal ellipsoid (50%) drawing of *meso*-[Pd₁₀(μ₁₀-β-carotene)₂][B(Ar^F)₄]₂ (**1-meso**). (**b**) Ball-stick drawing of **1-meso**. (**c**) Thermal ellipsoid (30%) drawing of *rac*-[Pd₁₀(μ₁₀-β-carotene)₂][B(Ar^F)₄]₂ (**1-rac**). (**d**) Ball-stick drawing of **1-rac**. (**e**) C–C bond lengths in **1-meso** and free β-carotene (CCDC-253816), determined by X-ray structural analyses. (**f**) A view of a part of an intermolecular backbone π–π stacking column of **1-meso** in the crystalline state. (**g**) A side view of a part of the π–π stacking column of **1-meso**. For **a-d,f,g**, B(Ar^F)₄ anions and non-coordinating solvent molecules were omitted for clarity.

(r.t.), contained four regioisomers (57:27:10:6) as shown in Supplementary Fig. 1.

We then found that the metal-deficient complexes of β-carotene can be obtained as a single product by demetalation from the Pd₁₀ complex **1-meso** with CO. Thus, the reaction of **1** with CO (1 atmosphere (atm)) at 5 °C for 1 day afforded [Pd₅(β-carotene)₂][B(Ar^F)₄]₂ (**2-meso**) in 74% yield, together with a significant amount of Pd black (Fig. 4a). ¹³C{¹H} NMR analysis as well as the single-crystal X-ray structure analysis of **2-meso** showed that the Pd₅ chain occupied the half-part of the bis-β-carotene framework (Fig. 5a). The ESI-MS monitoring experiments on the demetalation reaction with CO (1 atm) at 0 °C showed that the starting Pd₁₀ complex **1-meso** and the half-filled Pd₅ complex **2-meso** were present as the major MS-detectable species during the reaction (Fig. 4b). The prolonged reactions for 1 week resulted in gradual increase of the MS signal for the Pd₄ complex. The Pd₇ complex [Pd₇(β-carotene)₂][B(Ar^F)₄]₂ (**3-meso**) was also obtained by exposing **1-meso** to CO (1 atm) at 30 °C for 3 h in 20% yield (Fig. 4a). During this reaction, the major ESI-MS-detectable species were the starting Pd₁₀ complex and the Pd₇ complex (Supplementary Fig. 2), while the prolonged reactions resulted in further demetalation. The ¹³C{¹H} NMR analysis as well as the single-crystal X-ray structure analysis (Fig. 5b) showed that the Pd₇ chain was located in the bis-β-carotene framework. The demetalation of **3-meso** with CO (1 atm) at 0 °C occurred rapidly to afford **2-meso** with complete consumption of **3-meso** within 15 min. Thus, the pseudo-

superposed β-carotene stacking structure was preserved during the demetalation under a CO atmosphere, giving a single regioisomer of metal-deficient sandwich. The results of the ESI-MS monitoring experiments suggested that the loss of a Pd⁰ atom from the Pd₁₀ complex and from the Pd₅ or Pd₇ complex are relatively slow. The loss of Pd⁰ likely occurs from one end of the Pd chain, while it is not easy to explain the reason why the demetalation almost stopped at the Pd₅ species or the Pd₇ species in each reaction condition. There may be several factors that affect aggregation and dissociation of metals and organic ligands (for example, M–CO affinity³⁰, M–carotene bond dissociation and M–M bond dissociation). In the case of associative ligand exchange, the relatively slow release of Pd⁰ from the filled Pd₁₀ complex is probably due to the lower accessibility of the terminal Pd atoms, which are sterically hindered by the bulky terminal β-groups of the β-carotene ligands, to CO. Consistently, demetalation from the Pd₇ complex, where one end of the Pd₇ chain is less hindered by ligands, occurred much faster at 0 °C than that from the Pd₁₀ complex.

We next confirmed that metal-refilling reaction from isolated metal-deficient complexes proceeds smoothly by using the Pd₅ complex **2-meso**, that is, addition of Pd₂(dba)₃·C₆H₆ to **2-meso** in C₂D₄Cl₂ at 60 °C afforded the Pd₁₀ complex **1-meso** (41% yield). We then tested whether metal refilling of **2-meso** with Pt⁰ is possible. Thus, the bimetallic Pd₅Pt₃ chain complex [Pd₅Pt₃(β-carotene)₂][B(Ar^F)₄]₂ (**4-meso**) was formed by treatment of **2-meso** with Pt₂(dba)₃·CHCl₃ in the presence of ethylene (1 atm)

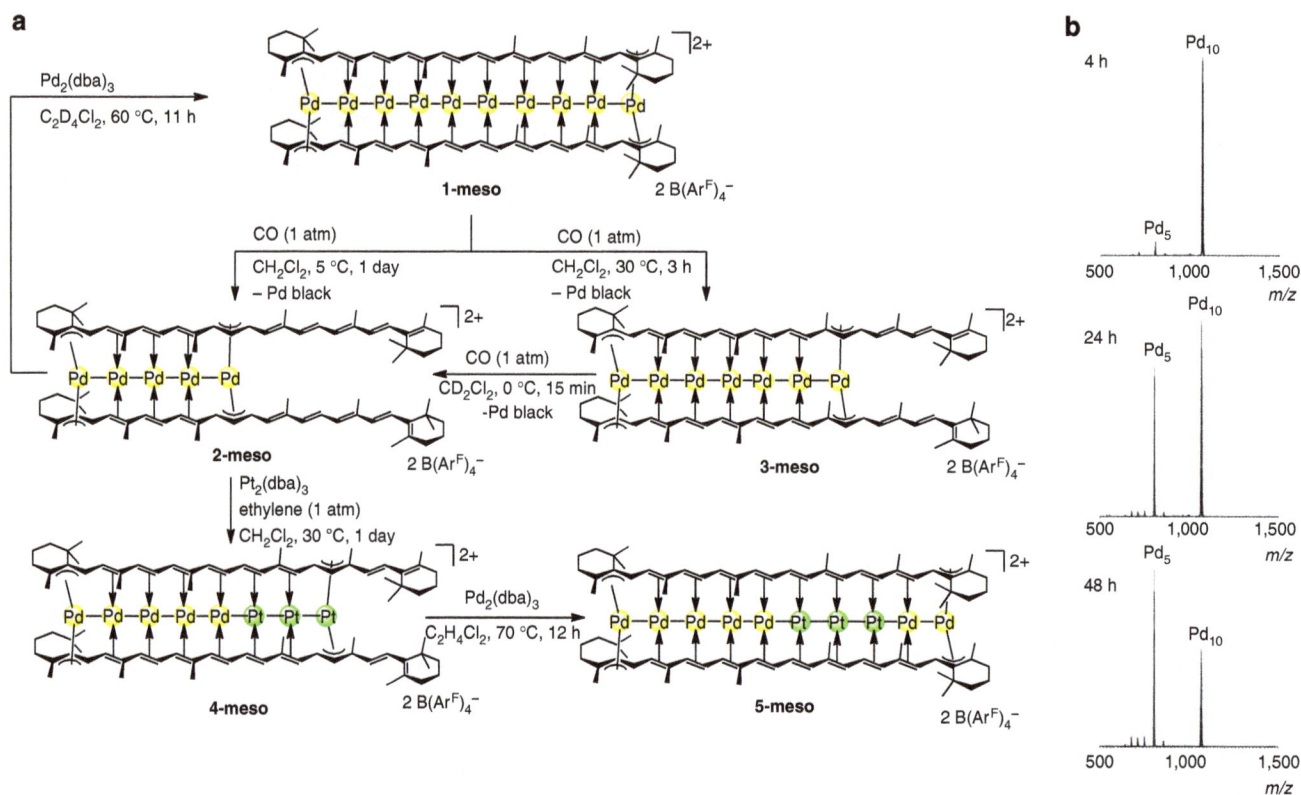

Figure 4 | Synthesis of bimetallic PdPt chain sandwich complexes of β-carotene. (a) Demetalation and metalation of bis-β-carotene framework. Demetalation of **1-meso** with CO afforded the metal-deficient complex $[Pd_5(\mu_5\text{-}\beta\text{-carotene})_2][B(Ar^F)_4]_2$ (**2-meso**) or $[Pd_7(\mu_7\text{-}\beta\text{-carotene})_2][B(Ar^F)_4]_2$ (**3-meso**). Subsequent metalation of **2-meso** with Pt^0 and then with Pd^0 gave a mixed metal complex $[Pd_5Pt_3(\mu_8\text{-}\beta\text{-carotene})_2][B(Ar^F)_4]_2$ (**4-meso**) and $[Pd_5Pt_3Pd_2(\mu_{10}\text{-}\beta\text{-carotene})_2][B(Ar^F)_4]_2$ (**5-meso**), respectively. **(b)** ESI-MS monitoring of the demetalation from **1-meso** at 0 °C under CO (1 atm) atmosphere.

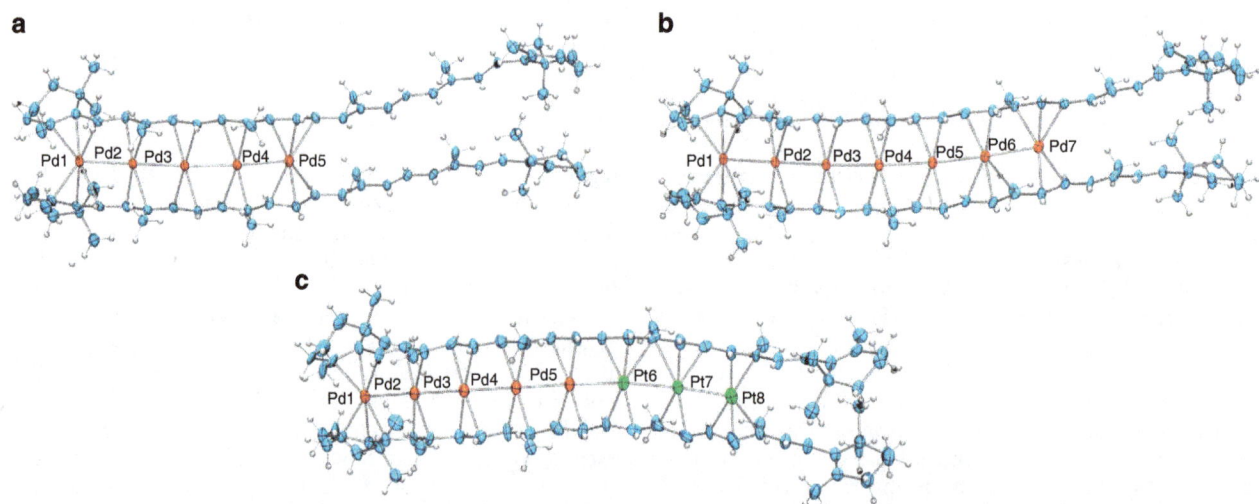

Figure 5 | Structures of metal-deficient sandwich complexes of β-carotene. (a) Thermal ellipsoid (30%) drawing of $[Pd_5(\mu_5\text{-}\beta\text{-carotene})_2][B(Ar^F)_4]_2$ (**2-meso**). **(b)** Thermal ellipsoid (30%) drawing of $[Pd_7(\mu_7\text{-}\beta\text{-carotene})_2][B(Ar^F)_4]_2$ (**3-meso**). **(c)** Thermal ellipsoid (30%) drawing of $[Pd_5Pt_3(\mu_8\text{-}\beta\text{-carotene})_2][B(Ar^F)_4]_2$ (**4-meso**). For **a**–**c**, $B(Ar^F)_4$ anions and non-coordinating solvent molecules were omitted for clarity. The coordination modes of two β-carotene ligands in the crystalline **3-meso** or **4-meso** are slightly different. Furthermore, in the crystal structures of the metal-deficient complexes, **2-meso**, **3-meso** and **4-meso**, the uncoordinated cyclohexenyl group are not superposed due to the rotation at C6–C7 bond. The NMR spectra of **2-meso**, **3-meso** or **4-meso** showed only a single set of β-carotene signals.

at 30 °C for 1 day (59% yield; Fig. 4a). In the absence of ethylene, the metalation with Pt$_2$(dba)$_3$ did not proceed at the present condition. Addition of ethylene generated Pt–ethylene complexes *in situ*, which might be more reactive and soluble than Pt$_2$(dba)$_3$

(ref. 31). The Pd$_5$ − Pt$_3$ mixed metal arrangement in **4-meso** was confirmed by X-ray crystallographic analysis (Fig. 5c). The Pd$_5$ chain (Pd–Pd = 2.731(2)–2.629(2) Å) and the Pt$_3$ chain (Pt–Pt = 2.6612(9) and 2.6864(10) Å) are connected through Pd–Pt bond (2.657(2) Å). The Pd_5Pt_3 arrangement was also confirmed by ^{13}C NMR analyses in CD_2Cl_2 where only one set of β-carotene signals was observed at 25 °C, and ^{13}C signals for Pt-bound carbons appeared at relatively higher field compared with those for Pd-bound carbons. Further metalation of **4-meso** with $Pd_2(dba)_3 \cdot C_6H_6$ at 70 °C gave decanuclear bimetallic chain complex $[Pd_5Pt_3Pd_2(\beta\text{-carotene})_2][B(Ar^F)_4]_2$ (**5-meso**) (Fig. 4a). The alternative metal arrangement, Pd_5–Pt_3–Pd_2 was confirmed by assignment of the upfield shifted Pt-bound carbons of the β-carotene ligands in $^{13}C\{^1H\}$ NMR analyses in CD_2Cl_2, that is, the substantial upfield shifts of the Pt-bound carbons of the β-carotene ligands in **5-meso** relative to those in **1-meso** ($\Delta\delta = 8$–13 ppm) were observed, while the chemical shifts of the Pd-bound carbons in **5-meso** are similar to those in **1-meso** ($\Delta\delta = 0$–2 ppm). Thus, β-carotene has the ability to bind bimetallic decanuclear chain, where stepwise demetalation-metalation sequence is useful in controlling the bimetal arrangement. Such reversible accommodation/liberation of multinuclear metal atoms has rarely been attained in metal cluster chemistry[32], representing a facile dynamic metal-binding feature derived from weakly coordinating olefin π-coordination[33,34] as well as M–M bonds in organometallic sandwich frameworks[35].

Absorption spectra of bis-(β-carotene) Pd complexes.

It is noted that $[Pd_n(\beta\text{-carotene})_2][B(Ar^F)_4]_2$ showed a nuclearity-dependent absorption profile, that is, the red-shift of maximum absorption bands was observed according to decrease of the number of Pd atoms (362 nm for **1-meso**, 373 nm for **3-meso** and 468 nm for **2-meso**) (Supplementary Fig. 3). The absorption spectra were well described by the time-dependent density functional theory calculations at the Coulomb-attenuated B3LYP level[36], suggesting that (i) the absorption bands originate mainly from the ligand-to-metal charge transfer (Supplementary Fig. 4 and Supplementary Table 2) and (ii) the observed red-shift with decreasing number of Pd atoms reflects the increasing stabilization of lower-lying unoccupied molecular orbitals that have antibonding character for $d\sigma(M)$-$d\sigma(M)$ and $d(M)$-$p(L)$ orbital interactions (Supplementary Figs 5–7 and Supplementary Tables 3–5). Visible-light irradiation resulted in the formation of **1** without heating, that is, the reaction similar to Fig. 2a with visible light irradiation (Xenon lamp, > 385 nm) yielded **1** at 20 °C in 24% yield (**1-meso:1-rac** = 7:3) (Supplementary Fig. 8 and Supplementary Methods).

Discussion

In this report, it has been proven that β-carotene, a naturally abundant and readily available unsaturated hydrocarbon pigment, has the ability to bind decanuclear homo- and heterometal chains through unprecedentedly large μ_{10}-bridging π-coordination. The present results showed that natural extended π-conjugated unsaturated hydrocarbons can be utilized as the multidentate π-scaffolds for the construction of giant metal clusters. Future studies will focus on the physical and chemical properties of the rod-like bis-carotene decametal chain sandwich complexes, such as self-assembling behaviour, multielectron redox behaviour and charge mobility.

Methods

Synthesis and characterization of compounds. All manipulations were conducted under a nitrogen atmosphere using standard Schlenk or drybox techniques. The β-carotene–metal complexes were characterized by elemental analyses, ESI-MS analyses and NMR. The assignment of each resonance in NMR analysis was made with aid of heteronuclear single-quantum correlation (HSQC) or heteronuclear multiple-quantum correlation (HMQC) and heteronuclear multiple-bond correlation (HMBC) techniques. Furthermore, the five complexes (**1-meso**, **1-rac**, **2-meso**, **3-meso** and **4-meso**) were structurally determined by X-ray crystallographic analyses (Supplementary Data 1–5).

Synthesis of $[Pd_{10}(\beta\text{-carotene})_2][B(Ar^F)_4]_2$ (1-meso) and (1-rac). To a suspension of β-carotene (679 mg, 1.27 mmol) in $ClCH_2CH_2Cl$ (200 ml) were added $Pd_2(dba)_3 \cdot (C_6H_6)$ (1.89 g, 1.90 mmol) and $[Pd_2(CH_3CN)_6][BF_4]_2$ (200 mg, 0.316 mmol) at r.t. The reaction mixture was stirred under nitrogen atmosphere at 60 °C for 1 day. The reaction mixture was filtered and the filtrate was dried *in vacuo*. The obtained brown powder and $NaB(Ar^F)_4$ (560 mg, 0.632 mmol) was added to CH_2Cl_2, and the mixture was stirred for 5 min at r.t. Et_2O was added to the solution and the mixture was filtered. The filtrate was dried *in vacuo* to yield a red powder. After washing with CH_3CN, $[Pd_{10}(\beta\text{-carotene})_2][B(Ar^F)_4]_2$ (**1-meso**) was isolated as an yellow powder (290 mg, 24%). $[Pd_{10}(\beta\text{-carotene})_2][B(Ar^F)_4]_2$ (**1-rac**) was obtained by recrystallization from the CH_3CN solution (107 mg, 9%). For **1-meso**: ^1H NMR (400 MHz, CD_2Cl_2, 25 °C): δ –0.27 (s, 12H), –0.15 (s, 12H), 0.33 (s, 12H), 1.54 (s, 12H), 1.64 (m, 4H), 1.78 (m, 4H), 1.99 (m, 4H), 2.08 (m, 4H), 2.10 (s, 12H), 2.65 (d, $J = 12$ Hz, 4H), 2.66 (d, $J = 12$ Hz, 4H), 2.86 (dd, $J = 3$ Hz, $J = 9$ Hz, 4H), 2.95 (d, $J = 12$ Hz, 4H), 3.04–3.11 (m, 8H), 3.28–3.45 (m, 8H), 3.47 (t, $J = 12$ Hz, 4H), 7.50 (s, 8H, p-$B(Ar^F)_4$), 7.65 (s, 16H, o-$B(Ar^F)_4$). ^{13}C NMR (100 MHz, CD_2Cl_2, 25 °C): δ 14.3, 15.0, 20.5, 23.7, 28.1, 31.8, 34.9, 36.8, 43.5, 74.1, 74.7, 80.7, 82.1, 83.0, 83.3, 84.5, 98.0, 103.3, 109.7, 110.1, 117.8 (p-$B(Ar^F)_4$), 125.4 (CF_3-$B(Ar^F)_4$), 129.2 (m-$B(Ar^F)_4$), 135.1 (o-$B(Ar^F)_4$), 162.1 (*ipso*-$B(Ar^F)_4$). MS (ESI) m/z calcd. for $[C_{80}H_{112}Pd_{10}]^{2+}$: 1,068.9601, found: 1,068.9647. Anal. calcd. For $C_{144}H_{136}B_2F_{48}Pd_{10}$: C, 44.76; H, 3.55, found: C, 44.77; H, 3.73. A single crystal suitable for X-ray crystallographic analysis was grown from a CH_2Cl_2-Toluene solution (Supplementary Methods). For **1-rac**: ^1H NMR (400 MHz, CD_2Cl_2, 25 °C): δ –0.72 (s, 12H), 0.20 (s, 12H), 0.38 (s, 12H), 1.40 (m, 4H), 1.70 (m, 4H), 1.73 (s, 12H), 1.80 (m, 4H), 1.83 (s, 12H), 2.08 (m, 4H), 2.77 (d, $J = 12$ Hz, 4H), 2.80 (d, $J = 12$ Hz, 4H), 2.85 (d, $J = 12$ Hz, 4H), 3.03 (d, $J = 12$ Hz, 4H), 3.07 (dd, $J = 3$ Hz, $J = 9$ Hz, 4H), 3.33 (m, 4H), 3.34 (t, $J = 12$ Hz, 4H), 3.46 (dd, $J = 3$ Hz, $J = 9$ Hz, 4H), 3.52 (m, 4H), 7.30 (s, 8H, p-$B(Ar^F)_4$), 7.65 (s, 16H, o-$B(Ar^F)_4$). ^{13}C NMR (100 MHz, CD_2Cl_2, 25 °C): δ 13.7, 13.8, 20.0, 24.4, 29.4, 29.7, 34.4, 36.5, 43.1, 69.3, 76.5, 79.5, 84.0, 87.5, 88.1, 89.3, 95.0, 96.1, 99.4, 108.1, 117.8 (p-$B(Ar^F)_4$), 125.4 (CF_3-$B(Ar^F)_4$), 129.2 (m-$B(Ar^F)_4$), 135.1 (o-$B(Ar^F)_4$), 162.1 (*ipso*-$B(Ar^F)_4$). MS (ESI) m/z calcd. for $[C_{80}H_{112}Pd_{10}]^{2+}$: 1,068.9601, found: 1,068.9432. Anal. calcd. For $C_{144}H_{136}B_2F_{48}Pd_{10} \cdot C_6H_{14}$: C, 45.60; H, 3.83, found: C, 45.48; H, 3.99. A single crystal suitable for X-ray crystallographic analysis was grown from a diethylether – hexane solution (Supplementary Methods).

Synthesis of $[Pd_5(\beta\text{-carotene})_2][B(Ar^F)_4]_2$ (2-meso). CO gas (1 atm) was bubbled in a CH_2Cl_2 solution (220 ml) of $[Pd_{10}(\beta\text{-carotene})_2][B(Ar^F)_4]_2$ (**1-meso**) (340 mg, 88.0 μmol) at 5 °C for 24 h. The reaction mixture was filtered and the filtrate was dried *in vacuo* to give a dark brown powder. After extraction with CH_3CN, the volatiles were removed in vacuo to give $[Pd_5(\beta\text{-carotene})_2][B(Ar^F)_4]_2$ (**2-meso**) as a dark brown powder (218 mg, 74%). ^1H NMR (400 MHz, CD_2Cl_2, 25 °C): δ –0.29 (s, 6H), 0.05 (s, 6H), 0.90 (s, 6H), 1.05 (s, 6H), 1.06 (s, 6H), 1.47 (m, 4H), 1.60 (s, 6H), 1.60–1.68 (m, 8H), 1.74 (s, 6H), 1.92–2.08 (m, 6H), 1.94 (s, 6H), 1.98 (s, 6H), 2.02 (d, $J = 12$ Hz, 2H), 2.06 (s, 6H), 2.12 (m, 2H), 2.39 (d, $J = 12$ Hz, 2H), 2.52 (d, $J = 12$ Hz, 2H), 2.92 (t, $J = 12$ Hz, 2H), 2.96 (t, $J = 12$ Hz, 2H), 3.3 (m, 4H), 3.67 (t, $J = 12$ Hz, 2H), 5.13 (t, $J = 12$ Hz, 2H), 5.85 (t, $J = 12$ Hz, 2H), 6.10 (d, $J = 12$ Hz, 2H), 6.15 (d, $J = 16$ Hz, 2H), 6.22 (d, $J = 12$ Hz, 2H), 6.26 (d, $J = 16$ Hz, 2H), 6.29 (d, $J = 16$ Hz, 2H), 6.91 (dd, $J = 12$ Hz, $J = 16$ Hz, 2H), 7.49 (s, 8H, p-$B(Ar^F)_4$), 7.64 (s, 16H, o-$B(Ar^F)_4$). ^{13}C NMR (100 MHz, CD_2Cl_2, 25 °C): δ 13.0, 13.3, 14.0, 14.2, 19.6, 19.9, 22.1, 24.0, 28.0, 29.2, 29.4, 31.6, 33.7, 34.6, 35.0, 36.8, 40.1, 42.7, 76.8, 78.9, 79.7, 89.2, 89.4, 89.6, 91.8, 92.5, 103.5, 111.8, 112.3, 117.8 (p-$B(Ar^F)_4$), 123.4, 125.5 (CF_3-$B(Ar^F)_4$), 126.7, 129.2 (m-$B(Ar^F)_4$), 129.3, 129.9, 130.4, 130.6, 135.1 (o-$B(Ar^F)_4$), 135.7, 137.6, 138.0, 140.3, 143.9, 162.1 (*ipso*-$B(Ar^F)_4$). MS (ESI) m/z calcd. for $[C_{80}H_{112}Pd_5]^{2+}$: 802.6996, found: 802.7014. Anal. calcd. For $C_{144}H_{136}B_2F_{48}Pd_5 \cdot C_6H_6$: C, 52.83; H, 4.20, found: C, 52.65; H, 4.35. A single crystal suitable for X-ray crystallographic analysis was grown from a diethylether–benzene solution (Supplementary Methods).

Synthesis of $[Pd_7(\beta\text{-carotene})_2][B(Ar^F)_4]_2$ (3-meso). CO gas (1 atm) was bubbled in a CH_2Cl_2 solution (100 ml) of $[Pd_{10}(\beta\text{-carotene})_2][B(Ar^F)_4]_2$ (**1-meso**) (107 mg, 27.8 μmol) at 30 °C for 3 h in the dark. The reaction mixture was filtered and the filtrate was dried *in vacuo* to give a dark red powder. The resultant powder was washed with CH_3CN and dried in vacuo. Et_2O was added and the mixture was filtered. The filtrate was dried *in vacuo* to give $[Pd_7(\beta\text{-carotene})_2][B(Ar^F)_4]_2$ (**3-meso**) as a red powder (20.0 mg, 20%). ^1H NMR (400 MHz, CD_2Cl_2, 25 °C): δ –0.27 (s, 6H), –0.04 (s, 6H), 0.07 (s, 6H), 0.67 (s, 6H), 1.12 (s, 6H), 1.13 (s, 6H), 1.51 (m, 4H), 1.59 (s, 6H), 1.61–1.70 (m, 6H), 1.79 (s, 6H), 1.97 (m, 2H), 2.06 (s, 6H), 2.06–2.15 (m, 8H), 2.23 (d, $J = 12$ Hz, 2H), 2.33 (d, $J = 12$ Hz, 2H), 2.34 (s, 6H), 2.40 (t, $J = 12$ Hz, 2H), 2.79 (d, $J = 12$ Hz, 2H), 2.85 (d, $J = 12$ Hz, 2H), 3.03 (d, $J = 12$ Hz, 2H), 3.27 (d, $J = 12$ Hz, 2H), 3.34 (m, 4H), 3.56 (t, $J = 12$ Hz, 2H), 3.65 (d, $J = 12$ Hz, 2H), 4.50 (d, $J = 12$ Hz, 2H), 6.08 (d, $J = 12$ Hz, 2H), 6.28 (d, $J = 16$ Hz, 2H), 6.44 (d, $J = 16$ Hz, 2H), 6.46 (t, $J = 12$ Hz, 2H), 7.48 (s, 8H, p-$B(Ar^F)_4$), 7.63 (s, 16H, o-$B(Ar^F)_4$). ^{13}C NMR (100 MHz, CD_2Cl_2, 25 °C): δ 13.6, 14.4, 14.6, 14.7, 19.6, 20.2, 22.2, 23.9, 28.0, 29.3, 29.5, 31.7, 34.0, 34.5, 34.9, 36.7, 40.2, 43.0, 75.5, 76.8, 79.0, 80.3, 80.6, 86.7, 87.9, 88.8, 89.2, 95.5, 96.1, 96.4, 100.9, 103.5, 111.2, 117.1, 117.8 (p-$B(Ar^F)_4$), 125.5 (CF_3-$B(Ar^F)_4$), 127.2, 129.2 (m-$B(Ar^F)_4$), 129.4, 131.5, 135.1 (o-$B(Ar^F)_4$), 136.6, 137.6, 140.4, 162.1 (*ipso*-$B(Ar^F)_4$). MS (ESI) m/z calcd. for $[C_{80}H_{112}Pd_7]^{2+}$: 909.1038, found: 909.1082. Anal. calcd. For $C_{144}H_{136}B_2F_{48}Pd_7 \cdot (C_6H_6)_2$: C, 50.62; H, 4.03, found: C, 50.73; H, 4.06. A single crystal suitable for X-ray crystallographic analysis was grown from a diethylether–benzene solution (Supplementary Methods).

Synthesis of [Pd$_5$Pt$_3$(β-carotene)$_2$][B(Ar^F)$_4$]$_2$ (4-meso).

Ethylene gas (1 atm) was bubbled in a CH$_2$Cl$_2$ solution (30 ml) of [Pd$_5$(β-carotene)$_2$][B(Ar^F)$_4$]$_2$ (2-meso) (54.0 mg, 16.2 μmol) and Pt$_2$(dba)$_3$·(CHCl$_3$) (100 mg, 82.5 μmol) at r.t. for 5 min. After the solution was stirred at 30 °C for 1 day, the colour turned dark brown. The reaction mixture was dried *in vacuo* to give a dark brown powder. After extraction with Et$_2$O and washing with C$_6$H$_6$, [Pd$_5$Pt$_3$(β-carotene)$_2$][B(Ar^F)$_4$]$_2$ (4-meso) was isolated as a yellow solid (37.6 mg, 59%). ^{1}H NMR (400 MHz, CD$_2$Cl$_2$, 25 °C): δ −0.28 (s, 6H), −0.11 (s, 6H), 0.50 (s, 12H), 1.17 (s, 6H), 1.21 (s, 6H), 1.56 (s, 6H), 1.56–1.59 (m, 4H), 1.60–1.72 (m, 8H), 1.84 (s, 6H), 1.98 (m, 2H), 2.01 (s, 6H), 2.07 (s, 6H), 2.09 (m, 2H), 2.13–2.18 (m, 4H), 2.34 (d, J = 12 Hz, 2H), 2.36 (d, J = 12 Hz, 2H), 2.43 (t, J = 12 Hz, 2H), 2.71 (d, J = 12 Hz, 2H), 2.76 (d, J = 14 Hz, 2H), 2.81 (d, J = 12 Hz, 2H), 3.00 (t, J = 11 Hz, 2H), 3.04 (d, J = 12 Hz, 2H), 3.11 (d, J = 12 Hz, 2H), 3.3 (m, 4H), 3.40 (m, 2H), 3.48 (t, J = 12 Hz, 2H), 3.79 (d, J = 11 Hz, 2H), 6.47 (d, J = 16 Hz, 2H), 6.68 (d, J = 16 Hz, 2H), 7.49 (s, 8H, *p*-B(Ar^F)$_4$), 7.65 (s, 16H, *o*-B(Ar^F)$_4$). ^{13}C NMR (100 MHz, CD$_2$Cl$_2$, 25 °C): δ 14.3, 14.6, 15.5, 17.5, 19.7, 20.3, 22.2, 23.8, 28.0, 29.2, 29.7, 31.7, 33.9, 34.7, 34.9, 36.8, 40.1, 43.2, 64.4, 67.3, 74.9, 75.1, 75.9, 78.1, 78.7, 79.9, 80.6, 85.2, 85.7, 87.6, 96.9, 98.4, 98.6, 103.5, 110.6, 114.4, 117.8 (*p*-B(Ar^F)$_4$), 125.5 (*CF$_3$*-B(Ar^F)$_4$), 129.1, 129.2 (*m*-B(Ar^F)$_4$), 132.3, 135.1 (*o*-B(Ar^F)$_4$), 136.0, 136.5, 162.1 (*ipso*-B(Ar^F)$_4$). MS (ESI) *m/z* calcd. for [C$_{80}$H$_{112}$Pd$_5$Pt$_3$]$^{2+}$: 1,095.1460, found: 1095.1291. Anal. calcd. For [C$_{144}$H$_{136}$B$_2$F$_{48}$Pd$_5$Pt$_3$·(C$_6$H$_6$)$_2$: C, 45.99; H, 3.66, found: C, 46.10; H, 3.86. A single crystal suitable for X-ray crystallographic analysis was grown from a dichloromethane–benzene solution (Supplementary Methods).

Synthesis of [Pd$_5$Pt$_3$Pd$_2$(β-carotene)$_2$][B(Ar^F)$_4$]$_2$ (5-meso).

To a solution of [Pd$_5$Pt$_3$(β-carotene)$_2$][B(Ar^F)$_4$]$_2$ (4-meso) (58.0 mg, 14.8 μmol) in ClCH$_2$CH$_2$Cl (50 ml) was added Pd$_2$(dba)$_3$·(C$_6$H$_6$) (200 mg, 201 μmol), and the reaction mixture was stirred under nitrogen atmosphere at 70 °C for 12 h. The mixture was filtered and the filtrate was dried *in vacuo*. After reprecipitation with CH$_2$Cl$_2$/hexane, the yellow powder was obtained. After drying *in vacuo*, the product was analysed by NMR, showing that [Pd$_5$Pt$_3$Pd$_2$(β-carotene)$_2$][B(Ar^F)$_4$]$_2$ (5-meso) was formed as a major product with an unidentified minor product (major:minor = 8:2; a mixture of two products: 43 mg). For 5-meso: ^{1}H NMR (600 MHz, CD$_2$Cl$_2$, 25 °C): δ −0.27 (s, 6H), −0.22 (s, 6H), −0.17 (s, 6H), 0.11 (s, 6H), 0.31 (s, 6H), 1.01 (s, 6H), 1.48 (s, 6H), 1.54 (s, 6H), 1.63 (m, 4H), 1.78 (m, 4H), 1.98 (m, 4H), 2.05 (s, 6H), 2.08 (m, 4H), 2.09 (s, 6H), 2.52 (d, J = 12 Hz, 2H), 2.56 (d, J = 12 Hz, 2H), 2.60 (t, J = 12 Hz, 2H), 2.62 (d, J = 12 Hz, 2H), 2.66 (d, J = 12 Hz, 2H), 2.74 (d, J = 12 Hz, 2H), 2.90 (d, J = 12 Hz, 4H), 2.91 (d, J = 12 Hz, 2H), 2.95 (d, J = 12 Hz, 2H), 3.07 (d, J = 12 Hz, 2H), 3.19 (t, J = 12 Hz, 2H), 3.30 (m, 8H), 3.40 (t, J = 12 Hz, 2H), 3.42 (t, J = 12 Hz, 2H), 7.51 (s, 8H, *p*-BAr$^F_4$), 7.66 (s, 16H, *o*-BAr$^F_4$). ^{13}C NMR (150 MHz, CD$_2$Cl$_2$, 25 °C): δ 14.1, 14.3, 14.8, 15.8, 20.4, 20.5, 23.7, 28.1, 28.2, 31.8, 31.9, 34.9, 36.8, 36.9, 43.4, 43.5, 70.0, 71.1, 71.2, 71.3, 71.8, 73.6, 74.1, 74.6, 74.8, 80.5, 81.8, 83.1, 83.4, 84.6, 89.9, 98.2, 103.5, 105.3, 109.8, 110.4, 111.4, 117.9 (*p*-BAr$^F_4$), 125.0 (*CF$_3$*-BAr$^F_4$), 129.2 (*m*-BAr$^F_4$), 135.2 (*o*-BAr$^F_4$), 162.1 (*ipso*-BAr$^F_4$). MS (ESI) *m/z* calcd. for [C$_{80}$H$_{112}$Pd$_7$Pt$_3$]$^{2+}$ found: 1,201.5553. Anal. calcd. For C$_{144}$H$_{136}$B$_2$F$_{48}$Pd$_7$Pt$_3$·(C$_6$H$_6$)$_2$: C, 42.81; H, 3.40, found: C, 42.87; H, 3.46. The sample suitable for elemental analysis was obtained from a dichloromethane-benzene solution. The minor product might be an isomer of 5-meso having a different metal arrangement (Supplementary Methods).

Synthesis of [Pd$_2$(1,10-diphenylpentaene)$_2$][B(Ar^F)$_4$]$_2$.

To a suspension of 1,10-diphenyl-1,3,5,7,9-decapentaene (80.0 mg, 0.28 mmol) in CH$_2$Cl$_2$ (30 ml) was added [Pd$_2$(1,4-diphenyl-1,3-butadiene)][B(Ar^F)$_4$]$_2$ (299.6 mg, 0.13 mmol). The mixture was stirred for 1 h at r.t. The reaction mixture was filtered and poured into hexane to give deep-green precipitation. [Pd$_2$(1,10-diphenylpentaene)$_2$][-B(Ar^F)$_4$]$_2$ was isolated as a mixture of four isomers by recrystallization (262.1 mg, 82%, isomer ratio 57:27:10:6). ^{1}H NMR (600 MHz, CD$_2$Cl$_2$, 25 °C) of vinyl protons in **A** (57% of products): δ 3.30 (dd, J = 12 Hz, J = 12 Hz, 1H), 3.38 (dd, J = 12 Hz, J = 12 Hz, 1H), 3.54 (dd, J = 12 Hz, J = 12 Hz, 1H), 3.75 (dd, J = 11 Hz, J = 12 Hz, 1H), 4.29 (dd, J = 11 Hz, J = 13 Hz, 1H), 4.33 (dd, J = 11 Hz, J = 12 Hz, 1H), 4.48 (dd, J = 12 Hz, J = 13 Hz, 1H), 5.07 (dd, J = 12 Hz, J = 14 Hz, 1H), 5.52 (dd, J = 12 Hz, J = 13 Hz, 1H), 5.77 (dd, J = 11 Hz, J = 13 Hz, 1H), 5.90 (dd, J = 11 Hz, J = 15 Hz, 1H), 6.05 (dd, J = 12 Hz, J = 14 Hz, 1H), 6.08 (dd, J = 11 Hz, J = 15 Hz, 1H), 6.14 (d, J = 14 Hz, 1H), 6.21 (dd, J = 11 Hz, J = 15 Hz, 1H), 6.40 (dd, J = 11 Hz, 1H), 6.47 (dd, J = 11 Hz, J = 14 Hz, 1H), 6.50 (d, J = 15 Hz, 1H), 6.69 (d, J = 15 Hz, 1H), 6.71 (dd, J = 15 Hz, 1H). **B** or **C** (27% of products): δ 3.34 (dd, J = 11 Hz, J = 11 Hz, 2H), 3.69 (dd, J = 11 Hz, J = 11 Hz, 2H), 4.32 (dd, J = 11 Hz, J = 13 Hz, 2H), 4.98 (dd, J = 11 Hz, J = 14 Hz, 2H), 5.46 (dd, J = 12 Hz, J = 14 Hz, 2H), 5.76 (dd, J = 12 Hz, 13 Hz, 2H), 6.21 (dd, J = 11 Hz, J = 14 Hz, 2H), 6.42 (d, J = 14 Hz, 2H), 6.61 (dd, J = 11 Hz, J = 15 Hz, 2H), 6.80 (dd, J = 15 Hz, 2H). **B** or **C** (10% of products): δ 3.30 (dd, J = 11 Hz, J = 11 Hz, 2H), 3.78 (dd, J = 11 Hz, J = 11 Hz, 2H), 4.23 (dd, J = 11 Hz, J = 13 Hz, 2H), 5.03 (dd, J = 11 Hz, J = 14 Hz, 2H), ca. 5.46 (2H), 6.13 (d, J = 14 Hz, 2H), 6.31 (dd, J = 11 Hz, J = 15 Hz, 2H), ca. 6.51 (2H), 6.56 (dd, J = 11 Hz, J = 14 Hz, 2H), 7.01 (dd, J = 15 Hz, 2H). **D** (6% of products): δ 3.48 (m, 4H), 4.49 (m, 4H), 6.01 (dd, J = 11 Hz, J = 13 Hz, 4H), 6.18 (dd, J = 11 Hz, J = 15 Hz, 4H), 6.69 (d, J = 15.6 Hz, 4H). Anal. calcd. For

C$_{108}$H$_{64}$B$_2$F$_{48}$Pd$_2$: C, 51.72; H, 2.57, found: C, 51.59; H, 2.58 (Supplementary Methods).

References

1. Polivka, T. & Sundstrom, V. Ultrafast dynamics of carotenoid excited states – from solution to natural and artificial systems. *Chem. Rev.* **104**, 2021–2072 (2004).
2. Polivka, T. & Frank, H. A. Molecular factors controlling photosynthetic light harvesting by carotenoids. *Acc. Chem. Res.* **43**, 1125–1134 (2010).
3. Takaichi, S., Mimuro, M. & Tomita, Y. *Carotenoids -Biological Functions and Diversity* (Shokabo, 2006).
4. Visoly-Fisher, I. *et al.* Conductance of a biomolecular wire. *Proc. Natl. Acad. Sci. USA* **103**, 8686–8690 (2006).
5. Ichikawa, M., Tsutsui, M. & Vohwinkel, F. Iron carbonyl complexes of β-carotene and lycopene. *Z. Naturforsch. B* **22**, 376–379 (1967).
6. Nakamura, A. & Tsutsui, M. π-Complexes with biologically significant materials. III. Vitamin A acetate iron tricarbonyl. *J. Med. Chem.* **7**, 335–337 (1964).
7. Birch, A. J., Fitton, H., Mason, R., Robertson, G. B. & Stangroom, J. E. Vitamin-A aldehyde iron tricarbonyl. *Chem. Commun.* 613–614 (1966).
8. Mason, R. & Robertson, G. B. Crystal and molecular structure of (vitamin-A aldehyde)tricarbonyliron. *J. Chem. Soc. A* 1229–1234 (1970).
9. Mashima, K. & Nakamura, A. Adaptable coordination modes of conjugated 1,3-diene: uniqueness of s-trans coordination. *J. Organomet. Chem.* **663**, 5–12 (2002).
10. Bera, J. K. & Dunbar, K. R. Chain compounds based on transition metal backbones: new life for an old topic. *Angew. Chem. Int. Ed.* **41**, 4453–4457 (2002).
11. Berry, J. F. Metal-metal bonds in chains of three or more metal atoms: from homometallic to heterometallic chains. *Struct. Bond* **136**, 1–28 (2010).
12. Zhao, Q., Harris, T. D. & Betley, T. A. [(^{H}L)$_2$Fe$_6$(NCMe)$_m$]$^{n+}$ (m = 0, 2, 4, 6; n = -1, 0, 1, 2, 3, 4, 6): An electron-transfer series featuring octahedral Fe$_6$ clusters supported by a hexaamide ligand platform. *J. Am. Chem. Soc.* **133**, 8293–8306 (2011).
13. Kanady, J. S., Tsui, E. Y., Day, M. W. & Agapie, T. A synthetic model of the Mn$_3$Ca subsite of the oxygen-evolving complex in photosystem II. *Science* **333**, 733–736 (2011).
14. Sunada, Y., Haige, R., Otsuka, K., Kyushin, S. & Nagashima, H. A ladder polysilane as a template for folding palladium nanosheets. *Nat. Commun.* **4**, 2014 (2013).
15. Ismayilov, R. H. *et al.* Two Linear undecanickel mixed-valence complexes: Increasing the size and the scope of the electronic properties of nickel metal strings. *Angew. Chem. Int. Ed.* **50**, 2045–2048 (2011).
16. Tanaka, K. *et al.* Programmable self-assembly of metal ions inside artificial DNA duplexes. *Nat. Nanotechnol.* **1**, 190–194 (2006).
17. Murahashi, T., Mochizuki, E., Kai, Y. & Kurosawa, H. Organometallic sandwich chains made of conjugated polyenes and metal-metal chains. *J. Am. Chem. Soc.* **121**, 10660–10661 (1999).
18. Tatsumi, Y., Shirato, K., Murahashi, T., Ogoshi, S. & Kurosawa, H. Sandwich complexes containing bent palladium chains. *Angew. Chem. Int. Ed.* **45**, 5799–5803 (2006).
19. Murahashi, T. *et al.* Discrete sandwich compounds of monolayer palladium sheets. *Science* **313**, 1104–1107 (2006).
20. Murahashi, T., Inoue, R., Usui, K. & Ogoshi, S. Square tetrapalladium sheet sandwich complexes: Cyclononatetraenyl as a versatile face-capping ligand. *J. Am. Chem. Soc.* **131**, 9888–9889 (2009).
21. Mednikov, E. G. & Dahl, L. F. Palladium: It forms unique nano-sized carbonyl clusters. *J. Chem. Edu.* **86**, 1135 (2009).
22. Murahashi, T. & Kurosawa, H. Organopalladium complexes containing palladium-palladium bonds. *Coord. Chem. Rev.* **231**, 207–228 (2002).
23. Murahashi, T., Nagai, T., Okuno, T., Matsutani, T. & Kurosawa, H. Synthesis and ligand substitution reactions of a homoleptic acetonitrile dipalladium(I) complex. *Chem. Commun.* 1689–1690 (2000).
24. Ukai, T., Kawazura, H., Ishii, Y., Bonnet, J. J. & Ibers, J. A. Chemistry of dibenzylideneacetone-palladium(0) complexes: I. Nobel tris(dibenzylideneacetone)dipalladium(solvent) complexes and their reactions with quinones. *J. Organomet. Chem.* **65**, 253–266 (1974).
25. Campbell, M. G. *et al.* Synthesis and structure of solution-stable one-dimensional palladium wires. *Nat. Chem.* **3**, 949–953 (2011).
26. Nakamae, K. *et al.* Self-alignment of low-valent octanuclear palladium atoms. *Angew. Chem. Int. Ed.* **54**, 1016–1021 (2015).
27. Sarkar, S. *et al.* Metals on graphene and carbon nanotube surfaces: From mobile atoms to atomtronics to bulk metals to clusters and catalysts. *Chem. Mater.* **26**, 184–195 (2014).
28. Chambers, A., Nemes, T., Rodriguez, N. M. & Baker, R. T. K. Catalytic behavior of graphite nanofiber supported nickel particles. 1. Comparison with other support media. *J. Phys. Chem. B* **102**, 2251–2258 (1998).
29. Tatsumi, Y., Nagai, T., Nakashima, H., Murahashi, T. & Kurosawa, H. Stepwise growth of polypalladium chains in 1,4-diphenyl-1,3-butadiene sandwich complexes. *Chem. Commun.* 1430–1431 (2004).

30. Diefenbach, A., Bickelhaupt, F. M. & Frenking, G. The nature of the transition metal-carbonyl bond and the question about the valence orbitals of transition metals. A bond-energy decomposition analysis of $TM(CO)_6^q$ ($TM^q = Hf^{2-}$, Ta^-, W, Re^+, Os^{2+}, Ir^{3+}). *J. Am. Chem. Soc.* **122**, 6449–6458 (2000).

31. Murahashi, T., Usui, K., Tachibana, Y., Kimura, S. & Ogoshi, S. Selective construction of Pd_2Pt and $PdPt_2$ triangles in a sandwich framework: carbocyclic ligands as scaffolds for a mixed-metal system. *Chem. Eur. J.* **18**, 8886–8890 (2012).

32. Takemoto, S. & Matsuzaka, H. Recent advances in the chemistry of ruthenium carbido complexes. *Coord. Chem. Rev.* **256**, 574–588 (2012).

33. Uddin, J., Dapprich, S., Frenking, G. & Yates, B. F. Nature of the metal-alkene bond in platinum complexes of strained olefins. *Organometallics* **18**, 457–465 (1999).

34. Stromberg, S., Svensson, M. & Zetterberg, K. Binding of ethylene to anionic, neutral, and cationic nickel(II), palladium(II), and platinum(II) cis/trans chloride ammonia complexes. A theoretical study. *Organometallics* **16**, 3165–3168 (1997).

35. Murahashi, T. *et al.* Redox-induced reversible metal assembly through translocation and reversible ligand coupling in tetranuclear metal sandwich frameworks. *Nat. Chem.* **4**, 52–58 (2012).

36. Yanai, T., Tew, D. P. & Handy, N. C. A new hybrid exchange-correlation functional using the Coulomb-attenuating method (CAM-B3LYP). *Chem. Phys. Lett.* **393**, 51–57 (2004).

Acknowledgements

This work was supported by JST through PRESTO programme, JSPS and MEXT through Grants-in-aid for Scientific Research and Institute for Molecular Science. We thank Professor H. Kurosawa and Professor S. Ogoshi for discussion.

Author contributions

The idea and plans of this research were made by T.Mu. Experiments and data analysis were performed by S.H., Y.T., M.Y., K.Y., K.M., K.T., T.Ma., S.K. and T.Mu. The theoretical calculations were performed by Y.K. and T.Y. The manuscript was co-written by T.Mu., Y.K. and T.Y. All authors discussed the results.

Additional information

Accession codes: The X-ray crystallographic coordinates for structures reported in this Article have been deposited at the Cambridge Crystallographic Data Centre (CCDC), under deposition numbers CCDC 1001723-1001727. These data can be obtained free of charge from The CCDC via www.ccdc.cam.ac.uk/data_request/cif.

19

General synthesis of complex nanotubes by gradient electrospinning and controlled pyrolysis

Chaojiang Niu[1,*], Jiashen Meng[1,*], Xuanpeng Wang[1,*], Chunhua Han[1], Mengyu Yan[1], Kangning Zhao[1], Xiaoming Xu[1], Wenhao Ren[1], Yunlong Zhao[1,2], Lin Xu[2], Qingjie Zhang[1], Dongyuan Zhao[1] & Liqiang Mai[1]

Nanowires and nanotubes have been the focus of considerable efforts in energy storage and solar energy conversion because of their unique properties. However, owing to the limitations of synthetic methods, most inorganic nanotubes, especially for multi-element oxides and binary-metal oxides, have been rarely fabricated. Here we design a gradient electrospinning and controlled pyrolysis method to synthesize various controllable 1D nanostructures, including mesoporous nanotubes, pea-like nanotubes and continuous nanowires. The key point of this method is the gradient distribution of low-/middle-/high-molecular-weight poly(vinyl alcohol) during the electrospinning process. This simple technique is extended to various inorganic multi-element oxides, binary-metal oxides and single-metal oxides. Among them, $Li_3V_2(PO_4)_3$, $Na_{0.7}Fe_{0.7}Mn_{0.3}O_2$ and Co_3O_4 mesoporous nanotubes exhibit ultrastable electrochemical performance when used in lithium-ion batteries, sodium-ion batteries and supercapacitors, respectively. We believe that a wide range of new materials available from our composition gradient electrospinning and pyrolysis methodology may lead to further developments in research on 1D systems.

[1] State Key Laboratory of Advanced Technology for Materials Synthesis and Processing, Wuhan University of Technology, Wuhan 430070, China. [2] Department of Chemistry and Chemical Biology, Harvard University, Cambridge, Massachusetts 02138, USA. * These authors contributed equally to this work. Correspondence and requests for materials should be addressed to L.X. (email: lxu@cmliris.harvard.edu) or to L.M. (email: mlq518@whut.edu.cn).

One-dimensional (1D) nanostructures, including nanowires and nanotubes, have been a focus of nanoscience and nanotechnology, due to the unique low-dimensional properties[1-6]. Various successful fabrication techniques, such as chemical/physical vapour deposition, hydrothermal, template based, electrochemical etching/deposition, laser ablation and electrospinning methods[7-15], have been developed for specific materials. However, owing to the restriction of applicable objects of each synthetic method and the difference of crystal growth orientation of different substances[16-18], 1D nanostructures are only achievable for some specific materials. Most inorganic nanotubes, especially for multi-element oxides and binary-metal oxides, have been scarcely obtained, which greatly restricts their further developments. Therefore, a universal technique is required, which can be used to fabricate nanotubes, as well as nanowires, for various inorganic materials with taking no account of the limitation of crystal orientation.

Electrospinning techniques have been studied for the fabrication of conductive polymer nanowires and a part of inorganic material nanowires[19-21]. This method along with different post-treatments has been applied to synthesize some interesting surface multilevel structures (branched nanowires and necklace-like nanowires) and inner multilevel structures (core/shell nanowires and multichannel microtubes)[22-26]. However, most of these structures have been confined to solid nanowires rather than nanotubes. Xia and co-workers have proposed a melt coaxial electrospinning method, with a coaxial spinneret, for fabricating core-shell nanowires and metal nanotubes[27-29]. Jiang and co-workers have developed a multifluidic compound-jet electrospinning technique for fabricating biomimic multi-channel microtubes[30]. Nevertheless, it has been rarely proposed to electrospin nanotubes for different kinds of inorganic oxides, especially for multi-element oxides and binary-metal oxides, with abundant materials diversity, low cost, good repeatability and high yield, which seriously limit their further applications.

Here we design a universal gradient electrospinning followed by controlled pyrolysis methodology to synthesize various types of mesoporous nanotubes and pea-like nanotubes (Fig. 1), including multi-element oxides, binary-metal oxides and single-metal oxides. This strategy is achieved through electrospinning with one ordinary syringe needle while modulating low-, middle- and high-molecular-weight poly(vinyl alcohol) (PVA) in the precursor. In this way, different nanotubes are obtained using controllable heat treatments. The resulting mesoporous nanotubes are composed of ultrathin carbon nanotubes (~ 5 nm in thickness and over 10 µm in length) and small nanoparticles (approximately 5–20 nm in diameter) on the tube walls. Pea-like nanotubes are composed of outer carbon nanotubes (~ 20 nm in thickness) and nanoparticles (approximately 100–300 nm in diameter) in the nanotubes. These structures have larger specific surface area and higher ionic–electronic conductivity compared with traditional nanowires, which exhibits great potential in energy storage fields. Therefore, $Li_3V_2(PO_4)_3$, $Na_{0.7}Fe_{0.7}Mn_{0.3}O_2$ and Co_3O_4 mesoporous nanotubes were selected as electroactive materials in lithium-ion batteries, sodium-ion batteries and supercapacitors, respectively.

Results

Gradient electrospinning and controlled pyrolysis mechanism.
First, the viscous homogeneous precursor solution was prepared by mixing low-, middle- and high-molecular-weight PVA (in a weight ratio of 3:2:1) and different needed inorganic materials. The precursor solution was then delivered into an ordinary metallic needle at a constant flow rate and electrospun at 20 kV. Under the strong electrostatic tension force, the low-,

middle- and high-molecular-weight PVA tend to be separated into three layers instead of mixing together. The reasons are as follows: at the same electrospinning conditions, the electric field (E), flow rate (Q), electric current (I), distance (D) between the injector nozzle and the receiver, and other operating characteristics of the low-, middle- and high-molecular-weight PVA are the same. According to the two equations proposed by Baumgarten and Rutledge[31,32] (equations. 1,2),

$$R = c \cdot \eta^{1/2} \tag{1}$$

$$R = c(I/Q)^{-2/3}\gamma^{1/3} \tag{2}$$

where R is the terminal jet radius, η is the viscosity, c is a constant, γ is the surface tension, Q is the flow rate and I is the electric current. The terminal jet radius (R) is directly proportional to the square root of the viscosity ($\eta^{1/2}$) and to the cube root of the surface tension ($\gamma^{1/3}$). The η values of the low-, middle- and high-molecular-weight PVA were measured as 0.0766, 0.5350 and 0.7685 dl g^{-1}, respectively, increasing gradually. The γ values of these three PVA were tested as 40.1, 41.6 and 51.4 mN m^{-1}, respectively (Supplementary Fig. 1a-c, g). Therefore, the high-weight PVA was distributed in the outer layer, the middle-weight PVA was located in the middle layer, and the low-weight PVA was concentrated in the centre in theory (Fig. 1a). To prove this important viewpoint, low-weight PVA was replaced by polyvinyl pyrrolidone (PVP), which has much smaller molecular weight, viscosity and surface tension (Fig. 2a, Supplementary Fig. 1g). After electrospinning low-, middle- and high-molecular-weight PVA composite polymer, the sample is solid nanowires, as shown in transmission electron microscope (TEM) image (Fig. 2b). After replacing the low-weight PVA by PVP, the sample is solid nanowires as well (Fig. 2c). Herein, PVP can be dissolved in trichloromethane ($CHCl_3$) solution, but PVA can not be dissolved in it. Therefore, the composite polymer nanowires were soaked in the $CHCl_3$ to remove PVP, getting middle-/high-molecular-weight PVA polymer nanotubes (Fig. 2d), which can clearly prove the layered distribution of low-, middle- and high-molecular-weight PVA. At the same time, the inorganic materials were homogeneously dispersed in all three layers.

After the electrospinning process, the composite nanowires were presintered in air. According to the thermogravimetric curves, the differentials of the mass loss (M) and temperature (T) (dM/dT) of the low-, middle- and high-weight PVA are -1.50, -0.95 and -0.58, respectively (Supplementary Fig. 1d,e). Consequently, the inner low-weight PVA first pyrolyzes and shrinks as the temperature is slowly increased, and moves towards the boundary of the low-/middle-weight PVA, carrying the inorganic materials simultaneously, thereby leading to the formation of nanotubes[33]. By this analogy, the middle-weight PVA then pyrolyzes and moves towards the middle-/high-weight PVA, together with the inorganic materials, thereby leading to the expansion of the inner diameter of nanotubes, as illustrated in Fig. 1a. Finally, all of the preliminarily decomposed PVA and inorganic materials converge together on the outer tubes. At last, inorganic mesoporous nanotubes can be obtained after sintering in air, which are only composed of tiny inorganic nanoparticles. Mesopores are formed via the decomposition of the inorganic materials and the partial pyrolysis of PVA. On the other hand, after high-temperature sintering under argon, all of the PVA carbonize, resulting in composite mesoporous nanotubes, which are composed of uniform inorganic nanoparticles and ultrathin carbon nanotubes.

For pea-like nanotubes, the preliminary electrospinning process of pea-like nanotubes is the same as that of the mesoporous nanotubes. After electrospinning, the composite

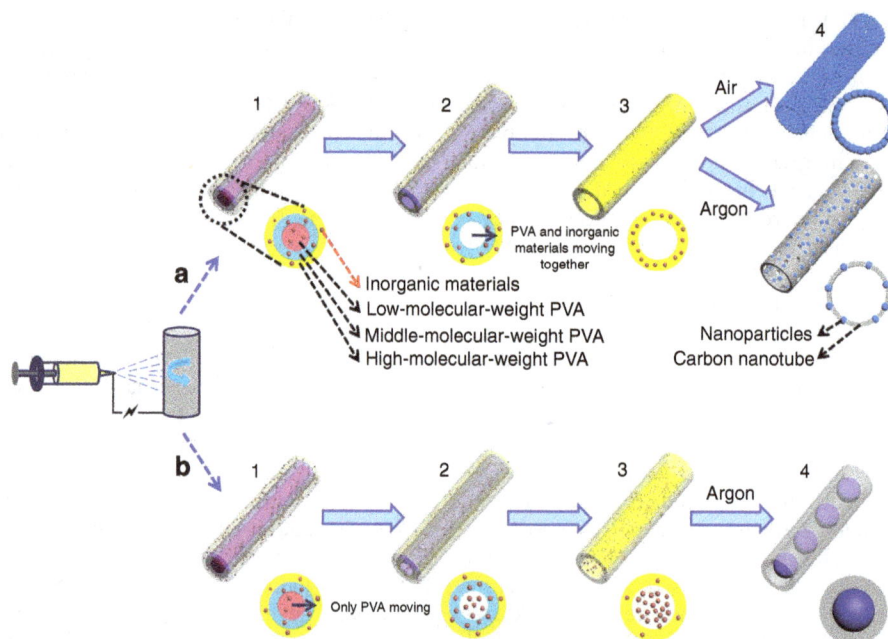

Figure 1 | Schematics of the gradient electrospinning and controlled pyrolysis method. (a) Preparation process of mesoporous nanotubes. (1) After the electrospinning process, the low-, middle- and high-molecular-weight PVA tend to be distributed into three layers in the radial direction of composite nanowires. (2) As the temperature is slowly increased, the inner low-weight PVA first pyrolyses and moves towards the boundary of the low-/middle-weight PVA, carrying the inorganic materials. Then the middle-weight PVA pyrolyses and moves towards the high-weight PVA as well. (3) All of the preliminary pyrolysed PVA and inorganic materials converge together in the tube walls. (4) After sintering in air, all of the PVA pyrolyse and uniform mesoporous nanotubes are obtained, which are composed of tiny inorganic nanoparticles. On the other hand, after high-temperature sintering under argon, PVA carbonize, uniform mesoporous nanotubes are also obtained, which consists of inorganic nanoparticles and carbon nanotubes. The mesopores result from the decomposition of the inorganic materials and a part of PVA polymers. **(b)** Preparation process of pea-like nanotubes. (1) After the electrospinning process, the composite nanowires are directly and immediately placed into a furnace in air, which is preheated to and maintained at 300 °C. (2, 3) All of the PVA decompose at the same time and rapidly move towards the outer high-weight PVA layer without carrying the inorganic materials, leaving them in the centre. (4) After high-temperature sintering under argon, the outer PVA carbonize and the inner inorganic materials develop into nanoparticles, forming pea-like nanotubes.

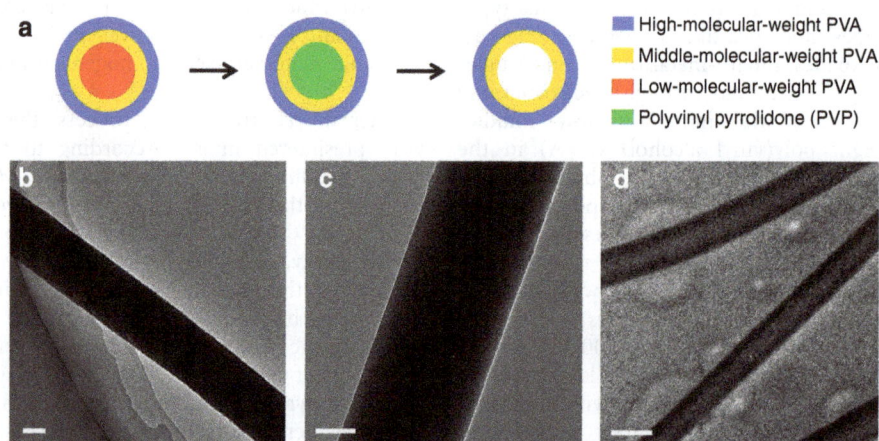

Figure 2 | Schematic and characterization of the gradient distribution of low-/middle-/high-molecular-weight PVA. (a) Schematic of the process of replacing low-molecular-weight PVA by PVP, then removing PVP with trichloromethane, to prove the layered distribution of low-, middle- and high-molecular-weight PVA. **(b)** TEM image of low-, middle- and high-molecular-weight PVA composite polymer nanowire after electrospinning with a scale bar at 100 nm. **(c)** TEM image of PVP and middle-/high-molecular-weight PVA composite polymer nanowire after electrospinning with a scale bar at 50 nm. Low-molecular-weight PVA is replaced by PVP. **(d)** TEM image of middle-/high-molecular-weight PVA polymer nanotubes with a scale bar at 500 nm, after removing PVP in the inner center using $CHCl_3$.

nanowires were directly and immediately placed into the furnace, which was preheated to 300 °C in air. At this temperature, all of the low-/middle-/high-weight PVA simultaneously decompose (Supplementary Fig. 1d) and quickly move towards the outer high-weight PVA layer without carrying the inorganic materials in the radial direction (Fig. 1b). Thus, the inorganic materials are

left *in situ* in the centre of the nanotubes. When the samples are annealed at high temperature under argon, the outer preliminary decomposed PVA carbonize and form carbon nanotubes, whereas the inner inorganic materials develop into nanoparticles, which are uniformly dispersed in the nanotubes. Eventually, pea-like nanotubes are obtained.

To confirm the mechanism of our gradient electrospinning and controlled pyrolysis method, various inorganic materials were electrospun into mesoporous nanotubes and pea-like nanotubes according to the aforementioned procedures (Fig. 3). First, multi-element oxides ($Li_3V_2(PO_4)_3$, $Na_3V_2(PO_4)_3$, $Na_{0.7}Fe_{0.7}Mn_{0.3}O_2$ and $LiNi_{1/3}Co_{1/3}Mn_{1/3}O_2$) were electrospun into uniform mesoporous nanotubes with a diameter of ~ 200 nm. Then, the binary-metal oxides ($LiMn_2O_4$, $LiCoO_2$, $NiCo_2O_4$ and LiV_3O_8) were electrospun into mesoporous nanotubes with a diameter of ~ 150 nm. For single-metal oxides (CuO, Co_3O_4, SnO_2 and MnO_2), mesoporous nanotubes with a smaller diameter of ~ 50 nm were fabricated. For the pea-like nanotubes, Co, $LiCoO_2$, $Li_3V_2(PO_4)_3$ and $Na_{0.7}Fe_{0.7}Mn_{0.3}O_2$ were selected from the different species, the outer layer was carbon and the inner particles were different inorganic salts. Additional scanning electron microscope (SEM) images and corresponding X-ray diffraction patterns of each sample are presented as well (Supplementary Figs 2 and 3). The detailed processes are clearly illustrated in the Methods section. Each precursor solution was electrospun at a constant flow rate of ~ 0.1 ml h^{-1} using one ordinary syringe needle (10 ml), and the corresponding production of inorganic materials was as high as ~ 0.6 mmol h^{-1} and ~ 900 cm^2 of aluminium foil for each run (Supplementary Fig. 1f), which was a large yield. And these products can also be collected in a parallel array around the rotating wheel, to improve the packing density.[34]

Electrochemical characterization. In energy-storage fields, most electrodes reported previously for batteries and supercapacitors suffer problems associated with a low conductivity, a small electrolyte/electrode surface area and self-aggregation during the charge/discharge process, leading to unsatisfactory performance, which greatly limits their applications[35–45]. One-dimensional nanomaterials have widely been investigated and applied in energy storage fields due to its unique low-dimensional properties. Remarkably, our complex nanotubes, especially mesoporous nanotubes, have the characteristics of large surface area, excellent stability and continuous carbon nanotubes with high conductivity and so on, which is expected to effectively improve the electrochemical performance of electrodes (Fig. 4). Therefore, to confirm it, $Li_3V_2(PO_4)_3$, $Na_{0.7}Fe_{0.7}Mn_{0.3}O_2$ and Co_3O_4 mesoporous nanotubes were selected and measured as typical examples of electroactive materials in lithium-ion batteries, sodium-ion batteries and supercapacitors, respectively.

First, the $Li_3V_2(PO_4)_3$ mesoporous nanotubes were further characterized, measured and compared with the $Li_3V_2(PO_4)_3$

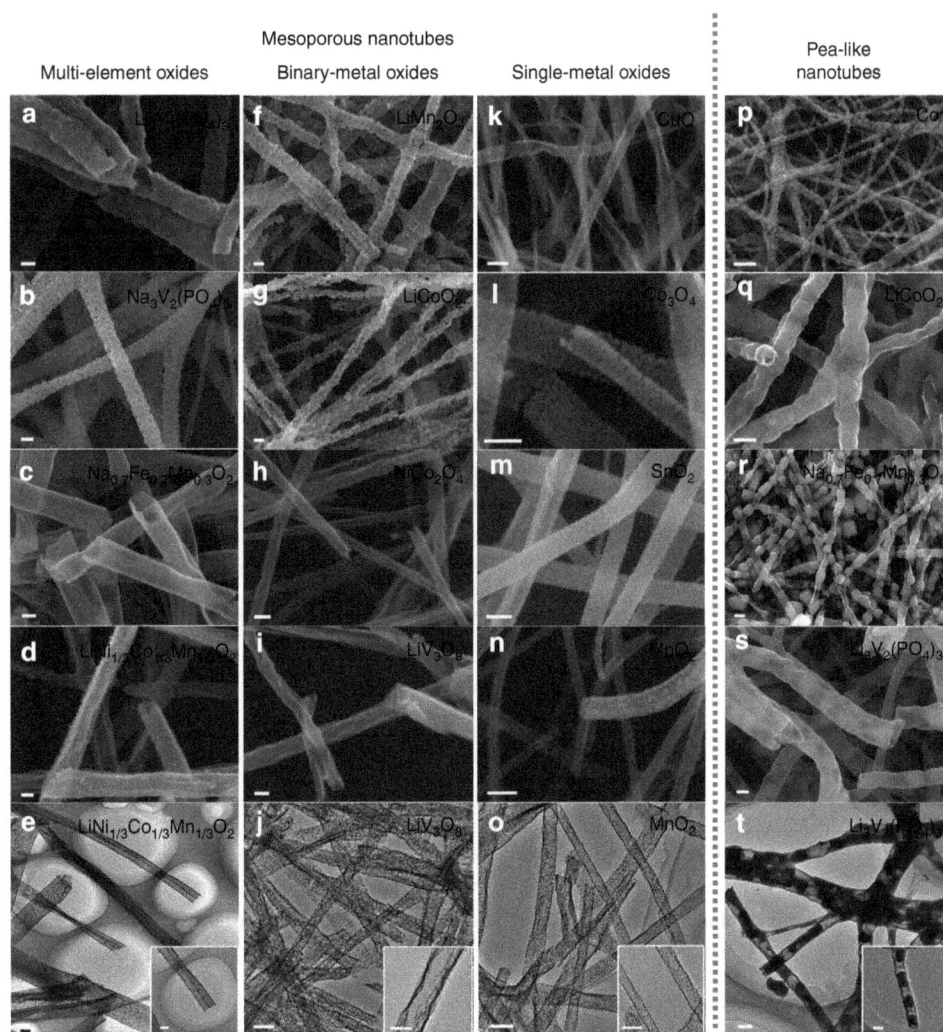

Figure 3 | Expansion of the gradient electrospinning and controlled pyrolysis method. (a–o) SEM and TEM images of multi-element oxides ($Li_3V_2(PO_4)_3$, $Na_3V_2(PO_4)_3$, $Na_{0.7}Fe_{0.7}Mn_{0.3}O_2$ and $LiNi_{1/3}Co_{1/3}Mn_{1/3}O_2$), binary-metal oxides ($LiMn_2O_4$, $LiCoO_2$, $NiCo_2O_4$ and LiV_3O_8) and single-metal oxides (CuO, Co_3O_4, SnO_2 and MnO_2) mesoporous nanotubes, respectively, scale bars, 100 nm. **(p–t)** SEM and TEM images of pea-like nanotubes (Co, $LiCoO_2$, $Li_3V_2(PO_4)_3$ and $Na_{0.7}Fe_{0.7}Mn_{0.3}O_2$) from different species with scale bars at 200 nm. The scale bars for the inset TEM images (**e, j, o, t**) are 100 nm.

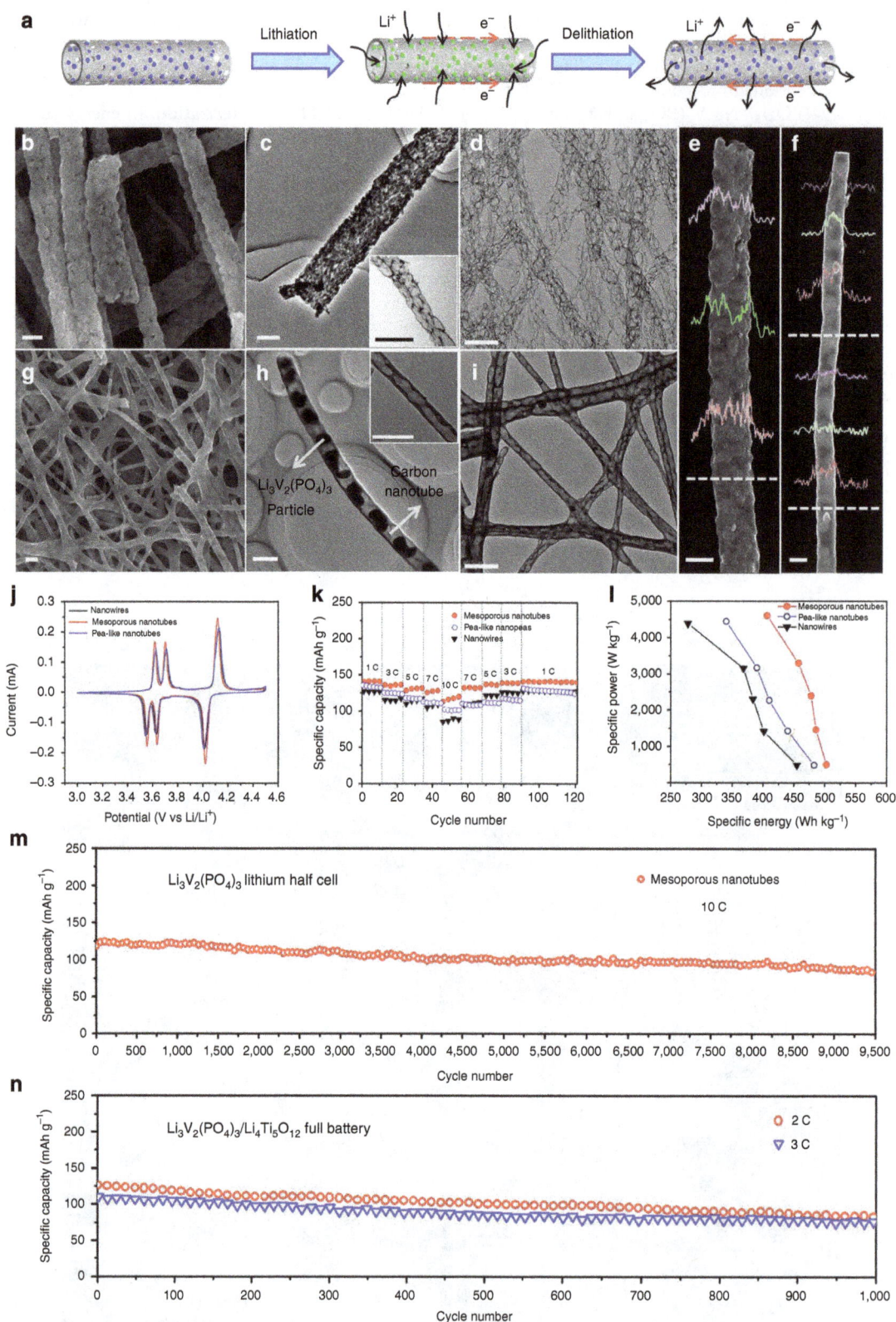

Figure 4 | Characterization and electrochemical performance in lithium-ion batteries. (a) Schematic of the lithiation and delithiation processes of mesoporous nanotubes. (**b,c**), SEM and TEM images of $Li_3V_2(PO_4)_3$ mesoporous nanotubes with scale bar at 200 nm. (**d**) and the inset of (**c**), TEM images of ultrathin carbon nanotubes after the $Li_3V_2(PO_4)_3$ was removed using hydrogen fluoride with scale bar at 500 nm. (**e,f**) Energy-dispersive X-ray spectra (EDS) line scans of $Li_3V_2(PO_4)_3$ mesoporous nanotubes (**e**) and pea-like nanotubes (**f**) with scale bar at 200 nm, respectively. (**g,h**) SEM and TEM images of $Li_3V_2(PO_4)_3$ pea-like nanotubes with scale bar at 500 nm. (**i**) and the inset of (**h**) TEM images of carbon nanotubes after removing $Li_3V_2(PO_4)_3$ with hydrogen fluoride with scale bar at 500 nm. (**j**) Cyclic voltammograms (CV) of the half cells collected at a sweep rate of 0.1 mV s^{-1} in the potential ranging from 3 to 4.5 V versus Li/Li$^+$. (**k,l**) Rate performance and the corresponding Ragone plots of these three cathodes measured at the rates of 1, 3, 5, 7 and 10 C, respectively. (**m**) Long cycling performance of $Li_3V_2(PO_4)_3$ mesoporous nanotubes measured at 10 C for a large number of 9,500 cycles in a lithium half cell. (**n**) Cycling performance of $Li_3V_2(PO_4)_3/Li_4Ti_5O_{12}$ lithium-ion full batteries measured at 2 and 3 C for 1,000 cycles (1 C equals to 133 mA g^{-1}).

pea-like nanotubes and solid nanowires (Supplementary Figs. 4-7). The $Li_3V_2(PO_4)_3$ mesoporous nanotubes consist of ultrathin, continuous, mesoporous carbon nanotubes (~ 200 nm in diameter) and $Li_3V_2(PO_4)_3$ nanoparticles (approximately 20–50 nm, uniformly dispersed on the tubes) (Fig. 4b–d), which is clearly confirmed by removal of the $Li_3V_2(PO_4)_3$ with hydrogen fluoride.

Moreover, the mesopore size can also be tuned by modulating the sintering temperature (Supplementary Fig. 5). In contrast, the $Li_3V_2(PO_4)_3$ pea-like nanotubes are composed of outer thick carbon nanotubes (~ 200 nm in diameter and ~ 20 nm in wall thickness) and $Li_3V_2(PO_4)_3$ nanoparticles in the nanotubes (~ 200 nm in diameter), which uniformly separate from each

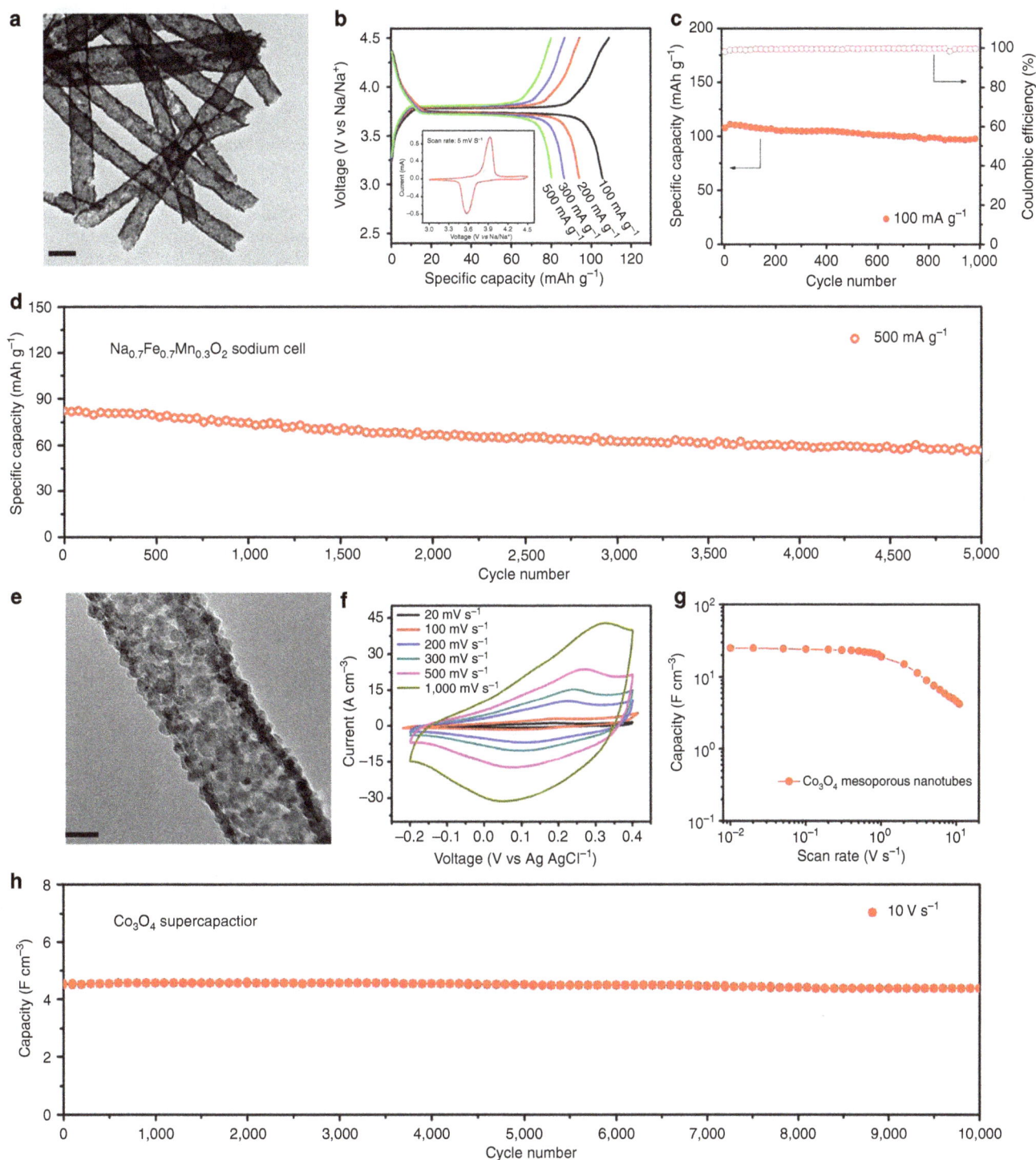

Figure 5 | Characterization and electrochemical performance in sodium-ion batteries and supercapacitors. (a) TEM image of the $Na_{0.7}Fe_{0.7}Mn_{0.3}O_2$ mesoporous nanotubes with a scale at 200 nm. (b) Charge–discharge curves of $Na_{0.7}Fe_{0.7}Mn_{0.3}O_2$ measured at 100, 200, 300 and 500 mA g^{-1}, respectively. The inset is the CV collected at a scan rate of 5 mV s^{-1} in the potential range 3.0-4.5 V. (c,d) Cycling performance of $Na_{0.7}Fe_{0.7}Mn_{0.3}O_2$ mesoporous nanotubes tested for 1,000 cycles at 100 mA g^{-1} and for 5,000 cycles at 500 mA g^{-1} (e) TEM image of Co_3O_4 mesoporous nanotubes with scale bar at 20 nm. (f) CV curves obtained at different scan rates from 20, 100, 300, 500 to 1,000 mV s^{-1}, respectively. (g) Stack capacitance of Co_3O_4 mesoporous nanotubes versus scan rate. (h) Long cycling performance of Co_3O_4 mesoporous nanotubes tested for 10,000 times at a high rate of 10 V s^{-1}.

other (Fig. 4g–i). The EDS line scans demonstrate the homogeneous distribution of V, P and C in mesoporous nanotubes, and display the difference between the hollow and solid parts in pea-like nanotubes (Fig. 4e,f). The carbon mass content of $Li_3V_2(PO_4)_3$ mesoporous nanotubes is only 7%, which is much smaller compared with pea-like nanotubes (17%) and nanowires (12%; Supplementary Fig. 4). $Li_3V_2(PO_4)_3$ mesoporous nanotubes exhibit higher capacity and better rate recovery (100%) than pea-like nanotubes (94%) and nanowires (98%) when tested at various rates ranging from 1, 3, 5, 7 to 10 C, demonstrating prominent rate performance (Fig. 4j–l). Particularly, $Li_3V_2(PO_4)_3$ mesoporous nanotubes can operate stably for as long as 9,500 cycles at the high rate of 10 C. The 9,500th discharge capacity is 86 mAh g^{-1}, which corresponds to a capacity retention of 80% and to a capacity fading of 0.0024% per cycle (Fig. 4m). However, the capacity retentions of $Li_3V_2(PO_4)_3$ pea-like nanotubes and solid nanowires are only 71% and 68%, respectively, after being cycled only 1,100 times at 10 C (Supplementary Fig. 6a). Then further electrochemical measurements of $Li_3V_2(PO_4)_3$ mesoporous nanotubes were implemented. When $Li_3V_2(PO_4)_3$ mesoporous nanotubes were measured at different temperatures of -20, 20 and 60 °C at a rate of 5 C, the corresponding initial capacities were 103, 131 and 136 mAh g^{-1}, respectively (Supplementary Fig. 6c,d). And the coulombic efficiency of $Li_3V_2(PO_4)_3$ mesoporous nanotubes can be kept on ~100% when tested at 1 C (133 mAh g^{-1}) for 500 cycles (Supplementary Fig. 6e). In the meantime, we assembled $Li_3V_2(PO_4)_3/Li_4Ti_5O_{12}$ lithium-ion full batteries. $Li_4Ti_5O_{12}$ was selected as the anode because of its stable charge/discharge plateau (Supplementary Fig. 7a–c). This full battery delivers capacities of 118, 93, 80 and 74 mAh g^{-1} at the rates of 2, 3, 4 and 5 C, respectively, and recovers 96% of its initial capacity when the rate is decreased back to 2 C (Supplementary Fig. 7d). Notably, the $Li_3V_2(PO_4)_3/Li_4Ti_5O_{12}$ battery is stably charged/discharged for over 1,000 cycles at rates of 2 and 3 C, with a capacity retention of 73% and 75%, respectively (Fig. 4n), demonstrating an excellent cycling performance and great commercial potential.

The superior performance is explained as follows: (1) $Li_3V_2(PO_4)_3$ mesoporous nanotubes contain ultrathin continuous carbon nanotubes. The charge transfer resistance (Rct) of $Li_3V_2(PO_4)_3$ mesoporous nanotubes is measured as 56 Ω, much smaller than that of nanowires (115 Ω), which can obviously improve the electronic conductivity[36] (Supplementary Fig. 6b). (2) $Li_3V_2(PO_4)_3$ mesoporous nanotubes exhibit the largest specific surface area (115 m^2 g^{-1}) among mesoporous nanotubes, pea-like nanotubes (33 m^2 g^{-1}) and nanowires (15 m^2 g^{-1}), which is beneficial to increase the electrode–electrolyte contact area and the active sites[39,40] (Supplementary Fig. 4). (3) These nanotubes have abundant mesopores on the tube walls, which can effectively buffer the stress induced during the charge/discharge process, and greatly improve the structural stability[40] (Fig. 4a).

The $Na_{0.7}Fe_{0.7}Mn_{0.3}O_2$ mesoporous nanotubes are also composed of ultrathin carbon nanotubes (~200 nm in diameter) and $Na_{0.7}Fe_{0.7}Mn_{0.3}O_2$ nanoparticles (~10 nm) on the tubes (Fig. 5a). And the X-ray diffraction pattern, ICP measurement and high-resolution TEM demonstrate the rhombohedral crystal structure of $Na_{0.7}Fe_{0.7}Mn_{0.3}O_2$ (Supplementary Fig. 8). Stable voltage plateaus are observed when measured at the current densities of 100, 200, 300 and 500 mA g^{-1}, respectively, corresponding to one pair of well-defined anodic (3.9 V) and cathodic (3.6 V) peaks (Fig. 5b). When $Na_{0.7}Fe_{0.7}Mn_{0.3}O_2$ mesoporous nanotubes are tested at a low density of 100 mA g^{-1} in the potential range 3–4.5 V, 90% of the initial capacity (109 mAh g^{-1}) is retained after 1,000 cycles (Fig. 5c). When measured at a high current density of 500 mA g^{-1}, 70% of the initial capacity (82 mAh g^{-1}) is maintained after cycling as long as 5,000 times, corresponding

to a capacity fading of 0.0071% per cycle (Fig. 5d). Compared with the conventional $Na_{0.7}Fe_{0.7}Mn_{0.3}O_2$ nanoparticles, which were synthesized as a control sample, our $Na_{0.7}Fe_{0.7}Mn_{0.3}O_2$ mesoporous nanotubes exhibit much higher specific capacity and better cycling performance (Supplementary Figs 8 and 9). Another comparison with those previously reported results for sodium-ion batteries reveals that our $Na_{0.7}Fe_{0.7}Mn_{0.3}O_2$ mesoporous nanotubes demonstrate superior electrochemical performance (Supplementary Fig. 8m).

The Co_3O_4 mesoporous nanotubes are very uniform with a diameter of ~50 nm, and only contain ~5 nm nanoparticles on the tubes (Fig. 5e, Supplementary Fig. 10). A micro-supercapacitor device based on these nanotubes was fabricated on a silicon wafer (The fabrication process of micro-supercapacitor device is clearly illustrated in Supplementary Fig. 10). These cyclic voltammograms reveal the exceptionally enhanced electrochemical performance of this material (Fig. 5f). 25.0, 24.0 and 18.9 F cm^{-3} for stack capacitance are obtained at the scan rates of 0.01, 0.1 and 1 V s^{-1}, respectively (Fig. 5g). When tested at a high rate of 10 V s^{-1}, it delivers a capacity of 4.5 F cm^{-3}, and 98% of the initial capacity is maintained after cycling as long as 10,000 times (Fig. 5h). All these unveil that Co_3O_4 mesoporous nanotubes demonstrate excellent electrochemical performance with respect to both high rate and long life, showing great application potential.

Discussion

In summary, our findings clearly indicate that the feasible gradient-electrospinning and controlled-pyrolysis methodology, with extensive material diversity, low cost, good repeatability and high yield, provides an efficient strategy to obtain controllable nanotubes for various inorganic materials, which can break through the limitation of the crystal growth orientation of each sample. On the basis of the formation mechanism, we successfully applied this technique to synthesize various multi-element oxides ($Li_3V_2(PO_4)_3$, $Na_3V_2(PO_4)_3$, $Na_{0.7}Fe_{0.7}Mn_{0.3}O_2$ and $LiNi_{1/3}Co_{1/3}Mn_{1/3}O_2$), binary-metal oxides ($LiMn_2O_4$, $LiCoO_2$, $NiCo_2O_4$ and LiV_3O_8) and single-metal oxides mesoporous nanotubes (CuO, Co_3O_4, SnO_2 and MnO_2). Therefore, we believe this technique could lead to rapid advancements in the development of 1D nanostructures. In addition, these novel mesoporous and pea-like nanotubes, which exhibit excellent electrochemical performance in lithium-ion batteries, sodium-ion batteries and supercapacitors, owing to their large surface area, high conductivity and robust structural stability, will have great potentials in not only the energy storage field, but also in numerous other frontiers.

Methods

Electrospinning $Li_3V_2(PO_4)_3$ mesoporous nanotubes. First, the uniform precursor solution for electrospinning was prepared with low-molecular-weight (98–99% hydrolysed), middle-molecular-weight (86–89% hydrolysed) and high-molecular-weight (98–99% hydrolysed) PVA in a weight ratio of 3:2:1 in 20 ml of deionized water. (This same uniform precursor solution was used to prepare the following samples.) $LiOH\bullet H_2O$, NH_4VO_3 and $NH_4H_2PO_4$ were also added in a molar ratio of 3:2:3. After the mixture was stirred at 50 °C for 5 h, a viscous, uniform, transparent yellow precursor solution was obtained. The concentration of the precursor solution was ~9.5 wt%, the average viscosity was measured as ~0.53 dl g^{-1}. The relative humidity was maintained at a constant of ~40% RH, and the temperature was maintained at ~20 °C. The precursor solution was subsequently electrospun at a constant flow rate of ~1 ml h^{-1} and at a high voltage of 20 kV (electrospinning equipment: SS-2534H from UCALERY Co., Beijing, China). The distance between the injector nozzle and the receiver was 15 cm. The composite nanowires were collected on revolving aluminium foil. After drying at 80 °C for 12 h, the composite nanowires were presintered at 300 °C (10 °C min^{-1}) in air for 3 h. The sample was then annealed at 800 °C (5 °C min^{-1}) under an argon atmosphere for 6 h to fully carbonize the decomposed PVA. Finally, uniform $Li_3V_2(PO_4)_3$ mesoporous nanotubes were obtained with a 7% carbon content (Supplementary Fig. 4).

Electrospinning Na$_3$V$_2$(PO$_4$)$_3$ mesoporous nanotubes. NH$_4$VO$_3$ and NaH$_2$PO$_4$•2H$_2$O were added in a molar ratio of 2:3 to the prepared PVA precursor. Then the precursor was electrospun at a high voltage of 20 kV. After the electrospinning process, the composite nanowires were sintered at 300 °C (10 °C min^{-1}) in air for 5 h and then annealed at 800 °C (5 °C min^{-1}) in an argon atmosphere for 5 h.

Electrospinning Na$_{0.7}$Fe$_{0.7}$Mn$_{0.3}$O$_2$ mesoporous nanotubes. NaNO$_3$, Fe(NO$_3$)$_3$•9H$_2$O, and Mn(CH$_3$COO)$_2$•4H$_2$O were added in a molar ratio of 7:7:3 to the prepared PVA precursor. Then the precursor was electrospun at a high voltage of 20 kV. After the electrospinning process, the composite nanowires were first sintered at 300 °C (10 °C min^{-1}) in air for 5 h and then annealed at 700 °C (5 °C min^{-1}) under argon for 8 h.

Electrospinning LiNi$_{1/3}$Co$_{1/3}$Mn$_{1/3}$O$_2$ mesoporous nanotubes. CH$_3$COOLi•2H$_2$O, Ni(CH$_3$COO)$_2$•4H$_2$O, Mn(CH$_3$COO)$_2$•4H$_2$O and Co(CH$_3$COO)$_2$•4H$_2$O were added in a molar ratio of 3:1:1:1. Then the precursor was electrospun at a high voltage of 20 kV. After the electrospinning process, the composite nanowires were annealed at 700 °C (5 °C min^{-1}) in air for 5 h.

Electrospinning LiMn$_2$O$_4$ mesoporous nanotubes. CH$_3$COOLi•2H$_2$O and Mn(CH$_3$COO)$_2$•4H$_2$O were added in a molar ratio of 1:2 to the prepared PVA precursor. The precursor was electrospun at a high voltage of 18 kV. After the electrospinning process, the composite nanowires were annealed at 700 °C (5 °C min^{-1}) in air for 5 h.

Electrospinning LiCoO$_2$ mesoporous nanotubes. CH$_3$COOLi•2H$_2$O and Co(CH$_3$COO)$_2$•4H$_2$O were added in a molar ratio of 1:1 to the prepared PVA precursor. The precursor was electrospun at a high voltage of 18 kV. After the electrospinning process, the composite nanowires were sintered at 500 °C (10 °C min^{-1}) in air for 5 h.

Electrospinning NiCo$_2$O$_4$ mesoporous nanotubes. Ni(CH$_3$COO)$_2$•4H$_2$O and Co(CH$_3$COO)$_2$•4H$_2$O were added in a molar ratio of 1:2 to the prepared PVA precursor. The precursor was electrospun at a high voltage of 18 kV. After the electrospinning process, the composite nanowires were sintered at 400 °C (10 °C min^{-1}) in air for 5 h.

Electrospinning LiV$_3$O$_8$ mesoporous nanotubes. LiOH•H$_2$O and NH$_4$VO$_3$ were added in a molar ratio of 1:3. The precursor was electrospun at a high voltage of 18 kV. After the electrospinning process, the composite nanowires were sintered at 450 °C (10 °C min^{-1}) in air for 5 h.

Electrospinning CuO mesoporous nanotubes. A moderate amount of Cu(NO$_3$)$_2$ was added to the prepared PVA precursor. The precursor was electrospun at a high voltage of 16 kV. After the electrospinning process, the composite nanowires were sintered at 350 °C (10 °C min^{-1}) in air for 5 h.

Electrospinning Co$_3$O$_4$ mesoporous nanotubes. A moderate amount of Co(CH$_3$COO)$_2$•4H$_2$O was added to the prepared PVA precursor. The precursor was electrospun at a high voltage of 16 kV. After electrospinning, the composite nanowires were sintered at 350 °C (10 °C min^{-1}) in air for 5 h. The carbon from PVA was removed via the sintering process.

Electrospinning MnO$_2$ mesoporous nanotubes. A moderate amount of Mn(CH$_3$COO)$_2$•4H$_2$O was added to the prepared PVA precursor. Then the precursor was electrospun at a high voltage of 16 kV. After the electrospinning process, the composite nanowires were sintered at 350 °C (10 °C min^{-1}) in air for 5 h.

Electrospinning SnO$_2$ mesoporous nanotubes. A moderate amount of SnCl$_4$ was added to the prepared PVA precursor. The precursor was electrospun at a high voltage of 16 kV. After the electrospinning process, the composite nanowires were sintered at 500 °C (10 °C min^{-1}) in air for 5 h.

Electrospinning Li$_3$V$_2$(PO$_4$)$_3$ pea-like nanotubes. The preliminary electrospinning process was the same as that used to prepare the mesoporous nanotubes. After drying at 80 °C for 12 h, the composite nanowires were directly and immediately heated for 1.5 h in a furnace preheated to 300 °C. After the composite nanowires were annealed at 800 °C (5 °C min^{-1}) under an argon atmosphere for 6 h, uniform pea-like Li$_3$V$_2$(PO$_4$)$_3$ nanotubes with 19% carbon content were obtained (Supplementary Fig. 4).

Electrospinning Co pea-like nanotubes. An amount of Co(CH$_3$COO)$_2$•4H$_2$O was added to the prepared PVA precursor. Then the precursor was electrospun at a

high voltage of 16 kV. After the electrospinning process, the composite nanowires were presintered in a muffle furnace at 300 °C for 1.5 h and then annealed at 600 °C under argon for 5 h.

Electrospinning LiCoO$_2$ pea-like nanotubes. CH$_3$COOLi•2H$_2$O and Co(CH$_3$COO)$_2$•4H$_2$O were added in a molar ratio of 1:1 to the prepared PVA precursor. Then the precursor was electrospun at a high voltage of 18 kV. After the electrospinning process, the composite nanowires were presintered in a muffle furnace at 300 °C for 1.5 h then annealed at 600 °C (5 °C min^{-1}) under argon for 5 h.

Electrospinning Na$_{0.7}$Fe$_{0.7}$Mn$_{0.3}$O$_2$ pea-like nanotubes. NaNO$_3$, Fe(NO$_3$)$_3$•9H$_2$O, and Mn(CH$_3$COO)$_2$•4H$_2$O were added in a molar ratio of 7:7:3. Then the precursor was electrospun at a high voltage of 20 kV. After the electrospinning process, the composite nanowires were presintered in a muffle furnace at 300 °C for 1.5 h and then annealed at 700 °C (5 °C min^{-1}) under argon for 8 h.

Morphology and structure characterizations. The crystallographic information of the final products was measured using a Bruker D8 Advance X-ray diffractometer equipped with a Cu Kα radiation source; the samples were scanned over the 2θ range from 10° to 80° at room temperature. SEM images were collected using a JEOL-7100F SEM, and TEM images were collected using a JEM-2100F TEM. The BET surface area was calculated from nitrogen adsorption isotherms measured at 77 K using a Tristar-3020 instrument. Energy-dispersive X-ray spectra were recorded using an Oxford IE250 system. thermogravimetric–DSC analyses were conducted using a STA-449C. X-ray photoelectron spectroscopy analysis was conducted on a VG Multilab 2000. Raman spectra were obtained using a Renishaw INVIA micro-Raman spectroscopy system. The surface tension was tested by an automatic surface tensiometer (CC2L202) from 2 to 200 N m^{-1}.

Preparation of Li$_3$V$_2$(PO$_4$)$_3$ lithium half cells. The 2016 coin cells were assembled in a glovebox filled with pure argon gas. Lithium foil was used as the anode, and a solution of LiPF$_6$ (1 M) in EC/DEC (1:1 vol/vol) was used as the electrolyte. The cathode was composed of a ground mixture of 70% Li$_3$V$_2$(PO$_4$)$_3$ active material, 20% acetylene black and 10% poly(tetrafluoroethylene). After coating onto aluminium foil, the cathode was cut into round slice with ∼0.36 cm^2 in area and ∼0.1 mm in thickness. And the loading rate of Li$_3$V$_2$(PO$_4$)$_3$ active material was ∼2 mg for one 2016 coin cell.

Preparation of Li$_3$V$_2$(PO$_4$)$_3$/Li$_4$Ti$_5$O$_{12}$ full batteries. The Li$_3$V$_2$(PO$_4$)$_3$/Li$_4$Ti$_5$O$_{12}$ battery was also assembled in a glovebox filled with pure argon gas. Li$_3$V$_2$(PO$_4$)$_3$ was coated onto aluminium foil as the cathode, and Li$_4$Ti$_5$O$_{12}$ was coated onto copper foil as the anode.

Preparation of Na$_{0.7}$Fe$_{0.7}$Mn$_{0.3}$O$_2$ sodium batteries. The 2016 coin cells were assembled in a glovebox filled with pure argon gas. Sodium foil was used as the anode, and 1 M NaClO$_4$ in a mixture of ethylene carbonate/dimethyl carbonate (1:2 in weight) with trace propylene carbonate (electrolyte additive) was used as the electrolyte. The prepared cathodes contained 70% Na$_{0.7}$Fe$_{0.7}$Mn$_{0.3}$O$_2$ active material, 20% acetylene black and 10% polyvinylidene difluoride (using N-methylpyrrolidone as the solvent). After coating onto aluminium foil, the cathode was cut into round slice with ∼0.36 cm^2 in area and ∼0.1 mm in thickness, and the loading rate of Na$_{0.7}$Fe$_{0.7}$Mn$_{0.3}$O$_2$ was ∼2.5 mg for one 2016 coin cell.

Preparation of Co$_3$O$_4$ supercapacitors. Electrochemical measurements were performed within the potential window of −0.2 to 0.4 V using an Autolab potentiostat/galvanostat (Autolab PGSTAT 302 N) in a three-electrode configuration with 1 M NaOH aqueous solution as the electrolyte; the measurements were performed at room temperature. A 1-cm^2 piece of silicon wafer on which the Co$_3$O$_4$ mesoporous nanotubes were deposited was directly used as the working electrode (Supplementary Fig. 10). The reference and counter electrodes were an Ag/KCl electrode and a platinum plate, respectively.

Electrochemical measurements. Galvanostatic charge/discharge measurements were performed using a multichannel battery testing system (LAND CT2001A). Cyclic voltammograms and electrochemical impedance spectra were collected at room temperature using an Autolab potentiostat/galvanostat.

References

1. Armand, M. & Tarascon, J.-M. Building better batteries. *Nature* **451**, 652–657 (2008).
2. Noorden, R. V. A better battery. *Nature* **507**, 26–28 (2014).
3. Wu, H. *et al.* Stable cycling of double-walled silicon nanotube battery anodes through solid–electrolyte interphase control. *Nat. Nanotechnol.* **7**, 310–315 (2012).

4. Zhao, Y. & Jiang, L. Hollow micro/nanomaterials with multilevel interior structures. *Adv. Mater.* **21**, 3621–3638 (2009).

5. Huang, Y., Duan, X. F., Wei, Q. Q. & Lieber, C. M. Directed assembly of one-dimensional nanostructures into functional networks. *Science* **291**, 630–633 (2001).

6. Jung, H. G., Jang, M. W., Hassoun, J., Sun, Y. K. & Scrosati, B. A high-rate long-life Li$_4$Ti$_5$O$_{12}$/Li[Ni$_{0.45}$Co$_{0.1}$Mn$_{1.45}$]O$_4$ lithium-ion battery. *Nat. Commun.* **2**, 516 (2011).

7. Yang, F. *et al.* Chirality-specific growth of single-walled carbon nanotubes on solid alloy catalysts. *Nature* **510**, 522–524 (2014).

8. Qin, Y., Lee, S.-M., Pan, A. L., Gösele, U. & Knez, M. Rayleigh-instability-induced metal nanoparticle chains encapsulated in nanotubes produced by atomic layer deposition. *Nano Lett.* **8**, 114–118 (2008).

9. Lee, S. W. *et al.* High-power lithium batteries from functionalized carbon-nanotube electrodes. *Nat. Nanotechnol.* **5**, 531–537 (2010).

10. Wang, X., Zhuang, J., Peng, Q. & Li, Y. D. A general strategy for nanocrystal synthesis. *Nature* **43**, 121–124 (2005).

11. Magasinski, A. *et al.* High-performance lithium-ion anodes using a hierarchical bottom-up approach. *Nat. Mater.* **9**, 353–358 (2010).

12. Deng, Y. H., Wei, J., Sun, Z. K. & Zhao, D. Y. Large-pore ordered mesoporous materials templated from non-Pluronic amphiphilic block copolymers. *Chem. Soc. Rev.* **42**, 4054–4070 (2013).

13. Yao, J., Yan, H. & Lieber, C. M. A nanoscale combing technique for the large-scale assembly of highly aligned nanowires. *Nat. Nanotechnol.* **8**, 329–335 (2013).

14. Chen, C. *et al.* Highly crystalline multimetallic nanoframes with three-dimensional electrocatalytic surfaces. *Science* **343**, 1339–1343 (2014).

15. Joshi, R. K. & Schneider, J. J. Assembly of one dimensional inorganic nanostructures into functional 2D and 3D architectures. Synthesis, arrangement and functionality. *Chem. Soc. Rev.* **41**, 5285–5312 (2012).

16. Yaman, M. *et al.* Arrays of indefinitely long uniform nanowires and nanotubes. *Nat. Mater.* **10**, 494–501 (2011).

17. Ren, Y., Ma, Z. & Bruce., P. G. Ordered mesoporous metal oxides: synthesis and applications. *Chem. Soc. Rev.* **41**, 4909–4927 (2012).

18. Xia, Y. N. *et al.* One-dimensional nanostructures: synthesis, characterization and applications. *Adv. Mater.* **15**, 353–389 (2003).

19. Li, D. & Xia, Y. N. Electrospinning of nanofibers: reinventing the wheel? *Adv. Mater.* **16**, 1151–1170 (2004).

20. Li, D., MaCann, J. T. & Xia, Y. N. Electrospinning: a simple and versatile technique for producing ceramic nanofibers and nanotubes. *J. Am. Ceram. Soc.* **89**, 1861–1869 (2006).

21. Yuan, C. Z., Wu, H. B., Xie, Y. & Lou, X. W. Mixed transition-metal oxides: design, synthesis, and energy-related applications. *Angew. Chem. Int. Ed.* **53**, 1488–1504 (2014).

22. Mai, L. Q. *et al.* Electrospun ultralong hierarchical vanadium oxide nanowires with high performance for lithium ion batteries. *Nano Lett.* **10**, 4750–4755 (2010).

23. Wu, J., Wang, N., Zhao, Y. & Jiang, L. Electrospinning of multilevel structured functional micro-/nanofibers and their applications. *J. Mater. Chem. A* **1**, 7290–7305 (2013).

24. Zussman, E., Yarin, A. L., Bazilevsky, A. V., Avrahami, R. & Feldman, M. Electrospun polyacrylonitrile/poly(methyl methacrylate)-derived turbostratic carbon micro-/nanotubes. *Adv. Mater.* **18**, 348–353 (2006).

25. Zhan, S. H., Chen, D. R., Jiao, X. L. & Tao, C. H. Long TiO$_2$ hollow fibers with mesoporous walls: sol-gel combined electrospun fabrication and photocatalytic properties. *J. Phys. Chem. B* **110**, 11199–11204 (2006).

26. Hou, H. L. *et al.* General strategy for fabricating thoroughly mesoporous nanofibers. *J. Am. Chem. Soc.* **136**, 16716–16719 (2014).

27. Sun, Y. G., Mayers, B. & Xia, Y. N. Metal nanostructures with hollow interiors. *Adv. Mater.* **15**, 641–646 (2003).

28. Li, D. & Xia, Y. N. Direct fabrication of composite and ceramic hollow nanofibers by electrospinning. *Nano Lett.* **4**, 933–938 (2004).

29. McCann, J. T., Marquez, M. & Xia, Y. N. Melt coaxial electrospinning: a versatile method for the encapsulation of solid materials and fabrication of phase change nanofibers. *Nano Lett.* **6**, 2868–2872 (2006).

30. Zhao, Y., Cao, X. Y. & Jiang, L. Bio-mimic multichannel microtubes by a facile method. *J. Am. Chem. Soc.* **129**, 764–765 (2007).

31. Shin, Y. M., Hohman, M. M., Brenner, M. P. & Rutledge, G. C. Experimental characterization of electrospinning: the electrically forced jet and instabilities. *Polymer (Guildf).* **42**, 9955–9967 (2001).

32. Fridrikh, S. V., Yu, J. H., Brenner, M. P. & Rutledge, G. C. Controlling the fiber diameter during electrospinning. *Phys. Rev. Lett.* **90**, 144502 (2003).

33. Wu, Z. X. *et al.* A general "surface-locking" approach toward fast assembly and processing of large-sized, ordered, mesoporous carbon microspheres. *Angew. Chem. Int. Ed.* **52**, 13764–13768 (2013).

34. Theron, A., Zussman, E. & Yarin, A. L. Electrostatic field-assisted alignment of electrospun nanofibres. *Nanotechonolgy* **12**, 384–390 (2001).

35. Liu, H. *et al.* Capturing metastable structures during high-rate cycling of LiFePO$_4$ nanoparticle electrodes. *Science* **344**, 1480 (2014).

36. Yang, P. D., Zhao, D. Y., Margolese, D. I., Chmelka, B. F. & Stucky, G. D. Generalized syntheses of large-pore mesoporous metal oxides with semicrystalline frameworks. *Nature* **396**, 152–155 (1998).

37. Kang, B. & Ceder, G. Battery materials for ultrafast charging and discharging. *Nature* **458**, 190–193 (2009).

38. Nishijima, M. *et al.* Accelerated discovery of cathode materials with prolonged cycle life for lithium-ion battery. *Nat. Commun.* **5**, 4553 (2015).

39. Mai, L. Q., Tian, X. C., Xu, X., Chang, L. & Xu, L. Nanowire electrodes for electrochemical energy storage devices. *Chem. Rev.* **114**, 11828–11862 (2014).

40. Masquelier, C. & Croguennec, L. Polyanionic (phosphates, silicates, sulfates) frameworks as electrode materials for rechargeable Li (or Na) batteries. *Chem. Rev.* **113**, 6552–6591 (2013).

41. Yabuuchi, N. *et al.* P2-type Na$_x$[Fe$_{1/2}$Mn$_{1/2}$]O$_2$ made from earth-abundant elements for rechargeable Na batteries. *Nat. Mater.* **11**, 512–517 (2012).

42. Kalluri, S. *et al.* Electrospun P2-type Na$_{2/3}$(Fe$_{1/2}$Mn$_{1/2}$)O$_2$ hierarchical nanofibers as cathode material for sodium-ion batteries. *ACS Appl. Mater. Interfaces* **6**, 8953–8958 (2014).

43. Xu, J. *et al.* Identifying the critical role of Li substitution in P2-Na$_x$[Li$_y$Ni$_z$Mn$_{1-y-z}$]O$_2$ ($0 < x, y, z < 1$) intercalation cathode materials for high-energy Na-ion batteries. *Chem. Mater.* **26**, 1260–1269 (2014).

44. Oh, S. M. *et al.* Advanced Na[Ni$_{0.25}$Fe$_{0.5}$Mn$_{0.25}$]O$_2$/C-Fe$_3$O$_4$ sodium-ion batteries using EMS electrolyte for energy storage. *Nano Lett.* **14**, 1620–1626 (2014).

45. Wu, Z. S., Parves, K., Feng, X. L. & Müllen, K. Graphene-based in-plane micro-supercapacitors with high power and energy densities. *Nat. Commun.* **4**, 2487 (2013).

Acknowledgements

This work was supported by the National Basic Research Program of China (2013CB934103, 2012CB933003), the International Science and Technology Cooperation Program of China (2013DFA50840), the National Natural Science Fund for Distinguished Young Scholars (51425204), the National Natural Science Foundation of China (51272197, 51302203), Hubei Province Natural Science Fund for Distinguished Young Scholars (2014CFA035) and the Fundamental Research Funds for the Central Universities (2014-VII-007, 2014-CL-A1-01). We thank Professor C. M. Lieber of Harvard University and Professor J. Liu of Pacific Northwest National Laboratory for useful discussions and assistance with the manuscript. We thank the Center for Materials Research and Analysis from Wuhan University of Technology.

Author contribution

L.Q.M. and C.J.N. conceived and designed the experiments, analysed the results and wrote the manuscript. J.S.M., X.P.W. and L.X. performed the experiments and analysed the results. All authors commented on the manuscript.

Additional information

20

Unique distal size selectivity with a digold catalyst during alkyne homocoupling

Antonio Leyva-Pérez[1], Antonio Doménech-Carbó[2] & Avelino Corma[1]

Metal-catalysed chemical reactions are often controlled by steric hindrance around the metal atom and it is rare that substituents far away of the reaction site could be differentiated during reaction, particularly if they are simple alkyl groups. Here we show that a gold catalyst is able to discriminate between linear carbon alkynes with 10 or 12 atoms in the chain during the oxidative homocoupling of alkynes: the former is fully reactive and the latter is practically unreactive. We present experimental evidences, which support that the distal size selectivity occurs by the impossibility of transmetallating two long alkyl chains in an A-framed, mixed-valence digold (I, III) acetylide complex. We also show that the reductive elimination of two alkyne molecules from a single Au(III) atom occurs extremely fast, in <1 min at $-78\,^\circ$C (turnover frequency $> 0.016\,\mathrm{s}^{-1}$).

[1] Instituto de Tecnología Química, Universidad Politécnica de Valencia-Consejo Superior de Investigaciones Científicas, Avda. de los Naranjos s/n, 46022 Valencia, Spain. [2] Departament de Química Analítica, Universitat de Valencia, Dr Moliner, 50, 46100 Burjassot, Valencia, Spain. Correspondence and requests for materials should be addressed to A.L.-P. (email: anleyva@itq.upv.es) or to A.C. (email: acorma@itq.upv.es).

Organic synthesis takes advantage of the predictable outcome of reactions where bulky groups are involved. Metal-catalysed carbon–carbon bond-forming reactions are not an exception since they are generally controlled by the steric hindrance around the catalytic site, which is given by the bulkiness of the ligands or the reactants, rather than for substituents far away from the reactive site. It is rare that substituents at more than six atoms distance from the reactive site can be sterically differentiated during catalysis[1,2]. Here we show a gold catalyst that is able to discriminate between linear chain alkynes only differing in one ethylene group at an eight-carbon distance of the catalytic site. This level of discrimination is extraordinary and difficult to achieve even for rigid microporous solid frameworks[3–5].

Gold catalysis has been intensively studied during the last decade and, in particular, the use of the redox pair Au(I)/Au(III) has arisen much interest among chemists as a catalyst for new carbon–carbon and carbon–heteroatom bond-forming reactions[6–10]. Despite the increasing number of new transformations reported, particularly with alkynes[11,12], fundamental studies covering the elemental steps of the gold-catalysed redox cycle remain scarce[13,14].

Here we dissect the fundamental steps during the oxidative coupling of alkynes catalysed by Au(I)/Au(III) to find a reasonable explanation to the distal size selectivity observed.

Results

Distal size selectivity of alkynes with gold catalysts. Fig. 1 shows the results for the homocoupling of 1-decyne **1a** and 1-dodecyne **1b** under gold-catalysed conditions. We found that 1-decyne **1a** (C_{10}) reacts smoothly under the gold-catalysed conditions, while 1-dodecyne **1b** is nearly unreactive for homocoupling (<5% yield after 24 h). Kinetic measurements by gas chromatography–mass spectrometry (GC–MS, see Supplementary Fig. 1) show that the initial rate of 1-decyne **1a** is 3.5 times higher than 1-dodecyne **1b,** and that while **1a** smoothly converts over the time to the homocoupling product **2a** 1-dodecyne **1b** rapidly stops converting to the homocoupled product **2b**. Notice that a minor non-catalysed polymerization of the alkyne occurs together with the homocoupling reaction.

Other linear terminal alkynes shorter than 1-decyne **1a** (C_6–C_9) react at a similar rate than **1a** under gold-catalysed conditions, with final product yields between 60–70%, while 1-undecyne (C_{11}) gives a low yield of homocoupling product (19%) and a longer linear terminal alkyne (C_{14}) is unreactive towards homocoupling. The use of freshly-prepared or just-open commercial bottles of Au catalyst and selectfluor assures reproducible results, otherwise lower yields can be obtained

although with still a similar distal size selectivity. This change based on the chain size is remarkable, when moving from 1-decyne **1a** (65–80%), then to 1-undecyne (19%) and finally to 1-dodecyne **1b** and longer linear alkynes (<5%). It must be highlighted that this distal size selectivity seems to be quite unique with gold, since when the homocoupling of alkynes was carried out under typical copper-mediated (Glaser coupling, that is, the most used protocol for the homocoupling of alkynes in organic synthesis)[15,16] or copper-catalysed (Glaser–Hay coupling, with O_2 as the final oxidant) conditions the reaction rate and final yields of 1-decyne **2a** and 1-dodecyne **2b** are very similar for both alkynes (see Fig. 1). These results suggest a profound difference in the mechanism of homocoupling of linear alkynes with gold or copper catalysts.

The nearly atomically precise selection of long linear alkynes by gold during the oxidative homocoupling of alkynes is perhaps more striking when we consider alkynes containing bulky substituents at the contiguous positions of the triple bond. For instance, *tert*-butylacetylene **1c** reacts smoothly (91% yield of **2c**) while cyclohexylacetylene **1d** gives only a moderate yield of homocoupling product **2d** (37% yield). These results clearly contrast with those obtained for copper: while *tert*-butylacetylene **1c** is poorly reactive in the Glaser and Hay couplings (11 and 34% yield after 24 h), cyclohexylacetylene **1d** shows a higher reactivity (46% yield after 4 days for the Glaser coupling and 53% yield after 24 h for the Hay coupling) closer to that of long linear alkynes **1a–b**. Overall, the results in Fig. 1 clearly show that the gold-catalysed oxidative homocoupling of alkynes occurs with size selectivity along the whole substrate rather than with the classical size selectivity around the reactive site of the substrate, the latter operating in the copper-catalysed reactions.

Study of the origin of the distal size selectivity with gold catalysts. The elemental steps of the homocoupling of alkynes on gold catalysts (supposing a mixed-valence gold complex) are shown in Fig. 2 and include oxidation of gold(I) to gold (III), formation of the digold complex (approach), transmetallation of the alkyne and reductive elimination[17,18].

To determine if the first step, that is, the oxidation by selectfluor of the *in situ* formed gold(I)-acetylide[17], is responsible for the distal size selectivity observed, the gold(I) acetylides of 1-decyne **1a** (**3a**) and 1-dodecyne **1b** (**3b**) were prepared and a stoichiometric amount of oxidant was added to each of them under the reaction conditions given in Fig. 3.

The results show that selectfluor is consumed within 45 min with an initial rate of 250 h^{-1} for acetylide **3a** and 220 h^{-1} for **3b** (measured from the initial slope of the curve and within the timing limitations of the ^{1}H NMR technique). The reaction rate is

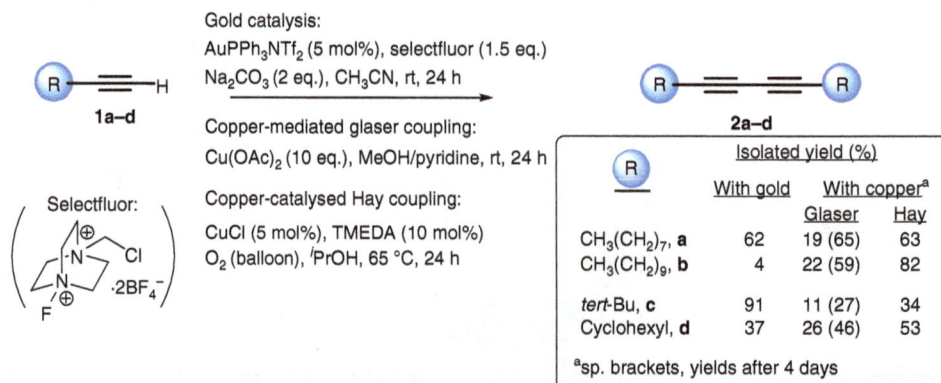

Gold catalysis:
AuPPh$_3$NTf$_2$ (5 mol%), selectfluor (1.5 eq.)
Na$_2$CO$_3$ (2 eq.), CH$_3$CN, rt, 24 h

Copper-mediated glaser coupling:
Cu(OAc)$_2$ (10 eq.), MeOH/pyridine, rt, 24 h

Copper-catalysed Hay coupling:
CuCl (5 mol%), TMEDA (10 mol%)
O$_2$ (balloon), ^{i}PrOH, 65 °C, 24 h

Selectfluor:

R	Isolated yield (%)		
	With gold	With copper[a]	
		Glaser	Hay
CH$_3$(CH$_2$)$_7$, **a**	62	19 (65)	63
CH$_3$(CH$_2$)$_9$, **b**	4	22 (59)	82
tert-Bu, **c**	91	11 (27)	34
Cyclohexyl, **d**	37	26 (46)	53

[a]sp. brackets, yields after 4 days

Figure 1 | Gold or copper-catalysed oxidative homocoupling of terminal alkynes. Notice the different reactivity of alkynes **1a–d** under the gold-catalysed conditions reported here and typical copper-catalysed conditions (Glaser or Hay conditions). Isolated yields are the average of two runs.

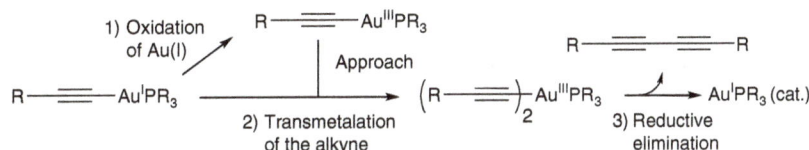

Figure 2 | Proposed reaction steps for the homocoupling of alkynes on gold catalysts. The reaction pathway includes oxidation of gold(I) to gold (III), formation of the digold complex (approach), transmetallation of the alkyne and reductive elimination[17].

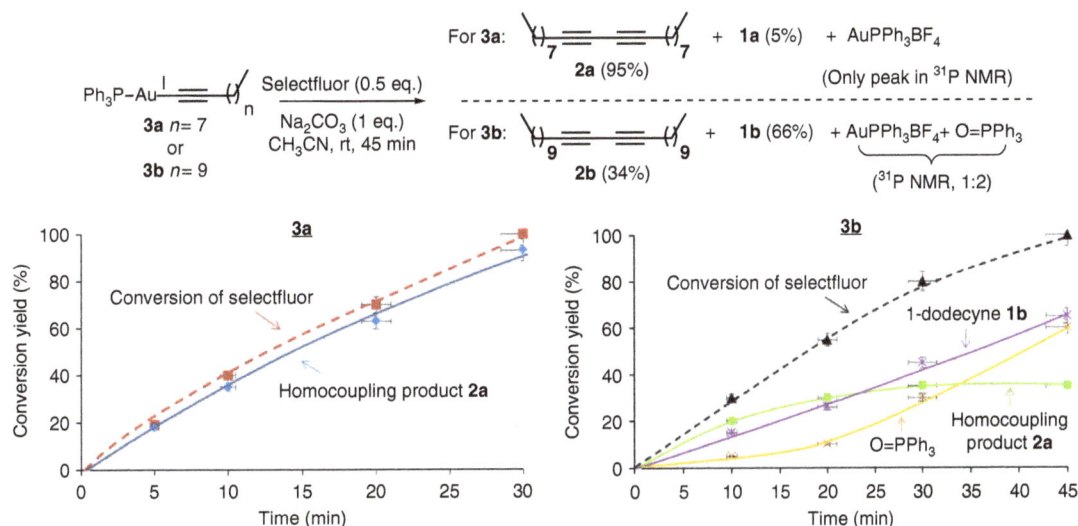

Figure 3 | Oxidation of gold(I) acetylides by selectfluor. Conversion and yields are calculated on the basis of ^{1}H and ^{19}F NMR spectroscopy for selectfluor, ^{1}H NMR spectroscopy and GC–MS (after extraction of the reaction mixture with n-hexane and addition of dodecane as an external standard) for alkynes and ^{31}P NMR spectroscopy for phosphines. Error bars represent an uncertainty of 5% in the value.

expressed in h^{-1} and increases with the concentration of selectfluor (Supplementary Fig. 2)[17]. While the consumption of selectfluor is similar for the two acetylides 3a–b, the appearance of the homocoupling product is not. For 3a (C_{10}), a new triplet at ~2.1 p.p.m. corresponding to the homocoupling product 2a appears in the ^{1}H NMR spectra, and this peak increases as selectfluor converts. In contrast, 3b (C_{12}) gives two new signals in the ^{1}H NMR spectra at ~2.3–2.1 p.p.m., and those signals correspond to the homocoupling product 2b and 1-dodecyne 1b. Extraction of the reaction mixture with n-hexane and analysis by GC–MS and also by ^{1}H and ^{13}C NMR spectroscopy confirms that the homocoupling product 2a is the only alkyne product after oxidation of the acetylide of 3a, and that the homocoupling product 2b and free 1-dodecyne 1b in a 1:2 molar ratio are the alkyne products from the acetylide of 3b. The appearance of free alkyne 1b from 3b prompted us to check carefully the spectra of 3a during oxidation, and small amounts (<5%) of free 1-decyne 1a were detected in solution by GC–MS. These results show that the free alkynes 1a or 1b are somehow formed during the oxidation of the Au(I) acetylide complexes 3a–b, respectively.

To check if the free alkyne is able or not to enter back the catalytic cycle, an equimolecular amount of 1-dodecyne 1b was added to the C_{10} acetylide 3a and then we proceed to oxidize with selectfluor. The results (Supplementary Fig. 3) show that the heterocoupling product C_{10}–C_{12} 4a was formed together with the homocoupling product C_{10}–C_{10} 2a. Complementary 1-decyne 1a was added during the oxidation of the C_{12} acetylide 3b, and we found again that the heterocoupling product C_{10}–C_{12} 4a was formed together with the corresponding homocoupling product C_{12}–C_{12} 2b. These results confirm the scrambling of alkynes during reaction. Notice that the homocoupling product of the free alkyne added externally is not obtained in any case, which is in

accordance with blank experiments that show that alkyne scrambling between the gold(I) acetylide and a second free alkyne does not occur without any oxidant in the reaction medium. When the two gold(I) acetylides 3a and 3b are combined in a single flask and selectfluor is added, the heterocoupling product 4a is formed in 37% yield, the C_{10}–C_{10} homocoupling product 2a is formed in 58% yield, the C_{12}–C_{12} 2b is formed in only 5% yield and the free alkyne 1b (50%) persists in solution. These results support that free alkyne is released after oxidation by selectfluor of the acetylide Au(I) complex, that the amount of free alkyne remaining in solution during the coupling is much higher for the C_{12} than for the C_{10} gold(I) acetylide, and that the released alkyne re-enters into the catalytic cycle. If anhydrous acetonitrile with a 0.5% (v:v) of D_2O is used as a reaction solvent (instead of non-dried acetonitrile) for the oxidation of 3b, much of the alkyne released (~80%) is d^1-1-dodecyne 1b-d^1, which suggests that the released alkyne takes the proton from the solvent.

^{31}P NMR spectroscopy also shows significant differences during the oxidation by selectfluor of 3a and 3b. For the former (C_{10}), the original peak of phosphine 3a at ~42 p.p.m. disappears as the oxidation converts and the generation of a single new peak at ~32 p.p.m. corresponding to PPh$_3$AuBF$_4$ is observed. In contrast, 3b (C_{10}) shows that the original peak (~42 p.p.m.) is transformed into PPh$_3$AuBF$_4$ (32 p.p.m.) and also into free O=PPh$_3$ (28 p.p.m.) in a 1:2 molar ratio. Despite homocoupling of the alkyne being incomplete, ^{1}H and ^{19}F spectroscopy clearly show the total consumption of selectfluor. Kinetic measurements (see Fig. 3 above) show that as soon as the formation of the homocoupling C_{12} product 2b stops during the oxidation of 3b with selectfluor, the formation of free O=PPh$_3$ increases. A new downshifted peak attributable to a phosphine oxide-gold complex was found during reaction by ^{31}P NMR spectroscopy

(Supplementary Fig. 3). No additional ^{31}P peaks were found during oxidation, which suggests that phosphine ligands are not detached during reaction, at least at a measurable rate, and that transmetallation (if occurs) and reductive elimination are much faster processes than gold(I) oxidation since no phosphine intermediates are detected. These results indicate that the oxidation of the phosphine follows the oxidation of gold(I) to gold(III), and when the former occurs the active acetylide gold species decompose to $O=PPh_3$, free terminal alkyne and gold[19,20].

In any case, the results showed above are consistent with the oxidation by selectfluor of the gold(I) acetylide complex independently of the size of the alkyne, with release of homocoupling product and also of free alkyne into the solution. This free alkyne is able to re-enter into the catalytic cycle provided the gold active species is still active. In the case of the

C_{10} acetylide **3a**, the amount of free alkyne **1a** detected in the solution is very low and the main alkyne product is the homocoupling product **2a**. In contrast, the C_{12} acetylide **3b** leaves significant amounts of 1-docecyne **1b** in solution which, at the end, does not form back productive acetylide gold complex for further homocoupling, at least at a rate that could compete with the destruction of the gold catalyst by oxidation of the metal ligand PPh_3 to $O=PPh_3$.

The results in Figs 1 and 3 suggest that the distal size selectivity observed for gold occurs after oxidation, when both species Au(I) and Au(III) are already present in the reaction medium. Previous mechanistic studies unveiled that the gold-catalysed oxidative homocoupling of alkynes in solution proceeds through a bimetallic Au(I)/Au(III) acetylide transition state[17], and density functional theory calculations confirm the feasibility of a digold intermediate species (Supplementary Fig. 4). The digold

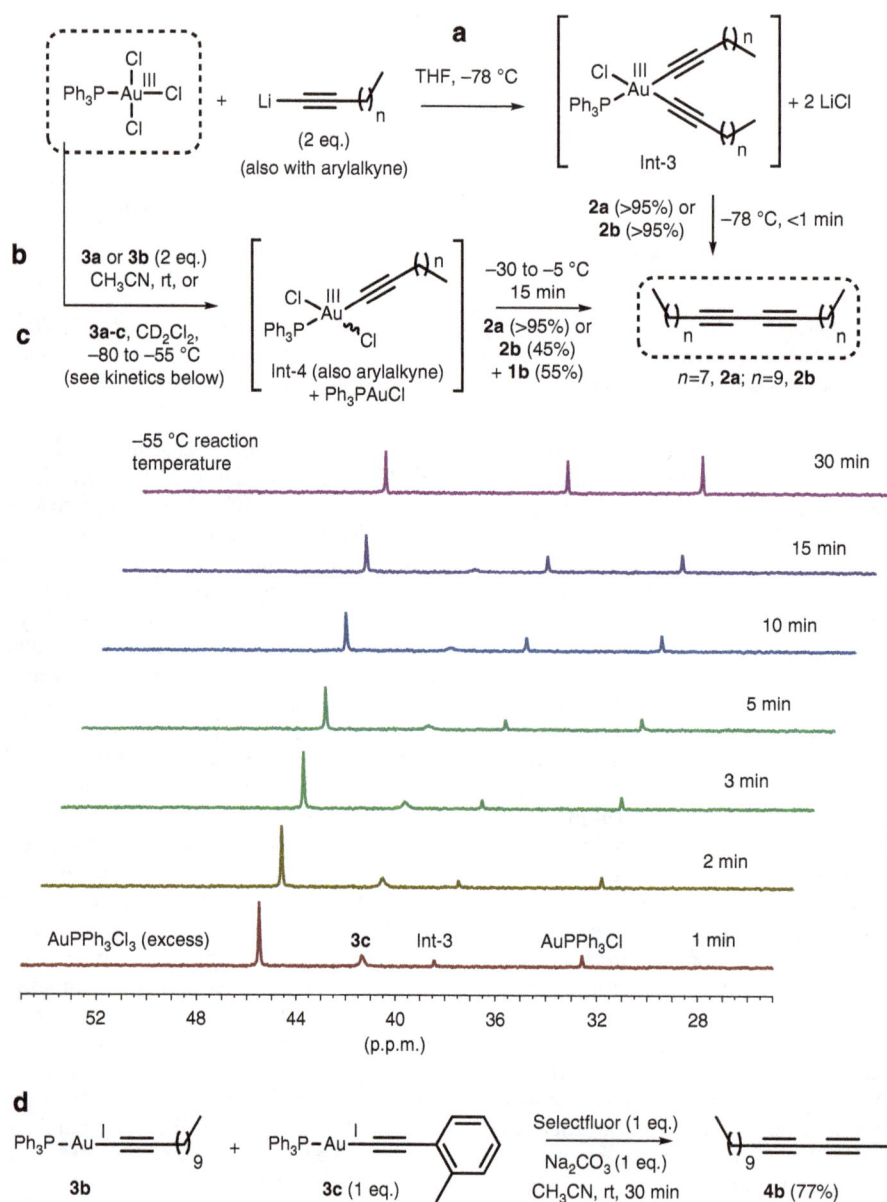

Figure 4 | Studies on the reductive elimination step. The top schemes (**a**–**c**) show the reductive elimination or transmetallation between gold and/or lithium acetylides in stoichiometric amounts. GC yields refer after extraction of the reaction mixture with n-hexane and addition of dodecane as an external standard. The kinetics at low-temperature of the transmetallation of tolylacetylene from **3c** to an excess (3 equiv.) of $AuPPh_3Cl_3$, followed by ^{31}P NMR, is also included. Spectra were recorded every 30 s after 1 scan (2.5 s), and a total of 60 measurements were taken. The in situ oxidation and transmetallation of gold(I) alkyl and aryl acetylides is shown at the bottom (**d**).

intermediate sets the Au(I)/Au(III) cations in position and, as a consequence, the alkynes are oriented in a *cisoid* conformation. If one compares this reaction intermediate for gold with that widely accepted for copper[21] consisting in a tetrahedrically coordinated dimeric Cu(II) acetylide species with the alkynes oriented in opposite directions, one can attempt to explain the different behaviour of gold and copper on the basis of the different geometries.

To assess if the spatial approaching of the two gold atoms has any influence on the reaction outcome, three different diphenylphosphine-bridged digold(I) complexes **5a–c** with one (methylene), two (ethylene) and three (propylene) methylene units in the tether were prepared (Supplementary Fig. 5 and Supplementary Discussion)[22,23]. X-ray Diffraction together with cyclic voltammetry studies showed that while **5a** presents aurophilic bonding between two gold atoms[14], **5c** has no aurophilic bonds at all[14] and **5b** has an intermediate behaviour, depending on the counteranion[24,25]. The results indicate that aurophilic bonding has a positive influence on the reaction rate but none on the distal size selectivity. Since previous kinetic studies have demonstrated that the rate-determining step of the gold-catalysed homocoupling of alkynes in solution is the oxidation of gold(I)[17], it is not surprising that aurophilic bonding improves the reaction rate by neighbouring gold atom-assisted oxidation of gold(I)[26,27]. To test if aurophilic bonding has also any influence in the distal size selectivity, we prepared the digold acetylide complexes **6a** [Au$_2$dppe(C$_{10}$)$_2$] and **6b** [Au$_2$dppe(C$_{12}$)$_2$], and the results after oxidation with selectfluor were compared with those obtained before for the corresponding monogold acetylides **3a–b**. The results were nearly identical, with the same amounts of homocoupling products **2a** (>95%) or **2b** (45%) obtained, and free alkyne 1-dodecyne **1b** (~50%) released into solution for the case of the dodecynyl acetylide **6b**. Cyclic voltammetry (Supplementary Figs 6 and 7 and Supplementary Discussion) confirmed that, in all the complexes tested, oxidation of Au(I) to Au(III) occurs without detecting any Au(II) species[9,16,23,28]. These results validate the mixed-valence digold intermediate **Int-1** and clearly show that aurophilic bonding improves the reaction rate but that it is not involved in the distal size selectivity.

The reductive elimination from Au(III) complexes to form new C$_{sp3}$–X and C$_{sp2}$–X bonds (X = heteroatom, C$_{sp3}$, C$_{sp2}$,…) is a well-known process[13,29–32] that has been recently reported to occur very rapidly (even at −52 °C)[33], faster than in palladium (II). Au(III) intermediates have been isolated and characterized[34–36]. Despite the great body of work reported in the literature for alkynes and gold, the reductive elimination of C$_{sp}$ bonds from Au(III) complexes has not been reported yet as far as we know[8], and very few examples of stable Au(III)-diacetylide complexes appear in the literature[37,38]. This lack of precedents for Au(III)-diacetylides might indicate that the reactivity of the alkynes once σ-bound to Au(III) is very high, including a potential reductive elimination. It has been proposed that the reductive elimination from a tetracoordinated planar Au(III) complex is favoured respect to other metals[33] since it leaves behind a two-coordinated Au(I) complex stabilized by relativistic effects[39–41]. To study the reductive elimination of alkynes from Au(III) and also if the distal size selectivity occurs at this stage, the Au(III) complex PPh$_3$AuCl$_3$ was prepared[42] and the corresponding Au(III) acetylide complex of 1-dodecyne **1b** was forced to be formed[43]. To do that, *n*-BuLi was added to 1-dodecyne **1b** in tetrahydrofurane (THF)-*d^8* at −78 °C and the resulting solution was added to PPh$_3$AuCl$_3$ suspended (not soluble) in THF-*d^8* at room temperature (2 equiv. of lithium acetylide with respect to gold). The result is shown in Fig. 4a. The bright-yellow complex PPh$_3$AuCl$_3$ became immediately

(<1 min) colourless and soluble. ^{31}P NMR spectroscopy of the solution just after addition of the acetylide solution shows the complete disappearance of the original signal of PPh$_3$AuCl$_3$ (42 p.p.m.) and the appearance of a new single peak at 33 p.p.m. that could correspond to AuPPh$_3$Cl. Addition of *n*-hexane to the THF solution precipitates a white solid that by isolation and characterization with ^{1}H and ^{31}P NMR spectroscopy was confirmed to be AuPPh$_3$Cl, the corresponding phosphine peak being observed during reaction. Analysis by GC–MS of the *n*-hexane supernatant shows that the homocoupling C$_{12}$ product **2b** is the only alkyne product of the reaction, no traces of free 1-dodecyne **1b** being present. A similar result is obtained when PPh$_3$AuCl$_3$ is treated with the lithium acetylide of 1-decyne **1a** since AuPPh$_3$Cl and the C$_{10}$ homocoupling product **2a** are the only products of the reaction. These results indicate that the reductive elimination of two acetylide fragments from a single Au(III) atom is very fast, in accordance with recent precedents in the literature that reports the efficient formation of C$_{sp2}$–C$_{sp2}$ bond with Au(III) via reductive elimination at −52 °C reaction temperature[30]. Thus, we must conclude that the distal size selectivity does not occur during the final reductive elimination but occurs in an intermediate step between oxidation and reductive elimination, since reductive elimination of the two alkyne molecules from Au(III) is extremely fast and does not give any distal size selectivity.

Gold is the most electronegative metal of the periodic table and, consequently, transmetallation of carbon ligands from Au(I)–C and Au(III)–C bonds to other metals is favoured and well-reported in the literature[44–46]. However, transmetallation from Au(I) to Au(III) is scarcely reported[29] although the migration of a carbon ligand from the more electronegative Au(I) to the more electropositive Au(III) cation should be favoured. Figure 4b shows the results obtained when the C$_{10}$ Au(I) acetylide **3a** was mixed with the Au(III) complex PPh$_3$AuCl$_3$ in acetonitrile. In this case, the homocoupling product **2a** is formed as only alkyne product in <5 min at room temperature. This result suggests that the transmetallation of the C$_{10}$ acetylide fragment of Au(I) complex **3a** to the Au(III) cation readily occurs. When the C$_{12}$ Au(I) acetylide **3b** was used, the homocoupling product **2b** was obtained together with free 1-dodecyne **1b**, as described before. To better study this step, we followed the reaction of PPh$_3$AuCl$_3$ with **3a** (C$_{10}$) or **3b** (C$_{12}$), and also with Au(I) arylacetylide **3c** (PPh$_3$Au-*ortho*-tolylacetylene), by means of low-temperature ^{31}P NMR spectroscopy (Fig. 4c and kinetics). For doing that, the NMR tube containing a solution of PPh$_3$AuCl$_3$ in CD$_2$Cl$_2$ was placed in liquid nitrogen (−196 °C) and then a solution of the corresponding Au(I) acetylide **3a–c** in CD$_2$Cl$_2$, cooled at −78 °C (acetone-dry ice), was slowly added into the NMR tube. After that, the tube was rapidly transferred to the NMR equipment at −80 °C, and measurements were started after shimming. It was found that **3a** or **3b** react with PPh$_3$AuCl$_3$ in <5 min at −80 °C to give 0.5 equiv. of PPh$_3$AuCl and 0.5 equiv. of an unknown phosphine product at 38 p.p.m. and that this intermediate (**Int-4**) is stable at least up to −30 °C. No differences between **3a** (C$_{10}$) and **3b** (C$_{12}$) were found here. When the NMR tube was left to warm, the only peak found by ^{31}P NMR was that corresponding to PPh$_3$AuCl. ^{1}H and ^{13}C NMR confirmed the formation of the homocoupling product **2a** from **3a** and the mixture of **1b** and **2b** from **3b**.

The reaction of PPh$_3$AuCl$_3$ with **3c** proceeds slower and a more accurate kinetics could be carried out at −55 °C. Again, PPh$_3$AuCl and the unknown peak at 38 p.p.m. were the only products formed in equimolecular amounts. *In situ* ^{1}H measurements of the reaction mixture at −55 °C showed a new peak at 2.40 p.p.m. downshifted −0.05 p.p.m. from the original methyl group of the Au(I) tolylacetylide **3c** (2.45) and also downshifted

(-0.11 p.p.m.) from the homocoupled product (2.51). It is perhaps more informative the result obtained by *in situ* ^{13}C NMR measurement at $-55\,^\circ$C. It was found that new peaks corresponding to the arene ring of the alkyne appear and that they present the typical J_{C-P} of the phosphine Au-acetylide complexes and, consistently, that the methyl peak is also shifted with respect to the starting material. The ^{13}C NMR spectrum strongly suggests that the unknown species at 38 p.p.m. is a new phosphine Au-acetylide that, according to the formation of 1 equiv. of AuPPh$_3$Cl per molecule of **3c** consumed, it might correspond to PPh$_3$Au(III)-tolylacetylene. A similar result is found when lithium tolylacetylide is used. The downshift of the methyl group in ^{1}H NMR points also in that direction. To further assess if **Int-4** is the result of one alkyne transmetallation from **3c** to PPh$_3$AuCl$_3$, the reaction was carried out with an excess (3 equiv.) of the latter to maximize monotransmetallation. Kinetics of the reaction in Fig. 4 gives an initial rate of 0.28 s^{-1}. *In situ* ^{1}H and ^{13}C measurements of the reaction mixture at $-55\,^\circ$C after consumption of the Au(I) acetylide **3c** gave neat spectra with the same phosphine intermediate found above. As for **3a** and **3b**, this intermediate is stable up to $-5\,^\circ$C and only at this temperature it starts to further convert into PPh$_3$AuCl. The kinetic results together with the stoichiometry of the reaction suggest that the gold(I)-acetylides **3a–c** transfer the first alkyne group to the Au(III) atom of PPh$_3$AuCl$_3$ at low temperature ($-80\,^\circ$C for alkyl, $-55\,^\circ$C for aryl) to give the intermediate **Int-4** PPh$_3$Au(III)-alkyne and that the second alkyne transfers at much higher temperature (-30 to $-5\,^\circ$C) to give **Int-3**. As soon as **Int-3** is formed, reductive elimination occurs and the homocoupling product is released. This means that the transmetallation of the second alkyne is much slower than the transmetallation of the first alkyne, since the former occurs at much higher temperature, and that the reductive elimination is in turn much faster than the second alkyne transmetallation, which correlates well with the results observed with lithium acetylides. In short, these results show that **Int-4** is an intermediate of the reaction and that it could very well occur that the mixture of Au(I) acetylide and **Int-4** is responsible for the distal size selectivity observed.

To mimic better the gold-catalysed system when having separated Au(III) and Au(I) acetylides, and to avoid the use of PPh$_3$AuCl$_3$, we made use of the different rates of oxidation with selectfluor of alkyl and aryl acetylides. It was shown above that the oxidation of the alkyl acetylide complexes **3a–b** (monogold) and **6a–b** (digold) by selectfluor proceeds in $\sim$30 min, and we

have previously shown that the same oxidation for PPh$_3$Au-*ortho*-tolylacetylene **3c** takes place in $\sim$4 h^{17}. Thus, in principle, the addition of selectfluor to a mixture of an Au(I) alkyl acetylide and an Au(I) aryl acetylide should oxidize first the alkyl Au(I) complex and form the corresponding Au(III) alkyl acetylide. Once the Au(III) alkyl acetylide is formed, the Au(I) aryl acetylide can transmetalate. It is true that while the oxidation with selectfluor occurs, the remaining Au(I) alkyl acetylide (not oxidized yet) could compete with the Au(I) aryl acetylide to transmetalate. To avoid the self-coupling of the alkyl acetylide, we employed **3b** (C$_{12}$), since the rate of homocoupling is the lowest in the alkyl series and it should not compete with the aryl acetylide. Thus, the Au(I) alkyl acetylide **3b** and PPh$_3$Au-*ortho*-tolylacetylene **3c** were mixed in the same flask and then selectfluor was added. The result in Fig. 4d shows that the heterocoupled product **4b** is the major product of the reaction with a 77% yield, indicating that the transmetallation from Au(I) aryl acetylides to Au(III) alkyl acetylides readily occurs.

With the hope that the digold intermediate of unreactive 1-dodecyne could be trapped and characterized, we carried out the oxidation of complex **6b** with selectfluor at $-78\,^\circ$C in a CD$_2$Cl$_2$/CD$_3$CN solvent mixture. The results (Supplementary Fig. 8) show the desymmetrization of the original H and P peaks of **6b** into two new signals, which points to the formation of the divalent-mixed Au(I,III) complex. Complex **6b** was independently prepared without one of the dodecylide fragments and its ^{31}P NMR signal fits well with the one detected after oxidation of **6b** by selectfluor, which supports that one of the dodecylide fragments of **6b** leaves as soon as one Au(I) atom is oxidized to Au(III).

With all the above results, we can suggest that the distal size selectivity occurs just before the transmetallation step of one alkyl acetylide fragment from Au(I) to Au(III), during the formation of the mixed-valence digold (I,III) complex. The tetracoordinated planar Au(III) site and the linear Au(I) site impose a parallel disposition of the aliphatic chains of each acetylene, which resembles the formation of a highly stabilized self-assembled monolayer, in this case just via two contiguous aliphatic chains. For longer chains, their interaction is enhanced thus blocking the coupling of the alkynes, while for shorter chains the lateral interaction is weak and the reaction readily occurs.

Mechanism of gold-catalysed distal-selective alkyne coupling. Figure 5 shows the proposed mechanism to explain the origin

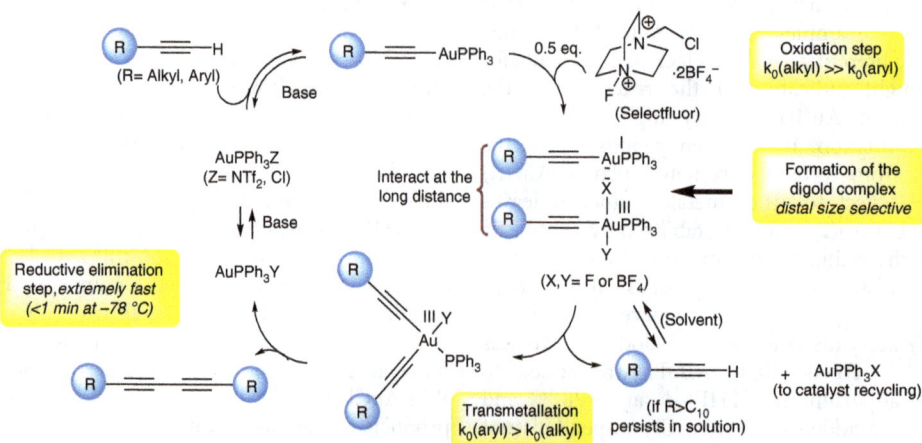

Figure 5 | Proposed mechanism for the gold-catalysed homocoupling of alkynes with distal size selectivity. The formation of a Au(I)/Au(III) digold intermediate is responsible for the distal steric differentiation.

of size selectivity during the gold-catalysed homocoupling of alkynes.

First, the linear Au(I) acetylide complex suffers the oxidation by selectfluor to give the corresponding Au(III) complex. This oxidation is favoured by a second neighbouring Au(I) acetylide molecule in solution but no size selectivity is found for this process, since selectfluor oxidation occurs at a similar rate for both C_{10} and C_{12} gold acetylides. At this point, we propose an Au(I)/Au(III) intermediate with a congested structure as responsible for the distal size selectivity observed, where the tetracoordinated planar Au(III) acetylide and the dicoordinated linear Au(I) acetylide present a parallel disposition of the aliphatic chains of each acetylene, which resembles the formation of a highly stabilized self-assembled monolayer. This intermediate is supported here experimentally by the similarity of the results obtained with the digold complexes 5a–b and the corresponding acetylides of 5b (complexes 6a–b), since these complexes have a locked cisoid conformation of acetylides and phosphines. Notice that this particular steric scenario, created by the interaction of a tetracoordinated planar metal complex and a linear metal complex, cannot be found in any metal redox pair of the periodic table but only in Au(I)/Au(III)[40]. In addition, this Au(I)/Au(III) intermediate suitably accommodates acetylides fully substituted in contiguous positions to the triple bond such as *tert*-butylacetylene 1c, which explains the feasibility of couplings with hindered alkynes around rather than along the triple bond. The experimental results also suggest that protodeauration competes with the transmetallation of the linear alkyl acetylide from Au(I) to Au(III), and that free acetylene is also released after decomposition of the fully oxidized gold acetylide. The alkyne can rebind to the Au(I) atom and enter back into the catalytic cycle provided the acetylide is still active.

Gold-catalysed heterocoupling of alkynes. The heterocoupling of terminal alkynes has been reported in the literature with different copper catalysts but, in all cases, a five-times amount of one of the alkynes must be added to achieve good yields of heterocoupling product[47–50]. Otherwise, with a one-to-one molar ratio of alkynes, a statistical 1:2:1 molar ratio of homo:hetero:homocoupled products, or a 1:1:1 molar ratio in some cases, is obtained[47–50]. In addition, the two alkynes must be differentiated either electronically (aryl or heteropropargyl versus alkyl alkynes, activated versus deactivated aryl alkynes) or sterically near the triple bond (linear versus bulky alkynes). The high excess of one of the alkyne reactants together with the need of electronic and/or steric differentiation of the terminal triple bond limits severely the application of these methodologies in organic synthesis. A much better methodology would consist in a catalytic system able to couple two different terminal alkynes regardless of the nature of the terminal triple bond and without the need of adding an excess of one of them. When we carried out the heterocoupling of different alkynes in equimolecular amounts under the present gold-catalysed conditions, we found that the heterocoupling products were formed in moderate to good yields and with high selectivity (Supplementary Fig. 9)[51,52].

One must not be surprised with these results when considering the high yield of heterocoupling obtained when we mix the C_{12} Au(I) alkyl acetylide 3b and PPh$_3$Au-*ortho*-tolylacetylene 3c (see Fig. 4d above). If we compare the initial rate of oxidation for the Au(I) alkyl acetylides (220 h^{-1}, see above) with that of the aryl acetylide (45 h^{-1})[17], a five-times relationship k_0(alkyl)/k_0(aryl) = k_{rel}(oxidation) ≈ 5 is found. However, the initial reaction rate for the homocoupling of 1-decyne 1a under the gold-catalysed reaction conditions indicated above (Fig. 1) is approximately three times lower than that for *ortho*-tolylacetylene

1e (ref. 17), k_0(alkyl)/k_0(aryl) = k_{rel}(homocoupling) = 0.36. Therefore, the rate of oxidation of Au(I)-acetylide and the rate of formation of homocoupling product are decoupled for alkyl and aryl alkynes, with an estimated value of k_{rel}(oxidation)/k_{rel}(homocoupling) ≈ 14. Since the last step, the reductive elimination from Au(III) is extremely fast (less than a minute at −78 °C) for both alkyl and aryl alkynes, the cross-coupling of aryl and alkyl terminal alkynes is favoured under the present gold-catalysed conditions. Not only that, but the cross-coupling between two different aryl alkynes and also between two different alkyl alkynes is also performed. The coupling of two different alkynes based on gold-catalysed conditions opens a reactivity window for carrying out oxidative heterocouplings of terminal alkynes in 1:1 molar ratio with yield and selectivity clearly beyond the statistical range[53,54].

Discussion

A near atomically precise distal size selectivity occurs during the oxidative coupling of terminal alkynes under gold-catalysed conditions. The formation of a crowded Au(I)/Au(III) digold intermediate is responsible for the distal steric differentiation. Oxidation by selectfluor, transmetallation and reductive elimination do not produce distal size selectivity. The reductive elimination of two alkynes from a single Au(III) atom is extremely fast and occurs in less than a minute at −78 °C. The subtle steric and electronic discrimination of alkynes by this gold-catalysed system allows the heterocoupling of two different alkynes in equimolecular amounts regardless of the nature of the terminal triple bond.

References

1. Leow, D., Li, G., Mei, T.-S. & Yu, J.-Q. Activation of remote meta-C–H bonds assisted by an end-on template. *Nature* **486**, 518–522 (2012).
2. Tang, R.-Y., Li, G. & Yu, J.-Q. Conformation–induced remote meta-C–H activation of amines. *Nature* **507**, 215–220 (2014).
3. Denayer, J. F. M. *et al.* Rotational entropy driven separation of alkane/isoalkane mixtures in zeolite cages. *Angew. Chem. Int. Ed.* **44**, 400–403 (2005).
4. Corma, A., Rey, F., Rius, J., Sabater, M. J. & Valencia, S. Supramolecular self-assembled molecules as organic directing agent for synthesis of zeolites. *Nature* **431**, 287–290 (2004).
5. Cantín, A. *et al.* Synthesis and structure of the bidimensional zeolite ITQ-32 with small and large pores. *J. Am. Chem. Soc.* **127**, 11560–11561 (2005).
6. Hashmi, A. S. K. Gold-catalysed organic reactions. *Chem. Rev.* **107**, 3180–3211 (2007).
7. de Haro, T. & Nevado, C. On gold-mediated C–H activation processes. *Synthesis* **16**, 2530–2539 (2011).
8. Engle, K. M., Mei, T.-S., Wang, X. & Yu, J.-Q. Bystanding F + oxidants enable selective reductive elimination from high-valent metal centers in catalysis. *Angew. Chem. Int. Ed.* **50**, 1478–1491 (2011).
9. Zhang, L. A Non-diazo approach to α-oxo gold carbenes via gold-catalysed alkyne oxidation. *Acc. Chem. Res.* **47**, 877–888 (2014).
10. Boronat, M., Leyva-Pérez, A. & Corma, A. Theoretical and experimental insights into the origin of the catalytic activity of subnanometric gold clusters: attempts to predict reactivity with clusters and nanoparticles of gold. *Acc. Chem. Res.* **47**, 834–844 (2014).
11. Corma, A., Leyva-Pérez, A. & Sabater, M. J. Gold-catalysed carbon–heteroatom bond-forming reactions. *Chem. Rev.* **111**, 1657–1712 (2011).
12. Hashmi, A. S. K. Dual gold catalysis. *Acc. Chem. Res.* **47**, 864–876 (2014).
13. Brenzovich, Jr W. E. *et al.* Gold-catalysed intramolecular aminoarylation of alkenes: C-C bond formation through bimolecular reductive elimination. *Angew. Chem. Int. Ed.* **49**, 5519–5522 (2010).
14. Tkatchouk, E. *et al.* Two metals are better than one in the gold catalysed oxidative heteroarylation of alkenes. *J. Am. Chem. Soc.* **133**, 14293–14300 (2011).
15. Siemsen, P., Livingston, R. C. & Diederich, F. Acetylenic coupling: A powerful tool in molecular construction. *Angew. Chem. Int. Ed.* **39**, 2632–2657 (2000).
16. Stefani, H. A., Guarezemini, A. S. & Cella, R. Homocoupling reactions of alkynes, alkenes and alkyl compounds. *Tetrahedron* **66**, 7871–7918 (2010).

17. Leyva-Pérez, A., Doménech, A., Al-Resayes, S. I. & Corma, A. Gold redox catalytic cycles for the oxidative coupling of alkynes. *ACS Catal.* **2**, 121–126 (2012).

18. Hopkinson, M. N., Ross, J. E., Giuffredi, G. T., Gee, A. D. & Gouverneur, V. Gold-catalysed cascade cyclization-oxidative alkynylation of allenoates. *Org. Lett.* **12**, 4904–4907 (2010).

19. Liu, L.-P., Xu, B., Mashuta, M. S. & Hammond, G. B. Synthesis and structural characterization of stable organogold(I) compounds. Evidence for the mechanism of gold-catalysed cyclizations. *J. Am. Chem. Soc.* **130**, 17642–17643 (2008).

20. Ball, L. T., Lloyd-Jones, G. C. & Russell, C. A. Gold-catalysed oxidative coupling of arylsilanes and arenes: origin of selectivity and improved precatalyst. *J. Am. Chem. Soc.* **136**, 254–264 (2014).

21. Kürti, L. & Czakó, B. *Strategic Applications of Named Reactions in Organic Synthesis* 186 (Elsevier Academic Press, 2005).

22. Berners-Price, S. J. & Sadler, P. J. Gold(I) complexes with bidentate tertiary phosphine ligands: formation of annular vs. tetrahedral chelated complexes. *Inorg. Chem.* **25**, 3822–3827 (1986).

23. Mirabelli, C. K. *et al.* Antitumor activity of bis(diphenylphosphino)alkanes, their gold(I) coordination complexes, and related compounds. *J. Med. Chem.* **30**, 2181–2190 (1987).

24. Li, D., Hang, X., Che, C.-M., Lo, W.-C. & Peng, S.-M. Luminescent gold(I) acetylide complexes. Photophysical and photoredox properties and crystal structure of [{Au(C≡CPh)}₂(μ-PPh₂CH₂CH₂PPh₂)]. *J. Chem. Soc. Dalton Trans.* **19**, 2929–2932 (1993).

25. Brandys, M.-C., Jennings, M. C. & Puddephatt, R. J. Luminescent gold(I) macrocycles with diphosphine and 4,4-bipyridyl ligands. *J. Chem. Soc. Dalton Trans.* **24**, 4601–4606 (2000).

26. Fackler, J. & John, P. Metal–metal bond formation in the oxidative addition to dinuclear gold(I) species. Implications from dinuclear and trinuclear gold chemistry for the oxidative addition process generally. *Polyhedron* **16**, 1–17 (1997).

27. Fackler, J. & John, P. Forty-five years of chemical discovery including a golden quarter-century. *Inorg. Chem.* **41**, 6959–6972 (2002).

28. Doménech, A., Leyva-Pérez, A., Al-Resayes, S. I. & Corma, A. Electrochemical monitoring of the oxidative coupling of alkynes catalysed by triphenylphosphine gold complexes. *Electrochem. Commun.* **19**, 145–148 (2012).

29. Cui, L., Zhang, G. & Zhang, L. Homogeneous gold-catalysed efficient oxidative dimerization of propargylic acetates. *Bioorg. Med. Chem. Lett.* **19**, 3884–3887 (2009).

30. Zhang, G., Peng, Y., Cui, L. & Zhang, L. Gold-catalysed homogeneous oxidative cross-coupling reactions. *Angew. Chem. Int. Ed.* **48**, 3112–3115 (2009).

31. Hopkinson, M. N. *et al.* Gold-catalysed intramolecular oxidative cross-coupling of nonactivated arenes. *Chem. Eur. J.* **16**, 4739–4743 (2010).

32. Zhang, G., Cui, L., Wang, Y. & Zhang, L. Homogeneous gold-catalysed oxidative carboheterofunctionalization of alkenes. *J. Am. Chem. Soc.* **132**, 1474–1475 (2010).

33. Wolf, W. J., Winston, M. S. & Toste, F. D. Exceptionally fast carbon–carbon bond reductive elimination from gold(III). *Nat. Chem.* **6**, 159–164 (2014).

34. Hashmi, A. S. K. Homogeneous gold catalysis beyond assumptions and proposals-characterized intermediates. *Angew. Chem. Int. Ed.* **49**, 5232–5241 (2010).

35. Hofer, M. & Nevado, C. Unexpected outcomes of the oxidation of (pentafluorophenyl)triphenylphosphanegold(I). *Eur. J. Inorg. Chem.* **9**, 1338–1341 (2012).

36. Hashmi, A. S. K. *et al.* Dual gold catalysis: σ,π-propyne acetylide and hydroxyl-bridged digold complexes as easy-to-prepare and easy-to-handle precatalysts. *Chemistry* **19**, 1058–1065 (2013).

37. Méndez, L. A., Jiménez, J., Cerrada, E., Mohr, F. & Laguna, M. A Family of alkynylgold(III) complexes [Auᴵ(μ-{CH₂}₂PPh₂)₂Auᴵᴵᴵ(C≡CR)₂] (R = Ph, tBu, Me₃Si): facile and reversible comproportionation of gold(I)/gold(III) to digold(II). *J. Am. Chem. Soc.* **127**, 852–853 (2005).

38. Au, V. K.-M., Wong, K. M.-C., Zhu, N. & Yam, V. W.-W. Luminescent cyclometalated dialkynylgold(III) complexes of 2-phenylpyridine-type derivatives with readily tunable emission properties. *Chem. Eur. J.* **17**, 130–142 (2011).

39. Schwerdtfeger, P. Relativistic effects in gold chemistry. 2. The stability of complex halides of gold(III). *J. Am. Chem. Soc.* **111**, 7261–7262 (1989).

40. Gorin, D. J. & Toste, F. D. Relativistic effects in homogeneous gold catalysis. *Nature* **446**, 395–403 (2007).

41. Leyva-Pérez, A. & Corma, A. Similarities and differences between Gold, Platinum and Mercury "relativistic" triad in catalysis. *Angew. Chem. Int. Ed.* **51**, 614–635 (2011).

42. Leyva, A., Zhang, X. & Corma, A. Chemoselective hydroboration of alkynes vs. alkenes over gold catalysts. *Chem. Commun.* **33**, 4897–5044 (2009).

43. Usón, R., Laguna, A. & Vicente, J. Novel anionic gold(I) and gold(III) organocomplexes. *J. Organomet. Chem.* **131**, 471–475 (1977).

44. Khairul, W. M. *et al.* Transition metal alkynyl complexes by transmetallation from Au(C≡CAr)(PPh₃) (Ar = C₆H₅ or C₆H₄Me-₄). *Dalton Trans.* **4**, 610–620 (2009).

45. Chen, Y., Chen, M. & Liu, Y. Gold-catalysed cyclization of 1,6-diyne-4-en-3-ols: stannyl transfer from 2-tributylstannylfuran through Au/Sn transmetallation. *Angew. Chem. Int. Ed.* **51**, 6181–6186 (2012).

46. Hofer, M., Gomez-Bengoa, E. & Nevado, C. A Neutral gold(III) − boron transmetallation. *Organometallics* **33**, 1328–1332 (2014).

47. Yin, W., He, C., Chen, M., Zhang, H. & Lei, A. Nickel-catalysed oxidative coupling reactions of two different terminal alkynes using O₂ as the oxidant at room temperature: facile syntheses of unsymmetric 1,3-diynes. *Org. Lett.* **11**, 709–712 (2009).

48. Balaraman, K. & Kesavan, V. Efficient copper(II) acetate catalysed homo and heterocoupling of terminal alkynes at ambient conditions. *Synthesis* **20**, 3461–3466 (2010).

49. Xiao, R., Yao, R. & Cai, M. Practical oxidative homo- and heterocoupling of terminal alkynes catalysed by immobilized copper in MCM-41. *Eur. J. Org. Chem.* **22**, 4178–4184 (2012).

50. Navale, B. S. & Bhat, R. G. Copper(I) iodide-DMAP catalysed homo- and heterocoupling of terminal alkynes. *RSC Adv.* **3**, 5220–5226 (2013).

51. Ohashi, K. *et al.* Indonesian medicinal plants. XXV.1) Cancer cell invasion inhibitory effects of chemical constituents in the parasitic plant scurrula atropurpurea (Loranthaceae). *Chem. Pharm. Bull.* **51**, 343–345 (2003).

52. Xu, Z., Byun, H.-S. & Bittman, R. Synthesis of photopolymerizable long-chain conjugated diacetylenic acids and alcohols from butadiyne synthons. *J. Org. Chem.* **56**, 7183–7186 (1991).

53. Lee, S., Lee, T., Lee, Y. M., Kim, D. & Kim, S. Solid-phase library synthesis of polyynes similar to natural products. *Angew. Chem. Int. Ed.* **46**, 8422–8425 (2007).

54. Liu, J., Lam, J. W. Y. & Tang, B. Z. Acetylenic polymers: syntheses, structures, and functions. *Chem. Rev.* **109**, 5799–5867 (2009).

Acknowledgements

Financial support by Consolider-Ingenio 2010 (proyecto MULTICAT) and Severo Ochoa programs from MCIINN and Prometeo program from Generalitat Valenciana is acknowledged. A. L.-P. thanks ITQ for the concession of a contract. We thank Dr J.A. Vidal for assistance with the low-temperature NMR experiments, and Dr M. Boronat for the DFT calculations.

Author contributions

A.L.-P. carried out the experiments and wrote the manuscript. A.D.-C. carried out the electrochemistry and wrote the conclusions thereof. A.C. wrote the manuscript.

Additional information

21

Deciphering the origin of giant magnetic anisotropy and fast quantum tunnelling in Rhenium(IV) single-molecule magnets

Saurabh Kumar Singh[1] & Gopalan Rajaraman[1]

Single-molecule magnets represent a promising route to achieve potential applications such as high-density information storage and spintronics devices. Among others, $4d/5d$ elements such as Re(IV) ion are found to exhibit very large magnetic anisotropy, and inclusion of this ion-aggregated clusters yields several attractive molecular magnets. Here, using *ab intio* calculations, we unravel the source of giant magnetic anisotropy associated with the Re(IV) ions by studying a series of mononuclear Re(IV) six coordinate complexes. The low-lying doublet states are found to be responsible for large magnetic anisotropy and the sign of the axial zero-field splitting parameter (D) can be categorically predicted based on the position of the ligand coordination. Large transverse anisotropy along with large hyperfine interactions opens up multiple relaxation channels leading to a fast quantum tunnelling of the magnetization (QTM) process. Enhancing the Re-ligand covalency is found to significantly quench the QTM process.

[1] Department of Chemistry, Indian Institute of Technology, Bombay Powai, Mumbai 400076, India. Correspondence and requests for materials should be addressed to G.R. (email: rajaraman@chem.iitb.ac.in).

In the quest of single-molecule magnets (SMMs)[1-6] with enhanced magnetic properties, magnetic anisotropy is found to be the most influential parameter, which governs the barrier height for slow relaxation of magnetization[7-11]. Owing to inherently large magnetic anisotropy, lanthanide-based complexes are promising candidate for single-ion magnets[2,3,5,12-17] and mononuclear SMMs based on transition metal ions are relatively scarce in the literature, as stronger ligand field interactions suppress the orbital contributions to the anisotropy and hence the barrier heights (U_{eff})[18-25]. In the past few years, late-transition metal ions have gained much attention in the area of SMMs. The diffused magnetic orbitals of the $4d/5d$ ions translate stronger magnetic exchange, whereas larger spin-orbit coupling constants (SOCs) exhibited by these ions[26], often lead to highly anisotropic ground state (highly anisotropic g-tensors with an unusually large zero-field splitting values (ZFS)). These two essential conditions along with a possibility of exhibiting anisotropic/anti-symmetric exchange makes this class of molecules ideal for observing SMM behaviour[27-33]. Owing to these advantages, these $4d/5d$ ions show better SMM behaviour compared with their $3d$ congeners at many occasions[27,28,31-33]. Among the $4d/5d$ ions, the chemistry of Re(IV) metal ion is very rich as they have been successfully used to isolate several single-chain magnets (SCMs)/SMMs with an attractive U_{eff} values[26,34-41]. Apart from rich magnetic studies, these Re(IV) complexes are also explored in the development of the new anticancer drugs[42].

Despite several years of comprehensive experimental efforts in designing Re(IV) ion-based SMMs/SCMs, the origin of giant magnetic anisotropy is not well understood[39]. As in most of the cases, the axial ZFS (D) values are extracted using magnetization measurements, which are known to be insensitive to the sign and strength of the D parameter[41,43-51]. On other hand, the most promising high frequency-electron paramagnetic resonance (HF-EPR) technique has its own limitation, where the sign can be accurately determined but such large magnitude of D are often difficult to estimate[39]. An alternate solution to resolve the ambiguities in the sign/magnitude of D value is to analyse the ZFS parameter using ab initio calculations[9,22,39,52-57], which has been widely used in this respect. Moreover, strategic designing of new generation SMMs based on Re(IV) ions requires a thorough understanding of the nature ZFS and how the magnitude and the sign of the D and E vary depending on the ligand field environment. As the magnitude of E and the hyperfine interactions are correlated to the quantum tunnelling of magnetization (QTM)[58], the possibility to fine tuning these values is of paramount importance in this area.

The goal of the present communication is to gain a thorough understanding of the magnetic anisotropy in six coordinate Re(IV) complexes using state-of-the-art ab initio calculations. By modelling structurally diverse 13 mononuclear six coordinate Re(IV) complexes[39,41,43-51], we aim to answer the following intriguing questions (i) What is the suitable theoretical methodology to compute ZFS parameters in $5d$ transition metal ions such as Re(IV) complexes? (ii) What is the origin of giant D values and is there a correlation between the nature of the donor atoms and the sign of the D values in these complexes? (iii) What is mechanism of magnetic relaxation in Re(IV) single-ion magnets and how this is influenced by the metal–ligand covalency?

Results

Magnetic anisotropy and spin-Hamiltonian. The free Re(IV) ion is a d^3 Kramers ion with a ^4F ground state term, which splits into three states $^4A_{2g}$, $^4T_{2g}$ and $^4T_{1g}$ with the $^4A_{2g}$ being the ground state in an octahedral environment. Due to perfect cubic symmetry, pure octahedral complexes do not possesses any ZFS,

however, any distortions from the octahedral geometry are expected to yield large D values via mixing of the subsequent excited states because of very large SOC ($\lambda \sim 1,000\ cm^{-1}$). To begin with, we have studied the homoleptic $[ReCl_6]^{2-}$ model complex in tetragonal environment to analyse the origin of ZFS. Abragam and Bleaney have proposed a qualitative equation to predict the sign of the D values of a tetragonally distorted d^3 ion.

$$D = -\frac{4\lambda^2}{\Delta_0} + \frac{4\lambda^2}{\Delta_1} \qquad (1)$$

where λ is the SOC and Δ_0, Δ_1 is the tetragonal distortion parameter. If $\Delta_0 < \Delta_1$, the sign is predicted to be negative, whereas for $\Delta_0 > \Delta_1$, the sign is predicted to be positive. To corroborate this qualitative analysis, we have performed ab initio calculations, using complete active self consistent field (CASSCF) and CASPT2 methods incorporating spin-orbit effects with the RASSI-SO module in MOLCAS[59,60]. A positive D value of $+19.3\ cm^{-1}$ has been obtained for axially elongated model, whereas a negative D value of $-24.3\ cm^{-1}$ has been observed for axially compressed D_{4h} $[ReCl_6]^{2-}$ model complex (see Supplementary Fig. 1 and Supplementary Table 1 for details). The energy splitting pattern of the first three same spin-free states ($^4T_{2g}(F)$) are arranged as expected based on the ligand field theory and the sign of the D values computed using CASPT2 are in line with the expected values based on the equation (1). Although CASSCF calculations predict a similar splitting pattern of first three same spin-free states, it fails to reproduce the correct sign of the D values compared with CASPT2 methods for both elongated and compressed geometries. This suggests that spin-flip states rather than same spin-free states govern the sign as well as magnitude of D values for $5d$ elements such as Re(IV) ion. Hence, here after all the results discussed are performed at CASPT2 level of theory (vide infra).

To further understand the nature of D and E values, we have selected 13 mononuclear Re(IV) complexes and classified into three categories type-I: $[ReX_4(L)]$ (where L = a bidentate ligand on the equatorial plane), type-II: $[ReX_4(L)_2]$ (where L = monodentate ligand in the axial positions) and type-III $[ReX_5(L)]^-$ (where L = monodentate ligand; see Figs 1 and 2 for details). Continuous symmetry measure analysis (SHAPE)[61] of the X-ray structures reveals that all the complexes are in the distorted octahedral geometry (see Supplementary Fig. 2 and Supplementary Tables 2 and 3 for details).

Sign and magnitude of ZFS parameter for 1–13. Calculations reveal that eight spin-free states corresponding to ^2G states are found to be low-lying and thus are expected to contribute significantly to the D values via spin-flip excitations in all complexes **1–13** studied (see Supplementary Figs 3 and 4, Supplementary Tables 4–16 and Supplementary Note 1 for details). The MS-CASPT2 + RASSI-SO computed D, E and the first spin-free excitation energies for all complexes are depicted in Table 1. For complexes **1–6**, large negative D and significantly large $|E/D|$ values, with D as high as $\pm 132\ cm^{-1}$ (for **4**) have been

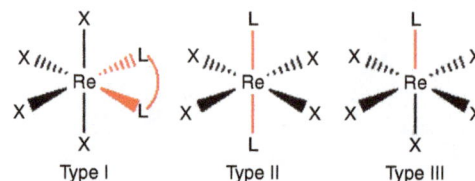

Figure 1 | Structural topology. Classifications of substituted hexa halo Re(IV) complexes (where X = Cl, Br and L = coordinating ligand).

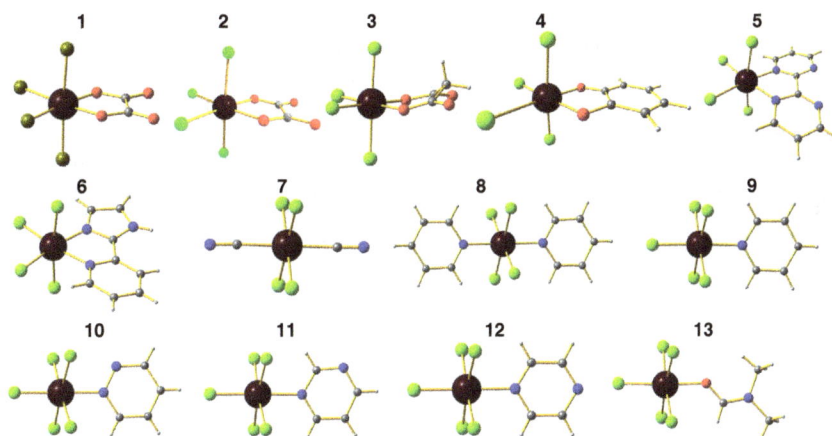

Figure 2 | X-ray crystal structures. Crystal structure of Re(IV) mononuclear complexes. Colour code: dark brown, Re; pale brown, Br; light green, Cl; blue, N; grey, C; white, H. $[ReBr_4(ox)]^{2-}$ (**1**); $[ReCl_4(ox)]^{2-}$ (**2**); $[ReCl_4(mal)]^{2-}$ (**3**); $[ReCl_4(cat)]^{2-}$ (**4**); $[ReCl_4(bpym)]$ (**5**); $[ReCl_4(pyim)]$ (**6**); $[ReCl_4(CN)_2]^{2-}$ (**7**); $[ReCl_4(py)_2]$ (**8**); $[ReCl_4(py)]^-$ (**9**); $[ReCl_5(pyz)]^-$ (**10**); $[ReCl_5(pyd)]^-$ (**11**); $[ReCl_5(pym)]^-$ (**12**); $[ReCl_4dmf]^-$ (**13**). bpym, 2,2'-bipyrimidine; cat, catechol; dmf, dimethyl formamide; mal, malonato; ox, oxalato; pyim, 2-(2'-pyridyl)imidazole; py, pyridine; pyd, pyridazine; pym, pyrimidine; pyz, pyrazine.

Table 1 | MS-CASPT2 + RASSI-SO computed D and $|E/D|$ value for all studied Re(IV) mononuclear complexes along with first spin-free excitation energy.

Complex	D_{cal}	$|E/D|_{cal}$	$D\ (|E/D|)_{exp}$	ΔE	References
1 $[ReBr_4(ox)]^{2-}$	− 93.0	0.18	− 73 (0.20)*	8,079.2	39
2 $[ReCl_4(ox)]^{2-}$	− 85.0	0.24	− 57 (0.26)*	8,873.5	39
3 $[ReCl_4(mal)]^{2-}$	− 61.6	0.15	55†	8,442.8	44
4 $[ReCl_4(cat)]^{2-}$	±132.6	0.30	95†	7,526.7	50
5 $[ReCl_4(bpym)]^{2-}$	− 47.9	0.23	−	7,616.5	46
6 $[ReCl_4(pyim)]^{2-}$	− 34.7	0.17	−	7,804.7	48
7 $[ReCl_4(CN)_2]^{2-}$	+ 16.2	0.23	+ 11 (0.29)*	7,377.3	41
8 $[ReCl_4(py)_2]$	+ 55.6	0.18	9.56†	5,950.5	43
9 $[ReCl_5(py)]^-$	±32.7	0.31	3.52†	7,841.7	51
10 $[ReCl_5(pyd)]^-$	+ 41.0	0.24	14.1†	7,385.9	47
11 $[ReCl_5(pym)]^-$	+ 24.6	0.17	6.2†	8,300.7	47
12 $[ReCl_5(pyz)]^-$	+ 32.5	0.20	9.4†	7,630.2	49
13 $[ReCl_5(dmf)]^-$	+ 18.5	0.18	10.1†	8,467.4	45

bpym, 2,2'-bipyrimidine; cat, catechol; dmf, dimethyl formamide; mal, malonato; ox, oxalato; pyim, 2-(2'-pyridyl)imidazole; py, pyridine; pyd, pyridazine; pym, pyrimidine; pyz, pyrazine.
All the D values and spin-free excitations energies are provided in cm^{-1}.
*HF-EPR reported values.
†Obtained from magnetic susceptibility measurements, no sign convention has been used.

witnessed. On other hand, complexes of type II and type III categories (**7–13**) found to posses positive ZFS parameter with D as high as $+ 55\ cm^{-1}$ has been noticed (for **8**; see Supplementary Fig. 5 and for the orientation of D tensor). The magnetic susceptibility and powder magnetization data computed for **1–13** reproduces nicely the experimental behaviour, adding confidence to the computed values (see Supplementary Figs 6–10 and Supplementary Note 2 for details). Simulation of HF-EPR spectra reported earlier[35,39] confirm the negative sign of the D with a large E/D values for complexes **1** and **2**, with the D estimated to be ca $- 73$ and $- 57\ cm^{-1}$, respectively. Calculations yield D value of $- 93$ and $- 85\ cm^{-1}$ for complexes **1** and **2**, where both the sign as well as the magnitude of the D values are correctly reproduced compared with the experimental values. More importantly, the magnitude of the $|E/D|$ values and g-tensors (both pseudo-spin 1/2 and 3/2), which are precisely estimated from the experiments are very well reproduced in our calculations (see Supplementary Tables 17 and 18 for details).

For complex **6**, the magnetization data[41] yield an estimate of D as $- 14.4\ cm^{-1}$, which is in agreement with CASSCF results

($- 21\ cm^{-1}$, see Supplementary Table 17) but in disagreement with MS-CASPT2 values ($+ 16.2\ cm^{-1}$, see Table 1). However, HF-EPR experiments performed lately[35], where both the magnitude as well as the sign of the D value is estimated accurately, places the D value to be $+ 11\ cm^{-1}$. This highlights the issue of obtaining the sign/magnitude of the D value from the magnetization data and also emphasise the need for CASPT2 approach and hence incorporation of dynamic correlation to correctly reproduce the sign of the D values[62,63]. Inclusion of dynamic correlation on CASSCF computed wavefunctions drastically stabilizes the doublet states compared with the computed CASSCF states, leading to a pronounced contribution from these states to the D values as discussed earlier (see Supplementary Fig. 3 for details). Moreover, either pseudo-spin or effective Hamiltonian approach[62] needs to be employed to extract the ZFS parameters as other theoretical methodologies found to yield ambiguous sign and magnitude of D values (see Supplementary Tables 19 and 20 for details).

The SINGLE_ANISO computed orientations of the main anisotropic axes (D_{XX}, D_{YY} and D_{ZZ}) and main magnetic axes (g_{XX}, g_{YY} and g_{ZZ}) for all the complexes **1–13** are provided in

Fig. 3 and Supplementary Fig. 5. It is evident from the figures that, the principal anisotropic axes (D_{ZZ}) are oriented towards the L_{ax}–R–L_{ax} (molecular –z axis) direction, however, a significant tilt from this axis is witnessed[64]. Moreover, the main magnetic axes (g_{XX}, g_{YY} and g_{ZZ}) and main anisotropic axes (D_{XX}, D_{YY} and D_{ZZ}) do not coincide with each other and such non-coincidence has been previously noticed by Askevold et al.[65] The orientation of the D_{ZZ} axis is tilted by 28.8°, 36° and 33.7° from their molecular –z axis for complexes 1, 7 and 10, respectively. Large structural distortions and the associated large |E/D| values are responsible for such deviations. Larger |E/D| values detected in complex 7 has larger tilt compared with complex 1, where smaller |E/D| value leads to smaller tilt. This trend is visible also for other structures. However, in the rhombic limit, the nature of the easy axis is ambiguous, as even the small structural distortions flip the Eigen values and hence the orientation of the easy axis of magnetization. This can be better visualized for complex 4, where presence of large |E/D| value of 0.30 causes the flipping of D_{ZZ} axis to the equatorial plane.

Origin of ZFS in complexes 1–13. To shed light on the sign of ZFS, we have analysed the molecular orbitals (MOs) of complexes 1 and 7 and 10 as a representative examples for type I, II and III

defined earlier (see Fig. 4 for details). For 1, presence of unsymmetrical ligand in the equatorial position leads to the d_{xz} orbital being the lowest lying in energy followed by degenerate d_{yz} and d_{xy} orbitals. The π^* orbitals of the oxygen donors interact rather strongly with the d_{yz} and d_{xy} orbitals because of acute ∠O-Re-O bite angle. The d_{xz} orbital on the other hand faces less repulsion leading to a slight stabilization. Besides, stronger σ^* interaction by the oxalate ligand lead to destabilization of the $d_{x^2-y^2}$ orbital compared with the d_{z^2} orbital. This orbital ordering has the following consequences to the D values: (i) all spin-conserved excitations from the d_{xy}, d_{xz} and d_{yz} orbitals to the vacant d_{z^2} orbital contribute to positive D values. This is affirmed by an additional calculation incorporating only the quartet states in the estimation of D and this yield a positive D value (+13 cm^{-1}). (ii) Spin-flip excitations from the d_{xz} to the d_{yz} orbitals contribute to negative D values (excitation between same |m$_l$| levels). As the gap between these two orbitals is very small (667 cm^{-1}), this transition governs both the sign and the magnitude of the D value for this complex. A similar pattern predicted for complexes 2–6 rationalize the observed negative sign for these complexes.

Strong π acceptor cyanide ligand stabilizes the d_{xz} and d_{yz} orbitals via $p\pi$–$d\pi$ interactions compared with the d_{xy} orbital. Stronger σ-donation along the axial directions destabilizes the d_{z^2}

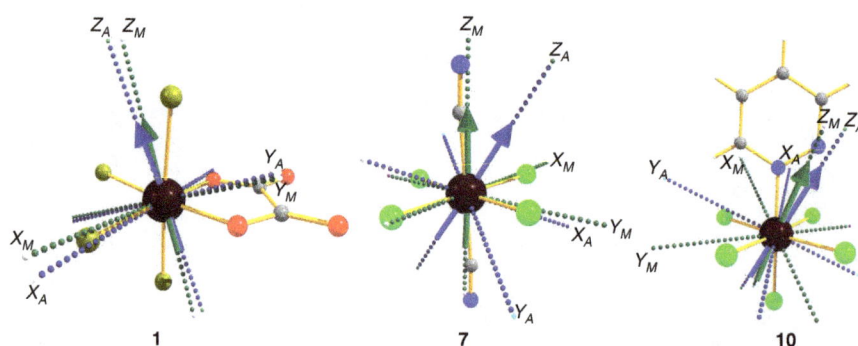

Figure 3 | Orientation of g and D tensors. SINGLE_ANISO computed main magnetic (X_M, Y_M and Z_M) axes representing g-tensors orientation and main anisotropic axes (X_A, Y_A and Z_A) representing D tensors orientation for complexes 1, 7 and 10.

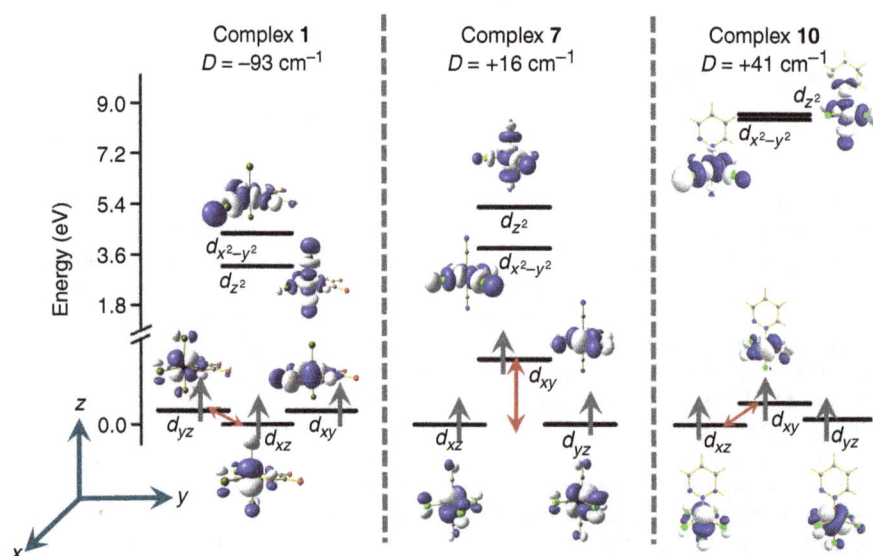

Figure 4 | Molecular orbital analysis and nature of excitations. Computed d-orbital ordering for complexes 1, 7 and 10. The iso-density surface plotted with the iso-value of 0.02 e$^-$/bohr^3. The double headed arrow represents the gap between the orbitals, which are contributing significantly to the D value. The orbitals which are appeared as degenerate in the figure are not strictly degenerate due to symmetry arguments.

orbital leading to different orbital arrangement compared with complex **1**. This orbital arrangement has the following consequences to the D value: (i) the spin-conserved $d_{xy} \to d_{x^2-y^2}$ transition contributes to the negative D value, affirmed by an additional calculation incorporating only the quartets in the estimation of D and this yields $-26\,cm^{-1}$ as the D value, (ii) the spin-flip excitations from the degenerate $d_{xz}-d_{yz}$ orbitals to the d_{xy} orbital contribute to the positive D value (the gap here is $3{,}123\,cm^{-1}$). Here also the later term dominates leading to an overall positive D value for **7**.

For complex **10**, positive D value is expected as the splitting pattern is found to be very similar to that of complex **7** (Fig. 4). However, the presence of weak π-donor pyridazine ligand on the axial position reduces the splitting between the $d_{xz}-d_{yz}$ orbital and the d_{xy} orbital ($1{,}044\,cm^{-1}$). This suggests that much less energy is required to flip the spin in complex **10** compared with complex **7**, thus the D value is expected to be large in this case. A similar splitting pattern is predicted for complexes **8**, **9** and **11–13** rationalize the observed positive D values for these complexes.

Rationale for the observed variation in the ZFS parameter. Among **1–6**, the equatorial positions are occupied by π-donor ligand, except in case of complex **5** where 2,2'-bipyrimidine (bpym) ligand serves as a weak π-acceptor ligand. Independent of the nature of the ligand (π-donor versus π-acceptor), in complexes **1–6**, the $d_{yz} \to d_{xz}$ transition dominates the D value over other transitions, leading to a large negative D values. To understand the large difference in the D values of complexes **1** and **2**, we have performed additional calculations on model complexes where Cl in complex **1** is modelled as Br maintaining Re–Br distance same as that of complex **2**. For this model, the D is estimated to be $-86\,cm^{-1}$ compared with $-85\,cm^{-1}$ for the Cl analogue and this suggest that apart from the spin-orbit coupling of Br, the structural distortion such as –cis angles play an important role in determining the strength of the D value (see Supplementary Table 3 for selected structural parameters of complexes **1** and **2**)[55].

The strength of the donor–acceptor abilities significantly affects the magnitude of the D values (see Fig. 5 and Supplementary Table 21 for further details)[20,62]. In complexes **1–6**, larger charge on the donor atoms are found to yield large D values (see Fig. 4 and Supplementary Figs 11 and 12 for quantitative charges computed). Larger charges on the donor atoms stabilizes the d_{xz} orbitals compared with the d_{xy}/d_{yz} orbitals leading to different transition energies (see Table 1 for details) and thus the computed charges are found to strongly correlated to the magnitude of the D values. This striking observation offers a rational approach to fine tune the magnitude of the negative D value in this set of complexes. Moreover, stabilization of d_{xz} orbital also affects the E values, as it increases the difference between the D_{XX} and D_{YY} contributions leading to a larger E with large charge on the ligand (see equation 2 and Supplementary Fig. 12 for details).

$$D = D_{zz} - \tfrac{1}{2}(D_{XX} + D_{YY}); E = (D_{XX} - D_{YY})$$
$$D_{ZZ} \sim \left[E(d_{xz}) - E(d_{yz})\right]^{-1}$$
$$D_{XX} \sim \left[E(d_{xz}) - E(d_{xy})\right]^{-1} \qquad (2)$$
$$D_{YY} \sim \left[E(d_{xz}) - E(d_{z^2})\right]^{-1}$$

In contrast to complex **7**, where axial positions are occupied by two strong π-acceptor ligands, complex **8** posses two weak π-donor (pyridine) ligands on the axial position and therefore the $d_{xz}-d_{yz}$ orbital to d_{xy} orbital is found to be ($1{,}360.7\,cm^{-1}$) much smaller than that of $3{,}123\,cm^{-1}$ observed in case of complex **7**. Moreover, the first spin-flip-excited state is found at $5{,}950.5\,cm^{-1}$ (in case of complex **8**), which is again much

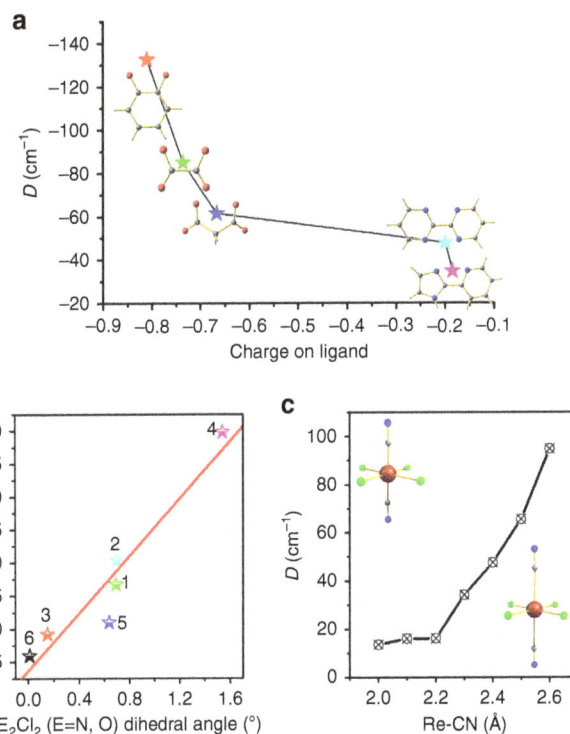

Figure 5 | Impact of structural distortions on magnetic anisotropy. (**a**) Plot of computed D value versus charge of the coordinated ligand atoms. (**b**) Plot of the computed $|E|$ value versus Re(O/N)$_2$Cl$_2$ dihedral angle; (**c**) magneto-structural correlation by varying Re–CN bond distances of complex **7**.

smaller than the $7{,}377.3\,cm^{-1}$ gap observed for complex **7** (see Table 1 for details). This leads to larger D value for complex **8** compared with complex **7**.

Complexes **9–13** possess positive D values ranging from $+18\,cm^{-1}$ (complex **13**) to $+41\,cm^{-1}$ (complex **10**). As the structural parameters across the series are very similar, the differences in the magnitude of the D values are expected to arise from the donor strength of the ligand. To affirm this point, we have analysed the donor–acceptor interactions using second-order perturbation theory natural bonding orbitals (NBO) analysis for complexes **10** and **13**. NBO analysis suggests a significant σ-donation from lone pair of nitrogen to Re d_{z^2} orbital and this strength is estimated to be $19.6\,kcal\,mol^{-1}$ for complex **10**, whereas $22.6\,kcal\,mol^{-1}$ for complex **13** (see Supplementary Fig. 13 for details). Larger σ-donation in complex **13** leads to smaller D value compared with complex **10**. A similar analogy can be drawn also for other complexes.

Here independent of the nature of π-donor/acceptor ligands, type I complexes found to yield negative D values, whereas type II and III complexes found to yield positive D values. This is in stark contrast to the earlier observations where lighter transition metal d^3 ions found to switch the sign of ZFS parameter by changing the nature of the ligand donor atoms (π-donor ligands found to yield $+D$ values, whereas π-acceptor ligands yield $-D$ values)[66]. This is essentially due to the fact that spin-flip doublet transitions are the dominating factor to the D values in Re(IV) complexes, whereas in lighter elements due to smaller crystal-field splitting, the spin-allowed transition dominates the D value.

Magneto-structural D-correlations. To probe how the axial bond length influences the D value, we have developed a magneto-structural D correlations on complex **7** (see Fig. 5c for details), where the axial -CN bonds are varied from 2.0 to 2.6 Å (compression and elongation of axial bonds). As $[ReCl_4(CN)_2]^{2-}$ unit has been employed as a building block for synthesis of polynuclear SMMs/SCMs[34-36,41], magneto-structural correlation developed on this model will serve the purpose of obtaining qualitative single-ion Re(IV) anisotropy in diverse polynuclear framework. This Re–C bond distance is found to vary significantly among structures, particularly when the -CN ligands are found to bind to other metal ions. In our correlation, the magnitude of the D value found to drastically increase (from $+13.6$ to $+94.7\,cm^{-1}$) as the Re-C bond length increases to 2.7 Å. As the metal–ligand interactions are weaker at longer distances, the transition energies are further lowered leading to larger D values for axially elongated structures[62,63]. Besides our results reveal that tetragonal distortions does not alter the sign of D values and this suggests that the $[ReCl_4(CN)_2]^{2-}$ unit unlikely to offer negative single-ion D value in any polynuclear framework.

Mechanism of magnetic relaxation. Complex **1** exhibits field-induced SMM behaviour with a barrier height of $9.6\,cm^{-1}$ at higher temperatures and $1.5\,cm^{-1}$ at lower temperatures. The experimental relaxation observed at higher temperature (up to 3.5 K) is unlikely due to Orbach process as the first excited Kramer's doublet (KD) is estimated to lie at $195\,cm^{-1}$. Thus, the relaxation is expected to be a multi-phonon Raman process. The fast relaxation observed at lower temperature is essentially due to QTM process, which is facilitated due to the presence of transverse anisotropy, hyperfine interactions and external perturbations such as internal magnetic field provided by surrounding molecules. To gain insights into the QTM process, we have analysed the wavefunction of the ground-state KD and our analysis suggests that the ground-state KD comprised of 44% of $|3/2, \pm 3/2\rangle$ and 47% of $|3/2, \pm 1/2\rangle$. As the D value is very large ($D >> kT$), the $|3/2, \pm 1/2\rangle$ KD will be completely depopulated and the ground state can be treated as a pseudo spin 1/2 system. The presence of large E term offers a strong mixing between the $|3/2, \pm 3/2\rangle$ and $|3/2, \pm 1/2\rangle$ components, which allows QTM to facilitate at low temperatures. To qualitatively analyse the mechanism of magnetic relaxation, we have computed the matrix elements between the connected KDs (see Supplementary Fig. 14 and Supplementary Note 3 for details)[67]. Our calculations predict very large tunnelling probability between the ground-state KDs and this is in line with the analysed wave function analysis. Such a prominent QTM process expected to quench the magnetization completely and this is consistent with the absence of zero-field SMM behaviour[56]. On the other hand, application of external d.c. field lifts the degeneracy and suppresses this fast relaxation. However, the QTM process cannot be ignored even under the applied field conditions as hyperfine interactions and intermolecular dipolar couplings facilitate this process. For diluted samples where intermolecular interactions are negligible, hyperfine interactions are the only factor which governs the resonant QTM process[58,68]. To gain further insights, we have computed hyperfine interactions of the Re(IV) ions as it has two dominant isotopes ^{185}Re and ^{187}Re ($I = 5/2$) with a significant natural abundance (see Supplementary Table 22). Particularly, the transverse component of this internal nuclear spin of Re(IV) (measured as $|A_X|$ and $|A_Y|$ hyperfine tensors) give rise to a small internal magnetic field inside the molecule facilitating the QTM in zero external field. The hyperfine interactions computed for complexes **1** and **2** are found to be significantly large (1,661, 1,668, 1,669 MHz for complex **1** and 1,962, 1,967 and 1,969 MHz for complex **2**) leading to fast QTM both in the presence and absence of magnetic field. A detailed experimental characterization on the diluted sample needs to be studied to verify the proposed mechanism and to gain further understanding on the relaxation process.

Role of metal–ligand covalency on magnetic anisotropy. With an aim to analyse the role of metal–ligand covalency on the D value of Re(IV) complexes, here we have modelled complexes, $[ReCl_4(E_2C_6H_4)]^{2-}$ (here E = O (complex **4**), S (**4a**) and Se (**4b**)). For the optimized geometries of **4a** and **4b**, the computed D values are $+112$ and $+114\,cm^{-1}$, respectively. Although the strength of the D values are only moderately affected, the sign can be predicted unambiguously here as there is a significant drop in the E/D values (0.30, 0.23 and 0.24 for **4**, **4a** and **4b**, respectively). Besides, the NBO analysis reveals that the Re–S/Se bond is more covalent than Re–O bond (for the Re–O, the Re contribution is 18.4%, whereas oxygen contribution is 81.6% and for Re–S(Se) bond, Re contribution is 32%(36%), whereas the S(Se) contribution is 67% (63%); see Supplementary Figs 15 and 16 and Supplementary Tables 23–25 for details). This difference in covalency lead to larger orbital splitting within t_{2g} sub-shell for S/Se analogues[22]. The larger orbital splitting are compensated by the large SOC associated with S and Se atoms leading to rigorous mixing of excited states with the ground state yielding similar strength of D values compared with the oxygen analogue. Interestingly, the π-interactions in **4a** and **4b** stabilize the d_{xy} and destabilize the d_{xz}/d_{yz} orbitals leading to a similar strength of D_{XX} and D_{YY} contributions. This lead to a decrease in the E values for complexes **4a** and **4b** compared with complex **4**.

Discussion

Here we have probed the origin of contrasting behaviour observed for Re(IV) SMMs where both the giant magnetic anisotropy and fast QTM found to co-exist. The low-lying doublet states are found to govern the sign and magnitude of ZFS parameters in this class of complexes. Our method assessment reveals that pseudo spin approach or effective Hamiltonian approach coupled with CASPT2 calculations needs to be employed to correctly reproduce the sign and magnitude of ZFS parameters. Quite interestingly, the sign of D values are found to be predictable based on the coordination mode of the ligands in these complexes where type I complexes found to possess larger negative D values, whereas type II and III found to possess a positive D values. Nature of the donor ligands as well as charge on the coordinated atoms found to influence only the strength but not the sign of D values. Very large hyperfine interactions (both transverse and axial) and rhombic anisotropy computed on these system found to govern the QTM process. By performing additional calculations and by developing magneto-structural correlations, we offer a way to enhance (diminish) the negative D ($|E|$) value in these classes of complexes. Some interesting observations are noted where the metal–ligand covalency found to govern the transverse anisotropy, offering a way to quench the inherent fast QTM process in this class of complexes.

Methods

Ab initio **calculations.** We have performed the *ab initio* calculations based of wave function theory approach to compute the ZFS in these set of mononuclear complexes. All the calculations have been performed using MOLCAS 7.8 suite of programme[69]. Here we have employed the state average-CASSCF method to compute the ZFS. The active space comprises of three active electrons in five active orbitals (CAS(3,5)). With this active space, we have computed all the 10 quartets and 40 doublet states in the configuration interaction procedure. On top of the converged CASSCF wave function, we have performed MSCASPT2 calculations to treat the dynamical correlations. We have employed ionization potential electron affinity (IPEA) shift of 0.25 to avoid the intruder states problem in CASPT2 calculations. The MS-CASPT2 computed states were further treated in RASSI-SO module, which explicitly computes the spin-orbit states. Furthermore, SINGLE_ANISO module has been utilized on top to compute the reliable spin-Hamiltonian (D and g-tensor, orientation of main magnetic axes and main anisotropic axes and local magnetic susceptibility) for each complex. The following ANO-RCC basis sets were used: [8s7p5d3f2g1h.] for Re, [ANO-RCC...5s4p2d.] for Cl, [ANORCC...6s5p3d.] for Br, [ANO-RCC...4s3p2d.] for O, C and N and [ANO-RCC...2s1p] for H during the calculations. The Cholesky decomposition for two electron integral is employed throughout the calculations to save the disk space. Moreover, additional ZFS calculations have been performed using two different techniques: (i) effective Hamiltonian approach and (ii) second-order perturbation method to check out the robustness of reported theoretical methods in predicting correct sign and magnitude of D values.

DFT calculations. Hyperfine interaction of the Re(IV) nuclei were computed within DFT framework, using electron paramagnetic resonance/nuclear magnetic resonance (EPR/NMR) module in the ORCA code[70]. We have employed meta-GGA TPSSH functional along with SARC basis set for the Re, which is much more flexible at core region to estimate all the components of the A-tensors (Fermi Fermi Contact, Spin-dipolar and Spin-orbit coupling) along $-x$, $-y$ and $-z$ directions. A very tight self consistent field (SCF) (1×10^{-8} E$_h$) has been kept throughout the calculations.

References

1. Sessoli, R., Gatteschi, D., Caneschi, A. & Novak, M. A. Magnetic bistability in a metal-ion cluster. *Nature* **365**, 141–143 (1993).
2. Blagg, R. J., Muryn, C. A., McInnes, E. J. L., Tuna, F. & Winpenny, R. E. P. Single pyramid magnets: Dy-5 pyramids with slow magnetic relaxation to 40 K. *Angew. Chem. Int. Ed.* **50**, 6530–6533 (2011).
3. Blagg, R. J. et al. Magnetic relaxation pathways in lanthanide single-molecule magnets. *Nat. Chem* **5**, 673–678 (2013).
4. Leuenberger, M. N. & Loss, D. Quantum computing in molecular magnets. *Nature* **410**, 789–793 (2001).
5. Woodruff, D. N., Winpenny, R. E. P. & Layfield, R. A. Lanthanide single-molecule magnets. *Chem. Rev.* **113**, 5110–5148 (2013).
6. Wernsdorfer, W. & Sessoli, R. Quantum phase interference and parity effects in magnetic molecular clusters. *Science* **284**, 133–135 (1999).
7. Neese, F. & Pantazis, D. A. What is not required to make a single molecule magnet. *Faraday Discuss.* **148**, 229–238 (2011).
8. Ruiz, E. et al. Can large magnetic anisotropy and high spin really coexist? *Chem. Commun.* 52–54 (2008).
9. Singh, S. K. & Rajaraman, G. Probing the origin of magnetic anisotropy in a dinuclear {MnIIICuII} single- molecule magnet: the role of exchange anisotropy. *Chem. Eur. J* **20**, 5214–5218 (2014).
10. Waldmann, O. A criterion for the anisotropy barrier in single-molecule magnets. *Inorg. Chem.* **46**, 10035–10037 (2007).
11. Cirera, J., Ruiz, E., Alvarez, S., Neese, F. & Kortus, J. How to build molecules with large magnetic anisotropy. *Chem. Eur. J* **15**, 4078–4087 (2009).
12. Langley, S. K. et al. A {(Cr2Dy2III)-Dy-III} single-molecule magnet: enhancing the blocking temperature through 3d magnetic exchange. *Angew. Chem. Int. Ed.* **52**, 12014–12019 (2013).
13. Liu, J. L. et al. A heterometallic Fe-II-Dy-III single-molecule magnet with a record anisotropy barrier. *Angew. Chem. Int. Ed.* **53**, 12966–12970 (2014).
14. Rinehart, J. D. & Long, J. R. Exploiting single-ion anisotropy in the design of f-element single-molecule magnets. *Chem. Sci* **2**, 2078–2085 (2011).
15. Meihaus, K. R. & Long, J. R. Magnetic blocking at 10 K and a dipolar-mediated avalanche in salts of the Bis(eta(8)-cyclooctatetraenide) complex [Er(COT)(2)](-). *J. Am. Chem. Soc.* **135**, 17952–17957 (2013).
16. Zhang, P. et al. Equatorially coordinated lanthanide single ion magnets. *J. Am. Chem. Soc.* **136**, 4484–4487 (2014).
17. Rinehart, J. D., Fang, M., Evans, W. J. & Long, J. R. Strong exchange and magnetic blocking in N-2(3-)-radical-bridged lanthanide complexes. *Nat. Chem* **3**, 538–542 (2011).
18. Chen, L. et al. Slow magnetic relaxation in a mononuclear eight-coordinate Cobalt(II) complex. *J. Am. Chem. Soc.* **136**, 12213–12216 (2014).
19. Colacio, E. et al. Slow magnetic relaxation in a Co-II-Y-III single-ion magnet with positive axial zero-field splitting. *Angew. Chem. Int. Ed.* **52**, 9130–9134 (2013).
20. Harman, W. H. et al. Slow magnetic relaxation in a family of trigonal pyramidal iron(II) pyrrolide complexes. *J. Am. Chem. Soc.* **132**, 18115–18126 (2010).
21. Jurca, T. et al. Single-molecule magnet behavior with a single metal center enhanced through peripheral ligand modifications. *J. Am. Chem. Soc.* **133**, 15814–15817 (2011).
22. Vaidya, S. et al. A synthetic strategy for switching the single ion anisotropy in tetrahedral Co(II) complexes. *Chem. Commun.* **51**, 3739–3742 (2015).
23. Zadrozny, J. M. & Long, J. R. Slow magnetic relaxation at zero field in the tetrahedral complex [Co(SPh)4]2-. *J. Am. Chem. Soc.* **133**, 20732–20734 (2011).

24. Zadrozny, J. M. et al. Magnetic blocking in a linear iron(I) complex. *Nat. Chem* **5**, 577–581 (2013).
25. Atanasov, M., Zadrozny, J. M., Long, J. R. & Neese, F. A theoretical analysis of chemical bonding, vibronic coupling, and magnetic anisotropy in linear iron(ii) complexes with single-molecule magnet behavior. *Chem. Sci* **4**, 139–156 (2013).
26. Wang, X. Y., Avendano, C. & Dunbar, K. R. Molecular magnetic materials based on 4d and 5d transition metals. *Chem. Soc. Rev.* **40**, 3213–3238 (2011).
27. Dreiser, J. et al. Three-axis anisotropic exchange coupling in the single-molecule magnets NEt4[MnIII2(5-Brsalen)2(MeOH)2MIII(CN)6] (M = Ru, Os). *Chem. Eur. J* **19**, 3693–3701 (2013).
28. Dreiser, J. et al. Frequency-domain fourier-transform terahertz spectroscopy of the single-molecule magnet (NEt4)[Mn-2(5-Brsalen)(2)(MeOH)(2)Cr(CN)(6)]. *Chem. Eur. J.* **17**, 7492–7498 (2011).
29. Magee, S. A. et al. Large zero-field splittings of the ground spin state arising from antisymmetric exchange effects in heterometallic triangles. *Angew. Chem. Int. Ed.* **53**, 5310–5313 (2014).
30. Mironov, V. S. Strong exchange anisotropy in orbitally degenerate complexes. A new possibilityfor designing single-molecule magnets with high blocking temperatures. *J. Magn. Magn. Mater.* **272**, E731–E733 (2004).
31. Mironov, V. S. Trigonal bipyramidal spin clusters with orbitally degenerate 5d cyano complexes [Os-III(CN)(6)](3-), prototypes of high-temperature single-molecule magnets. *Dokl. Phys. Chem* **415**, 199–204 (2007).
32. Pedersen, K. S. et al. Enhancing the blocking temperature in single-molecule magnets by incorporating 3d-5d exchange interactions. *Chem. Eur. J.* **16**, 13458–13464 (2010).
33. Singh, S. K. & Rajaraman, G. Can anisotropic exchange be reliably calculated using density functional methods? A case study on trinuclear Mn-III-M-III-Mn-III (M = Fe, Ru, and Os) cyanometalate single-molecule magnets. *Chem. Eur. J* **20**, 113–123 (2014).
34. Feng, X. W., Harris, T. D. & Long, J. R. Influence of structure on exchange strength and relaxation barrier in a series of (FeReIV)-Re-II(CN)(2) single-chain magnets. *Chem. Sci* **2**, 1688–1694 (2011).
35. Feng, X. W., Liu, J. J., Harris, T. D., Hill, S. & Long, J. R. Slow magnetic relaxation induced by a large transverse zero-field splitting in a (MnReIV)-Re-II(CN)(2) single-chain magnet. *J. Am. Chem. Soc.* **134**, 7521–7529 (2012).
36. Harris, T. D., Coulon, C., Clerac, R. & Long, J. R. Record ferromagnetic exchange through cyanide and elucidation of the magnetic phase diagram for a (CuReIV)-Re-II(CN)(2) chain compound. *J. Am. Chem. Soc.* **133**, 123–130 (2011).
37. Martinez-Lillo, J. et al. A heterotetranuclear [(NiRe3IV)-Re-II] single-molecule magnet. *J. Am. Chem. Soc.* **128**, 14218–14219 (2006).
38. Martinez-Lillo, J. et al. Metamagnetic behaviour in a new Cu(II)Re(IV) chain based on the hexachlororhenate(IV) anion. *Chem. Commun.* **50**, 5840–5842 (2014).
39. Martinez-Lillo, J. et al. Highly Anisotropic Rhenium(IV) Complexes: New Examples of Mononuclear Single-Molecule Magnets. *J. Am. Chem. Soc.* **135**, 13737–13748 (2013).
40. Pedersen, K. S. et al. [ReF6](2-): A robust module for the design of molecule-based magnetic materials. *Angew. Chem. Int. Ed.* **53**, 1351–1354 (2014).
41. Harris, T. D., Bennett, M. V., Clerac, R. & Long, J. R. [ReCl4(CN)(2)](2-): A high magnetic anisotropy building unit giving rise to the single-chain magnets (DMF)(4)MReCl4(CN)(2) (M = Mn, Fe, Co, Ni). *J. Am. Chem. Soc.* **132**, 3980–3988 (2010).
42. Martinez-Lillo, J. et al. Rhenium(IV) compounds inducing apoptosis in cancer cells. *Chem. Commun.* **47**, 5283–5285 (2011).
43. Mroziński, J., Kochel, A. & Lis, T. Synthesis, structure and magnetic properties of trans-tetrachloro-bis-(pyridine)-rhenium(IV). *J. Mol. Struct.* **610**, 53–58 (2002).
44. Cuevas, A. et al. Rhenium(IV)-copper(II) heterobimetallic complexes with a bridge malonato ligand. Synthesis, crystal structure, and magnetic properties. *Inorg. Chem.* **43**, 7823–7831 (2004).
45. Martinez-Lillo, J. et al. Ligand substitution in hexahalorhenate(IV) complexes: synthesis, crystal structures and magnetic properties of NBu4[ReX5(DMF)] (X = Cl and Br). *Inorg. Chim. Acta* **359**, 3291–3296 (2006).
46. Chiozzone, R. et al. A novel series of rhenium-bipyrimidine complexes: synthesis, crystal structure and electrochemical properties. *Dalton Trans.* 653–660 (2007).
47. Arizaga, L. et al. Synthesis, crystal structure, electrochemical and magnetic properties of (NBu4)[ReCl5(L)] with L = pyrimidine and pyridazine. *Polyhedron* **27**, 552–558 (2008).
48. Martinez-Lillo, J., Armentano, D., De Munno, G. & Faus, J. Magneto-structural study on a series of rhenium(IV) complexes containing biimH(2), pyim and bipy ligands. *Polyhedron* **27**, 1447–1454 (2008).
49. Martinez-Lillo, J. et al. Pentachloro(pyrazine)rhenate(IV) complex as precursor of heterobimetallic pyrazine-containing Re-IV M-2(II) (M = Ni, Cu) species: synthesis, crystal structures and magnetic properties. *Dalton Trans.* 4585–4594 (2008).
50. Cuevas, A. et al. Synthesis, molecular structure and magnetic properties of a rhenium(IV) compound with catechol. *J. Mol. Struct.* **921**, 80–84 (2009).

51. Kochel, A. Solvothermal synthesis, characterization and properties of [ReCl5(py)] − complex with T (Néel) of 5.5 K. *Trans.Met. Chem* **35,** 1–5 (2010).

52. Maganas, D., Sottini, S., Kyritsis, P., Groenen, E. J. J. & Neese, F. Theoretical analysis of the spin hamiltonian parameters in Co(II)S4 complexes, using density functional theory and correlated *ab initio* methods. *Inorg. Chem.* **50,** 8741–8754 (2011).

53. Zadrozny, J. M. *et al.* Slow magnetization dynamics in a series of two-coordinate iron(II) complexes. *Chem. Sci* **4,** 125–138 (2013).

54. Gomez-Coca, S., Cremades, E., Aliaga-Alcalde, N. & Ruiz, E. Mononuclear single-molecule magnets: tailoring the magnetic anisotropy of first-row transition-metal complexes. *J. Am. Chem. Soc.* **135,** 7010–7018 (2013).

55. Singh, S. K., Gupta, T., Badkur, P. & Rajaraman, G. Magnetic anisotropy of mononuclear Ni-II complexes: on the importance of structural diversity and the structural distortions. *Chem.Eur. J* **20,** 10305–10313 (2014).

56. Ungur, L., Thewissen, M., Costes, J. P., Wernsdorfer, W. & Chibotaru, L. F. Interplay of strongly anisotropic metal ions in magnetic blocking of complexes. *Inorg. Chem.* **52,** 6328–6337 (2013).

57. Ye, S. & Neese, F. How do heavier halide ligands affect the signs and magnitudes of the zero-field splittings in halogenonickel(II) scorpionate complexes? A theoretical investigation coupled to ligand-field analysis. *J. Chem. Theo. Comput* **8,** 2344–2351 (2012).

58. Wernsdorfer, W., Bhaduri, S., Boskovic, C., Christou, G. & Hendrickson, D. N. Spin-parity dependent tunneling of magnetization in single-molecule magnets. *Phys. Rev. B* **65,** 180403 (2002).

59. Chibotaru, L. F. & Ungur, L. Ab initio calculation of anisotropic magnetic properties of complexes. I. Unique definition of pseudospin Hamiltonians and their derivation. *J Chem Phys* **137,** 064112–064122 (2012).

60. Ungur, L. & Chibotaru, L. *Lanthanides and Actinides in Molecular Magnetism* (Wiley-VCH Verlag GmbH & Co. KGaA, 2015).

61. Pinsky, M. & Avnir, D. Continuous symmetry measures. 5. The classical polyhedra. *Inorg. Chem.* **37,** 5575–5582 (1998).

62. Maurice, R. *et al.* Universal theoretical approach to extract anisotropic spin hamiltonians. *J. Chem. Theo. Comp* **5,** 2977–2984 (2009).

63. Ruamps, R. *et al.* Giant ising-type magnetic anisotropy in trigonal bipyramidal Ni(II) complexes: experiment and theory. *J. Am. Chem. Soc.* **135,** 3017–3026 (2013).

64. Carmieli, R. *et al.* The catalytic Mn2 + sites in the enolase-inhibitor complex: crystallography, single-crystal EPR, and DFT calculations. *J. Am. Chem. Soc.* **129,** 4240–4252 (2007).

65. Askevold, B. *et al.* Square-planar ruthenium(II) complexes: control of spin state by pincer ligand functionalization. *Chem. Eur. J* **21,** 579–589 (2015).

66. Goswami, T. & Misra, A. Ligand effects toward the modulation of magnetic anisotropy and design of magnetic systems with desired anisotropy characteristics. *J. Phys. Chem. A* **116,** 5207–5215 (2012).

67. Ungur, L. & Chibotaru, L. F. Magnetic anisotropy in the excited states of low symmetry lanthanide complexes. *Phys. Chem. Chem. Phys.* **13,** 20086–20090 (2011).

68. Gomez-Coca, S. *et al.* Origin of slow magnetic relaxation in Kramers ions with non-uniaxial anisotropy. *Nat. Commun* **5,** 4300 (2014).

69. Aquilante, F. *et al.* MOLCAS—a software for multiconfigurational quantum chemistry calculations. *WIREs Comput. Mol. Sci* **3,** 143–149 (2013).

70. Neeese, F. The ORCA program system. *WIREs Comput. Mol. Sci* **2,** 73–78 (2012).

Acknowledgements

G.R. acknowledges DST (EMR/2014/000247), INSA, DST Nanomission (SR/NM/NS-1119/2011) for funding. S.K.S. thanks Department of Chemistry, IITB, for Research Associate position. The authors also thank the anonymous reviewer for his constructive comments.

Author contributions

S.K.S and G.R. designed the project and S.K.S performed all the calculations. S.K.S and G.R analysed the results and wrote the manuscript.

Additional Information

22

Atomically isolated nickel species anchored on graphitized carbon for efficient hydrogen evolution electrocatalysis

Lili Fan[1,2,*], Peng Fei Liu[3,*], Xuecheng Yan[1], Lin Gu[4], Zhen Zhong Yang[4], Hua Gui Yang[3], Shilun Qiu[2] & Xiangdong Yao[1]

Hydrogen production through electrochemical process is at the heart of key renewable energy technologies including water splitting and hydrogen fuel cells. Despite tremendous efforts, exploring cheap, efficient and durable electrocatalysts for hydrogen evolution still remains as a great challenge. Here we synthesize a nickel–carbon-based catalyst, from carbonization of metal-organic frameworks, to replace currently best-known platinum-based materials for electrocatalytic hydrogen evolution. This nickel-carbon-based catalyst can be activated to obtain isolated nickel atoms on the graphitic carbon support when applying electrochemical potential, exhibiting highly efficient hydrogen evolution performance with high exchange current density of $1.2\,mA\,cm^{-2}$ and impressive durability. This work may enable new opportunities for designing and tuning properties of electrocatalysts at atomic scale for large-scale water electrolysis.

[1] School of Natural Sciences, Queensland Micro- and Nanotechnology Centre, Griffith University, Brisbane, QLD 4111, Australia. [2] State Key Laboratory of Inorganic Synthesis and Preparative Chemistry, College of Chemistry, Jilin University, Changchun 130012, China. [3] Key Laboratory for Ultrafine Materials of Ministry of Education, School of Materials Science and Engineering, East China University of Science and Technology, Shanghai 200237, China. [4] Institute of Physics, Beijing National Laboratory for Condensed Matter Physics, Chinese Academy of Sciences, Beijing 100190, China. * These authors contributed equally to this work. Correspondence and requests for materials should be addressed to H.G.Y. (email: hgyang@ecust.edu.cn) or to S.Q. (email: sqiu@jlu.edu.cn) or to X.Y. (email: x.yao@griffith.edu.au).

Molecular hydrogen (H_2), when generated from carbon-neutral processes, plays a vital role in the sustainable energy systems and chemical industry[1-3]. Electrolysis has stood out among various hydrogen production technologies because of high energy-conversion efficiency and environmentally benign process, especially coupled with energy sources such as solar and wind[4-6]. In this rapidly developed field, finding active catalysts for hydrogen evolution reaction (HER) to lower large potentials is of paramount importance to promote water electrolysis application. It has been confirmed that platinum (Pt) is the most active and stable electrocatalyst for HER, with only small overpotentials for high reaction rate. However, its low abundance and consequently high cost limit its large-scale commercial applications. Hence, exploring cheap, efficient and durable alternatives of Pt is crucial to facilitate the global scalability of such potential clean-energy technology[7]. Therefore, a series of Pt-free electrocatalysts have been identified for HER in strong acids, such as metal compounds[8-17] (sulfides, nitrides, carbides, borides and phosphides) and carbon-based materials[18-20]. Moreover, remarkable advances in nanostructuring or stabilizing these electrocatalysts on carbon substrates have further exposed more active sites simultaneously with enhanced electrical conductivity[21-27]. Recently, design strategies like encapsulating 3d transition metals into carbon nanotubes and graphene nanoshells have been reported to improve the long-term activity[28-30]. Even so, compared with Pt, these reported materials show dwarfed in activity and durability.

Isolated metal atoms, owing to their low-coordination and unsaturated atoms functioned as active sites, have been demonstrated more catalytic active than nanometer-sized metal particles[31-35]. Supported isolated noble metals, such as palladium, Pt and gold, have been reported as highly active catalysts for hydrogenation reactions, CO oxidation reactions and water–gas shift reactions[33-35]. However, earth-abundant 3d transition metals, which have been investigated as promising alternatives in alkaline electrolytes[36,37], have not been downsized to single atoms to improve their activity for HER so far.

Herein, we show a nickel–carbon (Ni–C)-based catalyst, which can be tuned by electrochemical methods to obtain atomically isolated Ni species anchored on graphitized carbon, consequently displaying high activity and durability for HER. The activated-Ni–Carbon (A-Ni–C) catalyst exhibts overpotential for the current density of $10\,mA\,cm^{-2}$, $20\,mA\,cm^{-2}$ and $100\,mA\,cm^{-2}$ at $-34\,mV$, $-48\,mV$ and $-112\,mV$, respectively, a low Tafel slope of $41\,mV$ per decade and a large exchange current density of $1.2\,mA\,cm^{-2}$. More importantly, A-Ni–C could maintain high activity in $>25\,h$ in the chronoamperometric test. A variety of analytical techniques illustrate that Ni single atoms formed during the activation process, which could contribute greatly to HER performance. Compared with Pt and other Pt alternatives, A-Ni–C is composed of more abundant elements in the earth's crust. These findings in our work may help accelerate the large-scale application of proton exchange membrane (PEM) electrolysers and solar photoelectrochemical (PEC) water electrolysers.

Results

Sythesis and characterization of A-Ni–C catalyst. A-Ni–C catalyst was obtained from well-dispersed Ni metal in the graphitized carbon matrix. The synthesized Ni-based metal-organic framework (Ni-MOF) was used as the precursor (Fig. 1). Followed by carbonization at 700 °C in nitrogen (N_2) atmosphere, Ni nanoparticles encapsulated in graphene layers (Ni@C) were prepared. Hydrochloric acid (HCl) leaching treatment was used to remove the redundant Ni metal (HCl-Ni@C). After HCl treatment, hollow onion-like carbon nanoshells along with little Ni

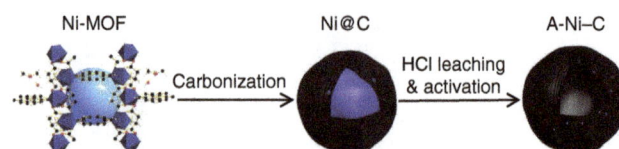

Figure 1 | Schematic diagram of synthesis and activation process of the Ni-C catalysts. The Ni-MOF used as a precursor consists of an orthorhombic crystal. Atoms are shown as follows: C, black; H, white; O, red; N, blue; Ni, royal blue. Carbonization of the synthesized Ni-MOF was at 700 °C in N_2 atmosphere to obtain Ni@C. HCl leaching treatment was repeated three times to sufficiently dissolve exposed Ni metal. Constant potential and CV treatments were performed to activate the catalysts until they reached the optimal performance and remained stable. During the activation process, Ni single atoms formed *in situ* anchored on the graphitized carbon.

nanoparticles protected by graphene layers remained. We then applied electrochemical cyclic-potential on the HCl-Ni@C catalysts-decorated electrode, unexpected activation process was observed. Similar activation processes were also reported in the literature[13,20]. To our surprise, direct constant potential on Ni@C can also activate the electrocatalyst, which was conducted until the electrocatalyst reached the optimized and stable activity. We found Ni single atoms formed during all these activation process, which is verified as discussed later.

The nanostructure of Ni@C was characterized by a transmission electron microscope (TEM). Low-magnification TEM image of Ni@C displays Ni metal inside carbon nanospheres (Supplementary Fig. 1a). The size distribution image (Supplementary Fig. 1b) suggests that the nanospheres have an average size around 10 nm. After HCl leaching, the high-resolution TEM image of HCl-Ni@C shows hollow onion-like nanoshells along with little Ni metal aggregates encapsulated in the graphene layers (Fig. 2a). To further study the structure changes of the catalyst before and after activation, a JEM-ARM200F scanning TEM (STEM) fitted with a double aberration-corrector for both probe-forming and the imaging lenses was used. In the STEM image of HCl-Ni@C, the measured Ni (111) d-spacing (0.204 nm) and Ni (200) d-spacing (0.176 nm) again confirm the existence of well crystalline Ni metal in the graphitic carbon nanoshells (inset of Fig. 2a). The thickness of graphene layer is uneven with exposure of Ni nanoparticles, which would provide possibilities for slow dissolution of Ni metal during the activation process in the acidic solution. The bright-field STEM image of A-Ni–C (Supplementary Fig. 2) illustrates that the measured (002) d-spacing of the graphitic carbon ranges from about 0.33 to 0.36 nm, indicating the carbon structure is somewhat disordered, rather than ideally graphitic, which was also evidenced by the X-ray diffraction (XRD) and Raman analysis. Individual metal atoms in practical catalysts can be discerned in the atomic resolution high-angle annular dark-field (HAADF) images[33-35]. For the sample A-Ni–C, the atomically isolated Ni species (marked by white circles) are dispersed on the partially graphitized carbon matrix (Fig. 2b). Examination of different regions (Supplementary Fig. 2) confirms abundance of Ni single atoms present in A-Ni–C. Considering that the elements of N, F and S have a smaller atomic number than Ni, these impuries can hardly exsit as bright tiny spots in the metal-carbon system. HAADF images of controled samples have also ruled out this possiblity (Supplementary Fig. 3). About 78% of the Ni species counted in the images were present as single atoms away from each other (inset of Fig. 2b). A minority of sub-nanometer Ni clusters present in the same sample did not have the packed Ni atom structure of Ni nanoparticles.

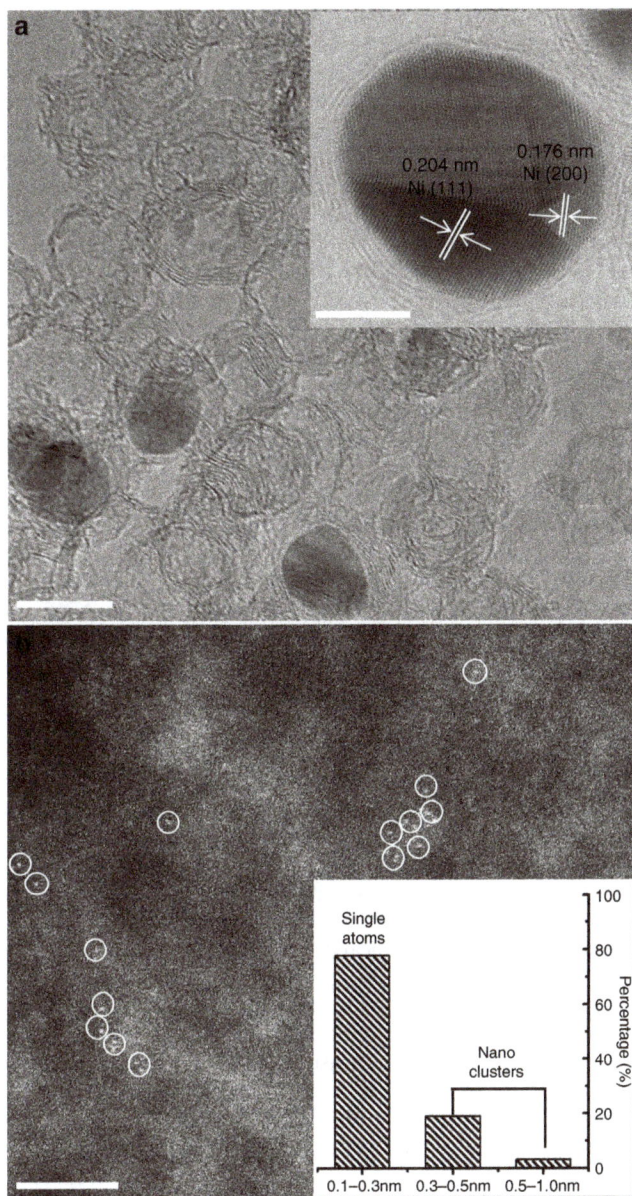

Figure 2 | Micrographs of samples HCl-Ni@C and A-Ni–C. (a) HRTEM image of HCl-Ni@C. Scale bar, 10 nm. Inset of **a**, BF STEM image of Ni metal encapsulated in graphene layers. Scale bar, 5 nm. The area of graphene with exposure of Ni nanoparticle provides the opportunity that Ni metal would slowly resolve during the activation process. (**b**) Subångström resolution HAADF STEM image of A-Ni–C. The circles are drawn around isolated Ni atoms. Scale bar, 3 nm. Inset of **b**, the size distribution figure, which is based on >120 observed Ni species counted from high-magnification images (recorded at 4 × original magnification).

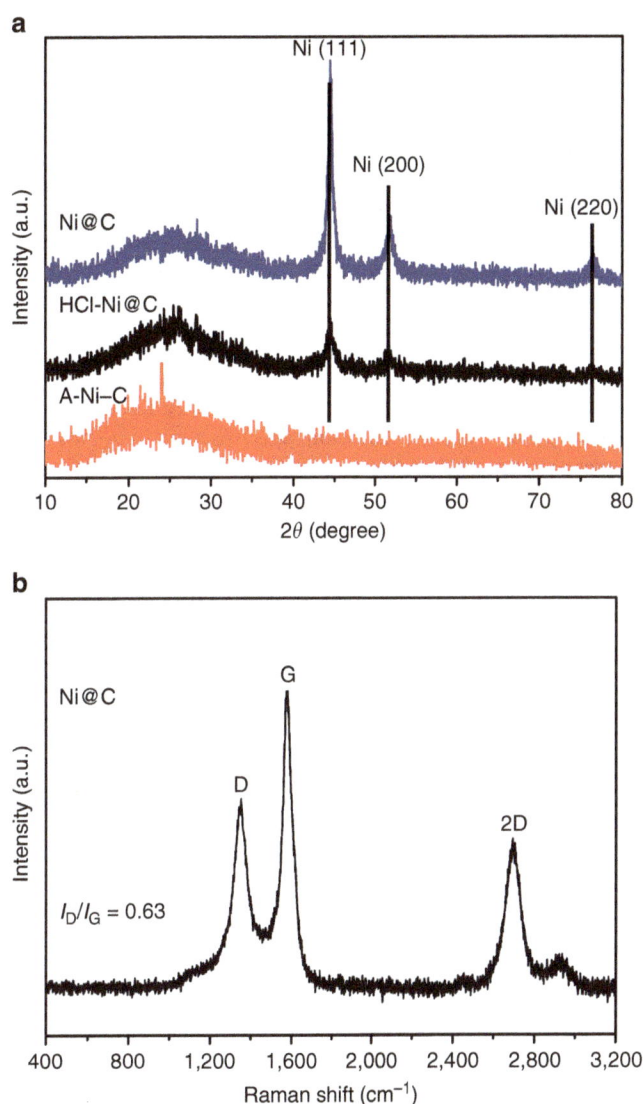

Figure 3 | Structure characterization of the Ni–C catalysts. (a) XRD patterns of Ni@C, HCl-Ni@C and A-Ni–C, respectively. The diffraction peaks of Ni in Ni@C and HCl-Ni@C confirm the existence of well crystalline Ni metal. The red line corresponding to A-Ni–C illustrates that no Ni-Containing crystal phase was detected after activation process. (**b**) Raman spectrum of Ni@C with $I_D/I_G = 0.63$, indicating the partial graphitization at 700 °C.

The multiple peaks on the XRD pattern of the synthesized Ni-MOF (Supplementary Fig. 4a) well match those of the typical crystalline structure of $[Ni_2(L-asp)_2(bpy)] \cdot CH_3OH \cdot H_2O$. Ni@C exhibits the cubic phase of Ni metal (JSPDS NO. 04-0850) with a broad shoulder peak in the range 20–30° (2θ), which originates from the partially graphitized carbon specie (Fig. 3a). The diffraction peaks of graphitic carbon become higher and sharper with the increased temperature (Supplementary Fig. 4b). Raman spectrum of Ni@C (Fig. 3b) further confirms that the Ni-MOF is partially graphitized at 700 °C. Among carbon-based catalysts, these graphitic carbon structures formed during the carbonization treatment may significantly enhance the electronic conductivity

and corrosion resistance, which would improve the electro-catalytic efficiency[29]. After sufficiently washed by HCl, little Ni nanoparticles encapsulated in the multi graphene layers remained. The XRD pattern of HCl-Ni@C proves the existence of the Ni metal. However, no Ni–Containing signal was detected in the sample A-Ni–C, primarily attributed to insensitivity of XRD to single atoms or small clusters.

The binding state was investigated by X-ray photoelectron spectroscopy (XPS). As shown in Supplementary Fig. 5, N–C bonds in the samples of Ni@C, HCl-Ni@C and A-Ni–C are in the form of pyridine (398.9 eV), pyrrole (400.4 eV) and graphite (401.3 eV) (refs 28,38). The quantitative analysis of the XPS results reveals that the atomic ratios of N/C in Ni@C, HCl-Ni@C and A-Ni–C are 4.62:100, 3.80:100 and 2.21:100, respectively. The existence of N dopants in the A-Ni–C would inherently increase the ability of interacting with reactants and render high positive charge density on their adjacent carbon atoms, which could also contribute to the high activity of A-Ni–C samples[28,29]. The Ni 2p XPS spectrum of Ni@C reveals the coexistence of Ni^0 (852.8 eV) and Ni^{2+} (853.7 eV), which can be ascribed to Ni-Ni and Ni-O bonding[39,40] (Fig. 4a). After HCl leaching, the Ni 2p XPS spectrum of HCl-Ni@C nearly remains the same as that of Ni@C, suggesting the residue of Ni metal. The existence of Ni-O bonding can be attributed to the slow oxidation of the exposed Ni metal[41–43]. We further collected samples of A-Ni–C for analysis. The XPS analysis of A-Ni–C (Supplementary Fig. 6, Supplementary Table 1) illustrates it contains Ni, C, N, F, S and O elements without any Pt impurity, eliminating the influence of Pt for HER performance. We should point out that F, S and O elements mainly resulted from the Nafion solution. To confirm that trace amount of Ni in the A-Ni–C sample contributes to the HER activity, we determined the Ni contents by indicatively coupled plasma atomic emission spectroscopy. As shown in Supplementary Table 2, the weight ratios of Ni in the Ni@C, HCl-Ni@C and A-Ni–C were 85.0%, 5.3% and 1.5%, respectively. We found that the Ni content decreased during the CV activation process. What is more, no detectable Pt impurities were found in all these samples. In addition, the Ni 2p XPS spectrum of A-Ni–C (Fig. 4b) has a sharp peak ~860.7 eV, which can be associated with the strong Ni–F binding. Specifically, the peak at 852.8 eV is assigned to Ni^0, reflecting Ni-Ni binding of the Ni clusters.

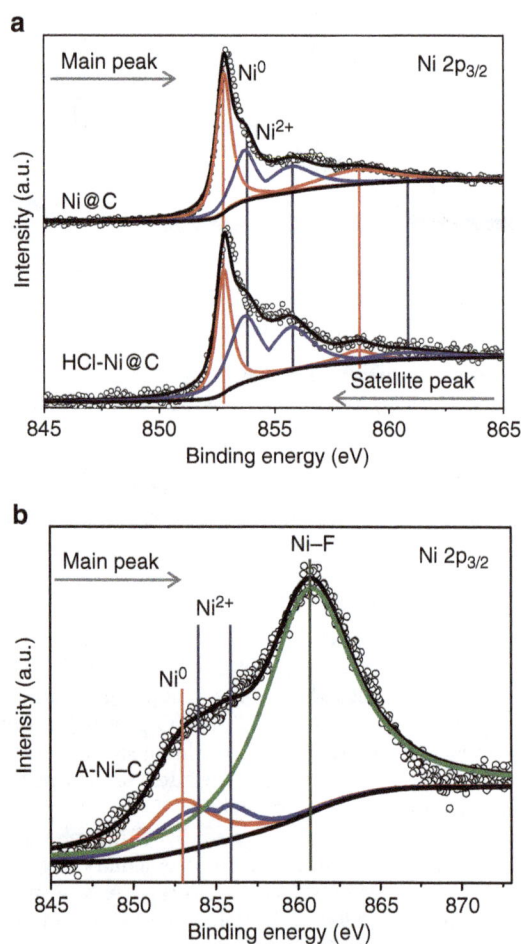

Electrocatalytic HER activity evaluation. The HER electrocatalytic activity was evaluated in 0.5 M H_2SO_4 using a typical three-electrode set-up. All tests were performed without iR compensation. The lineal sweep voltammetry (LSV) curves showing the normalized current density (j) versus voltage (j versus V) for A-Ni–C and HCl-Ni@C along with commercial Pt/C (5 wt% Pt on Vulcan carbon black), for comparison, are show in Fig. 5a. As expected, Pt/C catalyst exhibits excellent HER performance. HCl-Ni@C requires overpotential (η) of about − 440 mV to reach j of 10 mA cm^{-2}. In sharp contrast, A-Ni–C produces j of 10 mA cm^{-2}, 20 mA cm^{-2} and 100 mA cm^{-2} at η of − 34 mV, − 48 mV and − 112 mV, respectively. These overpotentials are the lowest among the reported acid-stable and earth-abundant HER electrocatalysts, including recently reported $Ni_2P^{[14]}$, $MoP \mid S^{[16]}$, $FeP^{[17]}$, MoS_2 (refs 8,22,24), WS_2 (refs 21,23,25), Mo_2C (refs 12,13) and chemically doped grapheme-based materials[18–20] (more details in Supplementary Table 3).

To understand the detailed underlying mechanism of HER activity, Tafel plots based on LSV curves were acquired (Fig. 5b). The linear regions of Tafel plots were fit to the Tafel equation ($\eta = b \log j + a$, where b is the Tafel slope), yielding Tafel slopes of ~34, 41, 194 mV per decade for Pt/C, A-Ni–C and HCl-Ni@C, respectively. The value for A-Ni–C does not correspond to one of the standard HER Tafel slopes (29, 38, 116 mV per decade), indicating the HER occurs through a Volmer–Heyrovsky mechanism and electrochemical recombination with an additional proton is the rate-limiting step[28]. Moreover, the exchange current density (j_0) was obtained by extrapolation of Tafel plots (Supplementary Fig. 7). The j_0 of A-Ni–C is about 1.2 mA cm^{-2} (the same magnitude as the value of 2.5 mA cm^{-2} for Pt/C), also highlighting the exceptional H_2 evolving efficiency of A-Ni–C catalyst.

The optimal activity of A-Ni–C catalyst would also be attributed to strong chemical and electronic coupling between graphitized carbon and Ni single atoms, permitting highly efficient electrical communication between the catalytic active sites and the underlying electrode substrate. Electrochemical impedance spectroscopy measurements were performed to confirm the hypothesis (Fig. 5c). The charge transfer resistance of A-Ni–C is similar to that of Pt/C, which is much lower than that of HCl-Ni@C. Thus, much faster electron transfer between the catalytic active sites of A-Ni–C and the electrode substrate is one of the key factors contributing to the superior HER kinetics.

Durability of A-Ni–C catalyst. The durability is a key concern of all catalysts. In this study, accelerated degradation measurements were adopted to evaluate the durability. Cyclic voltammetric (CV) sweeps between + 905 mV and − 95 mV versus the reversible hydrogen electrode potential (versus RHE) at 100 mV s^{-1} were applied to HCl-Ni@C-decorated glassy carbon (GC) electrode. Post-potential-cycling LSV curves were recorded (Supplementary Fig. 8). HER performance after 8,000 cycles retained almost the same to the test after 4,000 cycles. To further probe the durability of A-Ni–C in the acidic environment, continuous HER at a static overpotential was conducted (Fig. 5d). When operating a constant overpotential of − 45 mV, a continuous HER process occurred to generate molecular H_2 (different overpotentials were applied to generate H_2 shown in Supplementary Movie 1). During the first 11 h, the current density of ~10 mA cm^{-2} was observed, and then gradually increased to ~14 mA cm^{-2}, which might be caused by adequate activation of the catalyst. In addition, the current density levelled out at ~12 mA cm^{-2}. It should be mentioned that the high activity was maintained for >25 h in the chronoamperometric electrolysis (after CV activation, ≥4,000

Figure 4 | XPS spectra of the Ni-C catalysts. (a) XPS spectra of Ni 2p peaks of Ni@C and HCl-Ni@C. (b) XPS spectrum of Ni 2p peaks of A-Ni-C. The peaks at 852.8 eV and 853.7 eV are assigned to Ni^0 and Ni^{2+}, respectively. The sharp peak at 860.7 eV can be correlated with Ni-F binding, and the element F comes from Nafion solution.

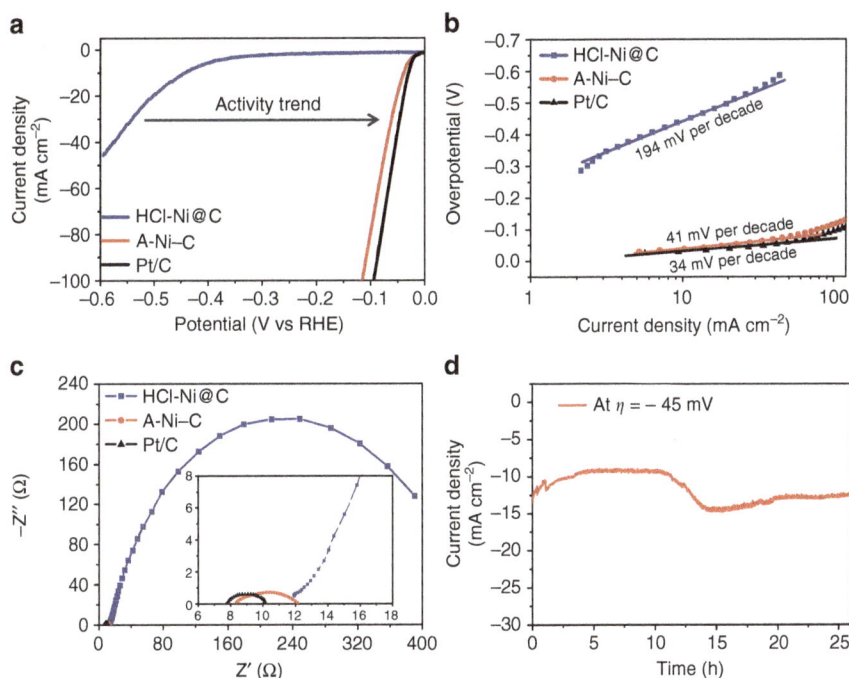

Figure 5 | HER electrocatalytic properties of the Ni–C catalysts. (**a**) LSV curves of HCl-Ni@C, A-Ni-C and Pt/C. (**b**) Tafel plots obtained from LSV curves in **a**, indicating significant improvement in activity for HER after activation. (**c**) EIS spectra of HCl-Ni@C, A-Ni-C and Pt/C at the overpotential of − 100 mV. Inset of **c** represents a close-up view of the spectra at high frequencies. A-Ni-C-modified GC electrode displays much lower impedance than HCl-Ni@C-modified GC electrode, similar to that of Pt/C-modified GC electrode. (**d**) Chronoamperometric curve of A-Ni-C at $\eta = -45$ mV, revealing nearly no apparent deactivation after 25 h. Note: all the electrochemical data of A-Ni-C catalysts were obtained after electrochemical cyclic-potential activation.

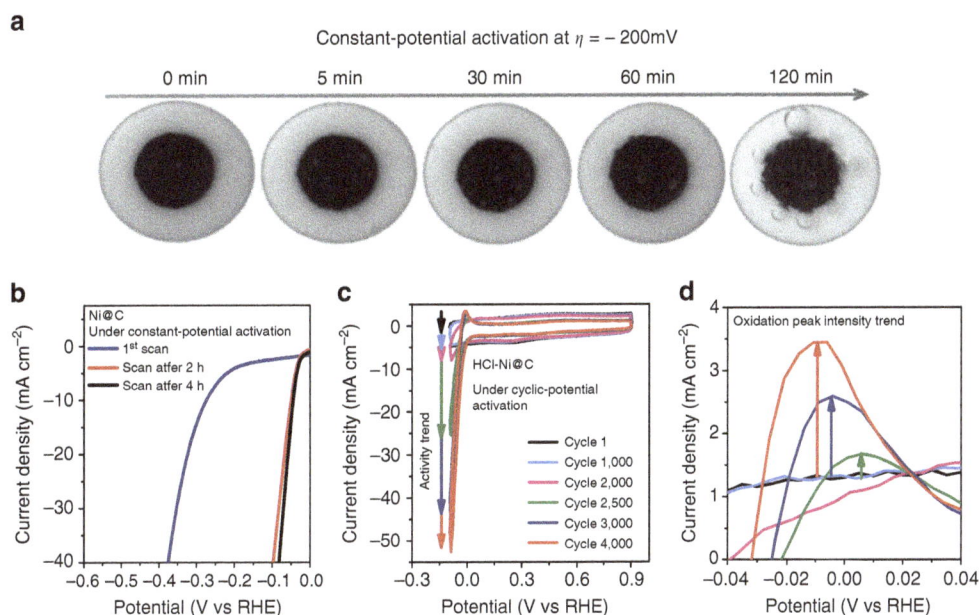

Figure 6 | Activation processes of the Ni–C catalysts. (**a**) Hydrogen bubbles driven at $\eta = -100$ mV at GC electrodes with different degrees of activation (constant-potential activation at $\eta = -200$ mV for different time). All the digital photos were taken when the cell turned on for 20 s. The diameter of glassy carbon electrode is 5 mm. (**b**) Initial and post-constant-potential activation LSV curves of Ni@C. The applied constant potential would activate Ni@C catalyst. (**c**) Cyclic voltammograms of HCl-Ni@C (cycle 1, cycle 1,000, cycle 2,000, cycle 2,500, cycle 3,000 and cycle 4,000 were selected). Catalytic activity trend corresponds to the improving current density. (**d**) Oxidation peaks of the catalyst in different cycles. The oxidation peak intensity increased and the peak potentials shifted in the negative direction in the anodic-going scan.

cycles). Furthermore, we did the galvanostatic electrolysis test to collect the evolved hydrogen gas to evaluate the Faradaic efficiency of the A-Ni–C catalyst. In Supplementary Fig. 9, the result demonstrates that A-Ni–C gives about 100% Faradaic yield during the electrolysis process. Overall, the high catalytic activity and outstanding durability illustrate the potential of A-Ni–C for cost-effective electrocatalytic hydrogen evolution in water electrolysis system and solar-driven hydrogen system.

Activation processes. Activation processes (constant-potential activation and cyclic-potential activation) were also recorded to investigate changes after activation. Hydrogen bubbles generated at $\eta = -100$ mV on Ni@C-decorated GC electrodes with different degrees of activation (Fig. 6a). Constant potential at η of -200 mV (Supplementary Fig. 10) was used to stabilize the catalyst to reach the best performance, simultaneously ruling out the possibility that the high j would result from the reaction between the unsecure Ni metal and the strongly acidic solution. The j gradually improved in the first 2 h and levelled out in the next 2 h. These activation treatments also ensured stable and reproducible results. LSV curves of Ni@C after different periods of constant-potential activation prove the activation process of Ni@C (Fig. 6b). During CV activation process, the catalytic j of HCl-Ni@C gradually increased with the consecutive CV scans and reached the maximum efficiency after about 4,000 cycles (Fig. 6c). Notably, the anodic peaks around 0 V (versus RHE) shifted to lower potentials during the process, indicating a more easily occurring oxidation reaction. Interestingly, the oxidation peak intensities were also enhanced during the CV treatment (Fig. 6d). These signs during CV activation reveal that A-Ni–C could lose electrons more easily in the anodic-going scan. The oxidation peak intensity gradually levelled out when the cathodic catalytic j nearly remained stable (Supplementary Fig. 11). These changes of the oxidation peak corresponding to valance changes of Ni coincide with the trend of the catalytic activity. In addition, we also collected the activated samples after different CV cycles for STEM analysis (Supplementary Figs 12 and 13). The downsizing of the Ni nanoparticles and the formation of single Ni atoms can be evident to be observed during the activation process. Throughout the detailed CV activation process, we summarize that active Ni single atoms would form during the activation process, leading to these changes of the oxidation peak and simultaneously contributing greatly to the improved activity. Furthermore, the HCl-Ni@C samples can also be activated on the fluorine-doped SnO$_2$ (FTO) glass substrate (Supplementary Figs 14 and 15).

Discussion

In summary, we have used carbonization of MOFs to synthesize Ni–C-based catalyst and applied electrochemical potential to activate Ni–C catalyst, which can reach outstanding performance during HER processes. The discovery of this A-Ni–C catalyst highlights a new area of tuning structure and functionality of metal-carbon-based catalyst at atomic scale by electrochemical methods, which would hold the promise for large-scale real-world water splitting electrolysers.

Methods

Chemicals. Nickel (II) carbonate basic hydrate (48–50% Ni, Fluka), L-aspartic acid (98%, Sigma-Aldrich), 4,4′-bipyridyl (98%, Alfa Aesar), Nafion-117 solution (5 wt. % in a mixture of lower aliphatic alcohols and water) (Sigma-Aldrich), sulphuric acid (98%, Merck) and HCl (32%, RCI labscan) were used as received without any further purification.

Synthesis of Ni(L-asp)(H$_2$O)$_2 \cdot$ H$_2$O. Measured amount of NiCO$_3 \cdot$ 2Ni(OH)$_2 \cdot \times$ H$_2$O (2.450 g) and L-aspartic acid (2.630 g) were dissolved in water (200 ml) under stirring and heating for 3 h. After filtering out all the remaining solid particles, the clear turquoise solution was then evaporated at 80 °C oven overnight to obtain the pale blue Ni(Asp) powder.

Synthesis of [Ni$_2$(L-asp)$_2$(bpy)] $\cdot$ CH$_3$OH $\cdot$ H$_2$O (Ni-MOF). Ni(L-asp)(H$_2$O)$_2 \cdot$ H$_2$O (0.219 g) and 4,4′-bipyridyl (0.312 g) were dispersed in a mixture containing water (6 mL) and methanol (4.742 g) under ultrasonic vibration. The final solution

was sealed into an autoclave and heated at 150 °C for 48 h. The product, [Ni$_2$(L-asp)$_2$(bpy)] $\cdot$ CH$_3$OH $\cdot$ H$_2$O, was filtered, washed with copious amounts of water and methanol and dried at 60 °C for 24 h.

Carbonization of Ni-MOF. Carbonization processes were carried out in N$_2$ atmosphere. The dried Ni-MOF powder was pretreated at 150 °C for 2 h and then carbonized at different temperatures for 5 h with a heating rate of 5 °C min^{-1}.

HCl leaching treatment. Excess concentrated HCl was slowly added into Ni@C hybrids and then the suspension was put under ultrasonic condition to dissolve Ni metal. The suspension was separated when being left to stand for 6 h. The entire process was repeated two additional times. Then the product was washed with copious amount of water until the pH of the solution was ~7 and then dried at 60 °C for 24 h.

Electrochemical measurements. All the electrochemical tests were performed in a conventional three-electrode system at an electrochemical station (CHI 660E), using Ag/AgCl (3.5 M KCl solution) electrode as the reference electrode, Pt mesh as the counter electrode and GC electrode as the working electrode. About 4 mg of sample and 80 μl of 5 wt% Nafion solution were dispersed in 1 ml of 4:1 v/v water/ethanol by at least 30-min sonication to form a homogeneous solution. Then 5 μl or 15 μl of the solution was loaded onto the GC electrode of 3 mm or 5 mm in diameter, respectively. The final loading for all catalysts and commercial Pt/C electrocatalysts on the GC electrodes is about 0.283 mg cm^{-2}. Similar preparation of titanium foil electrode was conducted with the same loading of the HCl-Ni@C samples and corresponding chronoamperometric curves of A-Ni–C was shown in Supplementary Fig. 16. LSV with scan rate of 5 mV s^{-1} was conducted in 0.5 M H$_2$SO$_4$ deaerated with argon at room temperature. The Ag/AgCl/3.5 M KCl reference electrode was calibrated with respect to RHE. The calibration was performed in the high purity hydrogen saturated electrolyte with Pt mesh as the working electrode and counter electrode. CVs were run at a scan rate of 1 mV s^{-1}, and the average of the two potentials at which the current crossed zero was taken to be the thermodynamic potential for the hydrogen electrode reactions (Supplementary Fig. 17). In 0.5 M H$_2$SO$_4$, $E_{RHE} = E_{Ag/AgCl} + 0.209$ V. AC impedance measurements were carried out in the same configuration when the working electrode was biased at the overpotential of 100 mV from 10^5 Hz to 10^{-1} Hz with an AC voltage of 5 mV. The CV measurements were performed by use of the scan rate at 100 mV s^{-1} between $+905$ mV and -95 mV (versus RHE) without accounting for uncompensated resistance. Chronoamperometric measurement ($\eta = -45$ mV) was performed to evaluate the long-term stability. The faradaic yield was calculated from the total amount of electrons passed through the cell during the galvanostatic electrolysis and the amount of the evolved H$_2$ gas collected by the water drainage method. The theoretically expected amount of H$_2$ was calculated by applying the Faraday law, which states that the passage of 96485.4 C causes 1 equiv. of reaction. The HER performance mentioned above was well reproducible in our laboratory. More than 10 A-Ni–C-decorated GC electrodes (after CV activation) were evaluated, giving similar activation phenomenon and negligible difference of HER performance.

Characterization. XRD patterns were acquired at room temperature using D/MAX 2550 VB/PC. Raman spectrum was recorded on a Rennishaw InVia spectrometer with a model 100 Ramascope optical fibre instrument. XPS data were collected from ESCALAB 250Xi, and the binding energy of the C 1 s peak at 284.8 eV was taken as an internal reference. TEM images were collected from TECNAI F-30 with an acceleration voltage of 300 kV. Subångström resolution STEM images were obtained on a JEM-ARM200F STEM fitted with a double aberration-corrector for both probe-forming and the imaging lenses.

References

1. Walter, M. G. *et al.* Solar water splitting cells. *Chem. Rev.* **110,** 6446–6473 (2010).
2. Cook, T. R. *et al.* Solar energy supply and storage for the legacy and nonlegacy worlds. *Chem. Rev.* **110,** 6474–6502 (2010).
3. Armaroli, N. & Balzani, V. The hydrogen issue. *ChemSusChem* **4,** 21–36 (2011).
4. U.S. Energy Information Administration. The impact of increased use of hydrogen on petroleum consumption and carbon dioxide emissions, http://www.eia.gov/oiaf/servicerpt/hydro/appendixc.html (2008).
5. U.S. Department of Energy Hydrogen Analysis Resource Center. Hydrogen Production, Worldwide Refinery Hydrogen, Production Capacities http://hydrogen.pnl.gov/cocoon/morf/hydrogen/article/706 (2012).
6. Carmo, M., Fritz, D. L., Mergel, J. & Stolten, D. A comprehensive review on PEM water electrolysis. *Int. J. Hydrogen Energy* **38,** 4901–4934 (2013).
7. Zou, X. & Zhang, Y. Noble metal-free hydrogen evolution catalysts for water splitting. *Chem. Soc. Rev.* **44,** 5148–5180 (2015).
8. Jaramillo, T. F. *et al.* Identification of active edge sites for electrochemical H$_2$ evolution from MoS$_2$ nanocatalysts. *Science* **317,** 100–102 (2007).

9. Karunadasa, H. I. *et al.* A molecular MoS$_2$ edge site mimic for catalytic hydrogen generation. *Science* **335**, 698–702 (2012).

10. Chen, W.-F. *et al.* Hydrogen-evolution catalysts based on non-noble metal nickel-molybdenum nitride nanosheets. *Angew. Chem. Int. Ed.* **51**, 6131–6135 (2012).

11. Cao, B., Veith, G. M., Neuefeind, J. C., Adzic, R. R. & Khalifah, P. G. Mixed close packed cobalt molybdenum nitrides as non-noble metal electrocatalysts for the hydrogen evolution reaction. *J. Am. Chem. Soc.* **135**, 19186–19192 (2013).

12. Vrubel, H. & Hu, X. Molybdenum boride and carbide catalyze hydrogen evolution in both acidic and basic solutions. *Angew. Chem. Int. Ed.* **51**, 12703–12706 (2012).

13. Xiao, P. *et al.* Novel molybdenum carbide-tungsten carbide composite nanowires and their electrochemical activation for efficient and stable hydrogen evolution. *Adv. Funct. Mater.* **25**, 1520–1526 (2015).

14. Popczun, E. J. *et al.* Nanostructured nickel phosphide as an electrocatalyst for the hydrogen evolution reaction. *J. Am. Chem. Soc.* **135**, 9267–9270 (2013).

15. Popczun, E. J., Read, C. G., Roske, C. W., Lewis, N. S. & Schaak, R. E. Highly active electrocatalysis of the hydrogen evolution reaction by cobalt phosphide nanoparticles. *Angew. Chem. Int. Ed.* **53**, 5427–5430 (2014).

16. Kibsgaard, J. & Jaramillo, T. F. Molybdenum phosphosulfide: an active, acid-stable, earth-abundant catalyst for the hydrogen evolution reaction. *Angew. Chem. Int. Ed.* **53**, 14433–14437 (2014).

17. Callejas, J. F. *et al.* Electrocatalytic and photocatalytic hydrogen production from acidic and neutral-pH aqueous solutions using iron phosphide nanoparticles. *ACS Nano* **8**, 11101–11107 (2014).

18. Zheng, Y. *et al.* Hydrogen evolution by a metal-free electrocatalyst. *Nat. Commun.* **5**, 3783 (2014).

19. Ito, Y., Cong, W., Fujita, T., Tang, Z. & Chen, M. High catalytic activity of nitrogen and sulfur co-doped nanoporous graphene in the hydrogen evolution reaction. *Angew. Chem. Int. Ed.* **54**, 2131–2136 (2015).

20. Das, R. K. *et al.* Extraordinary hydrogen evolution and oxidation reaction activity from carbon nanotubes and graphitic carbons. *ACS Nano* **8**, 8447–8456 (2014).

21. Voiry, D. *et al.* Enhanced catalytic activity in strained chemically exfoliated WS$_2$ nanosheets for hydrogen evolution. *Nat. Mater.* **12**, 850–855 (2013).

22. Kibsgaard, J., Chen, Z., Reinecke, B. N. & Jaramillo, T. F. Engineering the surface structure of MoS$_2$ to preferentially expose active edge sites for electrocatalysis. *Nat. Mater.* **11**, 963–969 (2012).

23. Cheng, L. *et al.* Ultrathin WS$_2$ nanoflakes as a high-performance electrocatalyst for the hydrogen evolution reaction. *Angew. Chem. Int. Ed.* **53**, 7860–7863 (2014).

24. Li, Y. *et al.* MoS$_2$ nanoparticles grown on graphene: an advanced catalyst for the hydrogen evolution reaction. *J. Am. Chem. Soc.* **131**, 7296–7299 (2011).

25. Yang, J. *et al.* Two-dimensional hybrid nanosheets of tungsten disulfide and reduced graphene oxide as catalysts for enhanced hydrogen evolution. *Angew. Chem. Int. Ed.* **52**, 13751–13754 (2013).

26. Li, D. J. *et al.* Molybdenum sulfide/N-doped CNT forest hybrid catalysts for high-performance hydrogen evolution reaction. *Nano Lett.* **14**, 1228–1233 (2014).

27. Youn, D. H. *et al.* Highly active and stable hydrogen evolution electrocatalysts based on molybdenum compounds on carbon nanotube–graphene hybrid support. *ACS Nano* **8**, 5164–5173 (2014).

28. Zou, X. *et al.* Cobalt-embedded nitrogen-rich carbon nanotubes efficiently catalyze hydrogen evolution reaction at all pH values. *Angew. Chem. Int. Ed.* **53**, 4372–4376 (2014).

29. Deng, J. *et al.* Highly active and durable non-precious-metal catalyst encapsulated in carbon nanotubes for hydrogen evolution reaction. *Energy Environ. Sci.* **7**, 1919–1923 (2014).

30. Deng, J., Ren, P., Deng, D. & Bao, X. Enhanced electron penetration through an ultrathin graphene layer for highly efficient catalysis of the hydrogen evolution reaction. *Angew. Chem. Int. Ed.* **54**, 2100–2104 (2015).

31. Vajda, S. *et al.* Subnanometre platinum clusters as highly active and selective catalysts for the oxidative dehydrogenation of propane. *Nat. Mater.* **8**, 213–216 (2009).

32. Lei, Y. *et al.* Increased silver activity for direct propylene epoxidation via subnanometer size effects. *Science* **328**, 224–228 (2010).

33. Qiao, B. *et al.* Single-atom catalysis of CO oxidation using Pt$_1$/FeO$_x$. *Nat. Chem.* **3**, 634–641 (2011).

34. Kyriakou, G. *et al.* Isolated metal atom geometries as a strategy for selective heterogeneous hydrogenations. *Science* **335**, 1209–1212 (2012).

35. Yang, M. *et al.* Catalytically active Au-O(OH)$_x$-species stabilized by alkali ions on zeolites and mesoporous oxides. *Science* **346**, 1498–1501 (2014).

36. Hall, D. E. Electrodes for alkaline water electrolysis. *J. Electrochem. Soc.* **128**, 740–746 (1981).

37. Döner, A., Karcı, İ. & Kardaş, G. Effect of C-felt supported Ni, Co and NiCo catalysts to produce hydrogen. *Int. J. Hydrogen Energy* **37**, 9470–9476 (2012).

38. Xia, W. *et al.* Well-defined carbon polyhedrons prepared from nano metal-organic frameworks for oxygen reduction. *J. Mater. Chem. A* **2**, 11606–11613 (2014).

39. Li, H. *et al.* XPS studies on surface electronic characteristics of Ni-B and Ni-P amorphous alloy and its correlation to their catalytic properties. *Appl. Surf. Sci.* **152**, 25–34 (1999).

40. Manders, J. R. *et al.* Solution-processed nickel oxide hole transport layers in high efficiency polymer photovoltaic cells. *Adv. Funct. Mater.* **23**, 2993–3001 (2013).

41. Du, J., Cheng, F., Wang, S., Zhang, T. & Chen, J. M (Salen)-derived nitrogen-doped M/C (M = Fe, Co, Ni) porous nanocomposites for electrocatalytic oxygen reduction. *Sci. Rep.* **4**, 4386 (2014).

42. Tang, C., Cheng, N., Pu, Z., Xing, W. & Sun, X. NiSe nanowire film supported on nickel foam: an efficient and stable 3D bifunctional electrode for full water splitting. *Angew. Chem. Int. Ed.* **54**, 9351–9355 (2015).

43. Carpenter, M. K., Moylan, T. E., Kukreja, R. S., Atwan, M. H. & Tessema, M. M. Solvothermal synthesis of platinum alloy nanoparticles for oxygen reduction electrocatalysis. *J. Am. Chem. Soc.* **134**, 8535–8542 (2012).

Author contributions

X.Y., S.Q. and H.G.Y. conceived the project and contributed to the design of the experiments and analysis of the data. L.F. and P.F.L. performed the catalyst preparation, characterizations and wrote the paper. Z.Z.Y. and L.G. conducted the STEM examination and contributed to writing the STEM section. L.F., P.F.L. and X.Y. collected the electrochemical data. All the authors discussed the results and commented on the manuscript.

Additional information

Competing financial interests: The authors declare no competing financial interests.

23

Cationic cluster formation versus disproportionation of low-valent indium and gallium complexes of 2,2'-bipyridine

Martin R. Lichtenthaler[1], Florian Stahl[1], Daniel Kratzert[1], Lorenz Heidinger[2], Erik Schleicher[2], Julian Hamann[1], Daniel Himmel[1], Stefan Weber[2] & Ingo Krossing[1]

Group 13 M^I compounds often disproportionate into M^0 and M^{III}. Here, however, we show that the reaction of the M^I salt of the weakly coordinating alkoxyaluminate $[Ga^I(C_6H_5F)_2]^+[Al(OR^F)_4]^-$ ($R^F = C(CF_3)_3$) with 2,2'-bipyridine (bipy) yields the para-magnetic and distorted octahedral $[Ga(bipy)_3]^{2+\bullet}\{[Al(OR^F)_4]^-\}_2$ complex salt. While the latter appears to be a Ga^{II} compound, both, EPR and DFT investigations assign a ligand-centred $[Ga^{III}\{(bipy)_3\}^\bullet]^{2+}$ radical dication. Surprisingly, the application of the heavier homologue $[In^I(C_6H_5F)_2]^+[Al(OR^F)_4]^-$ leads to aggregation and formation of the homo-nuclear cationic triangular and rhombic $[In_3(bipy)_6]^{3+}$, $[In_3(bipy)_5]^{3+}$ and $[In_4(bipy)_6]^{4+}$ metal atom clusters. Typically, such clusters are formed under strongly reductive conditions. Analysing the unexpected redox-neutral cationic cluster formation, DFT studies suggest a stepwise formation of the clusters, possibly via their triplet state and further investigations attribute the overall driving force of the reactions to the strong $In-In$ bonds and the high lattice enthalpies of the resultant ligand stabilized $[M_3]^{3+}\{[Al(OR^F)_4]^-\}_3$ and $[M_4]^{4+}\{[Al(OR^F)_4]^-\}_4$ salts.

[1]Institut für Anorganische und Analytische Chemie and Freiburger Materialforschungszentrum (FMF), Albert-Ludwigs-Universität Freiburg, Albertstr. 21 and Stefan-Meier Str. 21, 79104 Freiburg, Germany. [2]Institut für Physikalische Chemie and Freiburg Institute of Advanced Studies (FRIAS), Albert-Ludwigs-Universität Freiburg, Albertstr. 21 and Albertstr. 19, 79104 Freiburg, Germany. Correspondence and requests for materials should be addressed to I.K. (email: krossing@uni-freiburg.de).

I n 1966, F. A. Cotton defined the term metal atom cluster compound as 'those containing a finite group of metal atoms which are held together entirely, mainly, or at least to a significant extent, by bonds directly between the metal atoms even though some non-metal atoms may be associated intimately with the cluster'[1]. Meanwhile, the term cluster has been expanded, describing various ensembles of bonded atoms (both metal and non-metal) or molecules, thus including compounds such as the boranes and carboranes[2,3], Zintl-like phases[4], salt-like clusters[5] as well as metalloid clusters[6]. Among the many routes leading to metal atom cluster compounds, the reductive and anionic syntheses prevail: such clusters are typically electron deficient and have a strong demand for additional electrons that can be provided by reductants such as alkaline metals or through electron transfer reactions like disproportionations[6,7]. Alternatively, aggregation can be achieved by applying strong donor ligands: for example, the carbene-mediated formation of the neutral P_{12} non-metal cluster[8]. Though the different approaches have yielded a vast number of neutral and anionic cluster compounds, the redox-neutral synthesis of cationic clusters by low-valent cation aggregation has not been reported to our knowledge. Thus, univalent group 13 metal salts such as—partly hypothetic—M^+A^- (M = Al, Ga, In, Tl; A = Cl, [AsF_6], [$Al(OR^F)_4$] with $R^F = C(CF_3)_3$) could in principle aggregate, yielding the electron precise $[M_3]^{3+}(A^-)_3$ cluster salt (Fig. 1, right).

The simple M^+Cl^- salts however, are prone to disproportionate to elemental M^0 and $M^{3+}(Cl^-)_3$, due to the high and favourable lattice energies of MCl_3 (Fig. 1, left). Thus, AlCl is only known as a gas phase molecule[9], and a hypothetic salt Al^+Cl^- would disproportionate to elemental Al^0 and $AlCl_3$ (solid-state reaction enthalpy $\Delta_r H^\circ$(solid) = −396 kJ mol^{-1}, cf. Born-Haber-Fajans Cycle (BHFC), Supplementary Fig. 9). For the heavier element indium on the other hand, the oxidation state +1 is more favourable due to inert pair effects[10] and the known salt In^+Cl^- (refs 11,12) is stable towards disproportionation by +31 kJ mol^{-1} (BHFC, Supplementary Fig. 9). Yet, the alternative formation of the metal atom cluster $[M_3]^{3+}$ is hampered by the lower lattice energies of $[M_3]^{3+}(A^-)_3$ compared with $M^{3+}(A^-)_3$ and, most of all, the very distinct Coulomb repulsion (Fig. 1, right). Thus, a gaseous triangular $[M_3]^{3+}$ cluster would Coulomb-explode, releasing +1,544 kJ mol^{-1} Coulomb energy at a typical M − M distance of 270 pm. The cluster formation, however, becomes favoured for larger anions A^-: that is, the difference of the estimated lattice enthalpies ($\Delta_{latt}H^\circ$) of the $In^{3+}(A^-)_3$ and $[In_3]^{3+}(A^-)_3$ salts is +697 kJ mol^{-1} for Cl$^-$ ($V^- = 0.047$ nm^3), +281 kJ mol^{-1} for [AsF_6]$^-$ ($V^- = 0.110$ nm^3) but only +26 kJ mol^{-1} for [$Al(OR^F)_4$]$^-$ ($V^- = 0.758$ nm^3; Supplementary Tables 10 and 11). Overall, salts with very large anions like [$Al(OR^F)_4$]$^-$ and ligands, allowing for a delocalization and thus 'dilution' of the positive charge on $[M_3]^{3+}$, may support cationic cluster formation. Yet, the

question remains: what are suitable Ga^I/In^I sources to perform such a chemistry?

Stabilizing gallium in its low oxidation states (<III) has been an objective since the 1,930 s. (ref. 13) Subvalent gallium halides $Ga^I[Ga^{III}X_4]$ (X = Cl, Br and I)[14], 'GaI' (refs 15,16) and metastable, donor stabilized Ga^ICl solutions[9] are important milestones and to this day used as starting material for further Ga^I chemistry: for example, arene-[17], Cp^x-complexes (Cp^x: Cp = C_5H_5 and Cp* = C_5Me_5), (ref. 18) N-heterocyclic carbene (NHC) analogues[19,20] of Ga^I, or metalloidal gallium clusters[7]. In addition, the neutral Ga^I-Cp/Cp* complexes as well as the galla-NHC analogues have been applied as ligands for transition-metals[21]. Due to the strongly coordinating halide anions, however, $Ga^I[Ga^{III}X_4]$ and 'GaI' are prone to com- and disproportionation reactions[15,17]. Donor-free Ga^{I}Cl and related compounds are only accessible at very low temperatures using elaborated matrix isolation techniques[9]. Simple indium halides In^IX (X = Cl, Br and I)[11,12,22] are stable at ambient conditions (vide supra). In contrast to the Ga^I systems, the halide anions can be replaced for weakly coordinating anions (WCAs) of different sizes including [BF_4]$^-$ (ref. 23), [OTf]$^-$ (Tf = SO_2CF_3)[24], [PnF_6]$^-$ (Pn = P, As, Sb)[25] and [$Al(OR^F)_4$]$^-$ ($R^F = C(CF_3)_3$)[26]. Various In^I-Cp complexes[22] have also been used as catalysts in organic syntheses[27]. Our group introduced a simple route to univalent gallium and indium salts of the weakly coordinating [$Al(OR^F)_4$]$^-$ anion, by oxidizing elemental gallium and indium with $Ag^+[Al(OR^F)_4]^-$ in fluorobenzene (C_6H_5F)[28,29]. Under inert conditions, the $[M(C_6H_5F)_2]^+[Al(OR^F)_4]^-$ salts (M = Ga (1), In (2)) are stable, soluble in aromatic solvents (preferably fluorinated), and a potent starting material for further Ga^I and In^I chemistry: for example, phosphine[29], crown ether[30], carbene[31] and N-heterocyclic arene[32] complexes. In addition, the [Ga(arene)$_{1-2}$]$^+$ complexes (arene = C_6H_5F, mesitylene, diphenylethane, m-terphenyl) are highly efficient isobutylene polymerization catalysts[33,34].

Herein we report on the surprising reactions of $[M(C_6H_5F)_2]^+[Al(OR^F)_4]^-$ salts mainly with 2,2'-bipyridine (bipy), but also with the bipy-relative 1,10-phenanthroline (phen).

Results

Orienting quantum-chemical calculations. In contrast to the anionic chelating N-ligands for univalent gallium and indium, for example, guanidinates[35], diazabutadienes[19,20,36], β-diketiminates[20,37] or tris(pyrazolyl)hydroborates (Tp)[38], neutral N-ligands, like pyridine derivatives, were shown to only stabilize the +1 oxidation state of indium[39], but not of gallium[40,41]. Applying [$Ga^I(C_6H_5F)_2]^+[Al(OR^F)_4]^-$ (1), however, we were able to isolate the first gallium(I) complexes with simple N-ligands, such as pyrazine and di-tert-butylmethylpyridine[32]. To expand the scope to neutral chelating N-ligands, we investigated the thermodynamics of potential ligand exchange reactions of the $[M(C_6H_5F)_2]^+$ complexes (M = Ga, In) and 1 to 3 equivalents bipy. All turned out to be exothermic/exergonic by at least −146 kJ mol^{-1} (Supplementary Table 9). Thus, it appeared interesting to test the reactions.

Reaction of $[Ga^I(C_6H_5F)_2]^+[Al(OR^F)_4]^-$ with bipy. On mixing colourless solutions of **1** and bipy (2.00 eq) in ortho-difluor-obenzene (o-$C_6H_4F_2$), an unexpected distinct change in colour towards moss-green was observed (Supplementary Data 1–7). To promote crystallization, the reaction mixture was concentrated by slowly removing the volatiles under reduced pressure. During this process, the formation of a black precipitate was observed, while the colour of the solution remained green. Applying C_6H_5F as

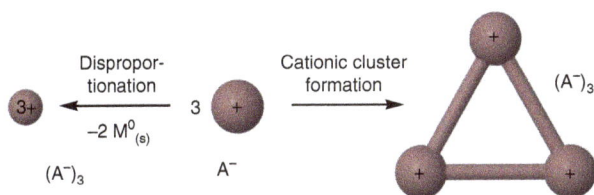

Figure 1 | Disproportionation versus cationic cluster formation of univalent group 13 metal salts M$^+$A$^-$. Generally, the disproportionation is favoured over the cluster formation due to the much higher lattice energies of the $M^{3+}(A^-)_3$ salt.

solvent, similar observations were made, though the precipitate formed without concentrating the solution. From the o-$C_6H_4F_2$ solutions, green platelet single crystals were repeatedly isolated and surprisingly characterized as $[Ga(bipy)_3]^{2+\bullet}\{[Al(OR^F)_4]^-\}_2$ (3) and not as the predicted $[Ga(bipy)_2]^+$ complex (cf. Supplementary Table 9). Though the $[Al(OR^F)_4]^-$ anions are crucial in terms of stabilizing the low-valent oxidation states of gallium and indium, all structures within this manuscript feature no Ga − F or In − F contact shorter than the sum of the corresponding van der Waals radii[42,43]. The structural discussions are therefore focused on the obtained cationic complexes and clusters. To our knowledge, the dicationic, paramagnetic and distorted octahedral $[Ga(bipy)_3]^{2+\bullet}$ complex (3^{2+}) is the first structurally characterized one of its kind, thus differing from the $[Ga(bipy)_3]^{3+}$ complex obtained, if 'GaI' is applied as gallium(I) starting material (Fig. 2)[40]. The synthesis of 3^{2+} is the first example, where 1 did not act as a pure Ga^I source but disproportionated (vide infra). With a total charge of 2 +, it is tempting to proclaim the successful stabilization of a monomeric, room-temperature stable Ga^{II} species, that is, a $[(Ga^{II})^\bullet(bipy)_3]^{2+}$ complex.

While earlier reported Ga^{II} compounds have proven to be mixed-valent species[14] or diamagnetic dimers[44], Aldridge et al. recently reported on a thermally robust monomeric Ga^{II} compound: $[Ga^{II}\{B(N(C_6H_3-2,3-iPr)CH)_2\}_2]$[45]. In this Ga^{II} molecule, the metal atom is coordinated in a bent fashion by two boryl ligands and over 70% of the unpaired spin density is located on the metal. The related $[Ga(dabab)_2]$ complex[46] (dabab = 1,4-di-tert-butyl-1,4-diazabutadiene) on the other hand, is correctly described as a Ga^{III} cation coordinated by one singly and one doubly reduced dabab ligand[36,47–50]. A related description could account for 3^{2+}, that is, a $[Ga^{III}\{(bipy)_3\}^\bullet]^{2+}$ complex, and therefore we conducted electron paramagnetic resonance (EPR) spectroscopy measurements of solutions of 3 in o-$C_6H_4F_2$ (Fig. 3). Both, the isotropic g-value ($g_{iso} = 2.0024$) being close to that of the free electron and the low g anisotropy[7a] speak in favour of a ligand-centred spin system. This assumption is further supported by the low hyperfine coupling constant $A(^{69}Ga) = 21$ MHz, thus being in good agreement with studies by Kaim et al.[47] and clearly differing from the above mentioned metal-centred spin system $[Ga^{II}\{B(N(C_6H_3-2,3-iPr)CH)_2\}_2]$ (cf. $A(^{69}Ga) = 670$ MHz)[45].

Furthermore, we applied density functional theory (DFT) calculations to compute the singly occupied molecular orbitals (Fig. 2, inset) and the spin density of 3^{2+} (Fig. 3, inset). We chose

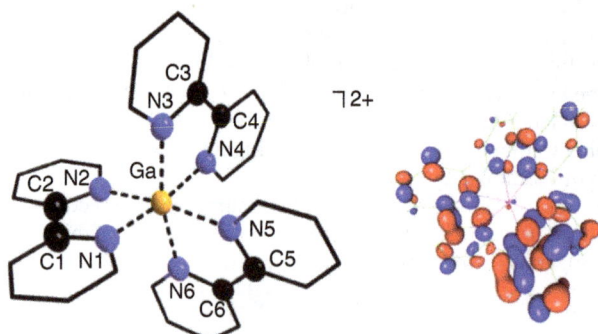

Figure 2 | Molecular structure of $[Ga(bipy)_3]^{2+\bullet}$ (3^{2+}). Selected bond lengths are given in pm: Ga − N1 = 202.6(4), Ga − N2 = 202.6(4), Ga − N3 = 205.8(4), Ga − N4 = 208.2(4), Ga − N5 = 208.2(4), Ga − N6 = 205.8(4), C1 − C2 = 149.2(12) and C3 − C4 = 148.6(7), C5 − C6 = 148.6(7). The earlier reported $[Ga(bipy)_3]^{3+}$ complex features a slightly longer average Ga − N bond length of 206.4 pm (cf. 205.5(4) pm for 3^{2+}) and a slightly shorter average $C_{1,3,5} - C_{2,4,6}$ bond length of 147.9 pm (cf. 148.8(9) pm for 3^{2+})[40]. The $[Al(OR^F)_4]^-$ anions and all of the hydrogen atoms were omitted for clarity. The thermal ellipsoids are set at 50% probability. An alternative presentation, featuring ellipsoids for all atoms, is included in Supplementary Fig. 6. As an inset, the figure also includes the B3LYP/SV(P) level frontier and singly occupied Kohn-Sham orbital (SOMO) of 3^{2+}, electron density cutoff at 0.04 a.u. The calculated distances of the latter are in good agreement with the experimental results (bond lengths in pm, similar label scheme): Ga − N1 = 206.4, Ga − N2 = 206.8, Ga − N3 = 211.6, Ga − N4 = 213.2, Ga − N5 = 211.1, Ga − N6 = 209.1, C1 − C2 = 145.7, C3 − C4 = 147.7 and C5 − C6 = 146.9. Yet, and more importantly, the SOMO is primarily located on the bipy ligands, thus speaking for a $[Ga^{III}\{(bipy)_3\}^\bullet]^{2+}$ rather than a $[(Ga^{II})^\bullet(bipy)_3]^{2+}$ complex. Applying the BHLYP/SV(P) level of theory, the SOMO is solely located on one bipy ligand, corresponding to a $[Ga^{3+}(bipy)_2(bipy)^{\bullet-}]^{2+}$ complex (Supplementary Fig. 7). The calculated distances of the latter are in inferior agreement with the experimental results and therefore 3^{2+} more likely corresponds to the $[Ga^{III}\{(bipy)_3\}^\bullet]^{2+}$ formulation.

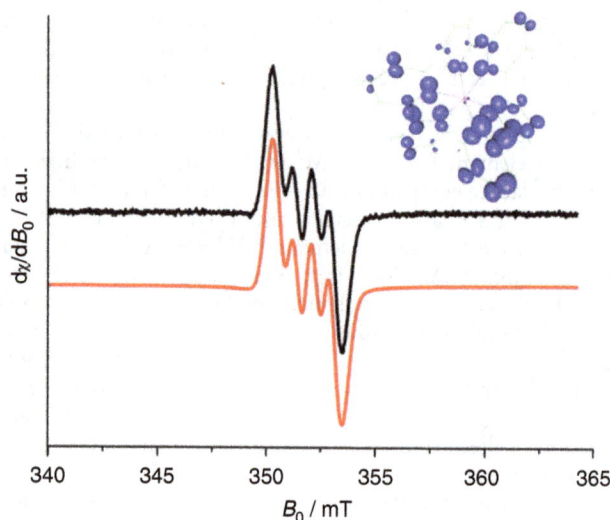

Figure 3 | EPR investigations of $[Ga(bipy)_3]^{2+\bullet}\{[Al(OR^F)_4]^-\}_2$ (3). X-band continuous wave EPR spectrum of a solution of 3 in o-$C_6H_4F_2$ (1mM) (black trace; recorded at room temperature, microwave frequency 9.861 GHz, magnetic-field modulation amplitude 0.1 mT, microwave power 2 mW) and the corresponding spectral simulation (red trace). The measurement and its spectral simulation exhibit an isotropic gallium dominated resonance with a g-value (g_{iso}) of 2.0024 and a hyperfine coupling constant of $A(^{69}Ga) = 21$ MHz, thus supporting an isotropic spin density of <1.0% on the gallium atom and a ligand-centred spin system, while the exact whereabouts of the radical in the ligand backbone could not be identified. The $A(^{69}Ga)$ value strongly deviates from the theoretical hyperfine coupling constant of an isolated ^{69}Ga atom ($A(^{69}Ga) = 12210$ MHz)[68,69]. In addition spectra were obtained from frozen solutions of 3 in C_6H_5F or o-$C_6H_4F_2$ and both show a weak g anisotropy (Supplementary Fig. 4). As an inset, the figure also includes the DFT calculated spin density of 3^{2+} (cutoff at 0.003 a.u., B3LYP/SV(P) level). With a distribution of 3.0% on the gallium atom and 97% on the bipy ligands, the latter corresponds to a $[Ga^{III}\{(bipy)_3\}^\bullet]^{2+}$ complex, thus being in very good agreement with the EPR results. An alternative modelling at the BHLYP/SV(P) level yielded a spin density distribution of 1.3% on the gallium atom and 98% on one bipy ligand (Supplementary Fig. 7), suggesting a formulation as $[Ga^{3+}(bipy)_2(bipy)^{\bullet-}]^{2+}$. Yet, a distribution of the spin density on all three ligands seems to be more likely as the calculated C − C and In − N bond lengths of the $[Ga^{III}\{(bipy)_3\}^\bullet]^{2+}$ complex are in much better agreement with the experimental results.

the hybrid B3LYP/SV(P) method, as it yielded good agreements between calculated and experimental bond lengths of 3^{2+} (Fig. 2), and the main absorption maximum (λ_{max}) in measured (302 nm) and simulated (296 nm) ultraviolet-visible spectra (Supplementary Fig. 5). These bands are reminiscent of the absorption of the related $[Ru^{3+}(bipy)_2(bipy)^{\bullet-})]^{2+}$ complex[51] (cf. $\lambda_{max} = 373$ nm). Beyond featuring a similarly distorted octahedral coordination mode as its single-crystal congener, the singly occupied molecular orbitals of the geometry-optimized 3^{2+} is solely located on the bipy ligands (Fig. 2) and only 3.0% of the spin is located at the gallium cation (Fig. 3). Overall, the experimental results (X-ray powder diffraction, EPR and ultraviolet-visible) and the DFT studies are in very good agreement and clearly assign a $[Ga^{III}\{(bipy)_3\}^{\bullet}]^{2+}$ complex as correct formulation of 3^{2+}. To our knowledge, 3 is the first reported single-crystal structure of a p-block metal complex, featuring a bipy radical anion as ligand (cf. the few examples of alkali-metal salts of bipy radicals and dianions of Goicoechea et al. as well as Wieghardt's extensive work on transition-metal complexes of bipy)[52,53].

Reaction of $[In^I(C_6H_5F)_2]^+[Al(OR^F)_4]^-$ with bipy. Due to the redox instability of 1, we additionally reacted the heavier, more redox-stable homologue $[In^I(C_6H_5F)_2]^+[Al(OR^F)_4]^-$ (2) with bipy. While the isolation of single crystals from highly concentrated, yellowish solutions of 2 and bipy (2.00 eq or 1.63 eq) in C_6H_5F was straightforward, the composition of the obtained crystals depended on the amount of bipy employed as well as the crystallization procedure. Overall, we were surprised not to isolate any of the predicted $[In(bipy)_{1-3}]^+$ complexes (cf. Supplementary Table 9), but the very first homonuclear **cationic** triangular or rhombic In^I clusters: $[In^I_3(bipy)_6]^{3+}\{[Al(OR^F)_4]^-\}_3$ (4), $[In^I_3(bipy)_5]^{3+}\{[Al(OR^F)_4]^-\}_3$ (5) and $[In^I_4(bipy)_6]^{4+}\{[Al(OR^F)_4]^-\}_4$ (6). As above, the direct interaction between the cationic clusters and the $[Al(OR^F)_4]^-$ anions is negligible and the latter are therefore not shown in Fig. 4. For the synthesis of 4, 2.04 equivalents of bipy were applied. In 4^{3+}, each In^I cation is coordinated in a distorted octahedral fashion, or in other words, three tetragonal pyramidal N-coordinated $[In^I(bipy)_2]^+$ fragments interact with each other, thus forming the observed triangular cationic In^I cluster. While the In1 – In2 and In1 – In3 bond lengths are very similar (266.07(4) and 266.98(5) pm, respectively),

the In2 – In3 distance is elongated by 11-12 pm. All three distances are well within the sum of the van der Waals radii[42,43] (386 pm) and among the shortest compared to the manifold of reported organometallic and inorganic compounds that feature In – In bonds (Table 1 and Supplementary Table 15).

Reducing the amount of bipy from 2.00 to 1.63 equivalents, we obtained compound 5. Though the molecular structure of 5^{3+} resembles the one of 4^{3+}, one bipy ligand now acts as bridging N-ligand between two In^I cations, thus resulting in a more twisted arrangement of the two pyridine rings. These findings are likely attributable to the reduced amount of bipy employed and correspond well with the stoichiometry of the reaction (cf. '(5 bipy ligands)/(3 In^I cations) ≈ 1.67'). Hence, only In2 is coordinated in a distorted octahedral fashion, while In1 and In3 are coordinated in distorted trigonal bipyramidal fashions. The very short In – In bond lengths (Table 1) are similar within 3 pm (av. In – In distance of 268.18(5) pm), thus resulting in an almost equilateral triangle. For 6, 2.00 equivalents of bipy were applied. Yet, and different to the synthesis of 4, the reaction mixture was additionally concentrated under reduced pressure, leading to the cationic, planar In^I rhomb 6^{4+}. While the coordination modes of In2 and In4 resemble the ones of the In^I cations in 4^{3+} and In2 in 5^{3+}, In1 and In3 are pentacoordinated, interacting with only one bipy ligand and featuring three In – In contacts. The peripheral In – In bond lengths only deviate by 5 pm (av. In – In distance = 277.99(14) pm) and are, with the exception of the In2 – In3 bond of 4^{3+}, 10 pm longer than the In – In distances of their triangular congeners. The bridging In1 – In3 distance on the other hand, is with 259.65(12) pm the shortest In – In bond that, to our knowledge, has been reported (the only shorter In – In bond derives from the structural relative seven, see below) (Table 1 and Supplementary Table 15).

Overall and despite the vast literature on compounds containing In – In bonds, the cationic molecular and univalent structures of 4, 5 and 6 are unique. Somewhat related to 6 is the $[In_4\{Cp_2Mo_2(CO)_4P_2\}_8]^{4+}\{[Al(OR^F)_4]^-\}_4$ salt reported by Scheer et al.[26] Herein, the In^I cations form a similar rhombic arrangement, but with intermetallic distances that are at least 60 pm longer (shortest In – In distance: 348.2 pm). These findings must be due to the different ligand system, as the group used the same $[Al(OR^F)_4]^-$ anion: that is, the interactions seem to be weakly dispersive rather than covalent. The remaining known

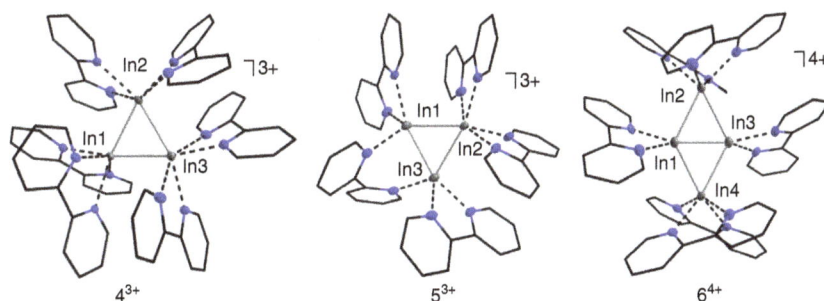

Figure 4 | Molecular structures of indium clusters. Molecular structures of $[In^I_3(bipy)_6]^{3+}$ (4^{3+}), $[In^I_3(bipy)_5]^{3+}$ (5^{3+}) and $[In^I_3(bipy)_6]^{4+}$ (6^{4+}). Selected bond lengths are given in pm and angles in °: 4^{3+}, In1 – In2 = 266.07(4), In1 – In3 = 266.98(5), In2 – In3 = 278.12(5), av. In1 – N = 236.9(4), av. In2 – N = 249.0(4), av. In3 – N = 246.5(4) and ∢(In – In – In) = 58.39(12) – 62.90(12); 5^{3+}, In1 – In2 = 267.89(5), In1 – In3 = 266.80(5), In2 – In3 = 269.84(5), av. In1 – N = 233.8(4), av. In2 – N = 234.0(4), av. In3 – N = 235.3(4) and ∢(In – In – In) = 59.49(14) – 60.62(14) (the asymmetric unit of 5 contains a second $[In^I_3(bipy)_5]^{3+}$ cluster featuring similar bonds lengths, Supplementary Table 4); 6^{4+}, In1 – In2 = 275.93(14), In1 – In3 = 259.65(12), In1 – In4 = 276.62(14), In2 – In3 = 280.83(14), In3 – In4 = 278.56(14), av. In1 – N = 232.40(10), av. In2 – N = 236.33(10), av. In3 – N = 233.60(10), av. In4 – N = 236.10(10) and ∢(In – In – In) = 55.59(3) – 63.16(4) (the asymmetric unit of 6 contains four $[In^I_4(bipy)_6]^{4+}$ clusters, featuring similar In – In bonds lengths, Supplementary Table 5). The $[Al(OR^F)_4]^-$ anions and all of the hydrogen atoms were omitted for clarity. The thermal ellipsoids are set at 50% probability. An alternative representation, featuring ellipsoids for all atoms, is included in Supplementary Fig. 6.

Table 1 | Selection of organometallic and inorganic indium compounds, featuring at least one In − In bond (d(In − In) in pm).

Compound	d(In − In) (pm)	Comment
$[In_4(phen)_6]^{4+}\{[Al(OR^F)_4]^-\}_4$ (**7**)	258.1–286.1*	this work
$[In_4(bipy)_6]^{4+}\{[Al(OR^F)_4]^-\}_4$ (**6**)	259.7–280.8*	this work
$In_5Mo_{18}O_{28}$ (In$_5$-moieties)	261.6–266.5	—†
$In_{11}Mo_{40}O_{62}$ (In$_5$- and In$_6$-moieties)	262.4–268.9	—†
$In_3(PO_4)_2$ (In$_2$-dimers)	263.0	—†
$[In\{C(Si(Me)_3)_3\}\{\{OC(C_6H_5)\}_2CH\}]_2$	264.6–279.3	—†
$[In\{(O_2CPh)C(SiMe_3)_3\}_2]_2$	265.4	—†
$[In\{N(C_6H_2-2,4,6-Me_3)CH\}_2O(SO_2)(CF_3)]_\infty$	265.6, 266.5	—†
$[In_3(bipy)_6]^{3+}\{[Al(OR^F)_4]^-\}_3$ (**4**)	266.1–278.1	this work
$[In_3(bipy)_5]^{3+}\{[Al(OR^F)_4]^-\}_3$ (**5**)	266.8–269.8	this work
In_5S_4 (In$_5$-moieties)	276.2–276.9	—†
In_4Se_3 (In$_3$-moieties)	275.6–277.6	—†
$InSe$ (In$_2$-dimers)	281.8	—†
$In_2O(PO_4)$ ($[In_2]^{4+}$ [In$_2$O$_2$(PO$_4$)$_2$]$^{4-}$)	286.2	—†
In − In (metal)	325.2, 337.7	—†
$[In_4\{Cp_2Mo_2(CO)_4P_2\}_8]^{4+}\{[Al(OR^F)_4]^-\}_4$	348.2–396.0	—†
InCl (distorted In$_4$-tetrahedra)	359.1–476.3	—†

The entries are ordered in terms of increasing d(In − In) values. Supplementary Table 15 contains a comprehensive compilation.
*The very short In − In bond lengths (258.1 and 259.7 pm) correspond to the bridging In1 − In3 distances within the rhombic 6^{4+} and 7^{4+}.
†The references are included with Supplementary Table 15.

cationic compounds are purely inorganic (cf. Supplementary Table 15) and while some of them feature similar In–In bond lengths, their chain-like substructures differ significantly. Though featuring a related geometry, the reported anionic triangular cyclogallanes differ in their electronic structures: that is, they are only accessible via reductive routes and feature delocalized electrons, resulting in a 2π-aromatic stabilization[54,55]. The dicationic rhombic tetraborane $[B_4H_2(\mu\text{-hpp})_4]^{2+}$ (hpp = 1,3,4,6,7,8-hexahydro-2H-pyrimido[1,2-a]pyrimidinate) on the other hand, is structurally and electronically related to **6** (ref. 56). Finally, the cationic In^I clusters are not consistent with the Wade-Mingos rules (polyhedral skeletal electron pair theory), which are often used to rationalize clusters from group 13 (refs 2,3,57).

Reaction of $[In^I(C_6H_5F)_2]^+[Al(OR^F)_4]^-$ with phen. While **3** and **4** reproduce the triangular $[In_3]^{3+}$ motif intrinsically, we additionally applied 1.49 equivalents of 1,10-phenanthroline (phen) to reproduce the rhombic $[In_4]^{4+}$ motif in **6**$^{4+}$. In doing so, we were able to isolate a structural relative of **6**, that is, the $[In^I_4(phen)_6]^{4+}\{[Al(OR^F)_4]^-\}_4$ salt (**7**) (Fig. 5). The structural parameters of the $[In^I_4(phen)_6]^{4+}$ complex (**7**$^{4+}$) resemble those of **6**$^{4+}$, despite small π-π-interactions within the phen ligands (distances in pm, angles in °, values for **6**$^{4+}$ are parenthesized): av. In − In = 279.55(11) (277.99(14)), In1 − In3 = 258.06(16) (259.65(12)), av. In − N = 234.15(80) (235.14(10)), ∢(In − In − In) = 54.92(3) − 65.11(3) (55.59(3) − 63.16(4)), no In − F contact to the $[Al(OR^F)_4]^-$ anions. However, with 258.06(16) pm the bridging In1 − In3 distance in **7**$^{4+}$ is the shortest In − In bond length reported to this day. Both, the short bridging In1 − In3 distances in **6**$^{4+}$ and **7**$^{4+}$, would be in agreement with the hypothetic interactions of a slightly trans-bent dicationic $[(N\text{-ligand})In = In(N\text{-ligand})]^{2+}$ fragment and two $[In(N\text{-ligand})_2]^+$ complexes (N-ligand = bipy, phen), thus resulting in two two-electron three-centre (2e3c) bonds (Fig. 5b).

Apart from **7**, the reaction mixture also contained small amounts of a second type of single crystals, which, being colourless and not yellowish, surprisingly corresponded to the $[(phen)_2In^I - Ag^I(C_6H_5F)]^{2+}\{[Al(OR^F)_4]^-\}_2$ salt (**8**). To our knowledge, this is the first monomeric dicationic In − Ag adduct. While the Ag^I cation likely derives from the originally used Ag$^+$ $[Al(OR^F)_4]^-$ salt (stemming from the synthesis of (**2**), the $[(phen)_2In^I - Ag^I(C_6H_5F)]^{2+}$ complex (**8**$^{2+}$) can be considered as an addition product of the Lewis basic, tetragonal pyramidal coordinated $[In^I(phen)_2]^+$ complex and the Lewis acidic, η^3-coordinated $[Ag^I(C_6H_5F)]^+$ complex (cf. $\Delta_rH°$(gas) = − 73/ − 84 kJ mol^{-1} and $\Delta_rG°$(gas) = − 24/ − 36 kJ mol^{-1} for the formation of this gaseous dication (!) at 298.15 K, 1.0 bar, BHLYP/SV(P) and B3LYP/SV(P) level). Yet, multi-charged species in the gas phase are subject to strong repulsion and Coulomb explosion (v.s.). Thus, the surprisingly favourable $\Delta_rH°$(gas) and $\Delta_rG°$(gas) values for the formation of **8**$^{2+}$ are probably attributable to the high Lewis basicity of the $[In^I(phen)_2]^+$ complex. Fig. 6 contains the single-crystal structure as well as the highest occupied molecular orbital of the converged calculated **8**$^{2+}$ structure.

Multinuclear solution NMR spectroscopy. All obtained single crystals were dissolved in o-$C_6H_4F_2$ and investigated by ^{1}H, ^{14}N, ^{19}F, ^{27}Al, ^{71}Ga and ^{115}In NMR spectroscopy. While the ^{14}N, ^{71}Ga and ^{115}In NMR spectra featured no resonances, thus being in good agreement with earlier reported σ-coordinated complexes of Ga^I and In^I [28,29,31,32], one singlet in the ^{19}F NMR and ^{27}Al NMR spectra at − 74.9 p.p.m. and + 33.8 pm, respectively, revealed the intactness of the $[Al(OR^F)_4]^-$ anions[58]. In the case of the mixed crystalline residue of **7** and **8**, the ^{19}F NMR spectrum additionally featured the triplet of a triplet at − 113 p.p.m. assigned to C_6H_5F, which likely derives from **8**$^{2+}$. Finally, the ^{1}H NMR spectra provided primary information on the stability of the obtained cationic complexes in solution: that is, the solution of **3** featured very weak and broad resonances due to the paramagnetic nature of **3**$^{2+}$ (cf. EPR studies). Solutions of **4**, **5** and **6** featured multiplets attributable to solvated bipy, thus speaking for a fragmentation of the cationic indium clusters in solution. The solution of the crystalline residue of **7** and **8** yielded a complex multiplet pattern in the aromatic region, which likely is attributable to different fragments of both sets of single crystals. From the multinuclear NMR studies we suggest that the cationic In^I clusters are unstable in solution, while the $[Al(OR^F)_4]^-$ anions stay intact. The dissociation of the cationic In^I clusters is probably attributable to the distinct Coulomb repulsion of the In − In bonded individual $[In^I(N\text{-Ligand})_{1,2}]^+$ units constituting **4**$^{3+}$, **5**$^{3+}$, **6**$^{4+}$ and **7**$^{4+}$ (cf. DFT studies below).

Disproportionation versus cationic cluster formation. To answer the question why **1** in the presence of bipy disproportionates and **2** functions as indium cluster source, as well as to elaborate a potential reaction pathway, we conducted further DFT calculations. From a retrosynthetic point of view, we chose the $[M^I(bipy)_2]^+$ complex (M = Ga, In) as starting point, as the latter seems to be a crucial building block during the disproportionation and cluster formations: that is, in **4**$^{3+}$, **5**$^{3+}$ and **6**$^{4+}$, six out of ten In^I cations are coordinated in such a manner. While this assumption is supported by the successful isolation of **7**$^{4+}$ and **8**$^{2+}$, the question remains, how the fragments interact with each other? (i) Via an ambiphilic route in which each $[M(bipy)_2]^+$ complex functions as a Lewis acid (empty 4p/5p orbitals) and as a Lewis base (occupied 4 s/5 s orbitals) (cf. the cyclopropane derivatives of the group 13 (ref. 59) and 14 homologues[60]) or ii) via a singlet-triplet route, that is, ligands such as o-quinones, N-hetero

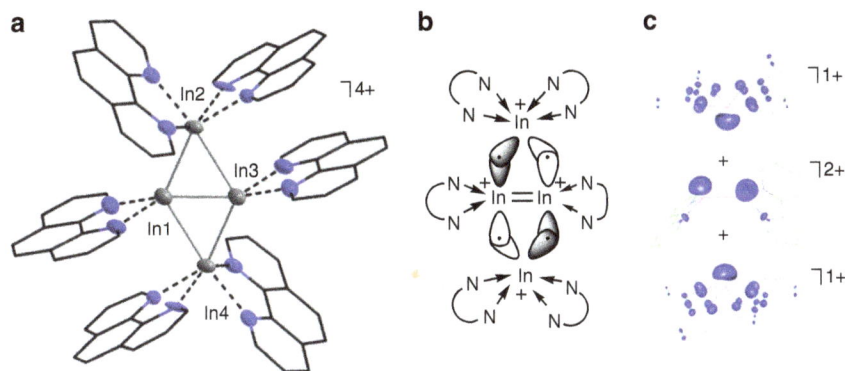

Figure 5 | Molecular structure and bonding description of $[In^I_4(phen)_6]^{4+}$ (7^{4+}) as well as spin density distributions of potential precursors.
(**a**) Molecular structure of 7^{4+}. Selected bond lengths are given in pm, angles in ° and the values for 6^{4+} are parenthesized: $In1 - In2 = 273.03(10)$ $(275.93(14))$, $In1 - In3 = 258.06(16)$ $(259.65(12))$, $In1 - In4 = 286.06(11)$ $(275.93(14))$, $In2 - In3 = 286.06(11)$ $(280.83(14))$, $In3 - In4 = 273.03(10)$ $(278.56(14))$, av. $In1 - N = 231.2(9)$ $(232.40(10))$, av. $In2 - N = 235.6(8)$ $(236.33(10))$, av. $In3 - N = 231.2(9)$ $(233.60(10))$, av. $In4 - N = 235.6(8)$ $(236.10(10))$ and $\sphericalangle(In - In - In) = 54.92(3) - 65.11(3)$ $(55.59(3) - 63.16(4))$. The $[Al(OR^F)_4]^-$ anions and all of the hydrogen atoms were omitted for clarity. (**b**) A possible hypothetic description of the bonding situation in 6^{4+} and 7^{4+}. The short bridging $In1 - In3$ values could originate from interactions between the π^*-orbitals of a dicationic, formally doubly bonded $[(N\text{-ligand})In = In(N\text{-ligand})]^{2+}$ fragment (N-ligand = bipy, phen) and singly occupied orbitals of two $[In(N\text{-ligand})_2]^+$ complexes, thus resulting in two two-electron three-centre bonds (2e3c), partially, but not fully, populating the antibonding π^*-orbital. The remaining small $In = In$ double-bonding contribution could account for the observed short $In - In$ separations in 6^{4+} and 7^{4+} and also the relatively long 2e3c $In - In$ bonds to the upper and lower $[In(N\text{-ligand})_2]^+$ moieties. **c**) Calculated spin density distributions of triplet state fragments that could interact to form the observed cationic clusters (spin density cutoff at 0.010 a.u., B3LYP/SV(P) level). In the calculated $[(bipy)In = In(bipy)]^{3+}$ fragment, the $In - In$ distance is 291.4 pm and the average spin density on each indium atom 77% (a planar dicationic fragment did not converge, even if the conductor-like screening model (COSMO)[65] was switched on and the permittivity was set to infinite $\varepsilon_r = \infty$). For triplet state of $[In(bipy)_2]^+$, the spin density on the indium atom is 35%.

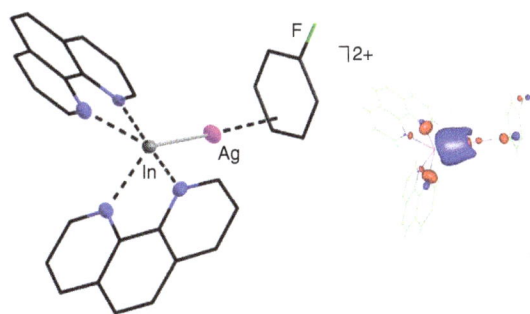

Figure 6 | Molecular structure of $[(phen)_2In^I - Ag^I(C_6H_5F)]^{2+}$ (8^{2+}).
Selected bond lengths are given in pm and angles in °: av. $In - N = 233.3(3)$, $In - Ag = 256.37(5)$, $Ag - cent = 239.4$ and $In - Ag - cent = 173.0$ (cent = centroid of the η^3-coordinating C_6H_5F; the linearity is in good agreement with earlier reported systems[70]. The $[Al(OR^F)_4]^-$ anions and all of the hydrogen atoms were omitted for clarity. The thermal ellipsoids are set at 50% probability. An alternate formulation as $[(phen)_2In^I - In^I(C_6H_5F)]^{2+}$ diindane is not conceivable, as a $[In^I(C_6H_5F)]^+$ complex would be η^6-coordinated. In addition, the X-ray diffraction refinement of the diindane formulation gave inferior R-factors and the DFT structure refinement did not converge. As an inset, the figure also contains the highest occupied molecular orbital (HOMO) of the converged 8^{2+} structure. The latter features a constructive interaction between the occupied 5s orbital of the $[In^I(phen)_2]^+$ complex and the unoccupied $5s/4d_{z^2}$ hybrid orbital[70] of the $[Ag^I(C_6H_5F)]^+$ complex. The calculated distances are in good agreement with experimental results (bond lengths in pm and angles in °): av. $In - N = 235.1$, $In - Ag = 262.2$, $Ag - cent = 253.4$ and $In - Ag - cent = 175.0$ (B3LYP/SV(P) level, electron density cutoff at 0.06 a.u.).

arenes and diazabutadienes have proven to be non-innocent[61], thus promoting electron transfer reactions and a more easy access to the triplet states of the complexes[62] (Fig. 7).

From an energetic point of view, the singlet-triplet route appears to be more conceivable, as the singlet-triplet gaps are distinctively smaller than the corresponding 4 s/5 s-4p/5p energy gaps (Table 2; cf. DFT studies by Macdonald et al.[63]). In addition and considering the distribution of spin densities, the triplet states of the $[M(bipy)_2]^+$ complexes offer important insights into the metal-dependent redox stabilities: that is, for gallium the tetrahedral $[Ga^{3+}\{(bipy)\bullet^-\}_2]^+$ complex forms and for indium the tetragonal pyramidal $[In^{2+}(bipy)(bipy)\bullet^-]^+$ complex. Hence, only the latter should be able to stepwise cyclotrimerize, while the former is labile and disproportionates. In this context, the choice of the redox-active ligand is crucial and for indium, bipy seems to be the perfect match as it enables reversible, single electron transfers between the metal centre and the ligand, thus making way for the stepwise cationic metal atom cluster formation. For gallium, this is not the case and the bipy-located electrons are prone to intermolecular rather than intramolecular transfer reactions, resulting in a disproportionation of the $[Ga(bipy)_2]^+$ complex[64].

Finally, we attempted to calculate the molecular structures of 4^{3+}, 5^{3+} and 6^{4+}. Though we implemented the conductor-like screening model[65] with an infinite permittivity and dispersive interactions (D3), the cationic In clusters fragmented due to the distinct Coulomb repulsion. However, we were able to calculate dicationic cluster fragments in their triplet state, such as $[(bipy)_2In - In(bipy)_2]^{2+}$ (Fig. 7, inset). With an average spin-density distribution of 32% at each indium atom, the latter could be seen as a reaction intermediate of the univalent indium clusters, thus supporting a stepwise cluster formation (for further dicationic cluster fragments see Supplementary Fig. 10). Furthermore, we assessed the gas-phase thermodynamics ($\Delta_r H°(gas)$, Table 3) of the formations of 4^{3+}, 5^{3+} and 6^{4+} from BHFCs and setting $\Delta_r H°(solid)$ in a worst-case scenario to ± 0 kJ mol^{-1} (Supplementary Fig. 8). The endothermic values are attributable to the above mentioned Coulomb repulsion and very well correspond to the large exothermicity of the Coulomb

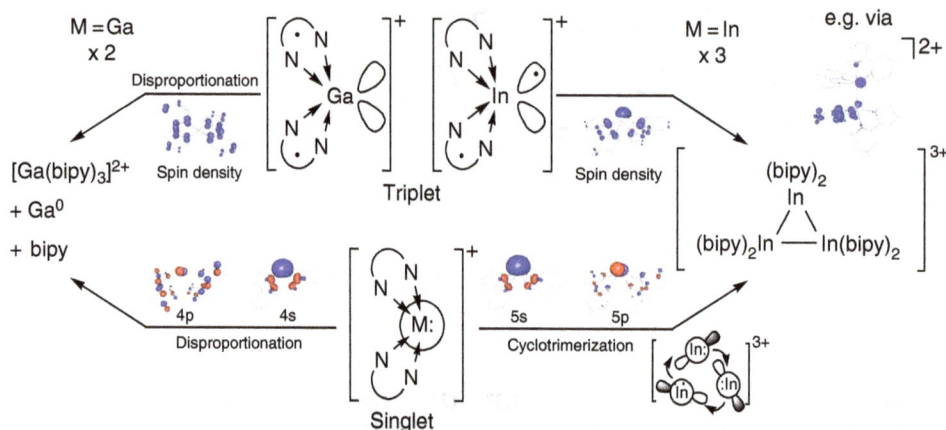

Figure 7 | Metal-dependent behaviour of $[M^I(C_6H_5F)_2]^+[Al(OR^F)_4]^-$ (M = Ga, In) in the presence of bipy. Disproportionation (for M = Ga) and cluster formation (for M = In) can occur via (i) an ambiphilic singlet route (bottom) or (ii) a triplet route (top). Regarding (i), the $[M^I(bipy)_2]^+$ complexes feature an occupied 4 s/5 s-orbital (HOMO) and empty 4p/5p-orbitals (LUMO + 6), thus being able to react via a disproportionation or a cyclotrimerization, yielding 3^{2+} and 4^{3+}, respectively. The conceivability of this reaction path directly correlates to the energy gap between the 4 s/5 s- and the 4p/5p-orbitals of the reactants (Table 2). Choosing reaction path (ii), the $[M^I(bipy)_2]^+$ complexes react via their triplet state. Dependent on the metal, the latter significantly differ in terms of geometries and localization of spin densities. Hence, the $[Ga(bipy)_2]^+$ complex changes its coordination mode from tetragonal pyramidal in the singlet to tetrahedral in the triplet state and both unpaired electrons are equally distributed over the bipy ligands (spin density at the gallium atom = 6.0%). These findings clearly speak for the tendency of the Ga^I cations to disproportionate and also explain the complex's inability to form a di-/oligo-gallane.[44] The $[In(bipy)_2]^+$ complex on the other hand, retains its tetragonal pyramidal coordination mode and 35% of the spin density is located on the indium atom, making a stepwise cyclotrimerization feasible (cf. the dicationic $[(bipy)_2In - In(bipy)_2]^{2+}$ cluster fragment shown). The conceivability of the reaction path directly correlates to the singlet-triplet gaps of the reactants (Table 2). The Kohn-Sham orbitals shown were selected on localization of the electron density on the metal cations and the lowest possible energy (electron and spin density cutoffs at 0.06 and 0.01 a.u., B3LYP/SV(P) level; as shown in Supplementary Fig. 10, the $[(bipy)_2In - In(bipy)_2]^{2+}$ cluster fragment only featured a spin density at the indium atoms, if the BHLYP functional was applied).

Table 2 | Energy gaps between the occupied 4 s/5 s- and unoccupied 4p/5p-orbitals as well as singlet-triplet gaps of $[M(C_6H_5F)_2]^+$, $[M(bipy)]^+$ and $[M(bipy)_2]^+$ in kJ mol^{-1} (M = Ga, In; gas phase, 298.15 K, 1.0 bar, values at BHLYP/SV(P)/B3LYP/SV(P) level).

	ΔE (kJ mol^{-1}) (M = Ga)		ΔE [kJ mol^{-1}] (M = In)	
	4 s/4p	singlet-triplet	5 s/5p	singlet-triplet
$[M(C_6H_5F)_2]^+$	853/653	311/-*	797/618	-*/-†
$[M(bipy)]^+$	582/386	196/-†	580/383	227/-†
$[M(bipy)_2]^+$	709/507	5‡/21‡	645/471	111§/96§

*Even after 500 iteration cycles, the geometry optimization of the triplet state did not converge.
†Though the geometry optimization converged, the electronic occupation of the triplet state is not correctly described.
‡The significant decrease of ΔE is accompanied by a geometry change of the $[Ga(bipy)_2]^+$ complex from a tetragonal pyramidal to a tetrahedral coordination mode in the singlet and triplet state, respectively (Supplementary Table 12).
§The $[In(bipy)_2]^+$ complex features a tetragonal pyramidal coordination mode both in the singlet and the triplet state (Supplementary Table 12).

Table 3 | Estimated $\Delta_r H^\circ$(gas) values for the formation of 4^{3+}, 5^{3+} and 6^{4+} as well as for the Coulomb explosion of 4^{3+} in kJ mol^{-1}.

Gas phase reaction	$\Delta_r H^\circ$(gas) [kJ mol^{-1}]
3 $[In(PhF)_2]^+ + 6$ bipy $\rightarrow [In_3(bipy)_6]^{3+} + 6$ PhF	+164
3 $[In(PhF)_2]^+ + 5$ bipy $\rightarrow [In_3(bipy)_5]^{3+} + 6$ PhF	+252
4 $[In(PhF)_2]^+ + 6$ bipy $\rightarrow [In_4(bipy)_6]^{4+} + 8$ PhF	+731
$[In_3(bipy)_6]^{3+} \rightarrow 3 [In(bipy)_2]^+$	−466

The $\Delta_r H^\circ$(solid) values for all BHFCs were deliberately set to ±0 kJ mol^{-1}, the $\Delta_{latt}H^\circ$ values calculated by applying the Jenkins generalized Kapustinskii equation[66] and all other enthalpies extrapolated from the given references (Supplementary Fig. 8).

possible through the application of matching ligands and ultimately is a solid-state-driven phenomenon. Both, bipy and phen lead to a pronounced decrease of the Coulomb repulsion within the clusters by diluting the positive charges on the In$^+$ cations to the ligand backbone, and contributing enough negative charge to yield ligand stabilized $[In_3]^{3+}(A^-)_3$ and $[In_4]^{4+}(A^-)_4$ salts with short In − In bonds. This corresponds to a ligand-to-metal charge transfer. The calculated high $\Delta_{latt}H^\circ$ values of −1,438 (**4**), −1,444 (**5**) and −2,266 kJ mol^{-1} (**6**) further stabilize the salts[66]. Together, the charge transfer leading to favourable metal-metal bond strengths, in combination with the lattice enthalpy gain are sufficient to overcome the strong Coulomb repulsion active in gaseous and solution phases. Last, it should be noted that for the central $[In_4]^{4+}$ cluster core with eight valence electrons deriving from four In^I cations, the stability of 6^{4+} and 7^{4+} would be in agreement with the Jellium model[67].

explosion of 4^{3+} with formation of three $[In(bipy)_2]^+$ monocations in the gas phase and assessed via a suitable BHFC as −466 kJ mol^{-1} (Table 3 and Supplementary Fig. 8).

This is in agreement with single-point DFT calculations on the frozen conformation of solid 4^{3+}, Coulomb exploding to give three $[In(bipy)_2]^+$ monocations cut out of this cyclic trimer solid-state structure. B3LYP and BHLYP suggest this gas phase process to be favoured by −684 and −705 kJ mol^{-1}, respectively. A non-ligand- supported triangular $[In_3]^{3+}$ cluster ($d_{In-In} = 270$ pm) was calculated to Coulomb explode at the same level with −1,447/−1,459 kJ mol^{-1}, Supplementary Table 13). Overall, the formation of the ligand supported $[In_3]^{3+}/[In_4]^{4+}$ clusters seems to be only

Discussion

The reaction of **1** and 2,2'-bipyridine resulted in a disproportionation of the former, thus yielding the monomeric and paramagnetic $[Ga(bipy)_3]^{2+\bullet}$ complex. Herein, the gallium cation is coordinated in a distorted octahedral fashion, and EPR and DFT studies reveal a ligand-centred radical: that is, a $[Ga^{III}\{(bipy)_3\}^\bullet]^{2+}$ complex. Applying the heavier homologue **2** on the other hand, we isolated the first homonuclear cationic triangular and rhombic clusters of univalent indium: $[In^I_3(bipy)_6]^{3+}$, $[In^I_3(bipy)_5]^{3+}$, $[In^I_4(bipy)_6]^{4+}$ and $[In^I_4(phen)_6]^{4+}$. Herein, the In^I cations are coordinated by one, 1.5 or two chelating bipy/phen ligands. To our knowledge, the In − In distances (258.1 and 259.7 pm) within the In − In bridges in the rhombic clusters are the shortest that have been reported so far. DFT studies suggest a stepwise formation of the clusters via their triplet state and an alternate ambiphilic route seems to be energetically less favourable. The general driving force for this cationic cluster formation is attributable to relatively strong In − In bonds, reduction of Coulomb repulsion by introduction of a suitable ligand and ligand-to-metal charge transfer in combination with the high lattice enthalpies of the resultant ligand stabilized $[M_3]^{3+}(A^-)_3$ and $[M_4]^{4+}(A^-)_4$ salts. We are convinced that this is a general phenomenon, which could be used as a pathway to cationic metal atom cluster formation of subvalent metal cations in combination with strong but sterically accessible (chelating?) ligands.

Methods

General experimental procedures. All manipulations were performed using Schlenk or glove box techniques in an argon atmosphere (H_2O and $O_2 < 1$ p.p.m.). o-$C_6H_4F_2$ and C_6H_5F were dried over CaH_2, distilled and had H_2O contents below 5 p.p.m. (Karl-Fischer titrations). Because the obtained compounds contain large amounts of fluorine in chemically very stable CF_3 groups, standard combustion analyses have proven to be incomplete. Characterizations of novel compounds were therefore done on the basis of single-crystal X-ray analysis and multinuclear NMR spectroscopy. As the highly symmetric and perfluorinated $[Al(OR^F)_4]^-$ anions are usually heavily disordered, prone to crystallize in superstructures (cf. compound **3**) and, due to their bulkiness, able to add up to very large unit cells (the unit cell sizes of **6** and the protein Viscotoxin B are comparable, Supplementary Table 8), processing of the crystal structure data was everything else than trivial. In this context, however, the quality of the data which were collected using a completely up to date crystallography is clearly well within the limits of accepted standards, thus allowing us to unambiguously identify all structures. Moreover, all compounds were reproduced from independent syntheses (apart from **5** and **6**) and **4**, **5**, **6** and **7** intrinsically confirm the central structural triangular $[In_3]^{3+}$ and rhombic $[In_4]^{4+}$ motifs. Therefore, access to international facilities for better quality X-ray diffraction data were not sought for. Further details are included in the Supporting Information. Moreover, it was several times attempted to obtain ESI-MS data of the reported systems. As the ionic compounds are very sensitive however, no meaningful spectra were obtained—presumably due to oxidation and/or hydrolysis on the way to the ionization chamber (as very frequently encountered with our sensitive systems). Since the other investigations strongly suggested these multiply charged cations to only exist in the solid state, this method was not further pursued. Compound **3** was additionally characterized by EPR and ultraviolet-visible measurements. The DFT calculations were performed at the BHLYP/SV(P) and B3LYP/SV(P) level of theory.

References

1. Cotton, F. A. Transition-metal compounds containing clusters of metal atoms. *Q. Rev. Chem. Soc.* **20**, 389–401 (1966).

2. Wade, K. The structural significance of the number of skeletal bonding electron-pairs in carboranes, the higher boranes and borane anions, and various transition-metal carbonyl cluster compounds. *J. Chem. Soc. D* 792–793 (1971).

3. Welch, A. J. The significance and impact of Wade's rules. *Chem. Commun.* **49**, 3615–3616 (2013).

4. Corbett, J. D. Polyanionic Clusters and Networks of the Early p-Element Metals in the Solid State: Beyond the Zintl Boundary. *Angew. Chem. Int. Ed.* **39**, 670–690 (2000).

5. Anson, C. E. *et al.* Synthesis and crystal structures of the ligand-stabilized silver chalcogenide clusters [Ag₁₅₄Se₇₇(dppxy)₁₈], [Ag₃₂₀(StBu)₆₀S₁₃₀(dppp)₁₂], [Ag₃₅₂S₁₂₈(StC₅H₁₁)₉₆], and [Ag₄₉₀S₁₈₈(StC₅H₁₁)₁₁₄]. *Angew. Chem. Int. Ed.* **47**, 1326–1331 (2008).

6. Schnöckel, H. & Schnepf, A. in *The Group 13 Metals Aluminium, Gallium, Indium and Thallium: Chemical Patterns and Peculiarities* 402–487 (John Wiley & Sons, Ltd, 2011).

7. Schnöckel, H. Structures and properties of metalloid al and ga clusters open our eyes to the diversity and complexity of fundamental chemical and physical processes during formation and dissolution of metals. *Chem. Rev.* **110**, 4125–4163 (2010).

8. Masuda, J. D., Schoeller, W. W., Donnadieu, B. & Bertrand, G. NHC-mediated aggregation of P₄: isolation of a P₁₂ cluster. *J. Am. Chem. Soc.* **129**, 14180–14181 (2007).

9. Dohmeier, C., Loos, D. & Schnöckel, H. Aluminum(I) and gallium(I) compounds: syntheses, structures, and reactions. *Angew. Chem. Int. Ed. Engl.* **35**, 129–149 (1996).

10. Jurca, T., Hiscock, L. K., Korobkov, I., Rowley, C. N. & Richeson, D. S. The tipping point of the inert pair effect: experimental and computational comparison of In(I) and Sn(II) bis(imino)pyridine complexes. *Dalton Trans.* **43**, 690–697 (2014).

11. Van Den Berg, J. M. The crystal structure of the room temperature modification of indium chloride, InCl. *Acta. Crystallogr.* **20**, 905–910 (1966).

12. van der Vorst, C. P. J. M., Verschoor, G. C. & Maaskant, W. J. A. The structures of yellow and red indium monochloride. *Acta Crystallogr., Sect. B: Struct. Sci.* **34**, 3333–3335 (1978).

13. Brukl, A. & Ortner, G. Die Sulfide des Galliums. *Monatsh. Chem.* **56**, 358–364 (1930).

14. Tuck, D. G. Gallium and indium dihalides: a classic structural problem. *Polyhedron* **9**, 377–386 (1990).

15. Baker, R. J. & Jones, C. 'GaI': a versatile reagent for the synthetic chemist. *Dalton Trans.* 1341–1348 (2005).

16. Malbrecht, B. J., Dube, J. W., Willans, M. J. & Ragogna, P. J. Addressing the chemical sorcery of 'GaI': benefits of solid-state analysis aiding in the synthesis of P→Ga coordination compounds. *Inorg. Chem.* **53**, 9644–9656 (2014).

17. Schmidbaur, H. Arene complexes of univalent gallium, indium, and thallium. *Angew. Chem. Int. Ed. Engl.* **24**, 893–904 (1985).

18. Schenk, C., Köppe, R., Schnöckel, H. & Schnepf, A. A convenient synthesis of cyclopentadienylgallium—the awakening of a sleeping beauty in organometallic chemistry. *Eur. J. Inorg. Chem.* **2011**, 3681–3685 (2011).

19. Asay, M., Jones, C. & Driess, M. N-Heterocyclic carbene analogues with low-valent group 13 and group 14 elements: syntheses, structures, and reactivities of a new generation of multitalented ligands. *Chem. Rev* **111**, 354–396 (2011).

20. Dange, D., Choong, S. L., Schenk, C., Stasch, A. & Jones, C. Synthesis and characterisation of anionic and neutral gallium(I) N-heterocyclic carbene analogues. *Dalton Trans.* **41**, 9304–9315 (2012).

21. Linti, G. & Schnöckel, H. Low valent aluminum and gallium compounds—structural variety and coordination modes to transition metal fragments. *Coord. Chem. Rev.* **206-207**, 285–319 (2000).

22. Pardoe, J. A. J. & Downs, A. J. Development of the chemistry of indium in formal oxidation states lower than +3. *Chem. Rev.* **107**, 2–45 (2007).

23. Fitz, H. & Müller, B. G. InBF₄, das erste komplexe Fluorid mit Indium(I). *Z. Anorg. Allg. Chem.* **623**, 579–582 (1997).

24. Macdonald, C. L. B., Corrente, A. M., Andrews, C. G., Taylor, A. & Ellis, B. D. Indium(I) trifluoromethanesulfonate and other soluble salts for univalent indium chemistry. *Chem. Commun.* 250–251 (2004).

25. Mazej, Z. Indium(I) hexafluoropnictates (InPnF₆; Pn = P, As, Sb). *Eur. J. Inorg. Chem.* **2005**, 3983–3987 (2005).

26. Welsch, S., Bodensteiner, M., Dušek, M., Sierka, M. & Scheer, M. A novel soluble In^I precursor for P$_n$ ligand coordination chemistry. *Chem. Eur. J.* **16**, 13041–13045 (2010).

27. Schneider, U. & Kobayashi, S. Low-oxidation state indium-catalyzed C–C Bond formation. *Acc. Chem. Res.* **45**, 1331–1344 (2012).

28. Slattery, J. M., Higelin, A., Bayer, T. & Krossing, I. A simple route to univalent gallium salts of weakly coordinating anions. *Angew. Chem. Int. Ed.* **49**, 3228–3231 (2010).

29. Higelin, A., Sachs, U., Keller, S. & Krossing, I. Univalent gallium and indium phosphane complexes: from pyramidal M(PPh₃)₃⁺ to carbene-analogous bent M(PtBu₃)₂⁺ (M = Ga, In) Complexes. *Chem. Eur. J.* **18**, 10029–10034 (2012).

30. Higelin, A., Haber, C., Meier, S. & Krossing, I. Isolated cationic crown ether complexes of gallium(I) and indium(I). *Dalton Trans.* **41**, 12011–12015 (2012).

31. Higelin, A., Keller, S., Göhringer, C., Jones, C. & Krossing, I. Unusual Tilted carbene coordination in carbene complexes of gallium(I) and indium(I). *Angew. Chem. Int. Ed.* **52**, 4941–4944 (2013).

32. Lichtenthaler, M. R. *et al.* σ- or π-coordination? Complexes of univalent gallium salts with aromatic nitrogen bases. *Eur. J. Inorg. Chem.* **2014**, 4335–4341 (2014).

33. Lichtenthaler, M. R. *et al.* Univalent gallium salts of weakly coordinating anions: effective initiators/catalysts for the synthesis of highly reactive polyisobutylene. *Organometallics* **32**, 6725–6735 (2013).

34. Lichtenthaler, M. R. *et al.* Univalent gallium complexes of simple and ansa-Arene ligands: effects on the polymerization of isobutylene. *Chem. Eur. J* **21**, 157–165 (2015).

35. Jones, C. Bulky Guanidinates for the stabilization of low oxidation state metallacycles. *Coord. Chem. Rev.* **254**, 1273–1289 (2010).

36. Baker, R. J., Farley, R. D., Jones, C., Kloth, M. & Murphy, D. M. The reactivity of diazabutadienes toward low oxidation state Group 13 iodides and the synthesis of a new gallium(I) carbene analogue. *J. Chem. Soc., Dalton Trans.* 3844–3850 (2002).

37. Tsai, Y.-C. The chemistry of univalent metal β-diketiminates. *Coord. Chem. Rev.* **256**, 722–758 (2012).

38. Reger, D. L. Poly(pyrazolyl)borate complexes of gallium and indium. *Coord. Chem. Rev.* **147**, 571–595 (1996).

39. Jurca, T., Lummiss, J., Burchell, T. J., Gorelsky, S. I. & Richeson, D. S. Capturing In⁺ monomers in a neutral weakly coordinating environment. *J. Am. Chem. Soc.* **131**, 4608–4609 (2009).

40. Baker, R. J., Jones, C., Kloth, M. & Mills, D. P. The reactivity of gallium(I) and indium(I) halides towards bipyridines, terpyridines, imino-substituted pyridines and bis(imino)acenaphthenes. *New J. Chem.* **28**, 207–213 (2004).

41. Jurca, T. *et al.* Disproportionation and radical formation in the coordination of 'GaI' with bis(imino)pyridines. *Dalton Trans.* **39**, 1266–1272 (2010).

42. Bondi, A. van der Waals volumes and Radii. *J. Phys. Chem.* **68**, 441–451 (1964).

43. Mantina, M., Chamberlin, A. C., Valero, R., Cramer, C. J. & Truhlar, D. G. Consistent van der Waals Radii for the whole main group. *J. Phys. Chem. A* **113**, 5806–5812 (2009).

44. Baker, R. J., Bettentrup, H. & Jones, C. The reactivity of primary and secondary amines, secondary phosphanes and N-heterocyclic carbenes towards Group-13 metal(I) halides. *Eur. J. Inorg. Chem.* **2003**, 2446–2451 (2003).

45. Protchenko, A. V. *et al.* Stable GaX₂, InX₂ and TlX₂ radicals. *Nat. Chem* **6**, 315–319 (2014).

46. Cloke, F. G. N., Hanson, G. R., Henderson, M. J., Hitchcock, P. B. & Raston, C. L. Synthesis and X-ray crystal structure of the first homoleptic main group diazadiene complex, bis(1,4-di-t-butyl-1,4-diazabuta-1,3-diene) gallium. *J. Chem. Soc., Chem. Commun.* 1002–1003 (1989).

47. Kaim, W. & Matheis, W. Bis(1,4-di-tert-butyl-1,4-diazabutadiene)gallium is not a gallium(II) compound. *J. Chem. Soc., Chem. Commun.* 597–598 (1991).

48. Baker, R. J. *et al.* An EPR and ENDOR Investigation of a Series of Diazabutadiene–Group 13 Complexes. *Chem. Eur. J.* **11**, 2972–2982 (2005).

49. Tuononen, H. M. & Armstrong, A. F. Theoretical investigation of paramagnetic diazabutadiene gallium(III) – pnictogen complexes: insights into the interpretation and simulation of electron paramagnetic resonance spectra. *Inorg. Chem.* **44**, 8277–8284 (2005).

50. Tuononen, H. M. & Armstrong, A. F. Theoretical investigation of paramagnetic group 13 diazabutadiene radicals: insights into the prediction and interpretation of EPR spectroscopy parameters. *Dalton Trans.* 1885–1894 (2006).

51. Zális̆, S. *et al.* Origin of electronic absorption spectra of MLCT-excited and one-electron reduced 2,2′-bipyridine and 1,10-phenanthroline complexes. *Inorg. Chim. Acta* **374**, 578–585 (2011).

52. Gore-Randall, E., Irwin, M., Denning, M. S. & Goicoechea, J. M. synthesis and characterization of alkali-metal salts of 2,2′- and 2,4′-bipyridyl radicals and dianions. *Inorg. Chem.* **48**, 8304–8316 (2009).

53. Wang, M., Weyhermueller, T., England, J. & Wieghardt, K. Molecular and Electronic structures of six-coordinate 'low-valent' [M(Mebpy)₃]⁰ (M = Ti, V, Cr, Mo) and [M(tpy)₂]⁰ (M = Ti, V, Cr), and seven-coordinate [MoF(Mebpy)₃](PF₆) and [MX(tpy)₂](PF₆) (M = Mo, X = Cl and M = W, X = F). *Inorg. Chem.* **52**, 12763–12776 (2013).

54. Li, X.-W., Pennington, W. T. & Robinson, G. H. Metallic system with aromatic character. synthesis and molecular structure of Na₂[[(2,4,6-Me₃C₆H₂)₂C₆H₃] Ga]₃ the first cyclogallane. *J. Am. Chem. Soc.* **117**, 7578–7579 (1995).

55. Wiberg, N. *et al.* On the Gallanyls R*₃Ga₂· and R*₄Ga₃ As Well As Gallanides R*₃Ga₂⁻ and R*₄Ga₃⁻ (R* = SitBu₃)—Syntheses, Characterization, Structures. *Eur. J. Inorg. Chem.* **2001**, 1719–1727 (2001).

56. Litters, S., Kaifer, E., Enders, M. & Himmel, H.-J. A boron–boron coupling reaction between two ethyl cation analogues. *Nat. Chem* **5**, 1029–1034 (2013).

57. Mingos, D. M. P. General theory for cluster and ring compounds of the main group and transition elements. *Nat. Phys. Sci* **236**, 99–102 (1972).

58. Krossing, I. The facile preparation of weakly coordinating anions: structure and characterisation of silver polyfluoroalkoxyaluminates AgAl(ORF)₄, calculation of the alkoxide ion affinity. *Chem. Eur. J* **7**, 490–502 (2001).

59. Power, P. P. in *Group 13 Chemistry I Vol. 103 Structure and Bonding*. Ch. 2 57–84 (Springer Berlin Heidelberg, 2002).

60. Driess, M. & Grützmacher, H. Main group element analogues of carbenes, olefins, and small rings. *Angew. Chem. Int. Ed. Engl.* **35**, 828–856 (1996).

61. Kaim, W. Manifestations of noninnocent ligand behavior. *Inorg. Chem.* **50**, 9752–9765 (2011).

62. Sundermann, A., Reiher, M. & Schoeller, W. W. Isoelectronic arduengo-type carbene analogues with the group IIIa elements boron, aluminum, gallium, and indium. *Eur. J. Inorg. Chem.* **1998**, 305–310 (1998).

63. Allan, C. J., Cooper, B. F. T., Cowley, H. J., Rawson, J. M. & Macdonald, C. L. B. Non-innocent ligand effects on low-oxidation-state indium complexes. *Chem. Eur. J.* **19**, 14470–14483 (2013).

64. Hellmann, K. W. *et al.* metal–ligand versus metal–metal redox chemistry: thallium(I)-Induced synthesis of 4,9-diaminoperylenequinone-3,10-diimine derivatives. *Angew. Chem. Int. Ed.* **37**, 1948–1952 (1998).

65. Klamt, A. & Schuurmann, G. COSMO: a new approach to dielectric screening in solvents with explicit expressions for the screening energy and its gradient. *J. Chem. Soc., Perkin Trans.* **2**, 799–805 (1993).

66. Jenkins, H. D. B., Roobottom, H. K., Passmore, J. & Glasser, L. Relationships among ionic lattice energies, molecular (formula unit) volumes, and thermochemical radii. *Inorg. Chem.* **38**, 3609–3620 (1999).

67. Brack, M. The physics of simple metal clusters: self-consistent jellium model and semiclassical approaches. *Rev. Mod. Phys.* **65**, 677–732 (1993).

68. Morton, J. R. & Preston, K. F. Atomic parameters for paramagnetic resonance data. *J. Magn. Reson.* **30**, 577–582 (1978).

69. Weil, J. A. & Bolton, J. R. in *Electron Paramagnetic Resonance* 583 John Wiley & Sons, Inc., 2006).

70. Dedieu, A. & Hoffmann, R. Platinum(0)-platinum(0) dimers. Bonding relationships in a d10-d10 system. *J. Am. Chem. Soc.* **100**, 2074–2079 (1978).

Acknowledgements

This work was supported by the Albert-Ludwigs-Universität Freiburg and by the DFG in the *Normalverfahren*. We would like to thank Fadime Bitgül and Dr Harald Scherer for the measurement of the NMR spectra, B.Sc. Boumahdi Benkmil for his support regarding single-crystal X-ray analysis, Dr Michael Schwarz and Prof. Dr Caroline Röhr for their support during the attempted preparation of single-crystalline EPR samples. This work is dedicated to the occasion of the 60th birthday of Prof. Dr. Manfred Scheer in Regensburg.

Author contributions

M.R.L. initiated and coordinated the project, conducted the synthesis of **4**, **5**, **6**, **7** and **8**, reproduced the synthesis of **3**, characterized the obtained compounds, performed the DFT calculations and wrote the manuscript with assistance of I.K.; during his apprenticeship and under the supervision of M.R.L., F.S. performed the synthesis of **1** and **3**; D.K. contributed valuable support in solving and refining the single-crystal structures; L.H., E.S. and S.W. conducted the EPR measurements of **3**, contributed the simulated spectra and drafted the EPR-related part of the manuscript. S.W. also revised the completed manuscript; During his bachelor thesis and under the supervision of M.R.L., J.H. performed the synthesis of **2**; D.H. provided important advice regarding the DFT calculations and revised the manuscript; I.K. directed the project, conceived the central ideas concerning the formation of the reported compounds and drafted parts of the manuscript.

Additional information

Accession codes: The X-ray crystallographic coordinates for structures reported in this study have been deposited at the Cambridge Crystallographic Data Centre (CCDC), under deposition numbers CCDC 1032681 (**3**), CCDC 1032680 (**4**), CCDC 1033040 (**5**), CCDC 1032732 (**6**), CCDC 1034231 (**7**) and CCDC 1034089 (**8**). These data can be obtained free of charge from The Cambridge Crystallographic Data Centre via www.ccdc.cam.ac.uk/data_request/cif.

Competing financial interests: The authors declare no competing financial interests.

24

A porous metal-organic framework with ultrahigh acetylene uptake capacity under ambient conditions

Jiandong Pang[1,2], Feilong Jiang[1], Mingyan Wu[1], Caiping Liu[1], Kongzhao Su[1,2], Weigang Lu[3], Daqiang Yuan[1] & Maochun Hong[1]

Acetylene, an important petrochemical raw material, is very difficult to store safely under compression because of its highly explosive nature. Here we present a porous metal-organic framework named **FJI-H8**, with both suitable pore space and rich open metal sites, for efficient storage of acetylene under ambient conditions. Compared with existing reports, **FJI-H8** shows a record-high gravimetric acetylene uptake of $224\,cm^3$ $(STP)\,g^{-1}$ and the second-highest volumetric uptake of $196\,cm^3$ $(STP)\,cm^{-3}$ at 295 K and 1 atm. Increasing the storage temperature to 308 K has only a small effect on its acetylene storage capacity ($\sim 200\,cm^3$ $(STP)\,g^{-1}$). Furthermore, **FJI-H8** exhibits an excellent repeatability with only 3.8% loss of its acetylene storage capacity after five cycles of adsorption–desorption tests. Grand canonical Monte Carlo simulation reveals that not only open metal sites but also the suitable pore space and geometry play key roles in its remarkable acetylene uptake.

[1] State Key Laboratory of Structure Chemistry, Fujian Institute of Research on the Structure of Matter, Chinese Academy of Sciences, Fuzhou, Fujian 350002, China. [2] University of Chinese Academy of Sciences, Beijing 100049, China. [3] Department of Chemistry, Texas A&M University, College Station, Texas 77843, USA. Correspondence and requests for materials should be addressed to M.W. (email: wumy@fjirsm.ac.cn) or to D.Y. (email: ydq@fjirsm.ac.cn).

A cetylene is a very important chemical feedstock for modern industry[1,2]. Many widely used polymer products such as polyurethane and polyester plastics are synthesized from acetylene. However, the safe storage and transportation of acetylene still remain challenging because of its explosiveness when compressed under pressures over 2 atm at room temperature[3,4]. Therefore, acetylene gas extensively used in industry so far has to be stored in special cylinders filled with acetone and porous materials suffering from lower acetylene purity and higher storage cost. Fortunately, the emergence of porous metal-organic frameworks (MOFs) brings promising solutions to the above problem due to their excellent performance for storage and separation of gases such as H_2, O_2, CH_4 and CO_2 (refs 5–17). MOFs have been recently studied for acetylene storage application[18–27]. For example, at 273 K ZJU-5 shows a high acetylene uptake of 290 cm^3 (STP) g^{-1}. However, the uptake drastically decreases to 193 cm^3 (STP) g^{-1} when the temperature rises to 295 K[18]. To effectively improve acetylene storage capacity at room temperature, Chen et al. explored a series of microporous MOFs with different structures and porosities and concluded that open Cu(II) sites and suitable pore space in MOFs played crucial roles for acetylene storage[18–22]. In addition, the dendritic multi-carboxylate ligands with m-benzenedicarboxylate moieties tend to form various polyhedral nanocages along with rich open Cu(II) sites, which has been demonstrated as an efficient approach to improve the gas uptakes[26–33].

Considering the previous studies, we designed and synthesized a new robust multi-carboxylate ligand 3,3',5,5'-tetra(3,5-dicarboxyphenyl)-4,4'-dimethoxy-biphenyl (H$_8$tddb, Supplementary Fig. 1). Reaction of H$_8$tddb and Cu(NO$_3$)$_2$ under solvothermal conditions resulted in a porous MOF ([Cu$_4$(tddb) · (H$_2$O)$_4$]$_n$ · (solvent)$_x$, abbreviated as **FJI-H8**) with both suitable pore space and open metal sites. At 295 K, **FJI-H8** exhibits a record-high acetylene uptake of 224 cm^3 (STP) g^{-1}, greatly exceeding the previous record of 201 cm^3 (STP) g^{-1} held by HKUST-1 (ref. 19). Increasing the storage temperature to 308 K has only small effect on its acetylene storage capacity ($\sim$200 cm^3 (STP) g^{-1}). Furthermore, the acetylene adsorption amount of **FJI-H8** at 295 K has no obvious loss after five cycles of adsorption–desorption test.

Results

Structure of FJI-H8. Single-crystal X-ray diffraction experiments revealed that **FJI-H8** crystallized in the tetragonal space group $P4_2/nnm$ (Supplementary Data 1). In the asymmetric unit, there are one-quarter of organic ligand and two kinds of Cu(II) ions both with the occupancies of 50%. One metal site is located on a mirror plane, while the other resides on a twofold axis. Further, the dinuclear core is centred about an inversion site. All the eight carboxyl groups are deprotonated in the organic ligand. Two inner phenyl rings of tddb are coplanar, while the four outer isophthalate groups are almost perpendicular to the diphenyl with a dihedral angle of 78.85° (Supplementary Fig. 2). Each tddb ligand coordinates to eight dicopper(II) paddlewheel secondary building units (SBUs) and each Cu$_2$ SBU links to four tddb ligands. As anticipated, there are three types of polyhedral nanocages in **FJI-H8** (Supplementary Fig. 3), that is, one regular cuboctahedron (Cage-A), one distorted octahedron (Cage-B) and one distorted cuboctahedron (Cage-C) (Fig. 1a). Cage-A is constructed by eight Cu$_2$ SBUs and four tddb ligands. The centres of the eight paddlewheels together with the centroids of the four tddb ligands are considered as the 12 four-connected vertices of the cage. Therefore, the pore diameter is around 15 Å, which is estimated through the separations of two opposite vertexes.

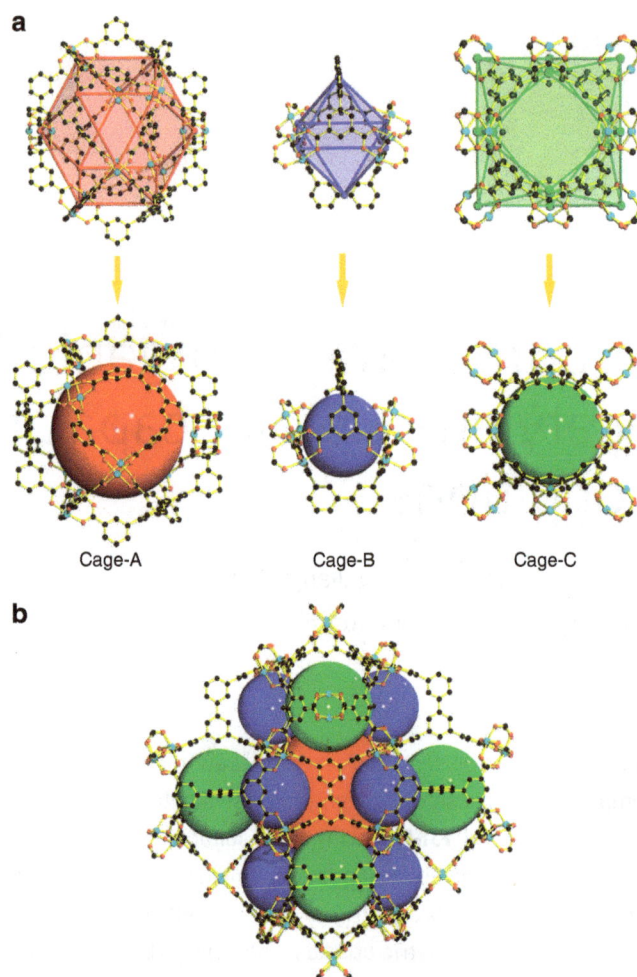

Figure 1 | Structural representations of FJI-H8 from X-ray diffraction data. (**a**) Three types of polyhedral nanocages observed in **FJI-H8**, that is, one regular cuboctahedral cage (Cage-A), one distorted octahedral cage (Cage-B) and one distorted cuboctahedral cage (Cage-C). (**b**) Combination of the three types of polyhedral nanocages. (The hydrogen atoms and hydroxymethyl groups are omitted for clarity.)

In addition, eight open Cu(II) sites point towards the centre of the cage and can interact directly with gas molecules residing inside, which can improve the gas adsorption ability[34–38]. Cage-B consists of four Cu$_2$ SBUs and two halves of tddb ligands, with a dimension of $\sim$8 Å. For Cage-C, its 12 vertices consist of the centroids of eight Cu$_2$ SBUs and four halves of tddb ligands, respectively. In addition, the dimension of this cage is estimated to be around 12 Å. On the whole, Cage-A is linked by six Cage-C through six rhombic faces and eight Cage-B through eight triangular faces (Supplementary Fig. 4). Similarly, Cage-C is linked by six Cage-A through six rhombic faces and eight Cage-B by sharing eight m-benzenedicarboxylate moieties (Supplementary Fig. 5). However, Cage-B is linked by four Cage-A through four triangular faces and four Cage-C by sharing four m-benzenedicarboxylate moieties (Supplementary Fig. 6). It should be noted that there are two kinds of Cu$_2$ SBUs in **FJI-H8** due to the fact that the distance between the paddlewheels and the centroids of the ligands are slightly different (8.96 and 9.45 Å, respectively). For the sake of clarity, if we simplify two types of Cu$_2$ SBUs as two kinds of four-connected nodes and the tddb ligands as eight-connected nodes, **FJI-H8** adopts the rare (4,4,8)-c URJ network with the topological point symbol of $4^{14}.6^{12}.8^2$ (Supplementary Fig. 7)[39].

Porosity and N_2 adsorption of FJI-H8. The solvent accessible volume in fully evacuated **FJI-H8** is 62.4% calculated by PLATON with a probe of 1.8 Å (ref. 40). To check the permanent porosity of **FJI-H8**, nitrogen adsorption isotherm was measured at 77 K and 1 atm. As demonstrated by powder X-ray diffraction (PXRD, Supplementary Fig. 8), the activated sample retained the crystallinity after being evacuated for 10 h under 80 °C. The N_2 sorption of **FJI-H8** exhibits a typical reversible type I isotherm with a saturated adsorption amount of 531 cm^3 (STP) g^{-1}, indicating the microporous nature of **FJI-H8**. The Brunauer–Emmett–Teller apparent surface area calculated from the N_2 adsorption data is 2025 ± 15 m^2 g^{-1} and is well consistent with theoretical one (1907 m^2 g^{-1}) calculated by Poreblazer[41], which demonstrates that the sample is fully activated. Accordingly, the total pore volume is 0.82 cm^3 g^{-1}. From the N_2 adsorption data, analysis by the non-local density functional theory (NLDFT) model confirms a narrow distribution of micropores around 12 Å (Fig. 2).

Acetylene adsorption property of FJI-H8. Considering the open metal sites and moderate pores in **FJI-H8**, its low-pressure acetylene uptake was measured under 1 atm. As expected, the C_2H_2 adsorption amount for **FJI-H8** reaches up to 277 cm^3 (STP) g^{-1} at 273 K and 1 atm, slightly less than the record of 290 cm^3 (STP) g^{-1} (ref. 18). In practice, C_2H_2 gas is stored at ambient temperatures. Therefore, the C_2H_2 adsorption experiment at room temperature (295 K) was carried out. Exhilaratingly, **FJI-H8** exhibits an adsorption amount of 224 cm^3 (STP) g^{-1} for acetylene at 295 K and 1 atm, which is greatly higher than those of the two famous MOFs, known as HKUST-1 (201 cm^3 (STP) g^{-1}) and CoMOF-74 (197 cm^3 (STP) g^{-1}) (Table 1). Surprisingly, **FJI-H8** exhibits an acetylene uptake of 206 cm^3 (STP) g^{-1} at 303 K and 1 atm, which prompts us to investigate its acetylene uptake at an even higher temperature. The most commendable aspect is that the acetylene adsorption amount of **FJI-H8** still reached 200 cm^3 (STP) g^{-1} even when the temperature increased to 308 K (Fig. 3a). This value is comparable to that of HKUST-1 at 295 K. In other words, the acetylene uptake capacity of **FJI-H8** decreases by a rate of 2.2 cm^3 g^{-1} K^{-1} with the experimental temperature increasing from 273 to 308 K, which is almost only half to that of ZJU-5 (3.9 cm^3 g^{-1} K^{-1}) from 273 K to 298 K (Supplementary Table 1). Therefore, **FJI-H8** is more suitable for practical applications over a wide temperature range around room temperature. In consideration of its practical application, we also tested the repeatability of **FJI-H8** for acetylene storage. About 100 mg of

desolvated sample was loaded onto an ASAP2020-M analyser and five cycles of acetylene adsorption at 295 K were recorded without the reactivation process between each cycle. For **FJI-H8**, there is only a 3.8% loss in absorbed quantity of C_2H_2 after five cycles, which indicates that **FJI-H8** is promising in refillable acetylene storage (Fig. 3b)[42].

The isosteric heat of acetylene adsorption for **FJI-H8** is calculated to be 32.0 kJ mol^{-1} based on the C_2H_2 adsorption isotherms at 273, 295, 303 and 308 K (Supplementary Fig. 9). This value is larger than that of HKUST-1 (30.4 kJ mol^{-1}). As has been reported previously, the high density of open metal sites within MOF materials may play crucial roles in the high acetylene storage capacities and high adsorption enthalpies. Supposing each open metal site binds one acetylene molecule, as established by Chen et al., the acetylene uptake by open Cu(II) sites in HKUST-1 accounts for c.a. 60% of the total acetylene uptake at room temperature. For CoMOF-74, almost 80% of the total acetylene uptake is contributed by open Co(II) sites assuming each Co(II) site binds one acetylene molecule. Moreover, for MgMOF-74, if all the open Mg(II) sites can be fully loaded with one acetylene molecule per one open Mg(II) site, the theoretical acetylene uptake by open metal sites is even larger than the experimentally measured value at room temperature. As listed in Table 1, for most MOF materials, acetylene uptakes by open metal sites account for almost half of the acetylene uptake or even higher. However, in **FJI-H8**, the open Cu(II) site density is 3.59 mmol g^{-1}, which is lower than those for the reported MOFs with high acetylene uptake capacities. On the whole, the open Cu(II) sites can only contribute 87 cm^3 of the total 224 cm^3 for the acetylene storage capacity at 295 K and 1 atm. Thus, the remaining acetylene storage capacity should come from the suitable pore space in **FJI-H8**. It should be noted that the acetylene uptake by the pore space accounts for >60% of the whole acetylene uptake in **FJI-H8**, which is rarely seen in other reported MOFs. More surprisingly, the whole acetylene uptake decreases to 200 cm^3 (STP) g^{-1} with a slight loss of 24 cm^3 (STP) g^{-1}, when the temperature rises from 295 to 308 K. The above result indicates relatively strong interactions between acetylene molecules and pore space in **FJI-H8**.

In general, there are two kinds of representations to measure the gas adsorption properties of adsorbent materials, that is, gravimetric capacity in the unit of cm^3 (STP) g^{-1} and volumetric

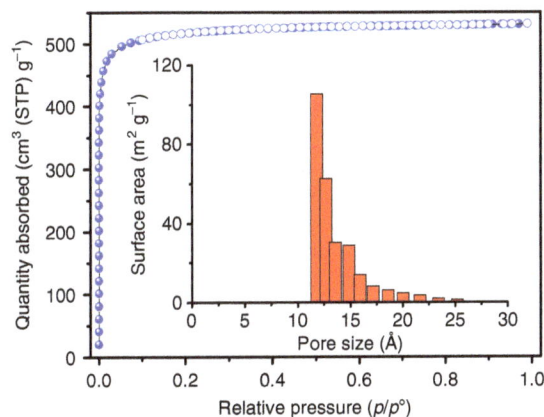

Figure 2 | N_2 sorption data for FJI-H8. N_2 sorption isotherm at 77 K (filled symbols: adsorption; open symbols: desorption); inset: pore size distribution analysed by NLDFT method.

Table 1 | Contributions of open metal sites (OMSs) and pore space in acetylene uptakes for selected MOFs at room temperature and 1 atm for gravimetric capacity in the unit of cm^3 (STP) g^{-1}*.

Material	OMS density (mmol g^{-1})	C_2H_2 uptake (cm^3 (STP) g^{-1})		
		By OMS	By pore space	Sum
FJI-H8	3.59	87	137	224
FJI-H8†	3.59	87	113	200
HKUST-1 (ref. 19)	4.96	120	81	201
CoMOF-74 (ref. 20)	6.41	155	42	197
ZJU-5 (ref. 18)	3.87	95	98	193
MgMOF-74 (ref. 20)	8.24	199‡	—	184
NOTT-101 (ref. 55)	3.44	84	100	184
ZJU-7 (ref. 56)	3.46	85	95	180
Cu-TDPAT (ref. 26)	3.74	91	87	178
PCN-16 (ref. 55)	4.19	102	74	176

*The OMS density of MOFs was calculated based on the crystal information files.
†Data for **FJI-H8** at 308 K.
‡The value of C_2H_2 uptake by OMS is larger than the sum value maybe because of the interaction between open Mg(II) sites and acetylene molecules are so weak that open Mg(II) sites cannot be fully loaded.

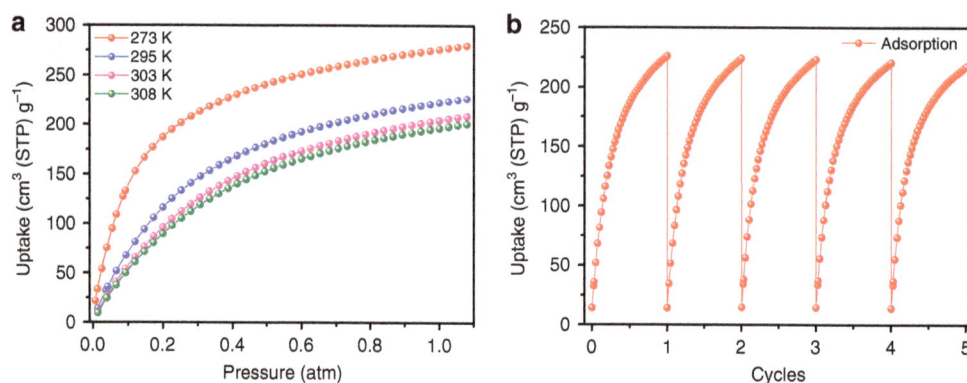

Figure 3 | C$_2$H$_2$ adsorption properties. (**a**) C$_2$H$_2$ adsorption isotherms of **FJI-H8** at 273, 295, 303 and 308 K. (**b**) Cycles of C$_2$H$_2$ adsorption for **FJI-H8** at 295 K.

Table 2 | Contributions of open metal sites (OMSs) and pore space in acetylene uptakes for selected MOFs at room temperature and 1 atm for volumetric capacity in the unit of cm^3 (STP) cm^{-3}*.

Material	Framework Density (g cm^{-3})[†]	OMS density (mmol cm^{-3})	C$_2$H$_2$ uptake (cm^3 (STP) cm^{-3})			Density[‡] (g cm^{-3})	P[§] (MPa)
			By OMS	By pore space	Sum		
FJI-H8	0.873	3.13	70	126	196	0.23	21.3
FJI-H8[‖]	0.873	3.13	70	105	175	0.20	19.9
HKUST-1 (ref. 19)	0.879	4.36	97	80	177	0.21	19.3
CoMOF-74 (ref. 20)	1.169	7.49	168	62	230	0.27	25.1
ZJU-5 (ref. 18)	0.598	2.31	52	63	115	0.13	12.5
MgMOF-74 (ref. 20)	0.909	7.49	168[¶]	—	167	0.19	18.2
NOTT-101 (ref. 55)	0.684	2.35	53	73	126	0.15	13.9
ZJU-7 (ref. 56)	0.750	2.60	58	77	135	0.16	14.7
Cu-TDPAT (ref. 26)	0.783	2.93	66	73	139	0.16	15.1
PCN-16 (ref. 55)	0.723	3.03	68	59	127	0.15	13.8

*The OMS density of MOFs was calculated based on the crystal information files.
†The framework density was calculated from single-crystal X-ray data.
‡Density of adsorbed C$_2$H$_2$ in bulk material.
§Pressure of C$_2$H$_2$ at 295 K corresponding to the calculated density of adsorbed C$_2$H$_2$ in bulk material.
‖Data for **FJI-H8** at 308 K.
¶The value of C$_2$H$_2$ uptake by OMS is larger than the sum value maybe because of the interaction between open Mg(II) sites and acetylene molecules are so weak that open Mg(II) sites cannot be fully loaded.

capacity in the unit of cm^3 (STP) cm^{-3}. The relationship between them is $V_{volumetric} = V_{gravimetric} \times \rho$. If we take the solvent-free crystal density into consideration, **FJI-H8** shows an acetylene uptake of 196 cm^3 (STP) cm^{-3} at 295 K and 1 atm, which is only inferior to the uptake of 230 cm^3 (STP) cm^{-3} for CoMOF-74 under the same conditions[20]. On the basis of the volumetric acetylene uptake, **FJI-H8** shows a safe acetylene storage density of 0.23 g cm^{-3} in bulk material at 295 K and 1 atm, which is equivalent to the value of an imaginary state of acetylene under 21.3 MPa at room temperature and is ~100 times of the compression limit for the safe storage of acetylene (0.2 MPa) at room temperature[3,43,44]. This value is also at the highest level for the reported MOFs, and is only slightly lower than that of CoMOF-74 (0.27 g cm^{-3}), which is consistent with the volumetric analysis (Table 2).

Similarly, supposing each open metal site binds one acetylene molecule, the acetylene uptake by open Cu(II) sites in HKUST-1 accounts for c.a. 54% of the volumetric acetylene uptake at room temperature. For CoMOF-74, 73% of the volumetric acetylene uptake is contributed by open Co(II) sites if each Co(II) site binds one acetylene molecule. Much like the gravimetric capacity, for the volumetric acetylene uptake of MgMOF-74 if all the open Mg(II) sites can be fully loaded with one acetylene molecule per each open Mg(II) site, the theoretical acetylene uptake by open metal sites is even larger than the experimental one at room

temperature. As listed in Table 2 for most MOF materials, the volumetric acetylene uptake capacities contributed by open metal sites account for almost half of the acetylene uptakes or even higher. However, in **FJI-H8**, the open Cu(II) sites can only contribute 70 cm^3 out of the total 196 cm^3 volumetric acetylene storage capacity at 295 K and 1 atm. Comparatively, the volumetric acetylene uptake by the suitable pore space in **FJI-H8** contributed to 64% of the whole amount, which is consistent with the gravimetric analysis aforementioned.

The GCMC simulation. Theoretically, the confirmation of acetylene adsorption sites within the MOF skeletons is very important for us to design new MOFs-based gas storage materials. The most intuitive method to get the confirmation of adsorbed acetylene molecules is single-crystal diffractions or the poly-crystal power diffractions[45,46]. However, the above methods usually require high quality crystalline samples with high stability, which is not always available. On the other hand, theoretical simulation is a powerful tool that can give us a lot of useful information[19,47,48]. To understand the acetylene-framework interactions, the acetylene adsorption property of **FJI-H8** was studied by grand canonical Monte Carlo (GCMC) simulations[45-48]. The calculated C$_2$H$_2$ adsorption isotherm is shown in Supplementary Fig. 10. As expected, the agreement between simulation and experiment of

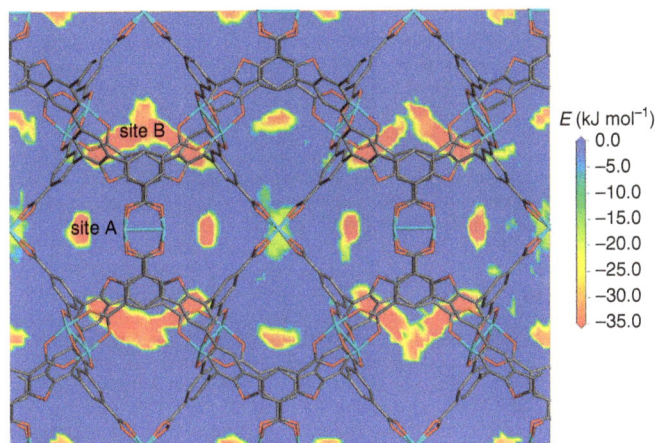

Figure 4 | Slice through the calculated potential field for acetylene in FJI-H8. The slice is viewed along the crystallographic *a* axis. The framework is displayed in capped sticks plot, and the hydrogen atoms are omitted for clarity.

acetylene adsorption is almost perfect at pressures below 0.3 atm, whereas the uptake at higher pressures is slightly overestimated with a deviation of 7%. Furthermore, the calculated adsorption heat of 28.7 kJ mol^{-1} is also comparable to the experimental one (Supplementary Fig. 11). Slice through the calculated potential field for acetylene is displayed in Fig. 4. Two preferential adsorption regions are showed in the field. As anticipated, the highest potential values are located around the unsaturated metal centres. Surprisingly, a strong increase of the interaction potential is visible in the small Cage-B, which is surrounded by eight benzene rings. However, it is noteworthy that the adsorption behaviour in Cage-B should be attributed to the interactions between the adsorbed acetylene molecules and the surrounding benzene rings rather than the four misaligned open copper sites. The adsorption in the octahedral pore is observed at the entrance windows to this pore, which is similar to HKUST-1 (refs 19,47). Therefore, the ultrahigh acetylene uptake of **FJI-H8** can be attributed to the suitable pore space together with open metal sites.

Discussion

To our knowledge, the open metal sites within MOFs materials usually play an important role in the high gas storage capacities due to the strong interactions between acetylene molecules and the metal sites. Supposing each open metal site binds one acetylene molecule, for most previously reported MOF materials the acetylene uptake by open metal sites theoretically accounts for over 60% of the whole acetylene uptake at room temperature or even higher. However, in **FJI-H8**, the open Cu(II) site density is low and the acetylene uptake by the pore space accounts for >60% of the whole amount, which is rarely seen in other reported MOFs. As is to be expected, the acetylene uptake by the pore space accounts for >60% of the whole amount even when the crystal density is taken into consideration (Table 2). Ideally, the perfect adsorbent materials should have both high gravimetric uptake and high volumetric capacity. In practice, very few adsorbents including MOFs can meet this requirement. For example, CoMOF-74 shows the highest volumetric acetylene uptake of 230 cm^3 (STP) cm^{-3} at 295 K and 1 atm; however, the gravimetric acetylene uptake of CoMOF-74 is only 197 cm^3 (STP) g^{-1} (ref. 20). Similarly, although the gravimetric acetylene capacity of HKUST-1 reaches up to 201 cm^3 (STP) g^{-1} at 295 K and 1 atm, it shows a volumetric acetylene uptake of 177 cm^3 (STP) cm^{-3} (ref. 19). In our case, compared with the reported

results measured at 295 K and 1 atm, **FJI-H8** exhibits the highest acetylene uptake of 224 cm^3 (STP) g^{-1} in gravimetric capacity, and also shows a high value of 196 cm^3 (STP) cm^{-3} in terms of volumetric capacity, which is only lower than that of CoMOF-74, due to its lower crystal density. More importantly, high acetylene uptakes around room temperature, low decrease rate of acetylene uptakes in the temperature range from 273 to 308 K and excellent repeatability make **FJI-H8** a suitable candidate for practical applications. Although MOF-based acetylene storage has not exhibited economical advantage to acetone-based one at this stage, the purity of acetylene stored in **FJI-H8** is higher than that in acetone. Furthermore, since acetone is an explosive pollutant, the storage of acetylene by recyclable **FJI-H8** is safer and cleaner. More importantly, the cost of MOF-based acetylene storage may decrease markedly if large-scale application is implemented. Therefore, the exploration of MOFs for safe and pure acetylene storage is significant not only for theoretical studies but also for practical applications.

In conclusion, **FJI-H8** is a promising candidate for acetylene storage at around room temperature. The high acetylene uptake of **FJI-H8** shows that not only open metal sites but also suitable pore space plays key roles in MOF-based acetylene storage. Our results shed light on the rational design and synthesis of new MOFs materials for acetylene storage based on the two factors above.

Methods

Materials and equipment. All reagents and solvents used in synthetic studies were commercially available and used as supplied without further purification. The ligand H$_8$tddb (H$_8$tddb = 3,3′,5,5′-tetra(3,5-dicarboxyphenyl)-4,4′-dimethoxy-biphenyl) was synthesized through the routine in the Supplementary Fig. 1.

Elemental analyses for C, H and N were carried out on a German Elementary Vario EL III instrument. ^{1}H NMR spectra were obtained on a Burker AVANCE 400 (400 MHz) for spectrometer. Mass-accurate match spectra were obtained using a DECAX-30000 LCQ Deca XP mass spectrometer with electro-spray ionization (ESI). The PXRD patterns were collected by a Rigaku Mini 600 X-ray diffractometer using Cu Kα radiation ($\lambda = 1.54$ Å). Simulations of the PXRD spectrum are carried out by the single-crystal data and diffraction-crystal module of the Mercury program available free of charge via internet at http://www.ccdc.cam.ac.uk/products/mercury/.

Synthesis of 3,3′,5,5′-tetrabromo-4,4′-biphenol (1). Bromine (13.8 ml, 268.6 mmol) was rapidly added to a solution of 4,4′-biphenol (10 g, 54 mol) in methanol (400 ml). After 1 h of stirring, the resulting precipitate was filtered and washed sequentially with aqueous solutions of NaHCO$_3$, Na$_2$SO$_3$ and water. The resulting white powder was dissolved in acetone and dried over anhydrous Na$_2$SO$_4$. Pure compound **1** was obtained by recrystallization in acetone (14.6 g, 54%). ^{1}H NMR (400 MHz, CDCl$_3$) δ 5.93 (s, 2H), 7.60 (s, 4H) p.p.m.

Synthesis of 3,3′,5,5′-tetrabromo-4,4′-dimethoxy-1,1′-biphenyl (2). Compound **1** (4.0 g, 8 mmol), iodomethane (6.8 g, 48 mmol) and K$_2$CO$_3$ (3.3 g, 24 mmol) were dissolved into acetonitrile (100 ml). The reaction mixture was heated at reflux for 18 h and then cooled to room temperature. Acetonitrile was removed using rotary evaporator, and the resulting mixture was poured into water and extracted with dichloromethane (3 × 100 ml). The combined organic layers were dried over anhydrous MgSO$_4$, and then the solvent was removed again using rotary evaporator. After purification by column chromatography on silica gel using hexane as eluent and evaporation of the fraction containing the product, compound **2** was obtained as a white powder. (2.88 g, 68%). ^{1}H NMR (400 MHz, CDCl$_3$) δ 3.94 (s, 6H), 7.65 (s, 4H) p.p.m.

Synthesis of 3,3′,5,5′-tetra(diethyl-3,5-dicarboxyphenyl)-4,4′-dimethoxy-biphenyl (3). Compound **2** (1.06 g, 2 mmol), diethyl 5-(4,4,5,5-tetramethyl-1,3,2-dioxaborolan-2-yl)-1,3-benzenedicarboxylate (4.18 g, 12 mmol), Cs$_2$CO$_3$ (11.8 g, 36 mmol) and tetrakis(triphenylphosphine)palladium (0.092 g, 0.08 mmol) were added to a 500-ml schlenk flask charged with stir bar. The flask was pumped under vacuum and refilled with N$_2$ for three times, and then 350 ml degassed 1,4-dioxane was transferred to the system and the reaction mixture was heated to 85 °C for 72 h under N$_2$ atmosphere. After the reaction mixture was cooled to room temperature, the organic solvent was removed using rotary evaporator, and the resulting mixture was poured into water and extracted with dichloromethane (3 × 100 ml). The combined organic layers were dried over anhydrous MgSO$_4$, and then the solvent was removed again using rotary evaporator. After purification by column chromatography on silica gel using ethyl acetate/hexane (1:5 v/v) as eluent and evaporation of the fraction containing the product, compound **3** was obtained as a pale yellow solid. (1.60 g, 73%). ^{1}H NMR (400 MHz, CDCl$_3$): δ 1.44 (t, 24H), 3.21 (s, 6H), 4.45 (q, 16H), 7.66 (s, 4H), 8.52 (s, 8H), 8.73 (s, 4H) p.p.m.

Synthesis of 3,3′,5,5′-tetra(3,5-dicarboxyphenyl)-4,4′-dimethoxy-biphenyl (H₈tddb). Compound **3** (2.19 g, 4 mmol) was dissolved in 20 ml of THF, to which 20 ml of 10 M NaOH aqueous solution was added. The mixture was stirred under reflux for 10 h, and then the organic solvent was removed using rotary evaporator. The aqueous phase was acidified to pH 2 using 6 M HCl aqueous solution. The resulting precipitate was collected via filtration, washed with water (200 ml) and dried under vacuum to afford **H₈tddb**. (1.58 g, 91%). ^{1}H NMR (400 MHz, DMSO-d_6) δ, 3.10 (s, 6H), 7.97 (s, 4H), 8.43 (s, 8H), 8.51 (s, 4H), 13.40 (s, 8H) p.p.m. ^{13}C NMR (100 MHz, DMSO-d_6) δ, 167.0, 154.5, 139.0, 136.4, 134.5, 134.4, 132.0, 130.1, 129.4, 61.1 p.p.m. ESI-MS (ESI$^-$ mode): calculated for $C_{46}H_{30}O_{18}$: 869.1. Found: 869.0.

Synthesis of FJI-H8. $Cu(NO_3)_2 \cdot 2H_2O$ (30 mg) and H₈tddb (10 mg) were dissolved in 1.5 ml of *N,N*-diethylformamide and 0.45 ml of water in a 2.5 ml pyrex vial, to which 25 μl of HCl were added. The mixture was heated in 85 °C oven for 12 h to yield 6 mg of blue–green crystals (yield: 46% based on H₈tddb). The crystals obtained were filtered and washed with *N,N*-diethylformamide. Elemental analyses calculated (%) for $C_{46}H_{42}O_{26}Cu_4$. (After activation and absorbed small amount of water, the crystal has a formula of $Cu_4(H_2O)_4(tddb) \cdot 6H_2O$): C 42.63, H 3.26; found: C 45.82, H 3.65.)

X-ray data collection and structure determination of FJI-H8. A blue–green block crystal of **FJI-H8** was taken directly from the mother liquor, transferred to oil and mounted into a loop. The crystal was kept at 100.0(1) K during data collection on a SuperNova diffractometer equipped with Cu-Kα radiation ($\lambda = 1.5418$ Å) using a ω scan mode. The crystal structure was solved by direct method and refined by full-matrix least squares on F^2 using SHELXTL package[49]. All non-hydrogen atoms were refined with anisotropic displacement parameters. The hydrogen atoms on the aromatic rings were located at geometrically calculated positions and refined by riding. However, the hydrogen atoms for the coordinated molecules cannot be found from the residual electron density peaks and the attempt of theoretical addition was not done. Therefore, the number of reported hydrogen atoms is more than the calculated one. The free solvent molecules are highly disordered in **FJI-H8**, and attempts to locate and refine the solvent peaks were unsuccessful. The diffused electron densities resulting from these solvent molecules were removed using the SQUEEZE routine of PLATON[40]; structures were then refined again using the data generated. Crystal data are summarized in Supplementary Table 2.

Low-pressure gas sorption measurements. Low-pressure (<1 bar) adsorption measurements were performed using a Micromeritics ASAP 2020-M surface area and pore size analyser. The fresh crystalline sample of **FJI-H8** was degassed under dynamic vacuum at 80 °C for 10 h after solvent exchange with methanol and then dichloromethane for 3 days each. A colour change from blue–green to deep purple–blue was observed during the activation process, which is attributed to the remove of terminal coordinated water of dicopper(II) paddlewheel SBUs, thus indicating the generation of open metal sites in the framework. Low-pressure N_2 adsorption isotherms were measured at 77 K in a liquid nitrogen bath. The specific surface areas were determined using the Brunauer–Emmett–Teller model from the N_2 sorption data. Low-pressure acetylene adsorption isotherms were measured at 273, 295, 303 and 308 K. The isosteric heat of adsorption was calculated through the Clausius–Clapeyron equation using the four sets of acetylene adsorption data collected.

Computational methods. To characterize the adsorption sites of acetylene molecules in **FJI-H8**, GCMC simulations were carried out using Sorption Module in Materials Studio[50]. The **FJI-H8** framework was fixed at the crystallographic data based on the single-crystal X-ray diffraction. Four unit cells of **FJI-H8** ($2 \times 2 \times 1$) were used to construct the simulation box of the GCMC run. Then, the structural parameters of simulation box are $a = b = 35.8514$ Å and $c = 28.0627$ Å, as well as $\alpha = \beta = \gamma = 90$.

DFT calculations. Density functional theory (DFT) calculations were performed to derive the charges to be used in the GCMC simulations to estimate the adsorption isotherms of acetylene in **FJI-H8**. The atomic coordinates were taken from the experimental crystallographic data. The cluster included building units (for example, metal ion nodes and the organic linker) representative of the unit cells. Details of structure and atom types of the **FJI-H8** clusters are shown in Supplementary Fig. 12. The DFT calculations were performed with the Gaussian 09 (ref. 51) software at the B3LYP/6-31G* level of theory. Partial atomic charges were extracted using the ChelpG method[52] by fitting them to reproduce the electrostatic potential generated by the DFT calculations. The charge of Cu was slightly adjusted to result in a neutral framework. Resulting partial charges for **FJI-H8** are given in Supplementary Table 3.

Acetylene model. The model of acetylene molecule was taken from the literature[47]. In this model, the acetylene molecule is a rigid structure where the C–C and C–H bond lengths are fixed at 1.2111 and 1.0712 Å, respectively (Supplementary Fig. 13). To account for the electrostatic interaction, point charges of 0.2780 *e* were assigned to H and C atoms of acetylene molecule, which were derived from DFT calculations using the ESP methods. Supplementary Table 4 shows bond lengths, ESP charge *q* and quadrupole moment θ for acetylene.

To represent van der Waals interactions, the acetylene molecule was treated as a two-site model, in which H atoms in acetylene molecule were represented as non-interacting atoms and the Lennard–Jones (LJ) positions located on the carbon atoms. The LJ parameters for *sp*-hybridized carbon was taken from the central CH groups of 2-butene in the literature[53].

Force-field parameters. LJ parameters for **FJI-H8** atoms were taken from the Universal force field[54]. LJ parameters for acetylene and LJ parameters representing the interaction of the acetylene molecules with the copper centres of **FJI-H8** were taken from the literature[47]. Supplementary Table 5 shows the LJ parameters for all atom types found in **FJI-H8** and acetylene.

GCMC simulation. These simulations were performed with the Sorption module of Materials Studio50. All GCMC simulations included a 4,000,000-cycle equilibration period followed by a 4,000,000-cycle production run. Atoms in **FJI-H8** were held fixed at their crystallographic positions. The van der Waals interactions were represented using a LJ potential, applying Lorentz–Berthelot mixing rules to calculate interactions between different atom types. An LJ cutoff distance of 12.5 Å was used for the simulations. The Ewald sum technique was used to compute the electrostatic interactions. Four unit cells of **FJI-H8** were used for the simulations. Acetylene isotherms were simulated at 295 K up to 1.0 bar. GCMC simulations reported the absolute adsorption data, which were then used to compute the excess adsorption data for comparison with experimental data using the relation

$$N_{\text{excess}} = N_{\text{total}} - V_{\text{p}} \times \rho,$$

where ρ is the bulk density of acetylene at simulation conditions. The density needed was calculated using the Peng–Robinson equation of state. V_{p} is the pore volume calculated by PLATON[40].

References

1. Stang, P. J. & Diederich, F. *Modern Acetylene Chemistry* (Wliey-VCH, 1995).
2. Chien, J. C. W. *Polyacetylene Chemistry, Physics, and Material Science* (Academic, 1984).
3. Matsuda, R. *et al.* Highly controlled acetylene accommodation in a metal-organic microporous material. *Nature* **436,** 238–241 (2005).
4. Budavari, S. *The merck Index.* 12th edn (Merck Research Laboratories, 1996).
5. Getman, R. B., Bae, Y. S., Wilmer, C. E. & Snurr, R. Q. Review and analysis of molecular simulations of methane, hydrogen, and acetylene storage in metal-organic frameworks. *Chem. Rev.* **112,** 703–723 (2012).
6. He, Y.-B., Zhou, W., Qian, G.-D. & Chen, B.-L. Methane storage in metal-organic frameworks. *Chem. Soc. Rev.* **43,** 5657–5678 (2014).
7. Furukawa, H., Cordova, K. E., O'Keeffe, M. & Yaghi, O. M. The chemistry and applications of metal-organic frameworks. *Science* **341,** 1230444 (2013).
8. Mason, J. A., Veenstra, M. & Long, J. R. Evaluating metal–organic frameworks for natural gas storage. *Chem. Sci.* **5,** 32–51 (2014).
9. Wilmer, C. E. *et al.* Gram-scale, high-yield synthesis of a robust metal–organic framework for storing methane and other gases. *Energy Environ. Sci.* **6,** 1158–1163 (2013).
10. Huck, J. M. *et al.* Evaluating different classes of porous materials for carbon capture. *Energy Environ. Sci.* **7,** 4132–4146 (2014).
11. Zhang, Z.-J., Yao, Z.-Z., Xiang, S.-C. & Chen, B.-L. Perspective of microporous metal–organic frameworks for CO_2 capture and separation. *Energy Environ. Sci.* **7,** 2868–2899 (2014).
12. Murray, L.-J., Dinca, M. & Long, J. R. Hydrogen storage in metal-organic frameworks. *Chem. Soc. Rev.* **38,** 1294–1314 (2009).
13. Suh, M. P., Park, H. J., Prasad, T. K. & Lim, D. W. Hydrogen storage in metal-organic frameworks. *Chem. Rev.* **112,** 782–835 (2012).
14. Zhuang, W.-J. *et al.* Robust metal–organic framework with an octatopic ligand for gas adsorption and separation: combined characterization by experiments and molecular simulation. *Chem. Mater.* **24,** 18–25 (2012).
15. Peng, Y. *et al.* Methane storage in metal-organic framework: current records, surprise findings, and challenges. *J. Am. Chem. Soc.* **135,** 11887–11894 (2013).
16. Decoste, J. B. *et al.* Metal-organic frameworks for oxygen storage. *Angew. Chem. Int. Ed.* **53,** 14092–14095 (2014).
17. Barin, G. *et al.* Isoreticular series of (3,24)-connected metal-organic frameworks: facile synthesis and high methane uptake properties. *Chem. Mater.* **26,** 1912–1917 (2014).
18. Rao, X.-T. *et al.* A microporous metal-organic framework with both open metal and Lewis basic pyridyl sites for high C_2H_2 and CH_4 storage at room temperature. *Chem. Commun.* **49,** 6719–6721 (2013).
19. Xiang, S.-C., Zhou, W., Gallegos, J. M., Liu, Y. & Chen, B.-L. Exceptionally high acetylene uptake in a microporous metal-organic framework with open metal sites. *J. Am. Chem. Soc.* **131,** 12415–12419 (2009).
20. Xiang, S.-C. *et al.* Open metal sites within isostructural metal-organic frameworks for differential recognition of acetylene and extraordinarily high acetylene storage capacity at room temperature. *Angew. Chem. Int. Ed.* **49,** 4615–4618 (2010).
21. Xu, H. *et al.* A microporous metal-organic framework with both open metal and Lewis basic pyridyl sites for highly selective C_2H_2/CH_4 and C_2H_2/CO_2 gas separation at room temperature. *J. Mater. Chem. A* **1,** 77–81 (2013).

22. Xu, H. *et al.* A cationic microporous metal–organic framework for highly selective separation of small hydrocarbons at room temperature. *J. Mater. Chem. A* **1**, 9916–9921 (2013).

23. Samsonenko, D. G. *et al.* Microporous magnesium and manganese formates for acetylene storage and separation. *Chem. Asian J.* **2**, 484–488 (2007).

24. Zhang, J.-P. & Chen, X.-M. Optimized acetylene/carbon dioxide sorption in a dynamic porous crystal. *J. Am. Chem. Soc.* **131**, 5516–5521 (2009).

25. Yang, S. *et al.* Supramolecular binding and separation of hydrocarbons within a functionalized porous metal–organic framework. *Nat. Chem.* **7**, 121–129 (2015).

26. Liu, K. *et al.* High storage capacity and separation selectivity for C_2 hydrocarbons over methane in the metal–organic framework Cu–TDPAT. *J. Mater. Chem. A* **2**, 15823–15828 (2014).

27. Xue, Y.-S. *et al.* A robust microporous metal–organic framework constructed from a flexible organic linker for acetylene storage at ambient temperature. *J. Mater. Chem.* **22**, 10195–10199 (2012).

28. Ma, L.-Q., Mihalcik, D. J. & Lin, W.-B. Highly porous and robust 4,8-connected metal–organic frameworks for hydrogen storage. *J. Am. Chem. Soc.* **131**, 4610–4612 (2009).

29. Yuan, D.-Q., Zhao, D., Sun, D.-F. & Zhou, H.-C. An isoreticular series of metal–organic frameworks with dendritic hexacarboxylate ligands and exceptionally high gas-uptake capacity. *Angew. Chem. Int. Ed.* **49**, 5357–5361 (2010).

30. Pang, J.-D. *et al.* Coexistence of cages and one-dimensional channels in a porous MOF with high H_2 and CH_4 uptakes. *Chem. Commun.* **50**, 2834–2836 (2014).

31. He, Y.-B., Li, B., O'Keeffe, M. & Chen, B.-L. Multifunctional metal-organic frameworks constructed from meta-benzenedicarboxylate units. *Chem. Soc. Rev.* **43**, 5618–5656 (2014).

32. Yan, Y., Yang, S., Blake, A. J. & Schroder, M. Studies on metal-organic frameworks of Cu(II) with isophthalate linkers for hydrogen storage. *Acc. Chem. Res.* **47**, 296–307 (2014).

33. Farha, O. K. *et al.* De novo synthesis of a metal-organic framework material featuring ultrahigh surface area and gas storage capacities. *Nat. Chem.* **2**, 944–948 (2010).

34. Lu, W.-G., Yuan, D.-Q., Makal, T. A., Li, J.-R. & Zhou, H.-C. A highly porous and robust (3,3,4)-connected metal-organic framework assembled with a 90 degrees bridging-angle embedded octacarboxylate ligand. *Angew. Chem. Int. Ed.* **51**, 1580–1584 (2012).

35. Wang, X.-S. *et al.* Enhancing H_2 uptake by "close-packing" alignment of open copper sites in metal-organic frameworks. *Angew. Chem. Int. Ed.* **47**, 7263–7266 (2008).

36. Ma, S.-Q., Simmons, J. M., Sun, D.-F., Yuan, D.-Q. & Zhou, H.-C. Porous metal-organic frameworks based on an anthracene derivative: syntheses, structure analysis, and hydrogen sorption studies. *Inorg. Chem.* **48**, 5263–5268 (2009).

37. Ma, S.-Q. *et al.* Metal-organic framework from an anthracene derivative containing nanoscopic cages exhibiting high methane uptake. *J. Am. Chem. Soc.* **130**, 1012–1016 (2008).

38. Li, J.-R. *et al.* Porous materials with pre-designed single-molecule traps for CO_2 selective adsorption. *Nat. Commun.* **4**, 1538 (2013).

39. Blatov, V. A., Shevchenko, A. P. & Proserpio, D. M. Applied topological analysis of crystal structures with the program package toposPro. *Cryst. Growth Des.* **14**, 3576–3586 (2014).

40. Spek, A. L. Single-crystal structure validation with the program PLATON. *J. Appl. Cryst.* **36**, 7–13 (2003).

41. Sarkisov, L. & Harrison, A. Computational structure characterisation tools in application to ordered and disordered porous materials. *Mol. Simul.* **37**, 1248–1257 (2011).

42. Qian, J.-J. *et al.* Unusual pore structure and sorption behaviour in a hexanodal zinc-organic framework material. *Chem. Commun.* **50**, 1678–1681 (2014).

43. He, Y.-B., Zhou, W., Krishna, R. & Chen, B.-L. Microporous metal-organic frameworks for storage and separation of small hydrocarbons. *Chem. Commun.* **48**, 11813–11831 (2012).

44. Chen, B.-L. Acetylene storage using metal-organic frameworks of the formula M_2(2,5-dihydroxyterephthalate). *US 8,664,419 B2* (2014).

45. Kubota, Y. *et al.* Metastable sorption state of a metal-organic porous material determined by in situ synchrotron powder diffraction. *Angew. Chem. Int. Ed.* **45**, 4932–4936 (2006).

46. Jang, S. B., Jeong, M. S. & Kim, Y. Crystal structure of an acetylene sorption complex of dehydrated fully Cd^{2+}-exchanged zeolite X. *Zeolites* **19**, 228–237 (1997).

47. Fischer, M., Hoffmann, F. & Froba, M. New microporous materials for acetylene storage and C_2H_2/CO_2 separation: insights from molecular simulations. *ChemPhysChem* **11**, 2220–2229 (2010).

48. Guo, H.-C., Shi, F., Ma, Z.-F. & Liu, X.-Q. Simulation of separation of C_2H_6 from CH_4 using zeolitic imidazolate frameworks. *Mol. Simul.* **40**, 349–360 (2013).

49. Sheldrick, G. M. A short history of SHELX. *Acta Cryst. A* **64**, 112–122 (2008).

50. Accelrys Materials Studio, version 6.0 (Accelrys Inc, San Diego, USA, 2012).

51. Frisch, M. J. *et al. GAUSSIAN 09 Program* (Gaussian Inc., 2009).

52. Breneman, C. M. & Wiberg, K. B. Determining atom-centered monopoles from molecular electrostatic potentials - the need for high sampling density in formamide conformational-analysis. *J. Comput. Chem.* **11**, 361–373 (1990).

53. Jorgensen, W. L., Madura, J. D. & Swenson, C. J. Optimized intermolecular potential functions for liquid hydrocarbons. *J. Am. Chem. Soc.* **106**, 6638–6646 (1984).

54. Rappe, A. K., Casewit, C. J., Colwell, K. S., Goddard, W. A. & Skiff, W. M. UFF, a full periodic table force field for molecular mechanics and molecular dynamics simulations. *J. Am. Chem. Soc.* **114**, 10024–10035 (1992).

55. He, Y.-B., Krishna, R. & Chen, B.-L. Metal–organic frameworks with potential for energy-efficient adsorptive separation of light hydrocarbons. *Energy Environ. Sci.* **5**, 9107–9120 (2012).

56. Cai, J.-F. *et al.* A NbO type microporous metal-organic framework constructed from a naphthalene derived ligand for CH_4 and C_2H_2 storage at room temperature. *RSC Adv.* **4**, 49457–49461 (2014).

Acknowledgements

This work was financially supported by the 973 Program (2011CBA00507 and 2011CB932504), National Nature Science Foundation of China (21131006, 21271172 and 21371169). W.L. acknowledges the Laboratory for Molecular Simulation of Texas A&M University for providing the Material Studio software.

Author contributions

D.Y., M.W., F.J. and M.H. conceived and designed the experiments and co-wrote the paper. J.P. performed most of experiments and analysed the data. C.L. and W.L. worked on all computational simulations. K.S. performed ligand synthesis. D.Y., M.W. and J.P. analysed the data and wrote the manuscript. All authors discussed the results and commented on the manuscript.

Additional information

Accession codes: The X-ray crystallographic coordinates for structure reported in this Article have been deposited at the Cambridge Crystallographic Data Centre (CCDC), under deposition number CCDC 1048132. These data can be obtained free of charge from The Cambridge Crystallographic Data Centre via www.ccdc.cam.ac.uk/data_request/cif.

25

Understanding silicate hydration from quantitative analyses of hydrating tricalcium silicates

Elizaveta Pustovgar[1], Rahul P. Sangodkar[2], Andrey S. Andreev[3], Marta Palacios[1], Bradley F. Chmelka[2], Robert J. Flatt[1] & Jean-Baptiste d'Espinose de Lacaillerie[1,3]

Silicate hydration is prevalent in natural and technological processes, such as, mineral weathering, glass alteration, zeolite syntheses and cement hydration. Tricalcium silicate (Ca_3SiO_5), the main constituent of Portland cement, is amongst the most reactive silicates in water. Despite its widespread industrial use, the reaction of Ca_3SiO_5 with water to form calcium-silicate-hydrates (C-S-H) still hosts many open questions. Here, we show that solid-state nuclear magnetic resonance measurements of ^{29}Si-enriched triclinic Ca_3SiO_5 enable the quantitative monitoring of the hydration process in terms of transient local molecular composition, extent of silicate hydration and polymerization. This provides insights on the relative influence of surface hydroxylation and hydrate precipitation on the hydration rate. When the rate drops, the amount of hydroxylated Ca_3SiO_5 decreases, thus demonstrating the partial passivation of the surface during the deceleration stage. Moreover, the relative quantities of monomers, dimers, pentamers and octamers in the C-S-H structure are measured.

[1] Institute for Building Materials, Department of Civil, Environmental and Geomatic Engineering, ETH Zürich 8093, Switzerland. [2] Department of Chemical Engineering, University of California, Santa Barbara, California 93106, USA. [3] Soft Matter Science and Engineering Laboratory, UMR CNRS 7615, ESPCI Paris, PSL Research University, 10 rue Vauquelin, Paris 75005, France. Correspondence and requests for materials should be addressed to J.-B.d.E.d.L. (email: jean-baptiste.despinose@espci.fr).

Since Le Chatelier[1], it is well understood that Portland cement hydration is initiated by the dissolution of calcium silicate monomers in water, followed by the precipitation of less soluble layered calcium-silicate-hydrates (C-S-H), in which silicate ions condense to form short chains. However, despite two centuries of widespread applications and a century of detailed study, the molecular mechanisms behind the kinetic stages of hydration (that is, induction, acceleration and deceleration) are still debated. Similar kinetic stages are observed in various heterogeneous hydration processes occurring during mineral weathering[2,3], glass alteration[4,5] and hydrothermal syntheses. For example, although hydrothermal zeolite syntheses under alkaline aqueous conditions proceeds over different timescales[6], the effective reaction rates in cementitious and zeolite systems exhibit similar distinct stages (induction, acceleration and deceleration), and are governed by several coupled parameters varying in space and time near the liquid–solid interface. This situation is thus extremely complex to describe accurately. An added difficulty is that for porous materials such as cement or zeolites, interfacial energy contributes to the stabilization of nanoscale intermediates, which are typically challenging to characterize. For Portland cement in particular, the lack of quantitative experimental data obtained with sufficient time resolution has precluded the validation of existing models aimed at explaining the complex kinetics of cement hydration.

Similar to the homogeneous versus heterogeneous pathways dichotomy in zeolite crystallization mechanisms[7], two landmark competing theories have been proposed to explain the early-age time dependence of the rate of tricalcium silicate (Ca_3SiO_5) hydration, the principal component in commercial Portland cements responsible for the development of mechanical strength[8–10]. The first theory proposes that early-age hydration products form a diffusion barrier on the surfaces of Ca_3SiO_5 particles, thus affecting subsequent reactions of the underlying non-hydrated core[11]. The second theory[12–14] suggests that the early-age time-dependence of the rate of hydration is determined by the rate of Ca_3SiO_5 dissolution and by a change in the associated rate limiting step from etch pit formation to step retreat, which is a mechanism also often invoked in the geochemical literature on natural weathering[15,16]. The relevance of these theories to silicate hydration can be examined by understanding the molecular compositions and structures of species at the solid–liquid interfaces during the early stages of hydration. Similar questions are raised in heterogeneous catalysis and geochemistry; however, Portland cement hydration faces the additional complexity that the main product, C-S-H, is not only poorly crystalline but also nanostructured with variable stoichiometry and silicate coordinations[17,18]. These challenges have been previously addressed partially through numerical modelling of hydration reactions at Ca_3SiO_5 surfaces[19,20] and of the local structure and disorder of the resulting hydration products[21]. Nevertheless, these models suffer from a lack of experimental support at the molecular level.

Here, solid-state NMR measurements of triclinic ^{29}Si-enriched Ca_3SiO_5 hydration are used to determine the transient molecular-level compositions at silicate surfaces and the interactions between silicate species, hydroxyl groups and water molecules, which influence the rates of hydration reactions. The isotopic ^{29}Si enrichment provides significantly enhanced NMR signal sensitivity that can be used to monitor the structures of the hydrates *in situ* during the hydration process, as a function of hydration time. In addition, ^{29}Si enrichment enabled two-dimensional (2D) through-bond (*J*-mediated) NMR measurements that are sensitive to ^{29}Si-O-^{29}Si covalent bonding. They are used to crucially provide detailed information on the local atomic-level compositions, structures and site connectivities in

hydrated silicate species, here C-S-H. These analyses shed new insights on the origin of rate limiting steps and the kinetics of silicate polymerization at the solid–liquid interface during Ca_3SiO_5 hydration.

Results

Experimental approach. To the seminal approach of ^{29}Si enrichment by Brough *et al.*[22], we added for the first time the sophistication of carefully controlled structure and granulometry of the Ca_3SiO_5 particles (see Supplementary Methods) and hydration reaction conditions (see Supplementary Notes 1 and 2). Indeed the surface structure and area of the Ca_3SiO_5 particles strongly affect their reactivity, which must be carefully controlled to ensure meaningful results[23]. For example, the high surface area of the synthesized ^{29}Si-enriched Ca_3SiO_5 ($4.4\,m^2\,g^{-1}$, see Supplementary Methods) allowed ~90% of the silicate hydration process to be monitored in 24 h of NMR spectrometer time, without external acceleration. In this way, subtle and unique quantitative information pertinent to hydration mechanisms can be obtained non-invasively and with a time resolution of 30 min

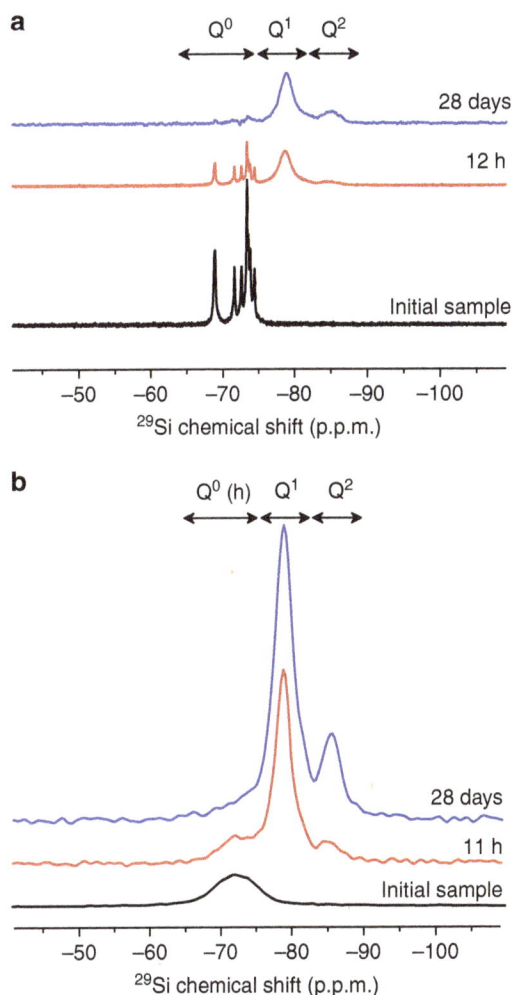

Figure 1 | Dynamics of silicate hydrates formation studied *in situ* by ^{29}Si NMR. (**a**) ^{29}Si MAS NMR and (**b**) {^{1}H}^{29}Si CPMAS NMR spectra of ^{29}Si-enriched triclinic Ca_3SiO_5 sample in its initial non-hydrated state (in black) and after hydration for 11 or 12 h (in red) and 28 days (in blue). ^{29}Si resonances from isolated silicate (Q^0) species in non-hydrated Ca_3SiO_5, hydoxylated surface Q^0 (Q^0(h)) species and polymerized calcium-silicate-hydrates (Q^1 and Q^2) are clearly resolved and can be quantified as a function of time.

(measurement time for the NMR spectra). Consequently, the progress of the hydration reaction could be accurately and quantitatively correlated to the corresponding ^{29}Si speciation. In addition, ^{29}Si enrichment allows NMR measurements to be performed on samples without the need for conventional water removal schemes for quenching the hydration process[24], which otherwise often disrupt the fragile microstructure of the C-S-H or may detrimentally alter chemical composition. Representative one-pulse ^{29}Si and ^{1}H{^{29}Si} cross-polarization (CP) magic-angle-spinning (MAS) NMR spectra are presented in Fig. 1a,b, respectively, for non-hydrated and hydrated Ca_3SiO_5. In anhydrous triclinic Ca_3SiO_5 which exhibits long-range crystalline order and well-defined local atomic ^{29}Si environments, eight distinct and narrow (<0.5 p.p.m. full-width at half maximum (FWHM)) ^{29}Si signals are resolved between -68 and -75 p.p.m. corresponding to anhydrous Q^0 species (Supplementary Fig. 2). In contrast, in hydration products, the ^{29}Si resonances are broad (3–4 p.p.m. FWHM) with signals centred at -72, -79 and -85 p.p.m. from silanol Q^0(h), hydrated Q^1 and hydrated Q^2 silicate species, respectively (Fig. 1). The last two species are associated with the C-S-H structure (Q^n refers to silicon atoms that are covalently bonded via bridging oxygen atoms to $0 \leq n \leq 4$ other silicon atoms[25]). These molecular-level insights of the local silicate structures in Ca_3SiO_5 hydration products (C-S-H) are consistent with previous ^{29}Si NMR (refs 26,27), ^{17}O NMR (ref. 28), X-ray and neutron scattering results[18] for C-S-H.

The degree of silicate hydration is determined by quantitative *in situ* ^{29}Si NMR analyses and forms the crux of our results, which are summarized in Fig. 2. These results are in close agreement with the degree of silicate hydration as established by independent isothermal calorimetric measurements, which reveal the successive stages of initial dissolution, induction, acceleration and deceleration (Fig. 2b) during the silicate hydration process. This comparison crucially establishes the accuracy of the quantitative ^{29}Si NMR results acquired during Ca_3SiO_5 hydration, and indicates that the hydration process is negligibly altered by factors such as the MAS conditions of the NMR experiment (see Supplementary Notes 1 and 2). This detailed

time-resolved, *in situ*, quantitative NMR analysis answers three central questions about Ca_3SiO_5 hydration: the molecular origin of the reduced apparent solubility of Ca_3SiO_5 during the induction period, the possible 'switch' from one type of hydration products to another between the acceleration and deceleration period, and the relative proportions of silica oligomers in the final C-S-H structure.

Induction period. The apparent solubility of Ca_3SiO_5 during the induction period of hydration has been reported to be lower compared with pristine anhydrous Ca_3SiO_5 (refs 11–13). This reduced apparent solubility has been proposed to arise from the deposition of a layer of hydration products (the metastable barrier hypothesis)[11] or from surface hydroxylation[12,13]. The molecular compositions at the Ca_3SiO_5 surface during this induction period

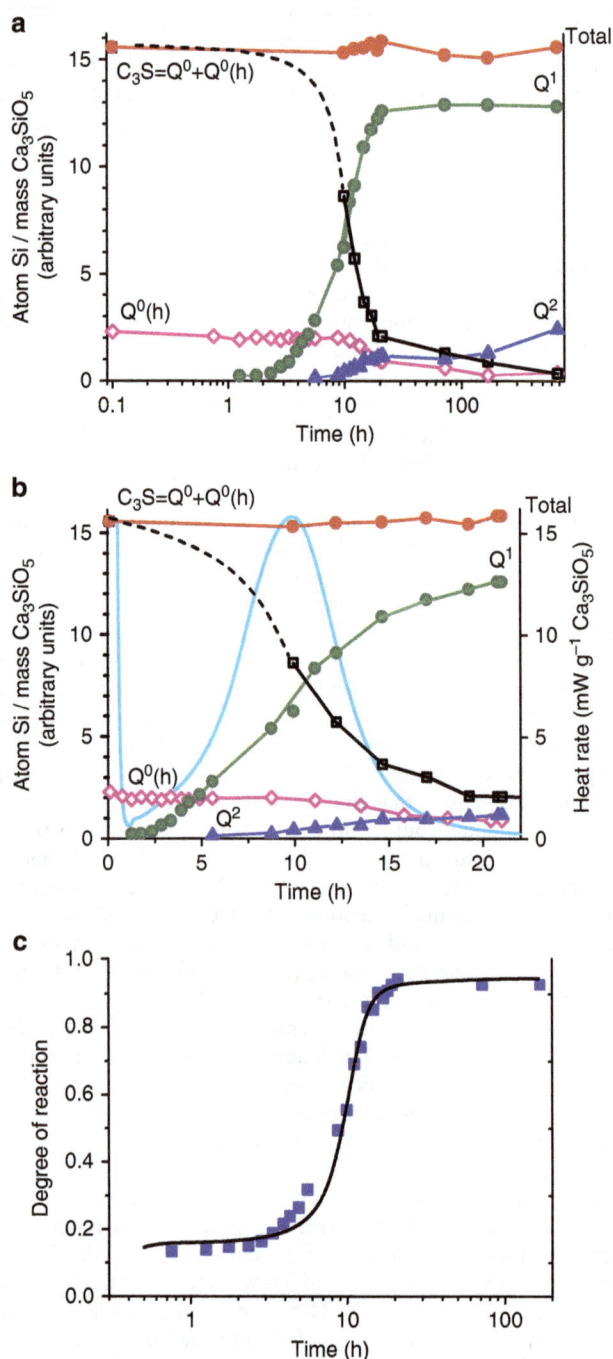

Figure 2 | Quantitative monitoring of silicate speciation during the hydration of ^{29}Si-enriched triclinic Ca_3SiO_5. (**a**) The quantities of different ^{29}Si silicate species as established by ^{29}Si MAS and {^{1}H}^{29}Si CPMAS NMR measurements for hydration times up to 28 days (see Supplementary Note 1). The quantities, normalized to the initial amount of Ca_3SiO_5, of anhydrous Q^0 (in black), hydroxylated Q^0(h) (in pink), hydrated Q^1 (in green), hydrated Q^2 (in blue) and total silicate species (in red) resulting from this analysis are as shown. (**b**) Comparison of the quantities of different ^{29}Si silicate species and the reaction heat flow rate determined by isothermal calorimetry (cyan line) for Ca_3SiO_5 up to 24 h of hydration. Based on the heat released in the calorimetry measurements, four stages in the hydration process can be identified: first a brief exothermic peak during the first few minutes (<15 min) corresponding to initial dissolution of Ca_3SiO_5, then a short (15 min–2 h) induction period during which no significant heat is released, followed by a peak corresponding to the acceleration period (2–10 h), and finally the deceleration period (>10 h) associated with decreasing rate of heat release (Supplementary Fig. 8). (**c**) Comparison of the degree of silicate hydration determined independently by ^{29}Si MAS and {^{1}H}^{29}Si CPMAS NMR quantitative analyses (squares) and isothermal calorimetry results (black line), which are in close agreement. The fact that the total amount of Si atoms remains constant, within the uncertainties of the measurements, over the entire hydration period (28 days) establishes the accuracy of the associated quantitative NMR methods and analyses. Details of these analyses are included in the Supplementary Note 1.

(as determined by the NMR analyses presented here) points towards the latter scenario. The ^{29}Si{^{1}H} CPMAS NMR measurements of the initial sample (that is, non-hydrated) (Fig. 1b) establish the presence of Q^0 silicate species in proximity to protons (henceforth labelled Q^0(h)) on Ca$_3$SiO$_5$ particle surfaces, even before contact with bulk water. Although previous studies have reported the presence of similar Q^0(h) silicate species at the surfaces of 'anhydrous' Ca$_3$SiO$_5$ particles[22,26], it has not been largely publicized nor quantitatively analysed. The 2D ^{29}Si{^{1}H} heteronuclear correlation (HETCOR) NMR spectrum of the same sample of non-hydrated Ca$_3$SiO$_5$ (Fig. 3) exhibits correlated intensities between the ^{29}Si signal at -72 p.p.m. from Q^0(h) species and unresolved ^{1}H signals around 1.3 and 0.9 p.p.m. from –SiOH and -CaOH moieties, thereby establishing the close molecular-level proximities of surface Q^0(h) species to at least one type of such ^{1}H moieties. In addition, the absence of resonances characteristic of polymerized hydration products (that is, Q^1 and Q^2 species), establishes that the reaction of surface silicate species in non-hydrated Ca$_3$SiO$_5$ with atmospheric moisture results solely in the formation of hydroxylated Q^0(h) species at particle surfaces, within the sensitivity limits of the measurement. In other words, no separate hydrate phase forms at this stage, it is solely the Ca$_3$SiO$_5$ particle near-surface which is hydroxylated.

From a crystal chemistry perspective, the Ca$_3$SiO$_5$ particle surface is unlikely to be inert when exposed to atmospheric water vapour. Specifically, Ca$_3$SiO$_5$ is an ionic crystal of Ca^{2+} cations with oxide and monomeric silicate anions $(3\text{Ca}^{2+} \cdot \text{O}^{2-} \cdot \text{SiO}_4{}^{4-})$ (refs 19,29). There is a strong ionization of the atoms ($+1.5$ on Ca^{2+} and -1.5 on O^{2-}) (ref. 19) and consequently Ca$_3$SiO$_5$ acts as a basic oxide that readily yields hydroxide ions when reacting with water,

$$\text{O}^{2-} + \text{H}_2\text{O} \rightarrow 2\,\text{OH}^- \qquad (1)$$

Therefore, one expects OH$^-$ to replace oxide ions on the particle surfaces. However, replacement of one O^{2-} by two OH$^-$ would yield a heterogeneous distribution of local atomic environments at the Ca$_3$SiO$_5$ surface, due to the different sizes and formal charges of these anions. Indeed the Q^0(h) ^{29}Si NMR resonance of the initial sample is very broad (Fig. 1b), reflecting a wide distribution of local ^{29}Si environments. In summary, the ^{29}Si

NMR analyses reveal that near-surface ^{29}Si species on Ca$_3$SiO$_5$ particles are predominantly hydroxylated and that negligible quantities of polymerized silicate hydration products form (within the sensitive detection limits of the measurements), a result consistent with previous force-field atomistic simulations[19]. Overall, hydroxylated Q^0 (h) species are predominant at particle surfaces during the induction period and expected to result in the reduced apparent solubility of Ca$_3$SiO$_5$, compared with pristine anhydrous Ca$_3$SiO$_5$ whose level of hydroxylation is lower.

Acceleration stage. With the progress of Ca$_3$SiO$_5$ hydration, the monomeric Q^0 silicate species polymerize to form oligomeric units of C-S-H. As shown in Fig. 2, while the population of hydroxylated Q^0(h) species remains constant, the populations of Q^1 species increase significantly during the acceleration stage (~ 2–10 h). Compared with the induction stage (<2 h), the ^{29}Si polymerization during the acceleration stage results predominantly in the formation of Q^1 species (dimers) at early times, and a combination of Q^1 and Q^2 species (for example, pentamers and octamers) at later time (10–20 h). In particular, the population of Q^1 species increases approximately linearly with the progress of hydration (Fig. 2b) across the entire acceleration stage, consistent with the formation of predominantly dimeric C-S-H units. No significant change nor in the silicon second coordination sphere of the hydration products nor in their rate of formation could be detected at this stage.

Deceleration stage. The data in Fig. 2 indicate that at the end of the acceleration stage (after ~ 10 h in the present case) greater quantities of long (>2 silicate tetrahedra) C-S-H chains containing Q^2 species are formed compared with dimeric C-S-H units (without Q^2). Although the amounts of Q^2 species increase progressively after the hydration peak (~ 20 h), the population of Q^1 species remains approximately constant, which indicates the formation of longer silicate chains besides the dimers. By comparison, the amount of Q^0(h) species remains constant for several hours (~ 10 h) during the induction and acceleration stages, it subsequently decreases just when, according to isothermal calorimetry, the Ca$_3$SiO$_5$ hydration slows down, that is during the so-called deceleration stage. This observation provides

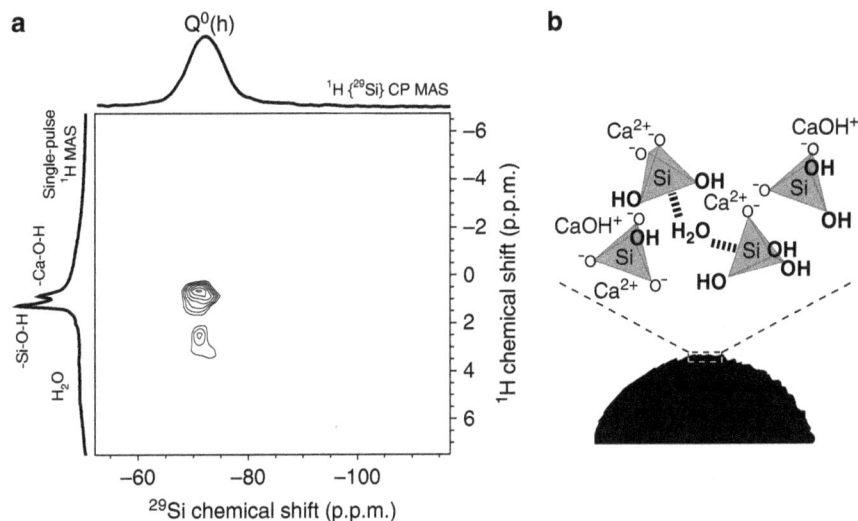

Figure 3 | Proton to silicon signal intensity correlations on the initial non-hydrated ^{29}Si-enriched triclinic Ca$_3$SiO$_5$. (**a**) The 2D {^{1}H} ^{29}Si HETCOR NMR spectrum shows intensity correlations between ^{29}Si and ^{1}H signals that result from molecular proximity between ^{29}Si and ^{1}H nuclei. ^{29}Si CPMAS and ^{1}H MAS 1D spectra are shown along the horizontal and vertical axis of the 2D spectrum. The chemical shift of ^{29}Si is detected (horizontal dimension), while chemical shift of ^{1}H is recorded in the indirect (vertical) dimension. (**b**) The right inset schematizes the protonated moieties detected on the Ca$_3$SiO$_5$ surfaces.

important insights regarding the debate on the origin of the deceleration period. While some previous studies suggest that the deceleration period results from coverage of Ca_3SiO_5 particles by hydration products[30], others claim that hydration initially results in products forming a low-density structure, the subsequent densification of which corresponds to the beginning of the deceleration stage[31,32]. Our analyses suggest that compared with the acceleration period that is associated with the formation of predominantly dimeric C-S-H units, the deceleration period corresponds to the formation of greater relative fractions of C-S-H units with longer chain lengths. Such increasing extents of silicate polymerization might possibly be accompanied by an increased density of the C-S-H that consequently would present a diffusion barrier for mass transport and, thus, slow the rate of hydration reaction, consistent with the deceleration stage. This alone is not conclusive as it could either support the view according to which the deceleration would be based indeed on the filling of an ultra-low-density gel[33] or the one based on an inhibition of hydration by hydrates themselves[34], impinging on each other's growth[35,36]. Nevertheless, the decrease of the amount of near-surface $Q^0(h)$ species population at the onset of the deceleration period reflects a proportional decrease of the average surface area available to drive hydration by silicate dissolution. The decrease of the particles surface area as revealed here by NMR supports strongly the conclusions of recent modelling studies[37], namely that the deceleration stage results from the reduction of the average particle surface area available for reaction due to increasing surface coverage of the Ca_3SiO_5 particles by hydration products. This conclusion is also supported by the fact that at 7 days 5% of the Ca_3SiO_5 has not yet hydrated, bringing support to a coverage and passivation of its surface by deposited hydrates. Moreover, the long period during which $Q^0(h)$ remains constant suggests that during dissolution, the surface decrease due to the reduction in particle size is compensated by roughening (opening of etch pits and step retreat)[38]. In other words dissolution does not simply proceed by shrinking of the core of the particles, but also by etching.

Final C-S-H structure. The atomic site interconnectivities of different silicate species can be used to elucidate the molecular structures and lengths of silicate chains in the C-S-H. Such detailed insights can be obtained by using solid-state 2D J-mediated $^{29}Si\{^{29}Si\}$ correlation NMR techniques[39] that probe J-coupled ^{29}Si-O-^{29}Si spin pairs and have been previously applied to establish silicate framework connectivities in a variety of heterogeneous materials[40–43]. Previously, Brunet et al.[44] have conducted 2D dipolar-mediated $^{29}Si\{^{29}Si\}$ NMR measurements that rely on through-space ^{29}Si-^{29}Si dipolar couplings and which yield information on the molecular-level proximities of different ^{29}Si moieties in synthetic C-S-H. However, such measurements cannot be used to directly establish the covalent connectivity among different ^{29}Si moieties in the C-S-H structure. In contrast, by relying on through-bond J-interactions associated with ^{29}Si-O-^{29}Si moieties (J-interactions between ^{29}Si spin pairs separated by more than two covalent bonds are negligibly small and consequently expected to be below the detection limits of the 2D J-mediated $^{29}Si\{^{29}Si\}$ NMR measurement.), 2D J-mediated $^{29}Si\{^{29}Si\}$ double-quantum (DQ) correlation NMR measurements provide detailed insights regarding the tetrahedral site connectivity in the C-S-H chains. Notably, the 2D J-mediated $^{29}Si\{^{29}Si\}$ NMR spectrum of hydrated ^{29}Si-labelled Ca_3SiO_5 shown in Fig. 4b provides significantly enhanced ^{29}Si resolution, compared with the single-pulse ^{29}Si MAS spectrum (Fig. 4a), and unambiguously establishes distinct ^{29}Si-O-^{29}Si covalent connectivities in the silicate chains.

The 2D J-mediated $^{29}Si\{^{29}Si\}$ NMR spectrum (Fig. 4b) exhibits three well separated regions of correlated intensities in the Q^1 (approximately -79 p.p.m.) and Q^2 (approximately -85 p.p.m.) chemical shift ranges along the single-quantum (SQ)–DQ $y = 2x$ line, and two pairs of cross-correlated peaks between the Q^1 and Q^2 chemical shifts ranges. The broad continuous distribution of correlated chemical shifts in the 2D $^{29}Si\{^{29}Si\}$ spectrum between signals at -82 and -87 p.p.m. in the ^{29}Si SQ dimension are attributed to different ^{29}Si-O-^{29}Si Q^2 moieties, consistent with the structural disorder of C-S-H. Interestingly, the spectrum reveals narrow (0.6 p.p.m. FWHM) ridges of intensity correlations that are parallel to the SQ–DQ line. Such features typically arise from structural disorder on length scales (>1 nm) that are larger than the distances between the ^{29}Si-^{29}Si spin pairs (or also due to anisotropy in the magnetic susceptibility)[45]. The presence of such poor long-range structural order is consistent with the broad distributions of local ^{29}Si environments that are associated with the heterogeneous nature of the C-S-H. Nevertheless, careful analysis of the 2D spectrum distinguishes discrete correlated signal intensities that are resolved to greater than a tenth of a p.p.m. Specifically, a strong correlated intensity (labelled i) between the ^{29}Si signals centred at -84.8 and -85.4 p.p.m. in the SQ dimension and at -170.2 p.p.m. in the DQ dimension (Supplementary Fig. 11) unambiguously establishes the presence of two chemically distinct Q^2 ^{29}Si species that are covalently bonded through a shared bridging oxygen atom. The different isotropic ^{29}Si chemical shifts of these distinct Q^2 species likely reflect differences in the number and types of species in the C-S-H interlayer (calcium ions or proton moieties such OH groups or water molecules) that are in close (<1 nm) molecular-level proximity to the non-bridging oxygen atoms of the four-coordinate silicate units. Indeed, the different electronegativities of Ca^{2+} and H^+ result in different ^{29}Si nuclear shielding, as shown by recent density functional theory calculations[46]. These molecular-level differences in the Q^2 species are shown in the schematic diagram (Fig. 4, inset) of a postulated structure of C-S-H that is consistent with the observed 2D NMR correlations (as well as previous experimental[28,18] and modelling analyses[17,47]). Although the Q^{2L} resonances (the four-coordinate Q^2 silicate units that are positioned away from the interlayer space between two C-S-H chains, as shown in the inset in Fig. 4) are not resolved in the spectrum, the external ridges of the Q^2 correlation spot correspond to correlated intensity between the ^{29}Si SQ signals of the two Q^2 silicate species at -85.4 and -84.8 p.p.m. with the ^{29}Si SQ signals from the Q^{2L} species to which they are, respectively, bound. Within this hypothesis and with the constraint that the DQ frequency must be the sum of the SQ frequencies, two additional correlations can be identified for the Q^2 species at SQ signals -85.4 and -84.8 p.p.m. at DQ signals approximately -168.9 p.p.m. (ii) and -168.1 p.p.m. (iii), respectively, thus establishing the presence of two distinct Q^{2L} species with SQ signals at -83.5 and -83.1 p.p.m. Furthermore, the same ^{29}Si SQ signals at -85.4 and -84.8 p.p.m. from the two Q^2 silicate species are also separately correlated with ^{29}Si signals centred around -79 p.p.m. (iv, v) (DQ $\simeq$ -164 p.p.m.) from Q^1 species, further corroborating that these Q^2 species are indeed chemically distinct. Therefore, analyses of the 2D J-mediated $^{29}Si\{^{29}Si\}$ spectrum establish the occurrence of oligomeric silicate units with two distinct Q^2 and two distinct Q^{2L} species in the C-S-H structure.

The partially resolved pair correlated intensities (ix–xii) in the range of -77 to -80 p.p.m. reveal the presence of different types of Q^1 silicate species associated with at least four distinct dimeric C-S-H units. These results are further corroborated by differences in the spin–spin (T_2) relaxation-time behaviours of the associated ^{29}Si Q^1 species, which were exploited to provide improved ^{29}Si resolution by using one-dimensional (1D) T_2-filtered ^{29}Si MAS measurements (Supplementary Fig. 12). In combination, the different pair correlated intensities establish the presence of dimeric units (*ix–xii*) and C-S-H chains that consist of two distinct Q^1-Q^2 (*iv, v*) and Q^2-Q^{2L} (*ii, iii*) connectivities and at least one Q^2-Q^2 (*i*) connectivity. To accommodate this diversity of atomic connectivity revealed by the 2D ^{29}Si{^{29}Si} NMR measurements, the C-S-H structure must contain a linear chain of at least eight four-coordinated silicate units (that is, an octamer). A similar analysis of pair correlated intensities *vi–viii* indicate the presence of pentameric C-S-H units, as discussed in the Supplementary Note 4. This result is supported by recent studies using density functional theory that have evaluated the relative stabilities of linear C-S-H units of different chain lengths and proposed the presence of stable octameric units[48], for which no direct experimental evidence has previously been available.

The relative populations of ^{29}Si silicate species associated with C-S-H units of different chain lengths (for example, dimers and octamers) are determined based on the enhanced ^{29}Si resolution afforded by the 2D ^{29}Si{^{29}Si} NMR spectrum. Specifically, the single-pulse ^{29}Si MAS spectrum shown in Fig. 5a can be simulated by using the peak positions of ^{29}Si signals as established by the 2D ^{29}Si{^{29}Si} NMR spectrum and the relative fractions of Q^1, Q^2 and Q^{2L} species associated with C-S-H units of different chain lengths (for example, $Q^2/Q^1 = 2$, $Q^2/Q^{2L} = 2$ for octamer as shown in Fig. 5b). Such an analysis yields estimates of 44, 7 and 42% ($\pm$ 4%) for the relative populations of ^{29}Si silicate engaged in octameric, pentameric and dimeric units, respectively. These values correspond to 20 mole% octamers, 5 mole% pentamers and 75 mole% dimers in the C-S-H. The salient result is, thus, that despite the fact that the average chain length is 5, pentamers are actually a minority feature. Such distributions of chain lengths are consistent with previous studies that have reported mean chain lengths for C-S-H, which suggest the presence of pentamers and octamers, in addition to dimers[22,49–51]. It must be understood that the high amount of octamers was obtained here in a relatively short hydration times (1.5 month) compared with what would be required in a usual cement paste. Specifically, the use of pure tricalcium silicate, the high surface area (4.4 m^2 g^{-1}) of the non-hydrated sample and the water-to-solids ratio (0.8) used in this study are expected to result in relatively fast hydration kinetics and a faster precipitation of C-S-H. The end result is a higher extent of hydration and silicate cross-linking. Interestingly, the analysis also indicates that small quantities of monomeric ^{29}Si

silicate species, such as hydroxylated Q^0(h) (5 $\pm$ 1%) and anhydrous Q^0 (2 $\pm$ 1%), are present even after hydration of Ca$_3$SiO$_5$ for 1.5 months at 25 °C. These monomers likely arise from remnants of surface hydroxylation of Ca$_3$SiO$_5$ particles or are components of the C-S-H structure, which is consistent with recent numerical modelling results[47].

Discussion

The carefully synthesized ^{29}Si-enriched sample enables, for the first time, 2D *J*-mediated (through ^{29}Si-O-^{29}Si bonds) ^{29}Si{^{29}Si} NMR measurements that provide detailed insights regarding the different silicate species, their respective site connectivities, and relative populations, especially for previously unidentified discrete silicate moieties in the C-S-H. Consequently, the lengths of C-S-H

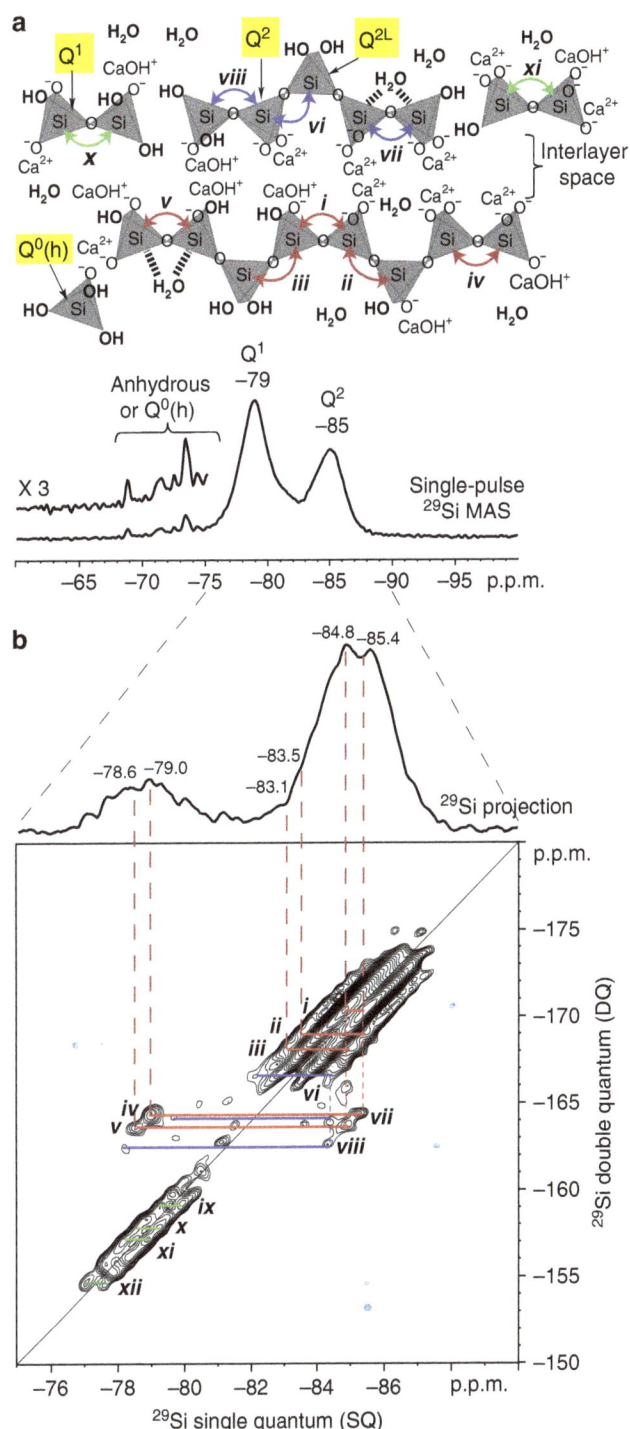

Figure 4 | Molecular structures and silicate site connectivities in partially polymerized calcium-silicate-hydrates. (a,b) Solid-state (**a**) 1D single-pulse ^{29}Si MAS and (**b**) 2D *J*-mediated ^{29}Si{^{29}Si} correlation NMR spectra of hydrated (1.5 month, 25 °C) ^{29}Si-enriched triclinic Ca$_3$SiO$_5$. The lowest contour lines in the 2D spectrum are 9% of the maximum signal intensity. The 'double-quantum' filter used to acquire the spectrum in **b** enables selective detection of pairs of signals (*i, j*) from distinct ^{29}Si nuclei that are covalently bonded. Consequently, the 2D spectrum exhibits intensity correlations between ^{29}Si signals at distinct frequencies (ω_i, ω_j) from ^{29}Si-O-^{29}Si spin pairs (*i, j*) in the horizontal SQ dimension (isotropic ^{29}Si chemical shifts) and at the sum of these frequencies ($\omega_i + \omega_j$) in the vertical DQ dimension. Therefore, correlated intensities at these specific positions in the 2D spectrum unambiguously establish the presence of covalently bonded ^{29}Si silicate species corresponding to the distinct isotropic ^{29}Si chemical shifts. The inset in **a** shows a schematic diagram of the different silicate moieties present in the calcium-silicate-hydrates with double-headed arrows indicating the *J*-interactions in ^{29}Si-O-^{29}Si species that are established by the intensity correlations in the 2D spectrum, specifically from dimeric (green), pentameric (blue) or octameric (red) units. For sake of clarity, the calcium layers are not represented.

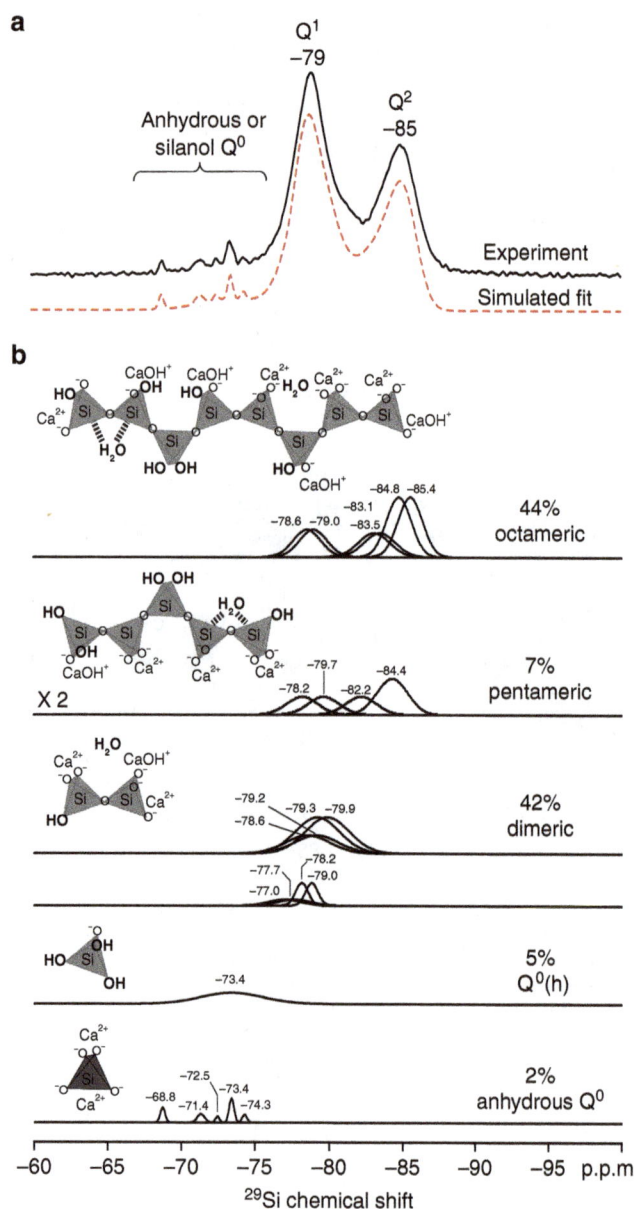

Figure 5 | Relative populations of ^{29}Si silicate species in hydrated triclinic Ca$_3$SiO$_5$. (a) Solid-state 1D single-pulse ^{29}Si MAS spectrum (black) of hydrated (1.5 month, 25 °C) ^{29}Si-enriched Ca$_3$SiO$_5$ and corresponding simulated fit (red) to the spectrum based on the signal decompositions shown in **b**. (**b**) Signal decompositions and relative populations of the different ^{29}Si moieties that comprise anhydrous Q^0, Q^0(h) and octameric, pentameric and dimeric C-S-H units, which contribute to the simulated fit (red) in **a**. Insets in **b** show schematic diagrams of the possible types of C-S-H units and the associated silicate moieties.

chains and the relative populations of associated silicate species are determined, which can be used to evaluate the validity of molecular models for Portland cement hydration that have been previously proposed in the literature[17,21,47]. This opens new perspective for understanding the complex molecular-level mechanical properties of C-S-H.

Solid-state ^{29}Si NMR measurements of ^{29}Si-enriched triclinic Ca$_3$SiO$_5$ also enable the transient silicate speciation and polymerization in the developing C-S-H structure to be monitored and quantified as a function of hydration time, especially during the crucial induction, acceleration and deceleration stages. Importantly, hydroxylated monomeric (Q^0(h)) silicate species can be detected

and quantified by using ^{29}Si{^{1}H} CPMAS NMR measurements to monitor changes in surface composition with the progress of hydration. The NMR results presented here establish that non-hydrated Ca$_3$SiO$_5$ particle surfaces predominantly consist of hydroxylated Q^0 silicate species with negligible quantities of Q^1 and Q^2 hydration products, including for the pre-induction and induction stages of the hydration process. Such detailed insights of silicate-water mixtures have heretofore been challenging and often infeasible to determine by other characterization techniques due to the low absolute quantities, complicated structures and poor long-range order of the hydroxylated surface species. Compared with the induction period, the onset of silicate polymerization (that is, Q^1 and/or Q^2 species) during hydration corresponds to the formation of dimeric units in C-S-H during the acceleration stage, consistent with previous cement literature. Interestingly, during the deceleration stage the hydration rate reduces (at a hydration level of 50%) before any significant reduction of the Q^0(h) populations are observed at the Ca$_3$SiO$_5$ surface. This corresponds to a relatively fast decrease in the reaction rate compared with the rate of reduction of the hydroxylated species available for reaction at the surface, which indicates that part of the surface is likely covered by C-S-H products. These results are consistent with previous studies that suggest that the rate of hydration is controlled by the surface coverage of C-S-H species during the deceleration stage[37]. Calculations based on a shrinking core model (hydration reaction slows down due to consumption of the particles) indicate that for monodispersed spherical particles, a decrease in volume by a factor of 0.5 would be accompanied by a decrease in surface area by a factor 0.63 ($2^{-2/3}$). Ca$_3$SiO$_5$ particles are neither spherical nor monodisperse but the present NMR results are definitely not compatible with a shrinking core model. Consequently, the surface area available for reaction is clearly modified by the surface roughness produced by dissolution driven etching of the surface[38].

The relations directly observed here for the first time between surface passivation and etching phenomena on the one hand and the succession of the induction, acceleration and deceleration stages of hydration of Ca$_3$SiO$_5$ on the other hand, provide new understanding for the occurrence of this complex kinetic behaviour actually observed in a variety of silicate systems. Ca$_3$SiO$_5$, because of its high reactivity, constitutes an interesting model for understanding long term silicate hydration processes occurring during geochemical weathering or hydrothermal synthesis[23].

Methods

NMR spectroscopy. The ^{1}H and ^{29}Si NMR isotropic chemical shifts were referenced to tetramethylsilane using tetrakis(trimethylsilyl)silane [((CH$_3$)$_3$Si$_4$)Si] as a secondary standard[52]. All measurements were performed using zirconia MAS rotors and at room temperature. Solid-state 1D ^{29}Si NMR experiments were carried out using a Bruker Avance-III 500 spectrometer (magnetic field 11.7 T). Magic-angle-spinning (MAS) spectra were measured using a Bruker MAS NMR probe with 4 mm rotors, at spinning frequencies of 7 kHz, and without decoupling. The single-pulse ^{29}Si MAS NMR spectra were acquired with a $\pi/2$ pulse length of 6 μs, a recycle delay of 1,000 or 100 s, and 64 or 16 scans for the ^{29}Si-enriched non-hydrated and hydrated Ca$_3$SiO$_5$ samples, respectively. {^{1}H}^{29}Si CPMAS NMR spectra were recorded using a ^{1}H rf power of 93 kHz, a contact time of 5 ms, and recycle delay of 10 s. The number of scans was 184 for hydrated Ca$_3$SiO$_5$ samples and 2,000 for non-hydrated sample. Hartmann–Hahn matching was ensured by a ramp on the ^{29}Si rf field intensity. 2D {^{1}H}-^{29}Si heteronuclear dipolar correlation (HETCOR) experiments were conducted on a Bruker Avance-700 (16.4 T) spectrometer at ambient temperature, under 4 kHz MAS conditions, with a 7 ms CP contact time, recycle delay of 10 s and 66 t_1 increments of 50 μs each. Solid-state 2D J-mediated ^{29}Si{^{29}Si} DQ correlation NMR experiments were conducted using the refocused-INADEQUATE technique[39] and a 18.8 T Bruker AVANCE-III NMR spectrometer. The experiments were conducted under conditions of 12.5 kHz MAS using a Bruker 3.2 mm H-X double resonance probehead. The 2D ^{29}Si{^{29}Si} spectrum was acquired using a 2.5 μs ^{1}H $\pi/2$ pulse, 3.5 ms contact time for ^{29}Si{^{1}H} CP, 6.0 μs ^{29}Si $\pi/2$ pulses, SPINAL-64 ^{1}H decoupling[53], 152 t_1 increments, an incremental step size of 80 μs, a recycle delay of 2 s and 3,072 scans for each t_1 increment, which corresponds to an experimental time of 260 h (~11 days).

Hydration experiments. Paste for *in situ* NMR measurements was prepared by mixing 0.3 g of non-hydrated ^{29}Si-enriched Ca$_3$SiO$_5$ and 0.24 g of ultrapure water in a cylindrical 2 ml plastic vial for 3 min using a vortex mixer (Analog, VWR) at 2,500 r.p.m. With the help of a syringe and needle, part this paste was introduced as such in the zirconia MAS rotor thus enabling the acquisition of the NMR spectra during the reaction and avoiding any possible microstructural changes caused by the commonly used drying techniques[24]. After 6 h of hydration, the paste was removed from the ZrO$_2$ rotor to prevent its hardening inside the rotor, and the NMR measurements were continued on the part of the sample previously set aside and stored in the closed vial at room temperature. The kinetics of ^{29}Si-enriched Ca$_3$SiO$_5$ hydration were measured by isothermal calorimetry using a TAM Air microcalorimeter at 23 °C. One gram of ^{29}Si-enriched Ca$_3$SiO$_5$ was mixed with 0.8 g of ultrapure water under identical conditions as for samples prepared for NMR measurements. The paste was immediately sealed in a glass ampoule and placed in the isothermal calorimeter. The degree of reaction of ^{29}Si-enriched Ca$_3$SiO$_5$ was calculated by dividing the cumulative heat released at a certain time by the enthalpy of the hydration reaction of Ca$_3$SiO$_5$ (-520 J g^{-1} Ca$_3$SiO$_5$) (refs 54,55). Additional details of synthesis, Ca$_3$SiO$_5$ characterization and NMR quantitative analysis are reported in the Supplementary Methods and Supplementary Note 1.

References

1. Le Chatelier, H. *Recherches Expérimentales sur la Constitution des Mortiers Hydrauliques* (Doctoral thesis, Faculté des Sciences de Paris, 1887).
2. Nugent, M. A., Brantley, S. L., Pantano, C. G. & Maurice, P. A. The influence of natural mineral coatings on feldspar weathering. *Nature* **395**, 588–591 (1998).
3. Casey, W. H., Westrich, H. R., Banfield, J. F., Ferruzzi, G. & Arnord, W. G. Leaching and reconstruction at the surface of dissolving chain-silicate minerals. *Nature* **366**, 253–256 (1993).
4. Conradt, R. Chemical durability of oxide glasses in aqueous solutions. *J. Am. Ceram. Soc.* **91**, 728–735 (2008).
5. Cailleteau, C. *et al.* Insight into silicate-glass corrosion mechanisms. *Nat. Mater.* **7**, 978–983 (2008).
6. Cundy, C. S. & Cox, P. A. The hydrothermal synthesis of zeolites: history and development from the earliest days to the present time. *Chem. Rev.* **103**, 663–702 (2003).
7. Serrano, D. & Van Grieken, R. Heterogenous events in the crystallization of zeolites. *J. Mater. Chem.* **11**, 2391–2407 (2001).
8. Bullard, J. W. *et al.* Mechanisms of cement hydration. *Cement Concrete Res.* **41**, 1208–1223 (2011).
9. Taylor, H. F. W. *Cement Chemistry* (Thomas Telford, 1997).
10. Bullard, J. W. & Flatt, R. J. New insights into the effect of calcium hydroxide precipitation on the kinetics of tricalcium silicate hydration. *J. Am. Ceram. Soc.* **93**, 1894–1903 (2010).
11. Gartner, E. M. & Jennings, H. M. Thermodynamics of calcium silicate hydrates and their solutions. *J. Am. Ceram. Soc.* **70**, 743–749 (1987).
12. Juilland, P., Gallucci, E., Flatt, R. J. & Scrivener, K. S. Dissolution theory applied to the induction period in alite hydration. *Cement Concrete Res.* **40**, 831–844 (2010).
13. Nicoleau, L., Nonat, A. & Perrey, D. The di- and tricalcium silicate dissolutions. *Cement Concrete Res.* **47**, 14–30 (2013).
14. Nicoleau, L., Schreiner, E. & Nonat, A. Ion-specific effects influencing the dissolution of tricalcium silicate. *Cement Concrete Res.* **59**, 118–138 (2014).
15. Lasaga, A. C. & Luttge, A. Variation of crystal dissolution rate based on a dissolution stepwave model. *Science* **291**, 2400–2404 (2001).
16. Arvidson, R. S., Ertan, I. E., Amonette, J. E. & Luttge, A. Variation in calcite dissolution rates: A fundamental problem? *Geochim. Cosmochim. Acta* **67**, 1623–1634 (2003).
17. Richardson, I. G. Tobermorite/jennite- and tobermorite/calcium hydroxide-based models for the structure of C-S-H: applicability to hardened pastes of tricalcium silicate, β-dicalcium silicate, Portland cement, and blends of Portland cement with blast-furnace slag, metakaolin, or silica fume. *Cement Concrete Res.* **34**, 1733–1777 (2004).
18. Allen, A. J., Thomas, J. J. & Jennings, H. M. Composition and density of nanoscale calcium-silicate-hydrate in cement. *Nat. Mater.* **6**, 311–316 (2007).
19. Mishra, R. K., Flatt, R. J. & Heinz, H. Force field for tricalcium silicate and insight into nanoscale properties: cleavage, initial hydration, and adsorption of organic molecules. *J. Phys. Chem. C* **117**, 10417–10432 (2013).
20. Thomas, J. J. *et al.* Modeling and simulation of cement hydration kinetics and microstructure development. *Cement Concrete Res.* **41**, 1257–1278 (2011).
21. Qomi, M. J. A. *et al.* Combinatory molecular optimization of cement hydrates. *Nat. Commun.* **5**, 4960 (2014).
22. Brough, A. R., Dobson, C. M., Richardson, I. G. & Groves, G. W. *In situ* solid-state NMR studies of Ca$_3$SiO$_5$: hydration at room temperature and at elevated temperatures using ^{29}Si enrichment. *J. Mater. Sci.* **29**, 3926–3940 (1994).

23. Fischer, C., Arvidson, R. S. & Lüttge, A. How predictable are dissolution rates of crystalline material? *Geochim. Cosmochim. Acta* **98**, 177–185 (2012).
24. Zhang, J. & Scherer, G. W. Comparison of methods for arresting hydration of cement. *Cement Concrete Res.* **41**, 1024–1036 (2011).
25. Engelhardt, G. & Michel, D. *High-Resolution Solid-State NMR of Silicates and Zeolites* (John Wiley & Sons, 1987).
26. Bellmann, F., Damidot, D., Möser, B. & Skibsted, J. Improved evidence for the existence of an intermediate phase during hydration of tricalcium silicate. *Cement Concrete Res.* **40**, 875–884 (2010).
27. Rawal, A. *et al.* Molecular silicate and aluminate species in anhydrous and hydrated cements. *J. Am. Chem. Soc.* **132**, 7321–7337 (2010).
28. Cong, X. & Kirkpatrick, R. J. 17O MAS NMR investigation of the structure of calcium silicate hydrate gel. *J. Am. Ceram. Soc.* **79**, 1585–1592 (1996).
29. Durgun, E., Manzano, H., Kumar, P. V. & Grossman, J. C. The characterization, stability, and reactivity of synthetic calcium silicate surfaces from first principles. *J. Phys. Chem. C* **118**, 15214–15219 (2014).
30. Garrault, S., Behr, T. & Nonat, A. Formation of the C-S-H layer during early hydration of tricalcium silicate grains with different sizes. *J. Phys. Chem. B* **110**, 270–275 (2006).
31. Kumar, A., Bishnoi, S. & Scrivener, K. L. Modelling early age hydration kinetics of alite. *Cement Concrete Res.* **42**, 903–918 (2012).
32. Gonzalez-Teresa, R., Dolado, J. S., Ayuela, A. & Gimel, J. C. Nanoscale texture development of C-S-H gel: a computational model for nucleation and growth. *Appl. Phys. Lett.* **103**, 234105 (2013).
33. Ioannidou, K., Pellenq, R. J. M. & Del Gado, E. Controlling local packing and growth in calcium-silicate-hydrate gels. *Soft Matter* **10**, 1121–1133 (2014).
34. Bishnoi, S. & Scrivener, K. L. Studying nucleation and growth kinetics of alite hydration using μic. *Cement Concrete Res.* **39**, 849–860 (2009).
35. Tzschichholz, F. & Zanni, H. Global hydration kinetics of tricalcium silicate cement. *Phys. Rev. E* **64**, 016115 (2001).
36. Garrault, S., Finot, E., Lesniewska, E. & Nonat, A. Study of C-S-H growth on C3S surface during its early hydration. *Mater. Struct.* **38**, 435–442 (2005).
37. Bullard, J. W., Scherer, G. W. & Thomas, J. J. Time dependent driving forces and the kinetics of tricalcium silicate hydration. *Cement Concrete Res.* **74**, 26–34 (2015).
38. Nicoleau, L. & Bertolim, M. A. Analytical model for the alite (C3S) dissolution topography. *J. Am. Ceram. Soc.* http://dx.doi.org/10.1111/jace.13647 (2015).
39. Lesage, A., Bardet, M. & Emsley, L. Through-bond carbon-carbon connectivities in disordered solids by NMR. *J. Am. Chem. Soc.* **121**, 10987–10993 (1999).
40. Fyfe, C. A. & Brouwer, D. H. Optimization, standardization, and testing of a new NMR method for the determination of zeolite host-organic guest crystal structures. *J. Am. Chem. Soc.* **128**, 11860–11871 (2006).
41. Cadars, S., Brouwer, D. H. & Chmelka, B. F. Probing local structures of siliceous zeolite frameworks by solid-state NMR and first-principles calculations of ^{29}Si-O-^{29}Si scalar couplings. *Phys. Chem. Chem. Phys.* **11**, 1825–1837 (2009).
42. Köster, T. K.-J. *et al.* Resolving the different silicon clusters in Li$_{12}$Si$_7$ by ^{29}Si and ^{6,7}Li solid-state NMR spectroscopy. *Angew. Chem. Int. Ed. Engl.* **50**, 12591–12594 (2011).
43. Shayib, R. M. *et al.* Structure-directing roles and interactions of fluoride and organocations with siliceous zeolite frameworks. *J. Am. Chem. Soc.* **133**, 18728–18741 (2011).
44. Brunet, F., Bertani, P., Charpentier, T., Nonat, A. & Virlet, J. Application of 29Si homonuclear and 1H-29Si heteronuclear NMR correlation to structural studies of calcium silicate hydrates. *J. Phys. Chem. B* **108**, 15494–15502 (2004).
45. Cadars, S., Lesage, A. & Emsley, L. Chemical shift correlations in disordered solids. *J. Am. Chem. Soc.* **127**, 4466–4476 (2005).
46. Rejmak, P., Dolado, J. S., Stott, M. J. & Ayuela, A. ^{29}Si NMR in cement: a theoretical study on calcium silicate hydrates. *J. Phys. Chem. C* **116**, 9755–9761 (2012).
47. Pellenq, R. *et al.* A realistic molecular model of cement hydrates. *Proc. Natl Acad. Sci. USA* **106**, 16102–16107 (2009).
48. Ayuela, A. *et al.* Silicate chain formation in the nanostructure of cement-based materials. *J. Phys. Chem.* **127**, 164710 (2007).
49. Chen, J. J. *et al.* Solubility and structure of calcium silicate hydrate. *Cement Concrete Res.* **34**, 1499–1519 (2004).
50. Kulik, D. A. Improving the structural consistency of C-S-H solid solution thermodynamic models. *Cement Concrete Res.* **41**, 477–495 (2011).
51. Richardson, I. G. Model structures for C-(A)-S-H (I). *Acta Crystallogr. B* **70**, 903–923 (2014).
52. Hayashi, S. & Hayamizu, K. Chemical shift standards in high-resolution solid-state NMR (1) 13C, 29Si, and 1H nuclei. *Bull. Chem. Soc. Jpn* **64**, 685–687 (1991).
53. Fung, B. M., Khitrin, A. K. & Ermolaev, K. An improved broadband decoupling sequence for liquid crystals and solids. *J. Magn. Reson.* **142**, 97–101 (2000).

54. Thomas, J. J., Jennings, H. M. & Chen, J. J. Influence of nucleation seeding on the hydration mechanisms of tricalcium silicate and cement. *J. Phys. Chem. C* **113,** 4327–4334 (2009).
55. Damidot, D. & Nonat, A. C3S hydration in diluted and stirred suspensions: (I) study of two kinetic steps. *Adv. Cem. Res.* **6,** 27–35 (1994).

Acknowledgements

This research was supported by the Commission for Technology and Innovation (CTI project number 15846.1), the US Federal Highway Administration (FHWA) under agreement No. DTFH61-12-H-00003 and by Halliburton, Inc. (Any opinions, findings and conclusions or recommendations expressed in this publication are ours and do not necessarily reflect the view of the US Federal Highway Administration). Funding was also provided by the program 'Germaine de Staël' and a mobility fellowship from the French Embassy in Bern. We thank Dr R. Verel (ETH Zürich), Dr M Plötze (ETH Zürich) and S. Mantellato (ETH Zürich) for their assistance in the 1D ^{29}Si NMR, XRD and BET surface area measurements, respectively. D. Marchon (ETH Zürich) is also thanked for her support in the development of the synthesis protocol of Ca_3SiO_5. The solid-state 2D *J*-mediated ^{29}Si{^{29}Si} correlation NMR measurements were conducted using the UCSB Materials Research Laboratory (MRL) Shared Experimental Facilities that are supported by the MRSEC Program of the US National Science Foundation under Award No. DMR 1121053; a member of the NSF-funded Materials Research Facilities Network (www.mrfn.org).

Authors contributions

E.P. was the main investigator. She developed the synthesis of the Ca_3SiO_5 samples, designed and carried out the characterization and calorimetry studies. J.-B.d.E.d.L., R.J.F. and M.P. designed the project. B.F.C. proposed the 2D J-mediated and relaxation NMR experiments. E.P., A.S.A., J.-B.d.E.d.L. and R.P.S. performed the NMR experiments. All authors contributed to the analyses of the results and the writing of the manuscript.

Additional information

Competing financial interests: The authors declare no competing financial interests.

26

One-dimensional Magnus-type platinum double salts

Christopher H. Hendon[1,2], Aron Walsh[1,3], Norinobu Akiyama[4,†], Yosuke Konno[4,†], Takashi Kajiwara[5], Tasuku Ito[6], Hiroshi Kitagawa[7] & Ken Sakai[8,9,10]

Interest in platinum-chain complexes arose from their unusual oxidation states and physical properties. Despite their compositional diversity, isolation of crystalline chains has remained challenging. Here we report a simple crystallization technique that yields a series of dimer-based 1D platinum chains. The colour of the Pt^{2+} compounds can be switched between yellow, orange and blue. Spontaneous oxidation in air is used to form black $Pt^{2.33+}$ needles. The loss of one electron per double salt results in a metallic d_{z^2} state, as supported by quantum chemical calculations, and displays conductivity of $11\,S\,cm^{-1}$ at room temperature. This behaviour may open up a new avenue for controllable platinum chemistry.

[1] Department of Chemistry, Centre for Sustainable Chemical Technologies, University of Bath, Claverton Down, Bath BA2 7AY, UK. [2] Department of Chemistry, Massachusetts Institute of Technology, Cambridge, Massachusetts 02139, USA. [3] Department of Materials Science and Engineering, Yonsei University, Seoul 03722, South Korea. [4] Faculty of Science, Department of Applied Chemistry, Science University of Tokyo, Kagurazaka, Shinjuku-ku, Tokyo 162-8601, Japan. [5] Faculty of Science, Department of Chemistry, Nara Women's University, Kitauoyanishi-machi, Nara 630-8506, Japan. [6] Department of Chemistry, Graduate School of Science, Tohoku University, Sendai 980-8578, Japan. [7] Division of Chemistry, Graduate School of Science, Kyoto University, Kitashirakawa-Oiwakecho, Sakyo-ku, Kyoto 606-8502, Japan. [8] Faculty of Science, Department of Chemistry, Kyushu University, Motooka 744, Nishi-ku, Fukuoka 819-0395, Japan. [9] International Institute for Carbon-Neutral Energy Research (WPI-I2CNER), Kyushu University, Motooka 744, Nishi-ku, Fukuoka 819-0395, Japan. [10] Center for Molecular Systems (CMS), Kyushu University, Motooka 744, Nishi-ku, Fukuoka 819-0395, Japan. † Present addresses: Fujisoft Incorporated, Kandaneribeicho 3, Chiyoda-ku, Tokyo 101-0022, Japan (N.A.); Nippon Kayaku CO., Ltd, Sanyo Onoda, Yamaguchi 757-8686, Japan (Y.K.). Correspondence and requests for materials should be addressed to A.W. (email: a.walsh@bath.ac.uk) or to K.S. (email: ksakai@chem.kyushu-univ.jp).

In 1828, Magnus[1] reported a one-dimensional (1D) Pt-chain compound that featured alternating $[PtCl_4]^{2-}$ and $[Pt(NH_3)_4]^{2+}$ building blocks ($d_{Pt-Pt} = 3.25$ Å), Magnus' green salt. However, it was not until the turn of the twentieth century that in-depth studies into the structure and reactivity of similar organometallic Pt chains began[2]. In 1908, the first Pt^{3+}-containing compound was discovered through the treatment of cis-PtCl$_2$(acetonitrile)$_2$ with a silver salt (for example, Ag$_2$SO$_4$)[3]. The material was described as 'platinblau' (platinum blue) because of its characteristic dark blue colour and indirect evidence suggested a polymeric/oligomeric structure involving Pt–Pt interactions[4,5]. Yet, despite the intriguing colour and nearly a century of enthusiastic efforts by chemists following the discovery, both the crystal and electronic structures of platinblau remain an enigma.

In the late 1960s, a series of conductive (0.1 S cm^{-1}) 1D Pt chains composed of either $[Pt(CN)_4]^{2-}$ or $[Pt(ox)_2]^{2-}$ (ox = oxalate; Fig. 1) were developed by Krogmann[6]. The properties of these materials were eventually realized to be a product of partial removal of electrons from the 1D Pt chain[7]. For the tetranuclear Pt-chain compounds (considered as partial structures of Krogmann's salts), the one-electron oxidized species (that is, $Pt_4^{2.25+}$) resulted in a blue compound reminiscent of the platinblau; other Pt-chain compounds with different nuclearity and oxidation states were also observed[8–12]. Then in 2006, Brédas and colleagues[13] revisited Magnus' green salt ($[Pt(NH_3)_4][PtCl_4]$) and showed that the material possesses interesting band properties for conductive applications, owing to the Pt–Pt interactions; later, Drew et al.[14] harnessed similarly highly coloured Pt-chain complexes as selective photochromic sensors.

In a Pt^{2+}-chain complex, the highest occupied state is constructed by an alternating array of antibonding Pt d_z^2 orbitals. Oxidation can result in either delocalization of the missing electron over many sites ($Pt^{[2+\delta]+}$)[15,16] or formal oxidation from $Pt^{2+} \rightarrow Pt^{3+}$ (ref. 17). In the past half century, a variety of highly coloured platinum complexes[18] have been isolated including red[19], orange[20], yellow[21,22], green[23,24] and blue[25] Pt-chain materials. In the case of tetranuclear platinum complexes, these colours arise from a variety of oxidation states including Pt^{2+} (Pt(II)$_4$), $Pt^{2.25+}$ (Pt(II)$_3$Pt(III)), $Pt^{2.5+}$ (Pt(II)$_2$Pt(III)$_2$) and Pt^{3+} (Pt(III)$_4$, although the fully oxidized systems tend to behave as Pt_2^{3+} dimers with two axial donors).

Around the same time as Krogmann's developments, Pt^{2+} complexes found application in medicinal chemistry, propelled by the anticancer treatment cis-platin[26–28]. These works provided an interesting avenue for platinum chemistry and subsequent focus shifted away from the unusual electronic properties associated with Pt-chain complexes towards molecular reactivity, with complexes often featuring relatively simple ligands similar to those shown in Fig. 2. However, fundamental research into ligand design continued alongside this more applied chemistry.

The amidate ligand (Fig. 1) is interesting, because it produces doubly bridged dimeric Pt$_2$ building blocks. A typical formula is $[Pt(II)_2(NH_3)_4(\mu\text{-amidate})_2]^{2+}$ (amidate can also be α-pyridonate, α-pyrrolidonate, 1-methyluracilate, 1-methylthyminate, acetamindate and so on)[8–12]. Tetranuclear complexes arise from a stack of two dimeric units where the dimer–dimer interaction is stabilized by ligand-mediated quadruple hydrogen bonds formed between the O(amidate) and N–H(amine) units. In such cases, the coordination manner of two amidates must be in a head-to-head (HH) arrangement, permitting a stack of N$_2$O$_2$-ligated Pt coordination planes. Moreover, the Pt–Pt bonding interactions within the tetranuclear units are reinforced by shortening the Pt–Pt distances as the oxidation state increases. Both the tetrenuclear and octanuclear species exhibit the characteristic partial oxidation associated with the colourful Pt-chain complexes[29,30].

With the recent progress in conductive metal-organic framework chemistry[31–36], interest in highly conductive hybrid materials has been reignited. Metal-organic materials typically do not have high intrinsic electrical conductivity, because the metals are spatially separated by organic ligands[37]. These 1D Pt chains pose interesting pathways to access highly ordered and conductive metal-organic wires[38]. There has only been one dimer-based conductive 1D Pt family[39]. The material was electrochemically grown in oxidizing conditions and produced a black Pt-chain complex ($[Pt_2(NH_3)_4(\mu\text{-acetate})_2]^{2+}$) that, similar to Krogmann's Pt salts, was susceptible to partial oxidation at the metal centres. This material demonstrated a moderate electrical conductivity of 0.001–0.01 S cm^{-1}, despite its relatively poor crystallinity.

Here we report the formation of a series of Pt^{2+} Magnus-type double salts: 1D Pt^{2+} chain composed of alternating cationic (+/+) and anionic (2-) Pt^{2+} building blocks. Our initial target was a stoichiometric mixture of $[HT\text{-}Pt(II)_2(bpy)_2(\mu\text{-pivalamidate})_2]^{2+}$ (1, HT indicates a head-to-tail arrangement of the pivalamidate ligands) and $[Pt(II)(ox)_2]^{2-}$ (2); however, the result is a new family of compounds. We detail the crystallization procedure, followed by characterization of their crystal structure and physical properties including colour and electrical conductivity.

Results

Crystal growth and characterization. We developed a simple kinetically controlled crystallization procedure to provide access to hitherto unknown Pt-chain systems. The petri-paper three-zone crystallization method (Fig. 2) isolates the reagents, located in zones 1 and 3, from the crystallization region, zone 2. In practice, this is achieved by using filter paper separating the zones, thereby slowing the rate of salt diffusion. Aqueous 1 and 2 were added to zones 1 and 3, respectively. After 2 days, bright yellow crystals of $[HT\text{-}Pt(II)_2(bpy)_2(\mu\text{-pivalamidate})_2][Pt(II)(ox)_2]$ (3), were isolated and were found to crystallize in a polymeric Pt^{2+}-chain complex (Fig. 3a).

As observed for the monomeric 1 (that is, the diplatinum cation $[HT\text{-}Pt(II)_2(bpy)_2(\mu\text{-pivalamidate})_2]^{2+}$)[40] (Supplementary Fig. 1), the yellow crystalline material 3 undergoes the HT → HH isomerization over 24 h through the gradual dissolution of 3 to form the compositionally identical 4

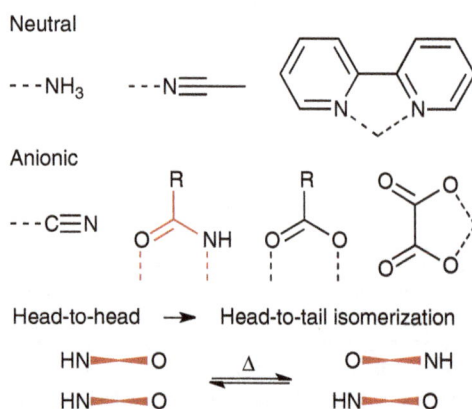

Neutral

Anionic

Head-to-head → Head-to-tail isomerization

Figure 1 | Ligands commonly employed in Pt-chain chemistry: Examples of familiar neutral and anionic ligands found in Pt-chain complexes. The bridging amidate ligand, shown in red, can form hydrogen bonds in the direction of the Pt chain, stabilizing the Pt centres and promoting oxidation.

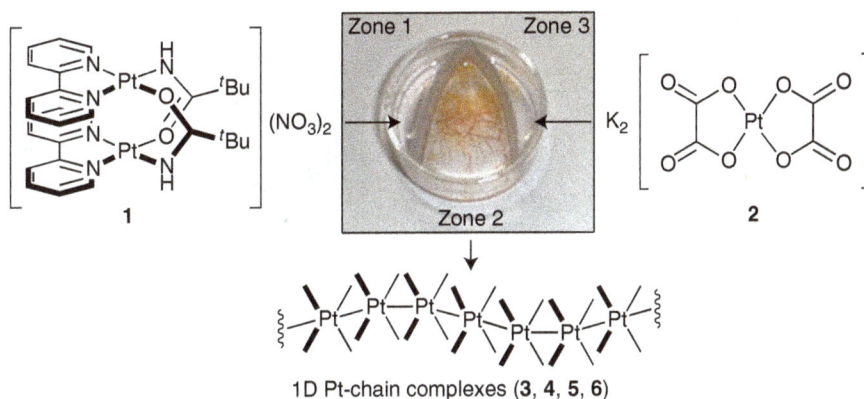

Figure 2 | The petri-paper three-zone crystallization procedure. Aqueous solutions of **1** and **2** are added to zones 1 and 3, respectively. Slow diffusion through the filter paper into the crystallization zone 2 yields large and highly crystalline **3**, with **4**, **5** and **6** resolved over time.

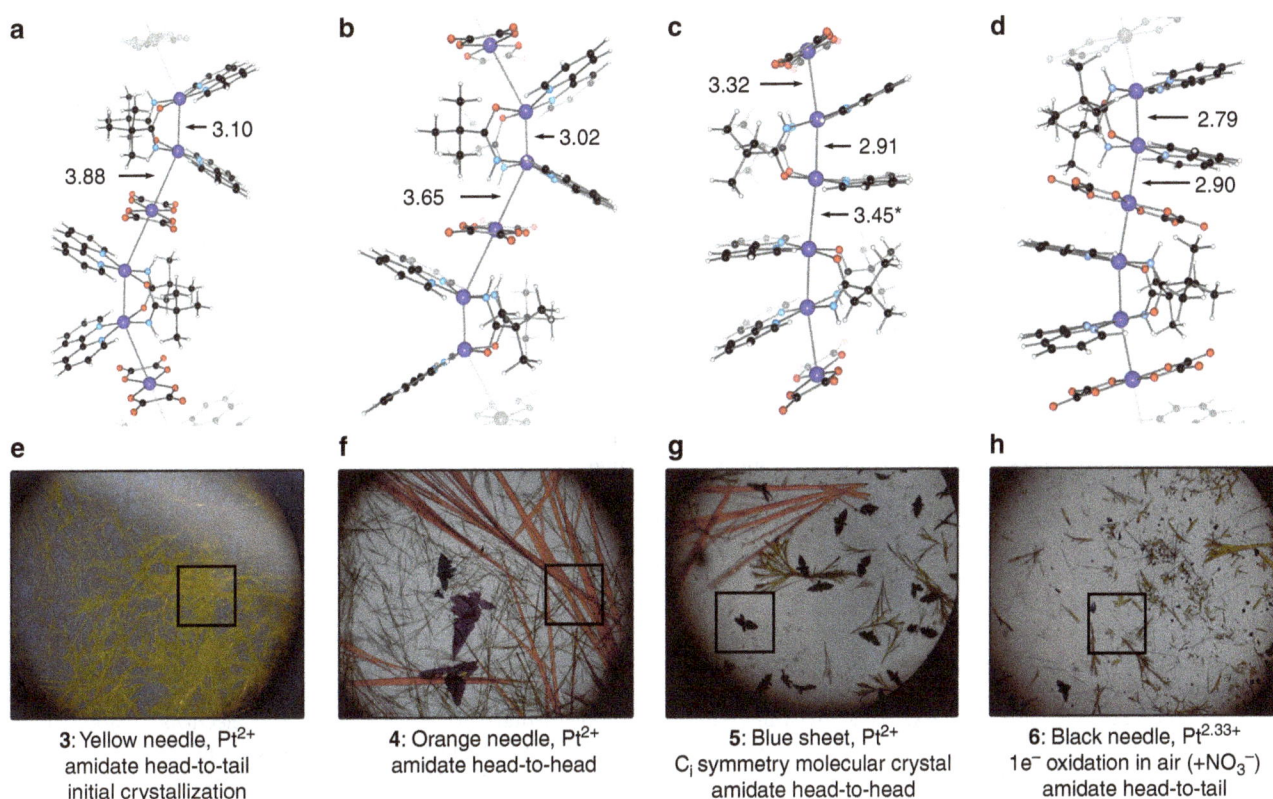

3: Yellow needle, Pt^{2+}
amidate head-to-tail
initial crystallization

4: Orange needle, Pt^{2+}
amidate head-to-head

5: Blue sheet, Pt^{2+}
C$_i$ symmetry molecular crystal
amidate head-to-head

6: Black needle, Pt$^{2.33+}$
1e$^-$ oxidation in air (+NO$_3^-$)
amidate head-to-tail

Figure 3 | Products isolated from the crystallization method: Sequential crystallization of Pt double salts. Structure **3**, shown in **a**, is a yellow needle (**e**) polymeric material. Over 24 h, **3** undergoes a HT→HH isomerization forming **4**, shown in **b**, orange needles. (**f**) Simultaneously, a discrete hexamer (**c**) with HH isomerism forms, **5** shown in **g**. Another purple sheet-like material is formed, shown by optical microscopy in **f**; it appears to be analogous to **5** with a different stacking manner or with different isomerism of the dimer unit, but the crystallinity proved insufficient for characterization. After exposure to air over several weeks, partially oxidized polymeric black needles of **6** with HT isomerism are formed (**d,h**). No HH mixed-valence compounds have been isolated. The Pt–Pt separations are in units of Å. In the case of **3**, **4** and **5**, H$_2$O is omitted for clarity. Both H$_2$O and NO$_3^-$ are omitted from the presentation of **6**.

([HH-Pt(II)$_2$(bpy)$_2$(μ-pivalamidate)$_2$][Pt(II)(ox)$_2$] · 5.5H$_2$O) (bright orange needles shown in Fig. 3b). Relative to **3**, the pivalamidate-bridged Pt–Pt distances decrease by ∼0.1 Å, which is consistent with the preference of HT-Pt(II)$_2$ dimers for a bridged Pt–Pt distance longer than the corresponding HH-Pt(II)$_2$ dimers, probably due to higher electrostatic repulsion given within a symmetric dimeric structure having two N$_3$O-ligated Pt geometries[24].

Remembering that the highest occupied states of Pt^{2+}-chain complexes are $\sigma^*(d_{z^2} - d_{z^2})$[41], the colours associated with Pt compounds are principally correlated with the lowest-energy optical transition, corresponding to the so-called metal-metal-to-ligand charge transfer transition. Compound **4** has shorter Pt–Pt bonds, leading to greater destabilization of the σ^*, and hence to the red shift in transition energy. In addition, the adjacent bipyridine ligands interact, lowering the lowest unoccupied molecular orbital energy proportional to inter-Pt distance (resulting in a red shift with HT→HH isomerization).

During the pivalamidate HT→HH isomerization and subsequent crystallization of **4**, a structurally dissimilar blue

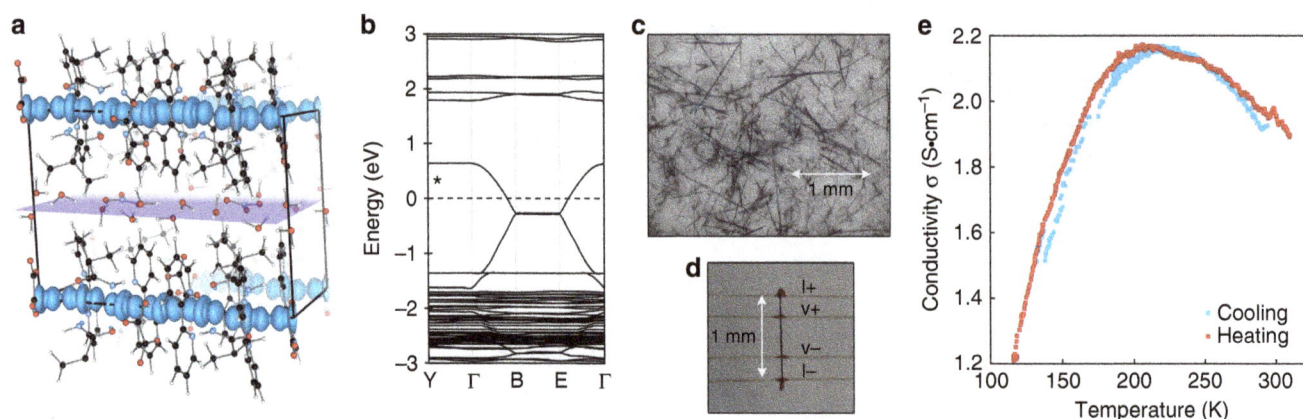

Figure 4 | Electronic properties of compound 6. [HT-Pt$_2$(bpy)$_2$(μ-pivalamidate)$_2$][Pt(ox)$_2$](NO$_3$)·7H$_2$O, **6**, where the calculated spin density (blue isosurface at 0.02 e/Å^3) is delocalized along the Pt chain in the <001> direction. The hydration layer solvates (NO$_3^-$)$_2$ in [200], shown as purple plane (**a**). The electronic band structure (**b**) depicts the metallic character in the direction of the delocalized Pt$^{2.33+}$ chain. The addition of 2e$^-$ (denoted by *) increases the Fermi level (dotted line) to fill the valence band, resulting in a semiconducting band gap analogous to **3**. The target synthesis of **6** resulted in large black needles (**c**), which could be mounted directly for conductivity measurements (**d**). Using gold paste contacts provided a peak conductivity of 11 S cm^{-1}, whereas the carbon paste used in the temperature sweeping measurement (**e**) produced a peak conductivity of 2.2 S cm^{-1} at ~210 K. O, H, N, C and Pt are depicted in red, white, blue, black and grey, respectively.

material simultaneously formed (shown in the optical microscope image; Fig. 3c and Supplementary Fig. 13). This structure was determined to be discrete hexamers of [HH-Pt(II)$_2$(bpy)$_2$ (μ-pivalamidate)$_2$]$_2$[Pt(II)(ox)$_2$]$_2$·8H$_2$O (**5**). From the immense amount of work on Pt-blue complexes, we assumed that this was a mixed-valence material that had partially oxidized in the presence of air[42]. However, the formulation of **5**, clearly solved by crystallography as a pair of **1** and **2**, allows us to conclude that **5** is a hexameric Pt^{2+} species with no paramagnetic character (confirmed by magnetic susceptibility measurements and a lack of electron spin resonance (ESR) response). Compound **5** is a rare example of a Pt^{2+}-blue compound and can be envisaged as the dimerization of two trimeric species. The resultant hexamer possesses a crystallographic inversion centre at which a stack of N$_2$O$_2$-coordinated Pt planes is realized in the same manner as established in various tetranuclear Pt-blue-related species.

The terminal Pt–Pt bonding observed in **5** provides one potential explanation for the blue chromophore (making **5** appear similar to the Pt$_4^{2.25+}$ blue species). The terminal Pt–Pt associations in the hexamer exhibit exceptionally shorter inter-Pt distances than any other inter-monomer–dimer associations. This indicates that dative bonding from the filled σ^* towards part of the vacant molecular orbitals (such as the Pt 6p$_z$ in [Pt(ox)$_2$]$^{2-}$) is dominant, leading to manifestation of net partial oxidation of the inner Pt$_4$ geometry.

Crystallization of Magnus-type double salts continued to evolve over 14 days: **3**, **4** and **5** dissolved and recrystallized into the terminal material, fine black needles, **6** (shown in Fig. 3d). Single crystal X-ray diffraction (XRD) confirmed the formation of [HT-Pt$_2$(bpy)$_2$(μ-pivalamidate)$_2$][Pt(ox)$_2$](NO$_3$)·7H$_2$O. This stoichiometry represents a 1e$^-$ partial oxidation per double salt in the presence of air (that is, delocalized Pt$^{2.33+}$). Compound **6** is structurally analogous to **3**, with the addition of charge balancing NO$_3^-$ solvated in the [200] disordered aqueous layer, as shown by the purple plane in Fig. 4a. Dramatic shortening occurs in both the bridged and non-bridged Pt–Pt distances, reflecting partial oxidation at all Pt centres including those ligated with oxalates. Only the HT was observed, in part due to crystal-lographic requirement: a crystallographic twofold axis is passing through the midpoint of the Pt–Pt bond within the dimeric unit, that is, this dimeric unit is considered to possess a

crystallographic HT isomerism. The HH analogue would result in anisotropic localization of the electrons (such as the blue species, **5** with discrete Pt centres). Following the classification of Robin and Day[43], these are class IIIB mixed-valence compounds with highly delocalized electrons.

Discussion

Unlike the colourful Pt-chain complexes, black compounds are interesting for electronic applications, owing to their potentially high conductivity and strong optical absorption. Before this, there has only been one Pt-black chain complex reported[39]. Quantum chemical calculations were used to elucidate the electronic structure of **6**. The unpaired valence electrons are delocalized along the Pt chain (Fig. 4a) in agreement with its diamagnetic character evidenced by electron spin resonance spectroscopy. Furthermore, the homogeneous charge distribution about the Pt nuclei is in agreement with X-ray photoemission spectroscopy measurements (Supplementary Figs 10 and 11d, and Supplementary Table 5). From the electronic band structure (Fig. 4b), we deduce that **6** is metallic with a partially occupied band along the Pt$^{2.33+}$ chain. In the reduced state (that is, electronically similar to compound **3**), the addition of two electrons per crystallographic unit cell (that is, one electron per Magnus stack) would increase the Fermi level (dotted line) to fill the valence band, forming a material with a finite band gap.

For electrical transport characterization, single crystal conductivity measurements were performed. The crystallinity and yield of **6**, as formed from the sequential crystallization, did not result in large enough crystals to perform such measurements. A revised target synthesis was designed, promoting partial oxidation of **2** to motivate the direct formation and isolation of **6** (yield = ca. 30%, >1 mm black needles; Fig. 4c). The conductivity was measured by mounting a single crystal of **6** (Fig. 4d) on four electrodes. The crystal was carefully attached using gold paste and a champion conductivity of 11 S cm^{-1} was obtained at room temperature. To investigate the temperature dependence of the electrical transport gold paste could not be used, because it resulted in cracking of crystal with sweeping temperature. Carbon paste results in a decrease in maximum conductivity to 2.2 S cm^{-1}, but the measurements identified a metal-semiconductor transition at ~210 K (Fig. 4e).

Our study has provided several advances in the synthetic chemistry and property control of Pt salts. First, we demonstrated that controlled crystallization using the petri-paper procedure provides a cheap and accessible method for the recovery of novel structures. Identification of a blue Pt^{2+} compound suggests that the elusive platinblau may not require non-integral oxidation states. We further showed that oxidation of this double salt to $Pt^{2.33+}$ results in a black highly conductive material, as supported by *ab initio* calculations that revealed a metallic state with delocalized electrons at the Fermi level. This provides an important design principle for developing conductive metal-organic networks, where redox reactions can be used to install conductivity postsynthetically. Our results highlight several important questions including whether we can harness the $Pt^{2+} \rightarrow Pt^{3+}$ oxidation for catalysis and whether these double salts can provide an alternative to molecular analogues for Pt^{2+}-based medicines? We anticipate that these findings will stimulate further interest in fundamental platinum chemistry.

Methods

Synthetic approach. [HT-Pt(II)$_2$(bpy)$_2$(μ-pivalamidate)$_2$](NO$_3$)$_2 \cdot$ 5H$_2$O, **1**, was prepared as previously described by Yokokawa and Sakai[40], and K$_2$[Pt(ox)$_2$] $\cdot$ 2H$_2$O, **2**, was prepared as previously described by Werner and Grebe[2]. Using the petri-paper three-zone method, aqueous solutions of **1** and **2** were added to zones 1 and 3, respectively. The crystals were then developed unsealed (that is, in air), at room temperature, within a few days, sequentially crystallizing as [HT-Pt(II)$_2$(bpy)$_2$(μ-pivalamidate)$_2$][Pt(II)(ox)$_2$] (**3**, yellow needles), [HH-Pt(II)$_2$(bpy)$_2$(μ-pivalamidate)$_2$][Pt(II)(ox)$_2$] $\cdot$ 5.5H$_2$O (**4**, orange needles) and [HH-Pt(II)$_2$(bpy)$_2$(μ-pivalamidate)$_2$]$_2$ [Pt(II)(ox)$_2$]$_2 \cdot$ 8H$_2$O (**5**, dark blue plates, hexaplatinum). Representative crystals were removed from the reaction mixture and physical property measurements were performed on them. Leaving of the mixture for 1–2 weeks led to the formation of [HT-Pt$_2$(bpy)$_2$ (μ-pivalamidate)$_2$][Pt(ox)$_2$](NO$_3$) $\cdot$ 7H$_2$O (**6**, fine black needles), a partially oxidized species. Although this sequential procedure resulted in very low yeids, a refined target synthesis was devised. The improved synthetic strategy involved the following: (i) oxidation by persulfate, (ii) ageing of crystals at higher temperature and (iii) manual separation of well-formed needles from the relatively small crystals deposited simultaneously. In this method, a solution of **2** (0.04 mmol) and K$_2$S$_2$O$_8$ (0.02 mmol) in water (12 ml) was heated at 65 °C for 1 h, followed by leaving it at room temperature for 1 h. To the solution, **1** (0.04 mmol, as solid) was added and the temperature was gradually raised to 65 °C over 1 h without stirring. The mixture was further left at 65 °C for 6 h to grow well-formed black metallic needles of **6**, crystalographically identical to the fine needles obtained in the other procedure. Finally, the solution was very slowly cooled down to room temperature over 6 h. As the crystals obtained possessed relatively wide distribution in size, only the needles longer than 180 μm were collected through tedious manual sieving. The resulting crystals were then washed with a minimum amount of water and dried in air (yield: *ca.* 30 %). Full characterization details, including single-crystal XRD, ^{195}Pt NMR, X-ray photoemission spectroscopy, thermogravimetric analysis (TGA), solid-state ultraviolet–visible and conductivity measurements are described in the Supplementary Figs 1–16, Supplemental Tables 1–5, Supplemental Notes 1–4 and Supplementary Data 1–4.

Computational approach. All calculations were performed within the Kohn–Sham density functional theory framework using periodic boundary conditions, to approximate the infinite salts, starting from the crystallographic unit cells determined from the XRD measurements. The Vienna *ab initio* simulation package[44], a planewave code (with PAW scalar relativistic core potentials)[45], was employed for all geometry optimizations and electronic calculations. For each system, the lattice vectors and internal atomic positions were relaxed with the semi-local Perdew–Burke–Ernzerhof exchange-correlation functional revised for solids[46]. For these calculations, a 500-eV plane-wave cutoff basis set was found to be suitable for convergence of the systems to within 0.01 eV per atom. To provide quantitative electronic information, non-local hybrid density functional theory calculations were performed using the HSE06 functional[47], with 25% of the short-range semi-local exchange functional replaced by the exact non-local Hartree–Fock exchange. Visualizations of the structures and orbitals were made using the codes VESTA[48].

References

1. Magnus, G. Ueber einige verbindungen des platinchlorrs. *Ann. Phys.* **90**, 239–242 (1828).
2. Werner, A. & Grebe, E. Beitrag zur konstitution anorganischer verbindungen. XIX. Mitteilung. Platinoxalatoverbindungen. *Z. Anorg. Chem.* **21**, 377–388 (1899).
3. Hofmann, K. A. & Bugge, G. Platinblau. *Ber. Dtsch. Chem. Ges.* **41**, 312–314 (1908).
4. Morgan, G. T. & Burstall, F. H. 201. Researches on residual affinity and coordination. Part XXXIV. 2:2'-Dipyridyl platinum salts. *J. Chem. Soc.* 965–971 (1934).
5. Gillard, R. D. & Wilkinson, G. Platinum blue and related compounds. *J. Chem. Soc.* 2835–2837 (1964).
6. Krogmann, K. Planar complexes containing metal-metal bonds. *Angew. Chem. Int. Ed.* **8**, 35–42 (1969).
7. Miller, J. S. *Extended Linear Chain Compounds* Vol. 3 (Springer Science & Business Media, 2012).
8. Barton, J. K., Best, S. A., Lippard, S. J. & Walton, R. A. Relationship of *cis*-diammineplatinum α-pyridone blue to other platinum blues. An X-ray photoelectron study. *J. Am. Chem. Soc.* **100**, 3785–3788 (1978).
9. Barton, J. K., Caravana, C. & Lippard, S. J. Chemical and spectroscopic characterization of *cis*-diammineplatinum α-pyridone blue in aqueous solution. comparison with other platinum blues. *J. Am. Chem. Soc.* **101**, 7269–7277 (1979).
10. Matsumoto, K. & Sakai, K. Structures and reactivities of platinum-blues and the related amidate-bridged platinum^{III} compounds. *Adv. Inorg. Chem.* **49**, 375–427 (1999).
11. Lippert, B. Impact of cisplatin on the recent development of Pt coordination chemistry: a case study. *Coord. Chem. Rev.* **182**, 263–295 (1999).
12. Zangrando, E., Pichierri, F., Randaccio, L. & Lippert, B. Structural aspects of Pt complexes containing model nucleobases. *Coord. Chem. Rev.* **156**, 275–332 (1996).
13. Kim, E. G. *et al.* Magnus' green salt revisited: impact of platinum-platinum interactions on electronic structure and carrier mobilities. *Adv. Mater.* **18**, 2039–2043 (2006).
14. Drew, S. M., Smith, L. I., McGee, K. A. & Mann, K. R. A platinum(II) extended linear chain material that selectively uptakes benzene. *Chem. Mater.* **21**, 3117–3124 (2009).
15. Barton, J. K., Rabinowitz, H. N., Szalda, D. J. & Lippard, S. J. Synthesis and crystal structure of *cis*-diammineplatinum α-pyridone blue. *J. Am. Chem. Soc.* **99**, 2827–2829 (1977).
16. Barton, J. K., Szalda, D. J., Rabinowitz, H. N., Waszczak, J. V. & Lippard, S. J. Solid state structure, magnetic susceptibility, and single crystal ESR properties of *cis*-diammineplatinum α-pyridone blue. *J. Am. Chem. Soc.* **101**, 1434–1441 (1979).
17. Uson, R. *et al.* The first mononuclear Pt^{III} complex. Molecular structures of (NBu$_4$)[Pt^{III}(C$_6$Cl$_5$)$_4$] and of its parent compound {NBu$_4$}$_2$[Pt^{II}(C$_6$Cl$_5$)$_4$].2CH$_2$Cl$_2$. *J. Chem. Soc. Chem. Commun.* 751–752 (1984).
18. Yan, X., Cook, T. R., Wang, P., Huang, F. & Stang, P. J. Highly emissive platinum(II) metallacages. *Nat. Chem.* **7**, 342–348 (2015).
19. Matsumoto, K. & Fuwa, K. *cis*-Diammineplatinum α-pyrrolidone tan, a structural analog of platinum blues. *J. Am. Chem. Soc.* **104**, 897–898 (1982).
20. Lippert, B., Schöllhorn, H. & Thewalt, U. Additive *trans*-influences of the axial ligand and metal-metal bond in diplatinum^{III} complex leading to an asymmetric structure with penta- and hexacoordination of the two metals. *J. Am. Chem. Soc.* **108**, 525–526 (1986).
21. Hollis, L. S. & Lippard, S. J. Synthesis, structure, and platinum-195 NMR studies of binuclear complexes of *cis*-diammineplatinum^{II} with bridging α-pyridonate ligands. *J. Am. Chem. Soc.* **105**, 3494–3503 (1983).
22. Matsumoto, K., Miyamae, H. & Moriyama, H. Crystal structure and carbon-13 and platinum-195 NMR spectra of an α-pyrrolidonate-bridged binuclear platinum^{II} complex, [Pt$_2$(NH$_3$)$_4$(C$_4$H$_6$NO)$_2$]$_2$(PF$_6$)$_3$(NO$_3$) $\cdot$ H$_2$O. *Inorg. Chem.* **28**, 2959–2964 (1989).
23. Mehran, F. & Scott, B. EPR evidence for metal-insulator transition in the "one-dimensional" platinum complex K$_2$[Pt(CN)$_4$]Br$_{1/3} \cdot$ 3H$_2$O. *Phys. Rev. Lett.* **31**, 1347 (1973).
24. Sakai, K. *et al.* New structural aspects of α-pyrrolidinonate- and α-pyridonate-bridged, homo- and mixed-valence, di- and tetranuclear *cis*-diammineplatinum complexes: eight new crystal structures, stoichiometric 1:1 mixture of Pt(2.25 +)$_4$ and Pt(2.5 +)$_4$, new quasi-one-dimensional halide-bridged [Pt(2.5 +)$_4$-Cl...]$_\infty$ system, and consideration of solution properties. *J. Am. Chem. Soc.* **120**, 8366–8379 (1998).
25. Ginsberg, A. P., O'Halloran, T. V., Fanwick, P. E., Hollis, L. S. & Lippard, S. J. Electronic structure and optical spectrum of *cis*-diammineplatinum α-pyridone blue: metal-metal bonding and charge transfer in a four-atom Pt(2.25 +) chain. *J. Am. Chem. Soc.* **106**, 5430–5439 (1984).

26. Rosenberg, B., Van Camp, L. & Krigas, T. Inhibition of cell division in *Escherichia coli* by electrolysis products from a platinum electrode. *Nature* **205,** 698–699 (1965).

27. Rosenberg, B. & Van Camp, L. Platinum compounds: a new class of potent antitumour agents. *Nature* **222,** 385–386 (1969).

28. Rosenberg, B. & Van Camp, L. The successful regression of large solid sarcoma 180 tumors by platinum compounds. *Cancer Res.* **30,** 1799–1802 (1970).

29. O'Halloran, T. V., Mascharak, P. K., Williams, I. D., Roberts, M. M. & Lippard, S. J. Synthesis, structure determination, and electronic structure characterization of two mixed-valence tetranuclear platinum blues with bridging α-pyridonate or 1-methyluracilate ligands. *Inorg. Chem.* **26,** 1261–1270 (1987).

30. Sakai, K. & Matsumoto, K. Mixed-valent octanuclear platinum acetamide complex, $[Pt_8(NH_3)_{16}(C_2H_4NO)^8]^{10+}$. *J. Am. Chem. Soc.* **111,** 3074–3075 (1989).

31. Allendorf, M. D., Schwartzberg, A., Stavila, V. & Talin, A. A. A roadmap to implementing metal-organic frameworks in electronic devices: challenges and critical directions. *Chem. Eur. J.* **17,** 11372–11388 (2011).

32. Hendon, C. H., Tiana, D. & Walsh, A. Conductive metal-organic frameworks and networks: fact or fantasy? *Phys. Chem. Chem. Phys.* **14,** 13120–13132 (2012).

33. Talin, A. A. *et al.* Tunable electrical conductivity in metal-organic framework thin-film devices. *Science* **343,** 66–69 (2014).

34. Butler, K. T., Hendon, C. H. & Walsh, A. Electronic chemical potentials of porous metal-organic frameworks. *J. Am. Chem. Soc.* **136,** 2703–2706 (2014).

35. Sun, L., Hendon, C. H., Minier, M. A., Walsh, A. & Dincă, M. Million-fold electrical conductivity enhancement in $Fe_2(DEBDC)$ versus $Mn_2(DEBDC)$ (E = S, O). *J. Am. Chem. Soc.* **137,** 6164–6167 (2015).

36. Park, S. S. *et al.* Cation-dependent intrinsic electrical conductivity in isostructural tetrathiafulvalene-based microporous metal-organic frameworks. *J. Am. Chem. Soc.* **137,** 1774–1777 (2015).

37. Hendon, C. H. & Walsh, A. Chemical principles underpinning the performance of the metal-organic framework HKUST-1. *Chem. Sci.* **6,** 3674–3683 (2015).

38. Tiana, D., Hendon, C. H. & Walsh, A. Ligand design for long-range magnetic order in metal-organic frameworks. *Chem. Commun.* **50,** 13990–13993 (2014).

39. Sakai, K., Ishigami, E., Konno, Y., Kajiwara, T. & Ito, T. New partially oxidized 1-D platinum chain complexes consisting of carboxylate-bridged *cis*-diammineplatinum dimer cations. *J. Am. Chem. Soc.* **124,** 12088–12089 (2002).

40. Yokokawa, K. & Sakai, K. Di-μ-pivalamidato-κ^4 *N*: *O*; *O*; *N*-bis[(2,2'-bipy-ridine-κ^2 *N*, *N*')(sulfato-κ*O*)platinum(III)] tetrahydrate in a head-to-tail isomerism. *Acta Crystallogr. C* **60,** m244–m247 (2004).

41. Whangbo, M.-H. & Hoffmann, R. The band structure of the tetracyanoplatinate chain. *J. Am. Chem. Soc.* **100,** 6093–6098 (1978).

42. Sakai, K., Tsubomura, T. & Matsumoto, K. Reaction of an α-pyrrolidonate-bridged *cis*-diammineplatinum^{II} dimer with molecular oxygen and its Application to the catalytic O_2 oxidation of hydroquinone. *Inorg. Chim. Acta* **234,** 157–161 (1995).

43. Robin, M. B. & Day, P. Mixed valence chemistry. A survey and classification. *Adv. Inorg. Chem. Radiochem.* **10,** 247–422 (1967).

44. Kresse, G. & Furthmüller, J. Efficient iterative schemes for *ab initio* total-energy calculations using a plane-wave basis set. *Phys. Rev. B* **54,** 11169 (1996).

45. Blöchl, P. Projector augmented-wave method. *Phys. Rev. B* **50,** 17953 (1994).

46. Perdew, J. P. *et al.* Restoring the density-gradient expansion for exchange in solids and surfaces. *Phys. Rev. Lett.* **100,** 136406 (2008).

47. Heyd, J., Scuseria, G. E. & Ernzerhof, M. Hybrid functionals based on a screened coulomb potential. *J. Chem. Phys.* **118,** 8207 (2003).

48. Momma, K. & Izumi, F. *VESTA3* for three-dimensional visualization of crystal, volumetric and morphology data. *J. Appl. Cryst.* **44,** 1272–1276 (2011).

Acknowledgements

This work was supported by a Grant-in-Aid for Scientific Research on Priority Areas ('Metal-assembled Complexes') from the Ministry of Education, Culture, Sports, Science and Technology of Japan. This was partly supported by the International Institute for Carbon Neutral Energy Research (WPI-I2CNER) sponsored by the World Premier International Research Center Initiative (WPI), MEXT, Japan. Computations benefited from access to the High Performance Computing Consortium, which is funded by EPSRC Grant EP/L000202. Additional support has been received from the Royal Society and the ERC (Grant 277757) and the NSF-funded XSEDE facilities (Grant ACI-1053575).

Author contributions

All compounds were prepared by Y.K. and K.S. The petri-paper method was invented by Y.K. and K.S.. Improved synthesis of compound **6** was developed by N.A. T.I., T.K. and K.S. performed the crystallographic measurements and structure refinement. Electrical conductivity measurements were carried out by N.A., K.S. and H.K.. C.H.H. and A.W. performed the density functional theory calculations. C.H.H., A.W. and K.S. wrote the paper.

Additional information

27

1,3,2,5-Diazadiborinine featuring nucleophilic and electrophilic boron centres

Di Wu[1], Lingbing Kong[1], Yongxin Li[2], Rakesh Ganguly[2] & Rei Kinjo[1]

The seminal discovery in 1865 by Kekulé that benzene nucleus exists with cyclic skeleton is considered to be the beginning of aromatic chemistry. Since then, a myriad of cyclic molecules displaying aromatic property have been synthesized. Meanwhile, borazine ($B_3N_3H_6$), despite the isostructural and isoelectronic relationships with benzene, exhibits little aromaticity. Herein, we report the synthesis of a 1,3,2,5-diazadiborinine ($B_2C_2N_2R_6$) derivative, a hybrid inorganic/organic benzene, and we present experimental and computational evidence for its aromaticity. In marked contrast to the reactivity of benzene, borazine, and even azaborinines previously reported, 1,3,2,5-diazadiborinine readily forms the adducts with methyl trifluoromethanesulfonate and phenylacetylene without any catalysts. Moreover, 1,3,2,5-diazadiborine activates carbon dioxide giving rise to a bicycle[2,2,2] product, and the binding process was found to be reversible. These results, thus, demonstrate that 1,3,2,5-diazadiborinine features both nucleophilic and electrophilic boron centres, with a formal B(+ I)/B(+ III) mixed valence system, in the aromatic six-membered $B_2C_2N_2$ ring.

[1] Division of Chemistry and Biological Chemistry, Nanyang Technological University, 21 Nanyang Link, 637371 Singapore, Singapore. [2] NTU-CBC Crystallography Facility, Nanyang Technological University, 21 Nanyang Link, 637371 Singapore, Singapore. Correspondence and requests for materials should be addressed to R.K. (email: rkinjo@ntu.edu.sg).

The concept of aromaticity has been of paramount importance in myriad fields of chemistry since the discovery in 1865 that benzene nucleus is cyclic[1]. Almost a century after the first identification of benzene by Faraday[2], borazine ($B_3N_3H_6$), also referred as inorganic benzene, was prepared[3]. Despite the isoelectronic and isosterism relationships between the $C=C$ and B–N units, however, $B_3N_3H_6$ displays a different electronic property from that of benzene[4,5], which is due to polarization of the B–N units arising from the variation of electronegativity between boron and nitrogen atoms. Thus, introduction of B–N units into the aromatic skeleton of isoelectronic organic counterparts leads to unique electronic structures, which indicate the potential to expand the diversity of aromatic molecules[6–8].

In 1958, Dewar and co-workers first reported the preparation of a polycyclic azaborinine[9]. Since then, the synthesis and full characterization of various mono-cyclic and ring-fused polycyclic azaborinine (BNC_4) derivatives, involving 1,2-azborinines, 1,3-azaborinines and 1,4-azaborinines, have been achieved[10]. Meanwhile, only a few diazadiborinine ($B_2N_2C_2$) derivatives have been structurally characterized so far[11,12]. Thermal stability as well as aromatic nature of both azaborinine and diazadiborinine more closely resembles those of benzene than $B_3N_3H_6$. Nevertheless, except for η^6-complexation with a chromium and ruthenium metals, the reactivities of the boron centre in these compounds are mainly associated with nucleophilic substitutions, which is in contrast to the reactivity of benzene where electrophilic substitution is archetypal. Thus, the boron in these compounds only acts as a classical electron pair acceptor. Recently, Bertrand and our groups[13–16] independently developed neutral tricoordinate organoboron species possessing a nucleophilic boron centre, which is formally in the +I oxidation state. We were interested in incorporating a nucleophilic boron centre into diazadiborinine skeleton because the resulting $B_2N_2C_2$ ring would involve both nucleophilic and electrophilic boron centres, which can be formally considered as a B(+I)/B(+III) mixed valence system[17–20]. Among extant non-metal-based mixed valence compounds[21,22], it has been reported that charge-neutral mixed valence system especially with closed-shell form, namely donor–acceptor system, exhibit high stability[23]. We reasoned that incorporation of the mixed valence system into aromatic ring would effectively lead to charge delocalization as found in mixed valence bimetallic compounds categorized into class II and III[24,25]. Meanwhile, preparation of such molecules with p-block heteroatoms is highly

challenging owing to the limitation of synthetic approach for their low oxidation state. Indeed, for heterobenzene featuring mixed valence systems of the p-block inorganic elements, only the valence isomers of hexasilabenzene[26] and tetraphosphabenzene[27] are known. No relevant species involving mixed valence system of boron atoms have been described thus far.

Herein, we report the synthesis, single-crystal X-ray diffraction analysis and computational studies of 1,3,2,5-diazadiborinine 4. We show that this compound possesses both nucleophilic and electrophilic boron centres with a formal B(+I)/B(+III) mixed valence system in the aromatic six-membered ring.

Results

Synthesis and characterization of 4. Oxazolinyl groups were introduced into a boron atom by treatment of two equivalents of 2-lithio-4,4'-dimethyl-2-oxazolide 1 with dichlorophenylborane (Fig. 1). Without further purification of the crude product 2, a subsequent reaction with one equivalent of dichlorophenylborane in toluene afforded a 2,5-dichloro-1,3,2,5-diazadiborinine derivative 3 (29% yield), which was fully characterized by standard spectroscopic methods, including a single-crystal X-ray diffraction study. Treatment of 3 with excess amounts of potassium graphite (KC_8) in toluene cleanly proceeded, and after workup 1,3,2,5-diazadiborinine derivative 4 was isolated as a white powder in 32% yield. In the ^{11}B nuclear magnetic resonance (NMR) spectrum of 4, a sharp signal for the boron atom between two carbon atoms appeared at $\delta = 7.3$ parts per million (p.p.m.) and a broad peak for the boron atom between two nitrogen atoms was observed at $\delta = 24.9$ p.p.m. Both signals shifted downfield compared with those of the precursor 3 ($\delta = -11.4$ and 3.5 p.p.m.). Compound 4 is thermally stable both in the solid state and in solutions, and it melts at 133 °C without decomposition.

Single crystals of 4 suitable for X-ray diffractometry were obtained by recrystallization from a benzene solution at room temperature, and crystallographic analysis revealed that the six-membered $B_2C_2N_2$ ring of 4 is nearly planar (Fig. 2a). Two boron atoms display trigonal-planar geometry (the sum of bond angles: B1 = 359.96° and B2 = 359.93°), which are characteristic for sp^2 hybridization. Phenyl ring at B2 and the $B_2C_2N_2$ six-membered ring are nearly perpendicular to each other with the twist angle of 89.1°, whereas phenyl group at B1 and the $B_2C_2N_2$ skeleton are slightly twisted by 11.9°. The B1–C5 (1.483(3) Å) and B2–N1 (1.443(3) Å) distances are significantly

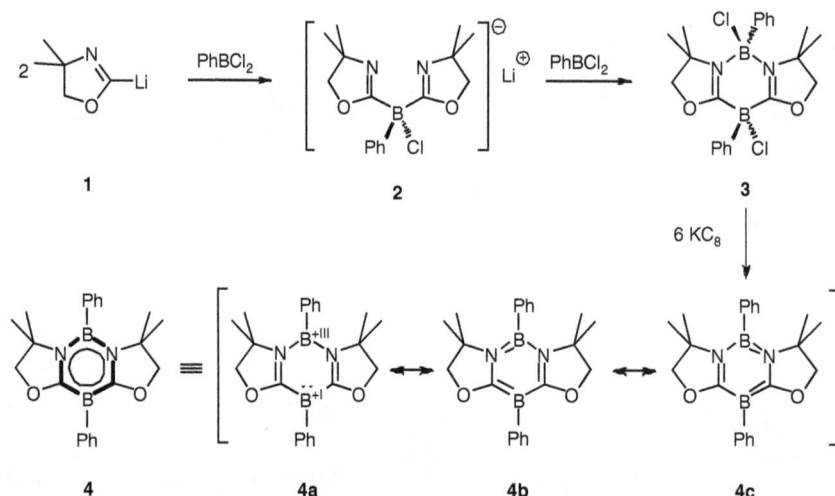

Figure 1 | Preparation of 1,3,2,4-diazadiborinine 4. 4a–c present the resonance forms.

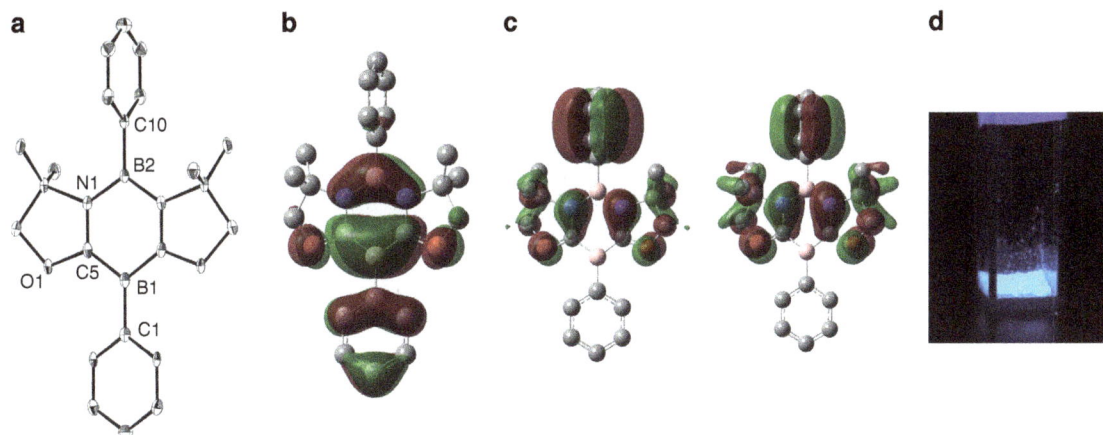

Figure 2 | Structural characterization and fluorescence property. (**a**) Solid state structure of **4**. Thermal ellipsoids are set at the 30% probability level. Hydrogen atoms are omitted for clarity. (**b**) Plot of the HOMO of **4**. (**c**) Plots of the HOMO-4 (left) and HOMO-5 (right) of **4**. Calculated at the B3LYP/6-311 + G(d,p) level of theory. Hydrogen atoms are omitted for clarity. (**d**) Photographic image of fluorescence emission in the solid state of **4** under irradiation of a ultraviolet lamp.

shorter than those (1.575(5)–1.589(5) Å and 1.573(4)–1.563(4) Å) in **3**, and lie between typical single and double-bond distances of boron–carbon and boron–nitrogen bonds, respectively. In contrast, the N1–C5 distance of 1.374(3) Å is longer than that (1.287(4)–1.292(4) Å) in **3**. These structural features suggest the delocalization of 6π-electrons over the six-membered $B_2C_2N_2$ ring in **4**, which can be represented by the average of the several canonical forms including **4a–c**.

Computational studies. To gain further insight into the electronic features of **4**, quantum chemical density functional theory calculation involving geometry optimization, natural bond orbital analysis and natural population analysis were performed at the B3LYP/6-311G + (d,p) level of theory. The optimized geometry of **4** was in good agreement with the structural parameters determined experimentally. Natural bond orbital analysis gave Wiberg bond index value for the boron–nitrogen bonds (0.96 for B2–N1). Meanwhile, Wiberg bond index values lager than 1 for the B1–C5 bonds (1.21) and the C5–N1 bonds (1.14) were obtained, thus suggesting the partial double-bond character of these bonds. Indeed, the HOMO of **4** displays a π-system over the six-membered $B_2C_2N_2$ ring featuring a node between the NBN and CBC π-unites, which exhibits anti-bonding conjugation with the π-orbital in the phenyl ring bounded to the B1 atom (Fig. 2b). π-Bonding interactions between the C5 atom and the N1 atom were confirmed in HOMO-4 and HOMO-5 (Fig. 2c). In the ultraviolet–visible absorption spectrum of **4** in a tetrahydrofuran (THF) solution, an absorption band was observed at wavelength (λ) of 275 nm, which is comparable to that ($\lambda = 277$ nm) of *para*-terphenyl (PhC$_6$H$_4$Ph) bearing all-carbon aromatic skeleton isoelectronic with **4** (ref. 28). Interestingly, we observed that **4** has little fluorescence in any solution, whereas it displays distinct light blue-fluorescence emission in the solid state under a ultraviolet lamp (Fig. 2d; refs 29,30). The broad emission peak of **4** at ($\lambda = 423$ nm) was bathochromically shifted by 80 nm in comparison with *para*-terphenyl ($\lambda = 343$ nm).

To evaluate the aromatic property of **4**, the nucleus-independent chemical-shift values NICS(0) and NICS(1) were calculated for parent 1,3,2,5-diazadiborinine **4′**, benzene (C_6H_6), 1,2-azaborine, 1,3-azaborine and $B_3N_3H_6$ (Fig. 3). The NICS values for **4′** are less negative than those of benzene and 1,3-azaborine, but comparable to those of 1,2-azaborine, and more negative with respect to those of $B_3N_3H_6$. Thus, it is predicted that 1,3,2,5-diazadiborinine features aromaticity, which seems smaller than those of benzene

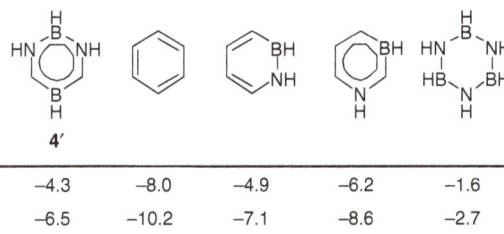

	4′				
NICS(0)	−4.3	−8.0	−4.9	−6.2	−1.6
NICS(1)	−6.5	−10.2	−7.1	−8.6	−2.7

Figure 3 | Results of density functional theory calculations. Calculated NICS(0) and NICS(1) values for **4′**, benzene, 1,2-azaborine, 1,3-azaborine and $B_3N_3H_6$. Calculated at the B3LYP/6-311 + G(d,p) level of theory.

and 1,3-azaborine, but greater than that of $B_3N_3H_6$. To estimate the resonance stabilization energy (RSE) of the parent 1,3,2, 5-diazadiborinine **4′**, we performed further computational analysis. The RSE value of **4′** is ~ 22.4 kcal mol^{-1} less than that of benzene (34.1 kcal mol^{-1}), which is smaller than those of 1,3-azaborinine (RSE = 29 kcal mol^{-1}) (ref. 31) and 1,2-azaborinine (RSE = 21 kcal mol^{-1}) (ref. 32).

Reactivity. To investigate the reactivity of **4**, we performed its reaction with methyl trifluoromethanesulfonate (MeOTf). A stoichiometric amount of MeOTf was added to an acetonitrile solution of **4** at ambient temperature. After removing the solvent under vacuum, **5** was obtained in 75% yield (Fig. 4a). An X-ray diffraction study confirmed that methyl group is attached to the boron atom between two carbons in the $B_2C_2N_2$ ring, whereas an oxygen atom of the triflate is bounded to the boron atom between two nitrogen atoms (Fig. 4b, left). This result, thus, demonstrates that **4** features both nucleophilic and electrophilic boron centres, thereby supporting the electronic property of the resonance structure **4a** (Fig. 1). The formal oxidation states of the B1 and the B2 atoms in **4a** are + I and + III, respectively. Thus, **4** presents a donor–acceptor mixed valence system.

Reactivity of neutral boron nucleophiles has seldom been explored thus far[13–16]. Based on the behaviours of **4** with MeOTf, we postulated that **4** would act as a frustrated Lewis pair, since it possesses both Lewis basic and acidic boron centres[33–37]. To bear out the hypothesis, we next investigated the reactivity of **4** towards non-activated alkynes. In a J-Young NMR tube, a stoichiometric amount of phenylacetylene was added to a C_6D_6 solution of **4**, and reaction was monitored by NMR

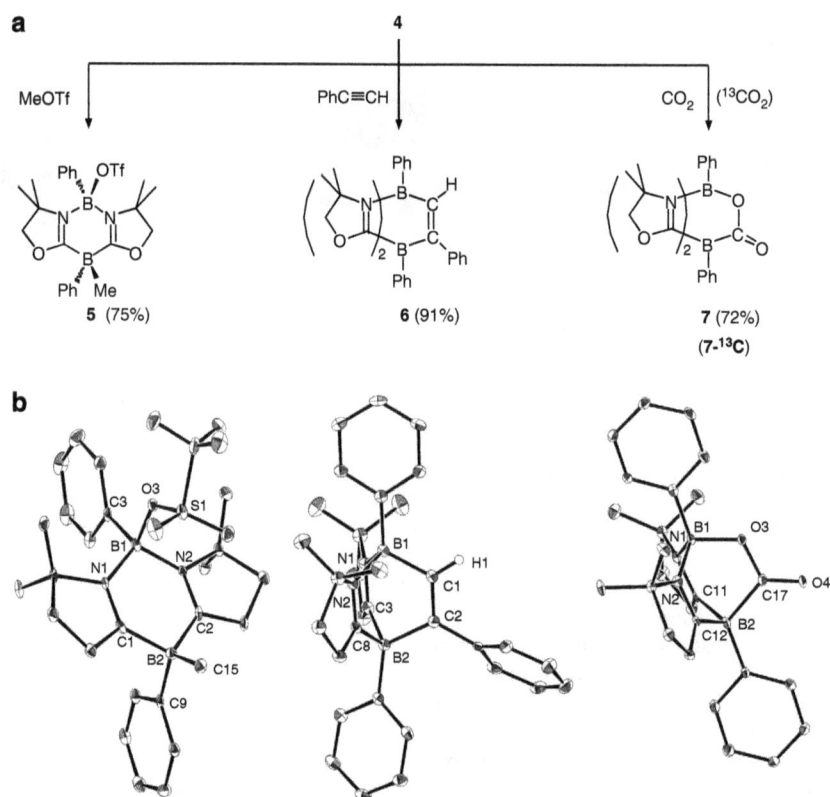

Figure 4 | Reactivity of 4. (a) Reactions of **4** with MeOTf, PhC≡CH and CO_2 ($^{13}CO_2$). **(b)** Solid state structures of **5** (left), **6** (middle) and **7** (right). Thermal ellipsoids are set at the 30% probability level. Hydrogen atoms except for H1 in **6** and solvent molecules are omitted for clarity.

spectroscopy. After 2 h at 70 °C, two new signals were detected at − 13.6 and 0.1 p.p.m. in the ^{11}B NMR spectrum. After workup, compound **6** was isolated as a white powder in 91% yield. The crystallographic study revealed the diazadiborabicycle[2.2.2] structure involving the C=N double bonds (N1–C3: 1.299(4) Å and N2–C8: 1.299(3) Å) (Fig. 4b, middle). Thus, **6** is a formal Diels–Alder product via a [4 + 2] cyclo-addition between **4** and the carbon–carbon triple bond of phenylacetylene. It is noteworthy to mention that Diels–Alder addition of alkynes to aromatic hydrocarbons normally is restricted to highly reactive aromatic compounds and activated alkynes under harsh conditions[38,39]. Moreover, **6** can be viewed as an analogue of bicycle[2.2.2]octatriene, also termed barrelene, which is inferred to be one of the Möbius aromatic compounds[39].

Next, we examined the carbon dioxide (CO_2) activation with **4** (ref. 40). CO_2 gas was introduced into a benzene solution of **4** at 1 bar, and the solution was heated at 70 °C. After 2 h, a white precipitate was filtered and dried under vacuum to afford compound **7** in 72% yield. We also carried out a ^{13}C-labelling study using $^{13}CO_2$, which produced **7**-^{13}C (80% yield). The ^{13}C NMR spectrum of **7**-^{13}C displayed a broad resonance at 191.1 p.p.m. In the ^{11}B NMR spectrum, a set of new peaks was observed at 3.1 p.p.m. as a broad singlet and − 16.7 p.p.m. as a broad doublet ($^1J_{B-C} = 42.4$ Hz), which is owing to the coupling with a ^{13}C carbon atom. These results indicate the presence of a bond between a boron and the carbon from CO_2 in **7**, which was decisively confirmed by X-ray diffractometry (Fig. 4b, right). One of the C=O double bonds of CO_2 was cleaved and new B–C and B–O bonds are formed through 1,4 addition, which is contrast to the behaviour of benzene that reacts with CO_2 to form benzoic acid only in the presence of Lewis acid catalysts[41]. We also found that the CO_2 activation process by **4** was reversible. Thus, treatment of **7** at 90 °C for 50 min reproduced **4**, quantitatively.

Discussion

One-hundred fifty years after the discovery of the cyclic structure of benzene, 1,3,2,5-diazadiborinine derivative **4** featuring a formal B(+ I)/B(+ III) mixed valence system has joined as a hetero-analogue into a class of 6π-Hückel aromatic molecules. In marked contrast to the chemical behaviour of benzene, $B_3N_3H_6$ or even other azaborinines, **4** exhibits unique optical property and reactivity like a boron–boron frustrated Lewis pair. Because the electronic property can be substantially modulated by varying the substituents on each boron atom, the isolation of this molecule paves the way for the discovery of new materials with useful photochemical properties. Moreover, the cooperative reactivity of the nucleophilic and electrophilic boron centres, and the reversible nature for activation of small molecules will be applicable to catalytic chemistry.

Methods

Materials. For details of spectroscopic analyses of compounds in this manuscript, see Supplementary Figs 1–25. For details of density functional theory calculations, see Supplementary Fig. 26, Supplementary Tables 3–5 and Supplementary Methods. For details of the synthetic procedures, see Supplementary Methods. For details of X-ray analysis, see Supplementary Data 1–5, Supplementary Tables 1 and 2, and Supplementary Methods.

General synthetic procedures. All reactions were performed under an atmosphere of dry argon using standard Schlenk or dry box techniques; solvents were dried over Na metal, K metal or CaH_2, and were distilled under nitrogen. Reagents were of analytical grade, obtained from commercial suppliers and were used without further purification. ^{1}H, ^{13}C, ^{11}B and ^{19}F NMR spectra were recorded on a Bruker AVIII 400 MHz or Bruker Avance 500 MHz AV500 spectrometers at 298 K. Chemical shifts (δ) are given in p.p.m. Coupling constants J are given in Hz. In the ^{13}C NMR spectra of compounds **3–7**, presumable owing to the coupling with boron atoms, signals for the carbon atoms directly bonding to boron atoms could not be observed. Electrospray ionization (ESI) mass spectra were obtained at the Mass Spectrometry Laboratory at the Division of Chemistry and Biological Chemistry, Nanyang Technological University. Melting points were measured with OptiMelt (Stanford Research System). Fourier-transform infrared (FT-IR) spectra were recorded on a SHIMADZU IRPrestige-21 spectrometer using solid compound. Ultraviolet and fluorescence spectra were recorded on Cary 100 UV-Vis and SHIMADZU RF-5301PC spectrofluorophotometer, respectively.

Synthesis of 3. A hexane solution (1.6 M) of n-BuLi (6.25 ml, 10.00 mmol) was added dropwise into a THF solution (50 ml) of 4,4-dimethyl-2-oxazoline (1.00 ml, 9.48 mmol) at −78 °C. After stirring for 1 h at −78 °C, dichlorophenylborane (0.62 ml, 4.72 mmol) was added into the solution. The reaction mixture was warmed to room temperature and stirred overnight. After the solvent was removed under vacuum, toluene (60 ml) was added and salts were filtered off, which was used directly for next step without further purification. Dichlorophenylborane (0.62 ml, 4.72 mmol) was added dropwise into the solution at −78 °C. The reaction mixture was left to stir for 1 h at −78 °C, and slowly warmed to room temperature and stirred overnight. The solvent was removed under vacuum, and the solid residue was recrystallized from benzene to afford colourless crystals of 3 (0.61 g, 29%). Melting point (Mp): 207 °C. ^{1}H NMR (400 MHz, C$_6$D$_6$): $\delta = 0.52$ (s, 6 H), 1.21 (s, 6 H), 3.06 (d, 2H, $J = 9.2$ Hz), 3.13 (d, 2 H, $J = 8.8$ Hz), 7.21−8.29 (m, 10 H); ^{13}C NMR (100 MHz, C$_6$D$_6$): $\delta = 24.2, 27.5, 68.3, 81.8, 127.6, 132.4, 133.1, 136.0$; DEPT−135 NMR (100 MHz, C$_6$D$_6$): $\delta = 24.2, 27.5, 127.6, 128.0, 128.3, 132.4, 133.1, 136.0$. ^{11}B NMR (76.8 MHz, C$_6$D$_6$): $\delta = −11.4$ (s), 3.5 (s); HRMS (ESI): m/z calculated for C$_{22}$H$_{26}$B$_2$N$_2$O$_2$Cl: 407.1869 [(M)]$^+$; found: 407.1866.

Synthesis of 4. Potassium graphite (1.49 g, 11.04 mmol) was slowly added to a solution of 3 (0.82 g, 1.84 mmol) in toluene (50 ml) and stirred overnight at room temperature. After filtration, the solvent was concentrated to 5 ml under vacuum. The solid residue was filtered off and washed with hexane (15 ml), and then dried under vacuum to afford 4 as a white powder (0.22 g, 32%). Mp: 133 °C. ^{1}H NMR (400 MHz, C$_6$D$_6$): $\delta = 0.79$ (s, 12 H), 3.63 (s, 4 H), 7.09−8.84 (m, 10 H); ^{13}C NMR (100 MHz, C$_6$D$_6$): $\delta = 27.5, 62.8, 80.9, 125.4, 127.7, 134.7, 135.1$; DEPT−135 NMR (100 MHz, C$_6$D$_6$): $\delta = 27.5, 125.4, 127.1, 127.7, 128.4, 134.7, 135.1$; ^{11}B NMR (76.8 MHz, C$_6$D$_6$): $\delta = 7.3$ (s), 24.9 (s); Ultraviolet–visible (THF): $\lambda = 275$ nm (ε: 5,740); HRMS (ESI): m/z calculated for C$_{22}$H$_{27}$B$_2$N$_2$O$_2$: 373.2259 [$(M + H)$]$^+$; found: 373.2271; IR (KBr, cm^{-1}): 1591.3, 1475.5, 1444.7, 1421.5, 1392.6, 1375.3, 1303.9, 1253.7, 1205.5, 1103.3, 1074.4, 1033.9, 1016.5, 954.8, 939.3.

Reaction of 4 with MeOTf. MeOTf (7.42 µl, 0.068 mmol) was added into an acetonitrile (6 ml) solution of 4 (0.021 g, 0.056 mmol) and stirred for 1 h at room temperature. After the solvent was removed under vacuum, the solid residue was recrystallized from a 2:1 mixture of fluorobenzene/hexane to afford colourless crystals of 5 (0.022 g, 75%). Mp: 151 °C (dec). ^{1}H NMR (400 MHz, THF-d$_8$): $\delta = 0.34$ (s, 3 H), 0.87 (s, 6 H), 1.38 (s, 6 H), 4.14 (d, 2 H, $J = 8.8$ Hz), 4.26 (d, 2 H, $J = 8.4$ Hz), 7.06−7.66 (m, 10 H); ^{13}C NMR (100 MHz, THF-d$_8$): $\delta = 22.4, 29.0, 67.3, 83.2, 126.1, 127.7, 128.1, 129.6, 130.8, 132.4$; ^{11}B NMR (76.8 MHz, THF-d$_8$): $\delta = −19.7$ (s), 4.2(s); ^{19}F NMR (225.6 MHz, THF-d$_8$): $\delta = −81.2$; HRMS (ESI): m/z calculated for C$_{24}$H$_{30}$B$_2$N$_2$O$_5$SF$_3$: 537.2014 [$(M + H)$]$^+$; found: 537.2006.

Reaction of 4 with phenylacetylene. Phenylacetylene (9.2 µl, 0.087 mmol) was added into a C$_6$D$_6$ (0.5 ml) solution of 4 (0.031 g, 0.083 mmol) and heated at 70 °C. Reaction was monitored by NMR spectroscopy. After 2 h, the solvent was removed under vacuum, and the solid residue was recrystallized from THF/hexane to afford colourless crystals of 6 (0.036 g, 91%). Mp: 120 °C. ^{1}H NMR (400 MHz, C$_6$D$_6$): $\delta = 0.52$ (s, 6 H), 0.87 (s, 6 H), 3.44 (d, 2 H, $J = 8.4$ Hz), 3.51 (d, 2 H, $J = 8.4$ Hz), 7.01−8.04 (m, 16 H); ^{13}C NMR (125 MHz, CDCl$_3$): $\delta = 25.0, 27.4, 65.9, 84.9, 124.7, 125.3, 126.7, 127.2, 127.6, 127.9, 128.5, 132.3, 135.2$; ^{11}B NMR (76.8 MHz, C$_6$D$_6$): $\delta = −13.6$ (s), 0.1 (s); HRMS (ESI): m/z calculated for C$_{30}$H$_{33}$B$_2$N$_2$O$_2$: 475.2728 [$(M + H)$]$^+$; found: 475.2733.

Reaction of 4 with CO$_2$. A benzene (15 ml) solution of compound 4 (0.061 g, 0.040 mmol) was degassed using a freeze-pump-thaw method, and then CO$_2$ (1 bar) was introduced into the schlenk tube at room temperature. After 2 h at 70 °C, a white precipitate was collected by filtration and dried under vacuum to afford 7 (0.049 g, 72%). Mp: 158 °C (dec). ^{1}H NMR (400 MHz, CD$_2$Cl$_2$): $\delta = 0.90$ (s, 6 H), 1.50 (s, 6 H), 4.36 (d, 2 H, $J = 9.2$ Hz), 4.39 (d, 2 H, $J = 9.2$ Hz), 7.22−7.95 (m, 10 H); ^{13}C NMR (100 MHz, CD$_2$Cl$_2$): $\delta = 25.9, 27.7, 66.5, 86.4, 126.9, 127.5, 129.1, 134.97, 134.99, 135.1$; ^{11}B NMR (76.8 MHz, CD$_2$Cl$_2$): $\delta = −16.7$ (s), 3.1 (s); HRMS (ESI): m/z calculated for C$_{23}$H$_{27}$B$_2$N$_2$O$_4$: 417.2157 [$(M + H)$]$^+$; found: 417.2161.

Reaction of 4 with ^{13}CO$_2$. Compound 4 (0.015 g, 0.040 mmol) was added to a sealable J-Young NMR tube and was dissolved in C$_6$D$_6$ (0.35 ml). The sample was degassed using a freeze-pump-thaw method. Labile ^{13}CO$_2$ was condensed into the NMR tube at 77 K, followed by warming to room temperature, and reacted for 2 h at 70 °C. A white precipitate was collected by filtration and dried under vacuum to afford 7-^{13}C (0.013 g, 80% yield). ^{13}C NMR (100 MHz, CD$_2$Cl$_2$): $\delta = 25.9, 27.7, 66.5, 86.4, 126.9, 127.5, 129.1, 134.97, 134.99, 135.1, 191.1$ (br d, $J_{CB} = 42.4$ Hz); ^{11}B NMR (76.8 MHz, CD$_2$Cl$_2$): $\delta = −16.7$ (br d, $J_{BC} = 42.4$ Hz), 3.1 (s).

Regeneration of 4 from 7. Compound 7 was dissolved in a 1:1 mixture of CD$_3$CN/THF-d$_8$, and heated at 90 °C. Reaction was monitored by ^{1}H and ^{11}B

NMR spectroscopy and after 50 min, a quantitative regeneration of 4 was confirmed.

References

1. Kekulé, A. Sur la constitution des substances aromatiques. *Bull. Soc. Chim. Fr* **3**, 98–110 (1865).
2. Faraday, M. On new compounds of carbon and hydrogen, and on certain other products obtained during the decomposition of oil by heat. *Phil. Trans. R. Soc. Lond.* **115**, 440–466 (1825).
3. Stock, A. & Pohland, E. Borwasserstoffe, IX.: B$_3$N$_3$H$_6$. *Ber. Dtsch. Chem. Ges. A/B* **59**, 2215–2223 (1926).
4. Bean, D. E. & Fowler, P. W. Effect on ring current of the Kekulé vibration in aromatic and antiaromatic rings. *J. Phys. Chem. A* **115**, 13649–13656 (2011).
5. Phukan, A. K., Guha, A. K. & Silvi, B. Is delocalization a prerequisite for stability of ring systems? A case study of some inorganic rings. *Dalton Trans.* **39**, 4126–4137 (2010).
6. Liu, Z. & Marder, T. B. B-N versus C-C: how similar are they? *Angew. Chem. Int. Ed.* **47**, 242–244 (2008).
7. Bosdet, M. J. D. & Piers, W. E. B-N as a C-C substitute in aromatic systems. *Can. J. Chem.* **87**, 8–29 (2009).
8. Ashe, III A. J. Aromatic borataheterocycles: surrogates for cyclopentadienyl in transition-metal complexes. *Organometallics* **28**, 4236–4248 (2009).
9. Dewar, M. J. S., Kubba, V. P. & Pettit, R. New heteroaromatic compounds. Part I. 9-Aza-10-boraphenanthrene. *J. Chem. Soc.* 3073–3076 (1958).
10. Campbell, P. G., Marwitz, A. J. V. & Liu, S.-Y. Recent advances in azaborine chemistry. *Angew. Chem. Int. Ed.* **51**, 6074–6092 (2012).
11. Jaska, C. A. *et al.* Triphenylene analogues with B$_2$N$_2$C$_2$ cores: synthesis, structure, redox behavior, and photophysical properties. *J. Am. Chem. Soc.* **128**, 10885–10896 (2006).
12. Forster, T. D. *et al.* A σ-donor with a planar six-π-electron B$_2$N$_2$C$_2$ framework: anionic N-heterocyclic carbene or heterocyclic terphenyl anion? *Angew. Chem. Int. Ed.* **45**, 6356–6359 (2006).
13. Kinjo, R., Donnadieu, B., Celik, M. A., Frenking, G. & Bertrand, G. Synthesis and characterization of a neutral tricoordinate organoboron isoelectronic with amines. *Science* **333**, 610–613 (2011).
14. Kong, L., Li, Y., Ganguly, R., Vidovic, D. & Kinjo, R. Isolation of a bis(oxazol-2-ylidene)-phenylborylene adduct and its reactivity as a boron-centered nucleophile. *Angew. Chem. Int. Ed.* **53**, 9280–9283 (2014).
15. Ruiz, D. A., Melaimi, M. & Bertrand, G. An efficient synthetic route to stable bis(carbene)borylenes [(L$_1$)(L$_2$)BH]. *Chem. Commun.* **50**, 7837–7839 (2014).
16. Kong, L., Ganguly, R., Li, Y. & Kinjo, R. Diverse reactivity of a tricoordinate organoboron L$_2$PhB: (L= oxazol-2-ylidene) towards alkali metal, group 9 metal, and coinage metal precursors. *Chem. Sci.* **6**, 2893–2902 (2015).
17. Creutz, C. & Taube, H. Direct approach to measuring the Franck-Condon barrier to electron transfer between metal ions. *J. Am. Chem. Soc.* **91**, 3988–3989 (1969).
18. Creutz, C. & Taube, H. Binuclear complexes of ruthenium ammines. *J. Am. Chem. Soc.* **95**, 1086–1094 (1973).
19. Robin, M. B. & Day, P. Mixed-valence chemistry: a survey and classification. *Adv. Inorg. Chem. Radiochem* **10**, 247–422 (1967).
20. Aguirre-Etcheverry, P. & O'Hare, D. Electronic communication through unsaturated hydrocarbon bridges in homobimetallic organometallic complexes. *Chem. Rev.* **110**, 4839–4864 (2010).
21. Hankache, J. & Wenger, O. S. Organic mixed valence. *Chem. Rev.* **111**, 5138–5178 (2011).
22. Abe, M. Diradicals. *Chem. Rev.* **113**, 7011–7088 (2013).
23. Heckmann, A. & Lambert, C. Organic mixed-valence compounds: a playground for electrons and holes. *Angew. Chem. Int. Ed.* **51**, 326–392 (2012).
24. Low, P. J. Twists and turns: studies of the complexes and properties of bimetallic complexes featuring phenylene ethynylene and related bridging ligands. *Coord. Chem. Rev.* **257**, 1507–1532 (2013).
25. Parthey, M. & Kaupp, M. Quantum-chemical insights into mixed-valence systems: within and beyond the Robin–Day scheme. *Chem. Soc. Rev.* **43**, 5067–5088 (2014).
26. Abersfelder, K., White, A. J. P., Rzepa, H. S. & Scheschkewitz, D. A tricyclic aromatic isomer of hexasilabenzene. *Science* **327**, 564–566 (2010).
27. Canac, Y. *et al.* Isolation of a benzene valence isomer with one-electron phosphorus–phosphorus bonds. *Science* **279**, 2080–2082 (1998).
28. Pistolis, G. Dual excimer emission of p-terphenyl induced by γ-cyclodextrin in aqueous solutions. *Chem. Phys. Lett.* **304**, 371–377 (1999).
29. Hong, Y., Lam, J. W. Y. & Tang, B. Z. Aggregation-induced emission. *Chem. Soc. Rev.* **40**, 5361–5388 (2011).
30. Hong, Y., Lam, J. W. Y. & Tang, B. Z. Aggregation-induced emission: phenomenon, mechanism and applications. *Chem. Commun.* **45**, 4332–4353 (2009).
31. Xu, S., Mikulas, T. C., Zakharov, L. N., Dixon, D. A. & Liu, S.-Y. Boron-substituted 1,3-dihydro-1,3-azaborines: synthesis, structure, and evaluation of aromaticity. *Angew. Chem. Int. Ed.* **52**, 7527–7531 (2013).

32. Marwitz, A. J. V., Matus, M. H., Zakharov, L. N., Dixon, D. A. & Liu, S.-Y. A hybrid organic/inorganic benzene. *Angew. Chem. Int. Ed.* **48**, 973–977 (2009).

33. Stephan, D. W. & Erker, G. Frustrated Lewis pairs: metal-free hydrogen activation and more. *Angew. Chem. Int. Ed.* **49**, 46–76 (2010).

34. Power, P. P. Main-group elements as transition metals. *Nature* **463**, 171–177 (2010).

35. Yao, S., Xiong, Y. & Driess, M. Zwitterionic and donor-stabilized N-heterocyclic silylenes (NHSis) for metal-free activation of small molecules. *Organometallics* **30**, 1748–1767 (2011).

36. Mandal, S. K. & Roesky, H. W. Group 14 hydrides with low valent elements for activation of small molecules. *Acc. Chem. Res.* **45**, 298–307 (2012).

37. Stephan, D. W. Frustrated Lewis pairs: from concept to catalysis. *Acc. Chem. Res.* **48**, 306–316 (2015).

38. Krespan, C. G., McKusick, B. C. & Cairns, T. L. Bis-(polyfluoroalkyl)-acetylenes. II. Bicyclooctatrienes through 1,4-addition of bis-(polyfluoroalkyl)-acetylenes to aromatic rings. *J. Am. Chem. Soc.* **83**, 3428–3432 (1961).

39. Zimmerman, H. E., Grunewald, G. L., Paufler, R. M. & Sherwin, M. A. Synthesis and physical properties of barrelene, a unique Moebius-like molecule. *J. Am. Chem. Soc.* **91**, 2330–2338 (1969).

40. Stephan, D. W. & Erker, G. Frustrated Lewis pair chemistry of carbon, nitrogen and sulfur oxides. *Chem. Sci.* **5**, 2625–2641 (2014).

41. Olah, G. A. *et al.* Efficient chemoselective carboxylation of aromatics to arylcarboxylic acids with a superelectrophilically activated carbon dioxide − Al_2Cl_6/Al System. *J. Am. Chem. Soc.* **124**, 11379–11391 (2002).

Acknowledgements

We gratefully acknowledge financial support from Nanyang Technological University (NTU; Singapore) and A*STAR (Agency for Science, Technology and Research, PSF/SERC 1321202066) of Singapore. We also thank Yanli Zhao and Pengyao Xing (NTU) for their assistance in fluorescence spectra analysis.

Author contributions

D.W. performed most of the synthetic experiments. L.K. conducted part of the syntheses and spectroscopic characterizations. Y.L. and R.G. performed the X-ray crystallographic measurements. R.K. conceived and supervised the study, and drafted the manuscript with the assistance from L.K.. All authors contributed to discussions.

Additional information

Accession codes: The X-ray crystallographic coordinates for structures reported in this article have been deposited at the Cambridge Crystallographic Data Centre (CCDC), under deposition numbers CCDC 1049275–1049279. These data can be obtained free of charge from The Cambridge Crystallographic Data Centre via www.ccdc.cam.ac.uk/data_request/cif.

Competing financial interests: The authors declare no competing financial interests.

28

Room temperature molecular up conversion in solution

Aline Nonat[1], Chi Fai Chan[2], Tao Liu[1,2], Carlos Platas-Iglesias[3], Zhenyu Liu[4], Wing-Tak Wong[4,5], Wai-Kwok Wong[2], Ka-Leung Wong[2] & Loïc J. Charbonnière[1]

Up conversion is an Anti-Stokes luminescent process by which photons of low energy are piled up to generate light at a higher energy. Here we show that the addition of fluoride anions to a D_2O solution of a macrocyclic erbium complex leads to the formation of a supramolecular $[(ErL)_2F]^+$ assembly in which fluoride is sandwiched between two complexes, held together by the synergistic interactions of the Er-F-Er bridging bond, four intercomplex hydrogen bonds and two aromatic stacking interactions. Room temperature excitation into the Er absorption bands at 980 nm of a solution of the complex in D_2O results in the observation of up converted emission at 525, 550 and 650 nm attributed to Er centred transitions *via* a two-step excitation. The up conversion signal is dramatically increased upon formation of the $[(ErL)_2F]^+$ dimer in the presence of 0.5 equivalents of fluoride anions.

[1] Laboratoire d'Ingénierie Moléculaire Appliquée à l'Analyse, IPHC, UMR 7178 CNRS, Université de Strasbourg, ECPM, Bât R1N0, 25 rue Becquerel, Strasbourg, 67087, France. [2] Department of Chemistry, Hong Kong Baptist University, Hong Kong SAR, Hong Kong. [3] Centro de Investigaciones Científicas Avanzadas, Departamento de Química Fundamental, Universidade da Coruña, Campus da Zapateira, Rúa da Fraga 10, A Coruña 15008, Spain. [4] State Key Laboratory of Chirosciences, Department of Applied Biology and Chemical Technology, Hong Kong Polytechnic University, Kowloon, Hong Kong. [5] Department of Applied Biology and Chemical Technology, Hong Kong Polytechnic University, Hong Kong SAR, Hong Kong. Correspondence and requests for materials should be addressed to W.-T.W. (e-mail: bcwtwong@polyu.edu.hk), or to K.-L.W. (e-mail: klwong@hkbu.edu.hk) or to L.J.C. (e-mail:l.charbonn@unistra.fr).

Conventional luminescence spectroscopy focuses on the emission of photons after absorption of photons of higher energy (that is, down shifting), but processes giving off emission of photons at higher energy than the incident beam and not related to thermal population of the excited states are far less common[1]. These processes gather aspects related to nonlinear optics, such as second and third harmonic generations or two photon absorption[2], and those associated to the cumulative effects of multiple first-order absorption phenomena, also named up conversion (UC)[3], among which are excited state absorption (ESA) and energy transfer up conversion (ETU)[1]. The last two became very popular in recent years, first because they are far more efficient than the others, affording observation of UC with more classical excitation sources, but also because of intensive research on the synthesis and preparation of UC materials, particularly in the field of lanthanide-doped phosphors[3] and more recently with lanthanide-based nanoparticles[4,5]. Lanthanide ions such as Er, Tm or Ho are particularly well-suited for UC processes as they gather two important properties: a ladder-like energy level profile spanning from the Near-infrared (NIR) to the visible region; and long-lived excited states due to Laporte forbidden electric dipole transitions[6].

UC is particularly appealing for biophotonic applications as the emitted signal is devoid of spurious signals such as auto-fluorescence of the sample or excitation bleed through, rendering an almost background-free measurement. Also, the NIR excitation is entirely dedicated to the excitation of the probe, as biological tissues absorb weakly in this region of the spectrum[7], and the penetration of excitation light into the tissues is thus deeper, allowing applications to bioimaging[8,9] or photodynamic therapy[10].

Despite these evident new opportunities, the development of UC devices is still confronted with a scientific challenge: the downscaling of the probes to the molecular scale. In bulk solids and nanoparticles, the phonon energy of the lattice plays a crucial role in luminescence quenching mechanisms. When phonons or their overtones are in resonance with the intermediate excited state levels of the emitting centres, non-radiative phonon-assisted deactivation pathways compete severely with the successive piling up of excited states necessary to reach the higher energy emitting level[11]. At the molecular level, the situation is even worse, where OH, NH and CH oscillators, commonly present in the ligand backbones of the Ln complexes or in their solvation shells, contribute as efficient non-radiative deactivation pathways[12]. As a result, even conventional NIR luminescence of such Ln complexes rarely exceeds quantum efficiencies of a few percent, in the best cases[13,14].

In spite of such apparently insurmountable obstacles, encouraging results were recently published. In 2011, Piguet, Hauser and co-workers reported on the observation of green erbium UC emission upon NIR chromium excitation in a heterotrinuclear triple-stranded helicate composed of two Cr sensitizers and a central Er acceptor in the solid state and solid solutions at room temperature and in frozen acetonitrile up to 150 K (refs 15,16). Unambiguous evidence for the UC process was brought forward by the quadratic dependence of the emitted intensity as a function of excitation power. The study of isostructural compounds of the spectroscopically silent Ga analogue[16] also allowed the authors to get a detailed overview of the kinetic parameters of their system. Within the frame of Förster's theory on energy transfer[17], the energy transfer efficiency is dependent on the inversed sixth power of the donor–acceptor distance and one can anticipate that the rather large distances between the Cr sensitizers and the Er acceptor (8.9 Å on the basis of the analogous Yb compound) may not lead to an efficient energy transfer[18]. In 2012, Faulkner, Beeby and Sorensen achieved multiphoton excitation of different lanthanide salts of Eu, Tb and Sm upon intense excitation of some concentrated ($c > 0.01$ M) triflate salts and complexes in d_6-dimethylsulfoxide solutions[19]. In the case of thulium, they observed a faint excitation band in the multiphoton excitation spectrum that was proposed to be a potential UC process, but lacked details to follow up with the work[20]. Very recently, Hyppänen and co-workers reported on the observation of UC in a system composed of an Er complex and a NIR dye photosensitizer in CDCl$_3$ at room temperature upon strong excitation (> 3.4 W cm^{-2}) at 808 nm (ref. 21).

In this contribution, we show that the macrocyclic complex of erbium with ligand L, [ErL(H$_2$O)]$^+$ can self-assemble in D$_2$O solution in the presence of fluoride anions to form a supramolecular [(ErL)$_2$F]$^+$ dimer (Fig. 1). Upon excitation into the NIR absorption band of Er at 980 nm in D$_2$O, a green emission is observed arising from an UC process, which is largely amplified with the formation of the Er dimer, representing, to the best of our knowledge, the first evidence of molecular UC in heavy water solution at room temperature.

Results

Synthesis and characterization of the complex. The [ErL(H$_2$O)] (NO$_3$) complex was prepared from ligand L and Er(NO$_3$)$_3$.5H$_2$O (ref. 22) and was fully characterized by elemental analysis, infrared spectroscopy, ^{1}H-NMR and electrospray mass spectrometry (Supplementary Figs 1–4). The presence of the nitrate anions was particularly well evidenced by infrared spectroscopy with new bands at 1,655 and 1,360 cm^{-1}, absent in the chloride complexes of Eu and Tb with L (ref. 22), and characteristic of nitrate anions[23]. In H$_2$O, the electrospray mass spectrum of the complex displayed a single intense peak with maximum at 714.19 m/z units, with an isotopic distribution perfectly fitting that calculated for a [ErL]$^+$ monocharged cation. A very minor species ($<2\%$) could be evidenced as a peak with maximum at 1,488.37 m/z, which was related to the presence of two monomers linked by a nitrate anion [(ErL)$_2$(NO$_3$)]$^+$ (calculated mass for C$_{56}$Er$_2$H$_{68}$N$_{17}$O$_{11}^+$: 1,488.39). Finally, the

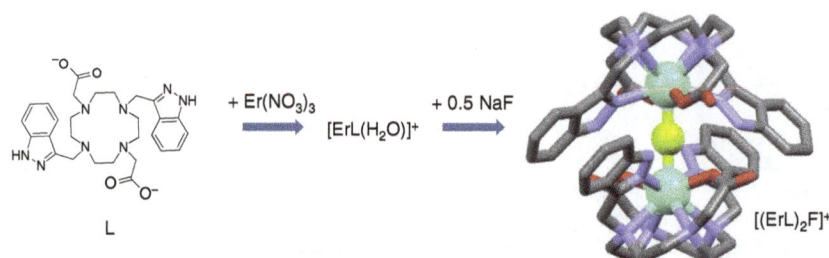

Figure 1 | Formation of the supramolecular erbium dimer. The ligand L react with Er^{3+} salts in water to form an [ErL(H$_2$O)]$^+$ complex. In the presence of half an equivalent of fluoride anions two complexes form an association in which the F$^-$ anion is sandwiched between the two complexes leading to the [(ErL)$_2$F]$^+$ dimer. The association is strengthened by two intercomplex stacking interactions and four hydrogen bonds.

complex was also characterized by ^{1}H-NMR spectrometry. Although at the limit of the solubility, and despite the paramagnetic contribution of the Er cation, it was possible to record the spectrum of the complex in D_2O. The spectrum of the paramagnetic complex displayed 15 broad peaks between ca -110 and $+110$ parts per million (p.p.m.), for 16 expected for a C_2 symmetry of the complex (Supplementary Figs 1 and 2). It is surmised that the missing peak is hidden under the peak of the partially deuterated water, HDO. Interestingly, minor peaks corresponding to a second paramagnetic species ($<6\%$) could also be observed in D_2O. Considering that the major species is in a square antiprismatic geometry, as observed for the parent Eu and Yb complex[22], the minor species may be attributed to the presence of some twisted square antiprismatic conformer, as observed in the Ln complexes of 1,4,7,10-tetraazacyclododecane-1,4,7,12-tetraacetic acid (DOTA)[24].

Spectroscopic characterization of the complex. The absorption spectrum of the Er complex in D_2O is composed of a broad absorption band with maximum at 294 nm ($\varepsilon = 9,850\,M^{-1}$

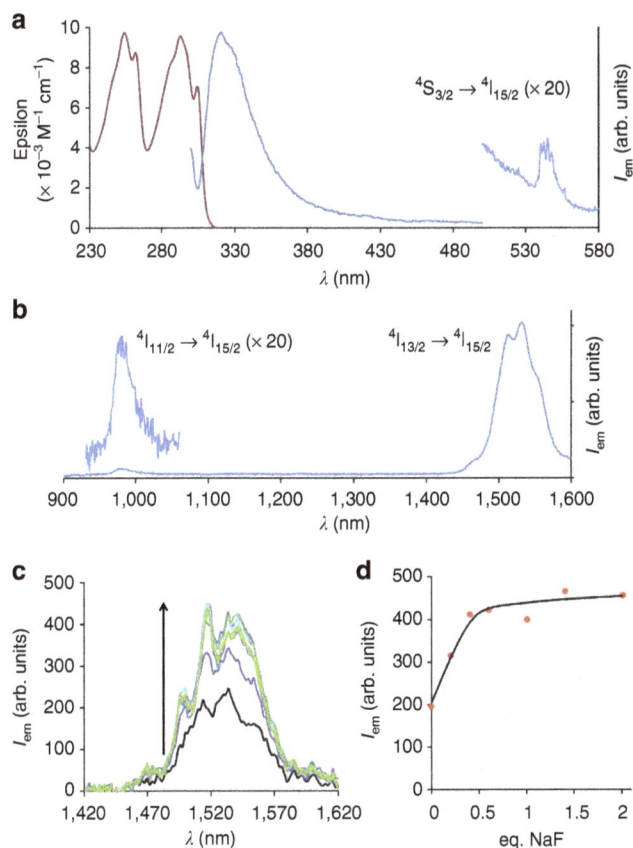

Figure 2 | Spectroscopic properties of the Er complexes. (a) The ultraviolet–visible absorption spectrum of the Er complex in D_2O (red, $c = 270\,\mu M$) displayed strong $\pi \to \pi^\star$ absorption bands. When excited into these bands ($\lambda_{exc} = 294\,nm$), the emission spectrum of the complex displayed a ligand centred emission band in the visible region (blue), together with a contribution attributed to the $^4S_{3/2} \to {}^4I_{15/2}$ transition of Er (recorded at 0.5 nm resolution). **(b)** Other Er centred transitions can be observed in the NIR region ($\lambda_{exc} = 294\,nm$, emission filter at 850 nm, D_2O). **(c)** The addition of fluoride anions into the D_2O solution of the Er complex resulted in an increase of the Er emission and **(d)** the intensity of the signal at 1,540 nm revealed a maximum increase for 0.5 equivalent (eq.) of fluoride anions, related to the formation of the dimeric $[(ErL)_2F]^+$ complex (the black curve corresponds to fitted data).

cm^{-1}, Fig. 2), which was attributed to $\pi \to \pi^\star$ transitions centred on the indazolyl moieties[25]. The transmittance spectrum was recorded in the solid state from 400 to 1,800 nm (Supplementary Fig. 5) and displayed numerous Er centred absorption bands attributed to transitions from the ground state to the $^4F_{7/2}$, $^2H_{11/2}$, $^4S_{3/2}$, $^4F_{9/2}$, $^4I_{11/2}$ and $^4I_{13/2}$, respectively at 488, 522, 541, 657, 980 and 1,514 nm.

Upon excitation into the ligand absorption bands (294 nm), the emission spectrum of the complex in D_2O (Fig. 2) displayed a broad emission band, with maximum at 318 nm, which is attributed to some ligand centred fluorescence from the singlet state. A close examination of the visible part of the spectrum also revealed the presence of a structured emission peak between 540 and 550 nm in the low-energy tail of the fluorescence band, which can be ascribed to the $^4S_{3/2} \to {}^4I_{15/2}$ transition of Er. In the NIR region of the spectrum, a weak and broad band can be found at 980 nm, attributed to the $^4I_{11/2} \to {}^4I_{15/2}$ transition of Er and a strong and structured emission band could be observed with maximum at 1,535 nm associated to the $^4I_{13/2} \to {}^4I_{15/2}$ transition of Er. These last emissions arising from Er centred bands point to the presence of a partial ligand to metal energy transfer[26]. Performing the same measurement in H_2O did not reveal any detectable Er emission in the NIR domain, nor in the 540–550 nm region. As previously observed for other lanthanides[22], the replacement of the water molecule in the first coordination sphere of the complex by a D_2O one has a dramatic influence on the luminescence properties, decreasing the non-radiative quenching of OH oscillators[27].

Addition of fluoride to the D_2O solution resulted in a ca 60% increase of the Er-centred luminescence intensity in the NIR region (Fig. 2c). At 1,540 nm, the maximum increase was observed at 0.5 equivalent of fluoride, corresponding to the formation of an Er_2F species (Fig. 2d). As it was previously observed for measurements in D_2O instead of H_2O, the coordination of fluoride resulted in a supplementary decrease of the non-radiative quenching processes with a concomitant increase of the luminescence. The addition of larger amounts of fluoride resulted in minor changes in the luminescence intensity, pointing to very similar metal-centred luminescence quantum yields for the species present in solution.

UC in D_2O solution. The possibility of an UC process with the Er complex in D_2O was investigated. A 1 mM solution of the complex in D_2O was irradiated at 980 nm into the Er $^4I_{15/2} \to {}^4I_{11/2}$ transition and the emission spectrum of the solution was recorded from 500 to 680 nm (Fig. 3). In the absence of fluoride, the UC spectrum displayed three weak emission bands at ca 525, 550 and 650 nm (Fig. 3a). These transitions were, respectively, assigned to the $^2H_{11/2} \to {}^4I_{15/2}$, $^4S_{3/2} \to {}^4I_{15/2}$ and $^4F_{9/2} \to {}^4I_{15/2}$ emission band from Er. To this solution, increasing amounts of fluoride anions were added in the form of NaF salt. After each addition, the mixture was agitated for 30 min to ensure the equilibration of the solution. As soon as fluoride was added, the intensity of the peaks strongly increased, with a maximum at 0.5 equivalent of added fluoride, and slowly decreased afterwards (up to 4 equivalents). Interestingly, the intensity of the peaks was clearly related to the formation of an Er_2F species, with a 7.7-fold increase at 0.5 equivalent compared with the fluoride-free solution (Fig. 3b).

To further characterize the UC process, the luminescence intensity of the $^2H_{11/2} \to {}^4I_{15/2}$ and $^4S_{3/2} \to {}^4I_{15/2}$ transitions from 505 to 580 nm was monitored as a function of the intensity of the incident beam at 980 nm (Fig. 3c) for a mixture containing 0.5 equivalent of fluoride anions (Supplementary Fig. 6)[22]. The calculated slope (2.14 ± 0.12, obtained as the average value of four

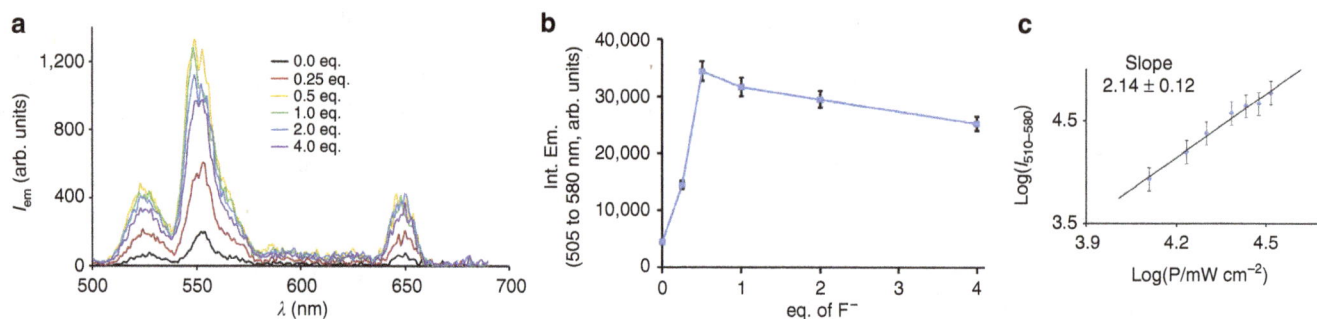

Figure 3 | UC emission spectroscopy. Excitation of a 1 mM solution of the Er complex in D_2O into the Er $^4I_{15/2} \rightarrow {}^4I_{11/2}$ absorption band at 980 nm revealed (**a**) a weak UC emission signal in the visible region associated to Er emission (black curve). Upon titration by fluoride anions, the UC emission strongly increased up to 0.5 equivalent (eq.) of fluoride (orange), as revealed (**b**) by the evolution of the intensity increase (integrated from 505 to 580 nm) as a function of added fluoride anions. (**c**) Plotting the UC emitted intensity (integrated from 505 to 580 nm) as a function of the incident pump intensity at 980 nm in a Log-Log format for the solution containing 0.5 eq. of fluoride anions gave a slope which corresponds to a two photon process (Errors were estimated assuming a Poisson's law for the intensities).

independent measurements) is slightly larger than the expected value of 2, in good agreement with UC processes[28], but such a slope may also be related with a possible ligand-based nonlinear process followed by sensitization of Er. To exclude this last possibility, the spectrum recorded at 0.5 equivalent of added fluoride was enlarged down to the ultraviolet region (from 350 to 700 nm, Supplementary Fig. 7). In these conditions, only the above-mentioned Er centred emission bands could be observed with no traces of ligand centred emission, as would be expected for a ligand-based nonlinear process followed by Er emission, similar to what is observed with one photon excitation of the ligand at 294 nm (Fig. 2). Also, the possibility of an emission at 545 nm resulting from a spurious signal arising from the $^5D_4 \rightarrow {}^7F_5$ transition of Tb impurities can be excluded by comparing the measured UC spectrum with that of the analogous Tb complex (Supplementary Fig. 8). Finally, the intensity of the UC signal at 545 nm was monitored by varying the excitation wavelength from 700 to 1,050 nm (Supplementary Fig. 9), showing a maximum of UC excitation at 980 nm for the $^4I_{15/2} \rightarrow {}^4I_{11/2}$ transition, thereby excluding the possibility of a nonlinear multiphoton process. Considering the relative weakness of the observed signal, an estimation of its quantum efficiency could be made by comparison with Er-doped UC nanoparticles pointing to an upper limit of 10^{-5}.

Additional titration experiments were performed to monitor the one photon emission spectra during Er titration with NaF with excitation into the ligand (294 nm). The visible part of the spectrum showed a weak increase of the ligand centred emission upon fluoride addition, but poor to negligible changes of intensity of the $^4S_{3/2} \rightarrow {}^4I_{15/2}$ transition of Er (at ca 550 nm, Supplementary Fig. 10). Following the NIR emission of the $^4I_{13/2} \rightarrow {}^4I_{15/2}$ transition of Er (1,450 to 1,600 nm, Supplementary Fig. 11) upon 980 nm excitation ($^4I_{15/2} \rightarrow {}^4I_{11/2}$) resulted in a maximum increase of ca 70% of the emission intensity, similar to what is obtained upon ligand centred excitation (Supplementary Fig. 12). Altogether, these results showed that the intensity increase of the one photon emission is not appreciably related to an improved ligand to metal energy transfer efficiency, but clearly to a better protection of the Er cation towards non-radiative processes. The weak intensity increase of the $^4S_{3/2} \rightarrow {}^4I_{15/2}$ transition compared with that of $^4I_{13/2} \rightarrow {}^4I_{15/2}$ may be related to important changes in the crystal field splitting of Er (ref. 29) with concomitant changes in the branching ratio.

Speciation of the species formed in solution. The formation of the new species was first monitored qualitatively by electrospray

mass spectrometry analysis of the mixtures. Addition of 0.5 equivalent of sodium fluoride to the complex (Supplementary Figs 13 and 14) led to a new peak at 1,449.46 m/z units, which was attributed to a $[(ErL)_2F]^+$ singly charged dimer, together with a doubly charged cation with a peak maximum at 736.23 m/z units, attributed to the sodium adduct $[(ErL)_2F + Na]^{2+}$. It is interesting to note that each complex is always present with two deuterium atoms, likely as a result of the H–D exchange of the imidazol protons in D_2O. At 1.0 equivalent of NaF (Supplementary Fig. 15), the peak of the singly charged dimer at 1,449.46 became intense in the spectrum, whereas the peak corresponding to the $[ErL]^+$ monomer almost vanished. The other important species were also attributed to the dimer and can be observed with maxima at m/z of 725.23 and 736.22, corresponding to the doubly charged dimer with a proton or a sodium cation, respectively. At two equivalents of fluoride (Supplementary Fig. 16), peaks corresponding to the singly charged and doubly charged dimer at 1,449.45 and 736.22 m/z units are still present, and the peak at 716.23, assigned to the $[ErL]^+$ complex became important. In these conditions, it is expected that the neutral [ErLF] monomer began to form in solution, and its ionization resulted in the breaking of the Er–F bond, releasing the $[ErL]^+$ species.

The formation of the $[(ErL)_2F]^+$ complex was also monitored by ultraviolet–visible absorption spectroscopy, following the evolution of the absorption spectra of the complex in the presence of increasing amounts of fluoride (Fig. 4) in H_2O solution.

The association of fluoride was fitted to the equilibria:

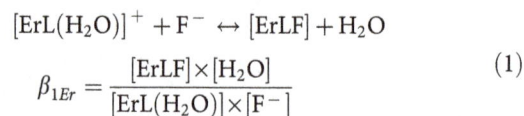

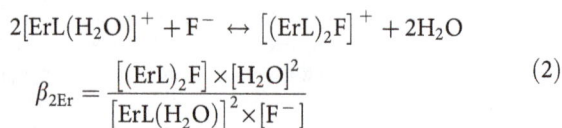

$$[ErL(H_2O)]^+ + F^- \leftrightarrow [ErLF] + H_2O$$

$$\beta_{1Er} = \frac{[ErLF] \times [H_2O]}{[ErL(H_2O)] \times [F^-]} \quad (1)$$

$$2[ErL(H_2O)]^+ + F^- \leftrightarrow [(ErL)_2F]^+ + 2H_2O$$

$$\beta_{2Er} = \frac{[(ErL)_2F] \times [H_2O]^2}{[ErL(H_2O)]^2 \times [F^-]} \quad (2)$$

Taking the water concentration as constant during the titration, analysis of the ultraviolet–visible absorption data by the Specfit software[30] using equations (1) and (2) afforded stability constants of 6.1(3) and 13.0(2) log units, respectively, for β_{1Er} and β_{2Er}. The results for the cumulative stability constant β_{2Er} are in excellent agreement with the one previously reported for other complexes (log $\beta_{2Ln} = 13.0(3)$, 12.5(1.0) and 12.6(1.0), respectively, for Eu, Tb and Yb)[22]. In contrast to what was

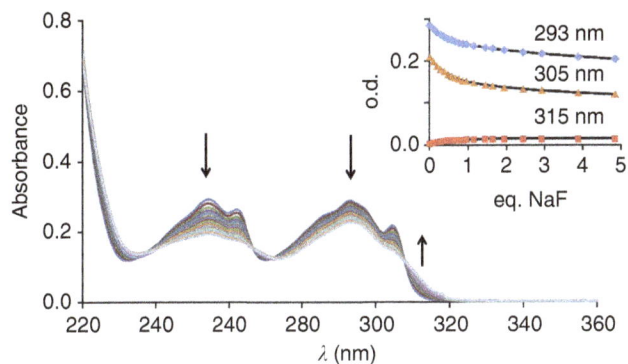

Figure 4 | Ultraviolet–visible titration of [ErL(H₂O)](NO₃). The progress of the dimerization of the erbium complex was followed by ultraviolet-visible absorption spectroscopy by addition of NaF (0–4.8 eq.) to a solution of $[ErL(H_2O)](NO_3)$ ($c = 3.15 \times 10^{-5}$ M) in H_2O. The evolution of the spectra can be fitted with equations (1) and (2) (see the text) and the comparisons between measured and calculated values (black lines) of the absorbances at 293, 305 and 315 nm are presented in the inset.

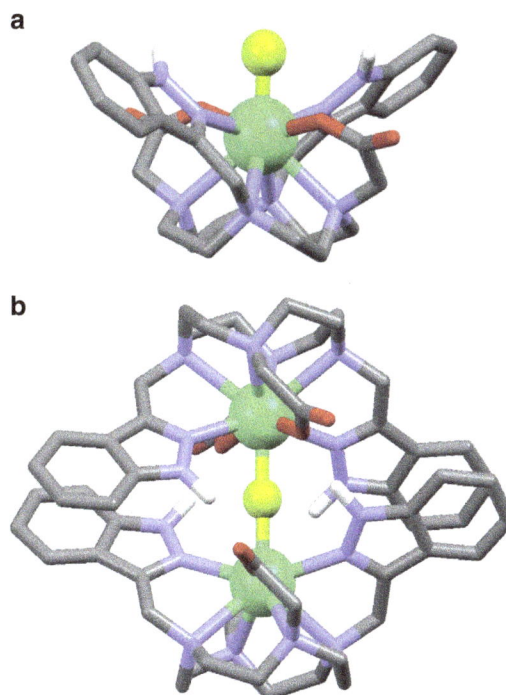

Figure 5 | DFT modelling. The [ErLF] monomer (a) and the $[(ErL)_2F]^+$ dimer (b) were modelled by DFT calculations and revealed the presence of strong π–π stacking interactions between indazolyl groups in the dimer, together with four hydrogen bonding interactions which, with the formation of Er-F-Er bridge, stabilize the supramolecular dimeric assembly.

observed for the other lanthanide cations, the titration of the Er complex allows the observation of the mononuclear [ErLF] intermediate complex. It is worth noting that the ratio of the stepwise association constants, $K_{2Er}/K_{1Er} = 6.3$ is larger than the value of 1.0 expected for a statistical behaviour (see Supplementary Methods for the calculation of the statistical value)[31,32], pointing to a weak positive cooperativity for the formation of the dimer. Also of interest is the value of β_{1Er}, which is larger than other stability constants measured for the coordination of fluoride anions to similar Ln complexes[29,33] (log $\beta_{1Eu} = 4.6$ was measured for pyridine containing cyclen analogue taking water concentration into account)[34]. The details of the calculated concentrations of the species formed during the titration and the recalculated ultraviolet–visible absorption spectra of the species are shown in Supplementary Figs 17 and 18.

Finally, the association was also monitored by ¹H-NMR spectroscopy (Supplementary Fig. 19). As previously stated, the ¹H-NMR spectrum of the Er complex in D_2O is composed of 16 broad peaks in the $+110/-110$ p.p.m. region. Addition of fluoride anions up to 0.5 equivalent immediately resulted in the observation of a second set of 16 signals, the two sets coexisting for F/Er ratios smaller than 0.5. This number of peaks is in agreement with the formation of a dimeric complex with D_2 symmetry and the observed variations of the chemical shifts are perfectly corresponding to the observations of Faulkner and co-workers[35] with a change of magnetic anisotropy associated to an anisotropic electronic distribution from prolate to oblate upon fluoride binding. For excess larger than 0.5 equivalent, only minor changes could be observed for the shapes of the signals. In the presence of very large excesses (>5 equivalent), a broadening of the peaks could be noticed. This observation may be related to very minor changes in the coordination of the ligand around Er in the $[(ErL)_2F]^+$ dimer compared with the [ErLF] monomer as expected on the basis of density functional theory (DFT) modelling (see below). Unfortunately, attempts to follow the titration by ¹⁹F-NMR proved unsuccessful. The strong paramagnetic contribution of the Er cation with the short Er–F distances in the complexes probably resulted in too broad peaks that we could not observe.

DFT modelling of the dimer. In the absence of available X-ray crystal structure, we turned our attention to DFT calculations for modelling the structure of the [ErLF] monomer and

$[(ErL)_2F]^+$ dimer (Fig. 5). DFT calculations performed at the M06/LCRECP/6-31G(d,p) level provided an optimized structure for $[(ErL)_2F]^+$ containing a linear Er-F-Er unit with a Er-F distance of 2.242 Å and a Er···Er distance of 4.484 Å. The Er-F distance is slightly shorter in the monomeric [ErLF] entity (2.104 Å). The dimeric structure is supported by π–π stacking interactions involving the indazolyl groups of the two (ErL) entities (distance between centroids 3.891 Å) and hydrogen bonds established between the NH groups of one indazolyl groups of a ErL entity and the oxygen atoms of the carboxylate groups coordinated to the second Er metal ion (N···O 2.751 Å, N-H···O 1.745 Å, N-H···O 162.4°). It is anticipated that the formation of hydrogen bonds and stacking interactions are at the origin of the positive cooperativity observed for the formation of the dimer during the ultraviolet–visible titration experiment (vide supra).

Discussion
Based on the stability constants determined above, it was possible to calculate the amounts of the three different species present in D_2O solution, that is, $[ErL(D_2O)]^+$, $[(ErL)_2F]^+$ and [ErLF], as a function of added F^-. Figure 6 represents the calculated concentrations of the different species as a function of the $[F^-]/[Er]$ concentration ratio. In the same graph, the evolution of the integrated intensity of the peak at 550 nm was reported, arbitrarily normalizing the emitted intensity at half an equivalent of added fluoride. The comparison clearly showed that the largest UC intensity increase is directly related to the formation of the dimer content. Interestingly, for larger amounts of fluoride for which the [ErLF] species is formed, the intensity decreased, pointing to a weaker UC process in the [ErLF] monomeric complex. It is expected that the coordination of fluoride anions in the monomeric complex will improve the metal-centred

luminescence quantum yield by displacing the non-radiative quenching processes associated to OD oscillators of the coordinated D_2O molecule oscillators of the coordinated heavy water molecule as observed by conventional fluorescence (Fig. 2b,c)[12,36]. This observation emphasizes the improvement of the UC process for the dimer compared with the two monomers, $[ErL(D_2O)]$ and $[ErLF]$.

Considering first-order processes, Fig. 7 summarizes the two possible mechanisms responsible of the UC process in the dimer. For both mechanisms, the first step is a ground-state absorption resulting in the formation of an $[Er^*FEr]$ excited state with population of the $^4I_{11/2}$ level of one Er atom. In the case of an ESA (red arrows), the absorption of a second photon by the excited Er atom led to population of the $^4F_{7/2}$ level, $[Er^{**}FEr]$ and relaxation to the $^2H_{11/2}$, $^4S_{3/2}$ and $^4F_{9/2}$ levels results in emission at 525, 550

and 650 nm, respectively. For the second process, the ETU (blue arrows), the ground-state absorption leading to $[Er^*FEr]$ is followed by the absorption of a second photon by the second Er atom, so that both Er atoms were in the $^4I_{11/2}$ excited state, forming the $[Er^*FEr^*]$ excited state. An intramolecular energy transfer from one Er to the other relaxed one Er to the ground state, whereas the second was promoted to a higher energy level such as $^4F_{7/2}$, $^2H_{11/2}$ or $^4S_{3/2}$, depending on whether the transfer process is a phonon-assisted one or not. Relaxation to the ground state also led to emission at 525, 550 and 650 nm.

A major difference in the ESA and ETU processes can be found in the risetime of the UC emitted signal. As ESA arise from two absorption phenomena, the risetime is very rapid. In contrast, the ETU process is slowed by the rate of energy transfer from the intermediate excited state and may lead to an observable risetime. Recording the risetime of the emission at 545 nm ($^4S_{3/2} \rightarrow {}^4I_{15/2}$) upon pulsed laser excitation at 980 nm (Supplementary Fig. 20) showed a ca 5 μs risetime in the emission, pointing to an ETU mechanism in the Er dimer.

The relationship between the UC efficiency and the concentration of the species in solution points to two observations; the efficiency is directly related to the presence of the Er dimer and decreases as the concentration of $[ErLF]$ monomer increases. Both observations are in favour of an ETU mechanism. If it proceeded via the ESA mechanism, the efficiency of the UC process should not decrease for larger concentrations of $[ErLF]$ monomer as the metal-centred luminescence quantum yield of the monomer was found to be almost similar to that of the dimer (Fig. 2). If one hypothesizes that the crystal field splitting and the transition probabilities of Er are similar for the monomer and the dimer (DFT models point to similar coordination environment around Er in the monomer and the dimer), the UC efficiency should also be the same, which is not the case. Although the UC signals observed for the monomers are probably related to an ESA mechanism, as observed by Faulkner and co-workers[19], the 7.7-fold increase of the UC efficiency associated with the formation of the dimer can hardly be explained only by the ESA mechanism, considering that the one photon luminescence quantum yield only increased by a factor 1.6. These observations,

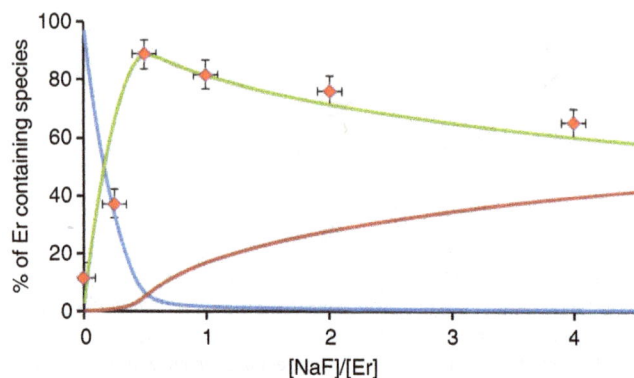

Figure 6 | Speciation in solution. The relative evolution of the percentages of the species present in solution ($[ErL(D_2O)]^+$, blue; $[(ErL)_2F]^+$, green; $[ErLF]$, red) can be recalculated from the knowledge of the stability constants and were compared with UC emission intensity (integrated from 505 to 580 nm, red squares) normalized on the concentration at 0.5 equivalent (errors correspond to those of a Poisson's law for the intensity and to ±5% for the relative concentrations). The speciation revealed a marked correlation between the UC intensity and the presence of the Er dimer.

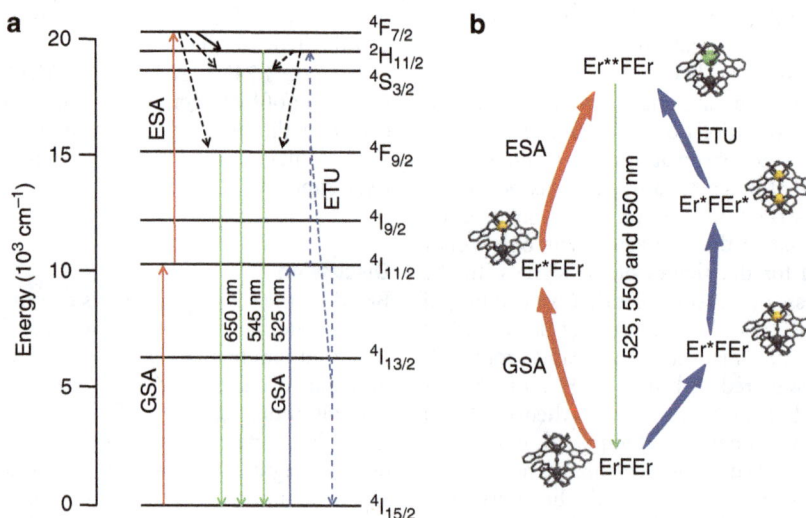

Figure 7 | Energy level diagram and mechanisms of the UC process. (**a**) The different mechanisms (ETU in blue and excited state absorption (ESA) in red) for the UC process are presented in the energy level diagram of the Er ion. In both cases, the first step is the excitation to the $^4I_{11/2}$ level upon excitation at 980 nm. For ESA, the erbium ion in the excited state absorbs a second photon to reach the upper levels and relaxation to the $^2H_{11/2}$, $^4S_{3/2}$ and $^4F_{9/2}$ precedes the emission of the UC signal (green arrows). For ETU, the higher excited state is reached by energy transfer from the second Er atom itself in the $^4I_{11/2}$ excited state. (**b**) Schematic representations of the two mechanisms (the vertical positions are not representative of the energy levels of the different excited states).

together with the evidence of a rising domain in the UC emitted signal, are fully consistent with an ETU mechanism, favoured by the spatial proximity of the two Er centres, as it has been observed in Er doped solids[37].

The coordination of fluoride anions with the $[ErL(H_2O)]^+$ complex resulted in the self-assembly of two complexes around a single fluoride anion. The assembly is characterized by a positive cooperative effect resulting from the formation of hydrogen bonds and aromatic stacking interactions. DFT models of the dimer showed that the intermetallic distance is very short (4.48 Å). In a D_2O solution at room temperature, excitation of the Er complex resulted in the observation of three UC emission bands at 525, 550 and 650 nm, respectively, attributed to the Er centred $^2H_{11/2} \rightarrow {}^4I_{15/2}$, $^4S_{3/2} \rightarrow {}^4I_{15/2}$ and $^4F_{9/2} \rightarrow {}^4I_{15/2}$ transitions, upon excitation in the Er $^4I_{15/2} \rightarrow {}^4I_{11/2}$ absorption band at 980 nm. Noticeably, addition of fluoride anions to the solution is accompanied by a very large 7.7-fold increase of the UC emission attributed to the formation of the dimer in a D_2O. These results unambiguously showed that UC can be achieved at the molecular and supramolecular level, in solution under ambient conditions, providing that the system is well-adapted. Although the observed UC may be related to an ESA mechanism in the monomeric complexes, the large increase observed for the dimer and the risetime observed for the UC emitted signal emphasize the importance of the association of the two cations of the dimer in the UC process and strongly indicate an excited state energy transfer mechanism. In this context, the very short intermetallic distance between the two Er cations is believed to be a crucial parameter.

These results are the first evidence of UC at the molecular level in deuterated aqueous solution at room temperature. We can expect that future developments on the structure and compositions of such complexes could lead to similar observations in non-deuterated water, hence opening a brand new window for luminescence tagging in numerous bio-analytical applications.

Methods

Materials and chemical analysis. Solvents and starting materials were purchased from Aldrich, Acros and Alfa Aesar and used without further purification. ^1H and ^{13}C NMR spectra were recorded on Bruker 500, 400 or 300 spectrometers. Chemical shifts are reported in p.p.m., with residual protonated solvent as internal reference[38]. Infrared spectra were recorded on a Perkin Elmer Spectrum One Spectrophotometer as solid samples and only the most significant absorption bands are given in cm^{-1}. Elemental analyses and routine mass spectrometry analysis were carried out by the Service Commun d'Analyses of the University of Strasbourg. High-resolution electrospray ionisation-time of flight (ESI-TOF) mass spectra were recorded using a micro-TOF-Q Applied Biosystems QSTAR Elite spectrometer in the positive mode.

Computational details. All calculations presented in this work were performed employing the Gaussian 09 package (Revision D.01)[39]. Full geometry optimizations of the $[(ErL)_2F]^+$ and $[ErLF]$ systems were performed in aqueous solution employing DFT within the hybrid meta generalized gradient approximation (hybrid meta-GGA), with the M06 exchange-correlation functional[40]. In these calculations, we used the large-core quasirelativistic effective core potential of Dolg and co. and its associated [5s4p3d]-GTO valence basis set for Er[41], whereas the ligand atoms were described by using the standard 6–31G(d,p) basis set. Solvent effects were included by using the CPCM variant of the polarizable continuum model, as implemented in Gaussian 09 (refs 42,43).

Spectroscopic characterizations. Spectroscopic measurements were performed with 10×10 mm^2 quartz suprasil certified cells (Helma Analytics). Ultraviolet-visible absorption spectra were recorded on a Perkin Elmer Lambda 950 spectrometer. Steady-state emission spectra were recorded on a FLS920P fluorescence spectrometer (Edimburgh Instrument) equipped with a 450-W continuous wavelength Xe lamp (range from 230 to 900 nm), using Hamamatsu R928 (visible) or R5509-72 (Vis and NIR range) photomultipliers. For steady-state emission in the NIR range, high-pass filters at 850 nm were used to remove second-order artefacts. For UC emission spectra, a 980-nm continuous wavelength laser diode (5 W, Laserwave, Beijing, China) was used as the light source and emission in the visible was detected with a photomultiplier tube (PMT, R928 Hamamatsu). Power

dependence of UC emission was measured under a focused excitation light with a spot size (average) of ≈ 0.07 cm^2, and laser power increased step by step (0.9–2.3 W). A focus lens (focal length is 10 cm) collimating the laser light (diameter of light cross section is about 5 mm) was placed 10 cm in front of the 980 nm laser, and 5 cm before the aperture. A power metre (FieldMaxII-TO laser power meter and PM30 probe) was placed in a position close to where the cuvette would be to measure the power. Laser power is measured both before and after spectral measurement of solution samples with the power probe placed in the pre-set position. Considering the response time of power metre probe (2 s), laser power is measured after allowing the laser to stabilize for 30 s. The error of the laser power and the power metre is < 1%, respectively. The UC excitation spectra and UC kinetic experiments were performed with an ultrafast laser system equipped with R928 Hamamatsu Photomultiplier Tube. Excitation light was generated by an infrared coherent pulsed laser (Ti:sappire compact laser system- 150 fs, with wavelength range 690–1,040 nm), excitation power at every point was calibrated to 750 mW and the UC emission intensity was monitored at 545 nm ($^4S_{3/2} + {}^2H_{11/2} \rightarrow {}^4I_{15/2}$).

Synthesis of the Er complex. $[ErL(H_2O)](NO_3)$ was synthesized by adaptation of previously published methods[22] using nitrate salts of Er instead of lanthanide chloride salts. Yield: 77%. ^1H NMR (D_2O, 400 MHz, 25 °C, Supplementary Figs 1 and 2): δ 110.16, 80.61, 45.98, 25.64, 21.47, 20.22, 15.50, 14.19, 12.12, 3.09, −1.52, −21.61, −82.97, −100.84, −110.54; infrared (attenuated total reflectance (ATR), Supplementary Fig. 3): 3,437 (br, m), 3,216 (br, m), 2,863 (w), 1,655 (m), 1,614 (s), 1,356 (s), 1,324 (s), 1,077 (m), 1,003 (w), 931 (w), 756 (m); ultraviolet-visible (H_2O): $\lambda_{max} = 294$ nm; electrospray mass spectrometry (ES-MS) (ESI$^+$, Supplementary Fig. 4): $[ErL]^+$ calculated for $C_{28}H_{34}ErN_8O_4$, 714.19; found, 714.20; analysis (calculated, found for $[(ErL)H_2O](NO_3).NaCl.4H_2O$ ($C_{28}H_{44}ClErN_9NaO_{12}$)): C (36.38, 36.49); H (4.80, 4.42); N (13.64, 13.62).

References

1. Auzel, F. Upconversion and Anti-Stokes Processes with f and d Ions in Solids. *Chem. Rev* **104**, 139–173 (2004).
2. Andraud, C. & Maury, O. Lanthanide complexes for nonlinear optics: from fundamental aspects to applications. *Eur. J. Inorg. Chem* 4357–4371 (2009).
3. Gamelin, D. R. & Güdel, H. U. Design of luminescent inorganic materials: new photophysical processes studied by optical spectroscopy. *Acc. Chem. Res* **33**, 235–242 (2000).
4. Wang, F. & Liu, X. Recent advances in the chemistry of lanthanide-doped upconversion nanocrystals. *Chem. Soc. Rev* **38**, 976–989 (2009).
5. Chen, G., Qiu, H., Prasad, P. N. & Chen, X. Upconversion nanoparticles: design, nanochemistry, and applications in theranostics. *Chem. Rev* **114**, 5161–5214 (2014).
6. Bünzli, J.-C.G. Lanthanide luminescence for biomedical analyses and imaging. *Chem. Rev* **110**, 2729–2755 (2010).
7. Weissleder, R. & Ntziachristos, V. Shedding light onto live molecular targets. *Nature Med* **9**, 123–128 (2003).
8. Zako, T. *et al.* Cyclic RGD peptide-labeled upconversion nanophosphors for tumor cell-targeted imaging. *Biochem. Biophys. Res. Comm* **381**, 54–58 (2009).
9. Zhang, T. *et al.* Water-soluble mitochondria-specific ytterbium complex with impressive NIR emission. *J. Am. Chem. Soc* **133**, 20120–20122 (2011).
10. Garaikoetxea Arguinzoniz, A. *et al.* Light harvesting and photoemission by nanoparticles for photodynamic therapy. *Part. Part. Syst. Charact* **31**, 46–75 (2014).
11. Haase, M. & Schäfer, H. Upconverting nanoparticles *Angew. Chem. Int. Ed* **50**, 5808–5829 (2011).
12. Beeby, A. *et al.* Non-radiative deactivation of the excited states of europium, terbium and ytterbium complexes by proximate energy-matched OH, NH and CH oscillators: an improved luminescence method for establishing solution hydration states. *J. Chem. Soc. Perkin Trans* **2**, 493–503 (1999).
13. Comby, S. & Bünzli, J.-C.G. in *Handbook on the Physics and Chemistry of Rare Earths 37* (eds Gschneidner, Jr K. A., Bünzli, J.-C. G. & Pecharsky, V. K. 217–470 (The Netherlands, 2007).
14. Doffek, C. *et al.* Understanding the quenching effects of aromatic C-H- and C-D oscillators in Near-IR lanthanoid luminescence. *J. Am. Chem. Soc.* **134**, 16413–16423 (2012).
15. Aboshyan-Sorgho, L. *et al.* Near-infrared to visible light upconversion in a molecular trinuclear d-f-d complex *Angew. Chem. Int. Ed* **50**, 4108–4113 (2011).
16. Zare, D. *et al.* Smaller than a nanoparticle with the design of discrete polynuclear molecular complexes displaying near-infrared to visible upconversion. *Dalton Trans* **44**, 2529–2540 (2015).
17. Förster, T. Transfer mechanisms of electronic excitation. *Disc. Faraday Soc* **27**, 7–17 (1959).

18. Andolina, C. M. & Morrow, J. R. Luminescence resonance energy transfer in heterodinuclear LnIII complexes for sensing biologically relevant anions. *Eur. J. Inorg. Chem* 154–164 (2011).
19. Sørensen, T. J., Blackburn, O. A., Tropiano, M. & Faulkner, S. Direct two-photon excitation of Sm^{3+}, Eu^{3+}, Tb^{3+}, Tb.DOTA$^-$, and Tb.propargylDO3A in solution. *Chem. Phys. Lett* **541**, 16–20 (2012).
20. Blackburn, O. A. *et al.* Luminescence and upconversion from thulium(III) species in solution. *Phys. Chem. Chem. Phys* **14**, 13378–13384 (2012).
21. Hyppanen, I. *et al.* Photon upconversion in a molecular lanthanide complex in anhydrous solution at room temperature. *ACS Photonics* **1**, 394–397 (2014).
22. Liu, T. *et al.* Supramolecular luminescent lanthanide dimers for fluoride sequestering and sensing. *Angew. Chem. Int. Ed.* **53**, 7259–7263 (2014).
23. Miller, F. A. & Wilkins, C. H. Infrared spectra and characteristic frequencies of inorganic ions. *Anal. Chem* **24**, 1253–1294 (1952).
24. Aime, S. *et al.* Conformational and coordination equilibria on DOTA complexes of lanthanide metal ions in aqueous solution studied by ^{1}H-NMR spectroscopy. *Inorg. Chem* **36**, 2059–2068 (1997).
25. Starck, M. *et al.* Towards libraries of luminescent lanthanide complexes and labels from generic synthons. *Chem. Eur. J* **17**, 9164–9179 (2011).
26. Nonat, A., Imbert, D., Pecaut, J., Giraud, M. & Mazzanti, M. Structural and photophysical studies of highly stable lanthanide complexes of tripodal 8-hydroxyquinolinate ligands based on 1,4,7-triazacyclononane. *Inorg. Chem* **48**, 4207–4218 (2009).
27. Bünzli, J.-C. G. On the design of highly luminescent lanthanide complexes. *Coord. Chem. Rev* **293-294**, 19–47 (2015).
28. Pollnau, M., Gamelin, D. R., Lüthi, S. R., Güdel, H. U. & Hehlen, M. P. Power dependence of upconversion luminescence in lanthanide and transition-metal-ion systems. *Phys. Rev* **B61**, 3337–3346 (2000).
29. Blackburn, O. *et al.* Spectroscopic and crystal field consequences of fluoride binding by [Yb.DTMA]$^{3+}$ in aqueous solution *Angew. Chem. Int. Ed* **54**, 10783 (2015).
30. Gampp, H., Maeder, M., Meyer, C. J. & Zuberbühler, A. D. Calculation of equilibrium constants from multiwavelenght spectroscopic data. *Talanta* **32**, 95–169 (1985).
31. Ercolani, G., Piguet, C., Borkovec, M. & Hamacek, J. Symmetry numbers and statistical factors in self-assembly and multivalency. *J. Phys. Chem. B* **111**, 12195–12203 (2007).
32. Ercolani, G. & Schiaffino, L. Allosteric, chelate and interannular cooperativity: A mise au point. *Angew. Chem. Int. Ed.* **50**, 1762–1768 (2011).
33. Lima, L. M. P. *et al.* Positively charged lanthanide complexes with cyclen-based ligands: synthesis, solid state and solution structure, and fluoride interaction. *Inorg. Chem* **24**, 12508–12521 (2011).
34. Tripier, R., Platas-Iglésias, C., Boos, A., Morfin, J.-F. & Charbonnière, L. Towards fluoride sensing with positively charged lanthanide complexes. *Eur. J. Inorg. Chem* 2735–2745 (2010).
35. Blackburn, O. A., Kenwright, A. M., Beer, P. D. & Faulkner, S. Axial fluoride binding by lanthanide DTMA complexes alters the local crystal field, resulting in dramatic spectroscopic changes. *Dalton Trans* **44**, 19509–19517 (2015).
36. Butler, S. J. Quantitative determination of fluoride in pure water using luminescent europium complexes. *Chem. Commun* **51**, 10879–10882 (2015).
37. Martín-Rodríguez, R. *et al.* Highly efficient IR to NIR upconversion in Gd_2O_2S: Er^{3+} for photovoltaic applications. *Chem. Mater* **25**, 1912–1921 (2013).
38. Gottlieb, H. E., Kottyar, K. & Nudelman, A. NMR chemical shifts of common laboratory solvents as trace impurities. *J. Org. Chem* **82**, 7512–7515 (1997).
39. Frisch, M. J. *et al.* Gaussian 09, Revision D.01 Gaussian, Inc. (2009).
40. Zhao, Y. & Truhlar, D. G. The M06 suite of density functionals for main group thermochemistry, thermochemical kinetics, noncovalent interactions, excited states, and transition elements: two new functionals and systematic testing of four M06-class functionals and 12 other functionals. *Theor. Chem. Acc* **120**, 215–241 (2008).
41. Dolg, M., Stoll, H., Savin, A. & Preuss, H. Energy-adjusted pseudopotentials for the rare earth elements. *Theor. Chim. Acta* **75**, 173–194 (1989).
42. Barone, V. & Cossi, M. Quantum calculation of molecular energies and energy gradients in solution by a conductor solvent model. *J. Phys. Chem. A* **102**, 1995–2001 (1998).
43. Cossi, M., Rega, N., Scalmani, G. & Barone, V. Energies, structures, and electronic properties of molecules in solution with the C-PCM solvation model. *J. Comp. Chem* **24**, 669–681 (2003).

Acknowledgements

L.J.C. thanks the French Ministère des Affaires Etrangères for a grant of the PROCORE program (no. 30707NM and F-HKBU201/13). Jean Marc Strub and Nicolas Busser are gratefully acknowledged for recording ES/MS spectra and for preparing high-quality images, respectively. We gratefully acknowledge the support of the University Research Facility on Chemical and Environmental Analysis (URFCE) of PolyU, HK Polytechnic University Area of Excellent Grants (1-ZVGG) and The Joint Research Program on Bioimaging, Hong Kong Polytechnic University and Hong Kong Baptist University. (RC-ICRS/15-16/02F-WKL02F-WKL).

Author contributions

L.J.C., K.-L.W. and W.-T.W. conceived and supervised the project. A.N. performed the spectroscopic titrations, characterized the complex and measured the spectroscopic properties in solution. T.L. performed the synthesis of the ligand and complex. C.P.-I. performed the DFT modelling. Z.L., C.F.C., W.K.W. and K.L.W. performed the UC experiments.

Additional information

Permissions

List of Contributors

Stafford W. Sheehan, Julianne M. Thomsen, Robert H. Crabtree, Gary W. Brudvig and Charles A. Schmuttenmaer
Department of Chemistry, Yale University, 225 Prospect Street, PO Box 208107, New Haven, Connecticut 06520-8107, USA

Ulrich Hintermair
Centre for Sustainable Chemical Technologies, University of Bath, Claverton Down BA2 7AY, UK

Cuiling Li
World Premier International (WPI) Research Center for Materials Nanoarchitectonics (MANA), National Institute for Materials Science (NIMS), 1-1 Namiki, Tsukuba, Ibaraki 305-0044, Japan

Ömer Dag
Department of Chemistry, Bilkent University, 06800 Ankara, Turkey

Thang Duy Dao
World Premier International (WPI) Research Center for Materials Nanoarchitectonics (MANA), National Institute for Materials Science (NIMS), 1-1 Namiki, Tsukuba, Ibaraki 305-0044, Japan
PRESTO and CREST, Japan Science and Technology Agency (JST), 4-1-8 Honcho, Kawaguchi, Saitama 332-0012, Japan
Graduate School of Materials Science, Nara Institute of Science and Technology, 8916-5 Takayama, Ikoma, Nara 630-0192, Japan

Tadaaki Nagao
World Premier International (WPI) Research Center for Materials Nanoarchitectonics (MANA), National Institute for Materials Science (NIMS), 1-1 Namiki, Tsukuba, Ibaraki 305-0044, Japan
PRESTO and CREST, Japan Science and Technology Agency (JST), 4-1-8 Honcho, Kawaguchi, Saitama 332-0012, Japan

Yasuhiro Sakamoto
PRESTO and CREST, Japan Science and Technology Agency (JST), 4-1-8 Honcho, Kawaguchi, Saitama 332-0012, Japan
Department of Physics, Graduate School of Science, Osaka University, 1-1 Machikaneyamacho, Toyonaka, Osaka 560-0043, Japan

Tatsuo Kimura
Advanced Manufacturing Research Institute, National Institute of Advanced Industrial Science and Technology (AIST), Shimoshidami, Moriyama, Nagoya 463-8560, Japan

Osamu Terasaki
Graduate School of EEWS (BK21Plus), KAIST, Daejeon 305-701, Korea
Department of Materials and Environmental Chemistry, EXSELENT, Stockholm University, 10691 Stockholm, Sweden

Yusuke Yamauchi
World Premier International (WPI) Research Center for Materials Nanoarchitectonics (MANA), National Institute for Materials Science (NIMS), 1-1 Namiki, Tsukuba, Ibaraki 305-0044, Japan
Department of Nanoscience and Nanoengineering, Faculty of Science and Engineering, Waseda University, 3-4-1 Okubo, Shinjuku, Tokyo 169-8555, Japan

Liang Wang, Xiangju Meng and Feng-Shou Xiao
Key Lab of Applied Chemistry of Zhejiang Province, Department of Chemistry, Zhejiang University, Hangzhou 310028, China

Yihan Zhu, Jianfeng Huang and Yu Han
Advanced Membranes and Porous Materials Center, Physical Sciences and Engineering Division, King Abdullah University of Science and Technology, Thuwal 23955-6900, Kingdom of Saudi Arabia

Jian-Qiang Wang
Key Laboratory of Interfacial Physics and Technology, Shanghai Institute of Applied Physics, Chinese Academy of Sciences, Shanghai 201800, China

Fudong Liu
Research Center for Eco-Enviromental Sciences, Chinese Academy of Sciences, Beijing 100085, China
The Materials Sciences Division, Lawrence Berkeley National Laboratory, Berkeley, California 94720, USA

Jean-Marie Basset
KAUST Catalysis Center, King Abdullah University of Science and Technology, Thuwal 23955-6900, Kingdom of Saudi Arabia

Alejo M. Lifschitz, Jose Mendez-Arroyo, Charlotte L. Stern, C Michael McGuirk and Chad A. Mirkin
Department of Chemistry and International Institute for Nanotechnology, Northwestern University, 145 Sheridan Road, Evanston, Illinois 60208, USA

Ryan M. Young and Michael R. Wasielewski
Department of Chemistry and International Institute for Nanotechnology, Northwestern University, 2145 Sheridan Road, Evanston, Illinois 60208, USA
Argonne-Northwestern Solar Energy Research (ANSER) Center, Northwestern University, Evanston, Illinois 60208, USA

Robert Hovden and Megan E. Holtz
School of Applied and Engineering Physics, Cornell University, Ithaca, New York 14853, USA

Stephan E. Wolf
Department of Materials Science and Engineering, Cornell University, Ithaca, New York 14853, USA
Department of Materials Science and Engineering, Institute of Glass and Ceramics, Friedrich-lexander-University Erlangen-Nürnberg, 91058 Erlangen, Germany

Frédéric Marin
UMR CNRS 6282 Biogéosciences, Université de Bourgogne Franche-Comté, 6 Boulevard Gabriel, 21000 Dijon, France

David A. Muller
School of Applied and Engineering Physics, Cornell University, Ithaca, New York 14853, USA
Kavli Institute at Cornell for Nanoscale Science, Ithaca, New York 14853, USA

Lara A. Estroff
Department of Materials Science and Engineering, Cornell University, Ithaca, New York 14853, USA
Kavli Institute at Cornell for Nanoscale Science, Ithaca, New York 14853, USA

Sohini Mukherjee, Manjistha Mukherjee and Abhishek Dey
Department of Inorganic Chemistry, Indian Association for the Cultivation of Science, 2A&2B Raja SC Mullick Road, Jadavpur Kolkata 700032, India

Arnab Mukherjee, Ambika Bhagi- Damodaran and Yi Lu
Department of Chemistry, University of Illinois at Urbana-Champaign, Champaign, Illinois 61801, USA

René-Chris Brachvogel, Frank Hampel and Max von Delius
Department of Chemistry and Pharmacy, Friedrich-Alexander-University Erlangen-Nürnberg (FAU), Henkestrasse 42, 91054 Erlangen, Germany

Prashant Kumar, Kumar Varoon Agrawal, Michael Tsapatsis and K Andre Mkhoyan
Department of Chemical Engineering and Materials Science, University of Minnesota, Minneapolis, Minnesota 55455, USA

Kai Wu, Kang Li, Ya-Jun Hou, Mei Pan, Lu-Yin Zhang and Ling Chen
MOE Laboratory of Bioinorganic and Synthetic Chemistry, State Key Laboratory of Optoelectronic Materials and Technologies, Lehn Institute of Functional Materials, School of Chemistry and Chemical Engineering, Sun Yat-Sen University, Guangzhou 510275, China

Cheng-Yong Su
MOE Laboratory of Bioinorganic and Synthetic Chemistry, State Key Laboratory of Optoelectronic Materials and Technologies, Lehn Institute of Functional Materials, School of Chemistry and Chemical Engineering, Sun Yat-Sen University, Guangzhou 510275, China
State Key Laboratory of Applied Organic Chemistry, Lanzhou University, Lanzhou 730000, China

Yuichi Takasaki and Satoshi Takamizawa
Graduate School of Nanobioscience, Yokohama City University, 22-2 Seto, Kanazawa-ku, Yokohama, Kanagawa 236-0027, Japan

Samantha K. Cary, Jared T. Stritzinger, Thomas D. Green, Kariem Diefenbach, Justin N. Cross, Kenneth L. Knappenberger, Mark A. Silver, A. Eugene DePrince, Matthew J. Polinski, Jane H. House, Alexandra A. Arico and Thomas E. Albrecht-Schmitt
Department of Chemistry and Biochemistry, Florida State University, Tallahassee, Florida 32306, USA

Monica Vasiliu and David A. Dixon
Department of Chemistry, The University of Alabama, Tuscaloosa, Alabama 35487, USA

Ryan E. Baumbach and Andrew Gallagher
National High Magnetic Field Laboratory, Tallahassee, Florida 32310, USA

Shelley M. Van Cleve
Nuclear Materials Processing Group, Oak Ridge National Laboratory, Oak Ridge, Tennessee 37830, USA

Naoki Kikugawa
National Institute for Materials Science, Tsukuba, Ibaraki 305-0047, Japan

Guokui Liu
Chemical Sciences and Engineering Division, Argonne National Laboratory, Argonne, Illinois 60439, USA

Dustin R. Cummins
Materials Physics and Applications (MPA-11), Los Alamos National Laboratory, Los Alamos, New Mexico 87545, USA
Chemical Engineering and Conn Center for Renewable Energy Research, University of Louisville, Louisville, Kentucky 40292, USA

Ulises Martinez, Aditya D. Mohite and Gautam Gupta
Materials Physics and Applications (MPA-11), Los Alamos National Laboratory, Los Alamos, New Mexico 87545, USA

Andriy Sherehiy, Alejandro Martinez-Garcia, Jacek Jasinski, Gamini Sumanasekera and Mahendra K. Sunkara
Chemical Engineering and Conn Center for Renewable Energy Research, University of Louisville, Louisville, Kentucky 40292, USA

Rajesh Kappera
Materials Physics and Applications (MPA-11), Los Alamos National Laboratory, Los Alamos, New Mexico 87545, USA
Materials Science and Engineering, Rutgers University, Piscataway, New Jersey 08854, USA

Roland K. Schulze
Materials Science and Technology (MST-6), Los Alamos National Laboratory, Los Alamos, New Mexico 87545, USA

Jing Zhang and Jun Lou
Materials Science and NanoEngineering, Rice University, Houston, Texas 77005, USA

Ram K. Gupta
Chemistry, Pittsburg State University, Pittsburg, Kansas 66762, USA

Manish Chhowalla
Materials Science and Engineering, Rutgers University, Piscataway, New Jersey 08854, USA

Teodora Miclăuş and Duncan S. Sutherland
Interdisciplinary Nanoscience Center (iNANO), Aarhus University, Gustav Wieds Vej 14, 8000 Aarhus, Denmark

Christiane Beer
Department of Public Health, Aarhus University, Bartholins Alle 2, 8000 Aarhus, Denmark

Jacques Chevallier
Department of Physics, Aarhus University, Ny Munkegade 120, 8000 Aarhus, Denmark

Carsten Scavenius and Jan J. Enghild
Department of Molecular Biology and Genetics, Aarhus University, Gustav Wieds Vej 10, 8000 Aarhus, Denmark

Vladimir E. Bochenkov
Interdisciplinary Nanoscience Center (iNANO), Aarhus University, Gustav Wieds Vej 14, 8000 Aarhus, Denmark
Department of Chemistry, Lomonosov Moscow State University, Leninskie gory 1/3, 119991 Moscow, Russia

Huajun He, Yuanjing Cui, Jiancan Yu, Yu Yang, Tao Song and Guodong Qian
State Key Laboratory of Silicon Materials, Cyrus Tang Center for Sensor Materials and Applications, School of Materials Science and Engineering, Zhejiang University, Hangzhou 310027, China

En Ma and Xueyuan Chen
Key Laboratory of Optoelectronic Materials Chemistry and Physics, Fujian Institute of Research on the Structure of Matter, Chinese Academy of Sciences, Fuzhou, Fujian 350002, China

Chuan-De Wu
Department of Chemistry, Zhejiang University, Hangzhou 310027, China

Banglin Chen
State Key Laboratory of Silicon Materials, Cyrus Tang Center for Sensor Materials and Applications, School of Materials Science and Engineering, Zhejiang University, Hangzhou 310027, China
Department of Chemistry, University of Texas at San Antonio, San Antonio, Texas 78249-0698, USA

Xu Peng, Huili Liu, Qin Yin, Junchi Wu, Pengzuo Chen, Guangzhao Zhang, Guangming Liu, Changzheng Wu and Yi Xie
Hefei National Laboratory for Physical Sciences at the Microscale, iChEM (Collaborative Innovation Centre of Chemistry for Energy Materials), Hefei Science Center (CAS), and CAS Key Laboratory of Mechanical Behavior and Design of Materials, University of Science and Technology of China, Hefei, Anhui 230026, China

Huixia Luo, Jason W. Krizan, Neel Haldolaarachchige, Weiwei Xie, Michael K. Fuccillo and Robert J. Cava
Department of Chemistry, Princeton University, Princeton, New Jersey 08544, USA

Lukas Muechler
Department of Chemistry, Princeton University, Princeton, New Jersey 08544, USA
Max-Planck-Institut für Chemische Physik Fester Stoffe, Dresden 01187, Germany

Tomasz Klimczuk
Faculty of Applied Physics and Mathematics, Gdansk University of Technology, Narutowicza 11/12, Gdansk 80-233, Poland

Claudia Felser
Max-Planck-Institut für Chemische Physik Fester Stoffe, Dresden 01187, Germany

Athanassios D. Katsenis, Cristina Mottillo, Patrick A. Julien and Tomislav Friščić
Department of Chemistry, McGill University, 801 Sherbrooke Street West, Montreal, Québec, Canada H3A 0B8

Andreas Puškarić, Krunoslav Užarević, Predrag Lazić and Ivan Halasz
RuXer Bošković Institute, Bijenička cesta 54, Zagreb, HR-10000, Croatia

Vjekoslav Štrukil
Department of Chemistry, McGill University, 801 Sherbrooke Street West, Montreal, Québec, Canada H3A 0B8
RuXer Bošković Institute, Bijenička cesta 54, Zagreb, HR-10000, Croatia

Minh-Hao Pham and Trong-On Do
Department of Chemical Engineering, Université Laval, Quebec City, Québec, Canada G1V 0A6

Simon A.J. Kimber
ESRF — The European Synchrotron, CS 40220, Grenoble 38043, France

Oxana Magdysyuk and Robert E. Dinnebier
Scientific Service Group X-ray Diffraction, Max Planck Institute for Solid State Research, Heisenbergstrasse 1, 70569 Stuttgart, Germany

Shinnosuke Horiuchi and Shiori Kawamata
Research Center of Integrative Molecular Systems (CIMoS), Institute for Molecular Science, National Institutes of Natural Sciences, Myodaiji, Okazaki, Aichi 444-8787, Japan

Yuki Tachibana
Research Center of Integrative Molecular Systems (CIMoS), Institute for Molecular Science, National Institutes of Natural Sciences, Myodaiji, Okazaki, Aichi 444-8787, Japan
Department of Applied Chemistry, Graduate School of Engineering, Osaka University, Suita, Osaka 565-0871, Japan

Mitsuki Yamashita, Koji Yamamoto and Kohei Masai
Research Center of Integrative Molecular Systems (CIMoS), Institute for Molecular Science, National Institutes of Natural Sciences, Myodaiji, Okazaki, Aichi 444-8787, Japan
Department of Structural Molecular Science, The Graduate University for Advanced Studies, Myodaiji, Okazaki, Aichi 444-8787, Japan

Kohei Takase and Teruo Matsutani
Department of Applied Chemistry, Graduate School of Engineering, Osaka University, Suita, Osaka 565-0871, Japan

Yuki Kurashige and Takeshi Yanai
Department of Theoretical and Computational Molecular Science, Institute for Molecular Science, National Institutes of Natural Sciences, Myodaiji, Okazaki, Aichi 444-8585, Japan
Department of Functional Molecular Science, The Graduate University for Advanced Studies, Myodaiji, Okazaki, Aichi 444-8585, Japan

Tetsuro Murahashi
Research Center of Integrative Molecular Systems (CIMoS), Institute for Molecular Science, National Institutes of Natural Sciences, Myodaiji, Okazaki, Aichi 444-8787, Japan
Department of Structural Molecular Science, The Graduate University for Advanced Studies, Myodaiji, Okazaki, Aichi 444-8787, Japan
PRESTO, Japan Science and Technology Agency (JST), Myodaiji, Okazaki, Aichi 444-8787, Japan

Chaojiang Niu, Jiashen Meng, Xuanpeng Wang, Chunhua Han, Mengyu Yan, Kangning Zhao, Xiaoming Xu, Wenhao Ren, Qingjie Zhang, Dongyuan Zhao and Liqiang Mai
State Key Laboratory of Advanced Technology for Materials Synthesis and Processing, Wuhan University of Technology, Wuhan 430070, China

Yunlong Zhao
State Key Laboratory of Advanced Technology for Materials Synthesis and Processing, Wuhan University

of Technology, Wuhan 430070, China
Department of Chemistry and Chemical Biology, Harvard University, Cambridge, Massachusetts 02138, USA

Lin Xu
Department of Chemistry and Chemical Biology, Harvard University, Cambridge, Massachusetts 02138, USA

Antonio Leyva-Pérez and Avelino Corma
Instituto de Tecnología Química, Universidad Politécnica de Valencia-Consejo Superior de Investigaciones Cient´ıficas, Avda. de los Naranjos s/n, 46022 Valencia, Spain

Antonio Doménech-Carbó
Departament de Química Analítica, Universitat de Valencia, Dr Moliner, 50, 46100 Burjassot, Valencia, Spain

Saurabh Kumar Singh and Gopalan Rajaraman
Department of Chemistry, Indian Institute of Technology, Bombay Powai, Mumbai 400076, India

Lili Fan
School of Natural Sciences, Queensland Micro- and Nanotechnology Centre, Griffith University, Brisbane, QLD 4111, Australia
State Key Laboratory of Inorganic Synthesis and Preparative Chemistry, College of Chemistry, Jilin University, Changchun 130012, China

Shilun Qiu
State Key Laboratory of Inorganic Synthesis and Preparative Chemistry, College of Chemistry, Jilin University, Changchun 130012, China

Peng Fei Liu and Hua Gui Yang
Key Laboratory for Ultrafine Materials of Ministry of Education, School of Materials Science and Engineering, East China University of Science and Technology, Shanghai 200237, China

Xuecheng Yan and Xiangdong Yao
School of Natural Sciences, Queensland Micro- and Nanotechnology Centre, Griffith University, Brisbane, QLD 4111, Australia

Lin Gu and Zhen Zhong Yang
Institute of Physics, Beijing National Laboratory for Condensed Matter Physics, Chinese Academy of Sciences, Beijing 100190, China

Martin R. Lichtenthaler, Florian Stahl, Daniel Kratzert, Julian Hamann1 Daniel Himmel and Ingo Krossing
Institut für Anorganische und Analytische Chemie and Freiburger Materialforschungszentrum (FMF), Albert-Ludwigs-Universität Freiburg, Albertstr. 21 and Stefan-Meier Str. 21, 79104 Freiburg, Germany

Lorenz Heidinger, Erik Schleicher and Stefan Weber
Institut für Physikalische Chemie and Freiburg Institute of Advanced Studies (FRIAS), Albert-Ludwigs-Universität Freiburg, Albertstr. 21 and Albertstr. 19, 79104 Freiburg, Germany

Jiandong Pang and Kongzhao Su
State Key Laboratory of Structure Chemistry, Fujian Institute of Research on the Structure of Matter, Chinese Academy of Sciences, Fuzhou, Fujian 350002, China
University of Chinese Academy of Sciences, Beijing 100049, China

Feilong Jiang, Mingyan Wu, Caiping Liu, Daqiang Yuan and Maochun Hong
State Key Laboratory of Structure Chemistry, Fujian Institute of Research on the Structure of Matter, Chinese Academy of Sciences, Fuzhou, Fujian 350002, China

Weigang Lu
Department of Chemistry, Texas A&M University, College Station, Texas 77843, USA

Elizaveta Pustovgar, Marta Palacios and Robert J. Flatt
Institute for Building Materials, Department of Civil, Environmental and Geomatic Engineering, ETH Zürich 8093, Switzerland

Rahul P. Sangodkar and Bradley F. Chmelka
Department of Chemical Engineering, University of California, Santa Barbara, California 93106, USA

Andrey S. Andreev
Soft Matter Science and Engineering Laboratory, UMR CNRS 7615, ESPCI Paris, PSL Research University, 10 rue Vauquelin, Paris 75005, France

Jean-Baptiste d'Espinose de Lacaillerie
Institute for Building Materials, Department of Civil, Environmental and Geomatic Engineering, ETH Zürich 8093, Switzerland
Soft Matter Science and Engineering Laboratory, UMR CNRS 7615, ESPCI Paris, PSL Research University, 10 rue Vauquelin, Paris 75005, France

Christopher H. Hendon
Department of Chemistry, Centre for Sustainable Chemical Technologies, University of Bath, Claverton Down, Bath BA2 7AY, UK
Department of Chemistry, Massachusetts Institute of Technology, Cambridge, Massachusetts 02139, USA

AronWalsh
Department of Chemistry, Centre for Sustainable Chemical Technologies, University of Bath, Claverton Down, Bath BA2 7AY, UK
Department of Materials Science and Engineering, Yonsei University, Seoul 03722, South Korea

Norinobu Akiyama and Yosuke Konno
Faculty of Science, Department of Applied Chemistry, Science University of Tokyo, Kagurazaka, Shinjuku-ku, Tokyo 162-8601, Japan
Fujisoft Incorporated, Kandaneribeicho 3, Chiyoda-ku, Tokyo 101-0022, Japan (N.A.); Nippon Kayaku CO., Ltd, Sanyo Onoda, Yamaguchi 757-8686, Japan (Y.K.)

Takashi Kajiwara
Faculty of Science, Department of Chemistry, Nara Women's University, Kitauoyanishi-machi, Nara 630-8506, Japan

Tasuku Ito
Department of Chemistry, Graduate School of Science, Tohoku University, Sendai 980-8578, Japan

Hiroshi Kitagawa
Division of Chemistry, Graduate School of Science, Kyoto University, Kitashirakawa-Oiwakecho, Sakyo-ku, Kyoto 606-8502, Japan

Ken Sakai
Faculty of Science, Department of Chemistry, Kyushu University, Motooka 744, Nishi-ku, Fukuoka 819-0395, Japan
International Institute for Carbon-Neutral Energy Research (WPI-I2CNER), Kyushu University, Motooka 744, Nishi-ku, Fukuoka 819-0395, Japan
Center for Molecular Systems (CMS), Kyushu University, Motooka 744, Nishi-ku, Fukuoka 819-0395, Japan

Di Wu, Lingbing Kong and Rei Kinjo
Division of Chemistry and Biological Chemistry, Nanyang Technological University, 21 Nanyang Link, 637371 Singapore, Singapore

Yongxin Li and Rakesh Ganguly
NTU-CBC Crystallography Facility, Nanyang Technological University, 21 Nanyang Link, 637371 Singapore, Singapore

Aline Nonat
Laboratoire d'Ingénierie Moléculaire Appliquée à l'Analyse, IPHC, UMR 7178 CNRS, Université de Strasbourg, ECPM, Bât R1N0, 25 rue Becquerel, Strasbourg, 67087, France

Chi Fai Chan, Wai-Kwok Wong and Ka-Leung Wong
Department of Chemistry, Hong Kong Baptist University, Hong Kong SAR, Hong Kong

Tao Liu
Laboratoire d'Ingénierie Moléculaire Appliquée à l'Analyse, IPHC, UMR 7178 CNRS, Université de Strasbourg, ECPM, Bât R1N0, 25 rue Becquerel, Strasbourg, 67087, France
Department of Chemistry, Hong Kong Baptist University, Hong Kong SAR, Hong Kong

Carlos Platas-Iglesias
Centro de Investigaciones Científicas Avanzadas, Departamento de Química Fundamental, Universidade da Coruña, Campus da Zapateira, Rúa da Fraga 10, A Coruña 15008, Spain

Zhenyu Liu
State Key Laboratory of Chirosciences, Department of Applied Biology and Chemical Technology, Hong Kong Polytechnic University, Kowloon, Hong Kong

Wing-Tak Wong
State Key Laboratory of Chirosciences, Department of Applied Biology and Chemical Technology, Hong Kong Polytechnic University, Kowloon, Hong Kong
Department of Applied Biology and Chemical Technology, Hong Kong Polytechnic University, Hong Kong SAR, Hong Kong

Loïc J. Charbonnière
Laboratoire d'Ingénierie Moléculaire Appliquée à l'Analyse, IPHC, UMR 7178 CNRS, Université de Strasbourg, ECPM, Bât R1N0, 25 rue Becquerel, Strasbourg, 67087, France

Index

9 781632 385482